Tutorials in Motor Neuroscience

NATO ASI Series

Advanced Science Institutes Series

A Series presenting the results of activities sponsored by the NATO Science Committee, which aims at the dissemination of advanced scientific and technological knowledge, with a view to strengthening links between scientific communities.

The Series is published by an international board of publishers in conjunction with the NATO Scientific Affairs Division

A Life Sciences **B Physics**	Plenum Publishing Corporation London and New York
C Mathematical and Physical Sciences **D Behavioural and Social Sciences** **E Applied Sciences**	Kluwer Academic Publishers Dordrecht, Boston and London
F Computer and Systems Sciences **G Ecological Sciences** **H Cell Biology** **I Global Environmental Change**	Springer-Verlag Berlin, Heidelberg, New York, London, Paris and Tokyo

NATO-PCO-DATA BASE

The electronic index to the NATO ASI Series provides full bibliographical references (with keywords and/or abstracts) to more than 30000 contributions from international scientists published in all sections of the NATO ASI Series.
Access to the NATO-PCO-DATA BASE is possible in two ways:

– via online FILE 128 (NATO-PCO-DATA BASE) hosted by ESRIN, Via Galileo Galilei, I-00044 Frascati, Italy.

– via CD-ROM "NATO-PCO-DATA BASE" with user-friendly retrieval software in English, French and German (© WTV GmbH and DATAWARE Technologies Inc. 1989).

The CD-ROM can be ordered through any member of the Board of Publishers or through NATO-PCO, Overijse, Belgium.

Series D: Behavioural and Social Sciences - Vol. 62

Tutorials in Motor Neuroscience

edited by

Jean Requin

Cognitive Neuroscience Unit,
Laboratory of Functional Neuroscience, C.N.R.S.,
Marseille, France

and

George E. Stelmach

Exercise and Sport Science Institute,
Psychobiology Section, Arizona State University,
Tempe, Arizona, U.S.A.

Kluwer Academic Publishers

Dordrecht / Boston / London

Published in cooperation with NATO Scientific Affairs Division

Proceedings of the NATO Advanced Study Institute on
Tutorials in Motor Neuroscience
Calcatoggio (Ajaccio), Corsica, France
15–24 September 1990

ISBN 0-7923-1385-2

Published by Kluwer Academic Publishers,
P.O. Box 17, 3300 AA Dordrecht, The Netherlands.

Kluwer Academic Publishers incorporates the publishing programmes of
D. Reidel, Martinus Nijhoff, Dr W. Junk and MTP Press.

Sold and distributed in the U.S.A. and Canada
by Kluwer Academic Publishers,
101 Philip Drive, Norwell, MA 02061, U.S.A.

In all other countries, sold and distributed
by Kluwer Academic Publishers Group,
P.O. Box 322, 3300 AH Dordrecht, The Netherlands.

Printed on acid-free paper

Printed in the Netherlands

TABLE OF CONTENTS

Section 1 : Stimulus-Response Coding

Section 2: Learning of Motor Actions

Section 6: Programming of Movement Parameters

Section 7: Planning of Movement Sequences

Section 8: Control of Movement Kinematics

Section 9: Control of Movement Dynamics

Section 10: Sensory Motor Integration

Section 11: Neural Plasticity in Motor Systems

PREFACE

This volume represents the proceedings of a NATO Advanced Study Institute (ASI) on the topic of "Motor Neuroscience" held at the Hotel San Bastiano, Calcatoggio (Corsica), September 15-24, 1990. The San Bastiano Hotel provided a beautiful setting for the ten day ASI in a resort on the west coast of Corsica, near the island's capital city of Ajaccio.

The motivation of this ASI originated from the success of an ASI that we organized eleven years ago at Senanque Abbey in the south of France. Our earlier meeting was successful in providing some coherence to a widely scattered literature while providing up to date knowledge on motor control and learning. Our goal for the second ASI was essentially the same. We wanted to appraise the main theoretical ideas that currently characterize the field by bringing together many of the internationally known scientists who are doing much of the contemporary work. It is our hope that these proceedings will provide some conceptual unification to an expanding and diverse literature on motor control.

The ASI organized in 1979, and the volume of proceedings which was published one year later (cf. Stelmach and Requin, 1980) may be considered retrospectively as having been well-timed for, at least, two reasons. At the end of the seventies, movement science was no longer the Cinderella of experimental psychology, but was developing very quickly, extending its interest to complex movement sequences. Secondly, most of the problems debated during the previous decade seemed to be, if not solved, at least on the way to being solved at both the psychological and physiological levels of analysis. Initial conceptions of motor programming have been amended so as to offer elegant answers to the recurrent criticisms: too much storage capacity, too much computation and not enough flexibility.

In the years since Senanque, the field has changed considerably. Interest in information processing, learning and memory has waned along with the strong alliance between experimental psychology and neuroscience. In their place has emerged a new interest in movement description, due in part to the availability of new recording techniques, aimed at obtaining a complete kinematic account of how a movement is executed. The field is now dominated by approaches which emphasize dynamics and kinematics and also utilize electrophysiological measures. Similarly, technological improvements in the field of neurophysiology have provided opportunities for scientists to examine the neural events preceding and accompanying behaviourally meaningful movements, which have fostered the development of cognitive neuroscience. The presentations at this ASI certainly reflected these changes.

The present volume was organized around eleven themes. While they are not comprehensive, these themes do reflect the structure of this motor neuroscience ASI. The chapters contained in each theme discuss many of the currently debated questions in the field concerning motor mechanisms and their implementation in motor control.

We thank all participants, lecturers and students who came from many different countries. Their presence and contributions to the discussions provided intense intellectual stimulation. These formal and informal discussions will surely have a lasting impact on all participants.

The ASI was supported by a grant from NATO Scientific Affairs Division. For this support we are truly grateful. In addition, support for the ASI was also provided by the Laboratory of Functional Neuroscience of the CNRS, University of Wisconsin, Arizona State University, the Assemblée de Corse, Northern Digital, and the Banque Populaire Provençale et Corse.

The smooth running of the ASI and the subsequent publication of the proceedings would not have been possible without the valuable assistance of Luce Moerman who kindly accepted the responsibility of the ASI secretariat.

Jean Requin and George E. Stelmach

Marseille

LIST OF PARTICIPANTS AND CONTRIBUTORS[1]

*M. AKAMATSU, Industrial Products Research Institute, 1-1-4, Higashi, Tsukuba, Ibaraki 305 *(Japan)*

J. ARMAND, Laboratoire de Neurobiologie, C.N.R.S., 31, Chemin Joseph Aiguier, 13402 Marseille Cedex 9 *(France)*

S. ATHENES, Unité de Neurosciences Cognitives, C.N.R.S. - LNF 1, 31 Chemin Joseph Aiguier, 13402 Marseille Cedex 9 *(France)*

S. BALKAN, Akdeniz Universitesi Tip Fakultesi, Hastanesi Bashekimi, Antalya *(Turkey)*

*C. BARD, Laboratoire des Sciences de l'Activité Physique, PEPS, Université Laval, Ste Foy, Quebec G1K 7P4 *(Canada)*

*P.J. BEEK, Free University, Faculty of Human Movement Sciences, Van der Boechorstraat 9, 1081 BT Amsterdam *(The Netherlands)*

*K.M. BENNETT, Department of Anatomy, Cambridge University, Downing Street, Cambridge CB2 3DY *(U.K.)*

A. BERTHOZ, Laboratoire de Physiologie Neurosensorielle, C.N.R.S., 15 Rue de l'Ecole de Médecine, 75270 Paris *(France)*

J. BLOEDEL, Division of Neurobiology, Barrow Neurological Institute, 350 West Thomas Road, Phoenix, AZ 85013 *(U.S.A.)*

*J. BLOUIN, Laboratoire des Sciences de l'Activité Physique, PEPS, Université Laval, Ste Foy, Quebec G1K 7P4 *(Canada)*

M. BONNARD, Cognition et Mouvement, I.B.H.O.P., Traverse Charles Susini, 13388 Marseille Cedex 13 *(France)*

M. BONNET, Unité de Neurosciences Cognitives, C.N.R.S. - LNF 1, 31 Chemin Joseph Aiguier, 13402 Marseille Cedex 9 *(France)*

*N.A. BORGHESE, I.F.C.N.-CNR, Via Mario Bianco 9, Milano 20131 *(Italy)*

I. BOTTEMANNE, Laboratoire de Neurophysiologie Sensorimotrice, Faculté de Médecine, UCL 5449, 54 Avenue Hippocrate, 1200 Bruxelles *(Belgium)*

*Y. BURNOD, Istituto di Fisiologia Umana, Facolta di Medicina e Chirurgia, Universita di Roma "La Sapienza", 00185 Roma *(Italy)*

[1] The names of contributors who did not attend the ASI are preceded by an asterisk

R. CAMINITI, Istituto di Fisiologia Umana, Facolta di Medicina e Chirurgia, Universita di Roma "La Sapienza", 00185 Roma *(Italy)*

*L.E. CARLSON, Department of Mechanical Engineering, University of Colorado, Boulder, CO 80302 *(U.S.A.),*

*M. CARROZZO, I.F.C.N.-CNR, Via Mario Bianco 9, Milano 20131 *(Italy)*

U. CASTIELLO, Istituto di Fisiologia Umana, Universita di Parma, Aspedale Maggiore, Via Gramsci 43100, Parma *(Italy)*

M. COLES, Cognitive Psychophysiology Laboratory, University of Illinois, Champaign, IL 61820 *(U.S.A.)*

*D. CORBETTA, Department of Psychology, Indiana University, Bloomington, IN 47405 *(U.S.A.)*

*A. DASZUTA, Unité de Neurochimie, C.N.R.S. - LNF 4, 31 Chemin Joseph Aiguier, 13402 Marseille Cedex 9 *(France)*

*A. DEAT, Unité de Neurosciences Intégratives, C.N.R.S. - LNF 3, 31 Chemin Joseph Aiguier, 13402 Marseille Cedex 9 *(France)*

J. DECETY, PET-Division, Department of Clinical Neurophysiology, Karolinska Hospital, Box 60 500, S-104 01 Stockholm *(Sweden)*

*G.C. DEGUZMAN, Florida Atlantic University, Center for Complex Systems, P.O. Box 3091, Boca Raton, FL 33421-0991 *(U.S.A.)*

H. DEUBEL, Max Planck Institut für Verhaltensphysiologie, Abteilung Mittelstaedt, D-8130 Seewiesen *(Germany)*

V. DIETZ, Department of Clinical Neurology and Neurophysiology, University of Freiburg, Hansastrasse 9, D-7800 Freiburg *(Germany)*

*E. DONCHIN, Cognitive Psychophysiology Laboratory, University of Illinois, Champaign, IL 61820 *(U.S.A.)*

M. FABRE-THORPE, Institut des Neurosciences, C.N.R.S. - U.P.M.C., Département de Neurophysiologie Comparée, 9, Quai St Bernard, 75230 Paris Cedex 05 *(France)*

J. FAGOT, Unité de Neurosciences Cognitives, C.N.R.S. - LNF 1, 31, Chemin Joseph Aiguier, 13402 Marseille Cedex 9 *(France)*

B. FARKIN, University of Stirling, Department of Psychology, Stirling FK9 4LA *(U.K.)*

*A.G. FELDMAN, Institute for Information Transmission Problems, Moscow *(U.S.S.R.)*

*S. FERRAINA, Istituto di Fisiologia Umana, Facolta di Medicina e Chirurgia, Universita di Roma "La Sapienza", 00185 Roma *(Italy)*

R. FLANAGAN, McGill University, Department of Psychology, Stewart Biological Sciences Building, 1205 Dr. Penfield Avenue, Montreal, Quebec 3HA 1B1 *(Canada)*

M. FLANDERS, Department of Physiology, University of Minnesota, 6-255 Millard Hall, 435 Delaware Street S.E., Minneapolis, MN 55455 *(U.S.A.)*

*M. FLEURY, Laboratoire des Sciences de l'Activité Physique, PEPS, Université Laval, Ste Foy, Quebec G1K 7P4 *(Canada)*

*W.J. GEHRING, Cognitive Psychophysiology Laboratory, University of Illinois, Champaign, IL 61820 *(U.S.A.)*

N. GOGGIN, Department of Kinesiology, University of North Texas, P.O. Box 13857, Denton, TX 76203-3857 *(U.S.A.)*

G. GOTTLIEB, Department of Physiology, Rush Medical College, Chicago, IL 60612 *(U.S.A.)*

*R. GOTTSDANKER, Psychology Department, University of California, Santa Barbara, CA 93106 *(U.S.A.)*

*G.M. GRAMMENS, Department of Mechanical Engineering, University of Colorado, Boulder, CO 80302 *(U.S.A.)*,

*G. GRATTON, Cognitive Psychophysiology Laboratory, University of Illinois, Champaign, IL 61820 *(U.S.A.)*

G. GROUIOS, B' Department of Neurology, AHEPA Hospital, Thessaloniki 540.06 *(Greece)*

Y. GUIARD, Unité de Neurosciences Cognitives, C.N.R.S. - LNF 1, 31 Chemin Joseph Aiguier, 13402 Marseille Cedex 9 *(France)*

D. GUITTON, Montreal Neurological Institute, McGill University, 3801 University Street, Montreal, Quebec H3A 2B4 *(Canada)*

P.A. HANCOCK , c/o 110 Cooke Hall, 1900 University Ave., S.E., University of Minnesota, Minneapolis, MN 55455 *(U.S.A.)*,

T. HASBROUCQ, Unité de Neurosciences Cognitives, C.N.R.S. - LNF 1, 31, Chemin Joseph Aiguier, 13402 Marseille Cedex 9 *(France)*

A. HAYASHI, Department of Neurology, University of Wisconsin, Clinical Sciences Center H6/546, 600 Highland Avenue, Madison, WI 53706 *(U.S.A.)*

*S.I. HELMS TILLERY, Division of Neurobiology, Barrow Neurological Institute, 350 West Thomas Road, Phoenix, AZ 85013 *(U.S.A.)*

M.C. HEPP-REYMOND, Institut für Hirnforschung der Universität Zürich, August-Forelstrasse 1, Postfach, CH 8029 Zurich *(Switzerland)*

C. von HOFSTEN, Department of Psychology, Umea University, S-90187 Umea *(Sweden)*

*G.A. HORSTMANN, Department of Clinical Neurology and Neurophysiology, University of Freiburg, Hansastrasse 9, D-7800 Freiburg *(Germany)*

J.C. HOUK, Department of Physiology, Northwestern University Medical School, Ward Building 5-319, 303 East Chicago Avenue, Chicago IL 60611 *(U.S.A.)*

*I. ISRAEL, Laboratoire de Physiologie Neurosensorielle, C.N.R.S., 15 Rue de l'Ecole de Médecine, 75270 Paris *(France)*

*H. JACOMY, Laboratoire de Neurobiologie, C.N.R.S., 31, Chemin Joseph Aiguier, 13402 Marseille Cedex 9 *(France)*

M. JEANNEROD, Laboratoire Vision et Motricité, INSERM U.94, 16 Avenue du Doyen Lépine, 69500 BRON *(France)*

*J.L. JENSEN, Department of Psychology, Indiana University, Bloomington, IN 47405 *(U.S.A.)*

*P.B. JOHNSON, Istituto di Fisiologia Umana, Facolta di Medicina e Chirurgia, Universita di Roma "La Sapienza", 00185 Roma *(Italy)*

*B. KABLY, Laboratoire de Neurobiologie, C.N.R.S., 31, Chemin Joseph Aiguier, 13402 Marseille Cedex 9 *(France)*

*K. KAMM, Department of Psychology, Indiana University, Bloomington, IN 47405 *(U.S.A.)*

Z. KATSAROU, Department of Neurology, AHEPA Hospital, Skolepou 3, Thessaloniki 546 36 *(Greece)*

*J.A.S. KELSO, Florida Atlantic University, Center for Complex Systems, P.O. Box 3091, Boca Raton, FL 33421 *(U.S.A.)*

L. KERKERIAN-LE GOFF, Unité de Neurochimie, C.N.R.S. - LNF 4, 31 Chemin Joseph Aiguier, 13402 Marseille Cedex 9 *(France)*

S. KORNBLUM, Mental Health Research Institute, University of Michigan, 205 Washtenaw Place, Ann Arbor, MI 48109 *(U.S.A.)*

D. KORNBROT, Division of Psychology, Hatfield Polytechnic, College Lane, Hatfield, Herts AL10 9AB *(U.K.)*

*P.N. KUGLER, Department of Kinesiology, University of Illinois, 906 South Goodwin Avenue, Urbana, IL 61801 *(U.S.A.)*

F. LACQUANITI, I.F.C.N.-CNR, Via Mario Bianco 9, 20131 Milano *(Italy)*

*Y. LAJOIE, Laboratoire des Sciences de l'Activité Physique, PEPS, Université Laval, Ste Foy, Quebec G1K 7P4 *(Canada)*

R. LEMON, Department of Anatomy, Cambridge University, Downing Street, Cambridge CB2 3DY *(U.K.)*

*F. LEVESQUE, Institut des Neurosciences, C.N.R.S. - U.P.M.C., Département de Neurophysiologie Comparée, 9, Quai St Bernard, 75230 Paris Cedex 05 *(France)*

W. MACKAY, Department of Physiology, University of Toronto, Toronto, Ontario M5S 1A8 *(Canada)*

C. MACKENZIE, Department of Kinesiology, University of Waterloo, Waterloo, Ontario N2L 3G1 *(Canada)*

R. MARTENIUK, Department of Kinesiology, University of Waterloo, Waterloo, Ontario N2L 3G1 *(Canada)*

O. MARTIN, Laboratoire RESACT Sport, Université J. Fourier, USRAPS BP 53 X, 38041 Grenoble Cedex *(France)*

J. MASSION, Unité de Neurosciences Intégratives, C.N.R.S. - LNF 3, 31 Chemin Joseph Aiguier, 13402 Marseille Cedex 9 *(France)*

*P.V. MCDONALD, Department of Kinesiology, University of Illinois, 906 South Goodwin Avenue, Urbana, IL 61801 *(U.S.A.)*

P. McKINLEY, School of Physical and Occupational Therapy, McGill University, 3654 Drummond, Montreal, Quebec H3G 1Y5 *(Canada)*

S. MELLAH, Unité de Neurosciences Intégratives, C.N.R.S. - LNF 3, 31 Chemin Joseph Aiguier, 13402 Marseille Cedex 9 *(France)*

M. MISSAL, Laboratoire de Neurophysiologie Sensorimotrice, Faculté de Médecine, UCL 5449, 54 Avenue Hippocrate, 1200 Bruxelles *(Belgium)*

P. MORASSO, Dipartimento di Informatica Sistematica e Telematica, Universita Genova, Via Opera Pia, 11 A, 16145 Genoa *(Italy)*

*I. MOURET, Unité de Neurosciences Cognitives , C.N.R.S. - LNF 1, 31 Chemin Joseph Aiguier, 13402 Marseille Cedex 9 *(France)*

F. MÜLLER, Neurology Department, University of Tübingen, Klinikum Schnarrenberg, Hoppe-Seyler-Str. 3, 7400 Tübingen *(Germany)*

*K.G. MUNHALL, Queen's University, Kingston, Ontario *(Canada)*

A. NAHOM, 914 East Lemon # 127, Tempe, AZ 85281 *(U.S.A.),*

K. NEWELL, Department of Kinesiology, University of Illinois, 906 South Goodwin Avenue, Urbana, IL 61801 *(U.S.A.)*

A. NIEOULLON, Unité de Neurochimie, C.N.R.S. - LNF 4, 31 Chemin Joseph Aiguier, 13402 Marseille Cedex 9 *(France)*

D. OSTRY, McGill University, Department of Psychology, Stewart Biological Sciences Building, 1205 Dr. Penfield Avenue, Montreal, Quebec 3HA 1B1 *(Canada)*

*J. PAILLARD, Unité de Neurosciences Comportementales, C.N.R.S. - LNF 2, 31 Chemin Joseph Aiguier, 13402 Marseille Cedex 9 *(France)*

C. PAUL, Depart. Psicologia , Instituto de Ciencias Biomédicas Abel Salazar, Universidade do Porto, Lg. Prof. Abel Salazar, 2, 4000 Porto *(Portugal)*

L. PELLAND, McGill University, School of Physical and Occupational Therapy, 3654 Drummond, Montreal, Quebec H3G 1Y5 *(Canada)*

*C.E. PEPER, Free University, Faculty of Human Movement Sciences, Van der Boechorstraat 9, 1081 BT Amsterdam *(The Netherlands)*

*C. PIERROT-DESEILLIGNY, INSERM U.289, Hôpital de la Salpêtrière, 47 Bd de l'Hôpital, 75013 Paris *(France)*

R. PLAMONDON, Ecole Polytechnique de Montréal, Département de Génie Electrique, P.O. Box 6079 Succ "A", Montreal, Quebec H3C 3A7 *(Canada)*

C. PRABLANC, Laboratoire Vision et Motricité, INSERM U.94, 16 Avenue du Doyen Lépine, 69500 BRON *(France)*

R. PROCTOR, Department of Psychological Sciences, Purdue University, West Lafayette, IN 47907 *(U.S.A.)*

T. G. REEVE, Motor Behavior Center - HHP, Auburn University, Auburn, AL 36849-5323 *(U.S.A.)*

J. REQUIN, Unité de Neurosciences Cognitives , C.N.R.S. - LNF 1, 31 Chemin Joseph Aiguier, 13402 Marseille Cedex 9 *(France)*

*A. RIEHLE, Unité de Neurosciences Cognitives , C.N.R.S. - LNF 1, 31 Chemin Joseph Aiguier, 13402 Marseille Cedex 9 *(France)*

*S. RIVAUD, INSERM U.289, Hôpital de la Salpêtrière, 47 Bd de l'Hôpital, 75013 Paris *(France)*

A. ROUCOUX, Laboratoire de Neurophysiologie, Université de Louvain, Faculté de Médecine, UCL 5449, B-1200 Bruxelles *(Belgium)*

*V. SANGUINETI, Dipartimento di Informatica Sistematica e Telematica, Universita Genova, Via Opera Pia, 11 A, 16145 Genoa *(Italy)*

G.J.P. SAVELSBERGH, Faculty of Human Movement Sciences, Vrije Universiteit Amsterdam, Van der Boechorstraat 7, 1081 BT Amsterdam *(The Netherlands)*

R. SCHMIDT, Motor Control Laboratory, Department of Kinesiology, University of California, Los Angeles, CA 90024 *(U.S.A.)*

*K. SCHNEIDER, Department of Kinesiology, University of California, Los Angeles, 405 Hilgard Avenue, Los Angeles, CA 93706 *(U.S.A.)*

J. SEAL, Unité de Neurosciences Cognitives, C.N.R.S. - LNF 1, 31 Chemin Joseph Aiguier, 13402 Marseille Cedex 9 *(France)*

A. SEMJEN, C.N.R.S. - LNF 1, 31 Chemin Joseph Aiguier, 13402 Marseille Cedex 9 *(France)*

H. SHAFFER, Department of Psychology, University of Exeter, Devon, EX4 4QJ *(U.K.)*

J. SHEA, Motor Behavior Laboratory, 139 White Building, Pennsylvania State University, University Park, PA 16802 *(U.S.A.)*

M. SHERIDAN, Department of Psychology, University of Hull, Cottingham Road, Hull, HU6 7RX *(U.K.)*

Y. SHINODA, Department of Physiology, School of Medicine, Tokyo Medical and Dental University, 1-5-45 Yushima, Bunkyo-ku, Tokyo, 113 *(Japan)*

J. SOECHTING, Department of Physiology, 6-255 Millard Hall, University of Minnesota, Minneapolis, MN 55455 *(U.S.A.)*

*M. SOLARI, Dipartimento di Informatica Sistematica e Telematica, Universita Genova, Via Opera Pia, 11 A, 16145 Genoa *(Italy)*

G.E. STELMACH, Department of Exercise Science, Arizona State University, Tempe, AZ 85287 *(U.S.A.)*

S. SWINNEN, Department of Physical Education, Group Biomedical Sciences, Tervuurse Vest 101, 3030 Leuven *(Belgium)*

N. TEASDALE, Laboratoire des Sciences de l'Activité Physique, PEPS, Université Laval, Ste Foy, Quebec G1K 7P4 *(Canada)*

H.L. TEULINGS, Psychology Laboratory, University of Nijmegen, Montessorilaan 3, P.O. Box 9104, 6500 HE Nijmegen *(The Netherlands)*

E. THELEN, Department of Psychology, Indiana University, Bloomington, IN 47405 *(U.S.A.)*

A. THOMASSEN, Department of Experimental Psychology, University of Nijmegen, Montessorilaan 3, P.O. Box 9104, 6500 HE Nijmegen *(The Netherlands)*

*H.J.C.M. TIBOSCH, Department of Experimental Psychology, University of Nijmegen, Montessorilaan 3, P.O. Box 9104, 6500 HE Nijmegen *(The Netherlands)*

P. VIVIANI, F.A.P.S.E., U.N.I. II, 24, rue du Général Dufour, 1211 Genève *(Switzerland)*

M. G. WADE, School of Physical Education and Recreation, 110 Cooke Hall, University of Minnesota, 1900 University Avenue S.E., Minneapolis, MN 55455 *(U.S.A.)*

S. WALLACE, Department of Kinesiology, University of Colorado, Boulder, CO 80302 *(U.S.A.),*

D. WEEKS, Ball State University, Ball Gymnasium, Muncie, IN 47360 *(U.S.A.)*

*W. WERNER, Department of Anatomy, Cambridge University, Downing Street, Cambridge CB2 3DY *(U.K.)*

H.T.A.. WHITING, Department of Psychology, University of York, Heslington, York Y01 5DD *(U.K.)*

P. van WIERINGEN, Free University, Faculty of Human Movement Sciences, Van der Boechorstraat 9, 1081 BT Amsterdam *(The Netherlands)*

C. WORRINGHAM, Department of Kinesiology, University of Michigan, 401 Washtenaw Avenue, Ann Arbor, MI 48109-2214 *(U.S.A.)*

P.G. ZANONE, Center for Complex Systems, Florida Atlantic University, 500 NW 20th Street, Boca Raton, FL 33431 *(U.S.A.)*

*R.F. ZERNICKE, Department of Kinesiology, University of California, Los Angeles, 405 Hilgard Avenue, Los Angeles, CA 93706 *(U.S.A.)*

SECTION 1

STIMULUS-RESPONSE CODING

STIMULUS–RESPONSE CODING IN FOUR CLASSES OF STIMULUS-RESPONSE ENSEMBLES

Sylvan Kornblum
Mental Health Research Institute
University of Michigan
205 Washtenaw Place
Ann Arbor, MI 48109-0720
U.S.A.

Abstract

Stimulus–Response Coding is treated in the framework of compatibility. A taxonomy is presented based on the dimensional overlap model (Kornblum, Hasbroucq, and Osman, 1990) that includes four different types of S-R ensembles. Some of the literature on human performance with each type of ensemble is reviewed, and new data are presented.

Introduction

If one were to take the title of this session seriously—Stimulus Response Coding—and put this title together with the length of the conference—eight and a half days—one might come away feeling reasonably optimistic that with luck, and lots of brilliant insights on the part of all the speakers, the conference might come close to adequately covering the high points of this one topic.

The reality that we face, however, is very different. Stimulus-response coding is just one of fifteen topics scheduled to be discussed here. And instead of eight and a half days we have a bare three and a half hours, of which I have been given a generous sixty minutes.

How will I use this time?

Thanks to a number of recently published reviews of the literature (Kornblum, Hasbroucq, and Osman, 1990; Proctor & Reeve, 1990; Welford, 1976), I will not have to be exhaustive in this talk. I can afford to be selective. In the next fifty minutes, therefore, I will try to use the concept of compatibility to address some of the fundamental issues of S-R coding, I will try to show how the concept of compatibility can be used to organize the material in this field, and how research on compatibility has generated experimental paradigms, theoretical models and data which, if they could all be made to fit together into a coherent whole, would give us a good grasp of what stimulus-response coding is, and how it works.

Let me start out by indicating what I think is the common sense meaning of the term "compatibility."

Consider two sets of lists, where the first set contains a list of letters and a list of letter names, and the second set contains the same set of letters and a list of male, English first names.

J. Requin and G. E. Stelmach (eds.), Tutorials in Motor Neuroscience, 3–15.

We may now speak of compatibility at two different levels: at the level of the lists themselves, and at the level of pairs, constructed of elements drawn from these lists. At the level of the lists themselves, we call the lists in the first set compatible, in that a list of letters and a list of letter names have a common referent and are conceptually similar. We call the lists in the second set non-compatible, in that a list of letters and a list of male, English first names have no common referents, and have nothing in common. Next, consider pairs of elements within each set, constructed by selecting one member of the pair from one of the lists in the set and the other member from the other list in the same set. If the elements are drawn from the lists in the first set they would consist of a letter and a letter name. These elements could either agree or conflict. If they agree, we call the members of the pair compatible; if they conflict, we call them incompatible. If members of the pairs are drawn from the second set, letters and male English first names, then no matter which letter was paired with which first name, the members of the pairs remain non-compatible. In a model that we have recently developed (Kornblum et al., 1990) we call the similarity of the lists themselves the dimensional overlap between the lists.

A number of interesting properties become clear by these terminological distinctions. First, note that the term "compatibility" is applicable to both the set of lists, and to the individual item pairs. Second, unless compatibility exists at the list level, the issue of compatibility at the item level is moot. Third, whether at the list, or at the individual item level, compatibility is a relational property that deals with the correspondence, similarity, or match between two or more entities. The consequences of compatibility at any level will, therefore, occur as interactions, not main effects.

By labeling the lists in our example appropriately we can generate conditions of stimulus–stimulus compatibility, stimulus–response compatibility, and response–response compatibility.

S-R Compatibility: The Origins

The problem of stimulus-response compatibility (SRC) was first identified by Fitts in two classic articles (Fitts & Deininger, 1934; Fitts & Seeger, 1953) published in the early 1950s. In the first (Fitts & Seeger, 1953), Fitts reports the results of an experiment in which he combined three different sets of spatially-arranged stimuli with three different sets of spatially-arranged responses to form nine different stimulus-response ensembles. The results indicate that for any one stimulus set, the fastest RT and the lowest error rate is obtained with a particular response set, and that this response set differs for different stimulus sets. Let us call the stimulus-response ensemble that yields the best performance, the most compatible ensemble. The most compatible ensembles turn out to be those for which the spatial arrangements of the stimuli and responses are most similar. This experiment illustrates the fact that set level, or list level, compatibility plays a significant role in determining performance.

In the second study (Fitts & Deininger, 1954) Fitts used a single set of eight spatially-defined responses together with four different stimulus sets to form four different stimulus-response ensembles. Within three of those ensembles he defined three separate stimulus-response pairings such that the correspondence between the elements in the pairs was either maximal, some systematic transformation of this maximal correspondence, or random. As was true of the effects of S-R compatibility at the set level, performance was best for pairings that had the greatest correspondence within each ensemble.

In drawing the theoretical implications from these two studies Fitts identified the stimulus–response coding process as the locus of the S–R compatibility problem: "Compatibility effects are conceived as resulting from hypothetical information transformation processes (encoding and/or decoding) that intervene between receptor and effector activity" (Fitts & Deininger, 1954, p. 483).

Irrelevant Dimensions

Next, I would like to turn to compatibility when it involves irrelevant dimensions.

When is a dimension irrelevant?

When it is uncorrelated with the response.

It is interesting to consider the question whether a dimension must be relevant in order to produce compatibility effects. The results of many experiments (e.g., Simon & Small, 1969; Wallace, 1971) indicate that when an irrelevant dimension is made to conflict, or correspond with the response, compatibility effects related to the irrelevant dimension may be observed in performance.

These studies, together with the two Fitts studies, provided the empirical underpinnings for the representational part of a model that we have recently developed (Kornblum et al., 1990), and on the basis of which we proposed a taxonomy of S-R ensembles in compatibility tasks. I now turn to the model.

Dimensional Overlap

We have just seen that there are at least two levels at which entities may be said to be compatible. The first is the global level, where sets of items are perceived as either corresponding or not. The second is the local level, where elements from these sets are selected to form pairs. The basis of the correspondence at the set level is what we have called the dimensional overlap between these sets. We define dimensional overlap as the degree to which two sets of items are physically or conceptually similar. Dimensional overlap is thus a characteristic of the way sets are represented, not of the physical properties of the sets themselves.

Given that two sets have dimensional overlap, a second level of compatibility can be identified at the element level. That is, pairs may be constructed out of elements from each set, so that the elements in the pairs either match, and are compatible—or conflict, and are incompatible. Note that unless the two sets from which the elements are chosen have dimensional overlap, their pairings are neither compatible, nor incompatible. Dimensional overlap is, therefore, a necessary condition for pair-wise compatibility.

Taxonomy

We have taken these two factors, dimensional overlap and dimensional relevance, and combined them to construct a general taxonomy of ensembles that generate compatibility effects in performance. The taxonomy includes four classes of ensembles that differ depending on whether the relevant, the irrelevant, neither, or both dimensions overlap.

Type 1 ensembles are characterized by the absence of any dimensional overlap on either the relevant or the irrelevant dimensions.

Type 2 ensembles are characterized by the presence of dimensional overlap on the relevant dimension, and its absence on the irrelevant dimension.

Type 3 ensembles are characterized by the presence of dimensional overlap on the irrelevant dimensions and its absence on the relevant dimension.

Type 4 ensembles are characterized by the presence of dimensional overlap on both the relevant and on the irrelevant dimensions.

This taxonomy will serve as the organizing framework for the remainder of my talk. For each class of ensembles I will discuss some of the major issues, data, and models as well as some of the experimental paradigms that have been used with the ensembles.

Ensembles of Type 1 and 2

We have developed a processing model, illustrated in Figure 1, for ensembles of type 1 and 2. The principal difference in the manner the model treats these two types of ensembles is in the automatic components that are postulated for type 2, and are absent for type 1 ensembles.

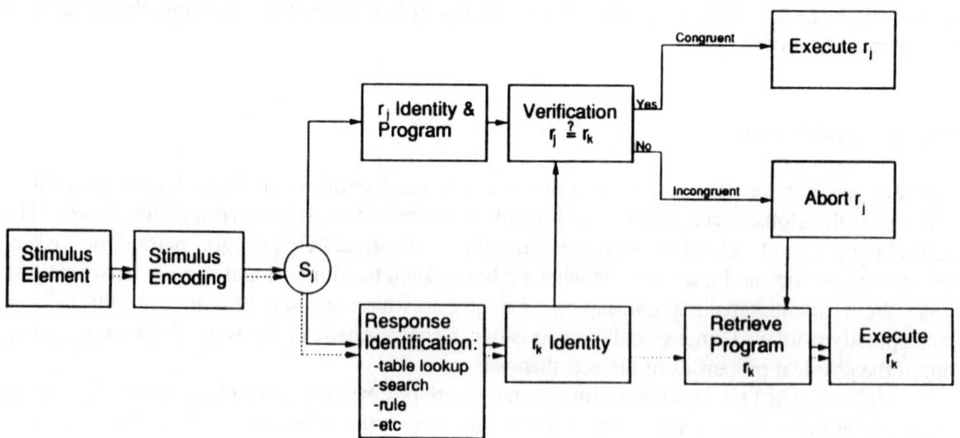

Figure 1. Block diagram of the major information-processing operations in stimulus–response (S–R) compatibility tasks with (solid lines) and with no (dotted lines) dimensional overlap. (The top branch of the solid line path illustrates the operations involved in the automatic activation of the congruent response for S–R ensembles with dimensional overlap. The bottom branch of the solid path illustrates the operations involved in the identification of the correct response.)

Consider ensembles of type 1, whose processing path is indicated by the dotted line on the bottom of figure 1. Here, the encoded form of the stimulus comprises the input to a response identification process. At the time the subject was given instructions for the task, a table of S–R pairs was set up and stored in the response identification process. The model postulates that for type 1 ensembles, the correct response is identified by searching through this table. Once identified, the response is programmed and subsequently executed. This is the classic, three-state information processing model.

Consider ensembles of type 2 next. The processing path for those ensembles is indicated by the solid lines in figure 1. As was true of ensembles of type 1, the encoded form of the stimulus constitutes the input to the response identification process. However, depending on the mapping instruction in the task (where mapping is the assignment of stimuli to responses), the correct response may be identified either by search (as in type 1 ensembles), or by rule. If by rule, it is either the identity rule, or some other rule. The model postulates that identification by search is the slowest, by identity rule the fastest, and by rules other than identity in-between. The response identification process thus constitutes a differential source of delay in response execution whose magnitude depends on the mapping instructions. In parallel with the response identification process, the stimulus automatically activates its corresponding response; this automatic activation is functionally equivalent to a response prime. These two responses, the one identified as correct and the automatically activated one, are now compared in the verification process. If, as a result of the mapping instructions, they are the same, then the mapping is said to be congruent and the response is executed rapidly and accurately. If, as a result of the mapping instructions, they are different, then the mapping is said to be incongruent and the automatically-activated, preprogrammed response must be eliminated by a time-consuming abort process. This generates a further delay beyond the one already caused by the response identification process.

Our model thus predicts that for ensembles with dimensional overlap, the fastest correct response will be the one produced with congruent mapping assignments, the next fastest will be the incongruent one that can be identified by rule, and the slowest will be the incongruent response whose identity is the result of a search. Furthermore, the model defines a neutral condition (appropriately constructed type 1 ensembles) with respect to which it predicts facilitation in the case of congruent mapping, and interference with respect to incongruent mapping.

SRC and the Effect of the Number of Alternatives

The model thus postulates the existence of three states for the response identification process. The particular state that the process is in depends on the mapping instructions. Response identification is either in a) the identity state, in which case it uses the identity rule to identify the correct response, in b) the rule state, in which case it uses a rule other than the identity rule to identify the correct response, or in c) the search state, in which case it use a serial search procedure though the table of S–R pairs to identify the correct response. The validity of this postulate can be verified from an examination of the joint effects of mapping and the number of alternatives on RT in SRC tasks. The argument is well summarized by Schvaneveldt and Staudenmeyer (1970): "Stimulus–response codes that involve a memory search to produce the appropriate response should show an effect of uncertainty (i.e., number of alternatives) on RT, while codes that produce the response directly though stimulus identification (i.e., the identity rule) or through a constant operation upon the result of the stimulus identification (i.e., other rules) should not" (p. 115)(the parenthetical remarks are ours). That is, when the response identification process is in the search state, the time to identify the correct response will depend on the number of items that are on the list being searched, and when the response identification process is in a rule state (whether identity or otherwise) the time to identify the correct response will be independent of the number of items on such a potential list. The difference in RT between the identity and non-identity rule states, according to the model, reflects the difference in complexity between the identity and other rules.

This prediction of the model is broadly supported by the data on figure 2 which represents a summary of the results from ten different experiments: six of them show RT increasing with the number of alternatives with roughly the same slope, and these presumably involved response identification through search; and the remaining four sets of data show almost no slope but are separated nevertheless.

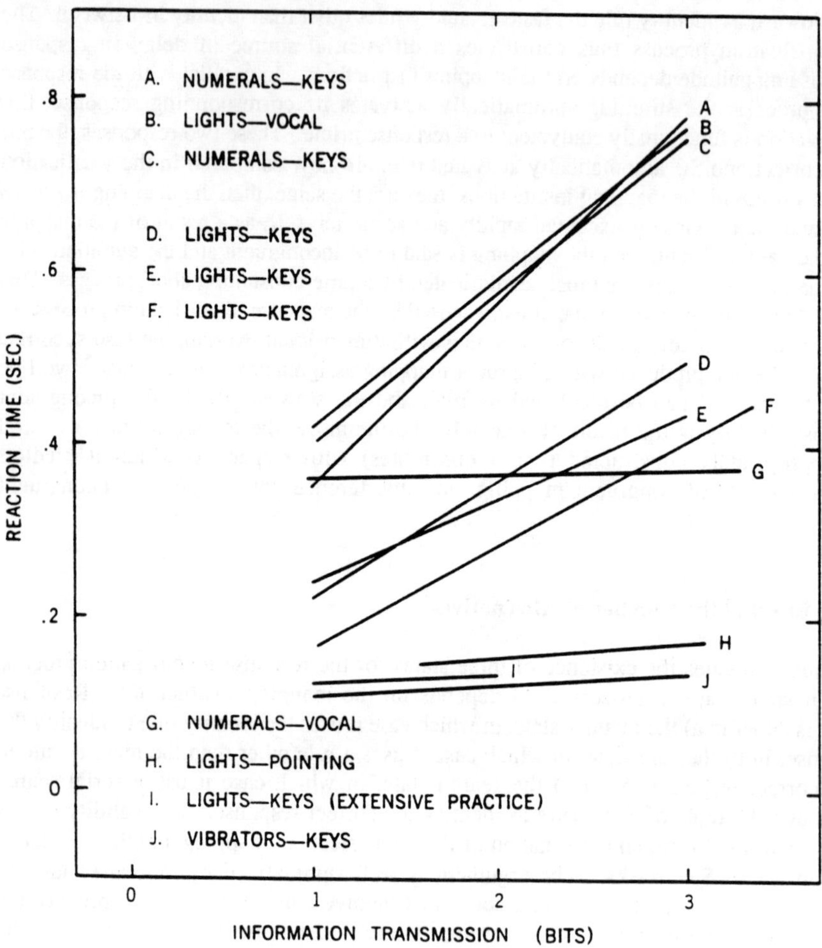

Figure 2. Reaction time as a function of uncertainty (number of alternatives) for a number of different S-R ensembles (reproduced from Fitts and Posner, 1967, in the public domain).

A study that bears more directly on this proposition was conducted by Schvaneveldt and Staudenmeyer (1970), who used digits as the stimuli in nine different experimental conditions that varied only in the response that was required. All nine experimental conditions included two-choice, four-choice and eight-choice subconditions. The data for four of these nine conditions are illustrated on figure 3A.

The fastest RT was obtained in the condition where subjects had to name the digits. This, of course, is the congruent condition, with the identity rule, and is expected to be the fastest. The slowest RT was in the random mapping condition where subjects had to produce a digit name that had been randomly chosen as the correct response for each stimulus digit. The in-between conditions were those in which response identification could be made by means of a rule. These results replicate the ordering that was found by Fitts and Deininger (1954), and are in accord with the three states whose existence we have postulated for the response identification process. Furthermore, the nature of these states is consistent with the fact that the congruent and rule-based conditions appear not to show any increase in RT as the number of alternatives increases from four to eight, whereas the RT in the random condition does increase with the number of alternatives. Thus far, the model does well. However, let us look at two other conditions in the Schvaneveldt and Staudenmeyer study (1970). In figure 3B, the data for two more rule-based conditions are displayed.

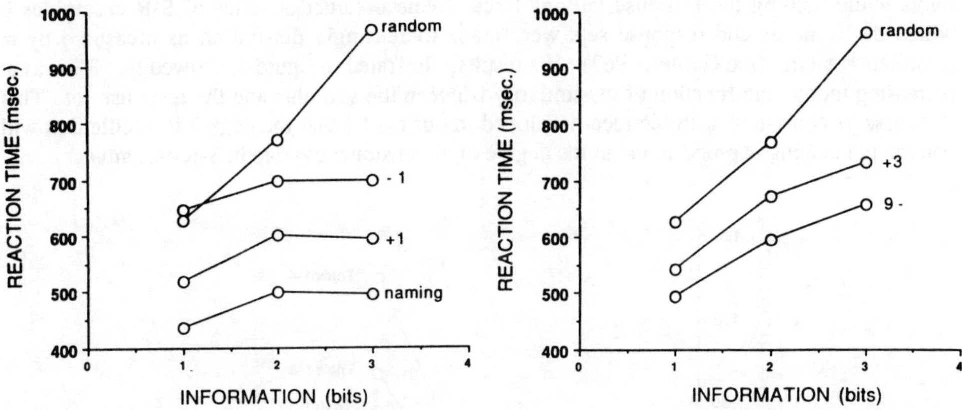

Figures 3A and 3B. Reaction time as a function of uncertainty (number of alternatives). The stimuli in all conditions were digits. Responses consisted either of naming the digit presented (naming), giving the name of randomly chosen digit (random), giving the name of the stimulus digit +1, -1, +3, or -9.

Like those on figure 3A, they also appear to be separated by a constant that may reflect the relative difficulty of these rules. However, even though these are rule-based conditions, the RT does increase with the number of alternatives. Furthermore, the slope of this function is less than that for the random mapping condition. Similar results were obtained by Smith (1978) in a key pressing task in which he used tactile stimuli to the finger tips. Clearly these data pose a problem for the three-state response identification process model. The continuous nature of the change in slope seems to indicate that either the three state response identification process is only one part of the story, or that one should not think of those three states as being discrete, or that some experimental conditions may in fact be a mixture of these states.

Spatial Properties

Even though SRC phenomena had been observed with many different types of stimuli and responses, it is relatively recently that the idea has been put forward that in spite of differences in stimulus and/or response modality, SRC phenomena may be a reflection of common, fundamental cognitive mechanisms. Much of the past work on SRC, therefore, gives a partitioned appearance because it occurs in connection with particular domains, or modalities, such as symbolic compatibility, or spatial compatibility (cf. Welford, 1976).

Some of the work that explored the spatial characteristics of stimuli and/or responses is particularly interesting because it not only cleared up a number of questions, and has come up with suggestions of general mechanisms, but also because spatial variables are ubiquitous and easily manipulable. Because of this, they continue to be widely used in current experimental paradigms. We have already seen that Fitts' original work made heavy use of spatially defined stimulus and response sets. Another early study that I find particularly interesting, for reasons that will be apparent, was done by Keele (1967). In that study Keele used eight stimulus lights and eight response buttons that were randomly, and identically, arranged in their own 16 x 16 matrices. By gradually, and systematically, distorting the spatial arrangements of the stimulus lights while holding the response buttons fixed, Keele obtained a series of S-R ensembles in which the stimulus and response sets were made increasingly dissimilar, as measured by an information metric (see Garner, 1962). His results, illustrated in figure 4, showed that RT was an increasing monotonic function of dissimilarity between the stimulus and the response set. This, of course, is consistent with the idea developed in our model that the degree of facilitation with congruent mapping is proportional to the degree of dimensional overlap in S-R ensembles.

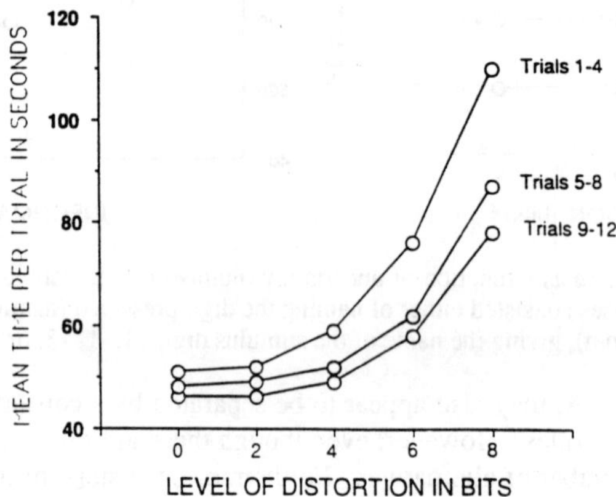

Figure 4. Reaction time as a function of dissimilarity between the stimulus and response sets (adapted from Keele, 1967).

A great deal of the work on "spatial compatibility" was, and continues to be, done with two-choice tasks. The early work focused on the question "what is the effective spatial property, or feature, underlying SRC effects?" This question took many different forms: For example, should the effective stimulus *left* and *right* be defined in terms of an ego-centric, or an external referent? Is congruence to be defined as the correspondence between the stimuli and the effectors, or between the stimuli and the manipulandum? I mention these only because they are important issues. However, the chapters by Umilta and Nicoletti, Heister et al., and Ladavas, in Proctor and Reeve's (1990) recent book make it unnecessary for me to review this work here.

An interesting development in the search for "the effective property" is the work of Proctor and Reeve (1990), whose studies include four-choice tasks in which spatial and non-spatial dimensions are often combined in a single S–R ensemble. By identifying what they call "salient features" in both the stimulus and the response sets, they have been able to show that these features function in very much the same way as points of reference do, in structuring the sets of stimuli and responses. (They are also hierarchically arranged). Performance with mapping instructions that preserve this structure is better than with mapping instructions that violate this structure. The structural homomorphisms that they observed are closely related to the dimensional overlap postulated in our model; and mappings that preserve, as against mappings that violate, this structure are close to what we have called congruent and incongruent mapping. By demonstrating the presence of such mapping effects in cross-modal tasks Proctor and Reeve's work extends, and identifies ways in which the notion of dimensional overlap may have to be elaborated in the future.

Type 3 Ensembles

Over twenty years ago, Simon published the first in a long series of studies in which he examined the role of irrelevant dimensions in choice RT (Simon & Small, 1969). His paradigmatic, experimental situations consisted of using a high and a low frequency tone as the stimuli, and mapping one tone to a left key press and the other tone to a right key press. By randomizing the ear to which these tones were presented he introduced ear as an irrelevant dimension that overlapped with the response set, and he showed that when the stimulated ear was on the same side as the response key RT was faster than when it was the opposite side. This effect was quite general and was accounted for by Simon in terms of a "natural tendency to respond towards the source of stimulation."

Based on the results of a large number of studies in the literature (cf. Proctor and Reeve, 1990), as well as results from experiments conducted in our own laboratory, we believe that the effects of irrelevant dimensions may be described more broadly. In particular, given type 3 ensembles, if the value of the overlapping irrelevant dimension corresponds with the value of the response, RT will be faster than when it conflicts.

The same processing model used to account for compatibility effects in type 2 ensembles is applicable to type 3 ensembles. In particular, whether an overlapping dimension is relevant or not, the model postulates that when a stimulus is presented with a value on that dimension it automatically activates its corresponding response. However, when the dimension is irrelevant, it plays no role in the response identification process which proceeds as it normally would with any other type 1 ensembles; i.e., by search. Since one of the responses in the response repertoire will have been activated by the irrelevant stimulus, the identity of this activated response must now be verified. If the activated response is also the correct response, its execution will be facilitated and the RT will be faster than when it is incorrect. If the irrelevant stimulus is neutral, then the RT will fall in between.

Type 4 Ensembles

We now come to type 4 ensembles in which a relevant and an irrelevant dimension both overlap. These dimensions may, but need not, be the same. The most frequent type of ensemble, however, and the one that I will focus on here, is the one in which these two dimensions *are* the same. This type of ensemble has prototypically been called the "Stroop task," or "Stroop-like" tasks.

Let me remind you very briefly of the principal elements in the generic Stroop task (Stroop, 1935). The stimuli consist of color names printed in different color ink. The task consists of naming the color that the words are printed in. The relevant stimuli are, therefore, the color of the ink used to print the words; the irrelevant stimuli are the words themselves, and the responses are color names. Clearly, when the color of the ink and that of the stimulus word itself match, naming the color of the ink does not present a problem. However, when the color of the ink and that of the color-word clash, then there is interference with the color-naming response.

		Irrelevant color name	
	S-----R	*"Red"*	*"Green"*
Relevant Ink Color	*Red--Red*	1	2
	Green--Green	4	3

Figure 5. Illustration of a standard, two-choice, Stroop task where the cells in the 2 x 2 matrix represent different types of trials. Interference normally occurs on trial types 2 and 4.

One of the critical issues raised is the site of this interference. In attentional terms, the question has been phrased: Does the attentional bottleneck occur early or late? In information processing terms the question becomes: Is the interference a stimulus effect or a response effect; does it arise at the perceptual analysis level or as the result of a response conflict? A cursory inspection of figure 5 makes it evident that this issue cannot be resolved by the standard experimental version of the Stroop task, because of the confounding that exists between the relevant stimuli and the responses. Two procedures may be used to eliminate this confounding: 1) one may choose a response set that does not overlap with either the relevant or the irrelevant dimensions; or 2) one may run the standard Stroop task with incongruent mapping instructions. Keele (1972) did the former, Green and Barber (1983) did the latter, and Simon and Sudalaimuthu (1979) did both. The results of these experiments strongly suggest that the site of the interference by the irrelevant dimension is before the response selection stage.

At this point we became interested in broadening the scope of our model to include extrinsic as well as intrinsic irrelevant dimensions. In particular, according to the model, whether a conflicting dimension or cue is presented as an intrinsic part of the stimulus, or as a physically separate stimulus object does not alter the fundamental processes that give rise to the facilitation and interference effects that are observed in performance.

We, therefore, performed a series of experiments with S-R ensembles of type 3 and 4 in which the visual displays consisted of digits or geometric shapes, as stimuli, flanked on both sides by digits or geometric shapes. The displays were similar in composition to those used by Eriksen,

Coles, and their colleagues (e.g., Coles, Gratton, Bashore, Eriksen & Donchin, 1985; Eriksen & Schultz, 1979) where we manipulated the dimensional overlap between the stimuli, the response and the noise so as to obtain ensembles of type 3 and type 4.

Type 4 Noise Ensemble

In this experiment the stimuli consisted of the digits 2 and 8; the responses consisted of the digit names "two" and "eight"; and the mapping was congruent and incongruent on different blocks. There were two flanking noise conditions: overlapping noise and neutral noise. The overlapping noise consisted of the digits 2 and 8; the neutral noise consisted of two geometric figures. When the noise overlapped, it produced a type 4 S-R ensemble that was formally equivalent to a Stroop-type ensemble. When the noise did not overlap, it produced a type 2 ensemble. The results, shown on table 1, indicate that with overlapping noise, when the noise conflicted with the stimulus, RT was longer than when it matched. This was true for both congruent and incongruent mapping conditions. When the noise did not overlap it did not interact with either stimuli or responses. The effects of noise are additives with the effects of mapping, and suggest strongly that the source of the conflict is at the stimulus end of the process.

Table 1. Reaction time with a type 4 (Stroop) S–R ensemble with extrinsic noise that overlaps (2/8) and does not overlap (H/T) with either stimuli or responses

	S–R Congruent			
	Noise		Noise	
S–R	*2*	*8*	H	T
2-2	402	457	406	403
8-8	445	401	393	398
	S–R Incongruent			
S–R	*2*	*8*	H	T
2-8	467	529	478	472
8-2	495	447	448	455

Type 3 Noise Ensemble

In this experiment the stimuli consisted of a trapezoid and a hexagon, and the responses consisted of the digit names "two" and "eight." Mapping was balanced. There were four flanking noise conditions: 1) where noise overlapped with the response (noise = 2, 8); 2) where noise overlapped with the stimuli (noise = trap, hex); 3) where noise overlapped with both (noise =

digits embedded in figures); and 4) where noise overlapped with neither (noise = asterisk, plus sign). The results, shown on table 2, were clear. Whatever the task variables that the noise overlapped with, it interacted with it so that the conflicting noise conditions were longer than the non-conflicting conditions. When the noise did not overlap with any of the task variables it did not interact.

Table 2. Response time (RT) in a two-choice task.

	noise = 2/8	noise = H/T
noise consistent	377	374
noise inconsistent	393	399
noise consistency effect	16*	25*

NOTE: Hexagon (H) and trapezoid (T) as stimuli, and "2" and "8" as responses. Flanking, irrelevant noise overlapped with either the stimuli (H/T), the responses (2/8) or neither (*/+). The data are the means for trials in which the noise was either consistent, or inconsistent with the task variables with which it overlapped. When it did not overlap (*/+) the mean RT was 370 ms.

* $p < .01$

The results of both experiments permit us to generalize the dimensional overlap model to include externally presented irrelevant dimensions.

Summary

In this paper we have examined various aspects of the stimulus-response coding problem in the framework of a model that we have recently developed for stimulus-response and stimulus-stimulus compatibility. This framework includes a taxonomy that identifies four broad classes of stimulus-response combinations each of which seemed to have its own characteristic processing consequences. This taxonomy enabled us to bring within a single framework many different types of studies, compare their properties and characteristics, and examine how they affect or contribute to the stimulus-response coding processes. Together with this taxonomy, which rested on a characterization of the representational aspects of coding, I presented a processing model that appears to work well for some ensembles, less well for others, and remains to be developed for others still. Some of the data that I presented appear to confirm certain aspects of the model, other data appear to identify problems that need to be worked on.

Acknowledgements

Supported by ONR grant number N00014-89-J-1557, and NIMH grant ROI MH 43287-01A1 to the author.

References

Coles, M. G. H., Gratton, G., Bashore, T. R., Eriksen, C. W., and Donchin, E. A. (1985). Psychophysiological investigation of the continuous flow model of human information processing. *Journal of Experimental Psychology: Human Perception and Performance, 11*, 529–553.

Eriksen, C. W., and Schultz, D. W. (1979). Information processing in visual search: A continuous flow conception and experimental results. *Perception & Psychophysics, 25*, 249–263.

Fitts, P. M., and Deininger, R. L. (1954). S–R compatibility: Correspondence among paired elements within stimulus and response codes. *Journal of Experimental Psychology, 48*, 498–492.

Fitts, P. M., and Posner, M. I. (1967). *Human performance.* Belmont, CA: Brooks/Cole.

Fitts, P. M., and Seeger, C. M. (1953). S–R compatibility: Spatial characteristics of stimulus and response codes. *Journal of Experimental Psychology, 46*, 199–210.

Garner, W. R. (1962). *Uncertainty and structure as psychological concepts.* New York: Wiley.

Green, E. J., and Barber, P. J. (1988). Interference effects in an auditory Stroop task: Congruence and correspondence. *Acta Psychologica, 53*, 183–194.

Keele, S W. (1967). Compatibility and time sharing in serial reaction time. *Journal of Experimental Psychology, 75*, 529–539.

Keele, S. W. (1972). Attention demands of memory retrieval. *J. Experimental Psychology, 93*, 245–248.

Kornblum, S., Hasbroucq, T., and Osman, A. (1990). Dimensional overlap: Cognitive basis for stimulus–response compatibility—A model and taxonomy. *Psychological Review, 97*, 253–270.

Proctor, R. W., and Reeve, G. T. (190). *Stimulus–response compatibility, An integrated perspective.* North-Holland: Amsterdam.

Schvaneveldt, R. W., and Staudenmayer, H. (1970). Mental arithmetic and the uncertainty effect in choice reaction time. *Journal of Experimental Psychology, 85*, 111–117.

Simon, J. R. and Small, A. M. (1969) Processing auditory information: Interference from an irrelevant cue. *Journal of Applied Psychology, 53*, 433–435.

Simon, J. R., and Sudalaimuthu, P. (1979). Effects of S–R mapping and response modality on performance in a Stroop task. *J. Experimental Psychology: Human Perception and Performance*, 5, 176–187.

Smith, G. A. (1978). Studies of compatibility and investigations of a model of reaction time. Unpublished Ph.D. dissertation. University of Adelaide.

Wallace, R. J. (1971). S–R compatibility and the idea of a response code. *Journal of Experimental Psychology, 88*, 354–360.

Welford, A. T. (1976). *Skilled performance: Perceptual and motor skills.* Glenview, IL: Foreman.

THE PREVALENCE OF SALIENT-FEATURES CODING IN CHOICE-REACTION TASKS

ROBERT W. PROCTOR
Department of Psychological Sciences
Purdue University
West Lafayette, Indiana 47907
U.S.A.

T. GILMOUR REEVE
Motor Behavior Center
Auburn University
Auburn, Alabama 36849
U.S.A.

ABSTRACT. Stimulus-response translation plays a prominent role in the performance of choice-reaction tasks. Our previous work with spatial-precuing and symbolic-cuing tasks has provided evidence that the codes used for translation are based on the salient features of the stimulus and response sets. Responding is fastest for situations in which the salient features for the respective sets correspond. In this chapter, phenomena previously attributed to motor-programming processes or to direct perceptual-motor interactions are shown to be due to translational coding. These coding effects are important cognitive factors in human performance.

1. Introduction

A variety of frameworks exist for the investigation of motor performance. Within the various frameworks, issues regarding the motor system often are examined by means of choice-reaction tasks, with the goal being to clarify the nature of the motor system (e.g., Klapp, 1977; Rosenbaum, 1983). The typical procedure is to require subjects to plan and execute movements following selection of a response from among two or more alternatives. The time to respond is interpreted as an indication of the characteristics of the motor system. However, because these tasks require choices between stimuli that vary along at least one dimension, the reaction times also can be affected by perceptual and cognitive factors.

Although the effects of these factors are widely acknowledged by researchers interested in motor performance (e.g., Klapp, 1981; Marteniuk & MacKenzie, 1981), the pervasive influence of perception and cognition in choice-reaction tasks is not fully appreciated. Of most concern is the cognitive factor of stimulus-response coding, because of its close relationship with motor programming. Such coding

J. Requin and G. E. Stelmach (eds.), Tutorials in Motor Neuroscience, 17–26.

occurs in what customarily is called the translation stage of information processing, in which a response to a stimulus is selected. Translation processes play an important role in any choice-reaction task for which the stimulus-response relations are not highly overlearned. Thus, stimulus-response coding necessarily will have a major influence on performance in most choice-reaction tasks.

2. Salient-Features Coding

Our work on spatial-precuing and symbolic-cuing tasks has clarified the nature of coding in the translation stage. Earlier views of translation in spatial-choice tasks (e.g., Teichner & Krebs, 1974) assumed that it was required when stimulus-response assignments lacked a direct spatial correspondence (i.e., incompatible assignments) but not when a direct correspondence existed (i.e., compatible assignments). This assumption was made by Miller (1982) when interpreting results from a four-choice spatial-precuing task in which pairs of responses from the index and middle fingers of each hand were cued. The task showed a pattern of differential precuing benefits in which pairs of fingers from the same hand benefited most from being precued. Because the procedure used a direct assignment of stimulus locations to response locations, Miller interpreted the advantage for the same-hand precues as a response-preparation effect that reflects a characteristic of the motor system.

Subsequently, we have reported several experiments showing that the pattern of differential precuing benefits is not a response-preparation effect (Proctor & Reeve, 1986, 1988; Reeve & Proctor, 1984, 1985). Rather, our research has indicated that the pattern arises from spatial stimulus and response codes used in the translation stage. Most important, by manipulating hand placement, we showed the pattern of precuing benefits to depend on the pairs of locations that are cued and not on the specific fingers assigned to the locations. This importance of spatial relations is consistent with the results obtained in studies of two-choice, spatial compatibility effects (Heister, Schroeder-Heister, & Ehrenstein, 1990; Umiltà & Nicoletti, 1990).

Similar coding effects were apparent in four-choice reaction tasks that used symbolic stimuli assigned to the keypress responses (Proctor & Reeve, 1985; Proctor, Reeve, Weeks, Dornier, & Van Zandt, in press). On any given trial, the assigned response was to be made to the stimulus, which was presented at a central location. The stimuli varied along two dimensions, one of which was more salient than the other. In several experiments, the stimulus set was composed of letters from two identities (O and Z) of two sizes (large and small). For this set, a preliminary study showed that two-choice discriminations between letter identities were faster than those between sizes, indicating that letter identity was more salient than size. In the primary study, a compatibility effect was obtained for different assignments of the stimuli to the four responses. Responding was faster when letter identity corresponded with the left-right feature of the response set (e.g., a left-to-right assignment of O, o, z, Z to the keys) than when it did not (e.g., a left-to-right assignment of O, z, o, Z). Additional evidence indicates that this symbolic-compatibility effect occurs because the salient letter-identity feature of the stimulus set corresponds with the salient left-right spatial feature of the response set in the former case but not in the latter.

Based on these and other findings with the spatial-precuing and symbolic-cuing tasks, we have proposed a **principle of salient-features coding** (Proctor & Reeve, 1985; Reeve & Proctor, 1990). This principle is that stimulus and response sets are coded in terms of the salient features of each, with responding being faster when the salient features of the two sets correspond than when they do not. Specific predictions derived from the salient-features coding principle have been confirmed in several studies (Cauraugh & Horrell, 1989; Reeve & Proctor, 1990).

Although most of the previous investigations have focused on the spatial-precuing and symbolic-cuing tasks, salient-features coding is intended as a general principle of stimulus-response translation. To test this notion, we have extended our investigations to other frameworks in which motor performance has been studied. Our intent has been to show the role of salient-features coding in phenomena that previously have been attributed to motor-programming processes or direct perceptual-motor interactions.

3. Reinterpretation of Motor-Programming Effects

The most frequent use of choice-reaction times is in the motor-programming framework. Within this framework, aimed movements to specific target locations often are investigated. Patterns of reaction time that vary as a function of the movement characteristics are interpreted as reflecting processes that operate during the motor-programming stage. However, as with discrete keypress responses, the time to initiate aimed movements can be influenced by stimulus-response coding. Two lines of research are described that illustrate the misinterpretation of cognitive effects as motor effects.

3.1. PRECUING LIMB MOVEMENTS

Serious concern with such misinterpretation developed in the study of limb movements. Rosenbaum (1980) obtained movement-precuing effects that he interpreted in terms of motor programming. However, Goodman and Kelso (1980) showed that the pattern of movement-precuing effects obtained by Rosenbaum was due to his use of symbolic precues. This pattern was not obtained by Goodman and Kelso when the precue information was specified by directly compatible location stimuli.

More recently, Larish (1986) conducted research investigating the role of translation processes in the movement-precuing task with spatial-location stimuli. He demonstrated that translation is evident for stimuli and responses that lack a direct spatial correspondence. This finding led Larish and Frekany (1985) to conclude that "if stimuli and responses do not maintain a direct spatial relationship, then a cognitive recoding process (S-R translation) is introduced into the processing sequence" (p. 169). This conclusion implies that translation is not a factor when the assignment of stimuli to responses is spatially compatible. Consequently, Larish and Frekany interpreted reaction-time differences between precuing conditions in spatially compatible situations as indicating that programming of limb-movement direction is required before that of the arm with which the movement is to be made.

Although Larish and Frekany (1985) based their conclusions about motor programming on compatibly assigned stimulus and response locations, the reaction-time differences still could be due to stimulus-response coding. This possibility arises from the fact that the spatial arrangements for the precues differed across conditions. Reaction times for programming arm were derived from precues that had a left-right spatial arrangement, whereas those for programming direction had an above-below arrangement.

To evaluate this possibility, Dornier and Reeve (1990) investigated situations in which the precues for all limb-movement parameters were derived from left-right spatially arranged stimuli (see Figure 1). No differences between programming arm ($\underline{M}$ = 345 msec) and direction ($\underline{M}$ = 338 msec) were apparent. This finding indicates that coding based on a left-right distinction allows for more efficient translation than does coding based on an above-below distinction, which is consistent with results obtained by Nicoletti and Umiltà (1984) for two-choice keypress tasks. When the confound of stimulus-response coding is eliminated, there is no evidence for a difference in the efficiency with which movement parameters can be programmed, a finding that is in agreement with that of Goodman and Kelso (1980).

3.2. ORTHOGONAL COMPATIBILITY EFFECTS

A second example in which the role of cognitive coding was underestimated involves stimulus-response compatibility effects for orthogonal spatial dimensions. Bauer and Miller (1982) had subjects perform a two-choice task for which the stimuli were oriented above and below a fixation point. The responses involved left or right movements of an index finger from a home key to a target location. In separate blocks of trials, the movements were made with either the

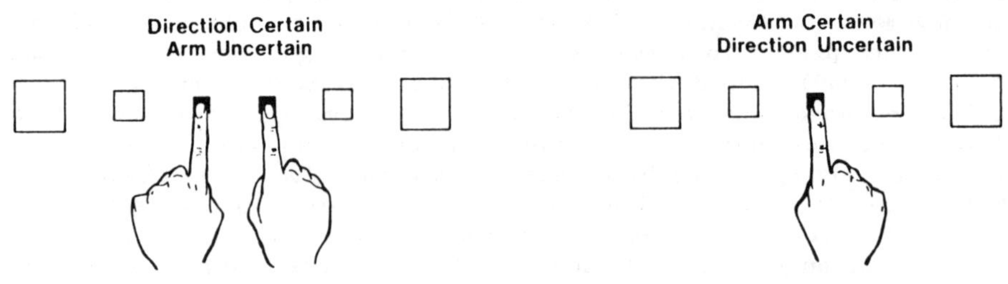

Figure 1. Response panel arrangement used by Dornier and Reeve (1990; Experiment 2).

left or right finger. The assignment of stimulus locations to response locations also was varied across blocks of trials (see Figure 2). Bauer and Miller obtained a compatibility effect for which the above-right/below-left assignment was faster than the reverse assignment. Because both assignments involved similar indirect relations between stimuli and responses, Bauer and Miller concluded that above-right/below-left advantage was due to characteristics of the motor system.

Weeks and Proctor (in press) have obtained evidence that this compatibility effect for orthogonal dimensions also has a cognitive basis. Numerous studies indicate that the polar referents "above"-"below" and "left"-"right" are asymmetric (Chase & Clark, 1971; Olson & Laxar, 1973). Thus, the findings of Bauer and Miller could be cognitive in nature. Because "above" is more salient than "below" and "right" more salient than "left," the salient features correspond for the above-right/below-left assignment, as do the nonsalient features. The consistent mapping of the features with this assignment should speed translation processes relative to the alternative assignment. In contrast to Bauer and Miller's motor-system account, this salient-features account predicts that the orthogonal compatibility effect should not be restricted to situations that involve the programming of aimed finger movements.

Weeks and Proctor (in press) confirmed this implication of the salient-features coding account in a series of experiments that used aimed features movements, keypress responses, and vocal responses with spatial-location and symbolic-arrow stimuli. For all of the stimulus and response sets examined, the above-right/below-left assignment was preferred. For example, "left"-"right" keypress responses to stimuli located above or below fixation were 42 msec faster with the above-right/below-left assignment than with the reverse assignment. Thus, the attribution of orthogonal compatibility effects to the motor system was premature and based on an underestimate of the role of stimulus-response coding.

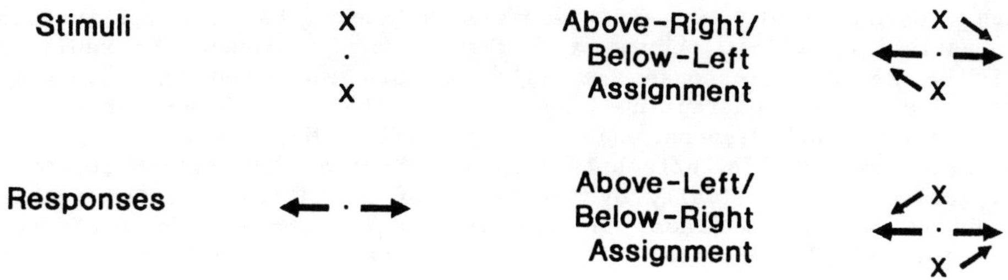

Figure 2. Stimuli, responses, and assignments used by Bauer and Miller (1982, Experiment 3).

4. Reinterpretation of Perceptual-Motor Effects

The perception-action framework provides an alternative to motor-programming investigations of human performance (Meijer & Roth, 1988). An implication of this framework is that a direct link exists between perception and action. Because of the emphasis on this direct link, the possibility exists in this framework for phenomena based on stimulus-response coding to be misinterpreted as perceptual-motor interactions.

4.1 MOVEMENT DESTINATION COMPATIBILITY EFFECTS

Michaels (1988) reported a compatibility effect for destination of apparent movement. The stimuli were two squares presented on a display screen, one to the left of fixation and one to the right (see Figure 3). The responses were forward joystick movements made with the left and right hands. After a delay, one of the squares appeared to move. The image either expanded symmetrically, to mimic movement directly toward the response location on the same side, or both expanded and shifted toward the center of the screen, to mimic movement toward the response location on the opposite side. When subjects were instructed to respond on the basis of the destination of the apparent movement rather than the origin, responses at the destination location were fastest.

Michaels (1988) interpreted this destination compatibility effect as inconsistent with stimulus-response coding accounts and consistent with an explanation in terms of environmental affordances. She proposed that the changes in the display information corresponded to those that afford a catching action at the destination location. According to her catching-affordance explanation, a catching response can be initiated faster when the response is consistent with the action afforded by the perceptual changes of the stimulus features.

In a series of experiments, Proctor, Van Zandt, Lu, and Weeks (1990) have evaluated implications of Michaels' (1988) catching-affordance account relative to those of a cognitive coding account. The catching-affordance account implies that the relation between the specific expansion properties of the stimulus and a "catching" action are crucial. Yet, our experiments have shown that the destination compatibility effect occurs when the catching movement is replaced with a keypress response, as well as when the expanding image is replaced with a contracting image. Thus, the specific properties of the stimuli and responses are not crucial. More interestingly, a similar compatibility effect is obtained when the destination location is indicated symbolically by the point of a stationary arrow rather than by movement. Thus, in the destination compatibility effect, movement simply provides the salient feature for coding relative location.

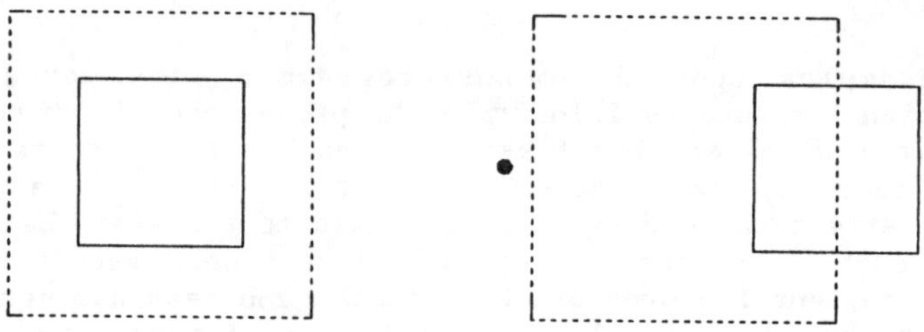

Figure 3. Display arrangement used by examined by Michaels (1988). The solid squares indicate the initial stimuli and the dashed squares the final positions for an ipsilaterally moving square (left) and a contralaterally moving square (right).

4.2. SPEECH COMPATIBILITY EFFECT

Another example comes from research on perceptual-motor interactions in speech by Gordon and Meyer (1984). They examined situations in which subjects responded with a consonant-vowel syllable to a spoken consonant-vowel stimulus. The stimuli and responses came from the set PUH, BUH, DUH, and TUH. These stimuli differ along two dimensions, voicing and place of articulation, with voicing being the most salient feature. The finding of interest is that responding was faster when the response shared a voicing feature with the stimulus than when it did not. For example, the unvoiced response "TUH" could be made to the unvoiced stimulus "PUH" faster than could the voiced response "DUH." To explain these results, Gordon (1990) proposed an interactive activation model in which features of speech stimuli automatically activate features of speech responses.

The two-dimensional set of speech stimuli used by Gordon and Meyer (1984) shares an obvious structure with the two-dimensional set of visual stimuli used previously by us in the symbolic-cuing task (Proctor & Reeve, 1985). This suggests that the speech compatibility effect may not depend on a special link between speech perception and production. Proctor, Weeks, and Kelly (1990) conducted an experiment in which two assignments of the speech stimuli to keypress responses were used, as in the symbolic-cuing task described previously. The assignment of the stimuli to the responses was manipulated so that the salient feature of the stimulus set (voicing) corresponded with the salient left-right spatial feature of the response set for one assignment, but not for the other. Reaction times were faster when the salient features corresponded ($\underline{M}$ = 785 msec) than they did not ($\underline{M}$ = 818 msec). In other experiments, spatial-location stimuli and two-dimensional visual stimuli were assigned to the speech response set. In both cases, responses again were faster when the salient features of the stimulus and response sets corresponded. Thus, these compatibility effects involving speech stimuli and/or responses apparently reflect salient-features coding in the translation stage, rather than direct activation of articulatory programs.

5. Summary

These examples from the motor-programming and perceptual-motor interaction literatures illustrate the prevalence of stimulus-response coding in choice-reaction tasks. In each case, effects that on the surface seem to have a basis at least in part in the motor system instead reflect the codings that are used to translate between stimuli and responses. Across a range of situations, responding is faster when the salient features of the stimulus and response sets correspond than when they do not. This finding is consistent with the view that salient-features coding is a fundamental principle of stimulus-response translation. Rather than regarding these coding effects as experimental nuisances to be controlled or eliminated, they should be regarded as important cognitive factors in human performance that are worthy of systematic investigation.

6. References

Bauer, D. W., & Miller, J. (1982). Stimulus-response compatibility and the motor system. Quarterly Journal of Experimental Psychology, 34A, 367-380.

Cauraugh, J. H., & Horrell, J. F. (1989). Advance preparation of discrete motor responses: Nonmotoric evidence. Acta Psychologica, 72, 117-138.

Chase, W. G., & Clark, H. H. (1971). Semantics in the perception of verticality. British Journal of Psychology, 63, 311-326.

Dornier, L. A., & Reeve, T. G. (1990). Evaluation of commpatibility effects in the precuing of arm and direction parameters. Research Quarterly for Exercise and Sport, 61, 37-49.

Goodman, D., & Kelso, J. A. S. (1980). Are movements prepared in parts? Not under compatible (nautralized) conditions. Journal of Experimental Psychology: General, 109, 475-495.

Gordon, P. C. (1990). Perceptual-motor processing in speech. In R. W. Proctor & T. G. Reeve (Eds.), Stimulus-response compatibility: An integrated perspective (pp. 343-362). Amsterdam: North-Holland.

Gordon, P. C. & Meyer, D. E. (1984). Perceptual-motor processing of phonetic features in speech. Journal of Experimental Psychology: Human Perception and Performance, 10, 153-178.

Heister, G., Schroeder-Heister, P., & Ehrenstein, W. H. (1990). Spatial coding and spatio-anatomical mapping: Evidence for a hierarchical model of spatial stimulus-response compatibility. In R. W. Proctor & T. G. Reeve (Eds.), Stimulus-response compatibility: An integrated perspective (pp. 117-143). Amsterdam: North-Holland.

Klapp, S. T. (1977). Reaction time analysis of programmed control. Exercise and Sport Sciences Review, 5, 231-253.

Klapp, S. T. (1981). Motor programming is not the only process which can influence RT: Some thoughts on the Marteniuk and MacKenzie analysis. Journal of Motor Behavior, 13, 320-328.

Larish, D. D. (1986). Influence of stimulus-response translations on response programming: Examining the relationship of arm, direction, and extent of movement. Acta Psychologica, 61, 53-70.

Larish, D. D., & Frekany, G. A. (1985). Planning and preparing expected and unexpected movements: Reexamining the relationships of arm, direction, and extent of movement. Journal of Motor Behavior, 17, 168-189.

Marteniuk, R. G., & MacKenzie, C. L. (1981). Methods in the study of motor programming: Is it just a matter of simple vs. choice reaction time? A comment on Klapp et al. (1979). Journal of Motor Behavior, 13, 313-319.

Meijer, O. G., & Roth, K. (1988). Complex movement behaviour: 'The' motor-action controversy. Amsterdam: North-Holland.

Michaels, C. F. (1988). S-R compatibility between response position and destination of apparent motion: Evidence of the detection of affordances. Journal of Experimental Psychology: Human Perception and Performance, 14, 231-240.

Miller, J. (1982). Discrete versus continuous stage models of human information processing: In search of partial output. Journal of Experimental Psychology: Human Perception and Performance, 8, 273-296.

Nicoletti, R., & Umiltà, C. (1984). Right-left prevalence in spatial compatibility. Perception & Psychophysics, 35, 333-343.

Olson, G. M., & Laxar, K. (1973). Asymmetries in processing the terms 'right' and 'left'. Journal of Experimental Psychology, 100, 284-290.

Proctor, R. W. & Reeve, T. G. (1985). Compatibility effects in the assignment of symbolic stimuli to discrete finger resonses. Journal of Experimental Psychology: Human Perception and Performance, 11, 623-639.

Proctor, R. W., & Reeve, T. G. (1986). Salient-feature coding operations in spatial precuing tasks. Journal of Experimental Psychology: Human Perception and Performance, 12, 277-285.

Proctor, R. W., & Reeve, T. G. (1988). The acquisition of task-specific productions and modification of declarative representations in spatial-precuing tasks. Journal of Experimental Psychology: General, 117, 182-196.

Proctor, R. W., Reeve, T. G., Weeks, D. J., Dornier, L. A., & Van Zandt, T. (in press). Acquisition, retention, and transfer of response-selection skill in choice-reaction tasks. Journal of Experimental Psychology: Learning, Memory, and Cognition.

Proctor, R. W., Van Zandt, T., Lu, C.-H., & Weeks, D. J. (1990, November). Stimulus-response compatibility for destination of apparent motion: Catching affordances or directional coding? Paper presented at the 31st annual meeting of the Psychonomic Society, New Orleans, LA.

Proctor, R. W., Weeks, D. J., & Kelly, P. (1990). Performance with consonant-vowel and spatial-location stimulus and response sets: A salient-features account. Manuscript in preparation.

Reeve, T. G., & Proctor, R. W. (1984). On the advance preparation of discrete finger responses. Journal of Experimental Psychology: Human Perception and Performance, 10, 541-553.

Reeve, T. G., & Proctor, R. W. (1985). Nonmotoric translation processes in the preparation of discrete finger responses: A rebuttal of Miller's (1985) analysis. Journal of Experimental Psychology: Human Perception and Performance, 11, 234-240.

Reeve, T. G., & Proctor, R. W. (1990). The salient-features coding princple for spatial- and symbolic-comaptibility effects. In R. W. Proctor & T. G. Reeve (Eds.), Stimulus-response compatibility: An integrated perspective (pp. 163-180). Amsterdam: North-Holland.

Rosenbaum, D. A. (1980). Human movement initiation: Specification of arm, direction, and extent. Journal of Experimental Psychology: General, 109, 444-474.

Rosenbaum, D. A. (1983). The movement precuing technique: Assumptions, applications, and extensions. In R. A. Magill (Ed.), Memory and control of action (pp. 231-274). Amsterdam: North-Holland.

Teichner, W. H., & Krebs, M. J. (1974). Laws of visual choice reaction time. Psychological Review, 81, 75-98.

Umiltà, C., Nicoletti, R. (1990). Spatial stimulus-response compatibility. In R. W Proctor & T. G. Reeve (Eds.), Stimulus-response compatibility: An integrated perspective (pp. 89-116). Amsterdam: North-Holland.

Weeks, D. J., & Proctor, R. W. (in press). Salient-features coding in the translation between orthogonal stimulus and response dimensions. Journal of Experimental Psychology: General.

STIMULUS-RESPONSE COMPATIBILITY AND PSYCHOPHYSIOLOGY

MICHAEL G. H. COLES, WILLIAM J. GEHRING,
GABRIELE GRATTON, & EMANUEL DONCHIN
Cognitive Psychophysiology Laboratory
University of Illinois
Champaign, Illinois 61820, USA

Psychophysiological measures can be used to illuminate the role of motor system activity in chronometric paradigms and, in particular, in studies of stimulus-response compatibility. These measures provide more or less direct access to those covert response processes that are proposed in various theories of stimulus-response compatibility (e.g., Kornblum, Hasbroucq, & Osman, 1990).

Our research has employed two-choice reaction time tasks, in which subjects must respond with either the left- or right-hand (or not at all) to stimuli that contain attributes that may be associated with both correct and incorrect response. For example, subjects have to respond to the center letter in a stimulus array and this letter is flanked by "noise" letters that are associated with either the same (correct) response or the other (incorrect) response (see Eriksen & Eriksen, 1974). Under conditions of stimulus-response incompatibility, when responses associated with target and noise letters are different, both responses may be activated even though the activation of the incorrect response does not usually reach the level of an overt behavioral response. Thus, Coles, Gratton, Bashore, Eriksen, and Donchin (1985) showed that incorrect *electromyographic* activity was observed more frequently for incompatible stimuli. Furthermore, when the incorrect response was activated, correct response activation appeared to be disrupted. Using a more central measure of response activation, *the lateralized readiness potential*, Gratton, Coles, Sirevaag, Eriksen, and Donchin (1988) showed that incorrect responses tended to be activated following the presentation of incompatible stimuli -- even when there was no evidence of peripheral response activation in the electromyogram.

In both these cases, incompatibility occurred because of a conflict between the response called for by the target letter and that associated with the noise letters. There was no "dimensional overlap" (Kornblum et al., 1990) between stimulus and response sets. However, because stimulus-response mapping was consistent, it seems likely that, after practice, particular responses would be "automatically" activated following presentation of the associated stimuli. Thus, in the Eriksen paradigm, the effects of incompatibility would be analogous to those observed in a situation in which there is dimensional overlap. Indeed, in a recent experiment by Osman and his colleagues (Osman, Bashore, Coles, Donchin & Meyer, 1990), in which there was dimensional overlap, the lateralized readiness potential suggested that the incorrect response was activated when the mapping between stimulus and response was incongruent. In this experiment, incongruent mapping occurred when the side of

J. Requin and G. E. Stelmach (eds.), Tutorials in Motor Neuroscience, 27–28.

stimulus presentation (to the left or right of fixation) and the side of the response hand (left or right) were different.

These data are consistent with the idea that stimuli can automatically activate particular responses (Kornblum et al., 1990). Such automatic activation occurs when there is dimensional overlap between stimulus and response sets, or following practice. Of course, if this activation process results in the activation of the incorrect response, the information processing system must make various adjustments if the correct response is to be activated. According to Kornblum et al., these adjustments involve a verification process, by which a representation of the activated response is checked against a representation of the response that should be activated and, if necessary, the incorrect response is aborted.

We (Gehring, Coles, Meyer, & Donchin, 1990) have recently identified a brain potential component that appears to be related to the response verification and abortion processes. This brain potential consists of a negative-going wave that is evident at about the time subjects initiate incorrect responses. Preliminary evidence suggests that it is larger when subjects operate under accuracy rather than speed emphasis conditions and that its amplitude is related to the probability that the incorrect response will be corrected. While these data strongly suggest that the potential is elicited when the verification-abort process is implemented, they do not, at present, permit a more detailed specification of its functional significance.

Coles, M. G. H., Gratton, G., Bashore, T. R., Eriksen, C. W., & Donchin, E. (1985). A psychophysiological investigation of the continuous flow model of human information processing. *Journal of Experimental Psychology: Human Perception and Performance*, *11*, 529-553.

Eriksen, B. A., & Eriksen, C. W. (1974). Effects of noise letters upon the identification of target letters in visual search. *Perception and Psychophysics*, *16*, 143-149.

Gehring, W. J., Coles, M. G. H., Meyer, D. E., & Donchin, E. (1990). The error-related negativity: An event-related brain potential accompanying errors. *Psychophysiology*, *27*, S34. (Abstract).

Gratton, G., Coles, M. G. H., Sirevaag, E., Eriksen, C. W., & Donchin, E. (1988). Pre- and post-stimulus activation of response channels: A psychophysiological analysis. *Journal of Experimental Psychology: Human Perception and Performance*, *14*, 331-344.

Kornblum, S., Hasbroucq, T., & Osman, A. (1990). Dimensional overlap: Cognitive basis for stimulus-response compatibility - A model and a taxonomy. *Psychological Review*, *97*, 253-270.

Osman, A., Bashore, T. R., Coles, M. G. H., Donchin, E., & Meyer, D. E. (1990). *On the transmission of partial information: Inferences from movement-related brain potentials*. Manuscript submitted for publication.

POSSIBLE NEURAL CORRELATES FOR THE MECHANISM OF STIMULUS-RESPONSE ASSOCIATION IN THE MONKEY.

Seal, J., Hasbroucq, T., Mouret, I., Akamatsu, M.[1] and Kornblum, S.[2], *Cognitive Neuroscience Unit, C.N.R.S., chemin J. Aiguier, 13402 Marseille Cedex 9, France,* [1] *Industrial Products Research Institute, 1-1-4, Higashi, Tsukuba, Ibaraki 305, Japan and* [2] *Mental Health Research Institute, University of Michigan, Ann Arbor, MI 48109-0720, U.S.A.*

ABSTRACT. We have attempted to describe the neuronal mechanisms of sensory to motor linking which can be considered to be a central component of any stimulus-response process. This mechanism was studied both at the level of a particular cortical area and within a sequence of cortical areas which can be considered to form the anatomical basis of a stimulus-response process. Single neuron recordings were made in monkeys performing reaction time tasks designed to identify sensory and motor properties. Descriptions of the patterns of change in neuronal activity and temporal analysis of these changes served to characterize three neuronal populations, one of which had the peculiarity of showing sensory and motor properties. The results suggested the presence of a continuum of function both within cortical areas and along a sequence of such areas.

1. Introduction

In the field of cognitive psychology, information processing has been broken down into a logical series of independent and functionally specialized processing stages (Sternberg 1969, Miller 1988). Likewise, the brain, and in particular the cerebral cortex, can be divided into an interconnected network of functionally distinct structures, with the correlation structure/function initially being based on data obtained in anatomical and lesion studies. Although these two points of view can be contested, together they provide a simple working hypothesis to study the neuronal mechanisms by which the sensory influx evoked by a cue for movement gives rise to the motor activity necessary for the execution of that movement.

A minimal model of information processing in reaction time tasks defines three stages; stimulus perception, stimulus-response translation and response elaboration (e.g. Proctor and Reeve 1986, Hasbroucq et al. 1989). The first and last stages have their correspondence in those cortical structures for which there exists a well-documented description of their sensory and motor functions. However, the neural correlates of the intermediate stage of stimulus-response translation remain poorly documented. This is perhaps because brain function has classically been described either in terms of sensory and motor mechanisms or in a wide range of purely descriptive and ill-defined terms. In our experiments, we have attempted to study the cortical mechanism of this hypothetical stage in which sensory input is linked to motor output. Our methods included not only qualitative descriptions of the patterns of change in neuronal activity but also

J. Requin and G. E. Stelmach (eds.), Tutorials in Motor Neuroscience, 29–39.

quantitative methods in an attempt to provide a comparable measure of sensory to motor linking in different cortical areas.

The site for this linkage is probably outside the primary sensory and motor areas of the cortex because the characteristics of neuronal discharge in these areas have been shown to be principally related to the physical features of the stimulus or the motor act (Mountcastle 1974, Evarts 1981). Considering its anatomical connections and the behavioural deficits following lesions (for review, see Mountcastle et al. 1975 and Hyvärinen 1982), the posterior parietal cortex (areas 5 and 7) is one of the possible sites for linking sensory and motor activity (see comments in Darian-Smith et al. 1979 and Wiesendanger 1981).

In experiments with behaving monkeys, several authors have described neurons in area 5 which modified their activity after a conditioned stimulus but before the onset of the conditioned movement (Mountcastle et al. 1975, MacKay et al. 1978, Bioulac and Lamarre 1979, Seal et al. 1982, Kalaska et al. 1983, Chapman et al. 1984). However, the interpretation of these results is difficult in that they may reflect sensory mechanisms, motor mechanisms or some higher brain function. The main debate has been about whether the early changes in activity of the posterior parietal cortex could best be described in terms of a sensory mechanism (Robinson et al. 1978) or as a form of motor command (Mountcastle et al. 1975) .

We have previously described stimulus- and movement-related changes in neuronal activity in area 5 and concluded that they may represent part of a mechanism by which the influx of neuronal activity evoked by a cue for movement is transformed into a form of motor activity (Seal and Commenges 1985). The aim of the two series of experiments presented in this paper was to describe, in greater detail, the characteristics of such a transformation mechanism. In a first series of experiments, we recorded single neuron activity in area 5 during the performance of an experimental protocol which allowed us to discriminate between sensory and motor mechanisms. We shall also briefly mention preliminary results of a second, on-going series of experiments in which we are attempting to follow the influx of neuronal activity evoked by the sensory cue along a sequence of cortical structures. The sequence under study may provide an anatomical basis for the mechanism linking sensory input to motor output.

2. Methods

The experimental protocol consisted of three different conditions: (i) stimulus followed by movement, (ii) stimulus alone and (iii) movement alone. Both sensory and motor mechanisms could be expressed in condition (i) but only sensory mechanisms in condition (ii), the movement being absent. In contrast, only motor mechanisms could be expressed in condition (iii), when the stimulus was absent. In addition, quantitative methods of data analysis were used to study the temporal organization of changes in neuronal activity and to indicate the degree to which a change in activity was related to the sensory and motor events of the experimental protocol. With this measure, the different neuronal populations could be placed on a hypothetical sensory to motor gradient from which it was possible to infer the role of each neuronal population in sensory to motor linking. The impetus to develop such a measure came, in part, from the comments of Mountcastle (1979) who proposed that there is nothing intrinsically motor about the motor cortex, nor sensory about the sensory cortex.

Single neuron activity was recorded in area 5 of a Rhesus monkey (Macaca mulatta) trained to perform in an experimental procedure which included the three conditions described above. The conditioned stimulus was an auditory cue of 400 or 1000 Hz (250 msec duration) and the limb movement was, respectively, extension or flexion of the arm about the elbow. A reward was given for movements in excess of 35°. Recordings in the different conditions were obtained in the following way. First, neuronal activity was recorded whilst the animal performed the condition (i). A decision was made to continue recording the activity of the neuron if examination of the peri-stimulus and peri-movement histograms of neuronal activity showed that the neuron modified its activity after the stimulus but before the onset of movement i.e. that it could be involved in the processes occurring between sensory input and motor output. Neurons which presented a modification of activity at or after the onset of movement were not studied further as it has been shown that these area 5 neurons are influenced by somaesthetic feedback (Seal et al. 1982). Condition (ii) was obtained by withholding the reward for correct trials and once the animal stopped making the limb movement, then the reward was re-established on a partial reinforcement schedule of one reward for three correct responses. The third condition of movement - no stimulus was obtained by increasing the inter-trial interval. Again, a partial reinforcement schedule was used. The animal performed well in this apparently contradictory experimental design and once the animal had experienced several complete experimental sessions, the change-over from one condition to another took only five to ten trials. It was as if the signal for the animal to switch to stimulus - no movement trials was the absence of the reward for a series of stimulus - movement trials. Given that the animal was well motivated to obtain the reward, absence of the stimulus and a prolonged inter-trial interval appeared to be taken as the trigger for initiating movement-only trials.

Standard procedures were used for preparing the animal for single neuron recordings and these have been described in detail elsewhere (Seal et al. 1982). The experimental events were controlled by a computer which was also used to store single neuron activity and arm position for each trial. The neuronal discharge data were visualized as peri-stimulus or peri-movement histograms and raster displays according to the experimental condition. The recorded neurons were classified into one of three groups - sensory, sensorimotor and motor - on the basis of the presence or absence of a change in neuronal activity in conditions (ii) and (iii). The detection of a change in neuronal activity, the estimation of the timing and duration of any change as well as a measure of where, on a sensory to motor gradient, the neuron could be placed were obtained from a trial by trial analysis of the data for each neuron. The methods used will be described only briefly here, further details being available elsewhere (Commenges and Seal 1985, Commenges et al. 1986a and b). The data for each neuron were analysed trial by trial to determine two time periods: onset of stimulus to change in neuronal activity (X) and change in neuronal activity to onset of movement (Y). For each neuron which had a significant change in neuronal activity, the variances of the time periods X and Y were calculated and the logarithm (variance Y / variance X) was used to place each neuron on a sensory to motor gradient.

3. Results

Taking into account the difficulties in defining the rostral and caudal limits of area 5 (Mountcastle et al. 1975, Jones et al. 1978), data are presented only for those neurons which were located within the middle part of the anterior bank of the intraparietal sulcus. Of the 254 neurons located in area 5, as defined in this way, 64 neurons presented changes in activity prior to movement in condition (i) and were thus selected for further study. These neurons were divided into three groups on the basis of the presence or absence of a change in neuronal activity in conditions (ii) and (iii). Sensory neurons were characterized by a change in neuronal activity in conditions (i) and (ii) and the absence of a change in condition (iii). This group represented 10% (n = 6) of the total population studied. Typically, the change occurred with a short latency (mean 23 msec) after the stimulus and consisted in a brief increase of discharge frequency which was difficult to detect in the trial by trial analysis. Motor neurons presented changes in activity in conditions (i) and (iii) but not in condition (ii). The majority of all neurons recorded were classified into this group (56%, n=36). The third group of neurons was the most interesting in that these neurons presented changes of activity in all three conditions. This pattern of change in activity was observed in 34% (n = 22) of the neurons studied. These neurons modified their discharge frequency after the stimulus alone (condition ii) and prior to the movement alone (condition iii) and appeared to have two components to the change of activity in the stimulus - movement condition, one related to the stimulus and another to the movement.

The temporal organization of the change in activity for the neurons of each group is shown in figure 1, with respect to the onset of the stimulus (top) and the onset of movement (bottom). In general, there was a sequential order of sensory neurons then sensorimotor neurons and finally the motor neurons. The overall means for the three groups, - 23 msec, 137 msec and 264 msec, respectively, - confirmed this order. With respect to the onset of movement, the overall means were 314 msec, 135 msec and 55 msec for the sensory, sensorimotor and motor groups of neurons, respectively. Again, the individual values for neurons in each group showed the general temporal organization of sensory, sensorimotor and motor neurons. A Student's t test showed that there was a significant difference ($p < 0.001$) between the time of change (with respect to either the stimulus or the movement) for the population of sensorimotor cells and the population of motor cells. From figure 1, it can be seen that the values of time of change in activity for the motor neurons appeared to be more closely grouped when aligned to movement onset (SEM ± 8) than when aligned to stimulus onset (SEM ± 29). The timing of the values for sensorimotor neurons, however, had a similar range with respect to the stimulus (SEM ± 17) and the movement (SEM ± 13). This suggests that the motor neurons were more motor than the sensorimotor neurons, and again implies a sequential organization for sensorimotor and motor neurons.

An indication as to the relative position of the three different groups of area 5 neurons within a hypothetical sensory to motor gradient can be obtained by studying the ratio of variance Y / variance X, as described above. The values of this ratio and its logarithm are given for sensorimotor and motor neurons in figure 2. For comparison, values for task-related neurons located in the motor cortex, area 4, (n = 8) and in an auditory area (around the caudal end of the Sylvian fissure, n = 12) are also indicated. The ratio of variance Y / variance X is

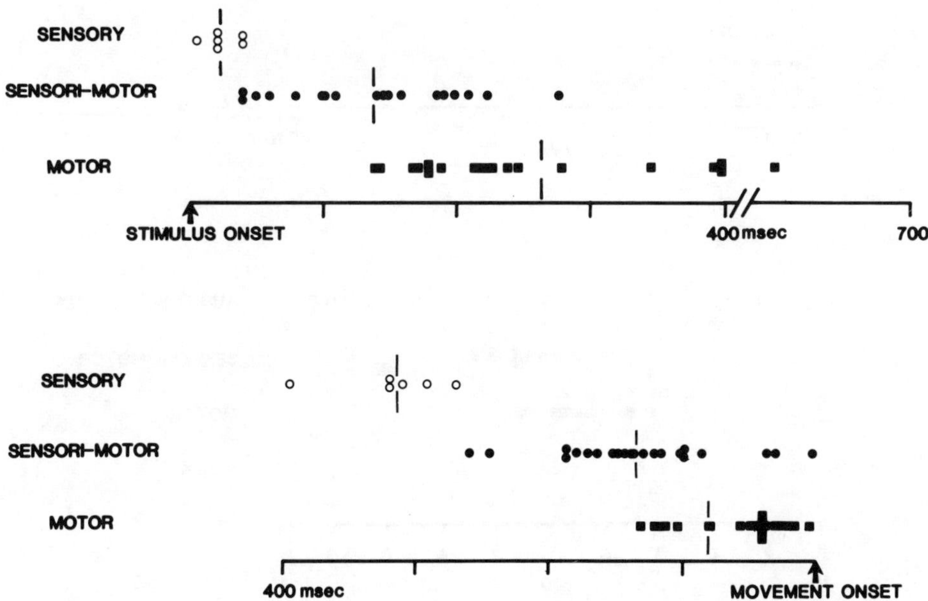

Figure 1. Timing of change in neuronal activity for the different neuronal populations in area 5. Mean values are given with respect to stimulus onset (top) and movement onset (bottom). The vertical bars indicate the overall mean of each group.

a geometric progression (figure 2, top) and so the relative position of the different neurons is best shown using a log scale (figure 2, bottom). It can be seen that the values of area 5 motor neurons are nearer to motor output, as represented by the values obtained for cells in the motor cortex, than the values of area 5 sensorimotor neurons. However, there is a certain degree of overlap between the two populations of area 5 cells. We did not have sufficient data to calculate variance Y / variance X for the sensory neurons because of the difficulty in detecting brief responses in the trial by trial analysis.

In a recent series of experiments, we have recorded single neuron activity in four cortical areas which are known to be interconnected and may form the anatomical basis of a stimulus-response process i.e. the primary somaesthetic cortex (S1), area 5 of the posterior parietal cortex, the premotor cortex of area 6 and the primary motor cortex (area 4). We have used a choice reaction time task in which the monkey received a brief tactile stimulus to either the fingers or the thumb and had to respond by a button press with the stimulated digit/s. Temporal analysis of the results obtained with single neuron recording showed the following sequential order for the time of change in activity of neurons in the four areas - S1 < area 5 < area 6 < area 4. The values of log (variance Y / variance X) for neurons in the different areas showed that there was a sensorimotor continuum such that cells which could be placed closest to sensory input were located in S1 followed by area 6, area 5 and then area 4, which was closest to motor output. Within each area, there appeared to be two distinct populations of cells without any overlap between the two, one closer to sensory input and the other closer to motor output. Furthermore, as observed in the experiments described above,

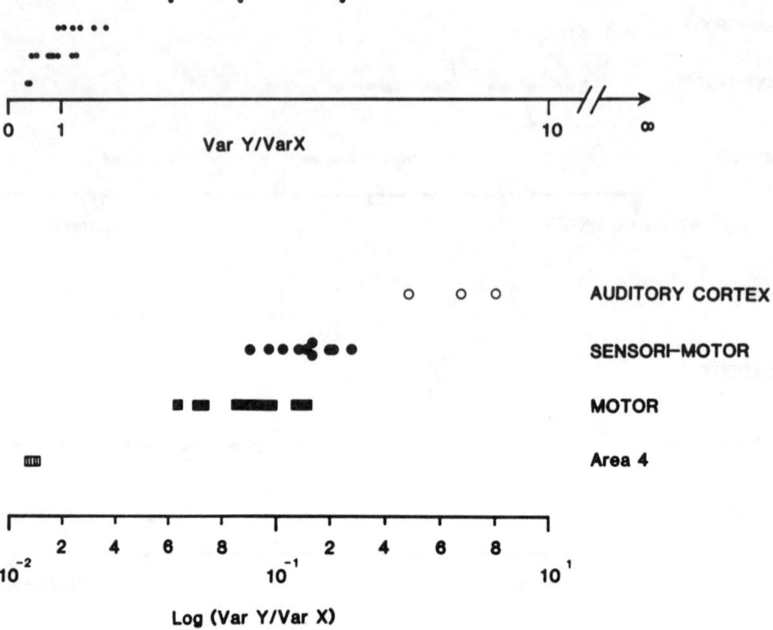

Figure 2. Representation of different neuronal populations on a hypothetical stimulus to response gradient. Data are shown as variance Y/ variance X (top) and the log of this value (bottom). The strength with which the different neuronal populations are influenced by sensory and motor events gradually changes from one population to another.

certain cells in area 6 presented two changes in activity, one related to the stimulus and a second related to and prior to movement onset. Within-burst analysis has shown that the change in activity of some area 5 cells occurs between the stimulus-related and the movement-related bursts in activity of area 6 cells.

The results presented in this paper suggest that, within area 5, there exists an ensemble of three neuronal populations which function together in the linking of behaviourally relevant sensory input with a form of motor output. In addition, this ensemble is itself inserted within a sequential organization of cortical areas consisting of S1, area 5, area 6 and area 4. From a comparison of the position of each neuron on a hypothetical sensory to motor gradient, it can be seen that there appears to be an intricate pattern for the passage of the sensory influx between S1, area 5 and area 6. The results of within-burst analysis suggest that the initial change in activity of area 6 cells following the S1 activation may be maintained at a high level by a rise in discharge frequency of certain area 5 cells.

Cells with a biphasic change in activity were observed in area 5 for the first series of experiments (simple reaction time task, arm movement) and in area 6 for the second, most recent series (choice reaction time task, digit presses). Comparison of histograms and within-burst analysis has shown that the first component of these responses is related to stimulus onset and the second component is related to, but precedes, the onset of movement. These populations of neurons may constitute the neural basis of a common cortical mechanism for the transformation of sensory input into motor output. For the results obtained in

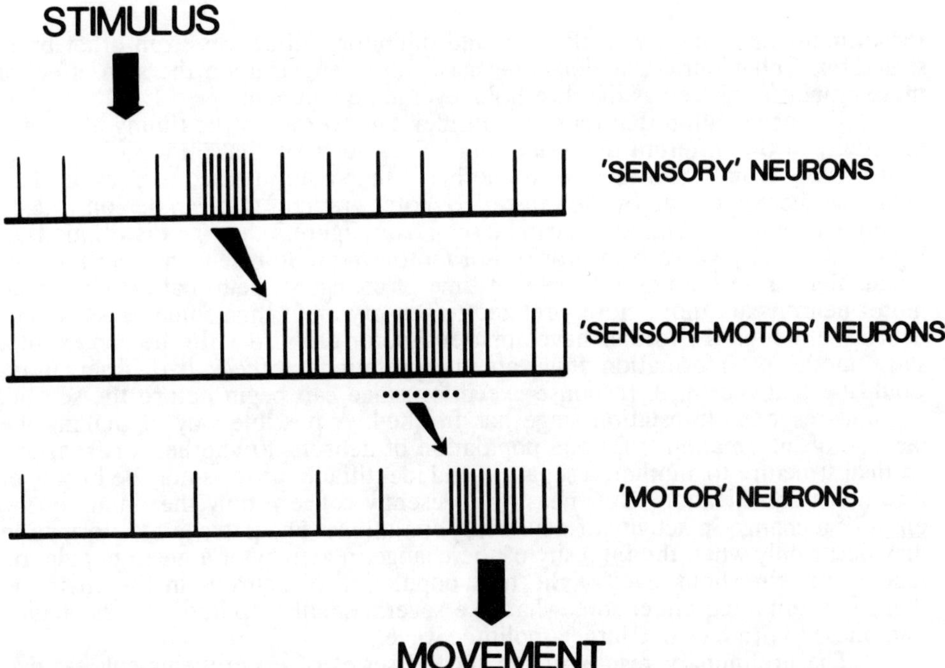

Figure 3. Model of sensory to motor transformation in area 5. Activation of the sensory neurons initiates the beginning of a change in activity (the sensory component) of the sensorimotor neurons. Under the effect of inhibitory and excitatory influences, the frequency of this initial discharge may increase until a threshold level (solid dots) is reached and the motor neurons are activated. The period of discharge at and beyond the threshold level represents the motor component of sensorimotor neurons.

area 5, it is possible to propose a simple model which accounts for the three populations of neurons we observed.

In figure 3, we show sequences of neuronal discharge which represent the activity of the three populations neurons described for area 5. The model assumes that the sensory neurons project to the sensorimotor neurons which, in turn, project to the motor neurons. We envisage that the initial sensory influx evoked at the level of the sensory receptors eventually arrives at area 5 and provokes a change in activity (say, excitation) of the sensory neurons. This excitation sets off a similar change in the activity of the sensorimotor neurons. We stipulate that before the activity in the sensorimotor neurons can influence the activity of the motor neurons, a certain threshold of activation must be attained. Once this level is reached, then the motor neurons are activated. Execution of the movement is conditional upon this activation of the motor neurons. Activation of motor neurons is, thus, governed by the level of activation attained by the sensorimotor neurons. The sensory component of the change in activity of the sensorimotor neurons may correspond to the initial activation due to the influence of the sensory neurons. The second, motor component may occur when the discharge frequency of sensorimotor neurons exceeds the threshold necessary for activation of the motor neurons which inevitably leads to movement. The linking of sensory input to motor output may be controlled by modulating the level of activity of sensorimotor neurons via facilitatory and inhibitory influences from other brain structures. Behavioural reaction time may, thus, depend upon the sum effect of these influences as well as the threshold level to be attained.

The observation that there is considerable overlap in the timing of change in activity of the different neuronal populations suggests that information passes continuously from one structure to another rather than in discrete packets. This idea is further supported by the sensory to motor gradient that can be constructed from the results we obtained. It can be seen from figure 2 that the distributions of log (variance X / variance Y) for sensorimotor and motor neurons overlap. This means that, from the point of view of time of change in neuronal activity, some motor neurons are more closely related to sensory input than some sensorimotor neurons although the former have apparent motor properties. In the context of a stage model of information flow, an interpretation of these two observations would be that the third, response execution stage can begin before the second, stimulus-response translation stage has finished. A possible way of making the passage of information from one population of neurons to another, or from one cortical structure to another, a separate and identifiable process may be by way of thresholds of activation. Our measures presently concern only the timing of the onset of a change in activity of neuronal populations. The passage of information may occur only when the intensity of the change in activity of a given population reaches the threshold level of the next population of neurons in the chain i.e. there is a summing effect somewhat like several small amplitude waves coming into phase to produce one large amplitude wave.

The preliminary results of the second series of experiments suggest that within the sequence of cortical areas that we have investigated, certain neurons of area 6 may be subject to a facilitatory influence from area 5. From the within-burst analysis, we have seen that certain area 6 cells have both sensory and motor components forming the change in neuronal activity and that the level of discharge frequency remains high during the time period between these two components. The activation of area 5 neurons can occur during this time interval and it is as if this activation serves to maintain the discharge frequency of the area 6 cells high. The sensory to motor gradient which was established with these results supports this hypothesis in that the majority of neurons in area 5 appeared to be less sensory than the neurons of area 6.

The results obtained from a temporal analysis of changes in neuronal activity within a cortical area (area 5) and within a sequence of cortical areas going from sensory input to motor output (S1, area 5, area 6 area 4) suggest that there is a continuum of function both within areas and within a sequence of areas. We postulate that this continuum serves to associate behaviourally relevant sensory input with the motor activity necessary for the execution of the movement called for by that sensory input.

Other authors have described results similar to ours but when recording in different cortical areas and with different stimuli and behavioural responses. Bruce and Goldberg (1984, 1985) have studied the activity of single neurons in the frontal eye fields (area 8a) using visual stimuli and eye saccades. The authors described cells which were essentially visual, cells which were essentially motor and visuomotor cells which showed a maintained increase in discharge frequency between the visual stimuli and onset of the saccade. These visuomotor cells were in a continuum with varying proportions of visual and motor activity. It was suggested that the different properties may reflect the transformation of sensory and central information into command signals for eye saccades. When discussing the results of studies on area 7 of the posterior parietal cortex, Motter and Mountcastle (1981) commented that there appeared to be a spectrum of cells ranging from those activated by light stimuli but unaffected by eye movement, through those with combined properties, to others active with saccades and insensitive to light stimuli. The authors suggested that cells with such a gradient change in apparent functional properties may be arranged in a sequential processing chain. If the measure of log (variance Y / variance X) can be taken as a reliable index of "function", then our results may provide evidence for the presence of one such processing chain within the sequence S1, area 5, area 6 and area 4.

Acknowledgements

This work was supported by the C.N.R.S. and The Office of Naval Research. We are greatly indebted to the following for their valuable help in providing the computer programs for data acquisition, transfer and analysis: D. Commenges, M. Coulmance and M. Culioli. Technical assistance was provided by N. Vitton and the experimental set up was built by B. Arnaud and R. Fayolle in the C.N.R.S. workshops.

References

Bioulac B, Lamarre Y (1979) Activity of postcentral cortical neurons of the monkey during conditioned movements of a deafferented limb. **Brain Res 172:** 427-437

Bruce CJ, Goldberg ME (1984) Physiology of the frontal eye fields. **Trends Neurosci 7:** 436-441

Bruce CJ, Goldberg ME (1985) Primate frontal eye fields. I. Single neurons discharging before saccades. **J Neurophysiol 53:** 603-635

Chapman CE, Spidalieri G, Lamarre Y (1984) Discharge properties of area 5 neurones during arm movements triggered by sensory stimuli in the monkey. **Brain Res 309:** 63-77

Commenges D, Pinatel F, Seal J (1986a) A program for analysing single neuron activity by methods based on estimation of a change-point. **Comp Meth Prog Biomed 23:** 123-132

Commenges D, Seal J, Pinatel F (1986b) Inference about a change-point in experimental neurophysiology. **Math Biosciences 79:** 1-28

Commenges D, Seal J (1985) The analysis of neuronal discharge sequences: change-point estimation and comparison of variances. **Stats Medicine 4:** 91-104

Darian-Smith I, Johnson KO, Goodwin AW (1979) Posterior parietal cortex: relations of unit activity to sensorimotor function. **Ann Rev Physiol 41:** 141-157

Evarts EV (1981) Role of motor cortex in voluntary movements in primates. In **Handbook of Physiology: The Nervous System.** American Physiological Society, Bethesda, Sect I, Vol II, pp 1083-1120

Hasbroucq T, Guiard Y, Kornblum S (1989) The additivity of stimulus-response compatibility with the effects of sensory and motor factors in a tactile choice reaction time task. **Acta Psychol 72**, 139-144

Hyvärinen J (1982) **The Parietal Cortex of Monkey and Man.** Studies of Brain Function, Vol 8. Springer, Berlin

Jones EG, Coulter JD, Hendry SHC (1978) Intracortical connectivity of architectonic fields in the somatic, motor and parietal cortex of monkeys. **J comp Neurol 181:** 291-348

Kalaska JF, Caminiti R, Georgopoulos AP (1983) Cortical mechanisms related to the direction of two-dimensional arm movements: relations in parietal area 5 and comparison with motor cortex. **Exp Brain Res 51:** 247-260

MacKay WA, Kwan MC, Murphy JT, Wong YC (1978) Responses to active and passive wrist rotation in area 5 of awake monkeys. **Neurosci Lett 10:** 235-239

Miller JO (1988) Discrete and continuous models of human information processing: Theoretical distinctions and empirical results. Acta Psychol 67, 191-257

Motter BC, Mountcastle VB (1981) The functional properties of the light-sensitive neurons of the posterior parietal cortex studied in waking monkeys: foveal sparing and opponent vector organization. **J Neurosci 1:** 3-26

Mountcastle VB (1974) Central nervous mechanisms in sensation. **Medical Physiology**, Part IV. Mosby, St. Louis

Mountcastle VB (1979) An organizing principle for cerebral function: the unit module and the distributed system. In Schmitt FO, Worden FG (eds) **The Neurosciences Fourth Study Program. MIT Press, Cambridge,** pp 21-42

Mountcastle VB, Lynch JC, Georgopoulos A, Sakata H, Acuna C (1975) Posterior parietal association cortex of the monkey: command functions for operations within extrapersonal space. **J Neurophysiol 38:** 871-908

Proctor RW, Reeve TG (1986) Salient-feature coding operations in spatial precuing tasks. **J Exp Psychol: Human Percept Perf 12**, 277-285

Robinson DL, Goldberg ME, Stanton GB (1978) Parietal association cortex in the primate: sensory mechanisms and behavioral modulations. **J Neurophysiol 41:** 910-932

Seal J, Commenges D (1985) A quantitative analysis of stimulus- and movement-related responses in the posterior parietal cortex of the monkey. **Exp Brain Res 58:** 144-153

Seal J, Gross C, Bioulac B (1982) Activity of neurons in area 5 during a simple arm movement in monkeys before and after deafferentation of the trained limb. **Brain Res 250:** 229-243

Sternberg S (1969) The discovery of processing stages: Extensions of Donders' method. In W.G. Koster (ed.), **Attention and Performance II**. Amsterdam: North-Holland, pp. 276-315

Wiesendanger M (1981) Organization of secondary motor areas of cerebral cortex. In **Handbook of Physiology: The Nervous System.** American Physiological Society, Bethesda, Sect I, Vol II, pp 1121-1147

SECTION 2

LEARNING OF MOTOR ACTIONS

INFANT MOTOR DEVELOPMENT: IMPLICATIONS FOR MOTOR NEUROSCIENCE

ESTHER THELEN, JODY L. JENSEN, KATHI KAMM, DANIELA CORBETTA
Department of Psychology
Indiana University
Bloomington IN 47405 USA

and

KLAUS SCHNEIDER, RONALD F. ZERNICKE
Department of Kinesiology
University of California, Los Angeles
405 Hilgard Avenue
Los Angeles CA 93706 USA

ABSTRACT.

Developmental studies can inform motor neuroscience by describing the **initial state** of the perception-action system, identifying **change** and the **emergence of new forms**, and by determining the **processes** which engender these changes. We describe a research program studying the limb dynamics of human infants using a combination of kinematics, kinetics, and EMG measures. The dynamics of **spontaneous kicking** reveal the self-organizing qualities of the neuromotor system. The **transition from spontaneous waving to reaching** suggests ways by which intention is mapped onto these self-organizing qualities. We conclude with a discussion of developmental processes of skill acquisition.

1. Introduction

The study of motor development is a small, but growing subsection of the discipline of motor behavior and neuroscience. In the last decade, researchers have explored issues of coordination and control in human infants and children from a variety of theoretical and disciplinary perspectives (see, for example, Bloch & Bertenthal, 1990; Kelso & Clark, 1982; Wade & Whiting, 1986; Woollacott & Shumway-Cook, 1989). Movement studies with young subjects present formidable problems beyond those normally encountered with adults. Infants and children must be presented with simple, age-appropriate tasks, they are easily fatigued and sometimes uncooperative, and their movements are less constrained and more variable. Nonetheless, the advances in movement analysis technology of the last few years have allowed us to begin to produce the kinematic and kinetic analyses of developing movement with

J. Requin and G. E. Stelmach (eds.), Tutorials in Motor Neuroscience, 43–57.

the same level of resolution as now possible in adult studies.

What can the field of motor neuroscience learn from developmental studies? Human infants and children acquire motor skills over a very protracted period of time. Studies of motor behavior spanning these years can contribute to our understanding of motor behavior in several ways. First, we can describe the **initial** state of the perception-action system as it generates the spontaneous and unskilled movements of early infancy. These spontaneous and often uncoordinated movements are important because they form the neuromuscular bases from which skills such as reaching, sitting, and walking are built. Second, knowing the initial state allows us to identify the relevant parameters that capture the **changes** in motor behavior that occur over the first months and years. And most importantly, we can then seek to determine the **processes** which engender these changes. Because the early acquisition of skills is both protracted and incremental, we have a window on the construction of **novel forms** that is not available with adults, who already have highly stable and practiced movement styles.

The period of infancy is perhaps the most instructive for these goals because we can see within the first year the entire spectrum of change from movements seemingly lacking in both coordination and control--and indeed newborns are unable even to counter the forces of gravity--to coordinated, goal-directed, and even graceful movements. In this chapter, we describe some results from our program of research designed to address these goals using a combination of kinematic, kinetic, electrophysiological, and behavioral studies on the functional limb and trunk movements of infants. We conclude with a discussion of the implications of this work for adult motor behavior.

1.1. TAKING BERNSTEIN SERIOUSLY

Traditional views of motor development have emphasized the dominant role of the maturation of the central nervous system and thus have tried to relate changes in the brain to their resultant changes in motor function. There are unquestionably large and dramatic changes in the size and organization of the brain during the first year when the basic motor skills are acquired (Goldman-Rakic, Isseroff, Schwartz, & Bugbee, 1983). Our goal, however, is to go beyond the correlational approach between brain anatomy and behavior to the more dynamic and interactionist perspective first associated with Bernstein (1967).

For developmentalists, Bernstein's great insight was that movement forms could never be **imposed** on the organism by an autonomously maturing brain, but had to be **sculpted** into the neuromuscular system strictly in interaction with the periphery. Bernstein's reasoning supporting this claim is now well-known. The brain cannot directly map functional movement forms on the limbs because of the indeterminancy between the brain and the effectors. The sources of this indeterminancy include 1) anatomy--there are multiple degrees of freedom in the multijointed segments, many modes of action in the muscles, and context-dependent agonist-antagonist relations; 2) the mechanical complexity of multisegmented kinematic changes leading to complex and varying reactive forces with each movement; and 3) normal variance in effector nerve impulses (Edelman, 1987). The means by which the CNS deals with this indeterminancy is, of course, the great puzzle facing our discipline, and although the solutions are not yet fully known, most agree that the construction of movement synergies, categories, or coordinative structures must be involved.

According to Bernstein, how the CNS forms these stable yet adaptive movement categories in the face of indeterminancy is primarily a problem of development. Movements evolve, develop, and react as they are performed. With each new stage in development, new problems arise to which the CNS must adapt. Consider, for example, infants' first attempts to stand supporting their weight. They must integrate inputs from muscle and tendon receptors with visual, tactile, and vestibular fields to produce the correct amount of tonic muscle stiffness to maintain stance and to produce accurately timed phasic contractions to restore upright balance. If Bernstein is correct, these ensemble responses could not have been anticipated by the CNS, but must evolve, in concert with the periphery, as a dynamic process.

1.2. CONSTRUCTING SYNERGIES IN A WORLD OF FORCES

Synergies of movement must be constructed, then, with reference to both the information and forces in the environment and the force fields generated by self-movement. Studies of adult movement alone cannot fully elucidate this constructive process because it is difficult to isolate the self-adjusting characteristics of the system from their intentional and often highly learned responses. Infants, in contrast, lack experience in the gravitational world and they must deal with gravity and the passive forces generated by their own movements long before they have much voluntary control or can guide their movements by vision. What, if any, adaptive characteristics does the newborn CNS have to deal with this world of forces? How does the acquisition of goal-directed and goal-corrected movements use or modulate these intrinsic properties of the system? What are the interactions between the acquisition of skill and the ability to both use and optimize forces? These are the questions we are addressing in our research program. In particular, here we contrast spontaneous kicking movements, spontaneous arm waving movements, and early intentional reaches.

2. Intersegmental Dynamics of Spontaneous Infant Kicking

The spontaneous kicking movements of young infants are an excellent window to begin answering these questions. Infant kicks are simple, but coordinated cycles of flexions and extensions of the three joints of the legs. All normal infants perform these movements when they are behaviorally aroused, and at least in the first months, they appear to have no immediate goal (Thelen & Fisher, 1983). Kicks are often forceful and rapid-- cycle durations of 600-800 ms are not uncommon.

When infants kick, therefore, the rapid movements of the segments of their legs create interactive forces that are transmitted among the segments. In addition, infants must lift their legs against the force of gravity, with the magnitude of the gravitational influence depending on the orientation of the limb. How do these inevitable and largely unpredictable forces influence the movement? How does the CNS modulate muscle contractions in the context of these dynamically changing forces? In short, what does the infant "come with" to deal with forces before they have significant voluntary control

2.1. INTERSEGMENTAL DYNAMICS OF INFANT MOVEMENT

We have adapted techniques of intersegmental dynamics to analyze infant movements, in concert with traditional kinematic and EMG measures. This method uses inverse dynamics and Newtonian equations of motion to partition the net torque acting at a given joint into components due to muscular (MUS), gravitational (GRA), and inertial and centripetal forces (called here motion-dependent torques, MDT) acting on the moving segments (e.g. Hoy & Zernicke, 1986; Schneider, Zernicke, Schmidt, & Hart, 1989). When the passive torques (GRA and MDT) are subtracted from the net torque, the result is a generalized muscle torque (active and passive muscle forces and passive viscoelastic forces arising from the soft connective tissues). It is this generalized muscle torque (MUS) which contains the only force component that is actively controlled by the CNS.

For our torque calculations, we derived accelerations from measured coordinate data as the limb segments moved freely through space. These kinematics were recorded by the WATSMART optical-electronic, computer-based, 3-D motion analysis system. For segmental moments of inertia, masses, and the location of segmental centers-of-mass, we developed a computational 17-segment model of the human infant based on detailed anthropometric measurements of each subject. (These procedures are reported in detail in Schneider, Zernicke, Ulrich, Jensen, & Thelen, in press.)

2.2. SPONTANEOUS KICKING IN THREE-MONTH OLD INFANTS

First, we consider spontaneous kicking in three-month old infants in the supine posture. The kinematics of an exemplar kick are shown in Figure 1 (a), expressed as the angular rotation of the hip, knee, and ankle joint. The hip and knee flex and extend nearly simultaneously. The ankle also flexes and extends during the kick, though slightly out of phase with the more proximal joints. Complex torque dynamics underlie this relatively simple movement, however. In Figure 1(b), we show the three torque components acting to produce the flexion (negative torques) and extension at the hip (positive torques). First, gravity acts throughout the kick to extend the leg. The motion-dependent-torques, here the summed influence from the movements of the linked segments, oscillate between flexor and extensor influences and oppose the direction of motion. Muscle torque, the only torque controllable by the nervous system, is modulated in response to these passive torques. At the initiation of the kick, MUS is flexor, counteracting the extensor influence of GRA and MDT. As the hip joint reverses from flexion to extension, MUS continues to exert a flexor influence. The flexor magnitude of MUS is diminished due to the flexor assistance provided by MDT. Note that muscle contraction does not explicitly generate the movement during extension, but is used primarily to brake and resist gravity, which pulls the leg down.

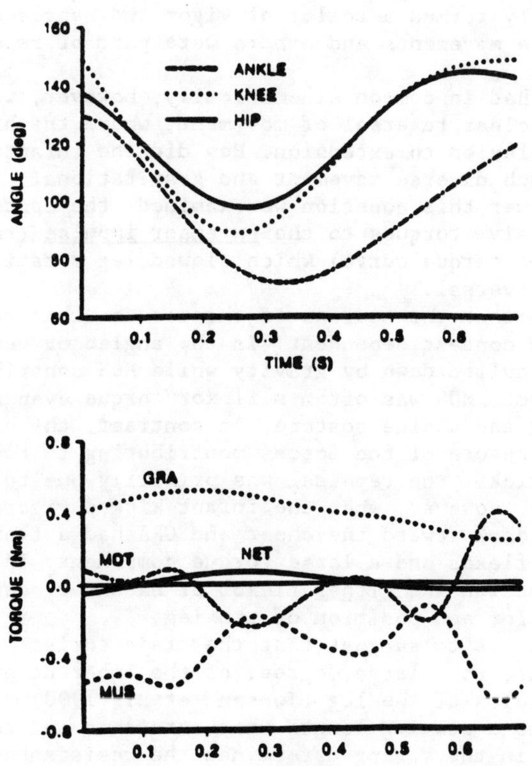

Figure 1. Kinematics and kinetics of exemplar infant kick 0 to 8 sec. (a.) Angular rotation of the hip, knee, and ankle joints. (b.) NET torque and the three contributing torque components for the hip joint. GRA: gravitational torque, MDT: motion-dependent torque, MUS: muscle torque. Positive values represent extensor influence, negative values represent flexor influence.

Since the movement trajectory is smooth, the muscles must work in concert with the inevitable passive forces acting on moving limbs. More muscle force is needed to initiate the kick because the muscles must overcome gravity and the MD torques to flex and lift the legs. Conversely, an _extensor_ torque may be unnecessary to reverse the leg because much of the work is done passively. The flexor influence does, however, keep the leg from extending too rapidly. The overall picture, therefore, is one of context-sensitive modulation of muscle activity (Schneider, et al., In press).

To further determine the context-sensitivity of this muscle-force loop, we also looked at a population of kicks from 3-month old infants (Jensen, Ulrich, Thelen, Schneider, & Zernicke, 1990). These kicks were recorded in three gravitational contexts--when the infant was supine, supported at 45 degrees, and held in a vertical position, and the kicks naturally formed a scalar of vigor and ranges of motion. Some kicks were single movements and others were part of repeating cyclic movements.

What these kicks had in common kinematically, however, was that they all contained a clear reversal of movement, where the hip joint motion changed from flexion to extension. How did the infant manage to reverse the leg in such diverse movement and gravitational environments? To answer this question we examined the contribution of the active and passive torques to the _extensor impulse_ (the integral under the net torque curve) which slowed leg rotation in preparation for hip reversal.

Again, we found that the sources of the torques that reversed the leg were entirely context dependent. In the angled or vertical posture, the leg was pulled down by gravity while MUS contributed a flexor impulse. In fact, MUS was often a flexor torque even when the leg was extending. In the supine posture, in contrast, the vigor of the kick changed the nature of the forces contributing to reversal. In supine non-vigorous kicks, the reversal was primarily due to GRA, as in the other postures. However, when the infant kicked vigorously, the leg was often pulled back toward the chest and GRA had a flexor effect. MDT was also flexor and a large torque component. In this case, the MUS contribution was _either_ flexor or extensor, depending on the combination of vigor and position of the leg.

Our analyses led us to suggest that the trajectories of infant kicks may be a product, to a large degree, of the inherent _spring-like_ properties of the muscles of the leg (Jensen, et al, 1990). Recall that springs return to a resting length when stretched and that the tension or stiffness in the spring determines the resistance offered to the stretch. Once muscle stiffness is set and maintained by the CNS, the purely mechanical properties of muscles give limbs this self-organizing quality, which we indeed observed in the smooth co-variation of spring force and joint angle in the phase plane diagrams of infant kicks (Jensen, et al., 1990). Thus, modulations in MUS during the kick may reflect-- at least at times-- these self-adjustments.

How then, do infants produce smooth and rapid kicks? We know from previous research (Thelen & Fisher, 1983), that infants initiate kicks by strong co-contraction of flexor and extensor muscles, changing the stiffness of the muscles around the joints. This results in an imbalance of forces, which initiates flexion. No such phasic bursts were detected at the reversal of direction, consistent with the present results showing that joint reversal is passive-- the result of lost momentum, the effects of gravity, and in some cases, the extensor bias of the limb shooting past the equilibrium point. Context-dependent modulations of MUS during the movement may also be a result of the spring effect. However, note that Jensen et al., (1990) also described kicks in which the muscle torque did not vary smoothly with the excursion of the leg, suggesting active intervention of the CNS to change stiffness and that the spring-model was, in itself, likely too simple.

In summary, we discovered that non-intentional leg movements in infants could be rapid, smooth, and coordinated from this combination of active and passive forces in a largely self-organizing manner, utilizing the springiness of muscles and their consequent sensitivity to a force-environment. Exploitation of the spring-like qualities of the neuromuscular system is an important part of the control mechanisms for voluntary behavior as well, but these muscle qualities must also be actively modulated and tuned to task demands. In the next section, we present a preliminary account of how this process may be accomplished during development.

3. Limb Dynamics and the Transition from Spontaneous to Voluntary Movement.

When infants learn to reach and grasp, they must acquire control of their rather uncoordinated waving arm movements to extend the arm and hand toward the desired object. Although directed arm extensions can be seen even in newborns under special postural and attentional conditions (Hofsten, 1982), voluntary directed reaches can be reliably elicited usually between 3 and 4 months of age (Hofsten, 1984). We have been observing the development of reaching in an intensive longitudinal study of four infants. We have tested these infants weekly from week 3 in several situations: toys presented by the parents, toys presented by an overhead apparatus (without the social dimension) and in social interaction with the parent (without an object present). The last condition was used to generate non-goal directed movements. At each session, we have recorded arm movements by the WATSMART Motion Analysis system, with IREDs at the shoulder, elbow, wrist, and knuckle to produce 3-D kinematics of both arms. We have also recorded concurrent EMG using surface electrodes on the biceps, triceps, deltoid, trapezius, and lower back muscles. Where kinematic data are sufficiently complete, we are also calculating intersegmental dynamics. For all trials, infants were seated in an armless infant-seat at 75 degrees inclination from the horizontal, and snugly strapped around the torso with a soft band.

The transition from spontaneous to voluntary movements is especially well-illustrated by trials in which the infants alternate between seemingly undirected waving or flapping movements and actual directed, and sometimes successful, attempts to touch the toy. Here we illustrate the emergence of directed reaches with an exemplar trial from a 15 week-old boy. In this 14-second trial, the parent presented the toy at midline, the infant flapped his arms several times, reached forward and up and finally lowered the arm toward the toy. Figure 2 shows the trajectory of his right wrist in the frontal plane during the portion of the trial that included a flapping action and reach by the right arm. The movement starts at the lower right-hand corner of the plot, commencing with one "flap" (9.9 s to 10.56 s) followed by the reach. The reach phase shows a raising and lowering of the arm combined with leftward movement (10.56 to 11.37 s). Figure 3(a) illustrates the corresponding 3-D joint angle rotations of the elbow and shoulder (declining angles represent flexion: shoulder flexion raises the arm). Again, the flap and reach can be seen in the alternate flexion and extension of the shoulder joint. At this age, infants reach primarily by flexion and horizontal adduction of the shoulder, with the elbow held relatively rigid. This brings the hand up and into midline. The associated angular velocities are presented in Figure 3 (b). In this movement, velocities associated with the flap and reach are comparable.

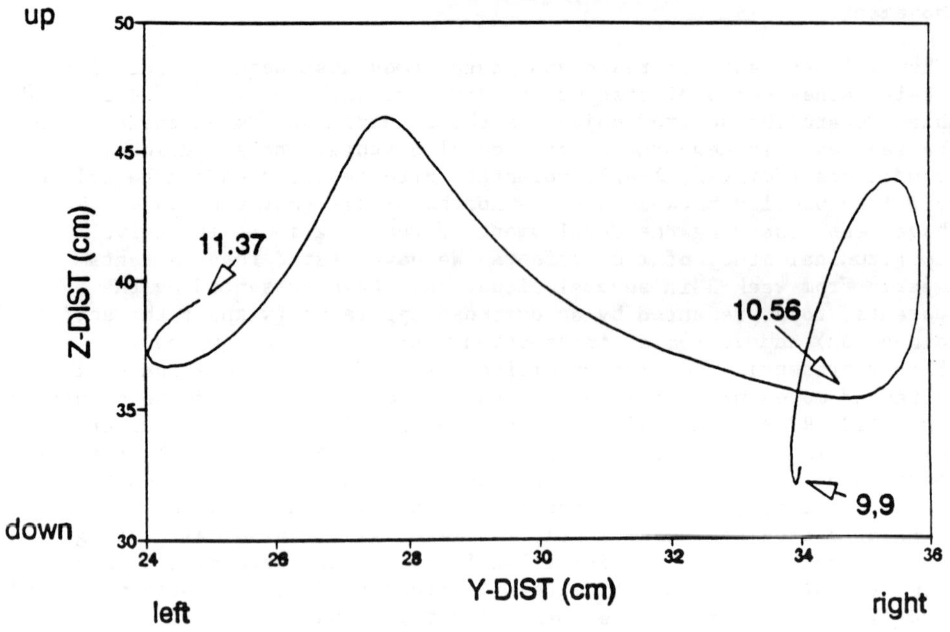

Figure 2. Endpoint trajectory of the right wrist viewed in the frontal plane for infant "flapping" and reach.

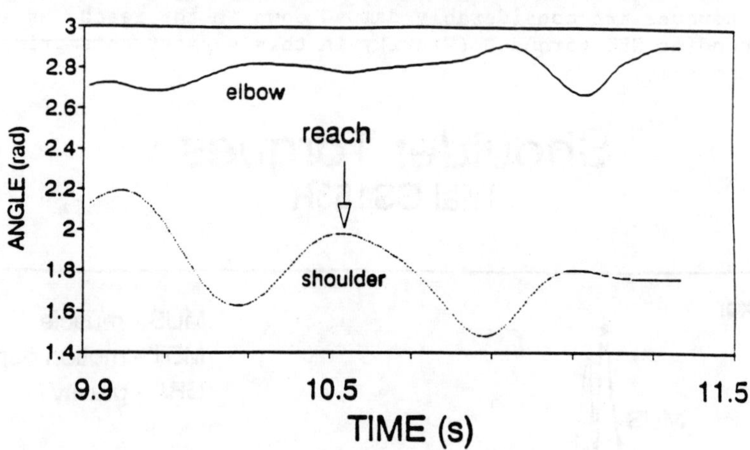

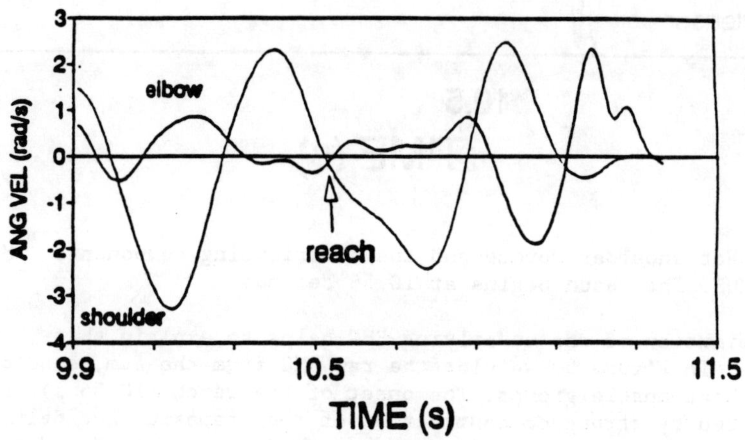

Figure 3. 3-D joint angle rotations, and joint angular velocities for shoulder and elbow. (a.) 3-D joint rotations for shoulder and elbow expressed in radians (rad). (b.) Angular velocities expressed in radians per seconds (rad/s). Positive values represent velocity toward extension, negative toward flexion.

The torque profiles associated with the spontaneous and directed movements are quite different, however, as can be seen in Figure 4. The spontaneous movement has high MDT balanced by high MUS torques, while both torques are considerably damped down in the reach, as are the corresponding NET torques. (Gravity in this segment acts primarily to extend the arm).

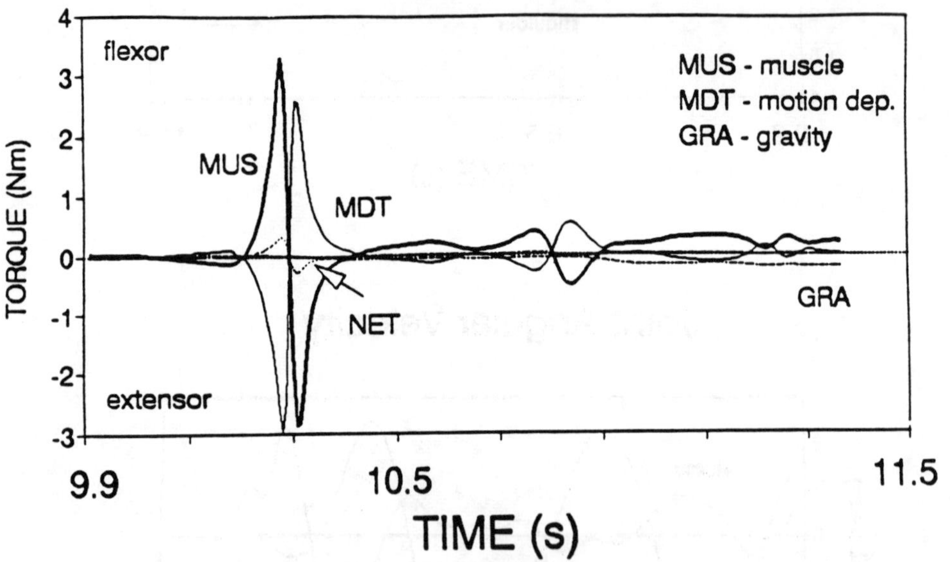

Figure 4. Net shoulder torque and the contributing components. MDT, GRA, and MUS. The reach begins at 10.56 seconds.

Examination of the underlying EMG helps to explain this difference. In Figure 5, we plot the raw EMG from the arm, shoulder, and lower back muscle groups. The onset of the reach (10.56 s) is characterized by strong co-contraction of the trapezius and deltoid muscles, with the triceps activated when the arm is raised and the elbow most fully extended. This pattern of co-contraction contrasts the more phasic and reciprocal muscle activity associated with the "flap" (eg. triceps and deltoid bursts between approx. 10.2 and 10.5 s). In this trial, the lower back muscles are largely inactive during this activity.

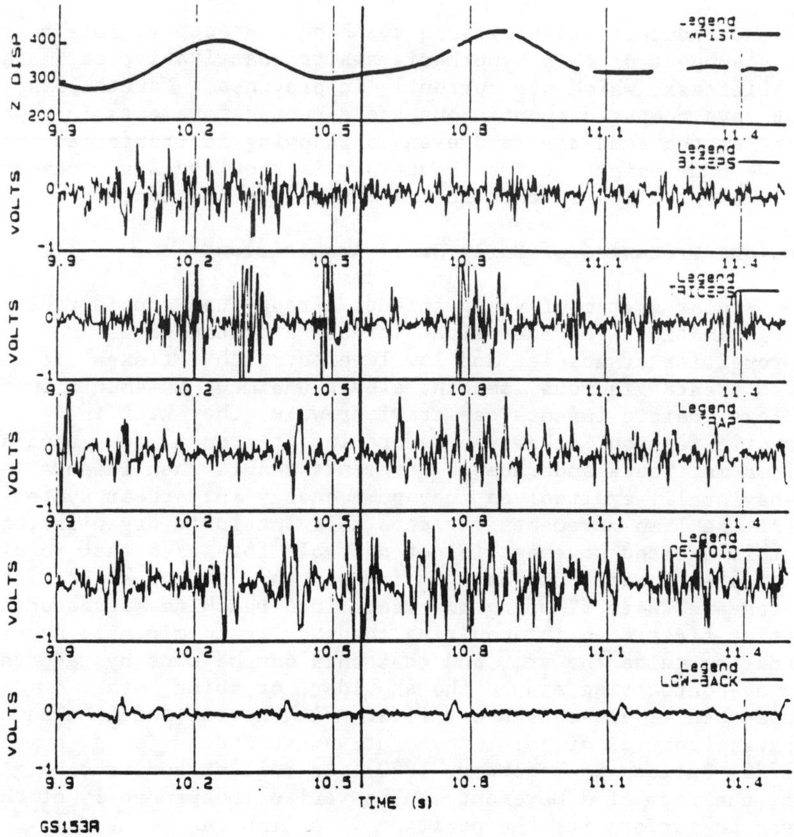

Figure 5. Raw EMG from biceps, triceps, upper trapezius, deltoid, and low-back (paraspinals 2cm above the level of the illiac crests). Movement of the wrist is represented in the Z-dimension (vertical displacement) for reference. Reach onset at 10.56 seconds.

Clearly, an uncontrolled flapping of the arm cannot be directed toward a precise target. What our preliminary evidence suggests is that infants can make this transition early in development by several strategies. In some infants and trials, the subjects attempt to reach using velocities considerably slower than those of their spontaneous movements. This has the effect of producing smaller motion-dependent forces with which to contend. Slower movements also allow time for error correction. Most common, however, is the extensive co-contraction we saw in Figure 5, which stiffens the shoulder joint and protects, so to speak, the shoulder against the MDTs generated by the lower arm segments. In this way, the arm can be swung toward the target with considerably greater velocity but with negligible effects of more distal MDTs. Accuracy may be increased because the increased stiffening about the shoulder joint contributes to proximal joint stability and reduces variability in the hand trajectory. Note that more direct support of this hypothesis awaits quantitative estimates of joint stiffness, which are currently in progress. Nonetheless, because we have compared spontaneous and directed movements in the same infant at the same age, and even as flapping is tranformed into reaching, we can begin to think about what is required for processes of change. We speculate on this in the concluding section.

4. Conclusion: Processes of Early Skill Acquisition.

The neuromuscular system of young infants already has considerable dynamic organization. The anatomical structure of the limbs, the elastic properties of muscle, and low-level stretch reflexes all work together to create vigorous, smooth, albeit unaimed, movements in the arms and legs. Before infants can crawl or walk, they kick in characteristic fashion. Likewise, before infants reach, they wave and flap their arms. These spontaneous movements result when infants become behaviorally excited: as they pump energy into their systems, the form of the limb movements reflects this intrinsic organization. However, this excited movement is not suitable for tasks that require accuracy or steadiness. Our data on infants who, at the sight of the toy, can convert their flapping movements into reaching suggested that one important first step in acquiring accuracy is to minimize excess MDT in order to guide the arm, and that this can be done by reducing velocity, co-contracting around the shoulder, or doing both.

These results are consistent first, with neural models that posit separate control of joint movement and stiffness (e. g. Humphrey & Reed, 1983; Bullock & Grossberg, 1990). As Bullock and Grossberg point out, the form of a movement can be varied independently of the energy used to perform it: the position of a limb can be held invariant despite great differences in the tension in it. Although infants' early movements are undirected, they do vary greatly in vigor, depending on the infant's level of behavioral excitation. Thus, from the earliest days, and before they have much successful directed movement, infants are experiencing modulation of their limb stiffness. Perhaps this modulation of limb stiffness through co-contraction, therefore, is the first strategy to be recruited for voluntary control.

Others have noted that, in general, unskilled movements are performed stiffly, with considerable co-contraction. Certainly, this is seen in new walkers (Okamoto & Goto, 1985). The combination of slow velocity and/or low compliance is a good way to assure steadiness and/or accuracy, depending on the needs of the task.

Since Bernstein, a number of theorists have also proposed that exploration of the task through multimodal feedback is one process whereby stable and efficient movements are established or selected from a larger universe of possibilities (e.g. Edelman, 1987; Grossberg & Kuperstein, 1989; Kugler & Turvey, 1977). If this hypothesis is true, then the infant (or adult performer) needs to "feel" success through all the perceptual modalities, including the proprioceptions of the appropriate level of stiffness, and the visual, haptic, etc.

consequences of completing the task. It seems likely that the first step in such exploration is to be able to do the task at all, which for infants, means breaking out of the cyclic waving movements of the arms or kicking of the legs. In a sense, they must interrupt the natural oscillatory dynamics by slowing or stiffening, and our early evidence suggests that they do.

Once the task can be accomplished, however clumsily, a ground is laid for the exploration-selective processes to work. As infants attempt to reach from different postures and starting positions, and at different energy parameterizations, and with their targets at different distances, sometimes moving and sometimes stationary, they are continually mapping these large perceptual spaces onto the movements they generate. By this manner they self-discover the task-specific dynamics. Studies documenting this self-discovery through multimodal exploration are a logical next step in our quest for understanding the acquisition of skill.

Finally, we may hypothesize that once accuracy is good and movement categories are stable and reliable, infants can rescale their speed and compliance to become more time and energy efficient (Bullock & Grossberg, 1990). That is, they can complete the Bernsteinian program by actually using passive forces rather than only resisting them. Adults are very adept at utilizing gravity, inertia, and the elastic properties of the system, and contracting muscles only when necessary to complement what comes "for free." Schneider, Zernicke, Schmidt, & Hart (1989) were able to demonstrate the more efficient use of passive dynamics with skill by documenting these changes after adults practiced novel, rapid arm movements. In infants, we would therefore predict reaching to use more passive dynamics and less muscle torque and co-contraction in particular and to have higher velocities, but only after they had attained a reliable degree of accuracy. At what stage in the development of reaching this transition will occur is unknown.

The use of these new kinetic measures in combination with kinematics and EMG is allowing us to pull apart processes of skill acquisition that were previously inaccessible. Developmental studies are an important complement to studies of control and coordination of practiced movements because they use natural experiments in processes of change. We believe it is essential now to go beyond superficial correlations between anatomy and behavior to address both what the infant has and what the infant does to become a skilled and graceful mover in a complex world.

Acknowledgement

This research was supported by grant HD 22830 from the National Institutes of Health and a Research Scientist Development Award to Esther Thelen from the National Institutes of Mental Health. We thank Michael Schoeny, Dexter Gormley, Deanna Berkoben, and Jürgen Konczak for their assistance with data collection and analysis. Address correspondence to Esther Thelen, Department of Psychology, Indiana University, Bloomington, IN 47405.

References

Bernstein, N. (1967). Coordination and regulation of movements. New York: Pergamon Press.

Bloch, H., & Bertenthal, B. I. (1990). Sensory-Motor organizations and development in infancy and early childhood. Dordrecht, Netherlands: Kluwer Academic Publishers.

Bullock, D., & Grossberg, S. (1990). Motor skill development and neural networks for position code invariance under speed and compliance rescaling. In H. Bloch, & B. I. Bertenthal (Eds.), Sensory-motor organizations and development in infancy and early childhood (pp. 1-22). Dordrecht, Netherlands: Kluwer Academic Publishers.

Edelman, G. M. (1987). Neural Darwinism. New York: Basic Books.

Goldman-Rakic, P. S., Isseroff, A., Schwartz, M. L., & Bugbee, N. M. (1983). The neurobiology of cognitive development. In P. H. Mussen, M. M. Haith & J. J. Campos (Eds.), Handbook of child psychology, 4th Ed., Volume II: Infancy and developmental psychobiology (pp. 281-344). New York: John Wiley.

Grossberg, S., & Kuperstein, M. (1989). Neural dynamics of adaptive sensory-motor control: Expanded edition. New York: Pergamon Press.

Hofsten, C. von (1982). Eye-hand coordination in newborns. Developmental Psychology, 18, 450-461.

Hofsten, C. von (1984). Developmental changes in the organization of prereaching movements. Developmental Psychology, 20, 378-388.

Hoy, M.B., & Zernicke, R.F.(1986). The role of intersegmental dynamics during rapid limb oscilations. Journal of Biomechanics, 19, 867-877.

Humphrey, D. R., & Reed, D. J. (1983). Separate cortical systems for control of joint movement and joint stiffness: Reciprocal activation and coactivation of antagonist muscles. In J.E. Desmedt (Ed.), Motor control mechanisms in health and disease (pp. 347-372). New York: Raven Press.

Jensen, J. L., Ulrich, B. D., Thelen, E., Schneider, K., & Zernicke, R. F. (1990). Posture-related limb dynamics in spontaneous infant kicking. Submitted for publication.

Kelso, J. A. S., & Clark, J. (1982). The development of movement control and coordination. New York: John Wiley.

Kugler, P. N., & Turvey, M. T. (1987). Information, natural law, and the self-assembly of rhythmic movement. Hillsdale, NJ: Erlbaum.

Okamoto, T., & Goto, Y. (1985). Human infant pre-independent and independent walking. In S. Kondo (Ed.), Primate morphophysiology, locomotor analyses and human bipedalism (pp. 25-45). Tokyo, Japan: University of Tokyo Press.

Schneider, K., Zernicke, R. A., Schmidt, R. A., & Hart, T. J. (1989). Changes in limb dynamics during practice of rapid arm movements. Journal of Biomechanics, 22, 805-817.

Schneider, K., Zernicke, R. F., Ulrich, B. D., Jensen, J. L., & Thelen, E. (in press). Understanding movement control in infants through the analysis of limb intersegmental dynamics. Journal of Motor Behavior.

Thelen, E., & Fisher, D. M. (1983). The organization of spontaneous leg movements in newborn infants. Journal of Motor Behavior, 15, 353-377.

Wade, M. G. & Whiting H. T. A. (1986). Motor skills acquisition. Dordrecht, Netherlands: Martinus Nijhoff Publishers.

Woollacott, M., & Shumway-Cook, A. (1989). The development of posture and gait across the lifespan. Columbia, S. C.: University of South Carolina Press.

FREQUENT AUGMENTED FEEDBACK CAN DEGRADE LEARNING: EVIDENCE AND INTERPRETATIONS

RICHARD A. SCHMIDT
Department of Psychology
University of California, Los Angeles
Los Angeles, CA 90024-1563
U.S.A.

ABSTRACT. The role of augmented information feedback for motor learning has been evaluated recently by an examination of its role on performance on transfer or retention tests. Several lines of evidence from various research paradigms show that, as compared to feedback provided frequently (after every trial), less frequent feedback provides benefits in learning as measured on tests of long-term retention. Such effects are of course contrary to most accounts of the learning process in human skills. In this paper, these lines of evidence are first briefly reviewed, and then several interpretations are provided in terms of the underlying processes that are degraded by frequent feedback. These decrements for frequent feedback seem to be caused by feedback's tendency to generate maladaptive short-term corrections, by a blockage of several sets of information processing activities, or by both of these factors in some combination.

1. Feedback For Skill Acquisition

It has been long understood that information about the outcome of an action, either in terms of meeting its environmental goal or in terms of the patterns of action that have led to the environmental goal, are critical for effective learning. Augmented information--that is, extrinsic information provided over and above any intrinsic information naturally available as a result of the "normal" conduct of the action--has been especially important not only in laboratory settings where it has been studied to determine feedback's role in learning, but also in practical settings where teachers or instructors provide it as a way to facilitate the learning process. Where no other sources of intrinsic feedback are available to the learner, failure to provide augmented information feedback has generally led to markedly degraded learning, or even to no learning at all in some cases where subjects cannot detect their own errors reliably. These properties of augmented feedback have been taken to mean that feedback is a critical variable for motor learning, and have fueled a healthy interest in its experimental study (for reviews, see Adams, 1971; Bilodeau, 1966; Newell, 1977; Salmoni, Schmidt, & Walter, 1984).

Most of the research, especially the early work, has used research paradigms where some variation in feedback (e.g., its frequency, its delay in time, etc.) was studied during the practice of some motor task, with different groups of subjects receiving different levels of this variable. Interpretations about the role of this variable for learning were made based on the relative proficiency of the groups during this practice phase, either in terms of the rate of gain in proficiency, in the final performance level, or both. With this kind of procedure, it was easy to show that nearly any variation that makes feedback more frequent, more immediate, more precise, or in general more "useful" produced benefits for performance during practice. It seemed logical to conclude that these effects were due to the added benefits of these feedback variations to the learning process, which further reinforced the beliefs that frequent feedback was a critical variable for learning. This generalization had numerous implications to practical situations such as the

J. Requin and G. E. Stelmach (eds.), Tutorials in Motor Neuroscience, 59–75.

design of training settings and simulators, and supported a class of theoretical positions about learning that were closely related to the basic ideas of Thorndike (1927; see Adams, 1971; Schmidt, 1975).

1.1. THE DISTINCTION BETWEEN LEARNING AND PERFORMANCE EFFECTS

The problem in the above line of thinking was that it ignored the well-known distinction between learning and performance, which can be traced to the writings of Tolman (1929; see also Hull, 1943; Guthrie, 1952). In this idea, a variable in practice can have two different kinds of effects on performance: (a) relatively permanent ones (learning effects) that survive a retention interval or a shift in conditions, and (b) temporary ones (performance effects) that do not, with their influences changing (or even vanishing) with time or a shift in conditions. Salmoni et al. (1984) pointed out in their review of this work that most of the feedback research had failed to separate the temporary effects from the relatively permanent ones, and thus most of the principles of feedback were limited to the effects of feedback on momentary performance, and not strictly to its role for learning.

1.2. THE ROLE OF FREQUENT FEEDBACK IN LEARNING

Over the past several years, our group at UCLA has pursued this notion in various ways, examining the role that several variations in augmented feedback have on performance during practice, but more importantly when subjects are given an immediate or delayed retention test. These retention tests are conducted under equated levels of the feedback variation, most typically with feedback withdrawn, allowing an analysis of the extent to which variations in augmented feedback influence the relatively permanent aspects of motor performance--i.e., learning.

These experiments, some of which are discussed later in this article, generally show some surprising and counterintuitive findings. I can summarize a number of effects shown in these studies as follows: Relative to a condition with frequent feedback (e.g., feedback after every practice trial), several variations in feedback scheduling that provide information less frequently produce decrements in performance during the acquisition phase, but produce benefits in performance in a retention phase when feedback has been withdrawn. That is, frequent feedback benefits performance during acquisition when feedback is present, but is detrimental to learning as it is measured on a retention test. Other experiments show that less frequent feedback benefits both performance in later practice and performance in retention. In both case, the evidence suggests that less frequent feedback is beneficial for learning. This is, of course, in violent contradiction to a number of relatively well accepted theoretical ideas which insist that learning can occur only if information about performance is provided quickly after each performance attempt (e.g., Adams, 1971; Schmidt, 1975; Thorndike, 1927).

In the sections which follow, I review briefly some of these empirical effects of feedback variations which document the above conclusion in various ways. Then, in the final section of the paper, I provide several suggestions or hypotheses about why frequent feedback might be detrimental for learning, and give some of the evidence for and against these propositions.

2. Frequent Feedback Degrades Learning: Empirical Evidence

In this section, I highlight three research paradigms that bear on the detrimental effects of feedback that is presented frequently and immediately during the learning process. These involve experiments on relative frequency of feedback, summary feedback, and delay of feedback. Several other paradigms related to these major divisions are included as well.

2.1. FEEDBACK FREQUENCY

In the literature, it has been useful to define absolute frequency as the total number of trials that receive feedback, and relative frequency as the proportion of the total trials having feedback provided. The prevailing view was that learning increased as a direct function of the absolute feedback frequency (see Bilodeau, Bilodeau, & Schumsky, 1959; Trowbridge & Cason, 1932), but that relative frequency was irrelevant for learning (Bilodeau & Bilodeau, 1958). Bilodeau and Bilodeau showed that when the total number of feedback trials was held constant, adding no-feedback trials contributed nothing to performance, suggesting that these no-feedback trials were in some way "neutral" for learning, neither helping nor hindering the process. However, these findings were based on performances when feedback was present and being manipulated, confounding the relatively permanent and temporary effects of feedback, and some of our more recent work with retention designs has contradicted this generalization.

2.1.1. *Relative-Frequency Effects.* In experiments by Winstein and Schmidt (1990; Winstein, 1988; see also Schmidt, Shapiro, Winstein, Young, & Swinnen, 1987), the total number of practice trials was held constant for various groups, while the number of feedback trials was varied between groups. Thus, the absolute and relative frequency of feedback were confounded. One experiment (Winstein & Schmidt, 1990, Experiment 2) used a spatial-temporal patterning task, in which the subjects learned to produce a pattern of horizontal limb movements with three reversals in direction that occupied 800 ms. The goal was to produce a goal pattern, and performance was scored as the RMS error between the goal pattern and the subject's movement. Subjects received feedback on a computer terminal, showing (a) the goal pattern with the subject's just-produced movement superimposed and (b) the computed RMS error for that trial. There were two experimental groups, one which received feedback after each of 196 trials across two days of practice (100% group), and another which received feedback on only 50% of the trials. This last group had feedback faded across practice, with high feedback frequencies in early practice and low frequency later (50%-faded group). We used this faded schedule in an attempt to improve learning by the 50% group, as we had a suspicion that frequent feedback would be needed for error reduction in early practice, but that subjects would become dependent on it in several ways if it were provided too frequently in later practice. (This idea is elaborated in the third section of the paper.) Then, learning was evaluated on retention tests given 10 min. and 2 days after the last acquisition day, with these trials being performed without feedback. Even though relative and absolute frequency of feedback are confounded here, this is actually an interesting experiment theoretically, as most extant theories of learning would predict that lowering the absolute frequency (and, hence the relative frequency here) would be detrimental for learning.

The results are shown in Figure 1, where the RMS errors are plotted for the two acquisition days on the left, and for the retention performances at the right. The feedback frequency did not have a strong effect in practice, with only a slight tendency for the 50%-faded group to perform with more error than the 100% group. But in the immediate retention test, there was a (nonsignificant) tendency for the 50% to have less error than the 100% group, and this effect was far larger and significant by the delayed retention test. We argue that the retention performances are indicants of the relative amount learning during the acquisition phase, uncontaminated by the temporary effects of feedback in the acquisition phase, and therefore that reduced relative (and absolute) frequency have facilitated learning relative to every-trial feedback. Reduced frequency improved performance in practice only slightly, but facilitated learning, contrary to the standard theoretical accounts of feedback functioning.

2.1.2. *The Role of Faded Feedback.* In the Winstein-Schmidt study, the relative frequency and the fading schedule were confounded, so it is not possible to separate the role of reduced frequency per se. Nicholson and Schmidt (1990a) examined this problem further with the same task, using four conditions of feedback scheduling. One was the 100% condition used before, and another (50%-faded) was essentially like that used in the Winstein-Schmidt study except for a minor variation in the schedule. Two other conditions with 50% relative frequency were also used, one which had feedback provided after every other trial, and another that had feedback provided in alternate blocks of 5 trials. There were two practice days as before, and retention tests without feedback after 10 min and 2 days.

The results are in Figure 2. As before, there were only small effects of feedback manipulations in the acquisition days, with only a tendency for the 100% group to perform best on the first day. There were negligible effects on the 10-min retention test, but on the 2-day test there was a clear effect for the 50% conditions to perform with less error than the 100% conditions, with no differences between the 50% groups. At face value, these results say that the main effect of the

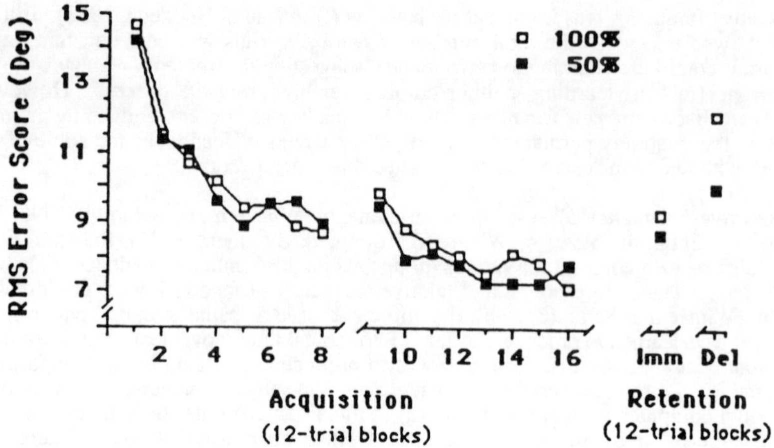

Figure 1. Average root mean squared (RMS) error in a limb-patterning task for the 100% and 50% conditions in two days of acquisition, and for immediate (10 min) and delayed (2 days) retention tests without feedback (from Winstein & Schmidt, 1990).

50%-faded schedule in Figure 1 was the reduced feedback frequency, and that the schedule had little further effect. But again, reducing the feedback frequency facilitated learning.

But there is an effect of scheduling after all, even with the relative frequency held constant, as shown by another experiment by Nicholson and Schmidt (1990a). Here, three groups all had 50% feedback schedules. One had a faded schedule as before (50%-faded), one had a uniform schedule with feedback after every other trial (50%-uniform), and a third group had a reversed-faded schedule, just the reverse of the 50%-faded group (50% reversed) where the feedback frequency was low in early practice and was gradually increased to 100% by the end. According to the arguments earlier, this last group should be relatively ineffective for learning, as it withholds feedback when the subject needs it in early practice to reduce errors, and provides it too frequently in later practice where the learner is likely to become dependent on it. This is essentially what happened, as in the 2-day retention test the 50%-reversed group performed poorest, with the 50% faded group being the best, and the 50%-uniform group being intermediate. Frequent feedback at the end of acquisition appeared to be detrimental for learning.

2.1.3. *Effect of Bandwidth Feedback.* In studies of so-called "bandwidth feedback," feedback is provided only if a learner's errors are outside of some predefined band of correctness; if the error is within the error band, no information is provided, which indicates to the learner that performance was acceptable. Relative to feedback after every trial, bandwidth feedback reduces the relative frequency, as feedback is withheld on trials for which the subject is correct. But it also tends to act as a fading variable. Early in practice, when errors are large, feedback is given frequently because the subject is seldom on target; but later in practice, when errors are smaller, feedback is produced less frequently. While the frequency of feedback is similar to those in fading experiments, this schedule differs because only those trials whose behavior is in error receive feedback, whereas in the fading schedule the no-feedback trials were determined a priori.

Sherwood (1983, 1988) has shown that, relative to every-trial feedback, bandwidth feedback tends to produce more stable behavior during the acquisition session, and to produce more effective performance when feedback is withdrawn in a retention test. The increased stability may be a product of the well known tendency for error information to induce performance change, with no information leading to relative stability. In addition, Lee and Carnahan (1990) used groups whose trial-to-trial feedback presentations were yoked to those of a group participating with bandwidth feedback. Of course, while the feedback frequency remains constant, one group (the bandwidth group) had feedback only after an error was made, whereas the yoked group had feedback after

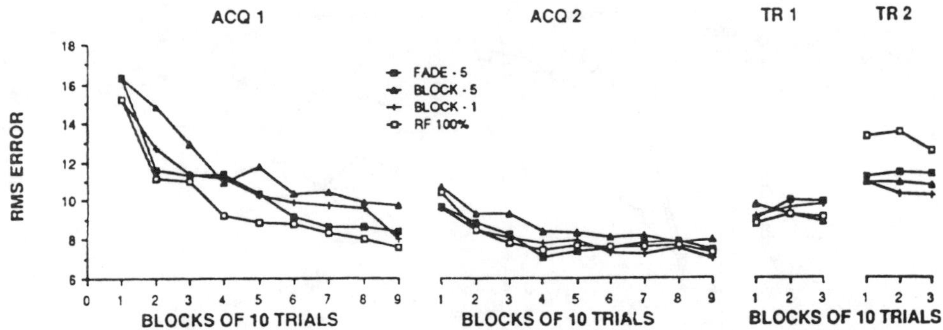

Figure 2. Average root mean squared (RMS) error in a limb-patterning task for four feedback conditions on two acquisition days, and for immediate (10 min, TR 1) and delayed (2 days, TR 2) retention tests without feedback (from Nicholson & Schmidt, 1990).

correct and error trials. They show that the bandwidth effects are over and above the effect of relative frequency, suggesting that interesting processes are occurring on error and no-error trials to affect learning differentially. Lee and Carnahan raise a number of additional issues here that deserve research attention.

2.1.4. *Summary of Frequency Effects.* Overall, even when the relative frequency is held constant, there is a tendency for schedules with reduced frequency of feedback at the end of practice to produce better retention performance, although this effect is a relatively small one as compared to the effect of reducing feedback frequency over all from 100% to 50%. Giving feedback in a bandwidth form reduces frequency to facilitate learning, but also has additional benefits probably because of the particular kinds of errors that receive feedback. One interpretation of all these effects is that frequent feedback, particularly if it is presented in later practice, tends to be detrimental for learning. Other interpretations of these data are possible, but this view is at least consistent with findings from several other experimental paradigms. We turn to the second of the paradigms next, that dealing with summary feedback.

2.2. SUMMARY AND AVERAGE FEEDBACK

In summary feedback schedules, originally studied by Lavery (1962; Lavery & Suddon, 1962), the learner performs for a set of trials (say, 10) without receiving feedback, and then receives feedback about each trial, usually in the form of a graph of performance against trials. Thus, formally the subject receives feedback on 100% of the trials, but with this schedule feedback is not delivered immediately after a performance where it could be effective in planning the next movement. Summary feedback is somewhat like reduced relative frequency, at least in terms of the learner's capability to use information after a trial. Of course, there are differences as well, such as the information provided in the summary graph, but as I argue later this is not the major contribution of summary feedback.

2.2.1. *Lavery's Pioneering Experiment.* Over 25 years ago, Lavery (1962; see also Lavery & Suddon, 1962) published experiments on summary feedback that were to have profound influence on our thinking about feedback's functioning. Using simple motor tasks, Lavery had subjects perform in 5 days of practice under three different feedback conditions. An Immediate condition had feedback after each trial, a Summary feedback condition had only summary feedback after the last trial in each 20-trial set, and a Both group had feedback after each trial and summary feedback after each set of 20 trials. Lavery then had subjects return for practice without feedback in several retention tests, as well as one month and three months later.

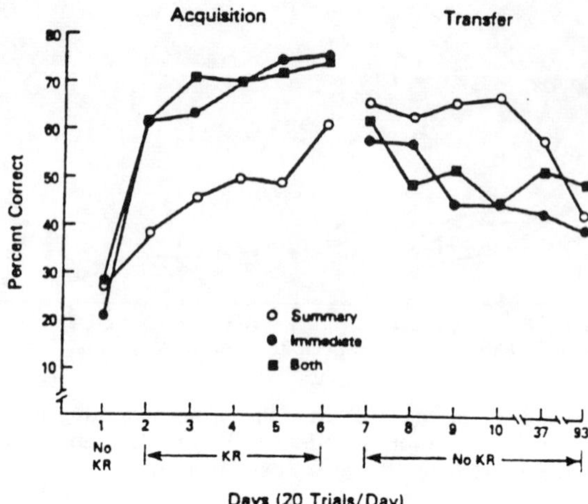

Figure 3. Average percentage correct responses for three feedback conditions in acquisition, and for in retention tests without feedback conducted on subsequent days (from Lavery, 1962).

As can be seen from the results in Figure 3, the Immediate and Both conditions performed far better than the Summary condition in acquisition, showing a faster rate of improvement and a higher asymptotic level near the last acquisition day. But in the retention tests, the order of conditions was reversed, such that the group performing best was the Summary condition, with the Immediate and Both conditions showing poorer performance at about the same level. These effects persisted through the one-month retention interval, and appeared to have been lost by 3 months. This experiment was surprising in that, relative to every-trial feedback, summary feedback degraded performance during practice, but enhanced learning as measured by performance in retention test.

It seems reasonable to ask why summary feedback was so effective for retention here. But, viewed in another way, this question does not seem to be reasonable after all. If summary feedback is so effective for learning, why did the Both group, which also had the benefits of summary feedback during acquisition, perform so poorly in retention? The common factor in the groups that performed poorly in retention was that every-trial feedback was given in acquisition. Thus it was not the summary feedback that aided retention, but rather the every-trial feedback that interfered with it. This is a radical suggestion when viewed in contrast to the earlier traditional concepts of feedback's operation, but it seems unavoidable given Lavery's data.

2.2.2. *Later Work With Summary Feedback.* In our group, we have followed up the intriguing findings from Lavery's work in several ways. In one experiment (Schmidt, Young, Swinnen, & Shapiro, 1989), we examined various summary lengths--the number of trials contained in the summary-feedback reports. We studied summary lengths of 1 (essentially every-trial feedback), 5, 10, and 15 trials in acquisition, with retention tests being provided after 10 min. and 2 days without feedback being provided. The task was a relatively simple arm movement task, where the goal was to make a movement pattern in 1000 msec.

Absolute constant errors for this experiment are shown in Figure 4. Increased summary length clearly degraded the performance in acquisition, slowing the rate of improvement and depressing the performance level, although there was some tendency for the groups to come together by the end of practice. On the 10-min retention test, there were no important performance differences. But on the delayed test, the order of treatment groups was just opposite to that in acquisition, with the 15-trial condition showing best performance and the 1-trial group being worst

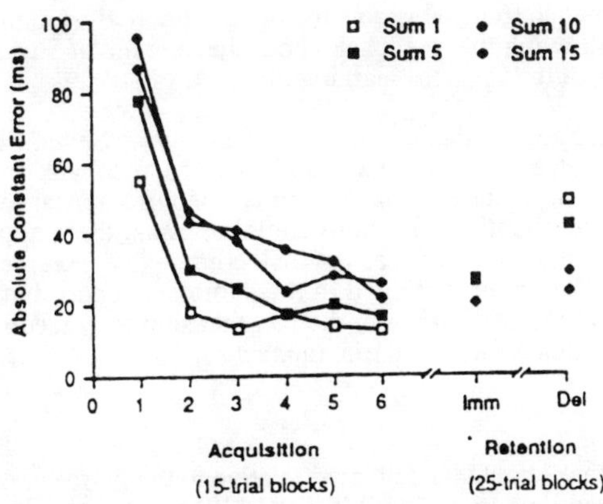

Figure 4. Average absolute constant error in a timing task for four summary feedback lengths in the acquisition phase, and for immediate (10 min) and delayed (2 days) retention tests without feedback (from Schmidt, Young, Swinnen, & Shapiro, 1989).

(see Guay, Salmoni, & McIlwain, 1990, for similar findings). Again, summary feedback degraded performance in acquisition, but facilitated learning as measured by performance in retention.

In another experiment (Schmidt, Lange, & Young, 1990, Experiment 1), we sought an optimal summary length for learning. The logic was based on the suspicion, mentioned earlier here, that frequent feedback had two opposite effects--guidance-like effects that directed the learner to the target, and dependency-producing effects that degraded learning--which operated together. If the summary length were too short, the learner would be guided effectively but the dependency-producing effects would be strong; and if the summary length were too long, the dependency-producing effects would be avoided, but there would not be sufficient guidance. This predicts an optimal summary length, where the benefits and decrements are matched. Such an optimal length was clearly beyond the 15-trial length in the earlier experiment (Figure 4), perhaps lying near the 20-trial value studied by Lavery (1962) in Figure 3. Would the optimal length be shorter if the task were more complex?

Using a more complex coincident-timing task, Schmidt et al. (1990) again studied summary lengths of 1 (every-trial feedback), 5, 10, and 15 trials. Again, the 1-trial summary length was best for performance in the acquisition phase when feedback was present. But in the retention tests without feedback, the 5-trial summary length was best, followed by the 1-, 10-, and 15-trial conditions. There was a clear inverted-U effect here, with the 5-trial condition being the best balance of the positive and negative effects of feedback.

2.2.3. *Kinematic Feedback.* Young and Schmidt (1990; Young, 1988) have also applied the notion of summary scheduling to kinematic feedback, where the information is about the pattern of action that led to the particular environmental outcome achieved (e.g., information that the arm is bent in a golf stroke). Here, though, we used average feedback, where instead of each of the trials being shown on the graph, only the average of them is provided as a single data point, but with no kinematic feedback after each trial. In many ways, this is like summary feedback, in that the learner must wait until the set of trials is completed until feedback is received, but the content of the feedback is of course different. Young and Schmidt showed that every-trial kinematic feedback about movement patterning in a coincident-timing task was less effective for learning than average feedback after 5-trial blocks. This tends to generalize the summary feedback work to information about movement patterning. To the extent that every-trial feedback is detrimental to learning as in Lavery (1962), then it might not make very much differences whether the feedback after a set of trials is about each of those trials or about their average, as the main feature would seem to be the avoidance of feedback after each trial. This conclusion is supported by data in a study by Proteau, Lee, and Schmidt (1990), showing that feedback about the average of 5 trials versus only the last one of the five trials was about the same for learning this task.

2.2.4. *Summary and Average Feedback Effects--A Summary.* Several experiments have not documented the beneficial effects of summary and average feedback for retention in several different kinds of tasks. In simpler tasks, longer summary lengths appear to degrade performance in acquisition but to enhance retention. In more complex tasks, there appears to be an optimal summary length, probably which decreases as the task complexity increases. Evidence suggests that it is not so much the information provided in the summary report that provides the benefits (although this may be a small factor), but rather the fact that summary feedback schedules prevent the learner from receiving feedback after each trial in practice.

2.3. FEEDBACK DELAYS

The question of when feedback should be delivered relative to the performance has been of interest for decades. Based on the notions from Thorndike's (1929) work, experimenter expected to find that separating the movement from its feedback in various ways would degrade learning, but this effect was almost never found in the skills literature (see, e.g., Bilodeau, 1966, and Salmoni et al., 1984, for reviews). Some kinds of separations do have this effect, such as inserting different tasks between a given performance and its feedback (e.g., Shea & Upton, 1976). But not all forms of separation have this effect, as seen in the work with summary feedback where numerous other trials of the same task can be inserted between a given performance and its feedback, which not only does not interfere with learning but seems to benefit it. However, in regard to "empty" feedback delays ranging from a few seconds to 1 min or so, there has never been any important effects of delay on learning.

Recently, Swinnen, Schmidt, Nicholson, and Shapiro (1990; see Swinnen, 1987) examined extremely short feedback delays ranging from literally 0 ms to about 240 ms. Their rationale was based on the idea, mentioned earlier here, that frequent feedback may have detrimental effects on learning because of its tendency to block certain information processing activities that occur just after a trial is completed. If so, then feedback presented instantaneously should degrade these processes further, and we should expect that instantaneously presented feedback should be detrimental for learning as compared to feedback that is delayed somewhat. In their Experiment 2, they used the coincident timing task mentioned earlier, and used feedback delays of 240 ms and 3.4 s in separate conditions for two days of acquisition, and then tested for retention after 10 min, 2 days, and 4 months.

The results are shown in Figure 5. There were no effects on feedback delay in the first acquisition day, but by the beginning of the second acquisition day there was an advantage for the delayed feedback condition. This advantage continued throughout the various no-feedback retention tests, and was even present but reduced somewhat on the 4-month retention test. We interpreted these findings to mean that instantaneous feedback blocked information-processing activities associated with subjective evaluation and error detection. When feedback is delayed, or when it is absent after a trial, these activities are supposedly carried out spontaneously, and they lead indirectly to the capability for the learners to detect their own errors, or perhaps to other benefits that contribute to performance on retention tests. In this sense, instantaneous feedback is the limiting value of every-trial feedback, which I have argued is detrimental to learning in the other paradigms mentioned in this paper to this point. Therefore, these experiments on instantaneous feedback contribute indirectly to our understanding of the ways in which frequent feedback can be considered as detrimental for learning. This is the major goal of the next major section of the paper.

3. Why Does Frequent Feedback Degrade Learning?

The most striking aspect of the data presented in the sections above is the tendency for feedback that is presented infrequently, and/or in ways that make it difficult for the learner to link the action with its feedback consequences, to enhance learning over every-trial feedback information. These findings are clearly contradictory to nearly every account about how feedback operates for learning,

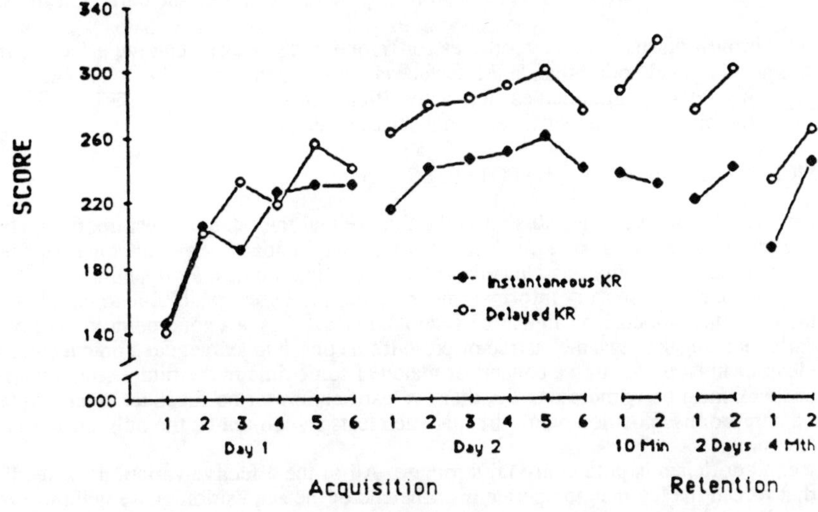

Figure 5. Average score (in arbitrary units) for a coincident-timing task for instantaneous versus delayed feedback in two days of acquisition, and for retention tests without feedback after 10 min, 2 days, and 4 months (from Swinnen, Schmidt, Nicholson, & Shapiro, 1990).

and thus demand additional concepts for us to understand the data available. In this final section of the paper , I attempt to bring these various findings together to contribute to an understanding, or the preliminaries to a theory, about how feedback operates to influence learning. In the sections that follow, I consider four major hypotheses about how frequent feedback operates, and consider the evidence for and against each of them.

3.1. A GUIDANCE HYPOTHESIS FOR FEEDBACK

First, however, I discuss the groundwork for the guidance hypothesis for feedback (Salmoni et al., 1984; Schmidt et al., 1989; Winstein & Schmidt, 1990) which has been mentioned briefly throughout the various sections so far. One way to view these data is to postulate two different kinds of feedback effects on learning--i.e., a two-factor theory of feedback operation. One set of effects is positive and contributes to learning. These have involved the processes recognized for decades, such as the tendency for feedback to facilitate motivation ("energizing" effects), but more importantly to provide information about errors so that the learner may correct them on the next trial, improving performance and maintaining it near the goal levels. We have termed these effects guidance effects, due to their similarity to physical and verbal guidance effects in the skills literature (e.g., Holding, 1976).

But we have also postulated a set of negative effects that detract from learning. We have not been able to be very specific about what these effects might actually be, however, and at this stage we have been able to list several likely possibilities. One is the general idea that the learner comes to depend on frequent feedback, almost as if the feedback became a "part of the task." If so , then if it is removed in a no-feedback retention test, performance suffers because of an acquired sensitivity to a particular kind of information which is no longer present. A second view is that frequent feedback, because it attracts attention so strongly, acts to interfere with or block important information processing activities which are important for learning, such as the processing of response-produced sensory information that leads to the development of error-detection capabilities. A third possibility is that frequent feedback encourages the learner to make too many corrections during practice, leading to the learner's failure to produce stable and consistent behavior. There are probably other hypotheses as well, and there is no reason that all could not be correct at the same time.

However, before I discuss these hypotheses for frequent feedback's negative effects, I must turn briefly to one additional notion that, at the surface at least, appears to be able to account for the feedback-scheduling effects mentioned earlier--a specificity, or similarity, hypothesis. There are several reasons why this view is not sufficient, as discussed next.

3.2. SPECIFICITY, OR SIMILARITY, HYPOTHESES

In the experiments discussed earlier, all shared the feature that the retention test upon which the estimates of learning were based were all conducted without feedback. Several concerns can be raised about this procedure. First, as Newell (1984) has pointed out, man lives in a world that contains abundant sources of sensory information, and it does not seem justifiable to base learning estimates only on tests that deprive subjects of such feedback. This is a good point. However, in many real-world teaching settings, the instructor provides feedback to learners in a practice session, but then the learner must perform in a concert or match at some time in the future, and where this instructor-feedback must be withdrawn. So, there is justification for no-feedback retention tests, and perhaps a compromise position would be that such tests should not be the only ones used for measures of learning.

But a second criticism is potentially far stronger. All of the effective variations of feedback were those that forced the learners to operate near the end of the acquisition phase without every-trial feedback. If so, then as compared to other conditions with every-trial feedback, the reduced feedback conditions are more "similar" to the no-feedback retention tests than are the every-trial feedback tests. A similarity view, a specificity view (Henry 1968), or an encoding specificity view (Tulving & Thomson, 1973) would argue that, for a given set of conditions at the retention test, those conditions in the acquisition phase that are most similar will produce the best retention performance (i.e., learning). Reduced relative frequency, summary feedback, and average feedback all have this feature. Thus, it could be the case that the only reason that retention performance was better for these conditions was that they resembled the acquisition conditions most strongly. There are several lines of evidence against this view, mentioned next.

First, several of the experiments in Winstein and Schmidt (1990) used retention tests that had feedback present on every trial. If the similarity view were correct, the conditions that had the highest frequency of feedback in acquisition would be best on these retention tests because they were the most similar. Yet in Winstein and Schmidt's Experiment 3, a 50%-fade condition was still more accurate than a 100% condition, with the difference being about the same as in the no-feedback retention test. In their Experiment 1, there was no interaction between the frequency of feedback in the acquisition phase (33 or 100%) and the frequency of feedback in the retention test (0, 33, 66, or 100%), with the 33% condition in acquisition always showing a slight advantage. Thus, in these experiments, there was no tendency for the similarity of conditions in acquisition and retention to influence retention performance.

Second, in Experiment 2 of the Schmidt et al. (1990) article on summary feedback, we used retention tests with feedback on every trial. There was no tendency for the 1-trial summary condition in acquisition (essentially every-trial feedback) to perform better than a group with 5-trial summaries, and some evidence that the 5-trial condition was better by the end of the retention test. Again, the conditions in acquisition most similar to the conditions at the retention test did not produce the best retention performance.

Finally, and perhaps most strongly, the logical extension of the similarity hypothesis predicts that a condition with no feedback at all in acquisition will be most effective for a no-feedback retention test--i.e., better than a group with some or all trials having feedback. There is ample evidence that this prediction fails (e.g., Bilodeau et al., 1959; Trowbridge & Cason, 1932), with the no-feedback conditions in the acquisition phase leading to very poor performance in retention, far poorer than a groups feedback on every trial.

For these various reasons, a specificity or similarity hypothesis cannot account for the findings provided earlier in this chapter. There may be a small sense in which the similarity, per se, of the acquisition conditions to the retention conditions influences retention performance, but it appears to be overshadowed by the feedback-scheduling variations. This hypothesis aside, another reasonable hypothesis--and the one promoted here--is that frequent feedback has in some way diminished the effectiveness of practice. We turn next to an elaboration of the views, mentioned earlier, that are designed to account for these frequent-feedback effects.

3.3. FEEDBACK BECOMES A "PART OF" THE TASK

Several authors have argued that feedback that is presented frequently, and particularly when it is useful for performance enhancement, begins to be relied on very strongly during acquisition. One by product of such a reliance is that the feedback stimuli become learned as a part of the collection of stimuli that comprise the task, just as the sounds of a car become a part of the information that contributes to driving. If the feedback does become a part of the task, then performance will suffer if the feedback conditions change, as they will in retention tests without feedback, for example. In this case, providing feedback less frequently in acquisition will not encourage such a dependency, because the learner must perform on his or her own resources in practice, and thus will not suffer decrements when the feedback conditions are removed. In other words, infrequent feedback in acquisition forces the subject to learn the task in a way that is independent of the feedback in acquisition.

A number of lines of evidence support this kind of hypothesis. Sanderson (1929) provided (or did not provide) instructions that subjects would be given a retention test to measure the learners' achievement. Performance on retention was generally facilitated by these instructions to retain. In Lavery's (1962) Experiment 2, the Immediate and Summary conditions that were used in Experiment 1 (see Figure 3 here) were used again, but now subjects were told that they would be given a no-feedback retention test. On this test, the Summary group was still more proficient than the Immediate group, but the differences were much smaller, suggesting that the instructions to retain in some way lessened the negative effects of the every-trial feedback conditions.

This kind of learner-dependency appears to operate with concurrent analog feedback, what might be called "on line" feedback, which is provided during the action for several kinds of tasks. For example, Annett (1969) gave or did not give concurrent visual information about the force being produced in a force estimation task. This information greatly facilitated performance over the no-vision condition, but produced drastic elevations in error on retention tests when the visual feedback was withdrawn. More recently, Proteau, Marteniuk, Girouard, and Dugas (1987) had subjects learn a rapid pointing task with vision either present or absent during practice, where vision facilitated performance. A retention test with no-vision produced a benefit for the group trained without vision, however. In both these examples, it appeared that the visual information became incorporated into the task as a part of the information supporting motor control, and when that part was removed in a retention test, performance suffered. Also, it is not clear that such processes will also occur with the augmented terminal (often verbal) feedback information under discussion here, as the information and the task performance are not so tightly intertwined as with concurrent feedback.

Against this view, however, is the evidence from Winstein and Schmidt (1990, Experiment 3), that reduced feedback had the same general facilitating effect on retention performance even when feedback was given after every trial. Here, if feedback had become a "part of the task," frequent feedback should have facilitated retention performance, not degraded it, because feedback was always present there. This is an interesting hypothesis, but it is not well supported by the evidence at present.

3.4. FREQUENT FEEDBACK MAY BLOCK OTHER PROCESSING ACTIVITIES

According to this general viewpoint, because augmented feedback is deemed to be so important for performance, and/or because learners are simply interested in how well they are performing, feedback attracts attention when it is given, and blocks or at least interferes with other important information processing activities that would otherwise occur after a trial. We have conceptualized two ways in which this interference might occur (others are also possible, of course), outlined below.

3.4.1. *Error-Detection Capabilities.* After a trial is completed, several lines of evidence suggest that the learner engages in an analysis of response-produced feedback in order to determine the effectiveness of performance on that trial. In one account of these processes, the learner compares the response-produced feedback from the previous action with a reference of correctness indicating how the correct action should have felt, with discrepancies indicating errors (e.g., Schmidt, 1975). With practice, the learner develops a sensitivity to response-produced feedback, leading to a learned capability to detect errors that provides the basis for recognition memory, or for an error-detection mechanism (Schmidt & White, 1972), that can facilitate performance in retention when augmented feedback may not be present. In terms of frequent feedback, this view is that feedback tends to block the processing of response-produced information (with frequent feedback blocking it more frequently). Therefore, frequent feedback produces less proficient performance in retention because of weaker error-detection capability developed during acquisition.

This was essentially the hypothesis under test in the instantaneous feedback study mentioned earlier (Nicholson & Schmidt, 1990a; see Figure 5). If feedback blocks such processing, then instantaneous feedback should block it most completely. After all, why should the learner go through the relatively effortful analysis of response produced feedback to obtain an estimate of performance proficiency when the experimenter provides that information "for free" in augmented feedback. And, this experimenter provided information is perfectly accurate, whereas the subjectively generated information is bound to be in error somewhat. Therefore, instantaneous feedback should lead to less effective learning, based on less effective error detection capabilities. Of course, we found that the instantaneous group showed less learning (in Figure 5), but we had no direct measures of error detection capabilities in this study, so it is not clear that the process by which less learning occurred involved these particular events.

However, in an experiment discussed earlier using summary feedback lengths of 1, 5, 10, and 15 trials in a coincident-timing task, where the 5-trial summary length was optimal for learning to perform (Schmidt et al., 1990, Experiment 1), we also gave a final unannounced retention test just after the no-feedback test of learning. In this final test, the learners were asked to perform the task as before, but to estimate their own performance score after each trial, and no feedback was given. We evaluated the strength of the error-detection capabilities by the within-subject correlation (over trials) between the estimated and actual performance score. The condition with the largest correlation was the 5-trial condition--the same condition that had the most effective performance, and the estimates of error-detection strength for the other conditions roughly paralleled the levels of proficiency as well. One interpretation is that the 5-trial summary condition led to the most proficient error detection capabilities, which translated into better performances on the retention test when feedback was withdrawn.

On the other hand, none of the groups in this experiment had correlations that were particularly strong (the largest was .43), suggesting that the error detection capabilities in this task were not particularly well learned; the corresponding correlations in the simple ballistic task of Schmidt and White (1972) ranged from .80 to .90. Therefore, this may have been a relatively weak mechanism in the learning of this task. At any rate, there is some support for this hypothesis, and it probably deserves additional study to determine the extent to which this is a major mechanism in frequent feedback's operation.

3.4.2. *Retrieval Processes.* Unlike the hypotheses discussed so far, which focus mainly on the interactions between the feedback and the preceding trial, it is also possible that the detrimental effects of frequent feedback occur because of the interactions with the next trial. This notion, borrowed from the literature on blocked and random practice effects (e.g., Magill & Hall, 1990), holds that frequent feedback makes the generation of the next action "too easy." In that literature, when practice is blocked, with the same task being produced on a series of trials, the learner needs to retrieve response information only on the first trial, and can merely modify the action slightly on successive trials. Random practice, on the other hand, requires the retrieval of response information on each trial. If these "retrieval processes" (Bjork, 1975) are important for learning, then blocked practice may be less effective for learning because it eliminates the need for the learner to practice retrieving response information. Because these retrieval operations are generally required on a retention test, it is easy to understand that retention performance (the measure of learning here) will suffer when learners practice under blocked as compared to random practice conditions.

How does this relate to the feedback literature? Because feedback directs and guides the next action, indicates what the errors were and how to correct them, and facilitates recall, various retrieval processes involved in response preparation are thought to be blocked when feedback is given too frequently. If it is also assumed that these retrieval processes are important for learning, then frequent feedback may interfere with learning because it prevents these retrieval processes from occurring. Less frequent feedback, or summary or average feedback, which has a less effective directive or guiding property, may require the subject to engage more actively in information processing activities necessary to produce the next response, thereby facilitating learning.

At present, there is no convincing evidence either for or against this hypothesis. But it is deserving of study, especially because it has the potential of tying together the literature on blocked vs. random practice conditions to the phenomenon of frequent feedback that is of interest here. It is possible that these two, rather distinct fields of study may have many common processes underlying them.

3.5. MALADAPTIVE SHORT-TERM CORRECTIONS

An old idea in the feedback literature is that when learners receive information about errors, it motivates a change in behavior; but when no information is provided after an action, the learner tends to repeat the previous action rather than to alter it (Bilodeau, 1966). This has led to the idea that frequent feedback leads to changes in the action on every trial, and thus it perhaps prevents the learner from producing stable behavior during practice. This may result in a practice sequence that is very effective for keeping the learner near the target, as the changes compensate for any tendency to drift off target, but may produce detrimental effects in the long term because the learner has not acquired a stable representation for producing the action at recall. In this way, these frequent trial-to-trial changes are maladaptive short-term corrections with respect to the main goal of generating effective retention performance. In short, then, this hypothesis says that frequent feedback is detrimental to learning because it produces too many modifications to the action during the acquisition phase.

Several of the feedback variations mentioned earlier have the feature that they tend to reduce the amount of trial-to-trial changes in acquisition. Of course, withholding feedback in the situations with reduced relative frequency has this effect, as does withdrawing feedback during the set of trials preceding the presentation of summary or average feedback. And, the bandwidth procedure withholds feedback on correct trials, which discourages the learner from making changes when the movement is already nearly error-free. With bandwidth feedback, the subject is in fact more consistent during the acquisition phase, and sometimes also in the retention phase (Sherwood, 1983, 1988; Lee & Carnahan, 1990). It is possible that all of these paradigms could operate in fundamentally the same way.

3.5.1. *Responding to Noise Processes.* One concern with this hypothesis is that it seems to contradict a large body of literature suggesting that variability in practice is effective for long term learning (Schmidt, 1975, 1988). But there is one important difference in these two situations. In the variable practice literature, the learner is typically given several different movement goals to produce, such as throwing an object to several different targets, or with several different speeds; this intentional variability in practice generally facilitates performance at a novel version of the task practiced in acquisition. However, in the feedback literature, the learner is generally asked to produce only a single version of the task (there are exceptions, e.g., Wulf & Schmidt, 1989); here, different outcomes on successive trials comes as a result of the learner's errors, caused by several sources of error. One of these sources of error is neuromuscular,"noise-like" processes associated with force and time variability in muscle contraction (as modeled by Meyer, Smith, Kornblum, Abrams, & Wright, 1990; Schmidt, Zelaznik, Hawkins, Frank, & Quinn, 1979), processes that are thought to be beyond the learner's control and fundamental to the neuromuscular system.

The problem is that feedback given from from trial to trial will contain this "noise" variability as a part of its variance. If the learner attempts to make a compensation for each deviation from the target, then some of the compensation will be directed to correcting for these noise processes. Two difficulties arise. First, the noise is inherently uncorrectable, so these compensations do not contribute to performance proficiency or learning. But second, because these noise processes are statistically independent of the subject's actual error, they should add to the size of the error reported as KR, and thus should increase the subject's tendency to make compensations as a result. If these compensations are maladaptive in the sense described above, then attempting to compensate for noise could actually degrade learning. And, this kind of process would be expected to be particularly strong in later practice, where the noisy processes are large relative to actual errors in performance. This is consistent with the fact that frequent feedback at the end of practice is particularly detrimental for learning, whereas it is probably beneficial for learning in early practice.

3.5.2. *Feedback Forces Compensations.* In addition to the idea that feedback gives the learner information which may then be used for compensations, there is the possibility that feedback operates even more strongly than this--perhaps even forcing compensations and disrupting behavior, almost as if the learner were responding to it involuntarily. Evidence for this suggestion comes from an experiment by Nicholson and Schmidt (1990b), patterned after an experiment by Bilodeau, Sulzer, and Levy (1962). In our experiment, we had subjects learn a simple task of pressing a lever for exactly 1600 ms. On certain trials during practice, we first gave (or did not give) the subject feedback about the errors on the previous trial, and then instructed the subject to repeat the action that was just done, and not to attempt to reduce the errors signaled by feedback. Thus, the subject was to ignore feedback, and simply repeat the previous trial. On these repeat trials, when the subject had received feedback prior to the repeat instruction there was far more error in repetition than when the subject did not receive feedback. It was as if this feedback information in some way interfered with the memory of the previous action, and thus prevented the learner from producing it again. Other experiments in this series are underway to determine if the feedback operates in the same general fashion when the instruction to repeat is given before feedback is received. If it does, it suggests that feedback might be very difficult to ignore, almost as if feedback invoked involuntary, undesired corrections to the action.

This idea, while being quite speculative at present, has the potential to account for why frequent feedback might be detrimental for learning. First, frequent feedback simply invites corrections, and provides a way to compensate for the error. But even if the learner were attempting to resist making changes, attempting to be consistent from trial to trial to stabilize behavior, it is still possible that feedback acts to interfere with the memory for the just-produced performance and makes it difficult, if not impossible, for the leaner to repeat the previous movement. If so, then feedback can be thought of as a kind of "distractor task," which during later learning seems to prevent the learner from acquiring the task in a way that would facilitate long-term retention. Clearly, much more work needs to be done on this idea before we can take it very seriously, however.

4. Summary

In the present paper, I have first indicated several lines of evidence that point to the conclusion that feedback presented frequently in practice, particularly at the end of practice when the learner is relatively proficient, can have detrimental effects on learning relative to providing feedback in a number of ways that produce less frequency. This tendency manifests itself mainly on performance of retention tests, particularly when the retention interval is longer than one day. This effect is not caused by the fact that the reduced-feedback schedules in acquisition tend to match the (often used) no-feedback retention tests, as there is evidence that they also operate in the same way for retention tests that have feedback after each trial. It seems unavoidable to conclude that frequent feedback degrades the establishment of long-term retention capabilities in some way.

This conclusion is inconsistent with a number of otherwise reasonably well supported theoretical viewpoints that account for feedback's operation for motor learning (Adams, 1971; Bilodeau, 1966; Newell, 1977; Schmidt, 1975). Each of these views demands that frequent feedback be beneficial for learning, as trials without feedback are regarded as "neutral," neither contributing to nor subtracting from long-term memory. Adams' (1971) view predicts that these blank trials are actually detrimental to learning, not beneficial to it as the data show. My view is that this accumulation of data now demands a reformulation of the account of how feedback operates for learning, as all present accounts of this process fail to account for one or more of the effects seen here.

Our group has attacked this problem by proposing a guidance hypothesis (Salmoni et al., 1984; Schmidt et al., 1989; Winstein & Schmidt, 1990), where feedback is argued to have two classes of effects on learning. The so-called positive effects have been known for decades, where feedback indicates errors, directs corrections, guides behavior to the target, motivates, "energizes," and generally facilitates performance. But we postulate several negative processes that are indicated here. Feedback could become a "part of the task" when presented frequently, raising problems when it is removed later in a retention test; this hypothesis does not account for why reduced feedback facilitates retention even in tests where feedback is given on every trial, however. Frequent feedback could block several important between-trial information processing activities, such as the development of error-detection capabilities, or the use of retrieval practice, both of which have strong implications for performance on the retention test. Finally, frequent feedback could encourage the learner to make too many trial-to-trial corrections that could degrade the capability to produce stable behavior on the retention test.

These views are not mutually exclusive, and two or more of these negative processes could be operating at the same time, interacting in reasonably complex ways, to facilitate long-term retention performance. One goal for future research in this area is to evaluate these various hypotheses on the way to the formulation of a viable theory of feedback's operation in motor learning.

5. References

Adams, J.A. (1971) 'A closed-loop theory of motor learning', Journal of Motor Behavior 3, 111-150.

Annett, J. (1969) Feedback and Human Behavior, Penguin, Middlesex, England.

Bilodeau, E.A., & Bilodeau, I.M. (1958) 'Variable frequency knowledge of results and the learning of simple skill', Journal of Experimental Psychology 55, 379-383.

Bilodeau, E.A., Bilodeau, I.M., & Schumsky, D.A. (1959) 'Some effects of introducing and withdrawing knowledge of results early and late in practice', Journal of Experimental Psychology 58, 142-144.

Bilodeau, E.A., Sulzer, J.L., & Levy, C.M. (1962) 'Theory and data on the interrelationships of three factors of memory', Psychological Monographs: General and Applied 76, 1-19.

Bilodeau, I.M. (1966) 'Information feedback,' in E.A. Bilodeau (ed.), Acquisition of Skill, Academic Press, New York, pp. 255-296.

Bjork, R.A. (1975) 'Retrieval practice', Unpublished manuscript, Department of Psychology, University of California, Los Angeles.

Guay, M., Salmoni, A.W., & McIlwain, J. (1990) 'Summary knowledge of results for skill acquisition: Testing the guidance hypothesis', Unpublished manuscript, Laurentian University.

Guthrie, E.R. (1952) The Psychology of Learning, Harper and Row, New York.

Henry, F.M. (1968) 'Specificity vs. generality in learning motor skill', in R.C. Brown and G.S. Kenyon (eds), Classical Studies on Physical Activity, Prentice-Hall, Englewood Cliffs, NJ, pp. 331-340. (Originally published in 1958)

Holding, D.H. (1976) 'An approximate transfer surface', Journal of Motor Behavior 8, 1-9.

Hull, C.L. (1943) Principles of Behavior, Appleton-Century-Crofts, New York.

Lavery, J.J. (1962) 'Retention of simple motor skills as a function of type of knowledge of results', Canadian Journal of Psychology 16, 300-311.

Lavery, J.J., & Suddon, F.H. (1962) 'Retention of simple motor skills as a function of the number of trials by which KR is delayed', Perceptual and Motor Skills 15, 231-237.

Lee, T.D., & Carnahan, H. (1990) 'Bandwidth knowledge of results and motor learning: More than just a relative frequency effect', Quarterly Journal of Experimental Psychology, in press.

Magill, R.A., & Hall, K.G. (1990) 'A review of the contextual interference effect in motor skill acquisition', Human Movement Science, in press.

Meyer, D.E., Smith, J.E.K., Kornblum, S., Abrams, R.A., & Wright, C.E. (1990) 'Speed-accuracy tradeoffs in aimed movements: Toward a theory of rapid voluntary action', in M. Jeannerod (ed.), Attention and Performance XIII, Erlbaum, Hillsdale, NJ, pp. 173-226.

Newell, K.M. (1977) 'Knowledge of results and motor learning', Exercise and Sport Sciences Reviews 4, 195-228.

Newell, K.M. (1984) Personal communication to R.A. Schmidt.

Nicholson, D.E., & Schmidt, R.A. (1990a) 'Feedback scheduling effects in acquisition', Unpublished manuscript, Motor Control Laboratory, University of California, Los Angeles.

Nicholson, D.E., & Schmidt, R.A. (1990b) 'Information feedback produces interference in reproducing movements', Unpublished manuscript, Motor Control Laboratory, University of California, Los Angeles.

Proteau, L., Lee, T.D., & Schmidt, R.A. (1990) 'Scheduling variations in summary feedback', Unpublished manuscript, University of Quebec.

Proteau, L., Marteniuk, R.G., Girouard, Y., & Dugas, C. (1987) 'On the type of information used to control and learn an aiming movement after moderate and extensive training', Human Movement Science 6, 181-199.

Salmoni, A.W., Schmidt, R.A., & Walter, C.B. (1984) 'Knowledge of results and motor learning: A review and critical reappraisal', Psychological Bulletin 95, 355-386.

Sanderson, S. (1929) 'Intention in motor learning', Journal of Experimental Psychology 12, 463-489.

Schmidt, R.A. (1975) 'A schema theory of discrete motor skill learning', Psychological Review 82, 225-260.

Schmidt, R.A. (1988) Motor Control and Learning: A Behavioral Emphasis (2nd ed), Human Kinetics Publishers, Champaign, IL.

Schmidt, R.A., Lange, C.A., & Young, D.E. (1990) 'Optimizing summary knowledge of results for skill learning', Human Movement Science, in press.

Schmidt, R.A., Shapiro, D.C., Winstein, C.J., Young, D.E., & Swinnen, S. (1987) Feedback and Motor Skill Training: Relative Frequency of KR and Summary KR. Technical Report No. 1/87, Motor Control Laboratory, University of California, Los Angeles.

Schmidt, R.A., & White, J.L. (1972) 'Evidence for an error detection mechanism in motor skills: A test of Adams' closed-loop theory', Journal of Motor Behavior 4, 143-153.

Schmidt, R.A., Young, D.E., Swinnen, S., & Shapiro, D.C. (1989) 'Summary knowledge of results for skill acquisition: Support for the guidance hypothesis', Journal of Experimental Psychology: Learning, Memory, and Cognition 15, 352-359.

Schmidt, R.A., Zelaznik, H.N., Hawkins, B., Frank, J.S., & Quinn, J.T. (1979) 'Motor-output variability: A theory for the accuracy of rapid motor acts', Psychological Review 86, 415-451.

Shea, J.B., & Upton, G. (1976) 'The effects on skill acquisition of an interpolated motor short-term memory task during the KR-delay interval', Journal of Motor Behavior 8, 277-281.

Sherwood, D.E. (1983) The Impulse Variability Model: Tests of Major Assumptions and Predictions, Doctoral dissertation, University of California, Los Angeles.

Sherwood, D.E. (1988) 'Effect of bandwidth knowledge of results on movement consistency', Perceptual and Motor Skills 66, 535-542.

Swinnen, S. (1987) Knowledge of Results Delay Activities and Motor Learning, Doctoral Dissertation, Katholieke Universiteit Leuven, Leuven, Belgium.

Swinnen, S., Schmidt, R.A., Nicholson, D.E., & Shapiro, D.C. (1990) 'Information feedback for skill acquisition: Instantaneous knowledge of results degrades learning', Journal of Experimental Psychology: Learning, Memory, and Cognition 16, 706-716.

Thorndike, E.L. (1927) 'The law of effect', American Journal of Psychology 39, 212-222.

Tolman, E.C. (1929) Purposive Behavior of Animals and Men, Century, New York.

Trowbridge, M.H., & Cason, H. (1932) 'An experimental study of Thorndike's theory of learning', Journal of General Psychology 7, 245-260.

Tulving, E., & Thomson, D.M. (1973) 'Encoding specificity and retrieval processes in episodic memory', Psychological Review 80, 352-373.

Winstein, C.J. (1988) Relative Frequency of Information Feedback in Motor Performance and Learning. Doctoral dissertation, University of California, Los Angeles.

Winstein, C.J., & Schmidt, R.A. (1990) 'Reduced frequency of knowledge of results enhances motor skill learning', Journal of Experimental Psychology: Learning, Memory, and Cognition 16, 677-691.

Wulf, G., & Schmidt, R.A. (1989) 'The learning of generalized motor programs: Reducing the relative frequency of knowledge of results enhances memory', Journal of Experimental Psychology: Learning, Memory, and Cognition 15, 748-757.

Young, D.E. (1988) Knowledge of Performance and Motor Learning, Doctoral dissertation, University of California, Los Angeles.

Young, D.E., & Schmidt, R.A. (1990) 'Augmented kinematic feedback for skill learning', Manuscript in preparation, Motor Control Laboratory, University of California, Los Angeles.

6. Acknowledgements

Much of the research cited in this chapter was conducted under Contract No. MDA903-85-K-0225 from the U.S. Army Research Institute, Basic Research, to R.A. Schmidt and D.C. Shapiro, Motor Control Laboratory, UCLA. Thanks to Diane Nicholson for technical help with this manuscript.

THERE MUST BE A CATCH IN IT SOMEWHERE!

H.T.A. WHITING
Department of Psychology
University of York
Heslington, York YO1 5DD
England

G.J.P. SAVELSBERGH
Department of Psychology
Faculty of Human Movement Sciences
1081 BT Amsterdam
The Netherlands

ABSTRACT. In a series of three experiments, the effect of training under restricted light conditions on one-handed catching performance is investigated. Training in the dark - with only a luminous ball visible - is shown to produce performances equivalent to those of subjects who train in full light provided that sessions of training in the dark are interspersed with trials in full-light. A proportion of the increase in performance, following on training under the restricted light condition, is demonstrated to be attributable to increases in spatial accuracy. The positive changeover effects - from training in the dark to performing in the light - and the negative effects - from performing in the light to training in the dark - are shown to be attributable to increases/decreases in spatial accuracy (additional increases/decreases in *temporal* accuracy also are not excluded - the methodology used, however, did not allow categorical statements in this respect to be made). These findings are discussed from the point of view of two contrasting theoretical frameworks - information-processing and ecological psychology.

1. Introduction

McLeod and Jenkins (1990) are of the opinion that the ability to make very accurate, *timed* responses, to coincide with the arrival of approaching objects at a very early age is a feature of the normal maturational process of a visual motor system developing in a world of moving objects which must be grasped or avoided. They invoke the studies of von Hofsten (1987) on the interception behaviour of 9 month-old infants to demonstrate how, even at such an early age, infants are able to demonstrate finely tuned perception-action couplings. They go on to suggest that normal childhood provides such massive opportunities for the practice of basic perceptual-motor skills - like timing actions to coincide with approaching objects - that there is relatively little further development, possible in adulthood, at least in terms of phylogenetic skills. This postulate should not, however, be extrapolated to ontogenetic skills which are, clearly, amenable to training although extended periods of practice - even in adulthood - may be necessary if optimal performance is to be achieved.. Whether or not such training effects relate to *timing* performance *per se* in, for example, the skill of one-handed catching to be addressed in this paper, is an open question.

Where does the difficulty for such people lie? In an early study, Alderson, Sully and Sully (1974) confirmed, using adult subjects, that the skill of one-handed catching - when the hand begins at the side of the body - involves serial organisation comprising:

- a gross orientation of the catching hand into the path of the ball, followed by
- a fine orientation of the catching hand, followed by
- a grasp and hold action.

J. Requin and G. E. Stelmach (eds.), Tutorials in Motor Neuroscience, 77–85.

2. Experiments

2.1 EXPERIMENT 1

On the basis of a series of (3)[1] fore-tests a group of 17 truly 'poor' catchers was assembled. They were required to catch tennis balls projected from a ball projection machine at a speed of 11.0 m/s at a distance of 5.60 m. Two groups of N=9 and N=8 subjects were randomly chosen. The N=9 group trained under normal lighting conditions (*environment light*) for a period of three days and the other group trained in the dark (*environment dark*) - only a fluorescent ball being visible in an otherwise totally dark room - for a similar period. Training each day comprised 3 sessions of 30 trials preceded by a pre-test and followed by a post-test under normal lighting conditions.

Design:

Environment light group:
Days 1,2 & 3
Pre-test (full light)
3x30 Training trials (full light)
Post-test (full light)

Environment dark group:
Days 1,2 & 3
Pre-test (full-light)
3x30 Training trials (in the dark with only the fluorescent ball visible)
Post-test (full light)

The rationale for the inclusion of these 'transfer' trials (pre-test to Training 1 and Training 3 to post-test) was to establish the practical relevance (more long-term effects) of the training, i.e., that any improvements noted when training in the dark did, in fact, carry over into the normal catching environment and were not just transient.

The results of this experiment (Fig. 1) demonstrated marked effects for the subjects who trained in the dark (with only a luminous ball visible) over a period of three days - at the end of the period, their catching performance *in the dark* was as good as that of the environment-light group of subjects *in the light*. Further, a positive and significant ($p<.05$) effect for the group that trained in the dark when transferred from the dark (training 3) to the light (post-test) and a negative effect - significant only at the 8% level - when transferred from the light (pre-test) to the dark (training 1).

Although the effects of training in the dark, are impressive, it was not possible, on the basis of the evidence collected, (*number of successful catches made* was the only dependent variable used) to specify whether or not these gains were due to increases in spatial and/or temporal efficiency. While, circumstantial evidence from the studies of Rosengren et al (1988) and Davids and Stratford (1989) points to improved *spatial* efficiency, this evidence is not conclusive. In a similar way, the positive and negative changeover effects - when subjects were transferred from dark to light and vice versa -could not *specifically* be assigned to increased/decreased spatial and/or temporal efficiency. Although the findings of Savelsbergh and Whiting (1988) point to *temporal* efficiency as being the key parameter, it has to be borne in mind that a changeover from dark to light not only provides a more structured environment but it also provides subjects with opportunity to view the hand. They were, after all, still relatively 'poor' catchers.

---

1 It has to be appreciated that selecting subjects on a single fore-test can lead to many misclassifications. This is confirmed when, on a subsequent test, many subjects initially placed in the 'poor' category revert to the 'good' category. It is necessary, therefore, by means of a series of tests to establish subjects who are persistently 'poor' or persistently 'good'.

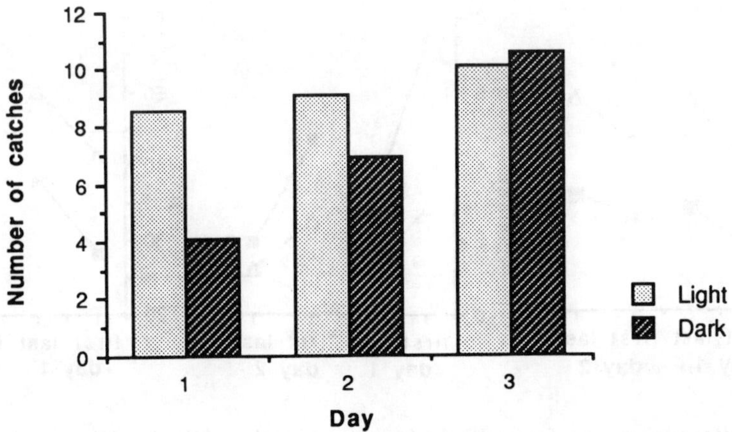

Figure 1. Mean catching performance over days for the *Environment Light* and *Environment* dark groups of subjects

In order to further clarify the *training in the dark* issue a second experiment was carried out. Clarification of the effects of transferring from dark to light, and vice versa, was left to a third experiment.

2.2 EXPERIMENT 2

The second experiment involved two groups of subjects, again, demonstrated to be (persistently) 'poor' catchers. Eight subjects were required to train under normal lighting conditions (serving as a control) over a period of two days and twenty-nine subjects (on whose performance interest was really focused) under similar dark conditions to that of the first experiment. Squash balls were used in this experiment (for pragmatic reasons as well as from the point of view of the generalisability of the results) instead of tennis balls. Each training day comprised 5 blocks each of 30 catching trials. In order to be able to specify the training effects more precisely, the number of dependent variables was increased to three:

- number of balls caught out of 30
- number of complete *misses* (per block of 30 trials) and,
- the total number of points - 2 points for a successful catch, 1 for a touch but not caught and 0 points for a complete miss - per block of 30 trials (following Rosengren et al. 1988).

Analysis of the results (Fig.2) of this experiment again showed a significant learning effect for both groups. The speculative conclusion, drawn on the basis of the results of Experiment 1, was confirmed. The significant improvement in catching behaviour, for subjects under the environment-dark condition could - at least in part - be attributed to more accurate spatial positioning of the hand.

This does not exclude, however, the possibility that there are also enhanced temporal effects (which could not be accurately ascertained on the basis of the experimental methodology used - no direct measure of temporal errors having been made). However, the significant increase in the *number of points* (Fig.2c) over trials, would not seem to be

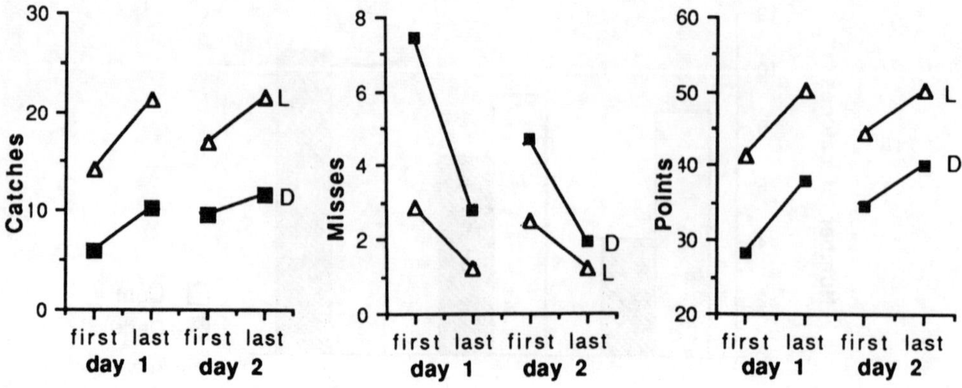

Figure 2. Mean performances for the dependent variables *Number of Catches, Misses* and *Points* over the two days of training.

accounted for by decreases in the number of spatial errors (Fig. 2b) alone. This experiment showed superior catching behaviour for those subjects who trained under normal lighting conditions suggesting that the changeover paradigm used in the first experiment has an important mediating effect on long-term training effects following on the two days training in the dark. The accuracy of this postulate is currently being investigated in a replication of Experiment 1 that includes a condition in which subjects receive their pre and post tests and train only in the dark.

2.3 EXPERIMENT 3

In a third experiment, an attempt was made to provide an adequate interpretation of the positive and negative effects demonstrated in Experiment 1 when subjects were transferred from the light to the dark and vice versa. There, it was speculated that the improvement in catching behaviour when transferring from performing in the dark (with only the ball visible) to performing in an environment that was not degraded could be attributed to the facilitatory effect that a structured environment has on the temporal accuracy of the catching action. It was, however, pointed out that transferring from the dark to the light also makes the hand visible - and vice versa - so that an alternative explanation of the catching improvement would be the availability of information about the position of the catching hand - especially for these, relatively ‘poor’ catching subjects.

To this end, an experiment was devised utilising the 29 subjects who had participated in Experiment 2. On the basis of their catching scores at the end of Day 2, they were divided into four (transfer) groups of similar mean catching ability: the so-called **H**(and) group (N=7); **V**(isual)**F**(rame)[2] group (N=7); **D**(ark) group (N=7) and **L**(ight) group (N=8). The single training day comprised 2 blocks of 30 trials in the dark room with only the fluorescent ball visible (training trials): the transfer trials comprised 3 blocks of 30 trials. The transfer conditions were as follows:

---

2 In the context of the present paper, this condition will not be expanded upon

- **H** group: ball and hand illuminated by UV light in an otherwise dark room
- **VF** group: ball and visual frame (a luminous frame surrounding the ball projection point and serving as a more structured environment than the conditions in which only the ball or ball and hand were illuminated)
- **L** group; normal lighting conditions
- **D** group, only ball visible in an, otherwise, dark room.

As the training consisted of 60 trials and the transfer task 90, scores were normalised out of 30. Analysis of the results (Fig. 3) for the dependent variable *number of catches* showed significant differences on the transfer trials between the **L** group and all other mean comparisons (**D,H,VF**: all fewer catches - p<.01). Also the performance of the **D** and **H** groups on the transfer task was significantly better than that of the **VF** group on all dependent variables (p<.05). A positive transfer effect from training in the dark to performing with vision of the hand and ball (**H**) or under the full light condition (**L**) was found for the dependent variable *misses* (Fig. 2b) implying an improvement in the accurate positioning of the hand following on practice in the dark. An apparent anomaly (which will not be pursued further here but which was the subject of further experimentation) was the decrement in performance (misses) when moving from the dark into the **VF** condition

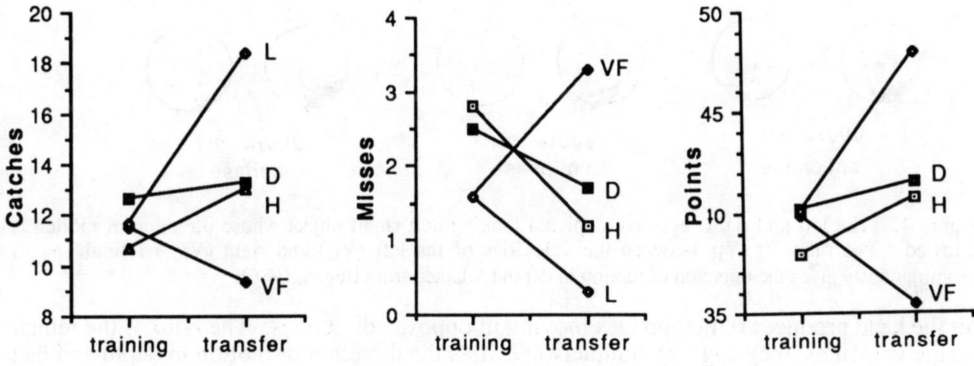

Figure 3. Mean performances for the dependent variables, *Number of Catches, Misses* and *Points* for the Training and Transfer trials.

for all three dependent variables. The **D** group, as expected (since they continued with the same condition) showed no significant differences when 'transferred'. The results for the dependent variable '*number of points*' (Fig. 3c) reflected the findings for the other two dependent variables.

2.4. Discussion

The significant positive effect for the dependent variable *number of misses* when transferring from training in the dark to performing with only the hand and ball visible (**H**) or to a full light condition (**L**) supports the idea that being able to see the hand, for subjects at this (relatively low) level of performance still has a positive effect on performance. The fact, however, that the effect was so much more pronounced for the group who transferred into full light (**L**) for the dependent variables *number of points* and *number of catches,* suggests that this is not the only advantage that being in the light contributes. This effect would not appear to be attributable to the modest, but significant, decrease in the *number of misses* alone, but suggests also (following on the findings of Savelsbergh & Whiting, 1988) an increment in *temporal accuracy*.

3. General Discussion

How can these positive effects of training in the dark be explained? From an information-processing point of view, Beverley and Regan (1973) have pointed out that the relative velocity of the two retinal images of an approaching object (Fig. 4) provides a precise cue as to its direction of motion in depth and suggest that the brain might be organised to take advantage of this geometrical fact. For example, a ball moving sideways produces retinal images that move at the same speeds in the same direction. A ball that will

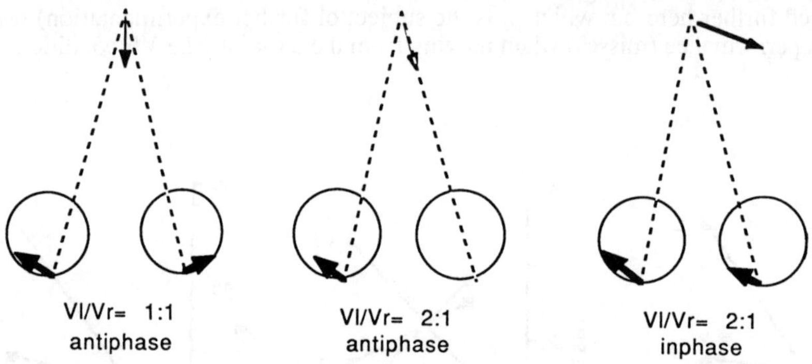

Figure 4. The left and right eyes are depicted looking at a small object whose direction of motion is arrowed. The ratio V_L/V_R between the velocities of the left (V_L) and right (V_R) retinal images unambiguously gives the direction of motion in depth (Adapted from Regan, 1986).

hit the head produces retinal images moving in opposite directions. The ratio of the retinal image velocities, they suggest, uniquely specifies the direction of motion in depth and this ratio is very sensitive when the trajectory passes near to the head (Regan, 1986). Thus, even in the environment dark training condition, subjects would have access to such information and, in principle, training could sensitize them to its usage.

From an ecological psychological point of view, an alternative potential source of predictive spatial information has been suggested by Fitch and Turvey (1977) and by Lee and Young (1986). These authors noted that an approaching ball, that will arrive at the point of observation, generates a stationary focus of expansion in the optic array. The displacing centre of expansion specifies the *direction* of the ball relative to the actor. Additional spatial information is - following Todd (1981) -provided by the ratio (the so-called Todd number) - of two time-to-contact components viz. time-to-contact with the vertical (longitudinal) plane and time-to-contact with the horizontal (sagittal) plane, both through the point of observation. When the Todd Number (Turvey & Carello, 1988) is varied, the following predictions can be made:

Tc vertical/Tc horizontal = 1 *ball will land at point of observation.*
Tc vertical/Tc horizontal > 1 *ball will land behind the point of observation*
Tc vertical/Tc horizontal < 1 *ball will land in front of the point of observation*

Thus, in principle both information processing and ecological psychological explanations can, account for the improved performance when training in the dark (with only the ball visible) - since both are dependent on the fact that all the necessary information is available in the approaching ball[3]. The explanation derived from Regan, however, is dependent upon binocular disparity/comparity whereas explanations based upon optical expansion information are not. Repeating the training in the dark experiment using monocular vision as well as binocular conditions would, therefore, help to clarify the explanation of the training effects shown. Such experiments are currently being carried out by the authors.

The positive and negative effects demonstrated in Experiment 1 (when subjects were transferred from the light to the dark and vice versa) - and clarified in Experiment 3 - can be attributed to the effect of the full light condition in enhancing both spatial (demonstrated) and temporal (postulated on the basis of the results) accuracy and the negative effects to the absence of such an enhancement. Explanations of these findings have been discussed by Savelsbergh, Whiting and Pijpers (1991) in ecological psychological terms. For example, transfer into a structured environment (dark to light) provides powerful information about the path of a ball based on different rates of gain (accretion) and loss (deletion) of optical texture outside the contour of the approaching ball. In these terms, increments in spatial accuracy following transfer need not be dependent only on the ability to see the hand. With respect to the postulated increase in temporal accuracy both information-processing interpretations (the presence of 3-D information enabling the computation of time to contact of the ball with the hand from velocity and acceleration information derived from the flight path of the ball) as well as ecological psychological approaches (based on the pick up of 'tau' the optical variable that *directly* specifies time to contact - Lee, 1975) can account for the findings. On the basis of parsimony, however, and on evidence for the primacy of *tau* in the control of the temporal aspects of one-handed catching, Savelsbergh, Whiting and Bootsma (1991) come down in favour of the ecological psychological explanation. .

In summary, the training of catching behaviour can be effective when subjects are required to practice in the dark with only a luminous ball visible. Such training, however, is only as effective as training in full-light if, following sessions of practice in the dark (with only the ball visible) subjects are allowed to perform under full-light conditions. Both an information-processing approach as well as an ecological psychological approach can account for the training effects reported in the series of studies. A future challenge - for proponents of one or other of these approaches - will be the designing of experiments that will demonstrate the greater explanatory power of the one as compared to the other. The proposed monocular viewing experiment will be a step in this direction.

Although hand-eye coordination (particularly in reaching behaviour) has been extensively studied in the motor control literature, the above studies suggest issues that, to date, do not appear to have been adequately addressed in such frameworks. For example, the fact that viewing the hand has a positive effect on catching performance has yet to be satisfactorily explained in motor control terms. Furthermore, the fact that 'good' catchers do not need to be able to see their hand for optimal catching performance brings into focus the often-cited - in the earlier literature on motor learning - phenomenon of a transfer (with learning) from visual to kinesthetic monitoring. It is interesting, therefore, with the increasing interest in theories of motor control, to see this putative phenomenon appear in a new guise. This has been in the context of mixed control models which imply an interplay between central planning and feedback processing. This latter type of control model has

---

3 In the case of the explanation provided by Beverley and Regan (1973) it would also be necessary to have knowledge of a scaling factor - such as the absolute size of the ball. The evidence provided by Judge and Bradford (1988) in relation to one-handed catching under a monocular condition suggests that such information might well be available.

been invoked by Proteau and his co-workers (Proteau and Cournoyer, 1990; Proteau, Marteniuk, Girouard & Dugas, 1987; Proteau, Marteniuk & Levesque, 1990; Marteniuk & Proteau, 1990) in the context of visual aiming movements. There experimental work led them to conclude that the earlier proposition suggesting that movement control goes from a reliance on visuo-spatial feedback to either a totally open-loop mode of control or a kinesthetic type of control should be rejected. To what extent the experiments on catching, carried out to date, question the generality of these findings awaits further explication and additional experimentation.

4. References

Alderson, G.J.K., Sully, D.L. and Sully, H.G. (1974) 'An operational analysis of a one-handed catching task using high speed photography', Journal of Motor Behavior, 6, 217-226.

Beverley, K.I. and Regan , D. (1973) 'Evidence for the existence of neural mechanisms selectively sensitive to the direction of movement in space', Journal of Physiology, 235, 17-29.

Carello, C., Grosofsky, A., Reichel, F.D., Solomon, H.Y. and Turvey, M.T. (1989) 'Visually perceiving what is reachable', Ecological Psychology, 1, 27-54.

Davids,K. and Stratford, R. (1989) 'Peripheral vision and simple catching the screen paradigm revisited', Journal of Sports Sciences, 7, 139-152.

Diggles, V.A., Grabiner, M.D. and Garhammer, J. (1987) 'Skill level and efficacy of effector visual feedback in ball catching', Perceptual and Motor Skills, 64, 987-993.

Fischman, M.G. and Schneider, T. (1985) 'Skill level, vision and proprioception in simple one-handed catching', Journal of Motor Behavior, 17, 219-229.

Fitch, H.L. and Turvey, M.T. (1977) 'On the control of activity: some remarks from an ecological point of view', in D. Landers and R. Christina (eds.), Psychology of Motor Behavior and Sport, Human Kinetics, Champaign, Illinois.

Hofsten, C von (1987) 'Catching', in H.Heuer and A.P. Sanders (eds.), Perspectives on perception and action, Lawrence Erlbaum, Hillsdale.

Judge, S.J. and Bradford, C.M. (1988) Adaptation to telestereoscopic viewing measured by one-handed ball-catching performance, Perception, 17, 783-802.

Lee, D.N. (1976) 'A theory of visual control of braking based on information about time-to-collision', Perception, 5, 437-459.

Lee, D.N. 1980) 'Visuo-motor coordination in space-time', in G.E. Stelmach and J. Requin (eds.), Tutorial in Motor Behaviour, North-Holland: Amsterdam.

Lee, D.N. and Young, D.S.(11986) 'Gearing action to the environment', Experimental Brain Rescarch Series 15, Springer-Verlag, Berlin.

Marteniuk, R.G. & Proteau, Luc. (1990). Evidence for feedback control of learned fast movements: Further support for specificity of learning. (submitted).

McLeod, P. and Jenkins, S. (1990) 'Timing accuracy and decision time in high-speed ball games', International Journal of Sport Psychology, (in press).

Proteau, L., Marteniuk, R.G., Girouard, Y. & Dugas, C. (1987). On the type of information used to control and learn an aiming movement after moderate and extensive training. Human Movement Science, 6, 181-199.

Proteau, L., Marteniuk, R.G. & Levesque, L. (1990). A sensorimotor basis for motor learning: evidence indicating specificity of practice. (submitted).

Proteau, L. & Cournoyer, J. (1990). Vision of the stylus in a manual aiming task: the effects of practice. Quarterly Journal of Experimental Psychology, (in press).

Regan, D.M. (1986) 'The eye in ball games: Hitting and catching', Proceedings of the conference on vision and sport, De Vriesenborch, Haarlem.

Rosengren, K.S., Pick, H.L. and Hofsten, C. von (1988) 'Role of visual information in

ball catching', Journal of Motor Behavior, 20, 150-164.
Savelsbergh, G.J.P.and Whiting, H.T.A. (1988) 'The effect of skill level,external frame of reference and environmental changes on one-handed catching', Ergonomics, 31, 1655-1663.
Savelsbergh, G.J.P., Whiting, H.T.A. and Pijpers, J.R. (1991) 'The acquisition and control of catching under restricted and textured environments', (submitted).
Savelsbergh, G.J.P., Whiting, H.T.A. and Bootsma, R.J. (1991) ''Grasping' TAU!', Journal of Experimental Psychology: Human Perception and Performance, (in press).
Smyth , M.M. and Marriott, A.M. (1982) 'Vision and proprioception in simple catching', Journal of Motor Behavior, 14, 143-152.
Todd, J.T. (1981) 'Visual information about moving objects', Journal of Experimental Psychology: Human Perception and Performance, 7, 795-810.

BIMANUAL MOVEMENT CONTROL: DISSOCIATING THE METRICAL AND STRUCTURAL SPECIFICATIONS OF UPPER-LIMB MOVEMENTS[1].

S. P. SWINNEN
Motor Control Laboratory ILO
Katholic University of Leuven
Tervuurse Vest 101
3001 Heverlee
Belgium

ABSTRACT. When performing upper-limb movements with different spatiotemporal features simultaneously, mutual synchronization effects occur that give rise to patterns of interference. Nevertheless, practice with appropriate information feedback can result in a reduction of these synchronization tendencies or in a dissociation of the bimanual movement patterns although individual differences are apparent. The present study attempted to obtain insight into the nature of interlimb dissociation. Findings revealed that success in bimanual skill was partly dependent on the capability to differentiate the neural activation levels for both limb movements through reduction of "neural crosstalk".

1. Introduction

Recently, we (Swinnen, Walter, & Shapiro, 1988) introduced a paradigm to investigate the human performer's capability to perform different movements in both upper-limbs simultaneously. The task consisted of performing a unidirectional flexion movement in the nondominant limb together with a flexion-extension-flexion movement in the

---

[1] Support for the present study was provided through a grant from the Research Council of K.U. Leuven, Belgium (Contract No. OT/89/26). Further support was made available through a Collaborative Research Grant from the NATO Scientific Affairs Division (Contract No. 86/732) for international collaboration between the Motor Control Lab, UCLA, Los Angeles, California, and those of ILO, K.U.Leuven, Belgium and University of Illinois at Chicago.

J. Requin and G. E. Stelmach (eds.), Tutorials in Motor Neuroscience, 87–94.

dominant limb (see Figure 1). Subjects were requested to initiate and terminate these movements simultaneously in the overall time of 600 ms but to perform them independently of each other. Consequently, the major goal of the task was to decouple or dissociate the upper-limbs (Swinnen, Walter, Beirinckx, & Meugens, 1990a).

In some of the experiments that we conducted, subjects were not informed about the patterns that were actually produced whereas in others, they were provided augmented (displacement) information feedback about both limb actions at regular intervals. In general, the findings revealed a moderate to strong tendency to synchronize otherwise unrelated movement patterns (for similar evidence, see Kelso, Southard, & Goodman, 1979 and Marteniuk, MacKenzie, & Baba, 1984). In order to evaluate success in accomplishing this goal, a distinction was made between metrical and structural dissociation. This distinction emanates from a categorization of movement features as promoted by action theorists in recent years (Kelso, 1981; Turvey, Shaw, & Mace, 1978): Essential or structural features hereby refer to the qualitative or structural aspects of a movement whereas nonessential or metrical features refer to quantitative, scalar changes that can be imposed on movement while leaving its underlying structure basically untouched. In our studies, structural dissociation was determined through cross correlating the angular acceleration patterns of the right and left limb movement. Metrical dissociation, on the other hand, was assessed through comparison of the amount of work produced in each of both upper limbs. Augmented information feedback was found to result in more successful interlimb dissociation as determined by means of the aforementioned structural and metrical coefficients. Nevertheless, some remnants of coupling persisted throughout acquisition. Moreover, there were marked individual differences: Some subjects were largely successful in dissociating the metrical and structural specifications of both limb movements whereas others were not. The present experiment sought to explore the phenomena underlying success in upper-limb dissociation.

2. Experiment

2.1. SUBJECTS

Subjects were 18-year-old students (N = 50) from the Katholic University of Leuven. All were right-handed and did not have prior experience with the task.

2.2. APPARATUS AND TASK

The apparatus consisted of two horizontal metal levers (43 cm long), attached to virtually frictionless vertical axles. An adjustable handle was located at the distal end of each lever. Shaft encoders (4096 bits per revolution) were mounted at the base of the axles to determine displacement, sampled at 500 Hz. The subject was seated behind the apparatus such that the front of the body was aligned between the lever axes. When the arms rested upon the levers, the elbows were nearly extended at the starting position. The end position was located in front of the subjects (85 deg from the starting position), just lateral of the body's midline. The elbow was positioned just above the axis of rotation of the lever.

Subjects were to perform two movements that differed in spatiotemporal features (Figure 1): an elbow flexion (hereafter referred to as the unidirectional movement) in the nondominant arm and an elbow flexion-extension-flexion movement in the dominant arm (hereafter referred to as the reversal movement). The latter movement was generally more forceful than the unidirectional movement because a greater total distance was required. Subjects were to reverse direction in each of two 3 cm wide vertical tárgets, located at 24 and 53 degrees from the starting position. Thus, the subject was to move the lever towards the second target zone (angular displacement = 53 degrees), reverse direction to the first target zone (angular displacement = 33 degrees), and then reverse direction again to move to the eindpoint (angular displcement = 65 degrees). Consequently, the reversal movement was divided into three segments that differed in amplitude and timing.

Both limb movements were to be produced in a time as close as possible to 600 msec. The subjects were instructed to initiate the movement shortly after the "go" signal, but reaction times were neither stressed nor measured.

2.3. PROCEDURE

There were two acquisition phases, administered on two consecutive days, and consisting of fifty trials each. Subjects performed the unidirectional movement in the left limb together with the reversal movement in the right limb at all times. Overall time and final amplitude was the same for both movements (MT = 600 ms; AMP = 85°). They were instructed "to perfom both movements simultaneously, that is, to initiate and terminate them at the same time". Subjects were informed about the deviations from the target movement times after every fourth trial for each of both limb movements (relative frequency KR schedule = 25%). Timing of the movement was initiated as soon as the subject left the starting position and stopped when peak displacement was reached at the end position (Figure 1).

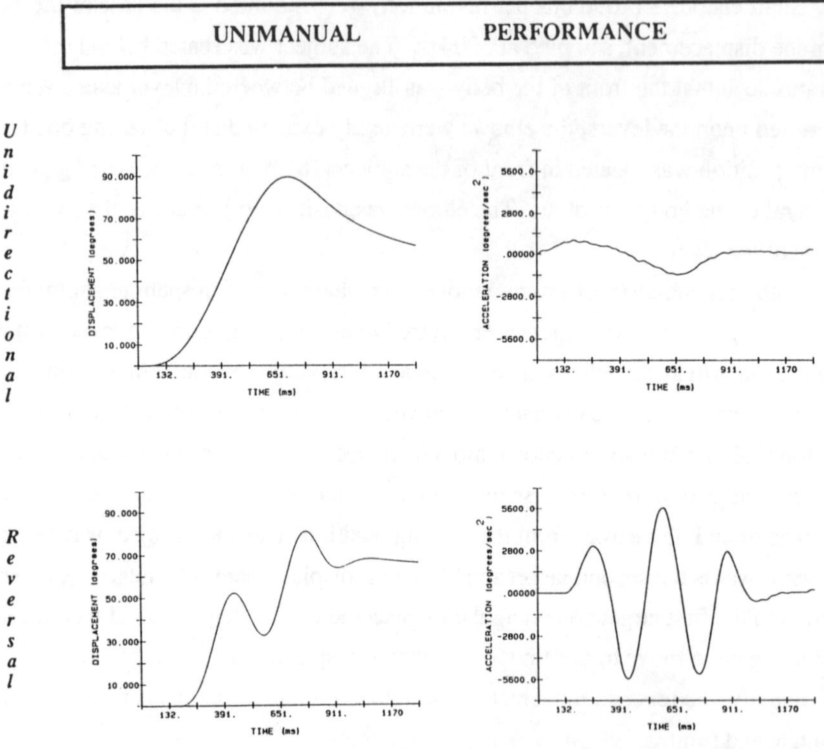

Figure 1. Illustration of the unidirectional and reversal movement as made in the nondominant and dominant limb, respectively.

All subjects were informed about the degree of coupling between the limbs through displacement information feedback (Knowledge of Performance, KP). Following every fourth trial, the computer terminal was turned to the subject and the displacement patterns of both limb movements (as a function of time) were shown, superimposed upon each other. The experimenter explained the patterns of interference that were mostly visible in the left limb movement. Interference was evident as an actual reversal, a short hesitation, or a small wave in the left displacement pattern that occurred at the time the reversal was made in the right limb.

At entering the testing room, the subject was to read preliminary information concerning the goal of the experiment, the movements to be performed, the number of practice trials during acquisition etc. It was stressed that independent limb movements were to be performed - that is a smooth arm flexion in the left limb and a reversal movement in the right limb. Moreover, subjects were told that both movements were to be made in the target time of 600 ms. Both these goals were repeated after the tenth and thirtieth trial during the first and second day of acquisition. A picture of an idealtypical bimanual trial was displayed in front of the subjects at all times.

In order to determine the degree of structural coupling between the limbs, the angular acceleration patterns of both movements were cross correlated in real time (Swinnen et al., 1988). Correlations were calculated per trial, transformed into Fisher's Z' scores, and then averaged across 10-trial blocks. On each trial, data points were taken from both acceleration traces every 10 ms as long as both limbs were being moved simultaneously, that is, from the time of the last initiated movement until the first completed movement. Previous research showed these cross correlation scores to approach zero when the movements were made independently of each other (Swinnen, Walter, Pauwels, Meugens, & Beirinckx, 1990b). No onset synchronization was applied prior to cross correlating the acceleration traces as it was intended to determine interactions between the limbs in real-time. On the basis of these cross correlations, two extreme groups were formed: Subjects of the group displaying high cross correlations were called Couplers (n = 10), those with low correlations were called Decouplers (n=10). Cross correlations of the Couplers were high throughout practice. In order, scores for Block 1 to 5 were, .75, .81, .83, .82, and .86, for the first day of acquisition and, .84, .87, .81, .82, and .82, for the second day. For the Decouplers, scores gradually decreased across practice: For the first day of acquisition, scores were .77, .59, .45, .35, and .2 for Blocks 1 to 5, respectively; for the second day, scores were .21, .12, .15, .08 and .1.

2.4. RESULTS

To better characterize the phenomena underlying success in interlimb dissociation, both groups were compared with respect to their performance on the following two dependent variables: overall movement timing and metrical dissociation.

Inspection of the overall timing data revealed that both groups showed progressively more temporal synchronization across practice. Indeed, average absolute time differences between the left and right limb movements decreased for both groups across practice, resulting in a significant effect for acquisition phase, $F(1, 18) = 23.35$, $p < .01$. Absolute time differences were larger for the Decoupling than for the Coupling Group, $F(1, 18) = 8.48$, $p < .01$. However, the differences between groups became smaller with increasing practice and no longer reached significance during the second acquisition phase ($p > .05$). Average time differences of the Decouplers were 259, 186, 159, 147, and 109 for the first day and 85, 77, 61, 80, and 74 ms for the second day. On the other hand, the Couplers' scores were 138, 92, 80, 43, and 37 ms for the first day and 46, 45, 56, 54, and 42, for the second day.

The question remains whether any differences in metrical dissociation could be observed between both groups that differed according to their level of structural dissociation. Metrical dissociation was assessed through division of the total amount of positive work[2] required for production of the reversal movement by the work required for production of the unidirectional movement (for an elaboration of this procedure see Swinnen et al., 1990a). The resulting ratio provided an indication of the differentiation of the intensity specifications for both limbs irrespective of their topological interactions. Previous studies have shown this ratio to approximate 8 when both limb movements are performed independently of each other (as under unimanual performance conditions). When the same intensity specifications are assigned to both limbs, the ratio tends towards 1.

The Decoupling Group's ratios were found to he higher than those of the Coupling Group during acquisition, $F(1, 18) = 8.35$, $p < .01$, indicative of a higher degree of differentiation of the intensity specifications for both limb actions. In order, scores for the Decouplers were 2.79, 3.32, 4.3, 4.34, and 4.6 for the first day and 4.6, 5, 5.18, 5.11 and 5.18, for the second day. For the Couplers, scores were 2.84, 2.7, 2.7, 2.73, and 2.63 for the first day, and 2.9, 3.2, 3.37, 3.15 and 3.31, for the second day of acquisition. Metrical dissociation increased with practice, resulting in an effect for acquisition phase, $F(1, 18) = 7.02$, $p < .01$. Whereas the Coupling Group only showed a small increase in metrical dissociation across practice, it was much larger for the Decoupling Group.

---

[2] Work was computed through integration of the positive values from the power-time profile. Muscle power is the product of the net muscle moment and angular velocity. Due to the proportionality between acceleration and force under the present task conditions, the power-time profile associated with the movement was obtained by multiplication of angular acceleration and velocity, expressed in deg/sec and deg/sec^2, respectively. Accordingly, the variable so obtained in the present study is proportional to power and its time integral is proportional to work.

To gain insight into the underlying causes of these changes in metrical dissociation, the amount of positive work generated in each limb was analyzed separately. It became evident that an excess of work was produced with respect to the unidirectional movement. Across practice, the Decoupling Group was capable to reduce this excess whereas the Coupling Group did not show marked changes across practice. For Blocks 1 to 5, scores for the Decouplers were 745, 727, 658, 593, and 523 units for the first day and 514, 479, 459, 464, and 456 units for the second day. For the Couplers, scores were 710, 891, 842, 740, and 801 for the first day and, 711, 716, 747, 712, and 698 units for the second day. The differences between groups almost reached significance, $F(1, 18) = 3.78$, $p = .06$. The decrease in work across both phases of acquisition was significant, $F(1, 18) = 6.68$, $p < .05$. With respect to the reversal movement, the Decouplers were found to generate a significantly higher amount of work than the Couplers during acquisition, $F(1, 18) = 4.22$, $p = .05$. Both groups showed increases in amount of positive work at the start of practice, followed by a small decrease. During the second day, scores stabilized for the Decoupling Group whereas the values for the Coupling Group further increased. Work scores with respect to the reversal movement for the Decouplers were 1840, 2186, 2377, 2268, and 2230 during the first day, and 2243, 2242, 2287, 2307, and 2283 units during the second day of acquisition. For the couplers, reversal scores were 1619, 1878, 1949, 1751, and 1728 during the first day, and 1682, 1904, 2011, 2008, and units 2014 during the second day.

3. Discussion

The findings of the present study demonstrated that subjects who were successful in overcoming the common organizational framework governing both limb actions (structural dissociation) were also more successful in dissociating the intensity specifications of the movements. This was accomplished through a reduction of the amount of excessive work that was generated for production of the unidirectional movement whereas the amount of work for the reversal movement was originally increased whereafter it stabilized. On the other hand, subjects not successful in structural dissociation showed no clear decreases in amount of work generated for the unidirectional movement. The amount of work for the reversal movement was increased across both days of practice. In contrast to the Decoupling Group, the Couplers showed a similar trend in the work scores for both limb movements, predominantly during the first practice day: When work for production of the reversal movement was increased, work for the unidirectional movement was increased as well. Consequently, it is hypothesized that success in the independent performance of these bimanual actions was at least partly dependent on the degree of neural output differentiation in the central nervous system. Apparently, less successful subjects had greater difficulty in preventing the "neural overflow" from the control centers of the right hand to those of the left hand and vice versa. As a consequence, intensity differentiation of the bimanual action patterns was less successfully accomplished. In addition, the cross correlations between the angular acceleration patterns of both limb movements remained very high across practice. The curtailing of this overflow effect or, what becomes evident at the behavioral level as a reduction of interference, is possibly accomplished through the recruitment of inhibitory neural networks serving to limit the spread of activity.

4. References

Kelso, J. A. S. (1981). Contrasting perspectives on order and regulation in movement. In J. Long & A. Baddeley (Eds.), Attention and performance (Vol 9) (pp. 437-457). Hillsdale, N.J.: Erlbaum.

Kelso, J. A. S., Southard, D. L., & Goodman D. (1979). On the coordination of two-handed movements. Journal of Experimental Psychology: Human Perception and Performance, 2, 229-238.

Marteniuk, R. G., MacKenzie, C. L., & Baba, D. M. (1984). Bimanual movement control: information processing and interaction effects. The Quarterly Journal of Experimental Psychology, 36A, 335-365.

Swinnen, S. P., Walter, C. B., Beirinckx, M. B., & Meugens, P. F. (1990a). Dissociating the structural and metrical specifications of bimanual movement. Manuscript accepted for publication in Journal of Motor behavior.

Swinnen, S. P., Walter, C. B., Pauwels, J. M., Meugens, P. F., & Beirinckx, M. B. (1990b). The dissociation of interlimb constraints. Human Performance, 3, 187-215.

Swinnen, S., Walter, C. B., & Shapiro, D. C. (1988). The coordination of limb movements with different kinematic patterns. Brain and Cognition, 8, 326-347.

Turvey, M. T., Shaw, R. E., & Mace, W. (1978). Issues in the theory of action: degrees of freedom, coordinative structures and coalitions. In J. Requin (Ed.), Attention and performance VII, Hillsdale, N.J.: Lawrence Erlbaum.

THE PERCEPTUAL-MOTOR WORKSPACE AND THE ACQUISITION OF SKILL

K. M. NEWELL, P. V. MCDONALD, & P. N. KUGLER
Department of Kinesiology
University of Illinois at Urbana-Champaign
Louise Freer Hall
906 S Goodwin Ave
Urbana, IL, 61801
USA

ABSTRACT. The acquisition of skill is examined with reference to the nature of the evolving perceptual-motor workspace. The evolution and dissolution of gradient and equilibrium regions within the workspace is discussed relative to the constraints on action. Some experimental data are reported which address the impact of some of the features of perceptual-motor workspaces on exploratory behavior, criterion performance, and transfer while learning to locate the minimum of an unknown function.

The process of acquiring a skill may be characterized as an exploratory search for a solution in a perceptual-motor workspace that continuously evolves from the confluence of constraints arising from the learner, the environment, and the task (Fowler & Turvey, 1978; Newell, Kugler, van Emmerik, & McDonald, 1989). The acquisition of skill in both phylogenetic and ontogenetic activities may be modeled in terms of the way subjects search and discover the gradient and equilibrium regions of this perceptual-motor work space. In this approach learning is synonymous with an increase in the coordination of the mapping of the perceptual kinematic informational properties with the kinetic motor properties in realization of the task demands (Shaw & Alley, 1985).

In this paper we continue to elaborate the search strategy approach to the study of skill acquisition and, in particular, emphasize with data based illustrations the way in which the structure of the perceptual-motor workspace provides information to channel the search in realization of the task demands. Initially we discuss how the constraints to action fashion the continually evolving and dissolving dynamics of the perceptual-motor work space. Subsequently, we show how the structure of this space, in terms of its equilibrium and gradient properties, can provide organization to the search strategy to realize the task goal. The protocol introduced by Krinskii and Shik (1964), whereby properties of this space can be manipulated apriori, is then elaborated as one experimental strategy to examine the influence of features of the perceptual-motor workspace on exploratory behavior during the the acquisition of skill. Some data are provided from recent experiments on adults learning to constrain two limbs in order to navigate through a computer generated perceptual environment (McDonald & Newell, 1990). The data illustrate how gradient, number of partial local minima[1], and symmetry of the perceptual-motor workspace all influence the acquisition and transfer of locating the minimum of an unknown function.

---

[1]We have termed these minima 'partial local' minima because these minima act in only one dimension of what is a 2 dimensional state space (with respect to anatomical degrees of freedom). An example of these minima are the 'corners 'of the inverted pyramid function illustrated in Figure 1.

J. Requin and G. E. Stelmach (eds.), Tutorials in Motor Neuroscience, 95–108.

Constraints on Action

Actions emerge from the set of constraints defined over the interaction of the organism, the environment, and the task. Traditionally, theories of learning have given emphasis to the constraints defined either as environmentally or organismically based. For example, the traditional tenets of behaviorism placed the emphasis external to the organism by virtue of seeking all the variance for learning in the environment. In contrast, current prescriptive approaches to skill acquisition emphasize the organism by virtue of having all the details necessary to execute a movement sequence explicitly represented within a motor program or action plan. The ecological approach to perception and action recognizes the reciprocity of the organism and the environment and consequently an account of perception and action is sought at the level of the organism-environment interaction.

In motor skill acquisition it is important to recognize a third class of constraints to action: those that arise from the task (Newell, 1986, 1989). Task constraints include the goal of the act and the rules that specify or constrain the movement sequence that may be produced in support of the action. Small quantitative variations in the task constraints can produce large scale qualitative changes in the movement sequence produced by the learner. For example, a small change in the distance that an object is to be projected can lead to a qualitative change in the action pattern used to realize the goal.

The constraints on action are open to description from several frames of reference. A typically chosen frame of reference has been one external to the learner, and and one example of this is efforts by experimenters and theoreticians of motor skills at task analysis. However, we emphasize it is the learner's *perceptions* of these constraints and their impact *relative* to the learner which are fundamental in defining the perceptual-motor workspace. This distinction is very important given that there is rarely an isomorphism between an experimenter's account and a learner's perception of the constraints to action. A property of some theoretical and operational importance is that task constraints viewed from the framework developed here can be characterized in the same terms used to describe the dynamics of the movement sequence (McGinnis & Newell, 1982; Saltzman & Kelso, 1987). Thus a unified description of the constraints for action, and the emergent properties of action, can be realized.

The Perceptual-Motor Workspace

The notion of the perceptual-motor workspace is drawn from the broader agenda of the ecological approach to perception and action (Gibson, 1979; Kugler & Turvey, 1987, Turvey & Kugler, 1984). Kugler and Turvey (1987) define the perceptual-motor workspace as that dynamic interface between the informational flows arising from perception and the kinetic flows arising from action. These fields are viewed as complementary in that not only can forces (kinetics) give rise to changes in flows (kinematics) but, in addition, flows can give rise to changes in the forces. In this view it is the increased coordination in the mapping of the perception and action fields that reflects learning (Shaw & Alley, 1985).

In order to clarify the concept of a perceptual-motor workspace it is necessary to say something about the features considered important in these workspaces. The physical basis of the workspace can be found in the concepts of dynamical systems. Consider a conservative pendulum set in motion by the addition of a squirt of energy. The pendulum begins to oscillate such that there is an exchange of potential and kinetic energy. At peak displacement, the potential energy is at a maximum and kinetic energy at a minimum; at the moment of peak kinetic energy the potential energy is at a minimum. Over the complete cycle energy is conserved. If left undisturbed the pendulum will continue to oscillate with exactly the same dynamics and there will be a conservation (of energy) over initial and final conditions. A system at equilibrium is said to be symmetric, and because this pendulum system is conservative the system is described as *maximally* symmetric. That is to say the system could be taken to any location in it's phase space and it's behavior defined at the initial conditions would continue unchanged.

In contrast consider a nonconservative pendulum. An isolated squirt of energy into a stationary, damped pendulum will also lead to oscillatory behavior, but because of the damping the energy added will be dissipated at some rate. This dissipation leads to a decrease in the energy defined over the pendulum cycle. The initial squirt of energy actually caused the system to move away from the point of equilibrium (zero motion). In moving away from equilibrium the pendulum is moved through a field defined with respect to energy (as one pushes the pendulum higher and higher the pendulum system gains potential energy). But in contrast to the conservative system there is a gradient in the energy field scaling to the level of dissipation. Thus the trajectory of the pendulum will slowly move back down the gradient to the point of equilibrium through the act of energy dissipation. The point of symmetry in this case has to be attained, and is done so through the action described in the second law of thermodynamics (cf. Kugler & Shaw, 1990). For a damped pendulum the equilibrium point at zero motion can be termed an attractor - that point at which the gradient region vanishes (the gradient is symmetrical). All trajectories of the pendulum system will converge to this point and as such this point is defined as the critical set for all trajectories. In addition, because this critical set is defined over an equilibrium 'point', the critical set is considered to be of (topological) dimension zero. Therefore the 'point' attractor is considered to be an attractor of topological dimension zero.

The physical concepts of symmetry and symmetry breaking, and the role of equilibrium and gradient dynamics (more generally attractor dynamics) have been proposed as appropriate for understanding the behavior of biological systems (Kay 1989; Kugler & Shaw, 1990; Kugler, Shaw, & Kinsella-Shaw, in press; Kugler & Turvey, 1987). However a sensient, intentional system negates an explanation of behavior based on current notions of existing physical laws (see Shaw, Kugler, & Kinsella-Shaw, 1990, for a discussion of this point). Generalization of these physical concepts to intentional systems demands the equilibrium set, or conservation, is defineable relative to both energetic and informational factors distributed across the task, the environment, and the organism. This equilibrium set is the critical set for the interacting perceptual (flow) and action (force) fields. Consistent with the physical example of the dissipative pendulum system the presence of an equilibrium set will give rise to a surrounding potential field of some gradient description. Consequently the biological behavior under investigation may be discussed relative to attractor dynamics.

We contend that behavior is organized relative to attractors, the attractor being defined over both physical variables and informational factors. These attractors (note the plural) act as building blocks for the perceptual-motor workspace, consequently biological systems are more often than not 'many attractor systems' (see Jackson, 1990). Interaction of attractors causes interesting features to emerge but this remains beyond the present discussion (see Abraham & Shaw, 1982, and Kugler & Turvey, 1987, for a more complete treatment). Skill acquisition may be considered as a search through this field of interacting attractors (the perceptual-motor workspace) with the aim being to realize a task-relevant mapping of perception and action. A point worthy of emphasis is that the layout of the perceptual-motor workspace is nonstationary. Although the task and environmental constraints may remain constant, the organismic constraints are constantly changing because the process of search/practice/activity acts to modify the learner in some way. Consequently the perceptual-motor workspace is also continually modified.

Gibson (1966, 1979) emphasized that it was the invariant features of the perceptual field that provided the informational support for action. In Gibson's view this information is movement relevant in that it affords certain actions or predisposes certain effectivities. For a learner who is naive with respect to accomplishing the task at hand many invariant perceptual features remain undiscovered. And, following the same line of reasoning, the naive actor has still to assemble many of the movement configurations available to a system of many degrees of freedom. The perceptual-motor workspace is in fact incompletely defined, or is defined relative to prior experience which may be inappropriate for the task at hand (consider this in light of the effects of negative or positive transfer). However, via exploratory behavior, which is guided by invariants recognized by the learner, additional information emerges which in turn channels the motor sequence. As a result, the mapping of perception and action in relation to the task goal becomes increasingly better defined. Of course this is not to say that 'all' practice is 'appropriate' practice (where 'appropriate' is seen to distinguish practice which aids in completing the task requirements effectively), but it certainly is the case that every moment of perceiving and acting causes the mapping to become more completely elucidated.

It should be recognized that features of the motor repertoire are often the source of information guiding the exploration in a perceptual-motor workspace. Boundaries can be defined with respect to many aspects of behavior including the continued maintenance of a certain 'mode' of coordination. In certain circumstances the rules of the task (a constraint on performance) may dictate crossing the boundary (changing the mode of coordination) to be inappropriate; in other situations operation beyond the boundary may preclude satisfactory performance of the task at hand. More severe are those boundaries which delimit regions of performance which threaten the integrity of the organism (an interesting issue is how much a performer knows of these boundaries, and what lies beyond, while never actually crossing the boundary!).

An additional problem facing experimenters and theorists alike is that the perceptual-motor workspace exists relative to a frame of reference different from that used for a behavioral description of the sources of constraint. A theoretical projection of this approach to motor skill acquisition is that the dimensionality of the perceptual-motor workspace will often be less than that of the biomechanical degrees of freedom inherent in the constraints to action. Bernstein (1967) directed attention to the degrees of freedom problem in motor control and suggested that skill acquisition is, in effect, the mastery of redundant degrees of freedom. One may consider this proposition in light of the concept of assembling a task specific device (cf. Bingham, 1988). Among other features, Bingham argued that a task specific device (TSD) is softly assembled. As well as reflecting the temporary nature of these assemblies Bingham (1988) stated "softness serves to emphasize the fact that TSD's are assembled out of the dynamic properties of anatomical structures as well as of the dynamic properties of objects associated with a task" (p. 245). Thus the dimensions for action become grounded in the functional requirements of behavior which by definition incorporates the anatomical and neural requirements. Consequently any efforts to develop a theory of perception and action must recognize the necessity of identifying a functional frame of reference. An elaboration of this general observation on the measurement problem remains a central issue to be addressed by future work in motor control.

A Method For Investigating Motor Learning

One of the orientations that we are pursuing in studying the acquisition of perceptual-motor skills involves the development of a protocol introduced by Krinskii and Shik (1964). Krinskii and Shik had subjects use motion at two joints in order to search for a minimum of some function, f(x,y), where x was the current angle of joint 1 and y was the current angle of joint 2. The subjects were provided continuous feedback with respect to their current status by way of an error value E, where E = f(x,y). Fluctuations in E would occur as the performer changed the joint angles and these fluctuations were continuously displayed for the performer. The minimum of the function was defined as some configuration of the two joint angles, so in essence the performers were required to find the combination of two joint angles which corresponded to the minimum of the function f(x,y). A representation of a mapping is illustrated in Figure 1 where,

$$f(x,y) = \left|\frac{\sqrt{2}}{2}x + \frac{\sqrt{2}}{2}y\right| + \left|\frac{\sqrt{2}}{2}y - \frac{\sqrt{2}}{2}x\right| \tag{1}$$

Krinskii and Shik (1964) actually used the function,

$$f(x,y) = |x - y - (a - b)| + \alpha|x - a| + \alpha|y - b| \tag{2}$$

where a, b, and α were held constant within a trial, and within experiments.

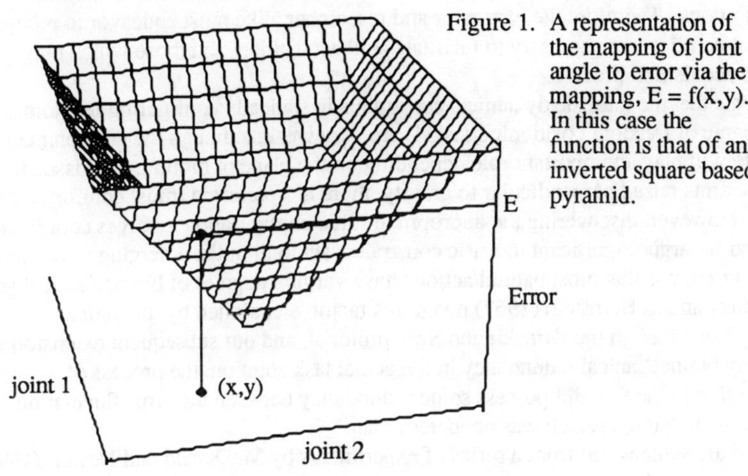

Figure 1. A representation of the mapping of joint angle to error via the mapping, E = f(x,y). In this case the function is that of an inverted square based pyramid.

Manipulation of two joints into some static configuration may seem like a fairly trivial movement task when compared to the intricate dynamic solutions that we all discover each day in tasks like walking, talking, eating, typing, etc. Nevertheless the Krinskii and Shik protocol possesses many of the facets we consider as crucial to understanding skill acquisition. First of all the task solution is represented as a minimum (an equilibrium point?) with a surrounding gradient field. Thus there is a symmetry defined at the solution and this symmetry is broken by being located away from the minimum. Second, the performer must engage in exploration via coordinated limb motion in order to locate the minimum, consequently the solution to the task is static but the process of achieving the goal state is dynamic. While the experimenter knows explicitly the invariant form of the mapping, f(x,y), between limb motions and fluctuations in E, the performer does not. The form of this mapping is revealed to the performer as limb movements are executed and the resulting fluctuations in E are monitored. Over time the performer establishes a mapping of fluctuations in E (information) with joint activity (action).

A useful property of the Krinskii and Shik protocol is that aspects of the perceptual space can be manipulated apriori to the running of the experiment, whereas the usual approach in motor control and skill acquisition has been to determine on a post hoc basis the nature of the perceptual space. Nevertheless, while the Krinskii and Shik protocol permits the experimenter to explicitly define a mapping of movements and error information this mapping should not be taken to represent the complete mapping of the performer's perceptual-motor workspace. Recall the workspace is defined over task, environment, and organismic constraints, but the function f(x,y) represents only the task constraints. The performer also brings to bear the influence of the organism's intrinsic dynamics. One consequence of this interaction is that the location of equilibrium regions in the externally defined function, f(x,y), and within the organism's intrinsically defined dynamics may not be consonant. Therefore the resulting critical set (that defined over the full complement of constraints) may be compromised in some manner. More specifically the performer may be diverted away from what is an intrinsically preferable solution by the form of the augmented information. This issue is relevant in a number of ways to skill learning, including individual differences, the necessity for providing appropriate augmented information, and the dynamic of this augmented information. Of course quite often skill acquisition necessarily involves transitions from 'comfortable', but relatively ineffective, task solutions. Therefore the instructor and researcher alike must endeavor to recognize which informational conflicts are necessary to facilitate skill acquisition, and those which may possibly hinder the acquisition process.

One may anticipate that the intrinsic dynamical constraints in statically configuring two joints (consider single degree of freedom positioning tasks!) are very weak, although the environmental constraint of gravity will play an important role depending on the plane of motion. The issue in fact is whether, with the arms raised perpendicular to gravity, there is a preferred (most comfortable) elbow joint angle? However, discovering the appropriate static configuration involves coordinated limb movement, and we argue significant intrinsic constraints act to limit the emerging movement patterns. One further point is that most natural actions have varying degrees of biomechanical and perceptual redundancy and as Bernstein (1967) noted this factor is exploited by the human movement system. Even though the Krinskii and Shik protocol, and our subsequent extension of it, did not permit any biomechanical redundancy in the actual task solution, the process of discovering the solution to the task did possess some redundancy between the error fluctuations and limb configurations such that the search was nondeterminate.

The data reported are synthesized from a series of experiments by McDonald and Newell (1990) in which subjects were tested under a protocol similar to that of Krinskii and Shik (1964). The subjects were required to minimize E as quickly as possible (where E = f(x,y)) by manipulating their elbow joint angles. Figure 2 is a schematic of the experimental setup. The experiments examined the effect of three geometrical manipulations: (a) the gain between elbow motion and error fluctuation; (b) the number of geometrically defined partial local minima (geometrically defined partial local minima act to cause certain patterns of error fluctuation); and (c) the symmetry of the gradient field around the minimum of the task function. The experimenter defined functions were stationary in that there was no change over time in the form of the function. The influence of these manipulations was assessed relative to performance on the task criterion (which was taken as the mean error over the final 3 seconds of the search), the form of the resulting exploratory behavior, and transfer performance between different functions.

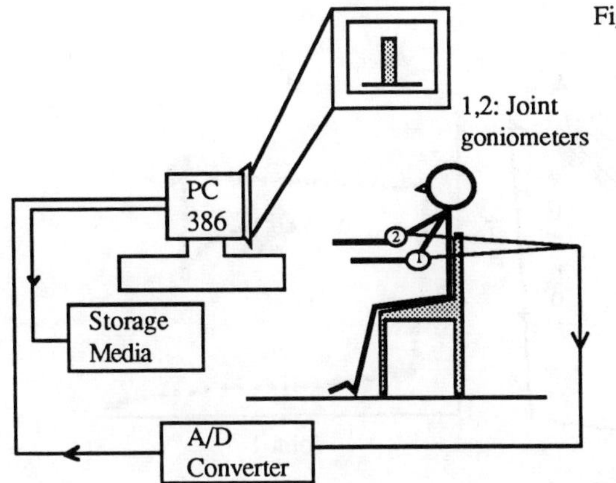

Figure 2. Joint angle displacements were monitored using two electrogoniometers; the signals were fed to the 386 via A/D converter. Changes in joint angles cause the error bar to fluctuate and this real-time feedback is used to search for the function's minimum.

The function used by Krinskii and Shik (1964), and those used by McDonald and Newell (1990), had a topological dimension of zero. However the general protocol is generalizable to functions of higher dimensions, and investigations of exploratory behavior and the acquisition of skill in such functions are currently underway. In addition the functions used by Krinskii and Shik (1964) and by McDonald and Newell (1990) all had a gradient which was continuously monotonic, radiating from the global minimum. There were no local minima which were separated from the global minimum by a local maximum. Hence it was possible for performers to take a trajectory which resulted in a continuous decrease in error all the way to the minimum. Of course rarely does learning proceed in such a manner. More importantly we contend that rarely would a workspace permit a trajectory which results in a monotonic approach to the goal state. Hence in future investigations we intend to examine the impact of nonmonotonic gradient functions on exploratory behavior and the acquisition of skill.

Figure 3 shows some key properties of the data that can be recorded in this experimental situation for a given perceptual-motor space function. Figure 3a provides a schematic of the potential function to be minimized where,

$$f(x,y) = \sqrt{x^2 + y^2} \tag{3}$$

Figure 3b shows the search behavior of the learner as reflected in the joint angle relations over the same trial. Figure 3c illustrates the error profile of a single subject over the 15 s of a single trial, and Figure 3d depicts the ratio of $\Delta E/E$ of each segment of the error profile over the same time series. The error ratio was calculated as the ratio of a local maximum in the error time series to the previous local maximum. Consequently an error ratio < 1 signifies an overall decrease in absolute error.

This experimental protocol provides for the potential derivation of many dependent variables beyond the three variables shown above. In this paper, we report data in a summary fashion and only as they relate to answers to specific questions about the influence of qualitative and quantitative manipulations of the perceptual-motor workspace on the acquisition of skill. The complete analysis may be found in McDonald and Newell (1990).

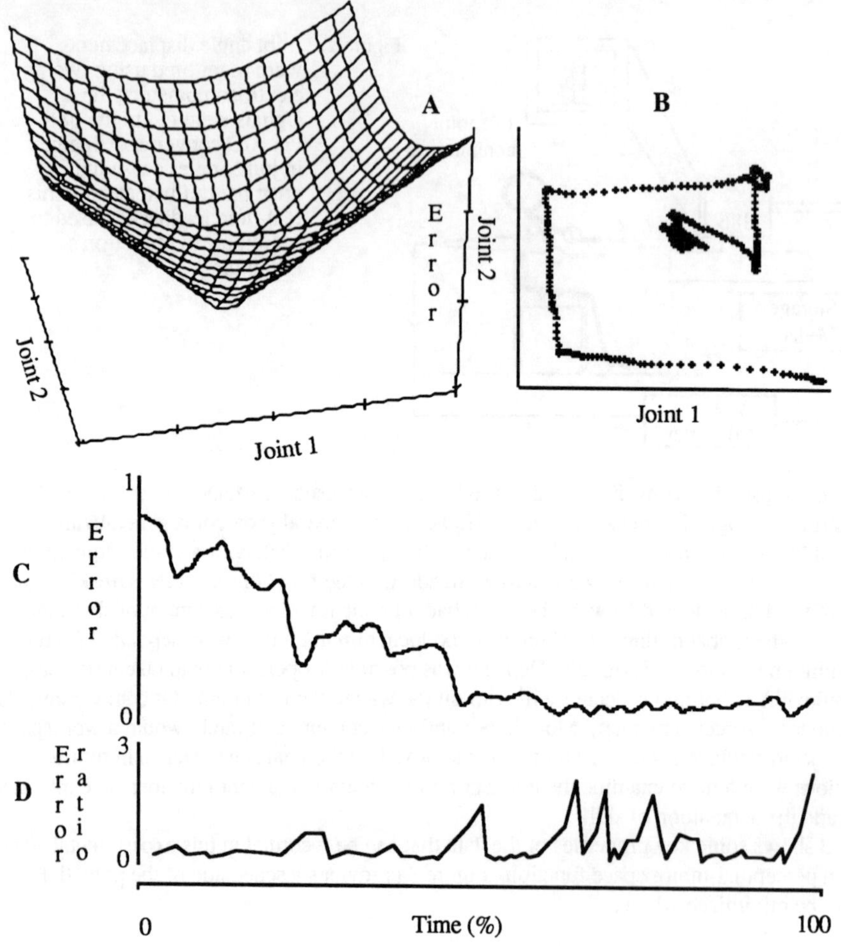

Figure 3. The cone function manifold (A), with a search trajectory performed in this space (B), and associated error time history (C), and error ratio time history (D). See text for details.

The Perceptual-Motor Workspace and Learning

The discussion will focus on the influence of three features of the perceptual space that channel the search and influence the level of performance achieved in the task. In the experiments conducted the main variables manipulated included the gradient, the number of partial local minima, and the symmetry of the search space. These features cannot always be manipulated independently because a change in one geometric property of the space can lead to changes in another geometric property. However, across the experiments conducted inferences can be drawn in relation to the relative influence of these variables on motor learning.

The gradient manipulation of the perceptual space is, in effect, a manipulation of the gain (or the ratio) between the change in error and the amount of motion at the elbow angles. It is analogous to manipulating the precision of KR in discrete tasks. The findings revealed a small but consistent influence of gradient on performance with the steeper gradient leading to less error in task performance. Thus, the greater gain led to less error in task outcome. This variable did not interact with degree of practice. Transfer between conditions with different gradients proved effective with subjects rapidly adjusting to the new gradient and hence the new gain between error information and elbow motion. The gradient or gain manipulated was not over the full range of the algorithm and therefore the facilitatory effect of gain may be limited to the range used.

CONSTRAINTS ON THE COORDINATION MODE

The manipulation of the perceptual space was confined to geometric aspects of the space. The stationarity of the space or the lack of time varying properties of the space would intuitively seem to place fewer constraints on the motor output. Thus, the intrinsic dynamics required of the motor output of this task would not seem to be so well defined as, for example, in rhythmical tasks.

However, there is considerably more structure in the motor output than may appear at first glance in Figure 3b. To examine this structure, the angle-angle relations were examined by segmenting the search into discrete subcomponents according to direction and length of the motion through the perceptual space. These subcomponents were then plotted on a graph with axes of the joint angles to provide a polar plot of the discrete motion segments as a function of coordination mode.

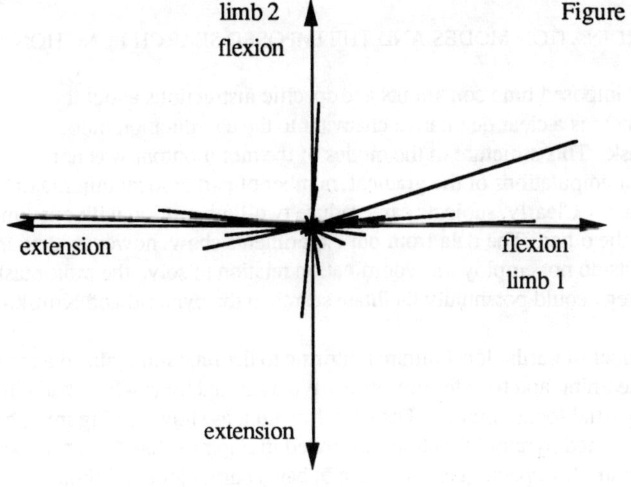

Figure 4. A polar plot of subcomponents of the search trajectory illustrating the limited range of coordination patterns.

Figure 4 shows an example polar plot of the search motion subsegments of a single trial as a function of angle-angle relations. This example, which is representative of most trials by most subjects, shows that not all angle-angle combinations were used for search steps beyond what may be considered the threshold of motion, which is defined by a small bandwidth around the center of the axes. In fact these data suggest, and further phase analysis confirmed that the motion of the two limbs can be characterized as being primarily of 5 modes: two limbs in-phase; two limbs out-of-phase; one limb (left); one limb (right); no motion. Subjects rarely, if ever, move the two limbs at different rates of motion in searching through the space. There is also a strong trend for the longer distance segments to be a consequence of two limb motion rather than one limb motion.

Most time during a trial was spent in the null movement mode, with single limb motion having more time than the dual limb motion. Within the dual limb mode most time was spent antiphase. There were also predominant patterns to the structure of the coordination mode transitions. That is, when in the null movement mode subjects predominantly switched to a single limb mode. When in the single limb mode there was about an equal probability of switching to either the null movement mode or a dual limb mode. Subjects rarely switched from one dual limb mode to another.

This tendency to utilize a small number of coordination modes, and furthermore for certain mode transition patterns to dominate, arose consistently across all the experimental manipulations performed. These trends appear to be exemplary of the performer's intrinsic dynamics. Kelso and colleagues have repeatedly demonstrated that the rhythmical coordination of two limbs varies in stability according to the relative phase of the two limbs. Indeed the inphase and antiphase patterns of motion are referred to as the system's intrinsic dynamics (see Kelso, 1990, for a summary). In addition Kelso, Scholz, and Schöner, 1988, provided empirical evidence that not all mode transitions are equally as easy (see also Schöner & Kelso, 1988, for a theoretical extension of this work). Even though the subjects in the search task could freely chose modes of coordination, and the search clearly involved periods of no motion (thus approaching discrete movement), the intrinsic dynamics emerge to be very consistent with those observed in rhythmical behavior. The coordination mode in effect not only provides boundaries to the nature of the current search, but it also seems to constrain the way in which the current status of the search is modified.

THE INTERACTION OF COORDINATION MODES AND THE IMPOSED SEARCH FUNCTION

Inspite of the lack of externally imposed time constraints and specific instructions about the coordination of the two limbs there is a clear qualitative character to the coordination modes employed by subjects in this task. This structure in the modes of the motor output was not significantly influenced by the manipulations of the gradient, number of partial local minima or symmetry of the perceptual space. Clearly, subjects can produce two-limb motion with one limb moving at a different rate than the other. The data from our experiments show, however, that under the conditions examined subjects do not employ this coordination relation to solve the motor task, inspite of the fact that this strategy could potentially facilitate search in the pyramid and Krinskii and Shik algorithms.

The manipulation of the number of partial local minima intrinsic to the function realized a strong effect on performance in both learning and transfer protocols. We used functions which had either 0, 4, or 6 intrinsically defined partial local minima. The cone function, as shown in Figure 3, has no partial local minima. The inverted pyramid function, illustrated in Figure 1, has 4 partial local minima. Finally, the Krinskii and Shik space, seen in Figure 5, has 6 partial local minima.

Increasing the number of partial local minima decreased performance level as reflected in a number of dependent variables, including error, distance travelled in the space, and percent of the space visited. These effects were most pronounced on transfering to a new space in that switching from the pyramid (4 partial local minima) to the cone (0 partial local minima) reduced error, whereas switching from the cone to the pyramid increased error (Figure 6). The Krinskii and Shik (1964) algorithm with 6 partial local minima proved the hardest space of all for subjects to learn, but it needs to be recognized that this manipulation is confounded with the asymmetry of the space.

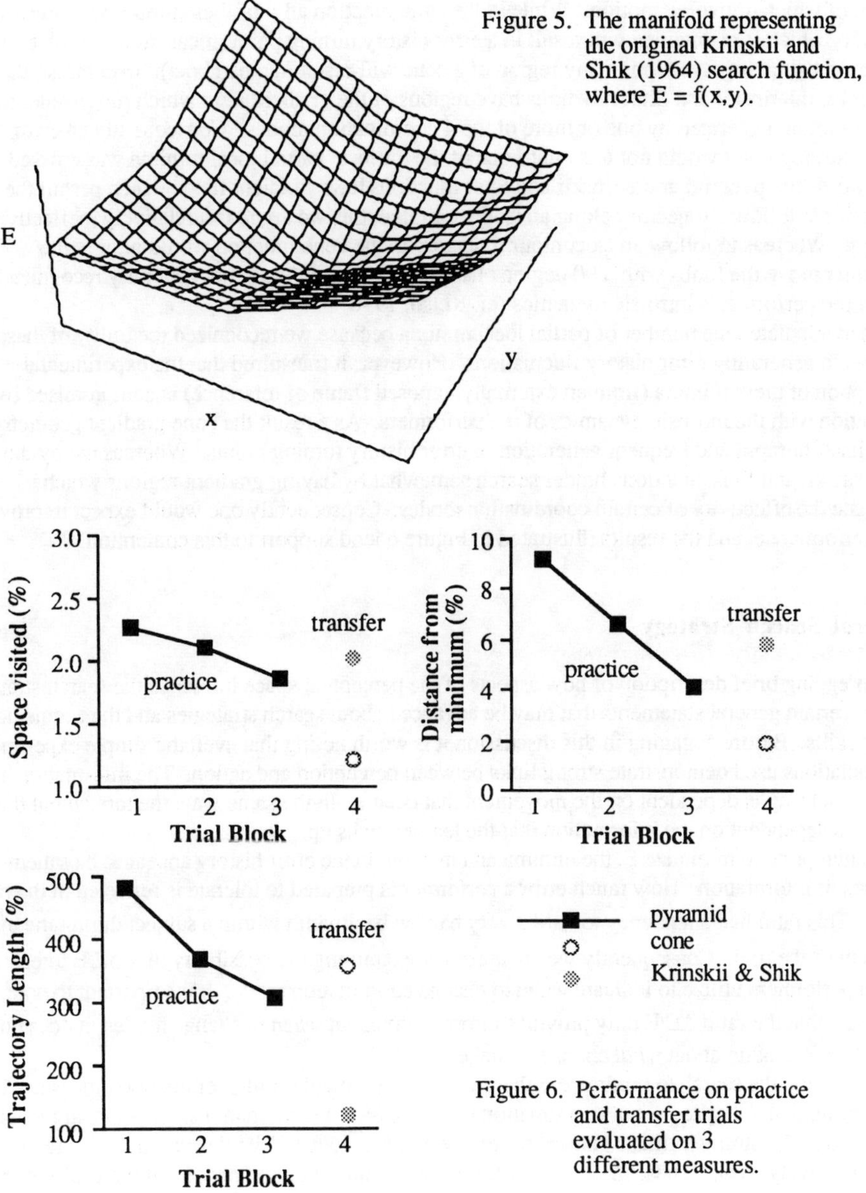

Figure 5. The manifold representing the original Krinskii and Shik (1964) search function, where E = f(x,y).

Figure 6. Performance on practice and transfer trials evaluated on 3 different measures.

The number of partial local minima intrinsic to the function is not necessarily isomorphic with the actor's perception of local minima or singularities in the perceptual space. The layout of the gradient interacts significantly with the coordination modes exhibited by the performers. To understand why, one must assume a major source of information in these tasks is the generation of a turning point in the error history. Recall that performers predominantly utilize 5 coordination modes, of which 4 involve motion. While in the cone function all 4 of these modes will allow a trajectory which if maintained will result in a error history turning point (recall from high school geometry that taking a section in any region of a cone will result in a parabola). In contrast, the pyramid and Krinskii and Shik functions have regions of the gradient field which run parallel to the axes of motion generated by one or more of the 4 coordination modes. Consequently an error history turning point would not arise until one of the intrinsic partial local minima was crossed. In other words the pyramid and Krinskii and Shik functions have gradient fields which permit the performer to follow a trajectory along an isocontour line, this we argue is a relatively ineffective strategy. Whereas to follow an isocontour trajectory in the cone function would require the performer move the limbs with a 90 degree phase lag. Such a mode is not generally recognized as part of the performer's intrinsic dynamics (cf. Kelso, 1990).

We manipulated the number of partial local minima because we recognized the utility of these features in generating error history fluctuations. However, it transpired that the experimental description of these minima (from an externally imposed frame of reference) is compromised by the interaction with the intrinsic dynamics of the performers. As a result the cone gradient geometry lends itself to rapid and frequent generation of error history turning points. Whereas the pyramid and Krinskii and Shik functions hinder search somewhat by having gradient regions which eliminate the effectivity of certain coordination modes. Consequently one would expect improved task performance, and the results illustrated in Figure 6 lend support to this contention.

General Search Strategy

The foregoing brief description of how aspects of the perceptual space influence the search strategy lead to certain general statements that may be advanced about search strategies and the acquisition of motor skills. Before engaging in this discussion it is worth noting that even the simple experimental manipulations used demonstrate strong links between perception and action. The information a learner picks up is dependent on the movement that occurs. In the same way, the movement that occurs is dependent on the information that the learner picks up.

In attempting to minimize E, the minima and maxima in the error history appear to be salient features of information. How much error a performer is prepared to tolerate is reflected in the ratio $\Delta E/E$. This ratio had a tendency toward a very narrow bandwidth within a subject during the initial segment of the trial. Consequently we are currently examining the possibility of a $\Delta E/E$ threshold which performers utilize to indicate when to change coordination mode. It is important to note, however, that the ratio $\Delta E/E$ only provides information about *when* to change modes, it does not provide information about *what* change to make.

In realizing the search to minimize E there was only a limited set of coordination modes used and, within that, a limited set of coordination mode transition combinations. This, we argue, is a reflection of the strong intrinsic constraints on the motor output despite the external constraints being relatively weak. These intrinsic constraints on the motor output prevented the subjects from picking up more information in certain search functions, which further emphasizes the strength of the intrinsic constraints. The data suggest that relative phase between limbs is a key variable even in tasks that do not demand it as a solution to the task constraint. One could argue that the performers do not take full advantage of the potential information inherent in the perceptual-motor space due to the constraints of the motor system. This limitation may be overcome with further practice, however the performers did perform relatively well even over the relatively short amount of practice allowed, and using only the limited number of coordination modes we observed.

The manipulation of properties of the perceptual space influence the acquisition and transfer of skill as reflected in discrete measures of the performance outcome. However, they do not seem to influence the information criterion for changing the coordination mode in the search or the probability of certain coordination modes or mode changes being employed. There appears to be some general rule employed about when to change coordination mode and this is accompanied by emergent relations in coordination mode transitions used to further minimize E.

Krinskii and Shik (1964) observed subjects to switch from a global to a local search strategy. Unfortunately it is difficult to relate the strategies we observed to this distinction because their strategies were defined only with respect to the motor output. Our preliminary analysis of the search strategies employed in these simple perceptual spaces suggests that subjects employ a general informational decision criterion with a probabilistic accompanying motor output. It is the form of the search through the space that is general, rather than the specific output of the motor system. Whether this consistency in strategy emerges as a function of the kinds of search spaces we used awaits further investigation.

References

Abraham, R. H., & Shaw, C. D. (1982). *Dynamics: The geometry of behavior. Part 1 - periodic behavior*. Santa Cruz, CA: Ariel Press.

Bernstein, N. (1967). *The co-ordination and regulation of movements*. New York: Pergammon Publishers.

Bingham, G. P. (1988). Task-specific devices and the perceptual bottleneck. *Human Movement Science, 7*, 225-264.

Fowler, C. A., & Turvey, M. T. (1978). Skill acquisition: An event approach with special reference to searching for the optimum of a function of several variables. In G. E. Stelmach (Ed.), *Information Processing in Motor Control and Learning* (pp. 1-40). New York: Academic Press

Gibson, J. J. (1966). *The senses considered as perceptual systems*. Boston: MA: Houghton Mifflin.

Gibson, J. J. (1979). *The ecological approach to visual perception.* Boston, MA: Houghton Mifflin.

Jackson, E. A. (1990) *On the control of complex dynamic systems*. Center for Complex Systems Research Tech. Report (CCSR-90-8). University of Illinois, Urbana, IL.

Kay, B. (1989). The dimensionality of movement trajectories and the degrees of freedom problem: A tutorial. *Human Movement Science, 7*, 343-364.

Kelso, J. A. S. (1990). Phase transitions: Foundations of behavior. In H. Haken & M. Stadler (Eds.), *Synergetics of Cognition* (pp. 249-268). Heidelberg, Germany: Springer-Verlag.

Kelso, J. A. S., Scholz, J. P., & Schöner, G. (1988). Dynamics governs switching among patterns of coordination in biological movement. *Physics Letters A*, *134*, 8-12.

Krinskii, V. I., & Shik, M. L. (1964). A simple motor task. *Biophysics*, *9*, 661-666.

Kugler, P. N, & Shaw, R. E. (1990). Symmetry and symmetry-breaking in thermodynamic and epistemic engines: A coupling of first and second laws. In H. Haken & M. Stadler (Eds.), *Synergetics of Cognition* (pp. 296-331). Heidelberg, Germany: Springer-Verlag.

Kugler, P. N, Shaw, R. E., & Kinsella-Shaw, J. (in press). The role of attractors in intentional systems. In D. S. Palermo & R. R. Hoffman (Eds.), *Cognition and the symbolic processes: Vol 3, Applied and ecological perspectives*. Hillsdale NJ: Erlbaum.

Kugler, P. N., & Turvey, M. T. (1987). *Information, natural law, and the self-assembly of rhythmic movement: Theoretical and experimental investigations*. Hillsdale, NJ: Erlbaum.

McDonald, P. V., & Newell, K. M. (1990). *Search strategies in perceptual-motor workspaces*. Manuscript submitted for publication.

McGinnis, P. M., & Newell, K. M. (1982). Topological dynamics: A framework for describing movement and its constraints. *Human Movement Science*, *1*, 289-305.

Newell, K. M. (1986). Constraints on the development of co-ordination. In M. G. Wade & H. T. A. Whiting (Eds.), *Motor Development in Children: Aspects of Coordination and Control* (pp. 341-360). Boston: Martinus-Nijhoff Publishers.

Newell, K. M. (1989). On task and theory specificity. *Journal of Motor Behavior*, *21*, 92-96.

Newell, K. M., Kugler, P. N., van Emmerik, R. E. A., & McDonald, P. V. (1989). Search strategies and the acquisition of coordination. In S. A. Wallace (Ed.), *Perspectives on the Coordination of Movement* (pp. 85-122). Amsterdam: North-Holland.

Saltzman, E. L., & Kelso, J. A. S. (1987). Skilled actions: A task dynamic approach. *Psychological Review*, *94*, 84-106.

Schöner, G., & Kelso, J. A. S. (1988). A dynamic theory of behavioral change. *Journal of Theoretical Biology*, *135*, 501-524.

Shaw, R. E., Kugler, P. N., & Kinsella-Shaw, J. (1990). Reciprocities of intentional systems. In R. Warren & A. H. Wertheim (Eds.), *Perception and Control of Self-motion* (pp. 579-619). Hillsdale, NJ: Erlbaum.

Shaw, R. E. and Alley, J. R. (1985). How to draw learning curves: Their use and justification. In T. D. Johnston & A. T. Pietrewicz (Eds.), *Issues in the Ecological Study of Learning* (pp. 275-304). Hillsdale, NJ: Erlbaum.

Turvey, M. T. and Kugler, P. N. (1984). An ecological approach to perception and action. In H. T. A. Whiting (Ed.), *Human Motor Actions: Bernstein Reassessed* (pp. 373-412). Amsterdam: North-Holland.

EFFECT OF PRACTICE ON THE KINEMATICS OF REACHING MOVEMENTS MADE TO MOVING TARGETS

JAMES R. BLOEDEL AND STEPHEN I. HELMS TILLERY
Division of Neurobiology
Barrow Neurological Institute
350 W. Thomas Rd.
Phoenix, Arizona 85013
USA

ABSTRACT. The purpose of these experiments was to examine the kinematic properties of the trajectories used by human subjects while reaching to moving targets and to determine whether these trajectories are modified with practice. Each subject was asked to reach for a target accelerated to a specified velocity along a horizontal track on twenty successive trials. Three different target speeds were used, and movements made by both the right and left hand were examined. Movements were assessed by determining the changes in the location of infrared emitting diodes located on the shoulder, elbow, wrist and index finger. The data show that movements made to targets of this type consist of several successive curved components and that these components do not vary qualitatively as the movement is practiced. However, when movements were made to faster moving targets and in some cases as movements were practiced, the curvature of the trajectory and the tangential velocity of the distal extremity both increased.

1. Introduction

Most previous studies which examined the characteristics of reaching movements focused on movements to stationary targets and usually employed paradigms in which the task was well practiced (Georgopoulos, 1986 for review). The purpose of this study was to assess the kinematics of movements made by human subjects while reaching to a moving target and to ascertain whether any changes in the strategy and/or kinematics of the movement occur as these movements are practiced.

The experiments were designed to test the hypothesis that the trajectories of limb movements directed towards moving targets are relatively invariant even when the movement is practiced repeatedly over several successive trials. Based on this view, the subject would not be expected to explore a variety of strategies over the first few trials in order to derive the one strategy that best achieved the movement's goal. Rather, the subject would execute a trajectory on the first trial similar to those performed after the movement had been practiced. The data presented below will show that the trajectories characterizing movements of a given subject consist of several curved components and that these components qualitatively remain unchanged despite changes in target speed and/or practice. However, increases in the magnitude of the curvature that paralleled increases in tangential velocity were observed as movements were practiced and when movements were made to faster-moving targets.

J. Requin and G. E. Stelmach (eds.), Tutorials in Motor Neuroscience, 109–120.

2. Methods

The experiments were performed in thirteen young adult human subjects that were normal neurologically. Each was comfortably seated in front of the apparatus on a chair. Four infrared-emitting diodes (IREDs) were attached to the extremity used to perform the movement. One diode each was attached to the shoulder, elbow, wrist, and index finger. These emitters were used together with a Watscope-Watsmart System in order to analyze several kinematic features characterizing the limb's trajectory. Although not presented in this manuscript, the subjects also were instrumented with surface EMG electrodes located over the biceps and triceps muscles.

The paradigm required that a reaching movement, triggered by a light signal, be made to a moving target consisting of a circle 26 mm. diameter moving along a fixed horizontal track. The subject was asked initially to place an index finger on an origin switch, a contact switch located 7.5 cm in front and 15.5 cm below the track of the target. Once the subject had contacted the switch for at least two seconds a red "ready" light was lit, and the movement of the target was initiated. The target moved slowly until it arrived at a point along the track unknown to the subject at which it was accelerated to one of three predetermined target velocities. As the target passed this point, a blue go-lamp was lit, indicating to the subject that a movement to the target could be made. Simultaneous to the illumination of this lamp, the target was accelerated to its specified velocity. Both velocity and size of target were selected so that the successful execution of the reaching movement was challenging to the subject, particularly at the highest speed. The subject was instructed to contact the target before it reached the end of its track. No instruction was given concerning how soon the subject was to contact the target or how rapidly the movement should be made.

In each experimental setting the subject was required to perform movements to the target in 20 consecutive trials for each of the three specified terminal velocities. Movements to targets moving to the right and to the left were assessed in the same sitting. In addition to analyzing the movement's trajectory, reaction time (RT) the movement time (MT) were also determined. Data were analyzed first by examining the stick figures representing the movements of the extremity of the target across all trials. In order to assess the extent to which the characteristics of the trajectory were modified as a function of practice, graphic methods were developed to compare two or more normalized trajectories generated by an individual subject. For this purpose the trajectories were normalized to the average length of the trajectories to be compared. The features of the trajectories then were compared qualitatively based on a visual inspection of the plots.

In addition, tangential velocity was calculated for the movement of the most distal IRED mounted on the index finger. This measurement was then compared over the succession of trials performed by the subject. The calculation was based on the changes in IRED position measured at a frequency of 100 Hz.

3. Results

3.1. CURVED NATURE OF THE TRAJECTORIES

The characteristics of the arm's trajectory to a target moving at 75 cm/sec is illustrated by the stick figures in Figure 1. This stick figure in A shows the characteristics of the movement as seen looking down on the subject from above with the target located at the top of the figure. In B and C two successive components of the same movement are shown in

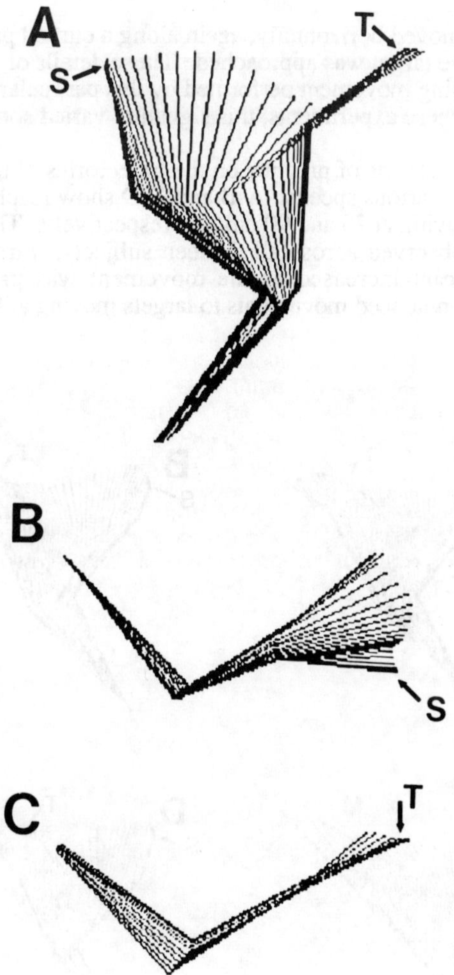

Figure 1.

views from the side of the subject with the target located on the right. A comparison of A and B demonstrate that the subject first moved upward from the start position (S) with a movement consisting of at least two curved components. There was an initial small curved movement almost in the vertical plain as the finger left the start position (S), after which the curvature of the movement reversed and the finger moved rapidly to the vertical-most position (B). Once maximum height of the movement was attained, the subject moved toward the target (T). Since the spacing between each stick figure is inversely proportional to the velocity of the limb, it is also apparent that the velocity of the movement increased substantially as the limb moved horizontally, again along a curved path. There was also an obvious deceleration as the target was approached. These details of the trajectory provided a "signature" for the reaching movement performed by this particular subject. Although not studied systematically in these experiments, this signature varied somewhat from subject to subject.

Figure 2 illustrates the effects of practice on the trajectories of these movements made towards targets moving at various speeds. A-B and C-D show reaches during the first and last trial to a target (T) moving at 75 and 90 cm/sec, respectively. These stick figures show two findings that were observed across all thirteen subjects. First the curvature of the trajectory near its midpoint increased as the movement was practiced. Second this parameter was greater for practiced movements to targets moving at faster speeds.

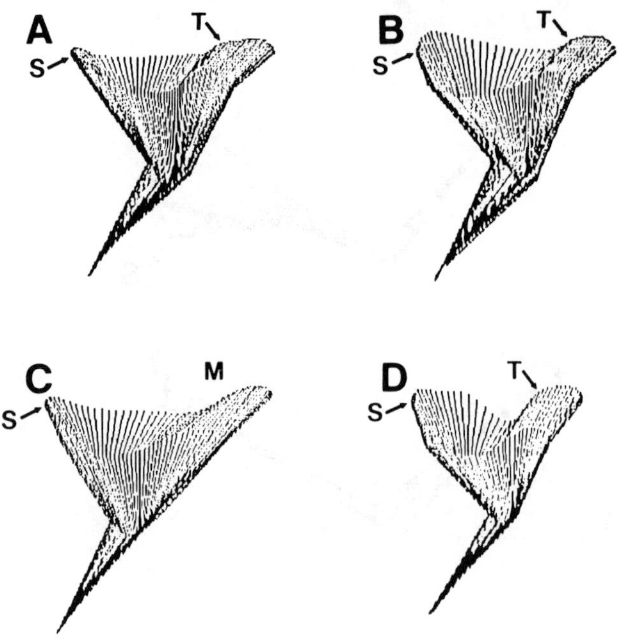

Figure 2.

Other more subtle characteristics of the specific trajectories in Figure 2 are also apparent. At the faster speed the subject decreased the distance moved horizontally before hitting the target. However, a progressive decrease in movement time with practice was not consistently observed, even within the trials of the same subject. This variability may have resulted from the fact that no instructions regarding movement speed were given. Also note that the subject missed the target on the first movement to the faster target (M in C). No differences have been found to date between trajectories made in successful and unsuccessful trials. As observed in Figure 1, the velocity of limb movement during the mid-portion of the trajectories increased as the movement was practiced.

3.2. TANGENTIAL VELOCITY PROFILES

The effects of practice on movement velocity is shown more clearly in the plots of Figure 3 illustrating the relationship between tangential velocity and absolute time for the IRED located on the index finger during movement execution. Zero on the abscissa is the time 100ms before the movement is initiated. Consequently the plot does not show reaction time. Consistent with the inferences drawn from Figure 2, it is apparent that

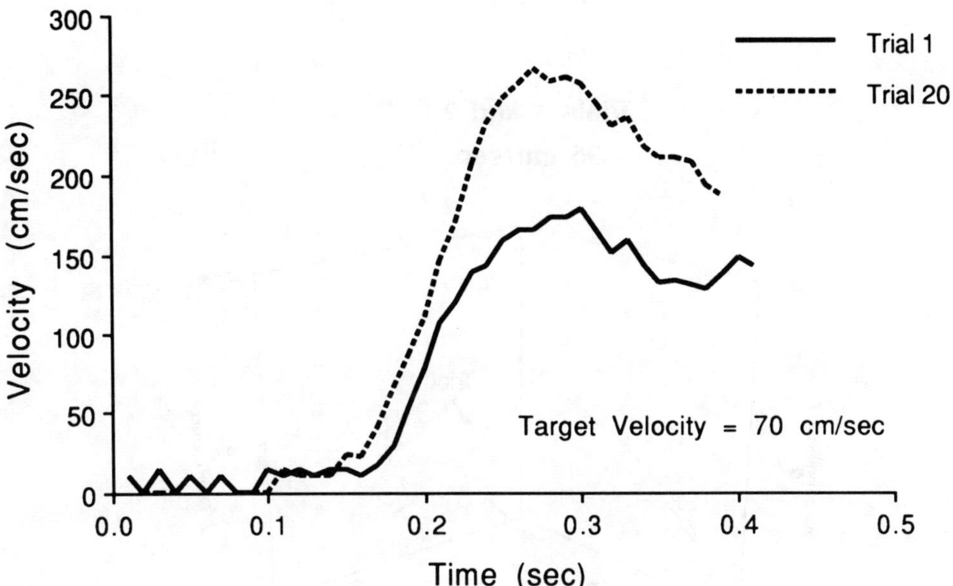

Figure 3.

the maximum tangential velocity was greater for this subject in the twentieth trial than in the first trial at this movement speed. In addition the velocity of the finger at the time of target contact also increased with practice. The contours of these two plots are rather similar. Note also that, for this subject and at this speed, the durations of the two movements were quite similar despite the 20 trials of practice. To reemphasize a point made previously, movement duration was not systematically altered with practice in this paradigm, one in which there were no instructions regarding the speed with which the movement was to be executed.

3.3. CHARACTERISTICS OF NORMALIZED TRAJECTORIES

To further examine whether a subject progressively modifies the basic features of the trajectory during practice or whether a rather consistent trajectory envelope was employed as the movement was progressively refined, trajectories made by each subject over a 20 trial session were compared after spatially normalizing them on the basis of trajectory length. Figure 4 compares two trajectories calculated for the IRED on the index finger during the first and twentieth movements to a target moving at 95 cm/sec. Displacement is plotted in millimeters along each axis. These are the same two trajectories shown in Figure 2, C and D. The flatter of the two trajectories occurred in the first trial. It is apparent in both the three-dimensional plot and its projection on the X-Y plane that practice over the 20 trials resulted in a decrease in the radius of curvature in at least two regions of the trajectory. Note also that, due to the fact that the subject missed the target on the first trial, there is a substantial difference in the end point of the trajectories on the Z axis. Despite the relatively poor accuracy of this movement, the contour of its trajectory was very similar to that observed for movements resulting in target contact.

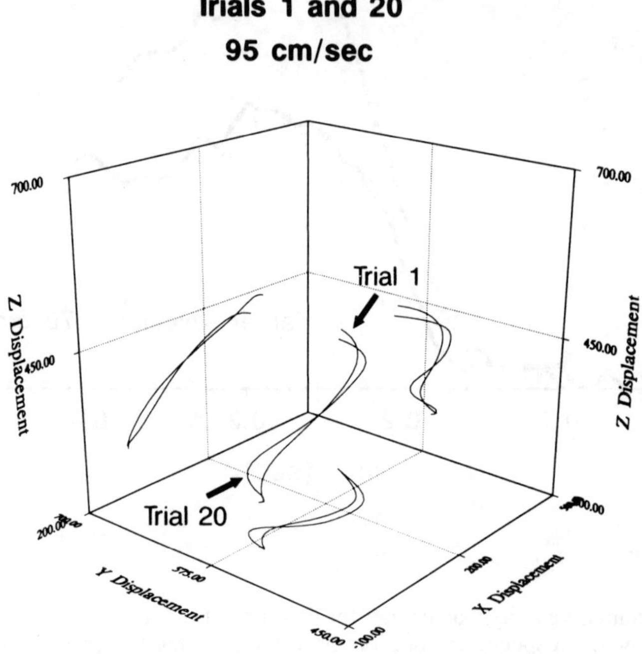

Figure 4.

Across all thirteen subjects the only variation from this principal occurred for initial movements made by some subjects to targets moving at the slowest velocity. The example in Figure 5 shows the movements in the first and last trial to a target moving at 70 cm/sec. Note that there is a much greater difference in the structure of these two trajectories than for the same subject's movements to a higher velocity target (Figure 4). Although there are similarities in these trajectories, the movement in trial one employs a much straighter trajectory during approximately the first 75% of the movement .

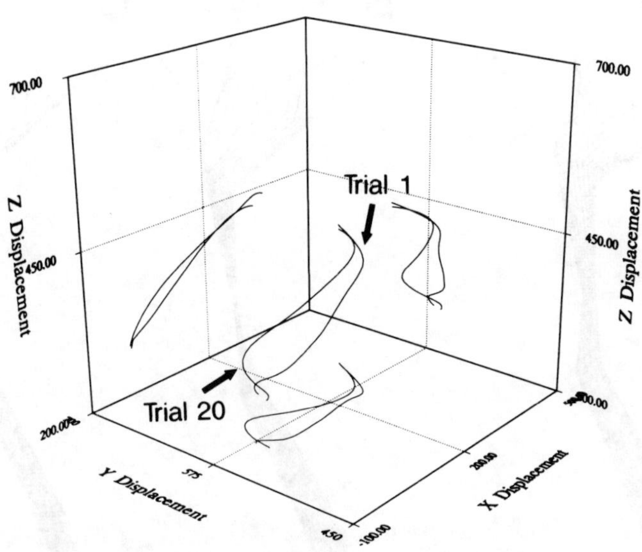

Figure 5.

3.4. COMPARISON OF MOVEMENTS TO MOVING AND STATIONARY TARGETS

In order to determine if the curved nature of these trajectories is due to the fact that the subjects are performing a movement to a moving target or whether this characteristic is dependent only on the spatial relationship between the start point and the target in the context of this paradigm (characteristics of target and start point and the distance and direction between them). If the latter were true, the trajectories would more likely be specified on the basis of the spatial relationship between the start point and the target rather the strategy required to hit a moving target. To address this issue trajectories generated by each of two subjects were compared under two conditions: (1) movements to a moving target as described above and (2) movements to a stationary target located at the same position as when contacted during the trials in which the target was moving.

Data from these experiments are shown in Figure 6, which illustrates the stick figures characterizing the arm movements to a stationary target (A) and to a moving target in the same location in two different trials (B and C). In this experiment 5 IREDs were placed on the subjects' extremity. In addition to the locations presented in Methods an IRED was

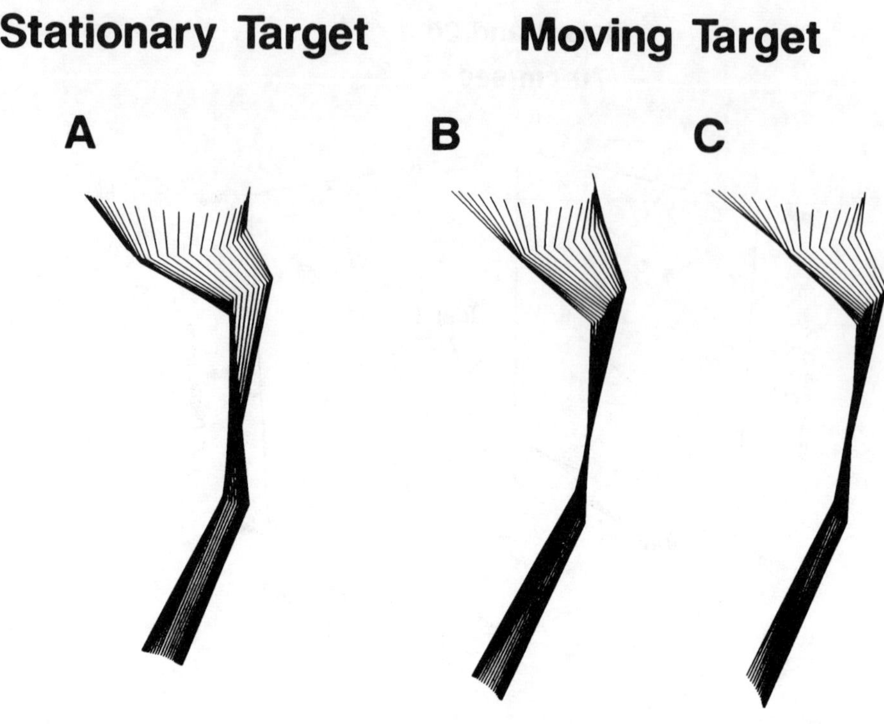

Figure 6.

placed on the knuckle just below the index finger. Notice the remarkable similarity in the hand paths made to the stationary (A) and moving targets (B and C). Notice again that not only do the curvatures of the hand paths occur at approximately the same location, but their magnitudes are virtually identical. Although not analyzed quantitatively, the stick figures also reveal modest changes in the trajectories of the elbow and shoulder despite the similarities in the hand paths.

3.5. AN ANECDOTE

The fact that movements of the type executed in this paradigm are inherently curved is further emphasized by several anecdotal observations obtained in a few subjects when they were instructed to contact the target as fast as possible. Under these circumstances the speed-accuracy tradeoff was clearly changed in the direction of increased movement speed, perhaps because the subjects already had completed a sequence of trials in which a high priority was placed on movement accuracy. In the movement shown in Figure 7 the subject failed to contact the target on the first attempt. The subject then made a second attempt at the target before it reached the end of the track. Notice that the second movement also was performed with a dramatically curved trajectory rather than a straight one. The

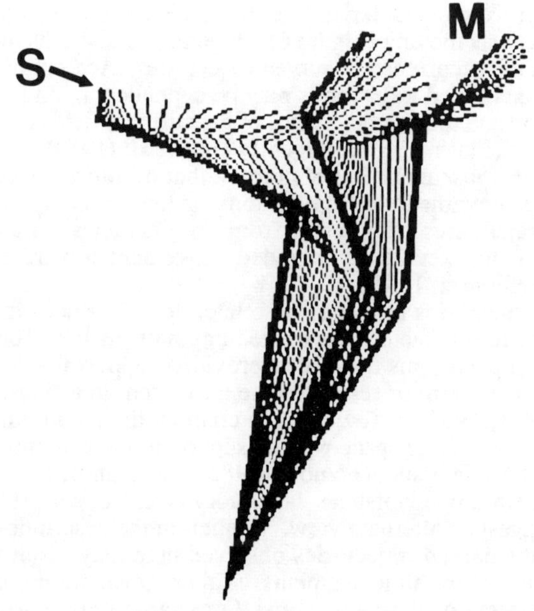

Figure 7.

subject chose to select a trajectory that resulted in a substantially greater movement distance rather than pursue the missed target with a movement along the board in a relatively straight line. These observations, although not a systematic inclusion in these experiments, further emphasize the curved nature of the trajectories characterizing the movement of the distal extremity in this paradigm.

4. Discussion

The trajectories characterizing the reaching movements in the paradigm employed in these experiments are clearly curved. Initially these data were somewhat surprising because of the predominance of straight trajectories in much of the reaching literature (Georgopoulos, 1986). However, there are now several reports of curved trajectories in reaching movements made under a variety of experimental and behavioral conditions. Curved trajectories were observed in movements to moving targets by Bairstow (1987) and von Hofsten (1979; see also Forsstrom and von Hofsten, 1982). Trajectories of this type also have been reported for reaching movements to stationary targets (Caminiti, et al., 1990). However even in paradigms resulting predominantly in straight trajectories, curved trajectories have been reported when the direction of movement is substantially modified (Flash, 1989; Lacquaniti, et. al., 1986; Uno, et. al., 1989).

The basis for the curvature in the trajectories reported here is a matter for speculation, since the studies did not address this issue specifically. It is apparent from the data that this feature is not due to the fact that subjects were reaching for moving targets. Movements to the same point in space were characterized by the same curved trajectories of the finger whether or not the target was moving (Figure 6). Several postulates have been presented in the literature to explain the occurrence of curved trajectories. According to the equilibrium trajectory hypothesis, the curved trajectories may be dependent upon the relation between the movement's direction and the characteristics of the stiffness fields measured at the hand at various limb positions (Flash, 1989). Recently Uno et al. (1989) proposed that curved trajectories could be the consequence of a strategy that minimizes torque change in the generation of reaching movements. Applications of tensor theory to limb movement suggest that curved trajectories may result from the characteristics of the coordinate transformation required to execute an intended movement toward the desired target (Bloedel et al., 1988; Pellionisz, 1985).

The nature of the trajectories also may be related to differences in the experimental paradigms used to study the characteristics of reaching movements. The studies reporting straight trajectories used paradigms that either provided support for the extremity with a device such as a draftsman's arm or restricted the movement to a single two dimensional plane (See Georgopoulos, 1986 for review). In contrast the paradigm employed in our studies required movements in free space with no support of the extremities.

The observations of Viviani and Terzuolo (1982), Lacquaniti, et al. (1983), Bairstow (1987) as well as those from von Hofsten's laboratory (von Hofsten, 1979; Forsstrom and von Hofsten, 1982) suggest an alternate view. Implicit in the description of their findings is the possibility that the curved trajectories observed here may result from the fact that these movements consist of multiple segments. This segmentation may actually reflect functionally distinct aspects of the movement. For example, Bairstow (1987) recently illustrated that the initial direction of a movement to a moving target is crudely related to the direction of the target, whereas subsequent changes in movement direction reflect more accurate estimates of the movement velocity. Speed of movement may also affect the extent to which a movement is segmented (Darling, et al., 1988).

The reason for curved trajectories in our paradigm has not yet been established. As a working hypothesis, we propose that this characteristic of the movement is task and target specific. Pertinent to this view, the increased curvature of the trajectory occurring with practice resulted in the terminal portion of the hand path becoming progressively more perpendicular to the target surface. This movement may reflect an implicit strategy for contacting targets using movements that optimize the activation of the switch signalling that the target was contacted. The curvature may also reflect a strategy that minimizes the complexity of torques and muscle activations required to generate the movement (Lacquaniti, et al., 1986). Whatever their bases, the data clearly demonstrate that subjects performing the specific task employed in these experiments utilized dramatically curved trajectories in performing the task and in attempting to improve their performance with practice.

One of the primary objectives of this study was to examine whether modifications in movement trajectories occurred with practice. Although on occasion there were modest changes in the trajectories over the twenty trials of practice at the slowest speeds (70-75 cm/sec), in general only small changes in curvature and speed were observed. With practice the peak tangential velocity of the movement increased, and the curvature of the trajectories increased.

The lack of a substantial change in the principal features of the trajectory across several successive trials has some extremely interesting implications. First of all the findings imply that a given subject uses a trajectory which is relatively fixed. In this specific paradigm all subjects appeared to select a strategy of target contact which resulted in terminal hand paths that approach the target perpendicular to its surface. The anecdotal finding reported in Figure 7 emphasizes this point by illustrating the second movement performed by a subject after missing the target the first time utilizes a dramatically curved trajectory despite the fact that the extremity must move over a substantially greater distance than if a straight line trajectory were selected. This characteristic of the trajectories is remarkably consistent across all subjects. Until additional studies can be performed, it is not possible to determine whether this is an inherent characteristic of movements of this type or whether it reflects an automatically adapted strategy due to the structural characteristics of the target. As described previously, the target consists of a disc that protrudes from the surface of the board.

In summary this study demonstrates that the spatial characteristics of a trajectory made by human subjects to moving targets do not vary substantially over the trials in which the subject optimizes the performance of the movement. Furthermore the fundamental spatial characteristic of the trajectory is that it is substantially curved and terminates with a movement approaching a line perpendicular to the surface of the disc-like target. These inherent spatial characteristics of the hand path associated with the trajectory are dependent upon only the start point and the target location and is not a consequence of pursuing a moving target.

This research was supported by NIH grant NS21958.

5. Bibliography

Bairstow, P.J., (1987) 'Analysis of hand movement to moving targets', Human Movement Science 6: 205-231.

Bloedel, J. R., Tillery, S. I. and Pellionisz, A. J. (1988) 'Effects of repetition on arm trajectories directed towards moving targets', Neurosci. Abstr. 14:952.

Caminiti, R. Johnson, P. B. and Urbano, A. (1990) 'Making arm movements within different parts of space: dynamic aspects of primate motor cortex', J. Neurosci. 10:2039-2058.

Georgopoulos, A. P. (1986) 'On reaching', Ann. Rev. Neurosci. 9:147-170.

Flash, T., (1989) 'Generation of reaching movements: plausibility and implications of the equilibrium trajectory hypothesis', Brain Behav. Evol. 33:63-68.

Forsstrom, A. and von Hofsten, C. (1982) 'Visually directed reaching of children with motor impairments', Develop. Med. and Child Neurol. 24: 653-661.

Lacquaniti, F., Soechting, J.F. and Terzuolo, C.A. (1986) 'Path constraints on point-to-point arm movements in three-dimensional space', Neuroscience 17: 313-324.

Lacquaniti, F., Terzuolo, C. and Viviani, P. (1983) 'The law relating the kinematic and figural aspects of drawing movements', Acta Psychol. 54: 115-130.

Pellionisz, A. J. (1985) 'Tensorial brain theory in cerebellar modelling', in J. R. Bloedel, J. Dichgans and W. Precht (eds.), Cerebellar Functions, Springer-Verlag, Berlin, pp. 201-229.

Uno, Y., Kawato, M., and Suzuki, R. (1989) 'Formation and control of optimal trajectory in human multijoint arm movement', Biol. Cybern. 61: 89-101.

Viviani, P. and Terzuolo, C. (1982) 'Trajectory determines movement dynamics', Neuroscience 7:431-437.

Von Hofsten, C. (1979) 'Development of visually directed reaching: the approach phase', J. Human Movement Studies 5:160-178.

EXPERIMENTAL STUDIES OF BEHAVIORAL ATTRACTORS AND THEIR EVOLUTION WITH LEARNING [1]

P.G. Zanone and J.A.S. Kelso
Program in Complex Systems and Brain Sciences
Center for Complex Systems
Florida Atlantic University
P.O. Box 3091
Boca Raton, Fla 33431 (USA)

ABSTRACT.

Only a few stable movement patterns may be produced initially in rhythmic bimanual coordination. Through a collective variable, relative phase, these patterns define attractors of the system's *intrinsic dynamics*. By visually requiring a relative phase, *behavioral information* is introduced into the dynamics, which attracts the system's behavior toward the required phasing pattern. Learning is the process by which such *environmental* behavioral information becomes *memorized*. The experimental rationale is to systematically probe the current collective variable dynamics while a phasing pattern is practiced in order to observe the evolution of the attractor layout as a new task is learned. Several dynamical processes associated with learning are identified. Pattern stability depends on whether behavioral information competes or cooperates with the intrinsic dynamics. A coordination pattern is learned to the extent that the initial dynamics are modified in the direction of the required pattern. If these practice-induced alterations of the dynamics are qualitative, learning takes the form of a *nonequilibrium phase transition*.

1 Introduction

As any adaptive mechanism of living things, learning involves durable and innovative modifications of behavior according to specific constraints. Such a process unfolds on a time scale allowing its experimental study, so that learning affords a convenient window for understanding the mechanisms and principles of behavioral adaptation. It is little wonder that most "historical" theories in psychology are theories of learning. In the field of perceptual-motor behavior — the subject of this paper —, a great deal of work has addressed the topic of skill acquisition (e.g., Adams, 1987; Marteniuk & Romanow, 1983; Newell, Kugler, van Emmerik & McDonald, 1989; Pew, 1974; Schmidt, 1987).

[1] This research was funded partly by NIMH grant MH42900, BRSG grant NSS 1-SO7-RR07258-01, and contract N00014-88-J-119 from the U.S. ONR. The first author was supported by the Swiss National Science Foundation, grant 8210-026064.

J. Requin and G. E. Stelmach (eds.), Tutorials in Motor Neuroscience, 121–133.

Among the numerous definitions of skill and skill acquisition, the following quotation from Fitts (1964) proves to be highly sensible and insightful:

> "[...] a skilled response is [...] highly organized, both spatially and temporally. The central problem for the study of skill learning is how such organization or patterning comes about" (p. 244).

Deep implications stem from this definition. First, skilled performance constitutes a spatiotemporal organization, that is, a pattern ordered in time and space. It is therefore essential that operational tools are available for identifying skilled behavior in terms of pattern. Ultimately, the question is how to distinguish between old and new and/or skilled and non-skilled behavior. Second, learning has to do with pattern formation, namely, the emergence of ordered behavior. However, a central issue that learning theorists have oftentimes raised, but have not been able to do very much about, is that organisms acquire new forms of skilled behavior on the background of already existing capacities. Thus, it is very likely that learning consists of the passage from an organized state of the system to another, rather than from plain disorder to order. For capturing such learning-related transitions experimentally, the following minimal set of problems are to be solved: a) to determine the behavioral repertoire that exists prior to exposure to the skill to be learned, and to assess its modifications with learning; b) to uncover the principles leading to behavioral change; and c) to identify the mechanisms governing behavior in response to environmental constraints.

The aim of this chapter is to present an ensemble of theory, methods, and results which may provide some insight into understanding perceptual-motor behavior and skill acquisition. The model situation that will be studied is learning a bimanual skill, namely, coordination to a visually-specified phase relationship. Human interlimb coordination proves to be rather limited, in that only few stable patterns of behavior are exhibited under fairly spontaneous and natural situations. The behavioral repertoire at hand is well-defined and sparse to start with, providing a window through which it becomes possible to trace the effects of learning a new task. Nevertheless, the principles of learning that are uncovered as well as the framework offerred here are general enough to tackle the formation of behavioral pattern and adaptive change more generally, that is, at other levels of description and on different time scales.

2 Theoretical framework

Understanding the emergence of patterns in complex systems, in which very many interacting components or degrees of freedom are involved, is a main challenge facing natural sciences (e.g. Haken, 1983a; Nicolis & Prigogine, 1989; Yates, 1987). Basically, the present conceptual framework derives from physical theories of self-organization in nonequilibrium systems, especially synergetics (see Haken, 1983a, 1983b, 1985). The operational tools are those of *nonlinear dynamical systems*. Such mathematical models have proven to be also highly relevant to collective behavior in biology, chemistry, sociology, and so forth, through the rather exotic concepts of bifurcations, catastrophes, fractals, and chaos (e.g., Gleick, 1987, for an easy overview; Glass & MacKey, 1988; Haken & Stadler, 1990; Kelso,

Mandel & Schlesinger, 1988; for more specialized readings). The fundamental and novel idea conveyed by such an approach is that a system, albeit determined by well-known and fairly simple dynamics (i.e., equations of motion), may exhibit very complex or totally unpredictable behavior (cf. May, 1976, for a simple but convincing example). Conversely, complicated and apparently random behavior does not imply absence of ordering principles. As we shall see, dynamical principles are also at work during learning, governing current behavior and underlying change.

The backbone of the study presented here is an entire body of work on biological systems, so-called "Dynamic Pattern Theory" (see Kelso & Schöner, 1987, 1988; Schöner & Kelso, 1988a, 1988b, 1988c), which tightly articulates modelling and experimental findings within a coherent framework. Following synergetics, the goal is to understand coordinated behavior in terms of pattern formation. The behavior of a system over time is to be characterized by *stable* collective states, that is, observable and reproducible *behavioral patterns.* The nature of the underlying dynamics is revealed near *phase transitions*, that is, when spontaneous (so-called self-organized) changes in the behavioral pattern occur, reflecting alterations of the system's dynamics. Typically, a reduction of the degrees of freedom occurs close to the transition, so that the system's dynamics can be characterized by a small number of collective variables or *order parameters.* These low-dimensional dynamics are defined by mapping the behavioral patterns onto *attractors* of the collective variables[2] and by studying their stability. Boundary conditions (the ensemble of extrinsic constraints impinging on the system) act as parameters on the dynamics, so that they may involve changes in the observed behavioral pattern. If these parameters are quite non-specific to the resulting behavioral pattern, they are coined *control parameters*, and the collective variables dynamics constitute the system's *intrinsic dynamics.* Intrinsic dynamics persist in absence of any specific requirement and determine the stable collective states toward which behavior is spontaneously attracted. In contrast, if the parameters are specific (i.e., they act directly on the collective variable), this constitutes *behavioral information* which attracts behavior toward a *required* behavioral pattern. Thus, behavioral information is expressed by the same collective variables that are used to characterize the behavioral patterns themselves. In particular, environmental constraints, intentional or purposeful needs, and tasks to be learned may operate as behavioral information.

The first evidence of phase transitions in biological systems was observed in bimanual finger movements by Kelso (1984). When requested to move homologous fingers at a common frequency, subjects exhibited only two stable patterns of coordination at various frequencies, namely, in-phase (simultaneous flexion/extension of both fingers) and anti-phase (the flexion of one finger coincides with the extension of the other). Nevertheless, with frequency increasing further (a non-specific parameter), a spontaneous switch from the anti-phase to the in-phase pattern was observed above a critical frequency (about 2.2 Hz). In contrast, the in-phase pattern remained equally stable at higher frequencies. Such a switch between behavioral patterns has been modelled as a nonequilibrium phase transition from a bistable to a monostable régime of the collective variable, relative phase, dynamics (Haken, Kelso, & Bunz, 1985; Schöner, Haken, & Kelso, 1986). Theoretically predicted features of phase transitions, such as enhancement of fluctuations and critical

[2]Attractors are asymptotically stable solutions of the equations of motion describing the system's behavior.

slowing-down, have been confirmed experimentally in the case of spontaneous transitions as well as voluntary switching (Kelso & Scholz, 1985; Kelso, Scholz, & Schöner, 1986, 1988; Scholz, Kelso, & Schöner, 1987; Scholz & Kelso, 1990; or see Kelso, 1990, for a review). Although they have not always been identified or analyzed as such, several instances of phase-transition-like phenomena in humans have been reported in interlimb coordination (e.g., Baldissera, Cavallari & Civaschi, 1982; Cohen, 1971), intralimb coordination (Kelso, Buchanan & Wallace, 1990), and perceptual-motor behavior (Kelso, DelColle & Schöner, 1990; Schmidt, Carello & Turvey, 1990; Tuller & Kelso, 1990),

3 Learning as a nonequilibrium phase transition

The dynamical perspective allows to capture learning theoretically through the joint concepts of intrinsic dynamics and behavioral information (see Schöner, 1989; Schöner & Kelso, 1988d, 1988e). The intrinsic dynamics define the stable behaviors to which the system relaxes spontaneously, independent of specific environmental constraints. More precisely, the layout of such behavioral attractors (or *phase diagram*[3] of the collective variables dynamics) determines the stable patterns adopted in absence of specific requirements, such as a learning task. The learning task itself affords behavioral information, which attracts the system's behavior to the required pattern. Learning is then the process through which *environmental* behavioral information defining a pattern to be learned becomes *memorized* behavioral information. Thus, a required pattern is learned to the extent that it modifies the intrinsic dynamics in the direction of the to-be-learned pattern. If any dramatic modification of the collective variable dynamics happens in that direction (i.e., the attractor layout undergoes qualitative alterations with practice of the required pattern), then *learning involves a phase transition.* A central methodological consequence is that the phase diagram allows one to know before any exposure to practice which pattern is to be learned "from scratch" and to know after learning to what extent this pattern has actually been learned. In other words, the phase diagram provides a probe of the modifications of the behavioral repertoire due to learning, beyond mere changes in performance.

The topic of learning has recently been investigated experimentally from the dynamical perspective (Zanone & Kelso, 1990). Our experiment was aimed at testing the theoretical prediction that learning may involve a phase transition. To test this idea, the individual phase diagram was systematically examined, while learning a required relative phase proceeded in time. The phase diagram probe consisted in scanning a large set of required phasing patterns. The rationale was that if no intrinsic pattern dynamics come into play, the produced relative phase would precisely match the required relative phase. On the

[3]The term "phase diagram" derives from thermodynamics. A phase diagram defines regions in parameter space that do not exhibit qualitative changes of the system's dynamics, as well as the boundaries across which such changes occur. It is important to note that a phase diagram may contain attractive, repelling, and saddle points that occupy basins and separatrices (e.g., Abraham & Shaw, 1982). In nonlinear dynamical systems, change can be continuous or abrupt depending on the region in parameter space occupied. We shall use the image "attractor layout" synonymously with phase diagram. Note the difference between the collective variable relative phase, ϕ, which is a relative timing variable expressing the coordination between active components, and the phase diagram, which is the attractor layout represented in parameter space.

contrary, systematic distortions with respect to the required performance may reflect the presence of stable patterns attracting the behavioral pattern elsewhere.

Practically speaking, subjects slipped their hands into a bimanual finger apparatus which allowed horizontal flexion-extension motion of the index fingers. A microcomputer-controlled "visual metronome" displayed different relative phases between the onsets of two light-emitting diodes (LEDs). Subjects were required to flex each finger in temporal coincidence with the onset of the respective left or right LED. No other constraint was imposed on motion except the instruction to execute as smooth and regular movements as possible. The metronome frequency was set at 1.75 Hz, below the threshold above which the 180-deg pattern becomes unstable. During an actual probe, the visual metronome displayed a required relative phase that was increased from 0 (simultaneous blink of the two LEDs) to 180 deg (alternate blink), by steps of 15 deg, hence constituting 13 plateaus lasting 20 s. For the sake of clarity, the results of only two subjects will be reported in the present paper, insofar as they are prototypical of key processes and principles pertaining to learning, but see Zanone & Kelso, 1990, for full details.

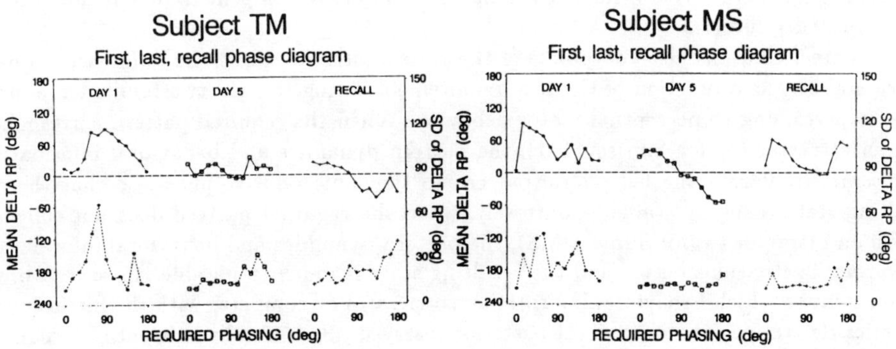

Figure 1: Individual phase diagrams before and after learning, and during the recall session. The upper solid graphs plot mean DELTA RP (produced minus required relative phase) as a function of the required phasing. The lower dotted graphs present the corresponding SD.

The phase diagrams of Subjects TM and MS are presented in the left and right Panels of Figure 1, respectively. The initial attractor layouts refer to the leftmost curves in both Panels. The upper (solid) curve in each Panel displays DELTA RP as a function of the required relative phase. DELTA RP is the difference between the produced and the required relative phase, so that when the required phasing is overestimated, DELTA RP has a positive value, and inversely. For Subject TM (left Panel of Figure 1), DELTA RP exhibits a humped curve as a function of the required relative phase, that is, the error is lowest when the required phasing is 0 or 180 deg. Moreover, DELTA RP exhibits a systematic bias below 180 deg: The produced relative phase departs from the required phasing in the direction of the 180-deg pattern, and such mismatch is roughly proportional to the difference between the required pattern and 180 deg. So, the negative slope of DELTA RP about 180 deg signifies that this pattern is an attractor for the nearby phasing patterns. A much smaller attraction to the 0-deg pattern is noticeable too. Accordingly, the lower (dotted) curves show that the SD of DELTA RP is lowest at 0 and 180 deg, while it increases markedly at intermediate values. The 0 and 180-deg patterns appear to be much more stable than the others, an interpretation that is consistent with demonstrations of loss of stability (cf. Section 2). Thus, the in-phase and anti-phase patterns constitute attractors of the collective variable dynamics. Such an attractor layout is expected based on previous work by Kelso (1984). Furthermore, a similar phase diagram was also shown by Tuller and Kelso (1989) and by Yamanishi, Kawato and Suzuki (1980), although they used different methods for probing the dynamics. In contrast, Subject MS (right Panel of Figure 1) presents a rather different attractor layout. Following the same scheme of interpretation, not only the 0 and 180-deg patterns act as attractors for the nearby phasing patterns (negative slope and/or low SD), but the 90-deg pattern is attractive as well. Thus, the collective variable dynamics for Subject MS appear to be tristable within the 0–180 deg interval.

Figure 1 contains also two points of theoretical interest. First, the differences in pattern stability as a function of the task requirements bring to light two fundamental processes pertaining to perceptual-motor behavior. When the required pattern corresponds to an attractor (as for Subject TM), the pattern dynamics and behavioral information *cooperate* to attract the behavioral pattern to the same relative phase, so that the resulting state is highly stable. Conversely, when the required pattern does not coincide with an attractor (as for Subject MS), the pattern dynamics and behavioral information *compete*, fluctuations occur, and the resulting state is much less stable. In other words, the extent to which behavioral information cooperates or competes with the intrinsic dynamics determines the behavioral patterns observed. Second, the existence of intrinsic dynamics that are qualitatively different initially between subjects emphasizes the fact that such dynamics encompass the soft-assembled, task-specific, and environmentally-attuned capacities that exist at the time a required pattern is to be learned. This makes the individual the relevant unit of observation, insofar as each subject may "bring in" different initial pattern dynamics depending on previous experience, volitional or perceptual factors, and so forth. Therefore, knowing the initial attractor layout is instrumental to the study of learning, in that it takes into account the various boundary conditions affecting the dynamics before any specific practice has occurred.

These two subjects were then exposed to a conventional learning task and procedure. Fifteen trials of practice, lasting 20 s each, were carried out per day, for five days in a row. The required phasing was set at 90 deg (right hand leading), that is, in-between the attractors at 0 and 180 deg. After every trial subjects received knowledge of results. One week after the last day of practice, two recall trials were carried out in order to test the persistence of learning effects. Subjects had to produce from memory the relative phase required in the practice sessions for one minute. Referring to the $\star$ curves, Figure 2 shows the performance of Subjects TM and MS across practice trials and the two recall trials

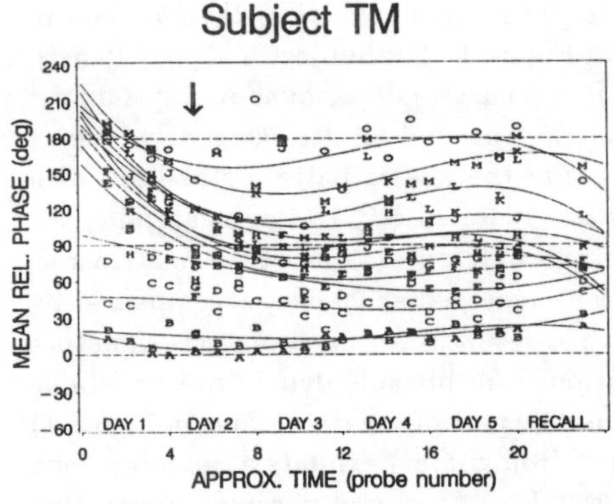

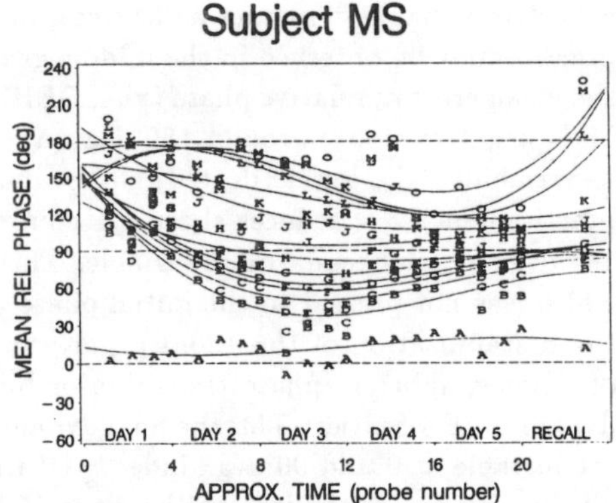

Figure 2: Evolution of performance with practice for Subjects TM and MS. The top solid curves plot the mean intratrial relative phase as a function of the practice trials. The lower dashed curves display the corresponding SD. Learning trials are coded by a $\star$ and probing trials by a $\square$.

On the first trials of Subject TM (left Panel of Figure 2), mean produced relative phase (upper solid graphs) was performed very steadily at about 180 deg, in spite of the task requirements. Then, performance tended gradually to 90 deg, until it reached the required relative phase at the beginning of the second day of practice. In this interval, the intratrial SD (lower dashed graphs) was rather large to begin with, suggesting a very variable pattern, but such fluctuations diminished substantially with practice within as well as across trials. The behavior of Subject TM in the first trials of practice is hardly surprising. Initially, no attractor exists at 90 deg, so that the behavioral pattern tend to relax to the 180-deg attractor of the intrinsic pattern dynamics. But competition between the task requirements and the intrinsic dynamics implies a highly variable performance except at 180 deg. In contrast, Subject MS (right Panel of Figure 2) performed the required relative phase in an accurate and fairly stable manner (compared to TM) just after the first trial. That Subject MS readily executed a 90-deg relative phase pattern is understandable with reference to the theory. The intrinsic dynamics, with an attractor at 90 deg, and behavioral information cooperate and tend to stabilize the produced pattern to the required phasing value. On the very first trial, Subject MS simply missed 90 deg, producing instead another pattern initially stable, namely, 180 deg. Note that in both subjects, more practice entailed a noticeable decrement in the variability of the produced pattern, and the various effects of 50 trials of practice are robust over the retention period of one week, a necessary condition to establish learning.

If one takes the idea of a close link between performance improvements and underlying pattern dynamics seriously, it is necessary to check out the predicted modifications in the collective variable dynamics as learning proceeds. The overall effects of the five days of practice may be observed in the middle graphs of Figure 1. For Subject TM (left Panel), the mean error in relative phase (viz., DELTA RP, top curves) does not show a systematic bias (negative slope) around 180 deg any longer, but around 90 deg instead. In terms of variability, the lower (dotted) curves indicate that the 90-deg pattern stabilized with practice, since SD reduces sharply for intermediate required RP. In the meanwhile, the 180-deg pattern became more variable. This ensemble of features signify that an attractor, which was not present in the initial phase diagram, emerged at 90 deg, accompanied by the destabilization of the 180-deg pattern. Such a result is an unambiguous signature of a nonequilibrium phase transition or bifurcation from bistable dynamics to tristable dynamics. For Subject TM, the final dynamics (middle part of the right Panel in Figure 1) are bistable at 0 and 90 deg. Indeed, DELTA RP (top curves) exhibits a negative slope all the way across the interval between 45 and 180 deg of required phasing, intersecting the zero axis at 90 deg, indicating a large basin of attraction. Meanwhile, SD is very low, comparable to that of the 0-deg pattern. Thus, the 90-deg pattern is a strong attractor, which "sucks in" the 180-deg pattern itself. From initially tristable dynamics, in which the 0-, 90-, and 180-deg patterns are attractors, a bistable régime emerges with learning, following the destabilization of the 180-deg attractor. Another important result is contained in Figure 1. As illustrated by the right graphs concerning the recall probe, the modifications of the phase diagrams associated with learning prevailed after the retention interval. Indeed, the destabilization of the 180-deg pattern persists in both subjects, and the newly created attractor at 90 deg is still present after one week. Such persistency suggests that the 90-deg pattern has been durably acquired, indicating that environmental behavioral information has become memorized behavioral information. Such a pattern was then available for recall upon request one week later, as shown in Figure 2.

The two learning-related phase transitions are rendered in a more flowing fashion in Figure 3. Consecutive probes of the pattern dynamics were performed periodically during the practice trials, as well as at the beginning and the end of each daily session. In Figure 3, the mean relative phase within each plateau of a probe are coded A to O as the required phasing increased from 0 to 180 deg. These scores are plotted as a function of the moment at which they were administered in the learning procedure, so that the time axis is roughly comparable to that of Figure 1. A cubic polynomial interpolation of these average values is drawn, such that the trend may be followed across consecutive probes. If there were no influence of the intrinsic dynamics, the produced relative phase would match each required relative phase on every probe. Thus, if no change in the dynamics occurred with practice, thirteen equidistant horizontal line would be plotted. On the contrary, the clustering of produced relative phase means is indicative of an attractor. Changes in the attractor layout may be then observed over the time scale of the experiment, such as drift of an attractor (fluctuation of a bunch of curves), broadening or narrowing of a basin of attraction (convergence or divergence of curves, respectively).

In the left Panel of Figure 3 (Subject TM), the salient phenomenon is the convergence of a large number of "iso-requirement" curves toward 90 deg across successive probes. This illustrates very nicely how the attraction of the 90-deg pattern with respect to its neighbors is progressively established. For the most part, curves belonging to the 180-deg basin fall into that centered at 90 deg. Episodically, the 90-deg attractor "sucks in"

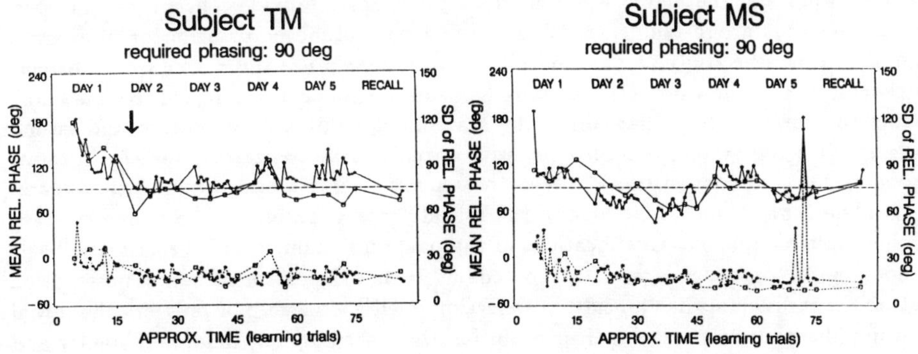

Figure 3: Evolution of the phase diagrams with learning.

the 180-deg pattern itself, as observed at the end of Day 2 and the beginning of Day 3. Afterwards, the relatively unstable nature of this pattern is still reflected by its very small basin of attraction. For Subject MS (right Panel of Figure 3), the initial tristable dynamics are indicated by the clusters of relative phase means around 0, 105, and 180 deg. With practice, all the iso-requirement curves except 0 deg phasing (coded A) converge gradually to 90 deg, as a huge basin of attraction is built up (see also Figure 1, middle). Finally, the 90-deg pattern absorbs the 180-deg pattern itself.

To complete the demonstration of a tight linkage between performance and the underlying pattern dynamics, a first test is that the substantial fluctuations across trials and days of the relative phase produced during the learning trials should correspond to comparable changes in performance when 90-deg RP is required in the probe trials. Recall that in Figure 1, the learning trials were coded by a $\star$. For the probe trials, coded by a $\square$, mean and SD produced relative phase within the 90-deg plateau are plotted for the 21 probes at the same position on the abscissa as the learning trial closest in time. Both Panels of Figure 1 show a very close covariation across trials and days between scores in the learning runs and the probes. In particular, the large jumps in produced RP between days in the learning task coincide with similar variations in the probes. In other words, within-day as well as across-day fluctuations, which unfold on two separate time scales, are captured coherently, suggesting that both tasks are tapping into the same underlying dynamics. A further validation of this interpretation is the fact that the moment at which Subject TM finally performed the 90-deg pattern stably in the practice trials coincides closely with the moment at which the 90-deg attractor appears first in the phase diagram, namely, in the middle of the second day (cf. the arrows in the left Panels of Figures 2 and 3). To sum up, there is indeed an intimate linkage between observed behavior and underlying dynamics.

4 Conclusion

The framework and the findings presented in this chapter provide conceptual and operational tools that are instrumental for describing perceptual-motor behavior in a lawful fashion. In turn, the study of coordination between perception and action from this perspective offers a window into the basic mechanisms of skill acquisition. On the one hand, probing the current phase diagram of the system's dynamics allows one to understand behavior when the system is confronted with specific task constraints, namely, in terms of cooperation and competition between the required pattern and the intrinsic dynamics. On the other hand, knowledge of the initial pattern dynamics allows one to trace learning "in real-time", as modifications of the pattern dynamics with practice. A basic methodological consequence, already pointed out, is that the pertinent unit of observation must be the individual (like for natural selection itself!). Indeed, not only are the initial dynamics idiosyncratic, but also their evolution with learning may unfold differently and along separate and varied time scales.

To sum up the results of the present study, learning effects have been clearly identified by changes in performance with practice in the direction of the task requirement, as would be the case were one to adopt conventional experimental methods. Nevertheless, the fundamental and demonstrable point of theory is that these behavioral changes result from the qualitative modifications of the current pattern dynamics toward the to-be-learned pattern. Such alterations accompanied by the destabilization of an initial attractor constitutes a typical nonequilibrium phase transition. Two types of transitions have been shown: A new behavioral attractor emerges in the phase diagram, and/or an old stable pattern vanishes from it.

There is no doubt that further investigations are needed to answer the numerous questions that arise from our results. In particular, one may wonder what makes learning take one or the other form of dynamical modifications. A candidate is the "distance" between the required pattern and existing attractors in parameter space, while another possibility is that only a limited number of behavioral attractors can exist simultaneously. Other learning-related problems may also be studied, such as the rate of learning as a function of the initial dynamics and the required pattern, the various time scales along which learning unfolds, forgetting, interference, and so forth. Fortunately, the methods used in the present study offer a promising way to operationalize these issues.

To conclude, our contention is that the processes and laws of learning identified in this chapter are not only those of perceptual-motor skill acquisition, but they may be paradigmatic of other, if not all, adaptive mechanisms. These mechanisms and principles are meant to be valid independent of the level of description ("micro" vs. "macroscopic") and the time-scale on which adaptation unfolds (e.g., developmental time) (cf. Schöner & Kelso, 1988a; Zanone & Kelso, 1990, in press). The notions of cooperation and competition, of stability and loss of stability may constitute the interpretative and operational tools for building a general theoretical model of adaptive change.

References

Abraham, R., & Shaw, C. (1982). *Dynamics – The geometry of behavior. Part 1: Periodic behavior.* Santa Cruz: Aerial.

Baldissera, F., Cavallari, P., & Civaschi, P. (1982). Preferential coupling between voluntary movements of ipsilateral limbs. *Neuroscience Letters, 34*, 95–100.

Cohen, L. (1971). Synchronous bimanual movements performed by homologous and non-homologous muscles. *Perceptual and Motor Skills, 32*, 639–644.

Fitts, P. M. (1964). Perceptual-motor skill learning. In A. W. Melton (Ed.), *Categories of human learning* (pp. 243–285). New York: Academic Press.

Glass, L., & MacKey, M. C. (1988). *From clocks to chaos – The rhythms of life.* Princeton, NJ: Princeton.

Gleick, J. (1987). *Chaos – Making a new science.* New York: Viking Penguin.

Haken, H. (1977/1983a). *Synergetics, an introduction:Non-equilibrium phase transitions and self-organization in physics, chemistry and biology.* Berlin: Springer.

Haken, H. (1983b). *Advanced synergetics: Instability hierarchies of self-organizing systems and devices.* Berlin: Springer.

Haken, H. (1985). *Complex systems: Operational approaches in neurobiology, physical systems and computers.* Berlin: Springer.

Haken, H., Kelso, J. A. S., & Bunz, H. (1985). A theoretical model of phase transitions in human hand movements. *Biological Cybernetics, 51*, 347–356.

Haken, H., & Stadler, M. (1990). *Synergetics of cognition.* Berlin: Springer.

Kelso, J. A. S. (1984). Phase transitions and critical behavior in human bimanual coordination. *American Journal of Physiology: Regulatory, Integrative and Comparative Physiology, 15*, R1000–R1004.

Kelso, J. A. S. (1990). Phase transitions: Foundations of behavior. In H. Haken (Ed.), *Synergetics of cognition* (pp. 249–268). Berlin: Springer.

Kelso, J. A. S., Buchanan, J. J., & Wallace, S. A. (1990). *Order parameters for the neural organization of single, multijoint limb movement patterns.* Manuscript submitted for publication.

Kelso, J. A. S., Delcolle, J. D., & Schöner, G. S. (1990). Action-perception as a pattern formation process. In M. Jeannerod (Ed.), *Attention and performance XIII* (pp. 139–169). Hillsdale, NJ: Erlbaum.

Kelso, J. A. S., Mandell, A. J., & Shlesinger, M. S. (1988). *Dynamic patterns in complex systems.* Singapore: World Scientific.

Kelso, J. A. S., & Scholz, J. P. (1985). Cooperative phenomena in biological motion. In H. Haken (Ed.), *Complex systems: Operational approaches in neurobiology, physical systems and computers* (pp. 124–149). Berlin: Springer.

Kelso, J. A. S., Scholz, J. P., & Schöner, G. S. (1988). Dynamics governs switching among patterns of coordination in biological movement. *Physics Letters, A134*(1), 8–12.

Kelso, J. A. S., & Schöner, G. S. (1987). Toward a physical (synergetic) theory of biological coordination. In R. Graham & A. Wunderlin (Eds.), *Lasers and synergetics* (pp. 224–237). Berlin: Springer.

Kelso, J. A. S., & Schöner, G. S. (1988). Self-organization of coordinative movement patterns. *Human Movement Science*, 7, 27–46.

May, R. (1976). Simple mathematical models with very complicated dynamics. *Nature, 261*, 459–467.

Nicolis, G., & Prigogine, I. (1989). *Exploring complexity: An introduction.* San Francisco: Freeman.

Schmidt, R. C., Carello, C., & Turvey, M. T. (1990). Phase transitions and critical fluctuations in the visual coordination of rhythmic movements between people. *Journal of Experimental Psychology: Human Perception and Performance, 16*(2), 227–247.

Scholz, J. P., & Kelso, J. A. S. (1990). Intentional switching between patterns of bimanual coordination is dependent on the intrinsic dynamics of the patterns. *Journal of Motor Behavior, 22*(1), 98–124.

Scholz, J. P., Kelso, J. A. S., & Schöner, G. S. (1987). Non-equilibrium phase transitions in coordinated biological motion: Critical slowing down and switching time. *Physics Letters, A123*, 390–394.

Schöner, G. S., Haken, H., & Kelso, J. A. S. (1986). A stochastic theory of phase transitions in human hand movement. *Biological Cybernetics, 53*, 442–452.

Schöner, G. S., & Kelso, J. A. S. (1988a). Dynamic pattern generation in behavioral and neural systems. *Science, 239*, 1513–1520.

Schöner, G. S., & Kelso, J. A. S. (1988b). Dynamic patterns in biological coordination: Theoretical strategy and new results. In J. A. S. Kelso, A. J. Mandell, & M. F. Shlesinger (Eds.), *Dynamic patterns in complex systems* (pp. 77–102). Singapore: World Scientific.

Schöner, G. S., & Kelso, J. A. S. (1988c). A synergetic theory of environmentally-specified and learned patterns of movement coordination. I. Relative phase dynamics. *Biological Cybernetics, 58*, 71–80.

Schöner, G. S., & Kelso, J. A. S. (1988d). A synergetic theory of environmentally-specified and learned patterns of movement coordination. II. Component oscillator dynamics. *Biological Cybernetics, 58*, 81–89.

Schöner, G. S., & Kelso, J. A. S. (1988e). A dynamic theory of behavioral change. *Journal of Theoretical Biology, 135*, 501–524.

Tuller, B., & Kelso, J. A. S. (1989). Environmentally-specified patterns of movement coordination in normal and split-brain subjects. *Experimental Brain Research*, *75*, 306–316.

Tuller, B., & Kelso, J. A. S. (1990). Phase transitions in speech production and their perceptual consequences. In M. Jeannerod (Ed.), *Attention and performance XIII* (pp. 429–452). Hillsdale, NJ: Erlbaum.

Yamanishi, T., Kawato, M., & Suzuki, R. (1980). Two coupled oscillators as model for the coordinated finger tapping by both hands. *Biological Cybernetics*, *37*, 219–225.

Yates, E. F. (1987). *Self-organizing systems. The emergence of order.* New York: Plenum.

Zanone, P. G., & Kelso, J. A. S. (1990). *The evolution of behavioral attractors with learning: Nonequilibrium phase transitions.* Manuscript submitted for publication.

Zanone, P. G., & Kelso, J. A. S. (in press). Learning and transfer as paradigms for behavioral change. In G. E. Stelmach & J. Requin (Eds.), *Tutorial in motor behavior II.* Amsterdam: North-Holland.

SECTION 3

COGNITIVE AND NEUROMOTOR IMPAIRMENTS

BASAL GANGLIA IMPAIRMENT AND FORCE CONTROL

George E. STELMACH
Departments of Exercise Science and Psychology
Arizona State University, Tempe
and Division of Neurobiology
Barrow Neurological Institute, Phoenix

ABSTRACT

Research on movement disorders can inform motor neuroscience by describing the effect of a given pathology on the perception-action cycle, by identifying structural-functional relationships, and by localizing processing impairments. A research program studying cognitive-motor impairments in Parkinson's disease is described which employs movement latencies, response dynamics and EMG measures. The data reviewed focus on the initiation, production and regulation of force and are grouped into five section headings: force control and muscular activation patterns, accuracy of force production, simultaneous force control, sequential force control and the amplitude and duration components of force. In addition to the more global findings of akinesia and bradykinesia, one of the most consistent impairments observed in Parkinson's disease patients is an inability to control force.

INTRODUCTION

Parkinson's disease (PD) is a neurodegenerative disease with predominant neuropathology of the basal ganglia. The main components of the basal ganglia are three large subcortical nuclei: the caudate, the putamen, and the globus pallidus. They are interconnected with two other subcortical nuclei, the subthalamic nucleus, and the substantia nigra. These five nuclei participate in the control of movement together with the cerebellum, the corticospinal system, and the brainstem (Cote and Crutcher, 1884). The damage associated with PD is primarily localized pharmacollogically, and appears to affect more than one neurotransmitter system; however, it most specifically interferes with dopaminergic functioning (i.e. a loss of the dopaminergic cells in the substantia nigra (Forno, 1982; Hornykiewicz, 1982). The dopamine levels are reduced in both the substantia nigra and the ventral tegmentum (Javoy-Agid, Ruberg, Taquet, Bohobza, Agid, Gasper, Berger, N'Guyen-Legros, Alvarez, Esourelle, Scatton & Rouquier, 1984).

In its advanced form, this disease is quite specific with four predominate symptoms: (1) akinesia, a general slowness in movement, (2) bradykinesia, a slowness in the execution of voluntary movement, (3) rigidity of the musculature and (4) resting tremor of the limbs (Phillips, Mueller & Stelmach, 1989). It is generally considered that the motor abnormalities indicative of basal ganglia dysfunction are akinesia and bradykinesia (Marsden, 1984; Wing and Miller, 1984).

The previously mentioned symptoms occur early in the illness and progress often as a function of the duration of the disease. These symptoms can be, however, significantly reduced following the introduction of drug therapy. This observation has led some

J. Requin and G. E. Stelmach (eds.), Tutorials in Motor Neuroscience, 137–148.

(Marsden, 1984; Stelmach, Worringham & Strand, 1986; Phillips et al., 1989; Wing, 1989) to postulate that the study of loss of motor function in PD patients, is perhaps the best way to learn of basal ganglia motor function, the primary pathology in humans. The task then, for research on movement disorders is to not only identify structural disturbances associated with the disease but to show how these disturbances influence the organization and control of movement. Brooks (1986) has postulated that basal ganglia dysfunction decomposes behavior into isolated motor acts and intended motor acts into movements of inappropriate amplitude.

The implication of these inferences is that the observed motor impairments in PD patients, the negative symptoms, may provide information about the role that the basal ganglia play in movement control. The basic assumption behind the research to be described is that by systematically varying the functional aspects of movement control, neuroscience will learn more about the abnormalities in the descending parameterization processes which result from structional disturbances caused by the disease. This notion predominates in this tutorial lecture; however, it is not given without some caution as the basal ganglia are interconnected with many other brain structures (Delong, Georgopoulos & Crutcher, 1985). Mountcastle (1978) suggested, and it is commonly accepted, that the CNS is a distributed system with function dispersed across its structures. Thus, the CNS may be thought of as having "redundancy of potential loci of command" (Mountcastle, 1978, p. 40).

Of the four cardinal symptoms, bradykinesia is the most interesting from a motor control standpoint as it is an observable deficit in the speed of movement execution. Bradykinesia is often associated with observations that PD patients make movements which have smaller amplitudes than required; e.g. reduced movement amplitude of the stride length in locomotion, and reduced amplitude of letter stroke size in handwriting. These observations suggest that PD patients may have difficulties initiating, regulating and maintaining the forces that underlie movement control. Force control in PD patients has been examined directly and indirectly by numerous investigators (see Phillips et al., 1989 for review). The sections that follow review the existing literature and suggest that basal ganglia pathology may be associated with force control (see Hefter, Homberg & Freund, 1989 for heightened dopamine effects).

FORCE CONTROL AND MUSCLE ACTIVATION PATTERNS

In normal subjects, it has been shown that discrete movements are typically made with single biphasic or triphasic patterns of EMG activation in agonist and antagonist muscles. For example, Brown and Cooke (1984), have shown that movements of greater amplitude are normally accomplished by increasing the size and duration of the EMG activation. In contrast to these findings, Hallett and Khosbin (1980) have postulated that PD patients do not activate their musculature in the typically observed biphasic or triphasic patterns. They have reported data which show that PD patients are unable to generate the appropriate amount of agonist EMG activation in producing a rapid elbow flexion movement. As reported, PD patients exhibited more EMG bursts in order to accomplish the requested movement. Draper and Johns (1964) also reported that in contrast to normal elderly, who show an increase in their peak velocity with increased movement amplitude, PD patients exhibited a nearly constant peak velocity across movements of different amplitude. Moreover, Hallet, Shahani & Young (1977) observed that in bradykinetic patients, additional cycles of alternating EMG bursts in the agonist and antagonist muscles are common.

These observations led Hallett and Khosbin (1980) to conclude that multiple EMG bursting is required because the amount of EMG activity that can be put into the initial burst is severly limited in PD patients. That is, PD patients fail to properly energize the agonist muscle. Subsequently Berardelli, Dick, Rothwell, Day & Marsden (1986) have shown that the first agonist burst does not completely saturate, and they suggested that in PD there is a breakdown of limb control between "perceptual appreciation" of the task goal and "delivery of the appropriate instruction" to the motor cortex.

In previous experiments, movement amplitude covaried with movement duration such that movements of longer amplitudes were also of longer durations. Thus it is not known whether the additional cycles of EMG activity previously observed in PD patients were the result of the larger movement amplitudes or of the longer durations associated with movements of larger amplitudes. Teasdale, Phillips & Stelmach (1990) performed a study which attempted to examine the degree to which PD patients can vary the duration of their movements, and to determine whether the increased cycles of EMG, previously observed, are associated with both slow and fast movements. Teasdale et al. (1990) examined EMG patterns and movement execution speed in PD patients using a discrete arm task. Subjects were required to make a discrete movement through a fixed twenty degree target barrier. Unlike previous tasks, there were no spatial accuracy constraints. In the baseline condition, subjects were instructed to produce ten self-selected maximum speed movements. In subsequent conditions, subjects were instructed to produce movements of varying speeds; ten percent faster, ten percent slower, thirty percent slower, and sixty percent slower.

The data revealed that while the PD patients were slower, they were able to vary their movement speed. PD patients were able to vary their movement time when requested (191, 225, 272 and 309 ms was observed for the four speed conditions, respectively). The temporal variability revealed that the PD patients were more variable than the elderly. For both groups, the variability increased with an increase in movement time (from 16 to 31 ms) for the elderly and (from 25 to 58 ms) for the PD patients.

To examine the Hallett and Khoshbin (1980) hypothesis, the integrated agonist EMG activity for the ten percent faster and the 60 percent slower movement time conditions were compared. The results showed that PD patients did not exhibit a saturation problem. As did the elderly, PD patients modulated the amount of EMG activity produced so that faster movements were produced with much larger initial EMG activation. However, despite that the PD patients could produce more EMG activity when required, they surprisingly showed multiple bursts of EMG. Further inspection showed that the multiple bursts of EMG were not observed in the faster conditions but were produced with the slower movements. The elderly generated a constant number of bursts across differing durations whereas PD patients needed more bursts and exhibited more bursting over longer durations. These data demonstrate that rather than producing movements of longer durations by increasing the length of their first agonist burst, seven of eight PD patients required additional cycles of EMG activity. Further there was no indication that PD patients had problems appropriately activating the agonist muscle. This latter finding may be the result of reducing the need for terminal control via antagonist muscle activity. Water and Strick (1981) found that the antagonist burst was considerably reduced and even abolished when ballistic responses were terminated by a mechanical stop.

Milner-Brown, Fisher & Wiener (1979) have provided some insight into perhaps why PD patients have difficulties with force and velocity control. They reported that even when PD patients made an effort to maintain voluntary force control, some motor units stopped firing for prolonged periods; some of the units also fired at abnormally low frequencies (2-3 per second). Their data also agrees with the Teasdale et al. (1990) data, in that they found more EMG abnormalities for slower movements. PD patients had difficulty in the recruitment of motor units at lower thresholds. Moreover, neurophysiological studies by Petajan (1983) and Dengler, Wolf, Schubert & Struppler (1986) indicate that PD patients have difficulty activating single motor units, and when activated, show an irregular firing pattern. Teulings and Stelmach (1991) use these findings to suggest that motor unit activation may be the cause of handwriting problems in Parkinsonians. Collectively, these data suggest that in PD patients the relationship between EMG activation and force control may be somewhat impaired due to the inability to recruit and regulate motor unit activity.

ACCURACY OF FORCE PRODUCTION

ACCURACY OF FORCE PRODUCTION

Stelmach & Worringham (1988a) examined force production in PD patients in a study of isometric contractions of the elbow flexors. Using force levels of 25, 50 and 75 percent of their maximum, they found PD patients to be as accurate as controls in achieving requested force levels; however, the patients took longer to reach peak force and to complete the isometric aiming task. Slower rates of force development and irregular force-time patterns were also observed, suggesting that PD patients are deficient in producing the desired force-time pattern rather than the desired force level. There was no evidence that the PD patients did not have an accurate "internal representation" of the requested force values.

Force production was considered in more detail in a subsequent experiment by Stelmach, Teasdale, Phillips and Worringham (1989) in which patients were asked to produce similar isometric force levels. The experiment focused on examining the relationship between the peak force produced and its variability; and determining, for a given force level, how it varied in PD patients. As previously reported, patients took longer to reach their peak forces; although they produced forces as accurately as controls. Patients exhibited more irregular force-time curves, characterized by jerkiness in force production. By examining the force-time curves, it was revealed that increased force production was associated with a negatively accelerating increase in variability, which was similar to that of controls. A similar pattern was found with representative point-to-point force variability.

In executing the force aiming task, PD patients showed a much slower rate of force development (an average of 62 vs 213 n/s) and needed nearly twice as much time to achieve peak force as did the elderly controls. In a related experiment, using oral facial muscles, and showing similar trends, Abbs, Hartman & Vishwanst (1987) have shown that when asked to sustain a given force level, PD patients exhibit greater variability. Sheridan, Flowers & Hurrel (1987) suggest that the increased variability of movement (both in time and space) exhibited by PD patients is related to their inherent variability in force production. They identify three potential sources of increased force variability: 1) an incorrect computation of the required force; 2) a defective memory for computed forces and 3) a noisy output system. While these statements are not all equally supported empirically, they do reflect trends observed in some data. Parkinsonians appear to be able to produce force levels required, but the manner in which they do it is somewhat atypical; the rate of force development is slow and the force-time curve is irregular.

SIMULTANEOUS FORCE CONTROL

Simultaneous activation and cessation of force is an essential aspect of skilled behavior since it is necessary to control the bodies many degrees of freedom for even the simplest skill (Bernstein, 1935). There have been many clinical reports that PD patients have difficulty in simultaneous actions; yet until only recently, there have been little data which have examined this ability. Benecke, Rothwell, Dick & Day (1986) examined performance of simultaneous movements in PD patients. Movements consisted of different combinations of (1) unilateral and bilateral flex and squeeze and (2) unilateral flex and cut. They found that unilateral squeeze and flex movements were significantly longer when performed together compared to the movement time of each movement performed alone. A different pattern was observed, however for the bilateral tasks, simultaneous movements showed only slight increases in movement time compared to when they were executed separately, while no change was observed in the controls. Benecke et al. interpreted these data as evidence that PD patients have an inability to carry out simultaneous movements.

Using a bimanual task which allows comparisons of initiation and termination synchrony, Stelmach and Worringham (1988b) compared PD and controls. Since bimanual arm movements to different targets have been shown to be often constrained by a temporal coupling, this task provides an excellent vehicle to examine simultaneous force control. Stelmach & Worringham (1988b) report that both control and PD patients took longer to initiate and execute asymmetrical movement compared to corresponding unimanual movements. PD patients were only shown to be slower. Unexpectedly, PD exhibited a pattern of reducing the movement durations of large amplitude movements when paired with smaller amplitude ones in the bilateral arm. The cumulative findings of Benecke et al. and Stelmach and Worringham suggest that PD does not impair the patient's ability to coordinate bimanual movement despite some reference to this in the PD literature.

Using discrete tasks designed to eliminate the temporal linkage between the limbs, as in the previous experiments, Lazarus and Stelmach (in preparation) utilized tasks which relied on force control in homologous muscle groups during simultaneous limb movement. At issue was whether this type of bimanual activity would bring more motor units closer to firing (Davis 1942; Wolff, Gunnoe & Cohen 1983; Todor and Lazarus 1986). Subjects performed an isometric contraction with one limb to a pre-specified submaximal force level and then initiated an isotonic movement with the other limb.

The attraction of this paradigm is that it avoids linkage via a temporal component and focuses on the potential for utilizing bilateral outflow to symetrical muscle groups. The subjects performed three tasks: a) an isotonic elbow flexion as rapidly as possible through an angle of 30 degrees, b) an isometric contraction of the flexor muscles at the elbow joint to 40 and 60 percent of maximal volitional force for a 5 second period, and c) an isometric contraction task for 5 seconds with one limb while initiating the isotonic flexion task with the contralateral limb at 2.5 seconds. Differential performance was observed in PD and control subjects.

The controls performed the isotonic movement faster and reached higher peak velocities when the movement was combined with an isometric contraction of the other limb as opposed to when it was performed alone. Conversely, PD patients performed the isotonic movement more slowly and some PD patients exhibited lower peak velocities when the movement was combined with an isometric contraction of the other limb. PD patients did not take advantage of the fact that they were using similar muscle groups (as did controls) and not only failed to enhance the isotonic movement, but demonstrated a differential slowing of the isotonic movement when performed in conjunction with the isometric contraction. As such, the patients exhibited an uncoupling or disengaging of the limbs from each other, requiring independent regulation. Consequently the tasks appeared to be more difficult for the PD patients than controls.

This finding suggests that decrements in bimanual force control may be limited to tasks which have few inherent temporal components; even when using homologous muscle groups. Benecke et al. (1986); Schwab, Chafety and Walker (1954); and Talland and Schwab (1964) using repetitive bimanual tasks have found slight impairments while using a discrete task which is known to exhibit temporal coupling. The exact nature of the decoupling observed by Lazarus and Stelmach must be further explored to determine whether the Parkinsonian difficulty can be attributed to the inability to time-share or shift attention between the bimanual tasks (Schwab et al., 1954; Talland and Schwab, 1964; Horstink, Berger Van Spaendonck & Van den Bercken, 1990).

SEQUENTIAL FORCE CONTROL

It has been known for sometime that PD patients have difficulties in executing movement sequences (see Phillips et al., 1989 for review). Most of the available data suggest that PD causes a gradual deterioration as the length and complexity of a movement-sequence increases. In a study which showed striking irregularities in movement-sequencing, Stelmach, Worringham and Strand (1986) found that PD patients did not organize a movement sequence as a total unit. That is, when asked to make a series of finger taps, PD patients did not exhibit a consistent inter-tap interval coupling across the movement sequence. The task employed consisted of making a series of 1, 2, 3, 4 or 5 taps as fast as possible when given an imperative response-signal. PD patients showed a striking dissociation between the first and second inter-tap interval regardless of sequence length. These findings were interpreted to suggest that PD patients may have difficulties releasing the force required to make the initial tap.

Subsequently, Stelmach, Garcia-Colera & Martin (1989) examined this issue further by requiring PD patients to make a series of 5 taps at either a slow or fast rate (200 or 600 ms inter-tap interval). Furthermore, the subjects were required to make a more forceful tap at each of the five tap locations, one through five. The subjects were either precued as to which tap was to be made more forceful or they were informed at the time of the imperative signal. This created a simple and choice RT situation. The data revealed that the production of a more forceful tap in the sequence created a dissociation in the inter-tap intervals that immediately followed the force tap. This was observed for both tap rate (200 or 600 ms) conditions and for all force tap locations. This part of the data suggests that when an additional force must be layered into a movement sequence the structure of the tap sequencing breaks at the point of where the additional force is to be introduced. In other words, it appears that the CNS breaks the sequence into before and after segments regardless of the location of the additional force application in the tap sequence.

When reaction times were compared for the tap conditions, there was additional evidence that PD patients have trouble adding a force transition component to a movement sequence. For the slow tap rate (600 ms) a RT difference was seen only for choice reaction situation. However, for the fast tapping rate (200 ms), where motor programming is required, both simple RT (precued) and choice RT rose in parallel while the no stress condition remained unchanged. These data suggest that having to make a more forceful tap in the programmed condition produced substantial delays in response latencies regardless of advance information.

Others have reported that PD patients have difficulties in making sequential responses. For example, Stern, Mayeux & Rosen (1984) found that using a tracing task, patients had difficulty linking parts of a movement sequence without exact guidelines. Furthermore, in a series of experimental conditions, Benecke et al. (1986) required PD patients to simultaneously flex their arm and squeeze a manipulandun with the same limb. Patients with PD had a problem of superimposing and sequencing movements. For patients unilateral squeeze and flex movements performed simultaneously were significantly longer when performed together compared to the movement times when performed alone. The data were interpreted by Benecke et al. as evidence that Parkinsonians are deficient in their ability to carry out simultaneous movements.

In a recent study which used a digitizer to record the handwriting trajectories produced while executing a series of e's and l's arranged in a sequence, Phillips, Teasdale, Stelmach & Meeuwsen (1991) found that PD patients had problems modulating their movement velocities and had difficulties producing sequences where more force alterations were required. These data were discussed in two ways: an impaired force control for individual strokes and an impaired force control between strokes when force levels are altered. On a relative basis, it appeared as though the most pronounced impairment was localized when switching from a small letter (e) to a larger letter (l) and the reverse. These data support the emerging picture that PD patients have a problem in dealing with changes in force amplitude.

Tremor has been suggested as one of the causes of jerkiness and irregularity in movement sequences. For example, Nakamura, Nagasaki and Narabayashi (1978) observed that sequential movements were intrained by a characteristic rhythm, approximately at the frequency of Parkinsonian tremor. Nakamura et al. had patients perform finger taps in synchrony with a periodic stimulus. Patients exhibited difficulties at certain frequencies: between 2.5 and 5 Hz, control of tapping deteriorated and between roughly 5 and 6 Hz, hastening of movement developed, independent of the stimulus. Such data suggest that tremor may be associated with the consistently observed problems on movement sequencing in Parkinsonians.

AMPLITUDE AND DURATION COMPONENTS OF FORCE

If a muscular force is to be effective in skilled movement, it must have an appropriate amplitude and duration. One area that has exploited these relationships is handwriting analyses. The analysis of stroke size and duration have proven to be an insightful way to examine force control in PD patients. Using such methods, Margolin and Wing (1984) performed handwriting analyses of micrographic patients to explore whether micrographia was caused by a primary change in applied force (measured by changes in the height of letter strokes) and/or reductions in the duration of individual letter strokes. Patients exhibited decreases in letter height with progressive decreases as a function of the number of strokes executed. Over consecutive letters, PD patients exhibited declining movement velocity and increased stroke duration. These data are supportive of the hypotheses that PD patients do not regulate or adjust force levels over sequence length.

In one of the seminal studies that attempted to gain insight into whether PD patients had a problem controlling force duration and/or force amplitude, Phillips, Stelmach & Teasdale (1989) required subjects to modulate their spatial or temporal aspects of handwriting. Subjects performed basic zig-zag writing strokes and were asked either to keep stroke length and duration constant, modulate the length of alternate strokes and keep stroke duration constant, or modulate duration of alternate strokes and keep amplitude constant. Phillips et al. found that PD patients were able to modulate stroke length while holding stroke duration constant. Both the control group and the PD patients had difficulty demonstrating control over their stroke durations compared to stroke length. These data are supportive of the earlier work of Teulings, Thomassen & Van Galen (1986) who found that signal-to-noise ratios (SNR) for stroke size were always higher than those of stroke duration of peak acceleration. This result was interpreted as evidence that stroke size is organized at a higher level in the control hierarchy than stroke information which is thought to be derived from movement information. The PD group produced consistently slower stroke durations.

Phillips et al. (1989) observed that PD patients had slower stroke velocities and that it was difficult to train them to execute faster. These data suggest that PD patients may be able to accurately control force, but that they take longer to prepare and execute these forces. Problems with the production of forces could impose limitations on the PD patient's ability to alter movement velocities, in as much as variable and irregular force pulses pose difficulties for the CNS to predict movement outcomes. An impairment of this nature could in turn have consequences for subsequent movements causing the often observed reduction in writing velocity and compensatory change in letter size.

Teulings and Stelmach (1991) also examined Parkinsonian handwriting and attempted to determine whether the control of stroke size or duration per stroke was substantially impaired in PD handwriting. The subject's task was to perform a specific writing pattern which contained long, short, up and down letter strokes. The subjects were requested to produce writing patterns of normal size and speed, slower, faster, and larger under various combinations of reduced tactile and visual feedback. Besides the trajectory profiles of acceleration, velocity, and duration, the primary analysis was the use of signal-to-noise ratios (SNR) of the writing pattern. The SNR expresses the ratio between the amount of modulation due to the invariant, programmed component (signal), and the amount of modulation due to the impairment (noise). Thus, the SNR is high if a movement pattern is reproduced accurately, and low if the pattern is variable due to motor impairments (Teulings et al., 1986).

The results fromTeulings and Stelmach (1991) showed that Parkinsonians, as well as normals, generally were able to modify stroke size, peak acceleration, force amplitude, and stroke duration. On the average, the PD subjects, wrote smaller and slower and did not exhibit any evidence of being influenced by the availability of feedback.

The signal-to-noise ratio analyses produced quite a different picture. For stroke size, the SNR was less in the PD patients than controls. In addition, the SNR for peak acceleration was also somewhat lower. In contrast to SNR's for stroke size and peak acceleration, the SNR for stroke duration yielded neither a significant subject group effect or a condition effect. These results offer no evidence that under conditions of reduced sensory feedback, PD patients produce their movement less accurately than do controls. Most importantly, stroke size and peak acceleration were impaired in PD patients, relative to controls, but stroke duration was not. Thus, the cumulative results suggest that handwriting in PD patients is relatively more impaired in force amplitude, rather than in duration. Teulings and Stelmach draw upon neurophysiological studies by Petajan, 1983; Petajan and Jarcho, 1975; and Dengler, wolf, Schubert & Struppler, 1986, to suggest that the observed handwriting impairments in Parkinsonians may be attributed to irregularities in motor unit activation patterns.

In order to produce legible handwriting, stroke shape and relative stroke size must be properly executed. These shapes and sizes are the result of net forces of appropriate amplitudes and durations at exact movements along two axes of movement. Using these concepts as a guide, Teulings and Stelmach (this volume) attempted to extend their previous findings by simulating distortions in horizontal and vertical force components of handwriting. In one simulation, force amplitude along the vertical axis of movement was lessened which created a stroke than planned, which also displaced all the following strokes. As a consequence, it was observed that the baseline, which is normally straight, became now more varied. In the second simulation, force duration was distorted in such a way that the vertical acceleration at the beginning of a stroke was delayed relative to the horizontal component. When then this was done, the sharp movement reversals required for such letters as i or u became rounded loops.

Based on the above simulations, it was possible to make predictions about the handwriting of PD patients. If the patient's primary force impairment was in amplitude, PD patients should have more irregular baselines; while if they have more loops in their letters, they would have more impairment in force duration. Of course, both force amplitude and force time could also be equally impaired.

These predictions were tested in PD patients and age-sex matched controls, where the subjects were required to perform a writing pattern repeatedly at normal speed and size, or larger or faster. The writing task was to execute a specific pattern containing upstrokes and downstrokes which were long and short and required sharp movement reversals as well as loops.

The results showed that the baseline in Parkinsonian handwriting was more irregular than that of control subjects. In contrast, relative to controls, PD patients did not make larger loops in their handwriting when letters with sharp movement reversals had to be performed. These observations are seen as supporting the notion that PD patients have precise control of force duration, but not force amplitude. Thus, these data support their previous finding (Teulings and Stelmach, 1991) which found that force amplitude control in handwriting was impaired by Parkinson's disease.

CONCLUDING COMMENT

The data reviewed in this lecture suggest that basal ganglia impairment, caused by Parkinson's disease, impairs the ability to precisely control muscular force. This loss of force control is seen in slower reaction times when an accurate force is required, in slower and more irregular force time curves, in movement sequences where force alterations are necessary, and in micrographic handwriting. Furthermore, preliminary data suggest that when applied force is decomposed into amplitude and duration components, both appear to be impaired; however, on a relative basis, force amplitude may be slightly more affected.

Cognitive-neuroscience has the task of elucidating the role(s) of the basal ganglia in motor control. Thus research must systematically examine the abilities of those individuals inflicted with Parkinson's disease. When information is transmitted from and through an impairment, research must carefully document how and which motor functions are disturbed. In the case of Parkinson's disease, this is best accomplished by combining graphic accounts, in terms of the brain function involved, in terms of the motor processing operations utilized and in terms of the pathology which propagated the perdition in the nervous system.

References

Abbs, J.H., Hartman, D.E. and Vishwanst, B. (1987) 'Orofacial motor control impairment in Parkinson's Disease', Neurology, 37, 394-398.

Benecke, R., Rothwell, J.C., Dick, J.P.R., Day, B.L. and Marsden, C.D. (1986) 'Performance of simultaneous movements in patients with Parkinson's disease', Brain, 109, 739-757.

Benecke, R., Rothwell, J.C., Dick, J.P.R., Day, B.L. and Marsden, C.D. (1987) 'Simple and complex movements off and on treatment in patients with Parkinson's disease', Journal of Neurology, Neurosurgery, and Psychiatry, 50, 296-303.

Berardelli, A., Accornero, N., Argenta, M., Meco, G. and Manfredi, M. (1986) 'Scaling of the size of the first agonist EMG burst during rapid wrist movements in patients with Parkinson's disease', Journal of Neurology, Neurosurgery and Psychiatry, 49, 1273-1279.

Berardelli, A., Dick, J.P.R., Rothwell, J.C., Day, B.L. and Marsden, C.D. (1986) 'Scaling of the size of the first agonist EMG burst during rapid wrist movements in patients Parkinson's disease', Journal of Neurology, Neurosurgery, and Psychiatry, 50, 1178-1183.

Bernstein, N. (1935) 'The problem of the interrelation of coordination and localization', Arch. Biol., 38, 15-59. Reprinted in Bernstein, N. (1967), 'The coordination and regulation of Movements', Oxford: Pergamon Press.

Brooks, V.B. (1986) 'The neural basis of motor control', New Jersey, Oxford University Press.

Brown, S.H. and Cooke, J.D. (1981) 'Amplitude and instruction dependent modulations of movement-related electromyogram activity in humans', Journal of Physiology, 316, 97-107.

Cote, L. and Crutcher, M.D. (1985) 'Motor functions of the basal ganglia and diseases of transmitter metabolism'. In E. Kandel and J. Schwartz (Eds), 'Principles of Neural Science', Amsterdam, Elsevier, pp. 524-535.

Davis, R.C. (1942) 'The pattern of muscular action in simple voluntary movement', Journal of Experimental Psychology, 315, 347-366.

DeLong, M.R., Georgopoulos, A.P., Crutcher, M.D., Mitchell, S.J. and Richardson, R.T. (1984) 'Role of basal ganglia in limb movement', Human Neurobiology, 2, 235-244.

Dengler, R.W., Wolf, M., Schubert, M. and Struppler, A. (1986) 'Discharge pattern of single motor units in basal ganglia disorders', Neurology, 36, 1061-1068.

Draper, I.T. and Johns, R.S. (1964) 'The disorders of movement in Parkinsonism and the effect of drug treatment', Bulletin of the John Hopkins Hospital, 115, 465-480.

Forno, L.S. (1982) 'Pathology of Parkinson's disease', In C.D. Marsden and Fahn (Eds.), 'Movement disorders', Butterworths, London, pp. 25-40.

Girotti, F., Carella, F., Grassi, M.P., Soliveri, P., Marano, R., Caraceni, T. (1986) 'Motor and cognitive performances of Parkinsonian patients in the on and off phase of the disease', Journal of Neurology, Neurosurgery and Psychiatry, 49, 657-660.

Girotti, F., Marano, R., Soliver, P., Geminiani, G. and Scigliano, G. (1988) 'Relationship between motor and cognitive deficits in Huntington's disease', Journal of Neurology, 235, 454-457.

Hallett, M. and Khosbin, S. (1980) 'A physiological mechanism of bradykinesia', Brain, 103, 301-314.

Hallett, M., Shahani, B.T. and Young, R.R. (1977) 'Analysis of stereotyped voluntary movements at the elbow in patients with Parkinson's disease', Journal of Neurology, Neurosurgery and Psychiatry, 40, 1129-1135.

Hefter, H., Homberg, V. and Freund, H.J. (1989) 'Quantitative analysis of voluntary and involuntary motor phenomena in Parkinson's disease', In H. Przuntel and P; Riederer

(Eds), 'Early diagnosis and preventive therapy in Parkinson's disease', Springer-Verlag, New York, pp. 66-73.

Hefter, J., Homberg, V., Lange, H.W. and Freund, H.J. (1987) 'Impairment of rapid movement in Huntington's disease', Brain, 110, 585-612.

Hornykiewicz, O. (1982) 'Brain neurotransmitter changes in Parkinson's disease'. In C.D. Marsden and S. Fahn (Eds.), 'Movement disorders', Butterworths, London, pp. 41-58.

Javoy-Agid, F., Ruberg, M., Taquet, H., Bokobza, B., Agid, Y., Gaspar, P., Berger, B., N'Guyen-Legros, J., Alvarez, C., Gray, F., Esourelle, R., Scatton, B. and Rouquier, L. (1984) 'Biomechanical neuropathology of Parkinson's disease. In G. Hassler and J.F. Chirst (Eds.), 'Advances in Neurology', Raven Press, New York, Volume 40, pp. 189-198.

Margolin, D.I. and Wing, A.M. (1983) 'Agraphia and Micrographia: Clinical manifestations of motor programming and performance disorders', Acta Psychologia, 54, 263-283.

Marsden, C.D. (1982) 'The mysterious motor functions of the basal ganglia: The Robert Wartenberg Lecture', Neurology, 32, 514-539.

Marsden, C.D. (1984) 'Function of the basal ganglia as revealed by cognitive and motor disorders in Parkinson's disease', Canadian Jounral of Neurological Science, 11, 129-135.

Mountcastle, V.B. (1978) 'An organizing principle for cerebral function: the unit module and the distributed system', In B.M. Edleman and V.B. Mountcastle, 'The Mindful Brain', MIT Press, Cambridge, P. 40.

Nakamura, R., Nagasaki, H. and Narabayashi, H. (1978) 'Distrubances of rhythm formation in patients with Parkinson's disease: Part I. Characteristics of tapping responses to the periodic signals', Perceptual and Motor Skills, 46, 63-75.

Petajan, J.H. (1983), 'Motor unit control in movement disorders', In E. Desmedt (Ed.), 'Motor Control Mechanisms in Health Disease', Raven Press, New York.

Petajan, J.H. and Jarcho, L.W. (1975), 'Motor unit control and Parkinson's disease and the influence of levodopa', Neurology, 25, 866-869.

Phillips, J.G., Mueller, F. and Stelmach, G.E. (1989), 'Movement disorders and the neural basis of motor control', In S. Wallace (Ed), 'Perspectives on the Coordination of Movement', North-Holland, Amsterdam, pp. 367-417.

Phillips, J. and Stelmach, G.E. (1991), 'The contribution of movement disorders research to theories of motor control and learning. In J. Summers (Ed.), 'Approaches to the Study of Motor Control and Learning', North-Holland Publishing, Amsterdam.

Phillips, J., Stelmach, G.E. and Teasdale, N. (1989) 'Preliminary assessment of spatio-temporal control of handwriting in Parkinsonians', In R. Plamondon, C.Y. Suen and M.L. Simner (Eds.), 'Computer Recognition and Human Production of Handwriting', World Scientific Publishing Company, Singapore, pp. 317-331.

Phillips, J.G., Stelmach, G.E. and Teasdale, N. (1991) 'What can indices of handwriting quality tell us about Parkinson's disease handwriting?', Journal of Human Movement Science (in press).

Phillips, J.G., Teasdale, N., Stelmach, G.E. and Meeuwsen, H. (1991) 'Movement sequencing in the handwriting of patients with Parkinson's disease' (in preparation).

Phillips, J.G., Teasdale, N. and Stelmach, G.E. (1991) 'Spatio-temporal control of handwriting in patients with Parkinson's disease' (in preparation).

Schwab, R.S., Chafetz, M.E. and Walker, S. (1954) 'Control of two simultaneous voluntary acts in normals and in Parkinsonians', Archives of Neurology and Psychiatry, 72, 591-598.

Sheridan, M.R., Flowers, K.A. and Hurrell, J. (1987) 'Programming and execution of movements in Parkinson's disease', Brain, 110, 1247-1271.

Stelmach, G.E., Garcia-Colera, A. and Martin, Z. (1989) 'Force transition control with a movement sequence in Parkinson's disease patients', Journal of Neurology, 236, 406-410.

Stelmach, G.E. and Phillips, J.G. (1990) 'Movement disorders: Limb movements and the basal ganglia', Journal of Physical Therapy, 21, 60-68.

Stelmach, G.E., Phillips, J. and Chau, A.W. (1989), 'Visuo-spatial information processing in Parkinsonians', Neuropsychologia, 27, 485-493.

Stelmach, G.E. and Teasdale, N. (1991), 'Force control in Parkinson's disease patients', Journal of Human Movement Science (in press).

Stelmach, G.E., Teasdale, N., Phillips, J. and Worringham, C.J. (1989), 'Force production characteristics in Parkinson's disease', Experimental Brain Research, 76, 165-172.

Stelmach, G.E., Teasdale, N. and Phillips, J. (1990), 'Initiation and accuracy of force production in Parkinson's disease', In 'Advances in Parkinson's Disease Research', Raven Press (in press). Also in M. Streifler (Ed.), 'Proceedings of the 9th International Symposium on Parkinson's Disease', Jerusalem, Israel, June 5-9, 1988 (in press).

Stelmach, G.E. and Worringham, C.J. (1988a), 'The control of bimanual aiming movements in Parkinson's disease', Journal of Neurology, Neurosurgery & Psychiatry, 51, 223-231.

Stelmach, G.E. and Worringham, C.J. (1988b), 'The preparation and production of isometric force in Parkinson's disease', Neuropsychologia, 26, 93-103.

Stelmach, G.E., Worringham, C.J. and Peters, H.H. (1989), 'Preparatory processes in a case of hemi-parkinsonism', Journal of Neurology, 236, 174-178.

Stelmach, G.E., Worringham, C.J. and Strand, E.A. (1986), 'Movement preparation in Parkinson's disease: The use of advanced information', Brain, 109, 1179-1194.

Stelmach, G.E., Worringham, C.J. and Strand, E.A. (1987), 'The programming and execution of movement sequences in Parkinson's disease', International Journal of Neuroscience, 36, 55-65.

Stern, Y. and Mayeux, R. (1986), 'Intellectual impairment in Parkinson's disease', In M.D. Yahr and K.J. Bergmann (Eds.), 'Advances in Neurology,' Vol. 45, Parkinson's disease, Raven Press, New York, pp. 405-408.

Talland, G.A. and Schwab, R.S. (1964), 'Performance with multiple tests in Parkinson's disease', Neuropsychologia, 2, 45-53.

Teasdale, N., Phillips, J. and Stelmach, G.E. (1990), 'Temporal movement control in patients with Parkinson's disease', Journal of Neurology, Neurosurgery, and Psychiatry, 53, 862-868.

Teulings, J.L., Thomassen, A.J.W.M., and van Galen, G.P. (1986), 'Invariances in handwriting: the information contained in a motor program', In H.S.R. Kao, G.P. van Galen and R. Hoosain (Eds.), 'Graphnomics: Contemporary Research in Handwriting', North-Holland, Amsterdam, pp. 305-315.

Teulings, J.H. and Stelmach, G.E. (1991), 'Force amplitude and force duration in Parkinsonian handwriting', In J. Requin and G.E. Stelmach (Eds.), 'Tutorials in Motor Neuroscience', Kluwer, Dordrecht, This volume.

Teulings, J.H. and Stelmach, G.E. (1991), 'Control of stroke size, peak acceleration, and stroke duration in parkinsonian handwriting', Journal of Human Movement Science (in press).

Todor, J.I. and Lazarus, J.C. (1986), 'Execution level and the intensity of associated movements', Developmental Medicine and Child Neurology, 28, 205-212.

Waters, P. and Strick, P.L. (1981), 'Influence of strategy on muscle activity during ballistic movements', Brain Research, 207, 189-194.

Wing, A.M. (1989), 'A comparison of the rate of pinch grip force increases and decreases in Parkinsonian bradykinesia', Neuropsychologia, 26, 479-482.

Wing, A.M. and Miller, E. (1984), 'Basal ganglia lesions and the psychological analyses of the control of voluntary movement', Ciba Foundation Symposium 107, 'Functions of the Basal Ganglia', Pitman, London, pp. 242-257.

Wolff, P.H., Gunnoe, C.E. and Cohen, C. (1983), 'Associated movements in man', In P.J. Vinken and G.W. Bruyn (Eds.), 'Handbook of Clinical Neurology, North Holland, Amsterdam, pp. 404-426.

FORCE AMPLITUDE AND FORCE DURATION IN PARKINSONIAN HANDWRITING

HANS-LEO TEULINGS (1) and GEORGE E. STELMACH (2) *

(1) Nijmegen Institute for Cognition Research and Information Technology (NICI), P.O. Box 9104, NL 6500 HE Nijmegen, The Netherlands.
(2) Exercise and Sport Science Institute, Psychobiology Section, Arizona State University, Tempe AZ 85287, USA.

ABSTRACT. Previous research has revealed that Parkinsonian handwriting may be impaired more in terms of force amplitude than force timing. These findings suggest that the source of the impairment may be located at the level of motor-unit recruitment. The present experiment demonstrates that an impairment of force amplitude or of force timing produce distinct handwriting distortions. To understand these distortions, it is assumed that pen movements may be decomposed into two movement axes. By simulations it can be shown that if force amplitude is manipulated the straight baseline of the handwriting pattern will become irregular. In contrast, if the force timing per axis is manipulated, sharp movement reversals (e.g., in cursive-script "u") become loops. Accordingly, it was found that Parkinsonians, in comparison to age-matched controls, exhibit no problems with sharp movement reversals but evince problems in maintaining a straight baseline. The effects observed were robust with respect to task load (i.e., writing "eluhule" interspersed with "eludule") and likewise, by alternating writing size (i.e., between normal and larger) and by writing speed (between normal and faster). Collectively, the results suggest a sequential organization of multi-joint movements in Parkinson's disease patients handwriting.

Introduction

Previous research has suggested that Parkinsonian handwriting is in comparison to age-matched controls impaired in terms of force amplitude and not in terms of force timing (Teulings & Stelmach, in press). Force amplitude refers to the strength of the force and force timing to force initiation and its duration. The impairment was quantified by the signal-to-noise ratio analyses (Teulings et al., 1986) of the force amplitude patterns versus those of stroke duration. These findings suggest that the source of the impairment may be located at the level of motor-unit recruitment (e.g., Dengler et al., 1986; 1988; 1989). The present experiment extends these findings by demonstrating that the force-amplitude and force-timing impairments produce distinct handwriting distortions.

* This research was supported by grants from the USA National Institute of Neurological Diseases and Stroke (NS 17421) and American Parkinson's Disease Association awarded to George E. Stelmach, and the Nijmegen University Research Pool (SW4/88). The researchers wish to thank the subjects from the Madison Area Parkinson Support Group for their motivated cooperation.

J. Requin and G. E. Stelmach (eds.), Tutorials in Motor Neuroscience, 149–160.

Force Amplitude

PD subjects have difficulty in producing the force amplitude necessary for large stroke size (McLennan et al., 1972; Margolin and Wing, 1983). However, PD patients can produce, isometrically, a target force as accurately as normals (Stelmach and Worringham, 1988b; Stelmach et al., 1989b).

Force Timing

PD subjects initiate their muscle contractions more slowly which seems to be a timing problem (Stelmach and Worringham, 1988b). In fact, although the rate of force development is slow in PD subjects, its release is even slower (Angel and Lewitt, 1978; Wing, 1988; Viallet et al., 1987). According to Phillips et al. (1989), the slowness of handwriting movements in PD subjects may be explained partially by temporal characteristics of muscle-force changes and, to a lesser degree, by low force amplitude. To achieve an acceptable writing size, PD subjects would need to produce longer stroke durations (Margolin and Wing, 1983). Phillips et al. (1989) hypothesized that the impaired force amplitude may contribute to a deliberate compensatory behavior for movement duration. Neither PD patients nor elderly controls were able to modify their stroke durations as proficiently as their stroke lengths.

When considering the timing of movement sequences, Stelmach et al. (1989a) have shown that Parkinsonians have difficulty producing alterations of force levels within a movement sequence. Moreover, their force pulses are more irregular than normal and not well sustained (Stelmach and Worringham, 1988b; Margolin and Wing, 1983; Stelmach et al., 1989b; Flowers and Sheridan, 1987).

A known timing problem in PD patients is the "hastening phenomenon" reported by Nagasaki and Nakamura (1982). PD subjects exhibited a hastening of movement when tapping at frequencies of 2.5 or 5 Hz. As the predominant frequencies in handwriting are close to 5 Hz (Teulings and Maarse, 1984), such a hastening effect could cause writing problems in PD patients trying to write at normal speeds. Freund (1989) found that serial hand movements faster than 2 Hz are only possible in PD patients if they are synchronized with the tremor. Such an impairment may underly the so-called 'hastening' phenomenon and lead to a disturbance of skills as handwriting or typing, which are usually performed at higher frequencies (Teulings and Maarse, 1984).

Benecke et al. (1987) and Goldenberg et al. (1986) found that sequential movements in PD patients are executed more slowly than movements executed in isolation. The second part of the sequence, in particular, is slowed and pauses between the first and second parts of the movement are increased. Even more difficulty is experienced when PD patients are asked to superimpose two separate motor programs (Benecke et al., 1986; Shimizu et al., 1987; Horstink et al., 1990). Also drawing lines of geometrical figures using both shoulder and elbow joints may be considered as superimposed tasks. Berardelli et al. (1986) found that PD patients executed lines straight but the time necessary to trace the geometric figures and the pauses at the vertices were prolonged relative to controls. They concluded that in Parkinson's disease the disability in generating two-joint ballistic movements depends on the difficulty in running motor programs for complex trajectories.

Description of Handwriting

Handwriting is the product of a sequence of well-practiced stroke shapes and relative stroke sizes. In order to produce these shapes and sizes the writer must produce preplanned net force amplitudes of appropriate durations and at appropriate moments along two axes of movements.

This description introduces three concepts which require some elaboration. First, there is the notion of a stroke. A stroke is defined as an elementary, ballistic movement having a unimodal absolute-velocity time curve during which on-line feedback cannot be processed. As such, relative minima of the absolute velocity-versus-time function form the segmentation points between strokes. In fluent handwriting the segmentation points are mostly at the bottoms and tops of the cursive-script characters where the curvature is high and the pen changes movement direction sharply (Thomassen and Teulings, 1985). However, in certain categories of subjects, pen movement is not fluent or ballistic (e.g., Maarse et al., 1987). Then, the absolute velocity-versus-time curve of a more or less straight trajectory shows two or more peaks. Therefore, a stroke will be defined as the segment between sharp changes of direction from up to down or vice versa, taking into consideration that there may be several peaks of the pen speed or even movement stops during that stroke. These points of sharp change in direction are referred to as 'segmentation points'.

The second concept concerns axes of movement. Many different systems of movement axes have been proposed (Teulings et al., 1989), such as the biomechanical degrees of freedom defined by flexions and extensions of the fingers and wrist, or the graphical axes of movement which are parallel to the baseline (horizontal) and perpendicular to it (vertical). The simulations of handwriting to be presented here do not depend upon any specific choice of movement axes, however. Moreover, it is assumed that each axis of movement is controlled by its own antagonistic muscle system. It is suggested that part of the impairment is located at the level of motor-unit control which is justified by assuming that impairments may exist in each of the axes of movement separately.

The third concept is muscle force as a function of time which, unfortunately, cannot be measured directly. Only the net muscle force can be approximately estimated, by assuming that it increases with the acceleration time function. It is currently the best available approximation. This estimate is correct to the extent that friction and viscosity forces are considerably smaller than inertial forces, which is obviously the case in small and fast handwriting under open-loop control (e.g., Denier van der Gon and Thuring, 1965).

Simulation of Impaired Force Amplitude

In order to understand the effects of impaired force amplitude and force timing upon the handwriting pattern, simulations were performed. Let us first consider to what extent the handwriting pattern is altered if force amplitude along the axes of movement is impaired. In Figure 1 a handwriting pattern is shown, together with the vertical position and acceleration time functions. In the example, the vertical force amplitude of one entire stroke has been decreased relative to the planned force amplitude, which generates a stroke shorter than planned, with all following strokes displaced. As a consequence, the baseline, which is implicitly straight, is now maintained less accurately. In fact, the acceleration integrated across the stroke had to be kept constant to prevent drift in the simulated trace. This was done by adding a small amount to the multiplied acceleration. In general, if many impairments occur, the baseline will be irregular. Note that the shapes and loops of all displaced strokes do not change. It should be noted that impairment of the horizontal component does not affect the base line. The same result would be produced for any other pair of main axes.

The extent of the baseline inaccuracy can be expressed by measuring the standard deviation of the vertical positions of the low segmentation points. As the standard deviation is proportional to the overall size of the handwriting pattern, the ratio of this standard deviation and the pattern's average stroke height forms a dimensionless measure for the inaccuracy of the baseline.

Simulation of Impaired Force Timing

For the second simulation excercise, let us consider to what extent the handwriting pattern changes if force timing is impaired. The sharp movement reversals at the tops of cursive-script "u" can only be produced if both movement axes simultaneously reach zero velocity no matter which axes of movement orientation is chosen. In Figure 2 a handwriting pattern is shown with the vertical acceleration at the beginning of a stroke delayed relative to the horizontal component. In this example, the first sharp movement reversal in the cursive-script "u" will be distorted into a loop. A sharp movement reversal may also become a blunt arc if the horizontal component is delayed relative to the vertical one (Schomaker et al., 1989). However, on average, the loop areas increase.

In order to return to the sequence of planned stroke shapes it is necessary for the horizontal component to phaselock with the vertical component. In order to keep the disruption of the remaining part of the pattern to a minimum, the horizontal component of the next stroke is delayed by the same amount such that none of the subsequent strokes are distorted. In fact, the first sample of the delay period and the first sample after the delay period are corrected in opposite directions such that velocity was zero after the first sample of the delay, which prevents drift in the simulated trace. Note that the relative baseline inaccuracy is virtually unaffected by the impairment of force timing.

Any timing inaccuracies, and therefore loop areas, may also be caused in higher speeds of handwriting. The extent of this effect will be tested by introducing a condition with faster than normal writing speed.

A measure for the impairment of force timing is the area of the loop between two subsequent strokes. As loop area increases with the square of the size of the entire pattern, the area of the loop divided by the heights of the two strokes enclosing the loop forms a dimensionless measure for loop area.

Summary

Two measures of distortion of handwriting have now been proposed: (1) distortion of the baseline which is a key measure of force-amplitude impairment, and (2) average loop area, at places were sharp movement reversals were to be performed (e.g., at the tops of cursive-script "u"), which is a key measure for force-timing impairment. On the basis of earlier research (Teulings and Stelmach, in press) we predict that Parkinsonian handwriting shows a less accurate baseline in comparison to controls but not larger relative loop sizes.

The writing task was performed in several conditions. The normal condition consisted of writing a number of times the same pattern. There are reasons to included other conditions as well. Writing the pattern alternately with another writing pattern or with the same pattern but at larger size or higher speed presents a substantial problem for Parkinsonians (e.g., Benecke et al., 1987, Yanagisawa et al., 1989). This suggests that in conditions of alternating tasks problems with force amplitude and timing control become more obvious. Therefore, one would predict that Parkinsonians may exhibit increased baseline inaccuracy and relative loop area in these conditions. Furthermore, in our previous research it was found that the controls tended to exhibit a low signal-

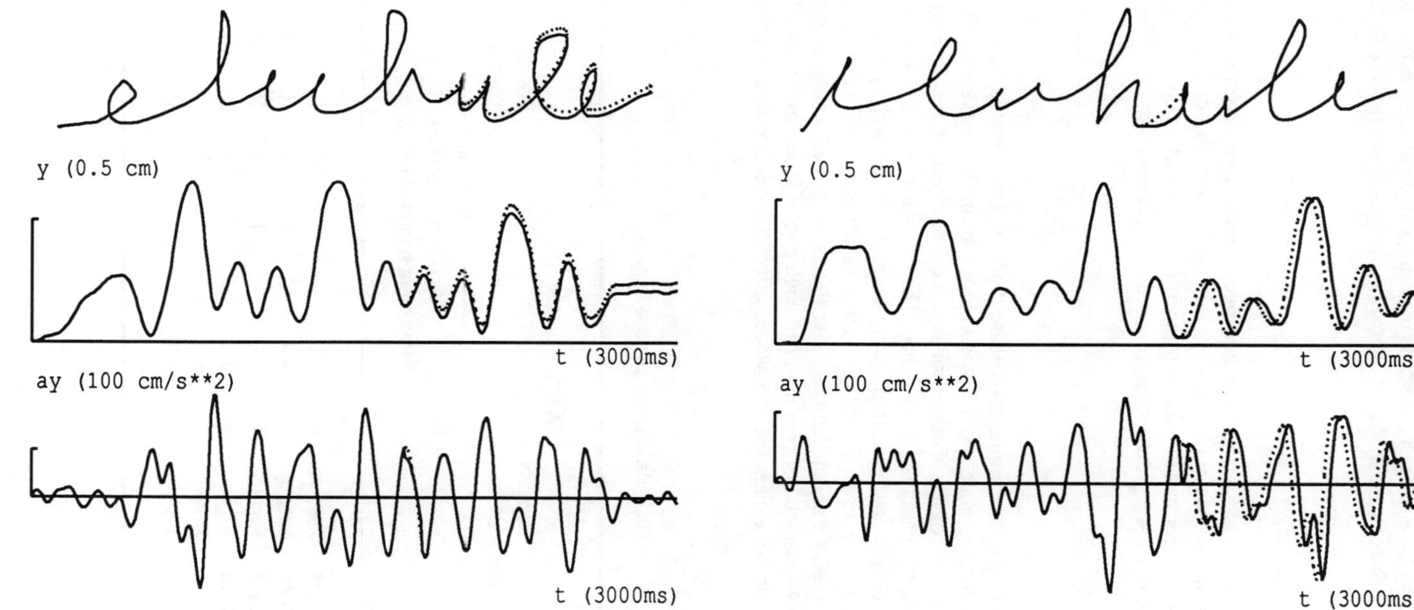

Figure 1. A trial by a control subject, together with the vertical position (y) and acceleration (ay) time functions, respectively. During stroke 13 the original vertical force amplitude (dotted) has been multiplied by 0.7. In the simulation, the relative inaccuracy of the baseline increases from 0.11 to 0.15, whereas the relative loop area between strokes 13 and 14 remains 0.27.

Figure 2. A trial by a PD subject, together with the vertical position (y) and acceleration (ay) time functions, respectively. At the beginning of stroke 13 the original vertical acceleration (dotted) has been delayed by 30 ms. At the end of the stroke, the horizontal acceleration has been delayed by the same amount. In this simulation. the relative loop area between strokes 13 and 14 increases from 0.20 to 0.85, whereas the relative inaccuracy of the baseline remains 0.13.

to-noise ratio in the write-larger and the write-faster conditions similar to that of the Parkinsonians. This suggests that controls may have increased baseline inaccuracy in these conditions. Finally, this research investigates whether these postulations are altered by changes in writing size or speed.

Experiment

To examine these predictions, PD subjects and age-matched control subjects performed a writing pattern, either at normal speed and size, or by executing it interspersed with a different, or a larger or a faster pattern. The subjects' task was to execute a specific writing pattern containing long and short, upstrokes and downstrokes, sharp movement reversals (as in cursive-script "u"), and loops (as in cursive-script "e").

Method

Subjects. Six subjects with idiopathic Parkinson's disease (5 males and 1 female; See Table 1) and six age-matched control subjects (1 male and 5 females) took part in the experiment. PD subjects were between 49.5 and 79.8 years of age. Control subjects were of a similar age, ranging from 1.0 to 4.2 years younger than the PD patients they matched. The controls reported having no signs of neurological disease. The only screening criteria were willingness and ability to perform the handwriting tasks. All subjects were proficient in cursive handwriting and were right-handed, except for one PD subject, who was left-handed. The PD subjects had a precise medication schedule, ranging from 2 to 4 Sinemet doses per day. Although gender is not matched well between both groups of subjects we do not expect that this would affect the baseline inaccuracy and loop size

Table 1 Parkinson's Disease Subjects.

Case	Age (y)	Gender	Handedness	Impaired side	Diagnosed onset (y)	Movement bandwidth (Hz)	Action tremor (Hz)	Tremor at rest	Micrographia	Rigidity	Bradykinesia
1JP	64.2	Female	Right	Bilateral	4	6.0	-	T	M	R	-
3RC	75.3	Male	Right	Left	4	4.4	-	T	-	R	B
4WF	79.8	Male	Right	Right	7	4.0	-	T	M	R	B
6JH	72.8	Male	Right	Right	5	2.2	5.6	T	M	R	B
7JR	53.0	Male	Left	Both	2	4.2	-	-	M	-	B
8PB	49.5	Male	Right	Both	1	4.5	-	T	M	R	-

- Absence of a symptom

Apparatus. The subjects wrote on a large, height-adjustable digitizer, (CalComp 91240) with its standard blunt-barreled pen. The digitizer provided horizontal and vertical coordinates of the pen tip at a frequency of 100 Hz, and an RMS accuracy of 0.2 mm. Non-simultaneous sampling of horizontal and vertical coordinate was supposed to have little effect in small handwriting. The subjects were asked to find their most comfortable writing posture. Once they chose the position and orientation of their writing, a sheet of paper with a 28-cm guide line was fixed upon the digitizer. This sheet was covered by a blank sheet transparent enough to see the guide line underneath. After each trial the covering paper was shifted by the subject such that they were able to write on this guide line while maintaining a constant writing posture. The forearm rested on the digitizer to minimize body movements and to dampen any tremor in the writing arm.

Procedure. Subjects were instructed to write in their normal style, size, and speed. At the beginning of each trial, the subject placed the pen on the paper in a position ready to write. When the subject was ready, the experimenter signaled the start of the recording with a beep, at which time the subject was free to choose when to initiate his/her handwriting. None of the subjects appeared to have initiation difficulties. The experimenter checked whether the correct letter sequence was produced and repeated trials where the letter sequence produced was clearly not the one requested. At any rate, no mistakes between e.g., cursive-script "u" and "el" were accepted in the data set.

Design and Condition. An experimental session consisted of the following tasks: (1) Write the sentence "They wrote eluhule." repeatedly for 2 trials. (2) Write the pseudo word "eluhule" at normal size and speed, for 10 trials. (3) Write "eluhule" but alternate it with writing "eludule" (consequently, yielding 20 trials). (4) Write "eluhule" but alternate it with the same word written larger than normal. (5) Write "eluhule" but alternate it with the same word written faster than normal. (The order of Tasks 3-5 was random). (6) Repeat as in (2). (7) Repeat as in (1). In Tasks 1 and 7 only the first 20.48 s per trial were recorded. In the other tasks the recording period was 4, 5, or 6 s per trial. However, in the slowest Parkinsonian the recording period was 8 or 10 s per trial and the number of trials per series was reduced to 10. The total duration of an experimental session was approximately 60 minutes. The trials of the pseudo word can be divided into 8 conditions: Task 2 (1 condition), Task 6 (1 condition), Task 3 (2 conditions), Task 4 (2 conditions), and Task 5 (2 conditions).

The purpose of Task 1 at the beginning and end of the experimental session was to estimate an individual's movement-frequency bandwidth and its constancy throughout the session. Task 2 was repeated at the end of a session in order to verify that the dependent variables did not change during the experimental session.

Data Analysis. Each subject's movement-frequency bandwidth was estimated in order to choose the appropriate lowpass cut-off frequency for calculating the velocity time function. The bandwidth was assumed to be equal to the frequency where the frequency spectrum dropped sharply. In order to estimate the frequency spectrum, each record of 2048 samples was divided into seven, equally spaced, overlapping segments of 512 samples; each segment was multiplied by a Hamming window. Next, a Fast Fourier Transform was applied to obtain the spectrum for position, a first time derivative taken by multiplication with angular frequency, and the absolute value taken, to obtain the amplitude spectrum for velocity. This spectrum was subsequently averaged across segments. The transition band of the lowpass filter was a half raised-cosine wave which extended to 3.7 times the frequency where the lowpass filter started suppressing (Teulings and Maarse, 1984).

The bandwidths of the writing apparatus in the control subjects were 5.4 to 6.3 Hz, and in the PD subjects, 2.2 to 6.0 Hz. The lowpass filtering frequencies (frequencies were the half raised cosine started) were 5.1 Hz for most subjects and 3.4 or 1.7 for the slower and the slowest subjects, respectively. In fact, the latter subject showed considerable action tremor with an average frequency of 5.6 Hz. Lowpass filtering with a transition band between 1.7 and 6.3 Hz was capable of reducing the effect of this action tremor sufficiently well and yielded a smooth movement pattern.

The writing patterns were first rotated to a horizontal baseline according to the orientation of the guide line, chosen by the subject. Subsequently, the data were lowpass filtered and differentiated and the absolute velocity was determined. Actually the absolute velocity was the weighted sum of the squared horizontal and vertical velocity component with weights 0.1 and 1.0, respectively. Then, the writing patterns were segmented into strokes. This was done by detecting all local minima of the absolute velocity (i.e., the lowest minimum within an interval of -50 and +50 ms). Only those minima were selected where a downward stroke, if not the first stroke, is followed by a significant upward movement (i.e., beyond 0.03 cm in vertical direction) and where an upward stroke is followed by a significant downward movement. Initial and trailing strokes shorter than 0.03 cm were also omitted. Ideally, the patterns of "eluhule" and "eludule" contain 21 strokes but the first two and the last three strokes were omitted, thus yielding 16 short and long up and down strokes. Four of the 12 subjects used to write "d" with an extra down and up stroke at the top of the c-shaped stroke. In order to enable comparison between subjects, these two strokes were discarded. Then the baseline was determined more precisely by the least-square error line through all adopted low segmentation points and the individual pattern was adjusted to horizontal orientation. A size-independent measure for the inaccuracy of the baseline was the standard deviation of these distances divided by the average stroke height. Finally, we estimated per stroke vertical size, duration, and loop area enclosed by the stroke and the previous one. A size-independent measure for the loop area is the area of the loop divided by the stroke heights of the two crossing strokes. Only those trials were accepted for continued analysis which showed the required pattern of short and long vertical strokes: The height of the longest short stroke should be less than the height of the shortest long stroke. Thus, 6 to 10 trials per condition per subject (average 8.9) remained for the controls, and, 2 to 10 (average 7.1) for the PD subjects. In the end, each trial yielded vertical stroke sizes, stroke durations, loop areas per stroke and an inaccuracy measure for the baseline. Medians were taken across replications in order to minimise effects of outlyers. Then, averages were taken across the strokes of a trial.

Results

Table 2 and Figure 3 summarize means across subjects of four types of data: vertical stroke size, stroke duration, inaccuracy of the baseline and loop area for Parkinsonians and controls and per condition. For each type of data a 2-groups x 6-subjects x 8-conditions analysis of variance was performed.

Vertical Stroke Size. PD subjects were found to write marginally smaller than controls ($F(1,10)=3.83$, $p=0.08$), which was to be expected (e.g., Lewitt, 1983). There is a significant condition effect ($F(7,70)=25.0$, $p<0.01$), which was caused by the instruction to write larger (See third panel of the first row of Figure 3). There was no interaction between group and condition ($F(7,70)=1.16$, $p>0.1$).

Table 2 The mean vertical stroke size, mean stroke duration, relative inaccuracy of the baseline, and the mean relative loop size of the tops of cursive-script "u", in the controls versus the PD subjects.

	Controls	Parkinsonians
Size (cm)	0.33	0.27
Time (s)	0.128	0.185
Base	0.073	0.098
Loop	0.77	0.42

Stroke Duration. PD subjects take only marginally more time to write than the controls (F(1,10)=3.44, p=0.09). There is a significant condition effect (F(7,70)=9.01, p<0.01) which was caused by the instructions to write larger and to write faster (See Panels 3 and 4 of the second row of Figure 3). There was no interaction between group and condition (F(7,70)=1.14, p>0.1).

Baseline Inaccuracy. PD subjects show greater relative inaccuracy of the baseline (F(1,10)=16.25, p<0.01). Condition does not show a main effect (F(7,70)=1.5, p>0.1) or interaction (F(7,70)=0.50, p>0.1).

Loop Area. PD subjects do not produce significantly larger relative loop sizes than controls (F(1,10)=2.86, p>0.1). They even seem to show a tendency to produce smaller loops. There is a marginal instruction effect (F(7,70)=2.05, p=0.06) but no interaction between group and condition (F(7,70)=1.27, p>0.1).

In summary, these data reveal that Parkinsonian handwriting shows a more irregular baseline but does not show larger loop areas at sharp movement reversals. The general absence of interactions indicates that Parkinsonians and controls do not behave differently in the various conditions with alternating patterns. Finally, differences between first and last normal series were only minor, indicating that learning nor fatiguing took place during the experimental session.

Discussion

PD patients and age-matched controls performed handwriting patterns under various natural conditions. It was found that the baseline in Parkinsonian handwriting is more irregular within a pattern than that of control subjects. However, relative to the controls, PD patients did not show larger relative loop areas in their handwriting when letters with sharp movement reversals had to be performed (See Figures 1 and 2 for examples of each). On the basis of simulations performed it is argued that these data support earlier findings (Teulings and Stelmach, in press) that PD patients show impaired control of force amplitude. but have an intact control of their force timing.

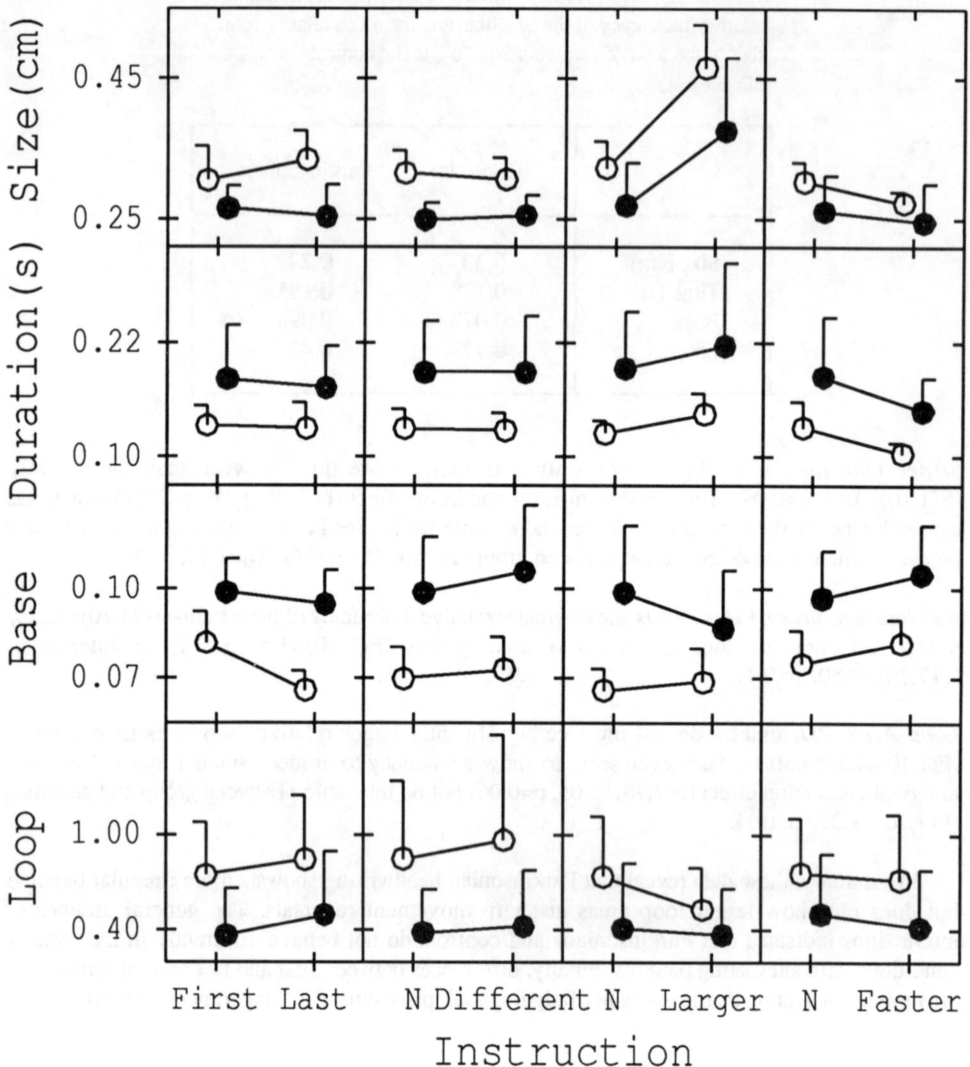

Figure 3. The mean vertical stroke size (top row), mean stroke duration (second row), relative inaccuracies of the baseline (third row), and the mean relative loop size of the tops of cursive-script "u" (bottom row) for the first versus last normal series (left column), for normal versus the different pattern (i.e., "eluhule" versus "eludule") (second column), for normal size versus larger (third column), and for normal speed versus faster (right column), in controls (open circles) and PD subjects (closed circles), respectively. Vertical bars indicate one side of the 5%-confidence interval.

An alternative explanation for the high degree of inaccuracy of the baseline in Parkinsonian handwriting is that Parkinsonians shifted their baseline back to the guide line only at casual moments when they processed visual feedback of the ongoing pattern. This is unlikely as it was found in previous research (Teulings and Stelmach, in press) that both Parkinsonians and controls do not alter their handwriting in the absence of visual feedback.

PD subjects show a tendency to produce even smaller relative loop areas at sharp movement reversals than controls. The larger average loop area in control subjects could be an artifact because they simply write faster. However, this can be rejected as no marked change in loop area was found in the condition with the speed instruction. An interesting explanation of the smaller relative loop sizes in Parkinsonian handwriting could follow from the recent findings by Flash (1990) that Parkinsonians do not tend to superimpose elementary motor programs in multi-joint movements, but rather finish one movement before starting the next. This deliberate putting into a disjunct sequence of movements, that could in principle be partly superimposed, may cause the very pointy and sharp movement reversals in characters such as the "u" observed in Parkinsonian handwriting. The lack of ability to superimpose elementary movement components is in line with similar observations by Berardelli et al. (1986), Benecke et al. (1986), Shimizu et al. (1987), and Horstink et al. (1990).

References

Angel, R. and Lewitt, P.A. (1978). Unloading and shortening in Parkinson's disease. *Journal of Neurology, Neurosurgery, and Psychiatry, 41*, 919-923.

Benecke, R. Rothwell, J.C. Dick, J.P. Day, B.L. and Marsden, C.D. (1986). Performance of simultaneous movements in patients with Parkinson's disease. *Brain, 109*, 739-757.

Benecke, R. Rothwell, J.C. Dick, J.P. Day, B.L. and Marsden, C.D. (1987). Disturbance of sequential movements in patients with Parkinson's disease. *Brain, 110*, 361-379.

Berardelli, A. Accornero, N. Argenta, M. Meco, G. and Manfredi, M. (1986). Fast complex arm movements in Parkinson's disease. *Journal of Neurology, Neurosurgery, and Psychiatry, 49*, 1146-1149.

Dengler, R. Elek, J. Hermans, R. and Wolf, W. (1988). [Double discharges of motor units and tremor strength]. *EEG EMG, 19*, 77-80.

Dengler, R. Gillespie, J. Argenta, M. Elek, J. Wolf, W. and Struppler, A. (1989). The impact of paired motor unit discharges on tremor. *Electromyography and Clinical Neurophysiology, 29*, 113-117.

Dengler, R. Wolf, W. Schubert M. and Struppler, A. (1986). Discharge pattern of single motor units in basal ganglia disorders. *Neurology, 36*, 1061-1066.

Denier van der Gon, J.J. and Thuring, J.Ph. (1965). The guiding of human writing movements. *Kybernetik, 2*, 145-148.

Flash, T. (1990). *Studies into the timing, sequencing, and modification of hand trajectories in healthy subjects and patients with basal ganglia disorders*. Paper presented at the NIAS Conference: Sequencing and timing of human movement. Wassenaar, The Netherlands, June.

Flowers, K.A. and Sheridan, M.R. (1987). *Movement variability and bradykinesia in Parkinson's disease*. E.S.P. Symposium on Motor Skills, Oxford, July.

Freund, H J. (1989). Motor dysfunctions in Parkinson's disease and premotor lesions. *European Neurology, 29*, 33-37.

Goldenberg, G. Wimmer, A. Auff, E. and Schnaberth, G. (1986). Impairment of motor planning in patients with Parkinson's disease: evidence from ideomotor apraxia testing. *Journal of Neurology, Neurosurgery, and Psychiatry, 49*, 1266-1272.

Horstink, M.W.I.M. Berger, H.J.C. Van Spaendonck, and Van den Bercken, J.H.L. (1990). Bimanual simultaneous motor performance and impaired ability to shift attention in Parkinson's disease. *Journal of Neurology, Neurosurgery, and Psychiatry, 53*, 685-690.

Lewitt, P.A. (1983). Micrographia as a focal sign of neurologic disease. *Journal of Neurology, Neurosurgery, and Psychiatry, 46*, 1152-1153.

Maarse, F.J. Meulenbroek, R.G.J. Teulings, H.L. and Thomassen, A.J.W.M. (1987). Computational measures for ballisticity in handwriting. In R. Plamondon, C.Y. Suen, J.-G. Deschenes and G. Poulin (Eds.), *Proceedings of the Third International Symposium on Handwriting and Computer Applications.* Ecole Polytechnique, Montreal, pp. 16-18.

Margolin, D.I. and Wing, A.M. (1983). Agraphia and micrographia: Clinical manifestations of motor programming and performance disorders. *Acta Psychologica, 54,* 263-284.

McLennan, J.E. Nakoma, K. Tyler, H.R. and Schwab, R.S. (1972). Micrographia in Parkinson's Disease. *Journal of Neurological Sciences, 15,* 141-152.

Nagasaki, H. and Nakamura, R. (1982). Rhythm formation and its disturbance; A study based upon periodic response of the motor output system. *Journal of Human Ergology, 11,* 127-142.

Phillips, J.G. Stelmach, G.E. and Teasdale, N. (1989). Preliminary assessment of spatio-temporal control of handwriting. In R. Plamondon, C.Y. Suen and M.L. Simner, (Eds.), *Computer and human recognition of handwriting.* World Scientific, Singapore, pp. 317-332.

Schomaker, L.R.B. Thomassen, A.J.W.M. and Teulings, H.L. (1989). A computational model of cursive handwriting. In R. Plamondon, C.Y. Suen and M. Simner (Eds.), *Computer recognition and human production of handwriting.* World Scientific, Singapore, pp. 153-177.

Shimizu, N. Yoshida, M. and Nagatsuka, Y. (1987). Disturbance of two simultaneous motor acts in patients with Parkinsonism and cerebellar ataxia. *Advances in Neurology, 45,* 367-370.

Stelmach, G.E. Garcia-Colera, A. and Martin, Z.E. (1989a). Force transition control within a movement sequence in Parkinson's Disease. *Journal of Neurology, 236,* 406-410.

Stelmach, G.E. Teasdale, N. Phillips, J.G. and Worringham, C.J. (1989b). Force production characteristics in Parkinson's disease. *Experimental Brain Research, 76,* 165-172.

Stelmach, G.E. and Worringham, C.J. (1988b). The preparation and production of isometric force in Parkinson's disease. *Neuropsychologia, 26,* 93-103.

Teulings, H.L. and Maarse, F.J. (1984). Digital recording and processing of handwriting movements. *Human Movement Science, 3,* 193-217.

Teulings, H.L. and Stelmach, G.E. (in press). Control of stroke size, peak acceleration, and stroke duration in Parkinsonian handwriting. *Human Movement Sciences.*

Teulings, H.L., Thomassen, A.J.W.M. and Maarse, F.J. (1989). A description of handwriting in terms of main axes. In R. Plamondon, C.Y. Suen, & M. Simner (Eds.), *Computer recognition and human production of handwriting.* World Scientific, Singapore, pp. 193-211.

Teulings, H.L, Thomassen A.J.W.M. and Van Galen, G.P. (1986). Invariances in handwriting: The information contained in a motor program. In H.S.R. Kao, G.P. Van Galen and R. Hoosain (Eds.), *Graphonomics, contemporary research in handwriting.* North-Holland Press, Amsterdam, pp. 305-315.

Thomassen, A.J.W.M. and Teulings, H.L. (1985). Time, size and shape in handwriting, exploring spatio-temporal relationships at different levels. In J.A. Michon and J.B. Jackson (Eds.), *Time, mind and behavior.* Springer-Verlag, Heidelberg, pp. 253-263.

Viallet, F. Massion, J. Massarino, R. and Khalil, R. (1987). Performance of a bimanual load-lifting task by Parkinsonian patients. *Journal of Neurology, Neurosurgery and Psychiatry, 50,* 1274-1283.

Yanagisawa, N. Fujimoto, S. and Tamaru, F. (1989). Bradykinesia in Parkinson's disease: Disorders of onset and execution of fast movement. *European Neurology, 29,* 19-28.

Young, P.T. (1931). Sex differences in handwriting. *Journal of Applied Psychology, 15,* 486-498.

Wing, A.M. (1988). A comparison of the rate of pinch grip force increases and decreases in Parkinsonian bradykinesia. *Neuropsychology 26,* 479-482.

SCALING PROBLEMS IN PARKINSON'S DISEASE

F. MÜLLER (1) and G. E. STELMACH (2)

(1) Department of Neurology
Klinikum Schnarrenberg
University of Tübingen
Hoppe-Seyler-Str. 3
D 7400 Tübingen
Germany

(2) Motor Behavior Laboratory
Psychobiology Section
Exercise and Sports Institute
Arizona State University
Tempe, Arizona, 85287
USA

ABSTRACT. Understanding functional impairments in a well defined movement disorder has drawn considerable interest from motor control researchers. Here Parkinson's disease (PD) serves as a model of deficient basal ganglia function. While several experiments implied that Parkinson's disease patients have problems intentionally scaling force, speed, spatial or temporal aspects of their movements and some authors even stated that patients could not intentionally vary the speed of their movements (Berardelli, Accornero, Argenta, Meco and Manfredi; 1986). The experiments presented here addressed the ability of PD patients to scale different movement parameters. Force pulses or levels are achieved by a pulse height control strategy. Experiment I examined the ability of patients to perform force step changes of different amplitudes and directions in finger pinch and elbow flexion. Experiment II analyzed reaching movements for different movement amplitudes and different instructional sets. Data from both experiments reject the hypothesis that Parkinson's disease patients cannot scale movement parameters according to task demands. However, the range of their scaling ability is limited.

INTRODUCTION

The study of movement disorders such as Parkinson's disease or cerebellar disorders uses the disruptive effect of defined lesions on movement parameters and profiles, to infer specific functions of the affected structures of the central nervous system on normal movements (Mai, Bolsinger, Avarello, Diener and Dichgans, 1988; Phillips, Müller and Stelmach, 1989; Brown, Hefter, Mertens and Freund, 1990). Parkinson's disease in this respect has quite often been studied as a model of basal ganglia dysfunction (Marsden, 1982). Brooks (1986) e.g. postulates, that basal ganglia dysfunction decomposes behavior into isolated motor acts and intended motor acts into movements of inappropriate amplitude. While some researchers demonstrated an

J. Requin and G. E. Stelmach (eds.), Tutorials in Motor Neuroscience, 161–174.

impairment in the performance of simultaneous movements (Benecke, Rothwell, Dick, Day and Marsden, 1986), this has not always been the case (Stelmach and Worringham, 1988). Similar contradictory results have been reported from studies that addressed the ability of patients to intentionally scale or modulate force, speed, spatial or temporal aspects of their movements.

In the chapter we present two experiments, that address the ability of Parkinson's disease patients to scale movement parameters according to task demands in various movement contexts (Teasdale and Stelmach, 1989) by using two different protocols, a paradigm of isometric force production and a paradigm of prehension production.

EXPERIMENT I: FORCE CHANGES

Control of force production is essential part of all motor acts and thus has served to study basic principles of movements in experiments where isometric muscle contractions had to be performed without overt movement. Gordon and Ghez (1987) e.g. have used isometric force production tasks to study the ability of normal subjects to perform amplitude scaling of force impulses and found that subjects regulated force rise time around a constant value and adopted a pulse height control strategy.

In recent years the control of isometric force production in Parkinson's disease has attracted considerable interest. It has been shown, that Parkinson's disease patients are able to produce accurate force levels, when expressed as a percentage of their maximum voluntary contraction (Stelmach and Worringham, 1988), although with a higher variability (Stelmach, Teasdale, Phillips and Worringham, 1989). Reaction time and rate of force increase are slowed in Parkinson's disease patients as shown in a variety of studies (Stelmach et al.; Benecke et al.; Viallet, Massion, Massarino and Khalil, 1987). Adequate force control however does not only require fast and accurate production of force (through isometric contractions of muscles) but demands fast and reliable termination of the movement itself as well. Stelmach, Worringham and Strand (1987) have noted in a repetitive finger-tapping sequence, that Parkinson's disease patients showed an abnormal prolongation of the first inter-tap interval, which suggests that PD patients may have problems with their control of force pulses. Similarly, Stelmach, Garcia Colera and Martin (1989) showed that when PDs had to intersperse a force among sequences they tended to inhibit the temporal sequencing. It was shown that PDs have difficulty initiating a sequence when they were requested to add a force to a finger tapping sequence. The sequence was also disrupted when the PD executed the adding force. Recently, Wing (1988) has compared increasing and decreasing step changes of pinch force in two untreated hemiparkinsonian patients (with the nondominant side being affected) and found a differential slowing of force decreases.

The experiment reported here compared smaller and larger isometric force increases with decreases of comparable amplitudes and assessed impairments of PD patients to perform force changes in a more comprehensive way. Everyday life poses different accuracy requirements on movements in different joints, affecting force control ability in different ways. We therefore considered it worthwhile, to examine the ability of Parkinson's disease patients to perform rapid force decreases and increases in finger pinch and elbow flexion independently.

Method

SUBJECTS. Eight subjects with idiopathic Parkinson's disease (7 males and 1 female; mean age 64,6 years) and eight age-matched healthy control subjects (7 males and 1 female) were studied with their dominant arm and hand. All patients had been selected as to show as little tremor as possible and had been screened by a neurologist for other neurological disorders. They were rated to be in stages II and III according to Hoehn and Yahr (1967) and were tested while on their normal medication.

APPARATUS AND PROCEDURE. Subjects were sitting in front of a table facing a CRT screen for presentation of stimuli and task feedback. Elbow flexion and thumb-index finger pinch forces were measured under near isometric conditions. During elbow flexion tasks upper arm and forearm of the subjects rested on an elevated padded lever, which was rigidly fixed to a heavy table; the arm was abducted at the shoulder joint, so that upper arm and forearm formed an angle of approximately ninety degrees in the horizontal plane. Subjects gripped a handle, so that any attempt to flex the elbow joint would cause the palmar surface of the hand to apply force to the transducer. For pinch force tasks subjects gripped a transducer with two parallel rods approximately 20 mm apart. The tips of thumb and index finger fitted neatly into two notches on the rods.
Pinch and elbow flexion tasks were studied separately. For each condition the data acquisition began with four trials of maximum voluntary force production, followed by one training session. Two blocks of data comprising six trials of each of the four different force step amplitudes were recorded for analysis. Control of the experiment and data sampling was performed by a PDP 11 computer with a sampling frequency of 250 Hz. Starting from a baseline level of 30 % of the maximum voluntary contraction the task required to increase , or decrease the existing force level by 15 respectively 30 % of mvc as fast and as accurately as possible; i.e. tasks required to change the actual force level by 50 or 100 %. A trial was initiated by an auditory start signal after the force level had staid for one second within the boundaries of +/- 10 % of the desired window, which was not visible to the subject. Feedback display of actual performance and display of the target window was given on the CRT screen after completion of each trial.

Unless indicated otherwise, data were analysed by ANOVAs with group, limb, direction and extent as independent factors.

Results

FORCE LEVELS. All subjects were very rapidly able to start a trial at the individual baseline level and did not require significant assistance to establish their baseline. Force levels however indicated quite different peak force levels for different subjects within and between groups. Average peak force levels differed significantly (t-test, $p<0.05$) between Parkinson's disease patients for peak pinch forces (mean 54.1 N for patients compared to 76.8 N for controls) and peak elbow flexion forces (mean 61.5 N for patients compared to 104.1 N for controls).

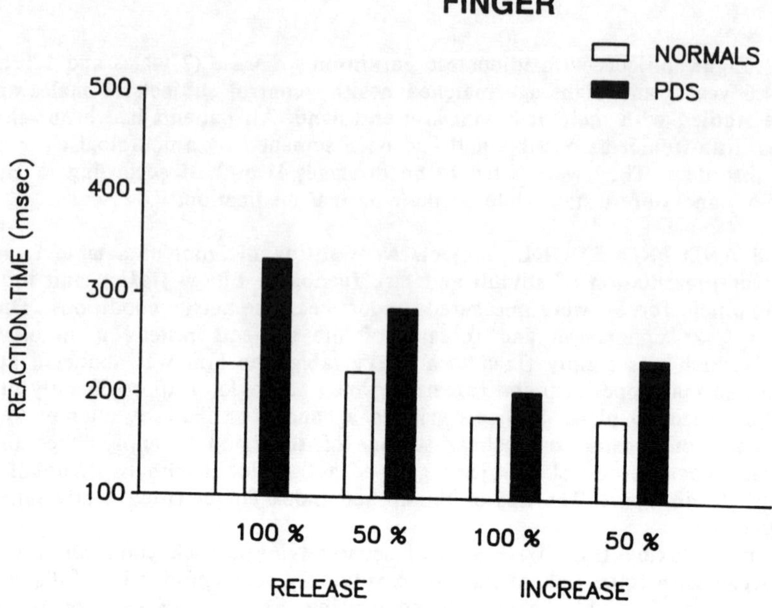

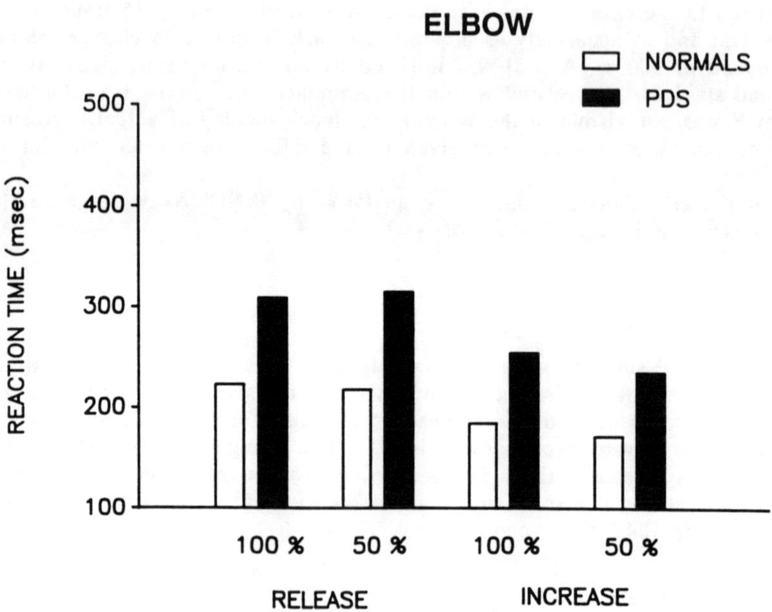

Figure 1. Illustration of group mean reaction times. Subjects had to perform step force changes, increasing or decreasing a baseline force level by 50 or 100 %.

REACTION TIMES: Reaction times were calculated from the onset of the imperative response stimulus to the first noticable change of force level in the required direction. Parkinson's disease patients were generally slower in response initiation with response times that were between 63 msec (pinch) and 79 msec (elbow flexion) longer ($p < 0.05$) than controls. Both groups took more time to prepare a decrease ($p < 0.01$), while this was unaffected by target level. No significant difference was found between limbs. Mean group reaction times for all conditions are shown in figure 1A and 1B.

MOVEMENT TIMES. Movement times were defined from onset of the step force change until the force trajectory reached the first boundary of the target window. Movement times confirmed the overall slowing of Parkinson's disease patients as shown in figure 2A and 2B ($p < 0.001$). Movement times showed longest duration for partial release, even longer than complete releases for both groups and both joints ($p < 0.001$), while changes of pinch force could generally be performed more rapidly than changes in elbow flexion forces ($p < 0.01$). The accuracy requirements of the task affected finger and elbow joints differentially. Normals were able to perform pinch force increases of different amplitudes within the same movement time duration, while this was not possible for elbow flexion tasks. However, Parkinson's disease patients were not able to scale their force rise trajectory in the same way. Higher force levels took 110 msec longer to be accomplished regardless of the effector muscles.

RATE OF FORCE CHANGES. A similar picture emerged when peak rate of force changes, i.e. first derivative of the force channel, were analyzed. Problems in force control of Parkinson's disease patients were confirmed by considerably smaller force pulse heights, averaging to only one third of pulse height of control subjects ($p < 0.001$). Even if force levels of patients are taken into account, as measured during the initial maximum voluntary contraction task, (40 % lower than peak forces of control subjects), pulse heights in patients are still profoundly impaired. Figure 3 shows mean pulse heights for all subjects in each of the different conditions of elbow flexion forces. For analysis purposes the pulse height of the small amplitude movement was divided by the amplitude of the large movement to create a measure of relative scaling, a ratio of 0.5 indicates a perfect amplitude scaling. The difference between the two joints was confirmed in this measure again. For pinch forces, normal controls adopt an impulse scaling strategy with ratios of 0.63 and 0.37 for increases, respectively decreases, while patients operate closer to the extremes, resulting in a ratio of 0.72 respectively 0.24 (figure 4). Thus the longer movement times in patients (for small decreases and large increases) is confirmed by comparable amplitudes of force impulses. This fact indicates that patients begin force increase movements of different amplitudes with nearly the same force pulses, and controls reach target levels by varying only the length of these pulses. For force releases, patients are able to use expanding pulses only for complete releases, however when required to decrease to a certain target level, they again use a small pulse amplitude and rather extend the duration of the force pulse until they reach the target level.

A different picture emerged for elbow flexion movements in controls; no interaction effect was found for direction and group. Thus, here normals had to use the same strategy of scaling force pulse duration that patients were using in both tasks. The reason for this difference is probably attributable to the accuracy requirement of our elbow flexion task.

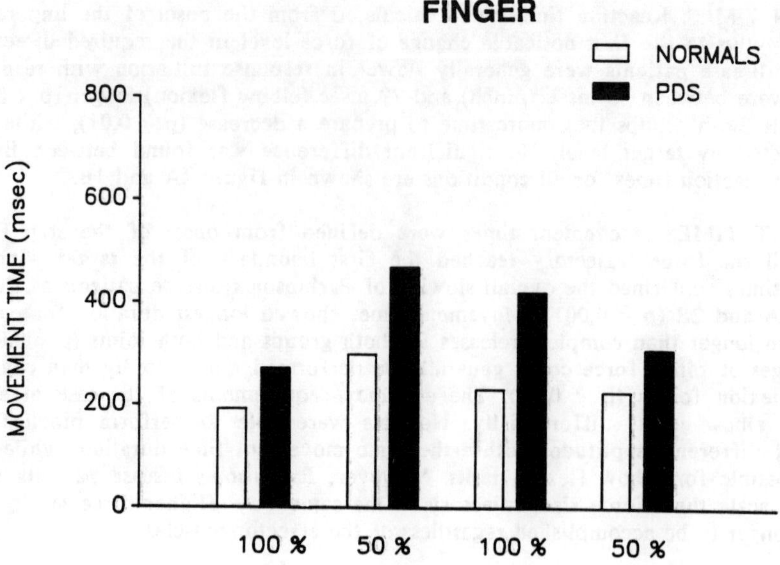

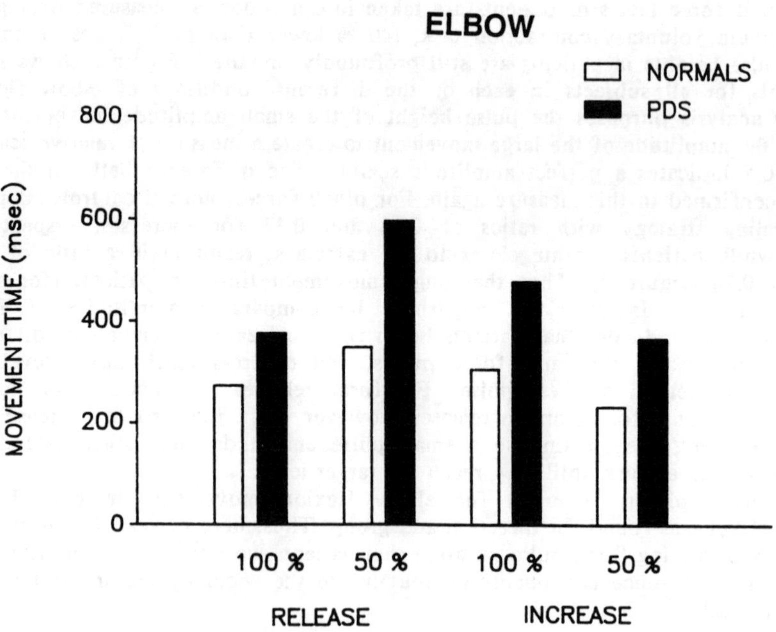

Figure 2. Illustration of group mean movement times. Subjects had to perform step force changes, increasing or decreasing a baseline force level by 50 or 100 %.

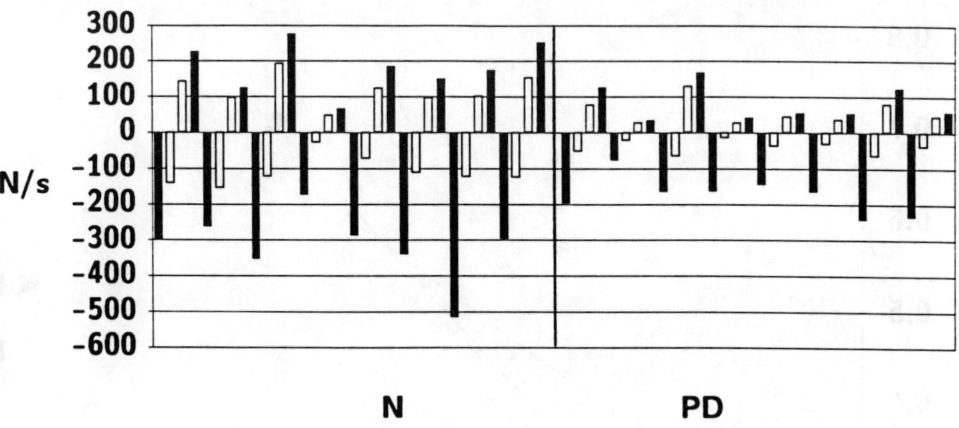

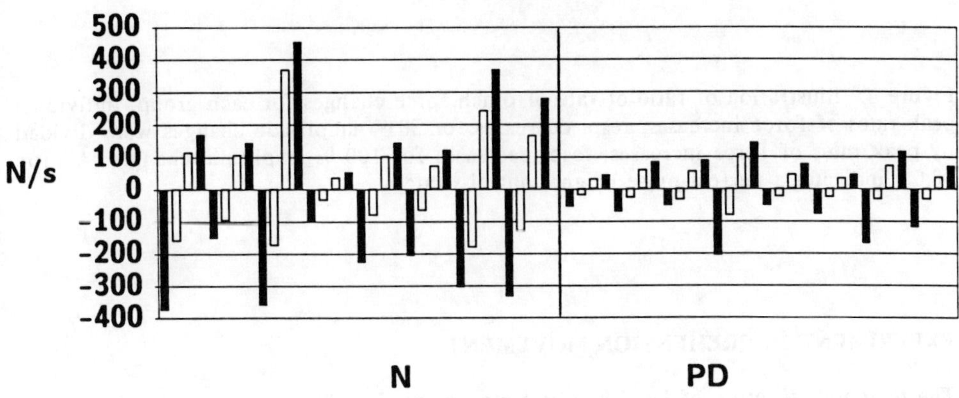

Figure 3. Mean force pulses of each individual subject aligned adjacent for each of the four conditions. Step force changes were performed in positive and negative direction by 100 % (▬) and by 50 % (▭) of a baseline force level.

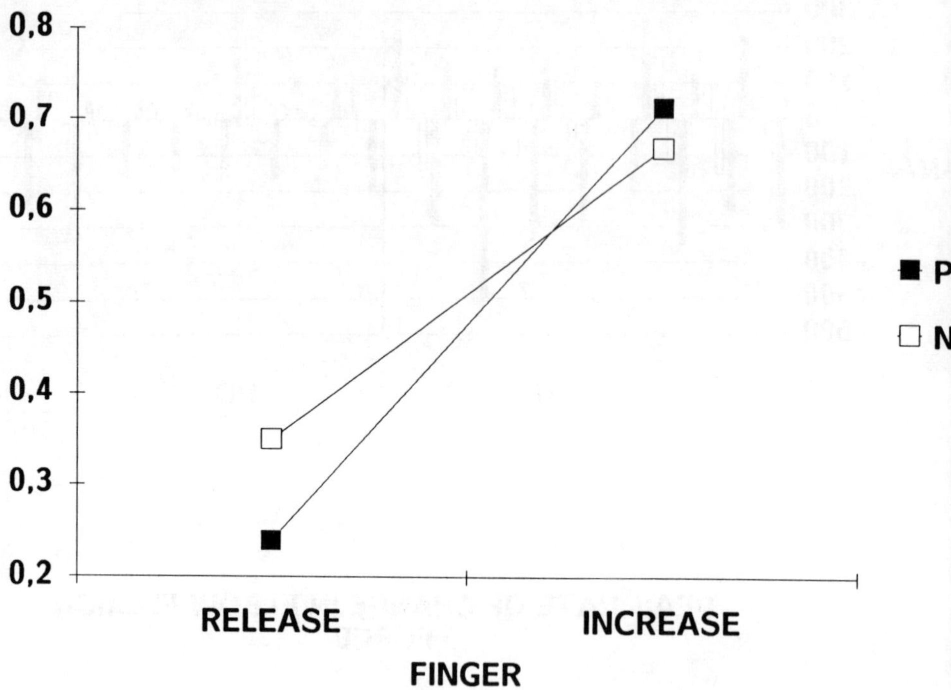

Figure 4. Illustration of ratio of rate of pinch force changes for each group. Individual peak rates of force increases, resp. decreases for 50 % amplitude changes were divided by peak rates of force increases, resp. decreases for 100 % amplitude changes. A ratio of 0.5 indicates a perfect pulse height control strategy.

EXPERIMENT II: PREHENSION MOVEMENTS

The temporal structure of hand and arm movements has drawn considerable interest in recent studies (Atkeson and Hollerbach, 1985; Soechting, 1984; Shapiro, 1986; Jeannerod, 1984; Wing, Turton and Fraser, 1986). Kinematic analyses show, that normal human arm movements share common organisational features, or as Bullock and Grossberg (1988) have put it, "produce velocity profiles whose global shape is remarkably invariant over a wide range of movement sizes and speeds". Berardelli,

Accornero, Argenta, Meco and Manfredi (1986) however stated, that patients suffering from Parkinson's disease, cannot intentionally vary the speed of their movements. Our second experiment addressed the ability of bradykinetic Parkinson's disease patients to scale reaching movements for different movement amplitudes and different instructions

Method

Movements were recorded with a WATSMART system of three cameras, allowing movements in three dimensions. Infrared light emitting diodes were attached to the radial bone at the wrist, the tip of thumb and index finger as well as on top of the target object. The starting position of the hand and the position of the grasped object were indicated by a small aluminum plate, which was connected to a force transducer thus allowing to register precisely lift-off and first touch at the object.

In this experiment the subjects had to reach for the dowel, grasp it and lift it. We studied the ability of Parkinson's disease patients to vary their movement control according to task demands. Task demands were varied by movement amplitude, object size, and instruction ("normal", "fast", "eyes closed"). Different accuracies of the reaching movement were required by using dowels of 2 different diameters (18 respectively 55 mm), which had to be grasped. 10 movements in each condition were analyzed.

4 bradykinetic Parkinson's disease patients and 4 age matched control subjects were studied. Subjects sat parallel to a table with their hand resting on the start plate. Upon an auditory stimulus they had to reach for the wooden dowel, grasp it and lift it.

Trajectorial velocity was computed off-line by the x-y-z coordinates of the marker on the radial bone at the wrist joint.

Hand shaping was analyzed by computing the three dimensional distance of the tips of thumb and index finger. First touch was indicated by a peak in the force signal of the target plate.

Results

MOVEMENT TIME. Data confirmed longer movement times and slower peak resultant velocities for Parkinson's disease patients overall ($p < 0.001$). Table 1 gives a summary of mean movement times for different conditions. Controls and patients were able to modulate movement time with movement amplitude and instruction. It took both groups 160 msecs longer to move double distance ($p < 0.001$). When instructed to move as fast as possible, controls could cut their movement time by half, while Parkinson's disease patients could reduce it only by one third (significant interaction group by instruction, $p < 0.05$). Closing the eyes did not affect movement time in normals, however expanded it in patients by 150 and 200 msecs respectively. Reaching for the smaller dowel took 50 msecs longer, however this difference was not statistically significant.

Table 1. A summary of group mean movement times (msec)of short and long amplitude movements for different instruction conditions ("normal", "fast","eyes closed").

instruction		movement short	long
normal	patients	1150	1340
	controls	840	1010
fast	patients	740	890
	controls	450	540
eyes closed	patients	1300	1540
	controls	800	1030

PEAK VELOCITY. Similar results were found for the amplitude of peak velocity. Peak velocity was faster in normals overall ($p < 0.001$), however for the "natural" movement there was only a small difference of 70 mm/second between groups, while this difference increased to 175 mm/sec, when subjects were asked to move as fast as possible (group by instruction interaction, $p < 0.05$). When eyes were closed Parkinson's disease patients again could not reach a similar peak velocity. The strongest difference of course stemmed from movement amplitude: peak velocity was twofold for twofold distance in normals and in patients ($p < 0.001$). Diameter of the dowel did not influence peak velocity at all.

Table 2. A summary of group mean peak velocities (mm/sec) of short and long amplitude movements for different instruction conditions ("normal", "fast","eyes closed").

instruction		movement short	long
normal	patients	244	448
	controls	304	525
fast	patients	347	670
	controls	483	883
eyes closed	patients	251	467
	controls	337	602

SYMMETRY RATIO. Marteniuk, MacKenzie, Jeannerod, Athenes and Dugas (1987) had shown a tremendous variability of velocity profiles produced by different task constraints. Data reported in the previous section had demonstrated the ability of parkinson's disease patients to scale peak velocity, however patients produced different velocity profiles. For purposes of statistical analysis the degree of movement symmetry was calculated by the ratio of acceleration duration divided by deceleration duration. Normal movements were characterised by a symmetry ratio of 0.8 in both groups as shown in figure 5, indicating a nearly symmetric profile with a slightly longer deceleration duration. Fast movements were again produced by parkinson's disease patients in a way that resulted in a symmetric velocity profile, while normals displayed considerably shorter deceleration phases, generating a symmetry ratio of 1.5. The opposite held with eyes closed, when the ratio changed only slightly in normals (0.69), whereas patients' movements were more asymmetric, the ratio being 0.54 ($p < 0.001$). Movement amplitude and diameter of the target object influenced the shape of the velocity profile in both groups, longer distance and smaller object increasing the proportion of the deceleration phase ($p < 0.001$).

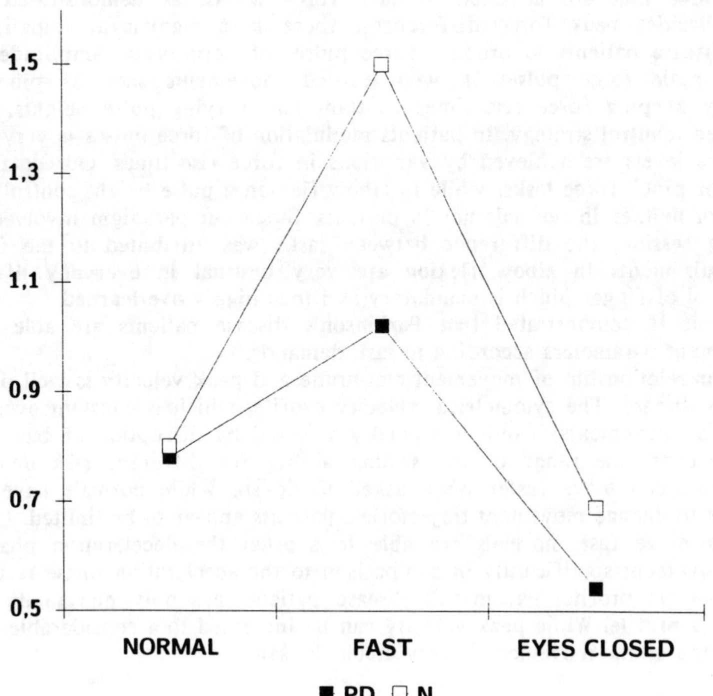

Figure 5. Symmetry ratio of total movement time: Duration of acceleration phase was divided by duration of deceleration phase. A perfectly bell-shape movement profile would result in a symmetry ratio of 1.0.

The absolute values of acceleration duration confirmed these observations: the differences varied in parallel with the variation in total movement duration. However, the results are different for the deceleration phase. The smaller diameter, resulting in a need for higher precision during homing in on the target, causes the deceleration phase to increase significantly by 60 msecs ($p < 0.05$). In addition, the instruction by group interaction is significant as well, showing a disproportionate increase in patients when the eyes are closed ($p < 0.05$). Together these data suggest, that the different effects of instruction on movement time and velocity profile arises mainly from the effect on the deceleration phase, where Parkinson's disease patients have more problems to deal with.

SUMMARY

Parkinson's disease patients have a restricted ability to produce high forces in several different muscles. Thus, experiments that try to analyze basic force production problems have to deal with that difference by either using only low force levels or by using amplitudes that are adjusted to each force levels, as demonstrated in this experiment. Besides peak force differences, there is a significant impairment in Parkinson's disease patients to produce force pulses of appropriate amplitudes. While normals can scale force pulses in well learned movements such as pinch force production by keeping force rise times constant and varying pulse heights, patients used a different control strategy. In patients modulation of force pulses is very limited, different force levels are achieved by variations in force rise times. Our data showed that clearly for pinch force tasks, while in elbow flexion a pulse height control strategy was not shown neither in normals nor in patients. Since our paradigm involved only a short training session, the difference between tasks was attributed to the fact that accuracy requirements in elbow flexion are very unusual in everyday life, while accurate control of finger pinch is mandatory and thus highly overlearned.

Experiment II demonstrated that Parkinson's disease patients are able to scale several movement parameters according to task demands.

The linear relationship of movement amplitude and peak velocity is well preserved in Parkinson's disease. The symmetrical velocity profile, which is constant over a wide class of normal movements is only marginally affected by disruption of basal ganglia function. However, the range of the scaling ability for different task demands is limited. Patients can move faster when asked to do so. While normals have a wide range of ways to change movement trajectories, patients appear to be limited. Upon the instruction to move fast, normals are able to shorten the deceleration phase of a prehension movement significantly in comparison to the acceleration phase resulting in a very asymmetric profile. Parkinson's disease patients are only marginally able to generate such a profile. While peak velocity can be increased to a considerable amount, the overall shape of the movement is very much the same.

REFERENCES

Atkeson, C. G., Hollerbach, J. M., (1985) 'Kinematic features of unrestrained vertical arm movements', Journal of Neuroscience, 5, 2318-2330.

Benecke, R., Rothwell, J. C., Day, B. L., Dick, J. P. R., Marsden, C. D., (1986) 'Motor strategies involved in the performance of sequential movements', Experimental Brain Research, 63, 585-595.

Benecke, R., Rothwell, J. C., Dick, J. P. R., Day, B. L., Marsden, C. D., (1986) 'Performance of simultaneous movements in patients with Parkinson's disease', Brain, 109, 739-757.

Berardelli, A., Accornero, N., Argenta, M., Meco, G., Manfredi, M., (1986) 'Fast complex arm movements in Parkinson's disease', Journal of Neurology, Neurosurgery and Psychiatry, 49, 1146-1149.

Brooks, V. B., (1986) 'The neurological basis of motor control', Oxford University Press, New York, 1986.

Brown, S. H., Hefter, H., Mertens, M., Freund, H. -J., (1990) 'Disturbances in human arm movement trajectory due to mild cerebellar dysfunction', Journal of Neurology, Neurosurgery and Psychiatry, 53, 306-313.

Bullock, D., Grossberg, S., (1988) 'Neural dynamics of planned arm movements: Emergent invariants and speed-accuracy properties during trajectory formation', Psychological Review, 95, 49-90.

Gordon, J., Ghez, C., (1987) 'Trajectory control in targeted force impulses. II. Pulse height control', Experimental Brain Research, 76, 241-252.

Hoehn, M. M., Yahr, M. D., (1967) 'Parkinsonism: onset, progression, and mortality', Neurology, 17, 427-442.

Jeannerod, M., (1984) 'The timing of natural prehension movements', Journal of Motor Behavior, 16, 235-254.

Mai, N., Bolsinger, P., Avarello, M., Diener, H. C., Dichgans, J., (1988) 'Control of isometric finger force in patients with cerebellar disease', Brain, 111, 973-998.

Marsden, C. D., (1982) 'The mysterious motor function of the basal ganglia: The Robert Wartenberg Lecture', Neurology, 32, 514-539.

Marteniuk, R. G., MacKenzie, C. L., Jeannerod, M., Athenes, S., Dugas, C., (1987) 'Constraints on human arm movement trajectories', Canadian Journal of Psycholology, 41, 365-378.

Phillips J. G., F., Müller G. E., Stelmach (1989). Movement disorders and the neural basis of motor control. In S. A. Wallace (Ed.), Perspectives on the Coordination of Movement, (pp. 367-413), Elsevier, Amsterdam

Shapiro, D. C., (1986) 'Rapid limb movements and their modulation with task demands', Experimental Brain Research Series, 15, 231-241.

Soechting, J. F., (1984) 'Effect of target size on spatial and temporal characteristics of a pointing movement in man.', Experimental Brain Research, 54, 121-132.

Stelmach, G. E., Garcia Colera, A., Martin, Z. E., (1989) 'Force transition control within a movement sequence in Parkinson's disease', Journal of Neurology, 236, 406-410.

Stelmach, G. E., Teasdale, N., Phillips, J., Worringham, C. J., (1989) 'Force production characteristics in Parkinson's disease', Experimental Brain Research, 76, 165-172.

Stelmach, G. E., Worringham, C. J., (1988) 'The preparation and production of isometric force in Parkinson's disease', Neuropsychologia, 26, 93-103.

Stelmach, G. E., Worringham, C. J., Strand, E. A., (1987) 'The programming and execution of movement sequences in Parkinson's disease', International Journal of Neuroscience, 36, 55-65.

Teasdale, N., Stelmach, G. E., (1988) 'Movement disorders: The importance of the movement context', Journal of Motor Behavior, 20, 186-191.

Viallet, F., Massion, J., Massarino, R., Khalil, R., (1987) 'Performance of a bimanual load-lifting task by parkinsonian patients', Journal of Neurology, Neurosurgery and Psychiatry, 50, 1274-1283.

Wing, A. M., (1988) 'A comparison of the rate of pinch grip force increases and decreases in parkinsonian bradykinesia', Neuropsychologia, 26, 479-482.

Wing, A. M., Turton, A., Fraser, C., (1986) 'Grasp size and accuracy of approach in reaching', Journal of Motor Behavior, 18, 245-260.

COORDINATION OF REACHING AND GRASPING IN PROSTHETIC AND NORMAL LIMBS

Gerald M. Grammens,* Stephen A. Wallace & Lawrence E. Carlson*
Departments of Kinesiology and Mechanical Engineering*
University of Colorado - Boulder
Boulder, Colorado 80309
U.S.A.

ABSTRACT. An experienced user of a body powered voluntary-close upper extremity prosthesis was studied while tending to the task of grasping a vertical cylindrical object. Trials of repeated grasps were carried out under three different speeds of action. Similar trials were run using the subject's normal right hand. Findings for the normal hand are consistent with previous studies by Fraser & Wing (1981) and Wing & Fraser (1983) in that the thumb tended to deviate significantly less from a straight line trajectory to the target than did the index finger. In the prosthetic hand, the moveable "thumb" maintained a straight line course toward the target while the fixed portion of the prehensor was moved normal to the trajectory in a manner which accounted for change in the aperture. Maximum aperture was greatest for the fast grasp in both prosthetic and normal hands. Other findings show significantly greater variability in most measures of the normal hand grasp compared to the prosthetic grasp. In general, the findings suggest multiple, but less variable, temporal patterns of coordination between transport and manipulation in the prosthetic compared to the normal hand.

1. INTRODUCTION

Of the estimated 50,000 upper extremity prosthesis wearers in the United States, a large majority opt to use a body powered (as opposed to myoelectrically controlled) prosthesis predominantly or exclusively (Muilenburg, & LeBlanc 1989). The body powered prostheses are of two major types -- voluntary opening and voluntary closing. In each case, the terminal device (prehensor) is operated by a cable attached to a harness which is typically anchored to the contralateral shoulder. The net result of combinations of muscle action on the trunk, shoulder girdle, and the upper extremity is a change in the tension of the cable. The resultant forces act to open or close the terminal device. Despite this apparently simple arrangement, the experienced user can display complex adaptive motor behavior to achieve function in the prosthetic device. It is anticipated that further understanding of specifics of this motor behavior can contribute to the design and evaluation of future prosthetic devices and control schemes, and to provide insight for the training in the use of prosthetic devices.

In the non-prosthetic subject, studies of upper extremity grasp have concentrated on observing and recognizing patterns of motor behavior. From a purely functional standpoint, the task of grasping an object can be broken into the independent components of transport (reach) and manipulation. Various investigations have sought out characteristic patterns of grasp, identified coupling of the two components in the integrated grasp, and observed manifestations of spatial and temporal constraints.

In a study of non-prosthetic grasping (Jeannerod, 1981), subjects were asked to grasp an object of variable size in the presence and absence of visual cues. The transport function was determined to be unaffected by test conditions, however the manipulation component or aperture (defined as the distance between the index finger and thumb) was shown to be influenced by these

J. Requin and G. E. Stelmach (eds.), Tutorials in Motor Neuroscience, 175–188.

same conditions. Specifically, maximum aperture was shown to increase with larger object size. Further work (Jeannerod, 1984) included the variables of speed and distance of transport. A fast phase and a slow phase of transport were described. Opening of the aperture, in preparation for grasp, occurred during the fast phase of transport while aperture closure appeared to correlate with the slow phase. A break point, the transition between fast and slow phase occurred at 70-80% of the transport--evidence of a spatial constraint (coupling) of the transport and manipulation functions.

Further studies on the relationship between manipulation and transport involved the examination of reach for a vertical dowel (Wing, Turton & Fraser, 1986). In observing the deviation of the thumb and index finger from a linear path to the target, the thumb was found to contribute less to closing of the aperture than the index finger. It was suggested from this study that the thumb was used as the visual guide for zeroing in on the target. An additional finding of this study was that maximum aperture increased with decreased total time of transport and with obscured vision. It was suggested that adjustments of aperture are made to compensate for errors in transport.

When a requirement of accuracy was added to the task of grasping (Wallace & Weeks, 1988), both transport and manipulation components were affected. In a series of experiments, subjects were asked to grasp an object without disturbing its position more than a prescribed amount. Increasing the allowable tolerance resulted in decreased transport time and maximum aperture size. Accuracy of manipulation was dependent on movement time (a temporal constraint), and the relative timing of maximum aperture remained constant over different movement times.

In a further study, Wallace, Weeks and Kelso (1990) provided evidence that initial starting conditions did not significantly affect the relative timing of final aperture closing to the target. Regardless of whether subjects started with aperture open or closed, the relative timing of final closing remained invariant across a wide range of transport movement times. This finding suggests only one preferred temporal pattern of coordination between transport and manipulation in non-amputee (normal) prehension.

The literature relative to prehension control in the upper extremity prosthetic user is limited. A case study (Fraser & Wing, 1981) of a 13 year old female with an Otto Bock voluntary opening upper extremity prosthesis reported similarities in grasp to the non-prosthetic function. This particular terminal device is designed to provide double-sided motion of the prehensor components. Even with the double-sided action, it was noted that the thumb closely tracked the axis of transport while the fingers deviated normal to this axis to accommodate changes in aperture. Further comparisons between the subject's prosthetic and natural hands showed similarity in transport, although slower in the prosthetic grasp.

In considering the action of grasping in the upper extremity prosthetic user, many questions arise relative to differences and similarities between the adaptive actions to those of the non-prosthetic individual. Clearly the body powered prosthesis mechanically couples the prehension activity with transport. Merely extending the arm results in cable pull to the terminal device. The questions to be addressed in this study center around the idea of spatial or temporal coupling of the components of grasp in a body powered prosthesis. Specifically addressed is the relative timing of final closure the same in the prosthetic and the normal hand.

2. METHOD

2.1. Subject

The subject (age = 41 years), a right handed adult male, had been a user of a left arm body powered below-the-elbow prosthesis for approximately 12 years. A terminal device with voluntary close action had been in use by the subject for the last five years (currently using the T.R.S. Grip III). The subject was naive to the potential findings of this study.

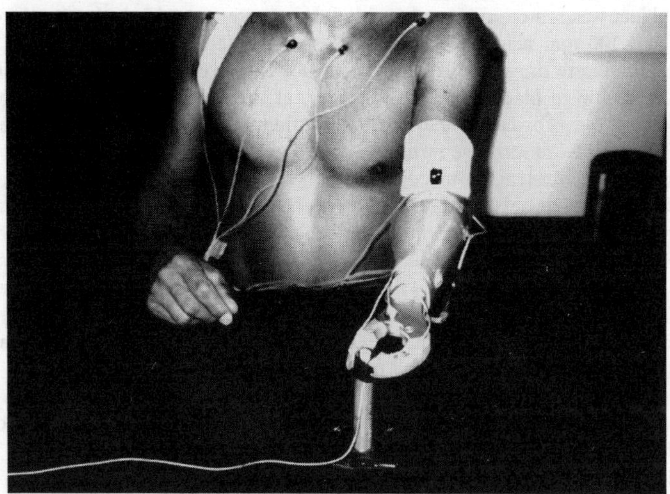

Figure 1. Test Setup. Subject with left below-the-elbow voluntary open prosthesis. A total of 9 infrared emitting diodes (IREDs) positioned one at each shoulder, the medial head of each clavicle, elbow, terminal device pivot, "thumb" and stationary component of terminal device, and the target. Front of harness worn on right shoulder can be seen.

2.2 Test Apparatus

A commercial motion study system (WATSMART, Northern Digital, Inc.) was employed to capture and analyze components of grasp under the prescribed test conditions. Infrared emitting diodes (IREDs) were attached to the subject and the target as demonstrated in Figure 1. The IREDs were pulsed sequentially at a rate of 100 pulses per second. Two infrared sensing cameras were positioned facing the subject so that all IREDs were unobstructed throughout the subject's motion (185 cm vertically from the floor and 142 cm apart). Data collection, processing, and analysis hardware and software were used to evaluate collected information.

A linear tracking device was positioned next to the subject. This was used to assist in pacing the speed of transport. In addition, the trials were videotaped for subsequent perusal.

2.3. Procedure

Appropriate precautions were taken to render the surrounding environment non-reflective of infrared light. External sources of infrared light were blocked.

The system was initialized by placing a calibration frame in the test space. The frame consists of 24 IREDs which are permanently fixed relative to one another at known distances. Cameras were aligned for optimal coverage of the test space. With the use of system supplied software, camera positions were derived by utilizing triangulation methods. This information was retained for conversion of camera data to three-dimensional coordinate location of each IRED in time. The reference coordinate system was established by the calibration frame.

The subject was seated in a comfortable fashion at a test table. A vertical cylinder of 18 mm diameter and 100 mm height was positioned approximately 50 cm directly in front of the shoulder corresponding to the side of reach. Eight IREDs were attached to the subject as shown in Figure 1. IREDs were placed on each shoulder, at the medial head of each clavicle, at the elbow, the pivot of the prosthesis, and at the "forefinger" and "thumb" of the prehensor. An additional IRED was placed atop the target.

The subject was instructed to begin with the prehensor (terminal device) in the closed position near his body. Trials were run at three speeds: slow, intermediate, and fast. A variable speed tracking device was placed adjacent to the subject to provide a gauge for rate of transport for slow and intermediate speeds. The subject was allowed to move as quickly as possible while maintaining control for the fast trials. The subject was instructed to grasp the target vertical cylinder with minimal movement of the target. Trials were executed in blocks of 15 for each speed. Practice runs were allowed at the beginning of each speed change to aid in adjustment. Block sequence was intermediate, slow, and fast.

After having completed the 45 trials with the prosthetic arm, the IREDs were positioned in analogous locations on the normal right arm and hand. Trials for the same speeds were then repeated for the non-prosthetic grasp.

In addition to the above, the subject was instructed to perform cyclic grasps in the absence of a target. The cycle began with the terminal device held in a closed position near the body, opened as transport began, then closed as if grasping the now displaced cylinder, and then returned in the reverse procedure. Five second segments of this continuous motion were sampled.

2.4 Data Analysis

Three and one-half seconds of data at 100 frames per second was collected for each trial. Using the parameters obtained from the calibration procedure, the raw data was converted into three-dimensional coordinates for each IRED. Each set of data was scanned to locate missing data points. Any gaps in the data were filled using cubic spline functions. Data was digitally low-pass filtered using a cutoff frequency of 5 Hz. to minimize noise. Other programs were utilized to calculate the aperture and to normalize the data with respect to time and distance variables. Other data manipulations were necessary to produce stick figure plots and average (or composite) profiles for the different test conditions.

3. RESULTS

The clustering of transport times for each of the three trial speeds can be seen along the abscissa in Figures 9 and 10. Transport time was determined to begin when the wrist had advanced .5 mm from the initial resting position followed by continuation of transport. Similarly, transport was declared complete when the wrist was within .5 mm of maximum transport. Mean times, along with the standard deviation, for each of the three speeds are shown in Table 1. In each case the mean and standard deviation are calculated from 15 trials.

TABLE 1. Mean and standard deviation (parentheses) of total transport times (seconds).

	Prosthesis	Natural hand
Slow	1.30 (.09)	1.24 (.07)
Intermediate	.91 (.08)	1.01 (.07)
Fast	.66 (.08)	.60 (.04)

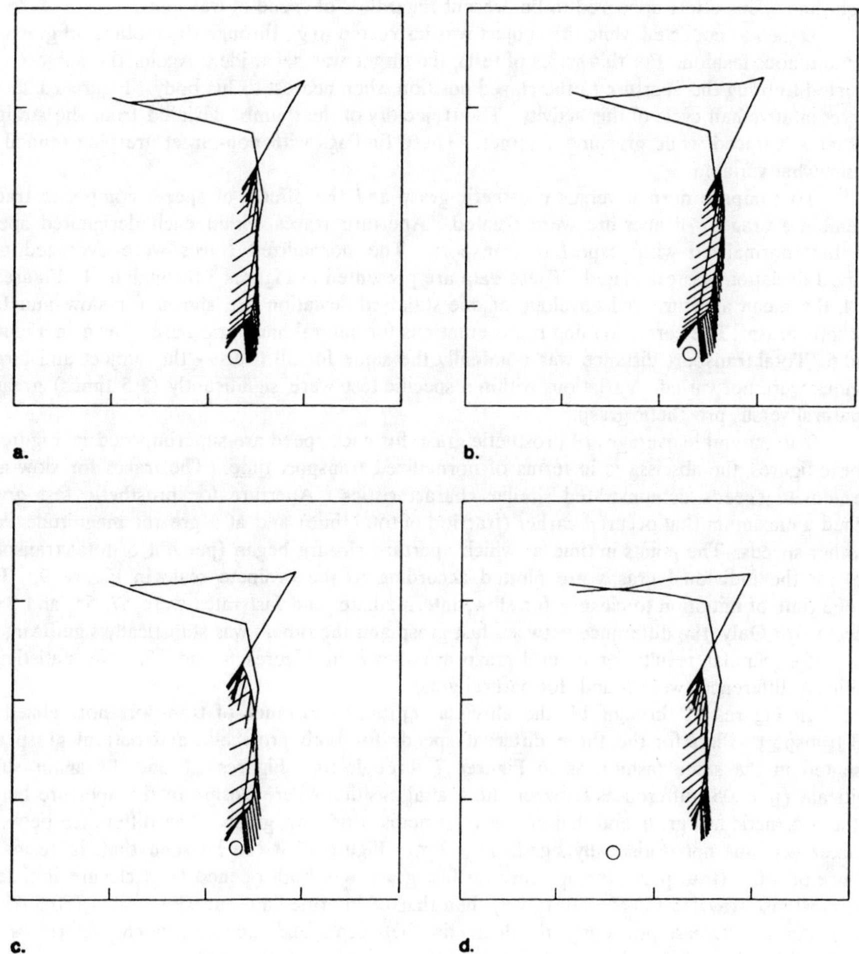

Figure 2. Stick figure representation for single grasp of target (circle) for a) intermediate, b) slow, and c) fast speed. Also, a representative half cycle d) of grasp with target removed. Full stick figure is shown only for first and last frame. Intermediate frames show position of prehensor pivot, "thumb" (medial), and fixed portion of terminal device at 50 msec. intervals. View is from overhead. Subject is using left prosthesis.

Figure 2 utilizes a modified stick figure representation of the complete grasp motion from a view looking down on the subject. The complete stick figure for all IREDs is drawn for the initial and final positions. The circle represents the target. The three IREDs which describe the prosthetic prehensor are drawn at 50 msec. intervals. A representative trial from each of the three speeds of transport is shown. The thumb component (movable) is located on the medial aspect of the terminal device.

The subject began with the aperture closed. As the transport progressed, the aperture widened with the release of tension on the control cable. The stationary component of the prehensor initially deviated laterally and then began to move medially at about the same time that the aperture began to close. The result was that the thumb position traced out as nearly a straight line. This effect appeared to be present regardless of speed of transport.

Data was collected while the subject was instructed to go through the motions of grasping in a continuous fashion. For this series of tests, the target was set aside. Again, the subject was instructed to bring the aperture to the closed position when nearest to his body. Figure 2d shows a representative half cycle of this activity. The trajectory of the thumb deviated from the straight line that was traced while grasping a target. These findings with non-target grasping tended to be somewhat variable.

To compare normal versus prosthetic grasp and the effects of speed, composite traces (ensemble averages) of aperture were created. Aperture traces within each designated speed were first normalized with respect to transport. The normalized traces were averaged and standard deviation was calculated. These data are presented in Figures 3 through 6. In Figures 3 and 4, the mean aperture and envelope of one standard deviation are shown for slow and fast prosthetic grasp. The corresponding representations for natural hand grasp are shown in Figures 5 and 6. Total transport distance was nominally the same for all trials -- the subject and target positions were not varied. Variations within a specific test were significantly (3-5 times) greater for natural versus prosthetic grasp.

The ensemble averages of prosthetic grasp for each speed are superimposed in Figure 7. In these figures, the abscissa is in terms of normalized transport <u>time</u>. The traces for slow and intermediate speeds demonstrated similar characteristics. Aperture for prosthetic fast grasp reached a maximum that occured earlier (fraction of total time) and at a greater magnitude than the other speeds. The points in time at which aperture closure began (percent of total transport time) for the individual grasps are plotted according to the ordinate scale in Figure 9. The average time of initiation to closure for slow, intermediate, and fast rates were 57, 54, and 38% respectively. Only the difference between fast grasp and the others was statistically significant ($p < .05$). Similar results for natural grasp are shown in Figures 8 and 10. No statistically significant differences were found for natural grasp.

In Figures 11 through 14, the abscissa represents distance of transport normalized to total <u>transport</u>. Plots for the three different speeds for both prosthetic and natural grasp are presented in the same fashion as in Figures 7 through 10. Figures 11 and 13 again show significant ($p < .05$) differences between the spatial position where closure of the aperture began for the prosthetic fast grasp and that of the intermediate or slow grasp. The difference between the later two was not statistically significant. From Figure 13 it can be seen that, in terms of distance of travel (transport), the aperture for fast grasp was both opened (and closure initiated) in less distance traveled (50-60% of travel) than that of intermediate and slow travel (80-90% of travel). As was the case previously, the distinction between speed and aperture characteristics for the normal hand are not in the range of statistical significance for this study.

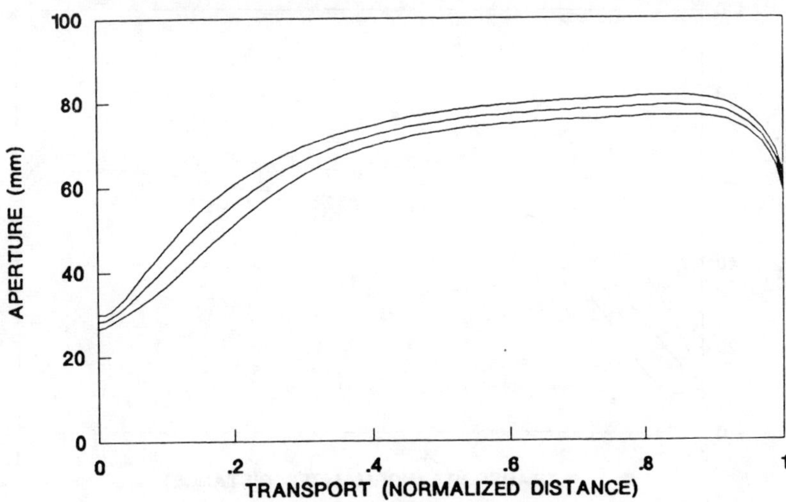

Figure 3. Prosthetic grasp: mean aperture of 15 trials within envelope designating + or - one standard deviation. Results of slow grasp are shown vs. normalized distance of total transport.

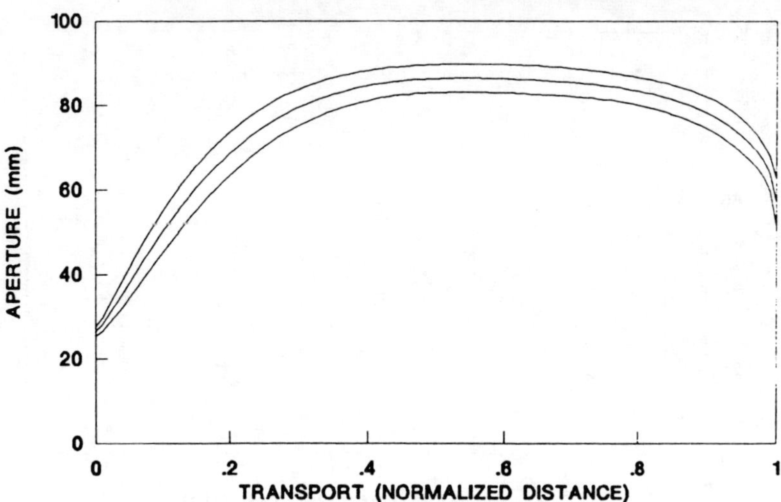

Figure 4. Same as in Figure 3, except for fast prosthetic grasp.

AVERAGE APERTURE FOR SLOW NATURAL GRASP
MEAN APERTURE +/- STANDARD DEVIATION

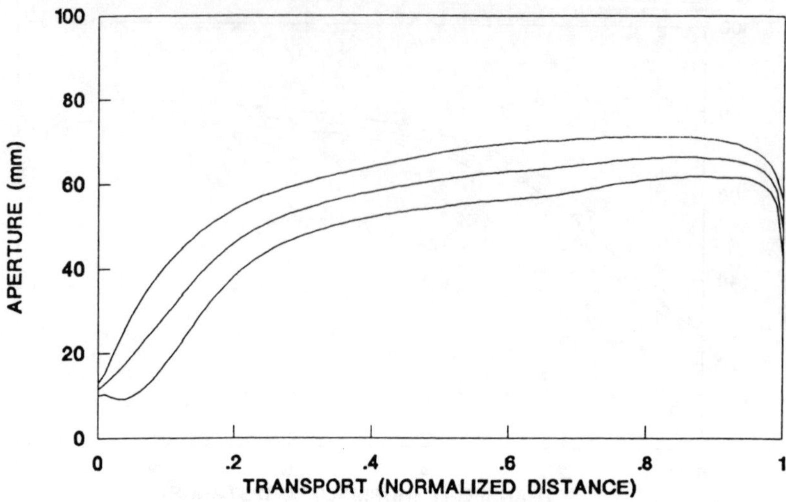

Figure 5. Natural grasp: mean aperture of 15 trials within envelope designating + & - one standard deviation. Results of slow grasp are shown vs. normalized distance of total support.

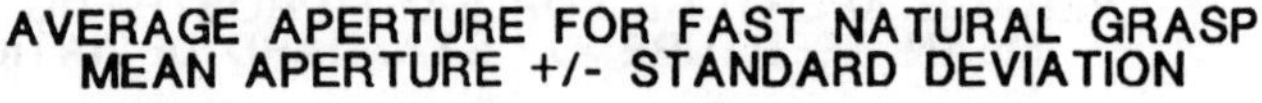

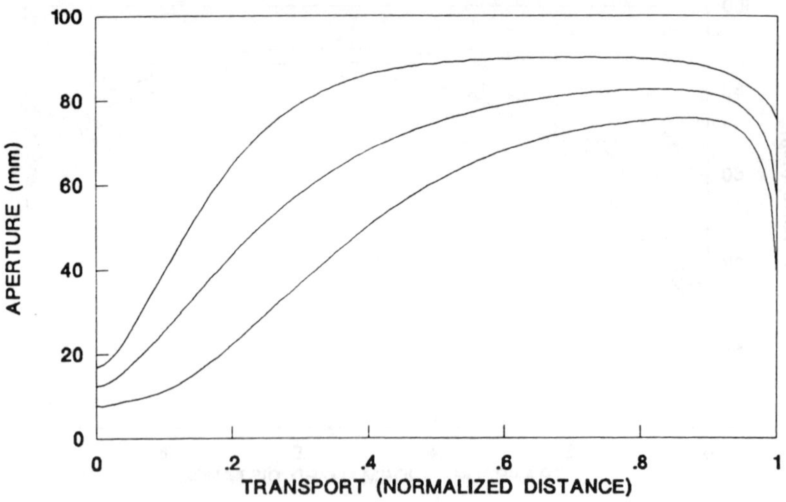

Figure 6. Same as for Figure 5, except for fast natural grasp.

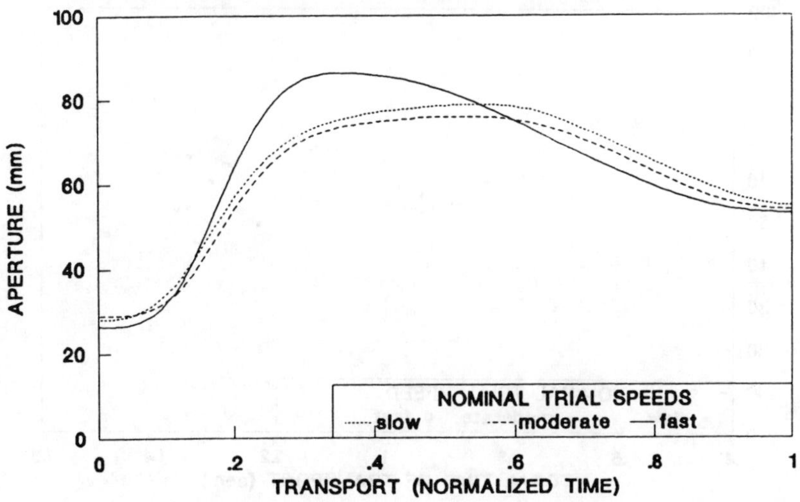

Figure 7. Ensemble average of 15 trials each for slow, intermediate, and fast prosthetic grasp. Aperture versus normalized time of total transport is shown.

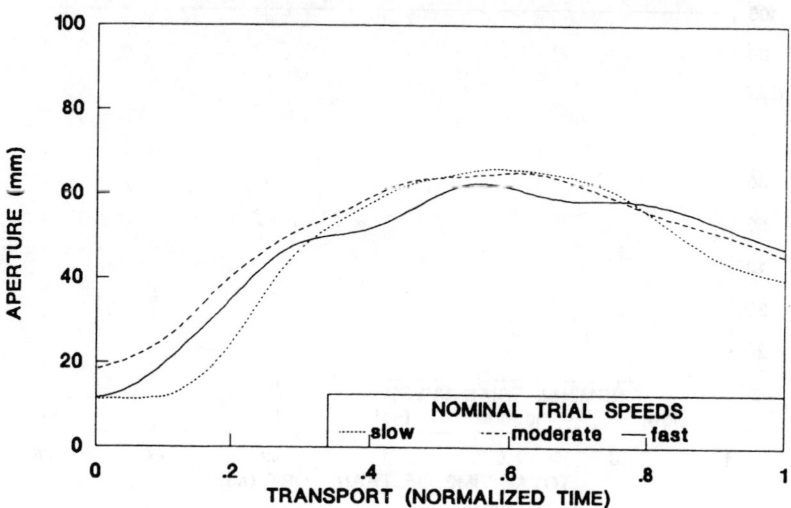

Figure 8. Ensemble average of 15 trial each for slow, intermediate, and fast grasp using the subject's natural right hand. Aperture versus normalized time of total transport is shown.

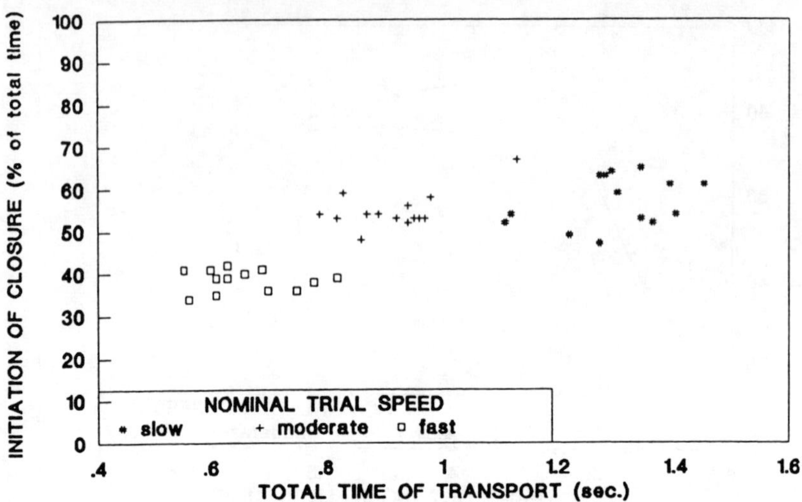

Figure 9. Initiation of aperture closure during grasp of a target. Each trial of prosthetic grasp is represented by total time of transport (abscissa) and the percent of that total transport time (ordinate) at which closure of aperture begins.

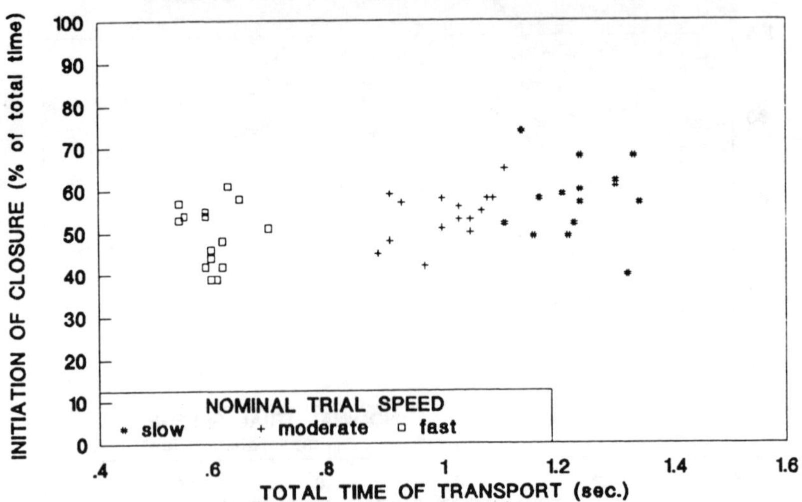

Figure 10. Initiation of aperture closure (as percent of total transport time) during grasp of a target by the natural right hand of the subject.

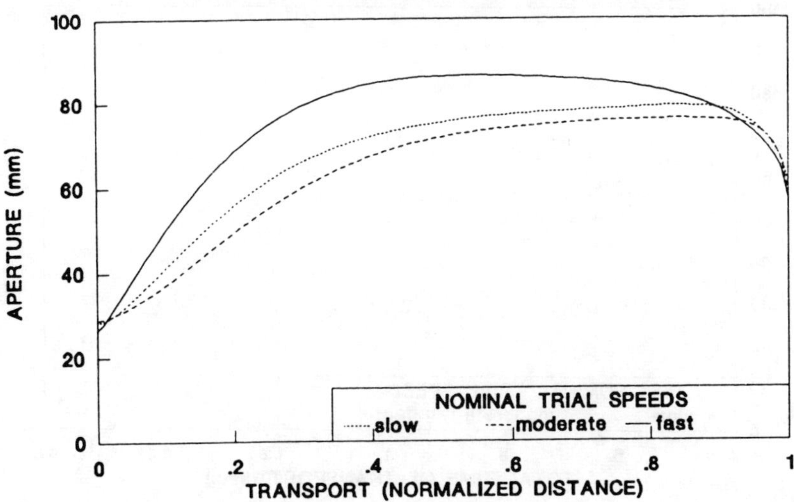

Figure 11. Ensemble average of 15 trials each for slow, intermediate, and fast prosthetic grasp. Aperture vs. normalized distance of transport is shown.

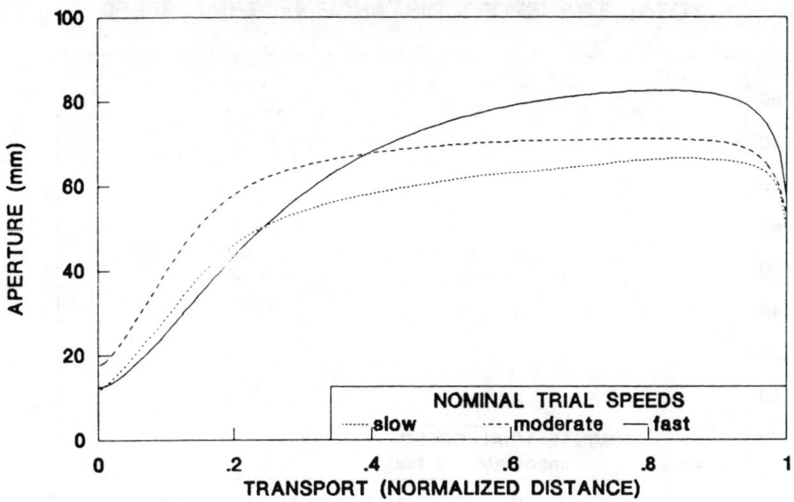

Figure 12. Ensemble average of 15 trials each for slow, intermediate, and fast grasp using the subject's natural right hand. Aperture vs. normalized distance of transport is shown.

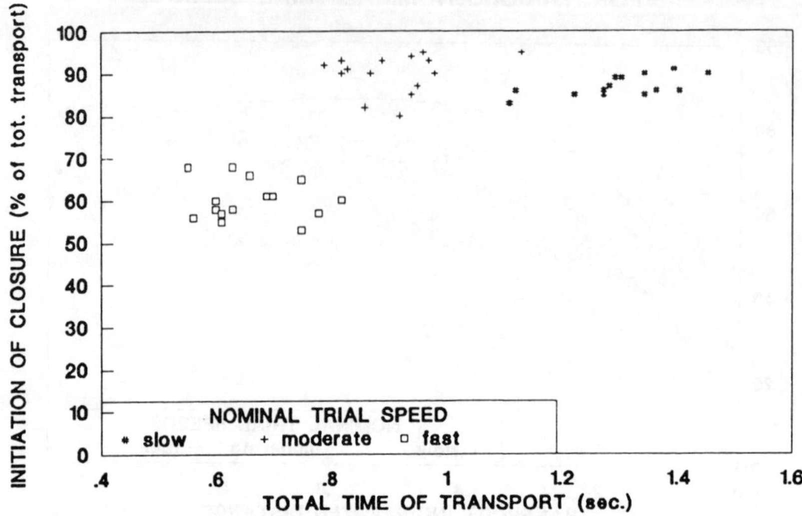

Figure 13. Initiation of aperture closure during grasp of a target. Each trial of prosthetic grasp is represented by total time of transport (abscissa) and the percent of that total transport distance (ordinate) at which closure of aperture begins.

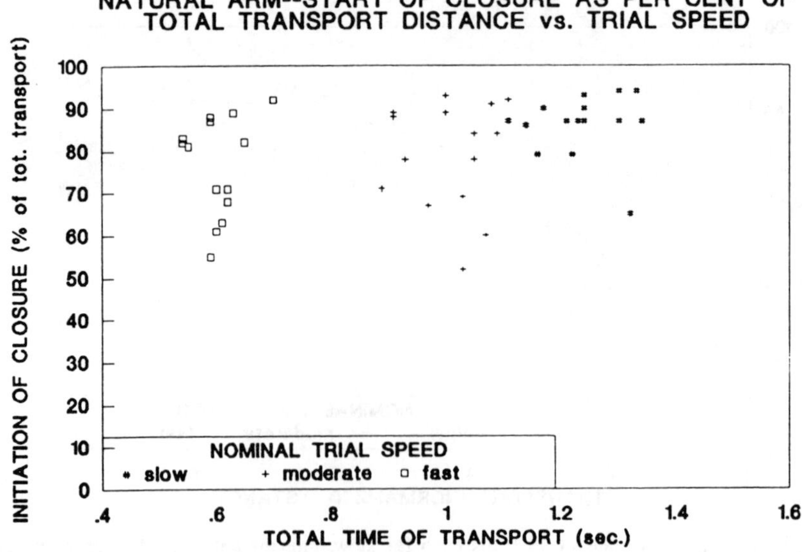

Figure 14. Initiation of aperture closure (as percent of total transport distance) during grasp of a target by the natural right hand of the subject.

3.1 Discussion of Results

The major results of this case study were: 1) the trajectory of the thumb (the movable component of the terminal device) traces nearly a straight line to the target while the stationary part of the prosthesis accommodates change in aperture size, 2) when compared to natural grasp in the same subject, significantly less variability from one grasp to the next occurs in the prosthetic hand, and 3) the initiation of closure of the aperture was dependent upon speed or total time of transport.

The observation of a straight line trajectory of the prosthetic thumb is consistent with findings in previous studies (Frazer & Wing, 1981; Wing & Frazer, 1983). However, in none of the previous studies was the thumb the only moveable component of the terminal device. In the natural hand, prehension using the thumb and index finger in pincerlike fashion can be seen to result predominantly from motion of the second metacarpophalangeal joint. It is possible that this anatomical consideration contributes to the motor pattern in which the thumb serves as the guide to the target.

A possible explanation for the observed pattern in this study is that control of the cable, which included elbow displacement laterally and medially, resulted in the hook approach of the stationary portion of the prosthesis. However, when the subject went through the motions of transport while closing the aperture at both ends of travel (in the absence of a target), the pattern was not reproduced. A second possibility is that the geometry of the prosthetic index finger necessitates a hook approach to facilitate effective and accurate grasp. A third possibility is that the thumb is an important visual guide which is incorporated into patterns of natural grasp, and that the behavior is carried over into adaptive prosthetic grasp. Future testing using different types of terminal devices will help to determine which of the above apply.

The significantly diminished variability in aperture for prosthetic grasp relative to natural grasp may be explained in terms of constraints. In the process of natural grasp, the subject can render virtually total uncoupling of the transport and manipulation components (at slow speeds of grasp). It is only when limits of speed and accuracy are reached that temporal and/or spatial coupling emerges. In contrast, the body powered prosthetic hand manipulation is mechanically coupled to transport. All else held motionless, the simple act of forward transport will cause the control cable to become taut and cause action on the terminal device. It is this very coupling, and therefore increased constraint, that renders prosthetic grasp less variable.

At the outset of this study, the possibility that little or no variability in prosthetic grasp would exist was contemplated. If the mechanical coupling was total (e.g. if the cable was anchored to the subject's chair), then change in aperture would be dependent solely on position of transport. The observation that initiation of closure of aperture began at 50-60% of travel for fast grasp and at 80-90% of travel for intermediate and slow rates, indicates that the mechanical coupling is not complete. Furthermore, that this change in character of aperture is a function of speed of transport suggests a temporal constraint in prosthetic grasp. More specifically, in the normal hand, only one temporal pattern of coordination (relative time to final closing) was observed, supporting previous work (Wallace & Weeks, 1988; Wallace et al. 1990). However, in the prosthetic, two patterns were observed--one pattern for slow and moderate speeds and a different one for the fast speed. The dynamics of pattern changes (e.g. Schoner & Kelso, 1989) will be addressed in future work.

Further studies are also needed to determine whether the above findings hold for a population of upper extremity prosthesis users and for other types of terminal devices. The fact that little difference was seen between the slow and intermediate speeds of transport indicates that reassessment of predetermined transport speeds is in order. Perhaps testing along a continuum of transport velocities will yield a point of instability beyond which distant initiation of aperture closure yields to more proximal initiation of closure. More detailed analysis of contributing motion of the shoulder and torso to prehensor control is also planned.

4. References

Fraser, C. and Wing, A.W. (1981). A case study of reaching by a user of a manually-operated hand. Prosthetics and Orthotics International, 5, 151-156.

Jeannerod, M. (1981). "Intersegmental coordination during reaching at natural visual objects." In: J. Long and A. Braddeley (eds.). Attention & Performance IX. Hillsdale, New Jersey: Erlbaum.

Jeannerod, M. (1984). The timing of natural prehension movements. Journal of Motor Behavior, 16, 235-254.

Muilenburg, A. & LeBlanc, M. (1989). Body-powered upper-limb components. In: D. Atkins and R. Meier (eds.), Comprehensive Management of the Upper-Limb Amputee, New York: Springer-Verlag.

Schoner, G., & Kelso, J.A.S. (1988). Dynamic pattern generation in behavioral and neural systems. Science, 239, 1513-1520.

Wallace, S.A. & Weeks, D.L. (1988). Temporal constraints in the control of prehensile movements. Journal of Motor Behavior, 20, 81-105.

Wallace, S.A., Weeks, D.L. & Kelso, J.A.S. (1990). Temporal constraints in reaching and grasping behaviors. Human Movement Science, 9, 69-93.

Wing, A.M. & Fraser, C. (1983). The contribution of the thumb to reaching movements. Quarterly Journal of Experimental Psychology, 35A, 297-309.

Wing, A.M., A. Turton, & C. Fraser. (1986). Grasp size and accuracy of approach in reaching. Journal of Motor Behavior, 18, 245-260.

SELF-INDUCED VERSUS REACTIVE TRIGGERING OF SYNCHRONOUS HAND AND HEEL MOVEMENT IN YOUNG AND OLD SUBJECTS.

C. Bard, J. Paillard*, N. Teasdale, M. Fleury & Y. Lajoie
Université Laval, Québec
Laboratoire de Performance Motrice Humaine
*C.N.R.S. Marseille
Laboratoire de Neurosciences Fonctionnelles

KEYWORDS / ABSTRACT : motor control / elderly / movement synchronization

When asked to synchronize simulnaneously ipsilateral hand and heel movements, subjects behave differently whether they have to react to an external trigger signal or to synchronize movements in a self-pacing mode. In the reactive mode, the central commands seem to be centrally coactivated. In contrast, in the self-pacing mode, the schedule of the two motor commands appears inverted. The present study addresses the problem of how this remarkable capacity of the nervous system evolves with age. Results show that elderly are not impaired significantly in either modes of control for simultaneity. However, a breaking point in the capacity of performing a dual task has been observed in the three oldest subjects, suggesting an impairment similar to trouble associated with Parkinson's disease.

Introduction

It is assumed that self-initiated motor commands are organized and scheduled according to the anticipated sensory consequences of a planned action. When asked to synchronize simultaneously the initiation of ipsilateral hand and foot movements, subjects behave differently whether they have to react with both limbs to an external triggering signal (reactive mode) or to self initiate both responses (projective mode). In the reactive mode, the commands for the two limbs are centrally coactivated. This mode of control yields a hand response that precedes the foot

J. Requin and G. E. Stelmach (eds.), Tutorials in Motor Neuroscience, 189–196.

response by a delay equivalent to the difference in the conduction time of the two efferent pathways. In contrast, for the self-paced mode, the schedule for the two motor responses is inverted: the foot movement is initiated first as if the synchronization of the reafferent sensory information coming from the two limbs is the goal to be reached. This early observation by Paillard (1948) has been recently replicated by Paillard, Bard & Fleury (1989). The present study addresses the problem of how this remarkable adaptive capacity of the nervous system in the time domain evolves with age.

Method

SUBJECTS

Right-handed subjects, eight young (mean age: 22.2 years) and five elderly (mean age 66.0 years), naive about the aim of the research, participated in the experiment on a voluntary basis. Three additional subjects (the oldest of elderly) were rejected from the study based on their incapacity to achieve the requirement of the task.

TASK

Two experimental conditions were presented:
Reactive Mode (RM). In response to an auditory stimulus, subjects were required to react as quickly as possible and simultaneously with the right hand and foot.
Projective Mode (PM). Subjects were instructed to activate simultaneously the ipsilateral hand and foot in a self-paced mode.

APPARATUS

Electrical contact switches allowed immediate rupture through extension of the index finger or heel raising. Muscle discharge patterns from the extensor digitorum longus and the internal gastrocnemius muscles were recorded with physiological amplifiers having a common mode of rejection of 87 db at 60 Hz. Pre-pasted Ag-AgCl surface electrodes (distant of 2.5 cm) were placed over the central portion of the muscles. EMG signals were pre-amplified at the source (35x), and filtered with a time constant of 2.5 ms. All signals were collected at 500 Hz. Onset of the EMG signals was determined with the help of numerical and interactive graphics procedures. An onset was marked when the EMG exceeded 10% of the peak amplitude. Trials for which no clear onset could be detected, because of background activity before movement onset, were not included in the analyses.

MEASURES

Two dependent measures were used and defined as follows: <u>Response Time Difference</u> (RTD): Delay between the rupture times of two electrical contact switches resulting from the foot and hand movements initiation.
<u>Pre-motor Time Difference</u> (PTD): Delay between the EMG onset of the gastrocnemius and the extensor digitorum longus.

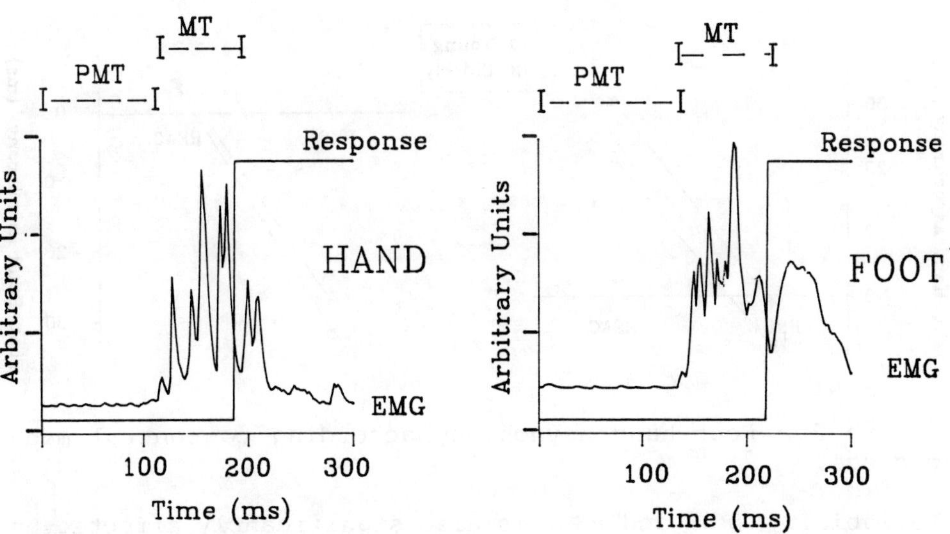

Figure 1. Temporal measurements. PMT: Pre-motor time; MT: Motor-time; EMG: Electromyogram.

PROCEDURES

Subjects participated in two experimental sessions. Thirty trials per session were given. The two conditions (RM and

PM) were counterbalanced between subjects. Each condition started with five practice trials.

Treatment of Data

RTD, PTD and their standard errors (RSE and PSE) were submitted to 2 x 2 (Projective/Reactive x Age) ANOVAs.

Results

When projective and reactive modes are compared, significant differences are found for RTD and PTD ($\underline{F}_{1,11}$ = 46.38, $\underline{p}$ <.001, and $\underline{F}_{1,11}$ = 43.49, $\underline{p}$ <.001, respectively), as shown in Figure 2. However, the main effect of Age and the Age x Mode interaction were not significant.

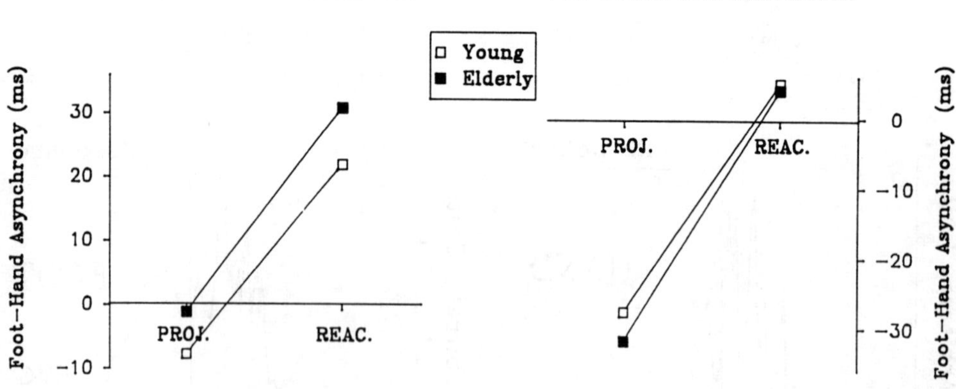

Figure 2. Foot-hand asynchrony according to control mode and age.

Variability (RSE and PSE) is also significantly affected by the experimental condition ($\underline{F}_{1,11}$ = 58.38, $\underline{p}$ <.001, and $\underline{F}_{1,11}$ = 51.52, respectively), subjects being more variable in the projective than in the reactive mode (see Table 1). Moreover, the effect of Age is significant for RSE ($\underline{F}_{1,11}$ = 6.60, $\underline{p}$ <.02, elderly being more variable than younger subjects.

TABLE 1. Mean and standard errors according to control mode and age

	PROJECTIVE MODE		REACTIVE MODE	
	Young	Elder	Young	Elder
RTD	- 7.74	- 1.20	21.97	30.81
RSE	3.60	5.80	2.10	3.10
PTD	-27.70	-31.84	4.97	4.02
PSE	4.00	4.50	2.02	2.90

Discussion

From the data, it can be hypothesized that execution of hand and foot movements, when externally triggered, are initiated at the central level by a simultaneous activation of two motor commands. The difference between EMG onset could be explained by the difference in the length of the efferent pathways. This hand over foot precession is enhanced at the response time level, motor times being longer for foot than for hand (see Table 1).

On the other hand, it has been found that heel movement initiation precedes finger extension by approximately 8 ms, whereas the pre-motor difference between foot and hand is 28 ms for young subjects and 32 ms for elderly, respectively. These results support the assumption that commands for goal directed movements are planned, organized and scheduled according to the anticipated consequences of the action, e.g., "an image motor achievement" (Teuber, 1974). Most theories assumed that these "goal images" are represented in terms of sensory events associated to motor achievement of the goal. In our experiment, motor achievement of the temporal goal can only be evaluated based on the sensory feedback linked to finger and heel movements. Therefore, the two motor commands for hand and foot must be delayed to produce a synchronous arrival of the two reafferent messages derived from the activation of both limbs.

The problem of where the evaluation takes place in the central nervous structure is an interesting one. The perceptual evaluation of simultaneity had been considered in an early work by Paillard (1948). This author showed that conscious perceptual evaluation is less accurate, the asynchrony between the two limbs must be at least 80 ms to be detected. A pilot study, performed on five subjects and using a hand/foot cutaneous stimulation, revealed the same tendencies. This lead to the assumption that the timing of motor commands disposes of exquisite computational resources regarding time measurement at a covert level.

Detection of coincidence has been recognized as a basic mechanism used by the nervous system in the processing of information (Fahle & Braitenberg, 1984). The cerebellum appears to be the best candidate for such a function (Braitenberg & Onesto, 1962; Ivry & Keele, 1989).

Another noticeable result of this experiment concerns the effect of aging on such basic capacities for evaluating simultaneity. Elderly seem to behave as efficiently as young subjects for the fine grain adjustment of the scheduling of their motor commands. In the projective mode, they account for the additional delays resulting from the slight changes in nerve conduction velocity, characterizing aging (Dorfman & Thomas, 1979), as much as for a slightly longer motor time. Nevertheless, a larger variability, specially in total response time, is worth mentioning in the elderly subjects.

Interestingly, the three oldest subjects of our group (aged 73, 74, and 80), were unable to move simultaneously the two limbs and thus, did not perform the task appropriately. They showed a general tendency to initiate movements in succession. This is obviously reminiscent of the impairment observed in Parkinson disease wherein patients have difficulty to plan complex movement involving simultaneous activation of different limbs (Marsden, 1982).

Conclusions

Self-paced execution of simultaneous movements of ipsilateral hand and foot (projective mode) induces a delay in the programing of the motor commands, resulting in an earlier foot movement initiation. It is hypothesized that the temporal target used for scheduling the onset of the two motor commands is the synchronous return of the reafferent information generated by each movement.

Simultaneous movements triggered by an external signal (reactive mode) results in a hand precession due to the shorter length of efferent pathways, suggesting a quasi simulataneous activation of the descending pathways at the central level. Most remarkable is the fact that, in both situations, subjects are unable to discriminate overtly and perceptually the order in which the hand or the foot was moving. In contrast, fine computational resources about time measurements seem available at a covert level.

Elderly are not impaired significantly in either modes of control for simultaneity. However, a breaking point in their capacity of performing a dual task has been observed in the three oldest subjects, suggesting an impairment similar to trouble associated with Parkinson's disease.

References

Braitenberg, V., & Onesto, N. (1962). The cerebellum cortex as a timing organ. Discussion of an hypothesis. Proceedings 1st International Conference of Medical Cybernetics, 1-19.

Dorfman, L.J., & Thomas, M.B. (1979). Age-related changes in peripheral and central nerve conduction in man. Neurology, 29, 38-44.

Fahle, M., & Braitenberg, V. (1984). Some quantitative aspects of cerebellar anatomy as a guide to speculation on cerebellar functions. In J. Bloedel, J. Dichgans, & W. Precht (Eds.), Cerebellar Functions. Berlin: Springer-Verlag.

Ivry, R.B., & Keele, S. (1989). Timing functions of the cerebellum. Journal of Cognitive Neuroscience, 1, 136-152.

Marsden, C.D. (1982). The mysterious motor function of the basal ganglia. The Robert Wartenburg Lecture. Neurology, 32, 514-519.

Paillard, J. (1948). Quelques données psychophysiologiques relatives au déclenchement de la commande motrice. Année Psychologique, 46-47, 28-47.

Paillard, J., Bard, C., & Fleury, M. (1989). Synchronizing hand and foot movement in a projective or reactive mode: two contrasting schedules of motor command. Society for Neurosciences Abstracts. Phoenix, p. 604.

Teuber, H.L. (1974). Key problems in the programming of movements. Brain Research, 71, 553-568.

SECTION 4

COORDINATION OF POSTURE AND MOVEMENT

TWO MODES OF COORDINATION BETWEEN MOVEMENT AND POSTURE

Jean MASSION and Annie DEAT
Laboratory of Functional Neurosciences, C.N.R.S.
31, chemin Joseph Aiguier
13402 MARSEILLE CEDEX 9, France.

ABSTRACT

Two modes of coordination between movement and posture are considered. The one deals with equilibrium control during movement performance. An example of this mode is provided by the axial upper trunk movements which are associated with hip and knee movements in the opposite direction. A distinction is made between strategy and synergy. The strategy is characterized by the kinematic changes which serve to regulate the center of mass position. The same strategy is observed under both normal gravity and microgravity conditions. The synergy or muscle pattern corresponds to the implementation of the strategy, and it depends on the biomechanical constraints. A change of synergy occurs under microgravity, as compared with earth conditions. The recovery of the initial synergy after return to earth occurs only after a few days, which indicates that a short term adaptation has taken place. The second mode of coordination serves to maintain the position of segments such as the head, trunk, or forearm which act as a reference frame for the organization of movements in the peripersonal space. Forearm position maintenance is an example involving this mode of coordination. The central organization of this coordination is analysed and the putative role of medial cortical frontal areas including the supplementary motor area in the gating of the anticipatory postural adjustments which prevent the forearm position from changing during unloading is discussed.

1. Introduction

Any motor activity belongs to one of two classes which subserve opposite functions.

Those in the first class involve maintaining of a reference value such as the position of one or several segments or that of the whole body posture. The reference value to be regulated can also be a more abstract parameter such as the projection onto the ground of the center of gravity (CG), which is relevant to equilibrium maintenance. The maintenance of these reference values does not depend only on a central command which provides the appropriate muscle contraction. It is usually regulated by a servomechanism, where error detecting inputs signal any mismatch with respect to the reference value and triggers a correction. In the case of equilibrium control, the correction is performed

J. Requin and G. E. Stelmach (eds.), Tutorials in Motor Neuroscience, 199–208.

through a continuous closed loop circuit, especially when the deviation from the reference position is slow. When the disturbance is fast, phasic postural reactions are induced, which depend on the support conditions and the body geometry at the time of the disturbance (see Massion and Dufossé, 1988). This type of organizational principle, which operates in the case of equilibrium control could also apply to single joint stabilization and other types of motor activity involving the maintenance of a reference value.

The second class of motor activities involves the performance of a movement which displaces one or several segments or the whole body toward a given goal. Here the initial posture is broken down and a new one built up which in turn will be stabilized.

During a given motor act, e.g. grasping an object while standing, the whole body is not involved in movement performance alone. At the same time, some of the body segments are involved in the maintenance of posture and equilibrium. The central nervous system therefore has to control both the movement and the maintenance of posture and equilibrium in parallel.

As given segments involved in movement and others dealing with postural and equilibrium maintenance belong to the same body, interactions between movement performance and posture and equilibrium maintenance are bound to take place. Some coordination between posture and movement occurs (Hess, 1943; Jung and Hassler, 1960). So-called anticipatory adjustments of posture can be observed which are associated with the onset of the voluntary movement, as first demonstrated by Belenkiy and Gurfinkel (1967). These constitute a typical feedforward command, which serve to minimize the disturbance of posture or equilibrium caused by the movement. A general diagram of this feedforward phasic control is shown in figure 1.

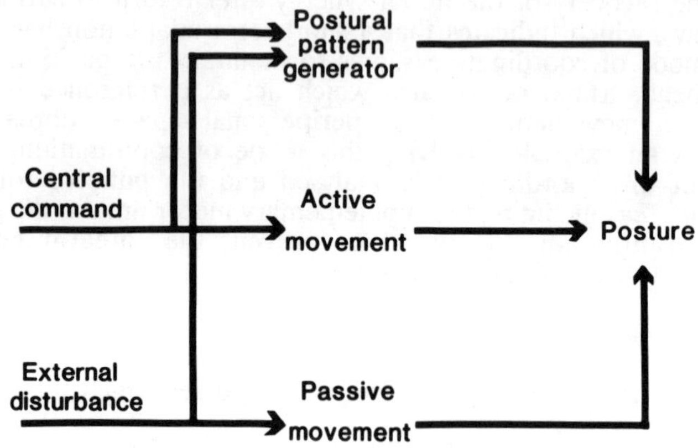

Figure 1. Diagram of the anticipatory and feedback control of posture. The anticipatory postural adjustment is associated with the central control of movement in a feedforward mode. The feedback postural adjustment is induced by a sensory signal which triggers a postural network providing the postural corrections.

2. Coordination aimed at equilibrium control

The first reason why equilibrium is disturbed during movement performance is the fact that the body is not comparable to a rigid block and that its geometry changes as the segments move during voluntary activity. Consequently, the projection of the CG onto the ground also changes and falling would occur if no associated postural adjustments were made.

Babinski (1899) was the first to illustrate this fact. When a subject is asked to bend the upper trunk backward, associated hip and knee movements in the opposite direction are observed which result in the maintenance of the projection of the CG at the same place. Babinski called the loss of these associated movements of the lower segments "asynergie" and assumed that it results from a cerebellar defect.

In a previous study (Crenna et al., 1987) the postural adjustments associated with upper trunk movements were investigated, using a kinematic recording (ELITE), a force platform recording and an EMG analysis. The investigation showed that upper trunk movements are associated with lower segment displacements in the opposite direction and that this results in minimizing the displacement of the CG onto the ground. The coordination between the displacements of upper trunk and lower segments is very efficient in terms of the stabilization of the CG position in the case of both forward and backward upper trunk movements. A characteristic EMG pattern is associated with this complex coordination (Fig. 2). In backward movements, for example, the prime mover is Erector Spinae which is a back extensor. A set of muscles located in the back of
the body, namely the hamstring and gastrocnemius, are activated at the same time. Activation of the hamstring results in flexing the knee and extending the hip. The gastrocnemius also contributes to flexing the knee, but its activation mainly leads to an ankle ventriflexion which gives rise to a ground reaction force pushing the body backward and accelerating the upper trunk movement. Here the distal muscle can be said to utilize the ground reaction forces in order to perform the movement faster.

The mechanisms whereby a set of muscles are activated simultaneously in order to displace the upper trunk and lower segments in opposite directions is still a matter of discussion. One hypothesis is that the muscle synergy at the origin of the coordination may a fixed, genetically predetermined pattern, which is selected by the central command in order to achieve the coordination. Alternatively, we may have a flexible synergy which can be variably modified as a function of the mechanical constraints or through training. The results of a study in trained gymnasts showed that the synchronous activation of the set of muscles in the back was replaced by a sequential disto-proximal activation of the same muscles in trained subjects (Pedotti et al., 1989). This indicates that the synergies are susceptible to training. In addition, under a specific constraint consisting of standing on a narrow support, the activation of the distal muscles during upper trunk movement disappears. This indicates that the synergy is flexible in the presence of a mechanical constraint. However, short term adaptation to support conditions has been observed only in trained subjects. This suggests that long term training not only results in a new EMG pattern which is more efficient in terms of movement velocity and equilibrium control, but also it facilitates short term adaptation to new support conditions.

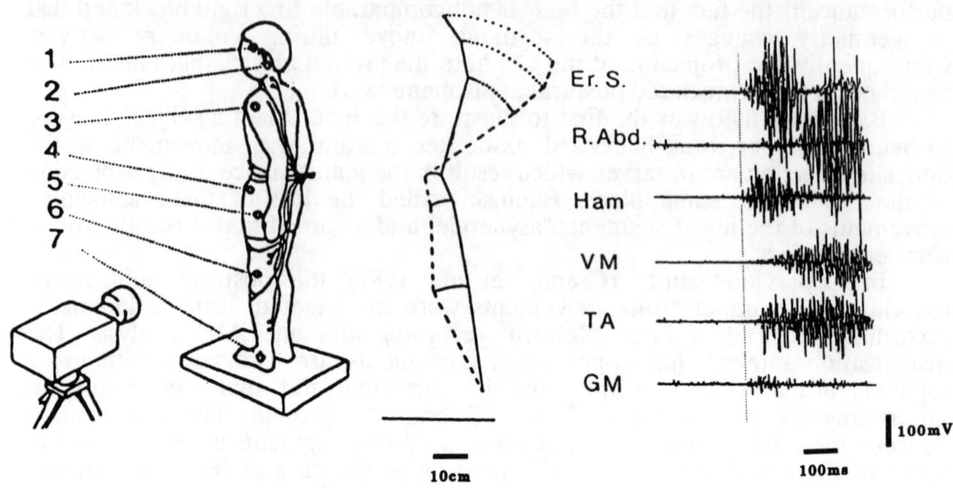

Figure 2. Anticipatory postural adjustments associated with upper trunk movement and served to control equilibrium during movement performance. On the left, recording of the position of the body segments (head, shoulder, iliac crest, trochanter, knee, ankle) with the ELITE system (Ferrigno et al., 1985) during upper trunk backward movements. Note the displacement of the upper and lower body segments in the opposite direction. On the right, EMG recording of a set of muscles of the trunk : Er Sp : erector spinae; R Abd : rectus abdominis, the thigh : hamstring, (Ham), vastus medialis (VM) and the leg : GM : gastrocnemius medius and TA : tibialis anterior in an untrained subject. Note the early activation of a set of muscles in the back of the body (Er Sp, Ham, Sol).

As the same kinematic changes are observed with different muscle patterns, it is tempting to make a distinction between strategy and synergy, as Horak and Nashner (1986) have done in the case of the postural reactions (see Macpherson, 1990). The strategy would consist of the kinematic changes utilized for maintaining the CG position during movement performance. These kinematic changes are the same in every subject and constitute the invariant part of the reaction.

The second aspect is the implementation of the strategy by the synergy i.e. the muscle pattern selected to carry out the strategy. This pattern varies depending on the subject's previous experience and the environmental constraints.

The hypothesis involving a distinction between strategy and synergy during upper trunk movements was explored under microgravity during the last French-Soviet flight (end of 1988) with a view to answering the following questions (Massion et al., 1991). Under normal gravity conditions, the hip and knee movements accompanying upper trunk movements are aimed at maintaining the CG projection inside the supporting surface in order to preserve the equilibrium. Under microgravity, with the feet attached to the "floor", there is no longer any need to keep the center of mass projection at the same place because there is no gravity. One might therefore expect these axial synergies to disappear under microgravity because they no longer serve any purpose.

Upper trunk movements performed under microgravity with the feet attached to the floor are still accompanied however by a displacement of the hip and knee in the opposite direction. Some of the associated movements were even enhanced. For example, the flexion of the knee increased in comparison with normal gravity conditions. This indicates that the position of the CG with respect to the feet is still regulated although it is not necessary for equilibrium maintenance. This means that at least in the case of flights lasting one month or more, the internal representation of the CG position with respect to the feet persists and is still regulated as a reference value which needs to be stabilized.

The second result concerns the implementation of the strategy by the muscle pattern. Under microgravity a clearcut change was observed in the EMG pattern associated with upper trunk movements. During backward upper trunk movements, the activation of the gastrocnemius was replaced by an activation of the tibialis, initiating the flexion of the knee. The activation of the hamstring persisted, causing flexion of the knee and extension of the hip. Interestingly, after return to Earth, the previous pattern was not immediately restored. The early TA activation still occurred 5 days after flight and was replaced by a gastrocnemius activation only at the 2nd recording, i.e. 7 days after landing. This indicates that a new adaptation had taken place.

To conclude, the general kinematics of upper trunk movements and the associated lower segment displacements in the opposite direction are maintained under microgravity. This indicates as proposed by Gurfinkel and Lestienne (1988) that the internal representation of the body segments is very stable and is not basically changed after the loss of the gravity forces. The muscle synergies which are responsible for the kinematic changes vary on the contrary as a function of the environmental constraints and/or training.

3. Coordination stabilizing the position of given segments

The coordination between posture and movement not only serve to maintain the equilibrium during movement performance.

The second goal of the coordination is to stabilize the position of given segments such as the head, trunk or limbs during ongoing movements.

In fact, the position of head, trunk or arm is used as a reference value (egocentric reference frame) for calculating the target position in the peripersonal space and the trajectory of the movement required to reach the target. As shown by Biguer et al. (1988), vibrating a neck muscle on one side to stimulate the Ia afferents from that muscle and mimic the stretching of that muscle induces an illusory displacement of a target in the dark and pointing movements are misdirected toward the illusory target. Stabilization of segments such as the head, neck or other segments is therefore a priority task for the nervous system in order to allow the movement to reach the target notwithstanding the postural disturbances due to movement performance. The movement results from internal forces generated by muscle contraction. These forces exerted on the moving segment are associated with reaction forces acting on the other body segments and causing changes in their positions. The anticipatory postural adjustements associated with the voluntary movement serve to minimizing these changes.

One of the most convincing examples of the anticipatory postural adjustments stabilizing the position of segments is provided by the bimanual load lifting task (Dufossé et al., 1982; 1985; Paulignan et al., 1989) (Fig. 3). Sitting subjects maintain their loaded forearm (postural forearm) in a horizontal position. Unloading is initiated either by the experimenter or by the voluntary movement of the subject's other hand (voluntary forearm). When the unloading is triggered by the subject himself, the forearm position is roughly maintained,

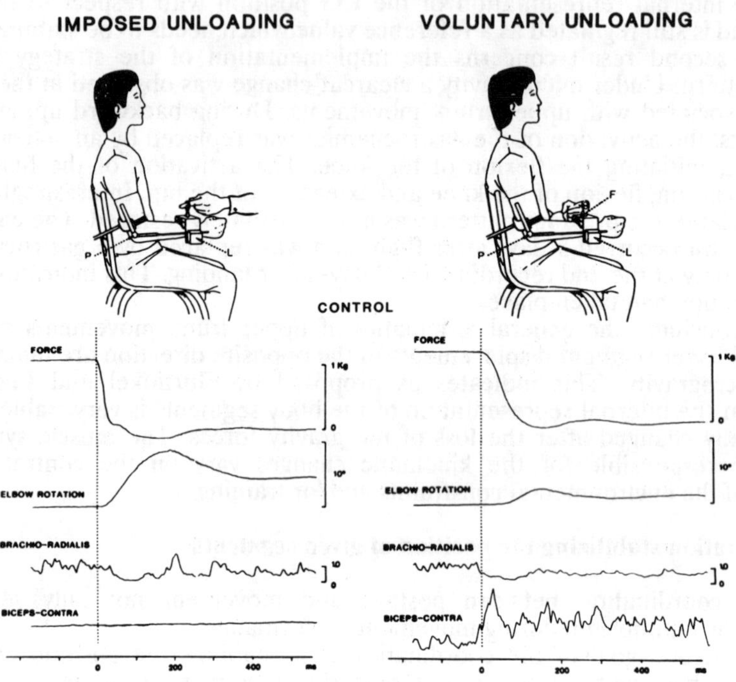

Figure 3. Comparison between imposed unloading and voluntary unloading. The sitting subject maintains a loaded forearm (1 kg) in a horizontal position. The load is lifted either by the experimenter (imposed unloading) or by the voluntary movement of the other hand (voluntary arm). Mean value of 20 trials. The vertical dashed line shows the onset of unloading. On the left, imposed unloading : the experimenter lifts the weight supported by the postural forearm. On the right, voluntary unloading : the subject himself lifts the weight with his other hand. Note that during voluntary unloading, the position of the postural forearm remained fairly stable and that a deactivation of the postural forearm flexors occurred before the onset of unloading.

nothwithstanding the disturbance due to the unloading. This is due to a phasic decrease in the forearm flexor activity which precedes the unloading and is more or less time locked with the activation of the biceps of the "voluntary arm". Thus, an anticipatory postural adjustment takes place which very efficiently minimizes the disturbance of the forearm position due to the unloading voluntary movement. This type of coordination between posture and movement is an old habit acquired during childhood, aimed at stabilizing the position of one forearm during the manipulation of an object with the other hand. It is one of the many movements which require a postural stabilization in order to be performed accurately because they are planned on the basis of a postural reference frame.

The anticipatory adjustments observed during a load lifting task differ in several respects from those associated with equilibrium control.

The common rule in both cases is the stabilization of a reference value. In the case of equilibrium control, the reference value is the CG projection onto the ground. In that of the bimanual task, it is the forearm position in space.

Contrary to equilibrium control, which necessitates the computation of the position of each body segment in order to evaluate the center of gravity position, the position control of a single segment is less complex, because the elbow joint is fixed and only one degree of freedom has to be regulated. It was observed that the anticipatory postural adjustment in the load lifting task still occurred in a patient with a section of the corpus callosum (Viallet et al., 1990) This indicates that the postural adjustments take place at subcortical level. As the onset of inhibition of the postural forearm flexors and the onset of activation of the voluntary forearm flexors are quite closely time locked, the anticipatory adjustment seems to be organized at subcortical level by collaterals of the pathways controlling the movement. This mode of organization is not comparable to that involved in equilibrium control, because in this case the onset of the postural adjustment precedes the onset of the prime mover activation by some 50-100 ms. This means that the command responsible for the movement onset is not at the origin of the anticipatory postural changes aimed at equilibrium control and that two parallel commmands (the one for the postural adjustment, and the other for the movement) are put into action.

Interesting pathological data have shed further light on the central organization of the coordination in bimanual load lifting (Fig. 4). It was observed that after unilateral lesion of the medial frontal cortex including the supplementary motor area (SMA), the anticipatory adjustment of the forearm position was reduced (Massion et al., 1990). This result is in agreement with the findings by Gurfinkel and Elner (1989) who noted a defective anticipatory postural adjustment in standing patients with lesions of the premotor and supplementary motor area when raising the arm. The defect was observed mainly when the postural forearm was contralateral to the lesion, however. In order to interpret this rather surprising result, indicating that the cortex contralateral to the postural arm plays a leading role in this co-ordinated action, we propose that the preparation for this bimanual load lifting task involves a hierarchical organization whereby the movement is planned on the basis of the forearm position taken as a reference. The priority in this task would be to keep the forearm posture stable. The role of the SMA area region together with other motor and premotor areas contralateral to the postural forearm might be to gate

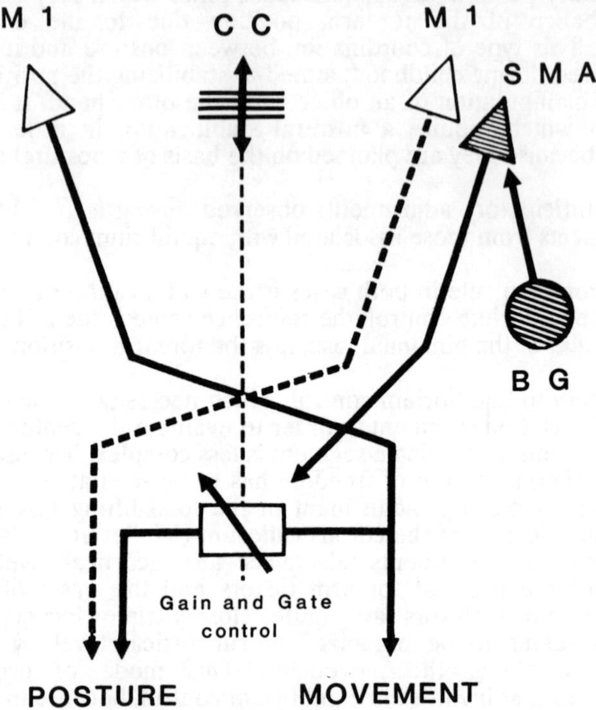

Figure 4. Scheme proposed to account for the central organization of coordination between posture and movement in the bimanual load lifting task. M1 : Primary motor cortex. SMA : Supplementary motor area region. In this scheme, the motor cortex on one side (M1) controls the load lifting movement (continuous line) whereas the motor cortex on the other side controls the postural maintenance of the postural arm (dashed line). The coordination between the 2 controls is not performed through the corpus callosum (CC). The movement control pathway gives off collaterals at subcorticxal level towards the postural arm, which are responsible for the anticipatory postural adjustment. The possibility of using this collateral pathway (gate) and its gain depend on a supraspinal control from the SMA area contralateral to the postural arm and also from the contralateral basal ganglia (BG).

on the phasic learned postural stabilizing circuits which are appropriate for the movement to be performed. The feedforward control of these circuits by collaterals on the movement control pathways might minimize the postural perturbation arising from the movement performance and increase the movement accuracy. As the same coordination deficit has been observed in parkinsonian patients (Wise and Strick, 1984), it seems likely that the basal ganglia and the SMA have a common function, namely the control of posture during movement performance.

4. References

Babinski, J. (1899) De l'asynergie cérébelleuse, Rev. Neurol., 7, 806-816.

Belenkiy, V.E., Gurfinkel, V.S. and Paltsev, E.I. (1967) On elements of control of voluntary movements (in Russian), Biofizika, 12, 135-141.

Biguer, B., Donaldson, I.M.L., Hein, A. and Jeannerod, M. (1988) Neck muscle vibration modifies the representation of visual motion and direction in man, Brain, 111, 1405-1424.

Crenna, P., Frigo, C., Massion, J. and Pedotti, A. (1987) Forward and backward axial synergies in man, Exp. Brain Res., 65, 538-548.

Dufossé, M., Hugon, M. and Massion, J. (1985) Postural forearm changes induced by predictable in time or voluntary triggered unloading in man, Exp. Brain Res., 60, 330-334.

Dufossé, M., Macpherson, J. and Massion, J. (1982) Biomechanical and electromyographical comparison of two postural supporting mechanisms in the cat, Exp. Brain Res., 45, 38-44.

Ferrigno, G. and Pedotti, A. (1985) ELITE : a digital dedicated hardware system for movement analysis via real-time TV signal processing. I.E.E.E. Trans. Biomed. Eng., 32, 46-62.

Gurfinkel, V.S. and Elner, A.M. (1988) Participation of secondary motor area of the frontal lobe in organization of postural components of voluntary movements in man, Neurophysiology, 20, 7-14.

Hess, W.R. (1943) Teleokinetisches und ereismatisches Kräftesystem in der Biomotorik, Hel. Physiol. Pharmacol. Acta 1, C62-C63.

Horak, F.B. and Nashner, L.M. (1986) Central programming of postural movements : adaptation to altered support surface configurations, J. Neurophysiol., 55, 1369-1381.

Jung, R. and Hassler, R. (1960) The extrapyramidal motor system, in J. Field, H.W. Magoun and V.E. Hall (eds.), Handbook of Physiology, Sect. 1, Vol. 2, Amer. Physiol. Soc.

Lestienne, F. and Gurfinkel, V.S. (1988) Postural control in weightlessness : a dual process underlying adaptation to an unusual environment, TINS, 11, 359-363.

Macpherson, J.M. (1990) How flexible are muscle synergies? in D.R. Humphrey and H.J. Freund (eds.), Motor control : concepts and issues. Dahlem Konferenzen, Chichester, John Wiley & Sons Ltd, in press.

Massion, J. and Dufossé, M. (1988) Coordination between posture and movement : why and how? NIPS 3, 88-93.

Massion, J., Déat, A., Gurfinkel, V., Lipshits, M., Popov, K. (1991) Axial synergies under microgravity conditions. Exp. Brain Res. (Submitted).

Paulignan, Y., Dufossé, M., Hugon, M. and Massion, J. (1989) Acquisition of co-ordination between posture and movement in a bimanual task, Exp. Brain Res., 77, 337-348.

Pedotti, A., Crenna, P., Déat, A., Frigo, C. and Massion, J. (1989) Postural synergies in axial movements : short and long-term adaptation, Exp. Brain Res., 74, 3-10.

Viallet, F., Massion, J., Massarino, R. and Khalil, R. (1990) Coordination between posture and movement in a bimanual load lifting task : putative role of a mesial region including the supplementary motor area, Exp. Brain Res., in press.

Wise, S.P. and Strick, P.L. (1984) Anatomical and physiological organization of the non-primary motor cortex, TINS, 7, 442-447.

AFFERENT CONTROL OF POSTURE

V. DIETZ and G.A. HORSTMANN
Department of Clinical Neurology and Neurophysiology
University of Freiburg
Hansastr. 9
7800 Freiburg (F.R.G.)

Abstract. The visual, vestibular and muscle proprioceptive systems have all been shown to contribute to sway stabilization. Nevertheless, an additional receptor system is needed to signal the position of the body's centre of gravity relative to the feet. This receptor system should be "gravity" dependent. The properties of this receptor system were evaluated during water immersion. An approximately linear relationship was found between contact force and impulse directed e.m.g. response amplitude in the leg muscles. Out of water, loading of the subjects resulted in no further increase of the response amplitude. A gain control mechanism for postural reflexes which is dependent on body weight was demonstrated. In a further experiment it could be shown that the receptors for this mechanism are distributed along the vertical axis of the body. It is suggested that these force dependent receptors are pressure receptors within the joints and the vertebral column.

1. Introduction

Common sense tells us that, on a trivial level, upright stance (and, to a lesser extent, bipedal gait) is largely a matter of keeping the body's centre of gravity over the feet. Since the human body may, in a very simplified way, be considered as an inverted pendulum, the projection of the centre of gravity has to meet the support surface within the area of the feet, in order to prevent falling. From these considerations, several questions arise: what type of mechanism tells us that the centre of gravity projects within the required range? What known receptor type could satisfy the requirements and, if such a receptor or receptor system exists then, how does it interact with the known postural control mechanisms? Finally, what kind of experiment can help to reveal its properties?

Several mechanisms have been discussed in the literature during the last years to be responsible for the control and stabilization of human posture. Among these, the influence of proprioceptive (Diener et al. 1984; Berger et al. 1984; Dietz et al. 1987), visual (Berthoz et al. 1979) and vestibular (Allum et al. 1986) inputs has been demonstrated by various experimental approaches. Nevertheless, according to recent experiments (Dietz et al. 1989b; Gollhofer et al. 1989), these mechanisms cannot, alone, account for postural stabilization. An additional posture control mechanism was suggested which regulates the position of the centre of gravity with respect to the feet. An additional receptor system was, therefore, proposed, seemingly contact force dependent and related to the joint receptors.

In order to evaluate the properties of these hypothetical contact force dependent

J. Requin and G. E. Stelmach (eds.), Tutorials in Motor Neuroscience, 209–222.

receptors, the following experiment was designed to test their influence on the e.m.g. responses to destabilizing impulses applied on a translation platform.

While there is no way to reduce gravity for more than a few seconds (parabolic flight) on earth, buoyancy in a water filled pool was used to simulate the effect of weightlessness. Such an experimental approach has the advantage, compared to postural reactions studied in weightlessness (Clément et al. 1984), of leaving vestibular function unaltered and allowing the possibility of manipulating body's contact forces. A possible disadvantage could arise from the viscosity of water. If a contact force dependence exists, than some effect on the responses to destabilizing platform impulses should be seen.

In order to evaluate whether contact force dependence is related to receptors within skeletal connections or joints, an experiment with partial loading and unloading out of water was designed. A subject unloaded by hanging in a parachute harness with a counterweight attached should produce similar results following destabilising impulses standing on a translational platform as those seen during buoyancy under water. If the subject is then simultaneously reloaded, then an increase of corrective responses should occur, despite the fact that total body weight remains unaltered, if a summation of the output of the proposed force or pressure dependent joint receptors has occurred.

2. Mechanisms involved in the control of posture

Displacement of the supportive surface during upright stance in man evokes early, functionally directed e.m.g. responses in the lower and upper leg musculature (Berger et al. 1984; Dietz et al. 1988). These compensatory reactions adapt quickly to a given mode of perturbation (Forssberg and Nashner 1982; Nashner 1976; Nashner et al. 1979; Quintern et al. 1985). For the compensatory e.m.g. responses in such a task the follwing three different control mechanisms have been considered.

2.1 Visual input

Results of earlier studies (Dietz et al. 1989) and those of the immersion experiments (Dietz et al. 1989a) indicate that visual input plays only a minor role: all the results obtained during water immersion and out of water were qualitatively and quantitatively similar if the subject was tested with the eyes opened or closed. An influence of a feedforward control mechanism, as described by Nashner and Berthoz (1978), can thus be excluded for the compensatory mechanism involved during translational displacements of the feet. The importance of vision for the correction of body tilt (Allum and Pfaltz 1985) or head tilt (Horstmann and Dietz 1988) has been demonstrated in patients with bilateral vestibular deficits, but it is not essential in subjects with normal vestibular function.

2.2 Vestibular input

Vestibular input has been shown to play a role for compensation during certain conditions of stance perturbations: the influence of vestibulo-spinal reflexes was demonstrated in toe-up- (Allum and Pfaltz 1985) and head acceleration experiments (Horstmann and Dietz 1988). Tibialis anterior e.m.g. responses produced following platform rotational toe-up tilt has been attributed to a vestibulo-spinal mechanism (Allum and Pfaltz, 1985), whereas the compensatory e.m.g. responses seem to arise predominantly as the result of a spinal proprioceptive mechanism following horizontal platform displacements (Dietz et al. 1984). A more recent study has shown that vestibulo-spinal reflex activity induced by a head acceleration device (without primarily stretching the leg muscles) - produced only 20% of the e.m.g. response activity induced by a comparable platform translation (Horstmann and Dietz 1988). Several observations (Mauritz and Dietz 1980; Berger et al. 1987; Nashner et al. 1979) suggested that the vestibulo-spinal reflex system plays a major role in the compensation of slow body sway. In experiments producing rapid perturbations of stance during water immersion the influence of vestibular input on the e.m.g. responses of the primarily stretched muscles appeared to be negligible for the following reasons: head accelerations occurring before the onset of leg muscle e.m.g. responses showed no significant differences, neither when comparing the different loads, nor when comparing the immersion and out of water experiments.

The usually stronger background e.m.g. activity in tibialis anterior m. during water immersion, compared to the out of water condition, could best be explained as being due to the loss of the static moment of force acting on the ankle. This effect was first observed by Clément et al. (1984) in microgravitation during spaceflight. The decrease in the e.m.g. activity reported during microgravity may, on the other hand, not necessarily be a short-term effect of microgravity since similar effects were observed after three days of water immersion (Christova et al. 1989) and after 20 hours of antiortostatic hypokinesia (Belyaeva et al. 1989).

2.3 Proprioceptive input

Proprioceptive input has been shown to play the major role in early compensatory e.m.g. responses of the leg musculature if destabilization of posture is induced in upright standing man by platform translation. This was suggested on the basis of the close correlation between impulse velocity and impulse amplitude, and the slope and amplitude of the compensatory e.m.g. responses when elicited during gait. It was suggested that muscle proprioceptive input is incorporated in a more complex, centrally programmed pattern in order to generate the compensatory e.m.g. responses (Dietz et al. 1987).

3. Postural reflexes depend on the presence of contact forces

The visual, vestibular and proprioceptive systems have all been shown to contribute to sway stabilization. Nevertheless an additional receptor system is needed to signal the position of the body's centre of gravity relative to the feet. The properties of this receptor system was evaluated during water immersion.

3.1 Influence of body's mass on postural response

The experiments concerning the influence of actual body weight on the postural responses were performed during water immersion and out of water. Under water, the subjects were standing with their feet 1.8 m below surface, breathing by means of a snorkel. Water temperature was 24 ^{0}C. The subjects were asked to stand in an upright "rigid" position, with their shoes attached to a pneumatically moveable platform (fig. 1). The platform was randomly and unexpectedly displaced forwards and backwards by means of two pneumatic cylinders controlled by an electrical valve.

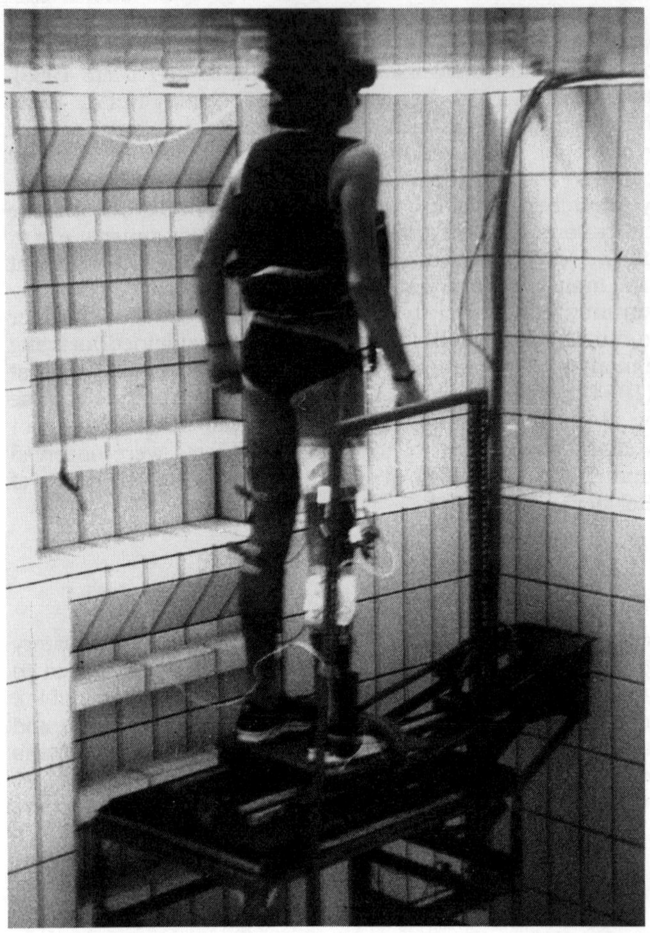

Fig. 1: Subject standing under water on a pneumatically movable platform. The subject breathed using a snorkel. Buoyancy was compensated for by a lead vest with different loads.

The e.m.g. responses of gastrocnemius med., tibialis anterior, rectus femoris and biceps femoris muscles were recorded following these acceleration impulses, using surface electrodes isolated from the water by means of a special adhesive film. Head acceleration was measured by a water resistant piezo-electric accelerometer mounted at the back of the subject's head. Ankle, knee and hip joint angles were indicated by mechanical goniometers.

The e.m.g. signals were amplified, rectified and, together with biomechanical parameters, were transferred on-line via an A/D converter to a microcomputer system. An average was then made from ten trials at each of the movement parameters (see below). Further processing was carried out on the averaged data with the SPSS-PC^{TM} statistic package.

Two different types of acceleration impulses (forwards or backwards) were applied under water in a 3.5 m deep pool with 4 different loads (with 10, 20, 40 or 60 kg lead vests attached) (figure 1) and out of water with 4 different loads (without load and 20, 30 or 40 kg lead vests). The acceleration rate was fixed for all experiments at 24 m/s^2, the duration of the displacement at 100 ms and the displacement amplitude at +/- 0.12 m. Background e.m.g. activity and the e.m.g. responses were calculated by integration. From the biomechanical parameters the peak angular velocities were calculated for the hip, knee and ankle joints, as well as the initial amplitude of head acceleration. Since fixed additional loads but different body weights had to be considered, a relative e.m.g. response was calculated by dividing the integrated e.m.g. response by body weight (multiplied by a factor to set the out of water - no load condition = 1.0).

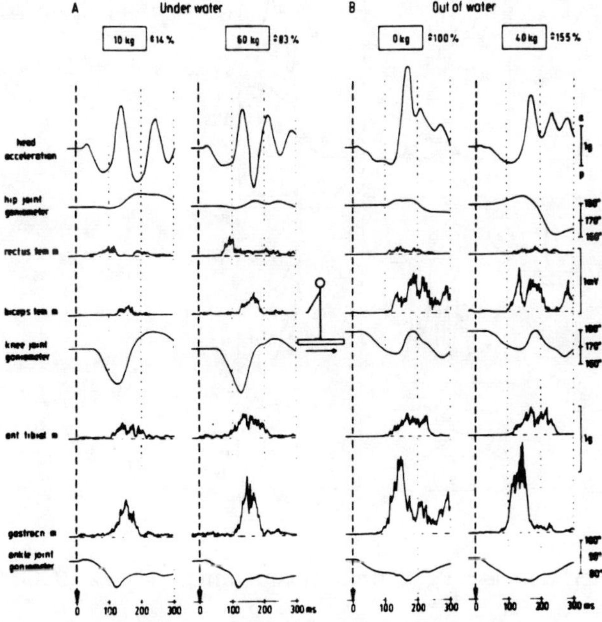

Fig. 2: The averaged biomechanical and e.m.g. (rectified) data of one subject during ten backward accelerations; (A) under water with a 10kg (= 14 % of body weight) and 60 kg (= 83 % of body weight) lead vest attached; (B) out of water without (= 100 % of body weight) or with 40 kg lead vest attached (= 155 % of body weight) (from Horstmann and Dietz 1990).

Figure 2 shows the results for one subject during immersion (A) with a load of 10 and 60 kg and out of water (B) without load and with a load of 40 kg, following backward displacement. There was a gradual increase in the gastrocnemius e.m.g. response with increasing load. Onset latency of the gastrocnemius e.m.g. response ranged from 60 to 80 ms after the beginning of ankle joint flexion. No significant difference was found between the e.m.g. response with a 60 kg load during immersion (which resulted in a relative body weight of 83%), and the response out of water, either without load or with an additional load of 40 kg. Loading the subject with 60 kg during immersion produced considerably larger e.m.g. responses in the primarily stretched gastrocnemius muscle than loading with 10 kg. No substantial change occurred in the activity of the antagonistic muscle. Also in the upper leg, the biceps femoris m. showed a stronger e.m.g. activity with increasing load. Very early (50 ms) but tiny e.m.g. responses were seen in the rectus femoris m., which showed no dependence on load.

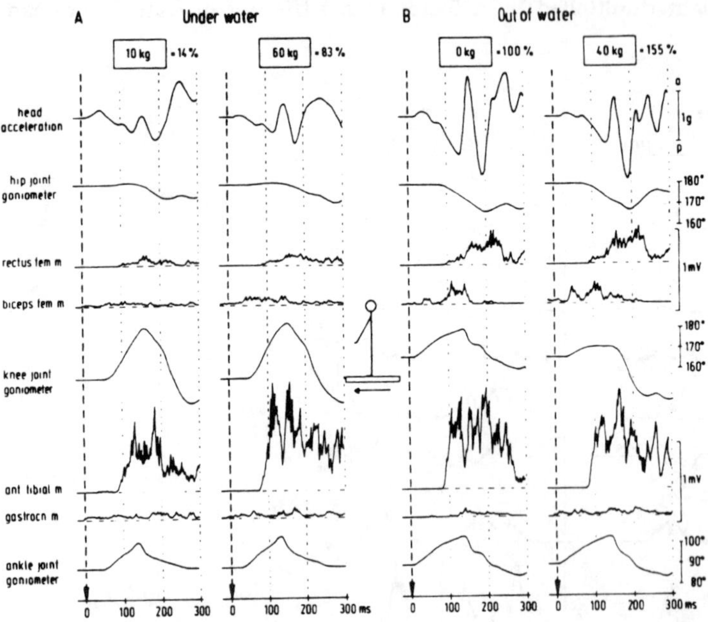

Fig. 3: Data for forward acceleration. (see fig. 2) (from Horstmann and Dietz 1990).

Figure 3 shows the comparable responses following forward acceleration impulses. These were followed by large responses in the primarily stretched tibialis anterior m., with latencies of about 70 ms and with amplitudes which were dependent on the attached load. These responses were followed by a very small gastrocnemius activation. The e.m.g. activity in the antagonistic gastrocnemius m. was not load dependent. The amplitude of rectus femoris m. activity in this case was, however, load dependent. The latencies were about 90 ms from the primarily knee joint extension, while the biceps femoris m. during immersion showed no activity discernible from the background activity.

Figure 4 shows the results obtained from 10 subjects (integrated e.m.g. responses +/- SD) in the gastrocnemius m. (A) for backward acceleration and in the tibialis anterior m. (B) for forward acceleration, with different loads during immersion and out

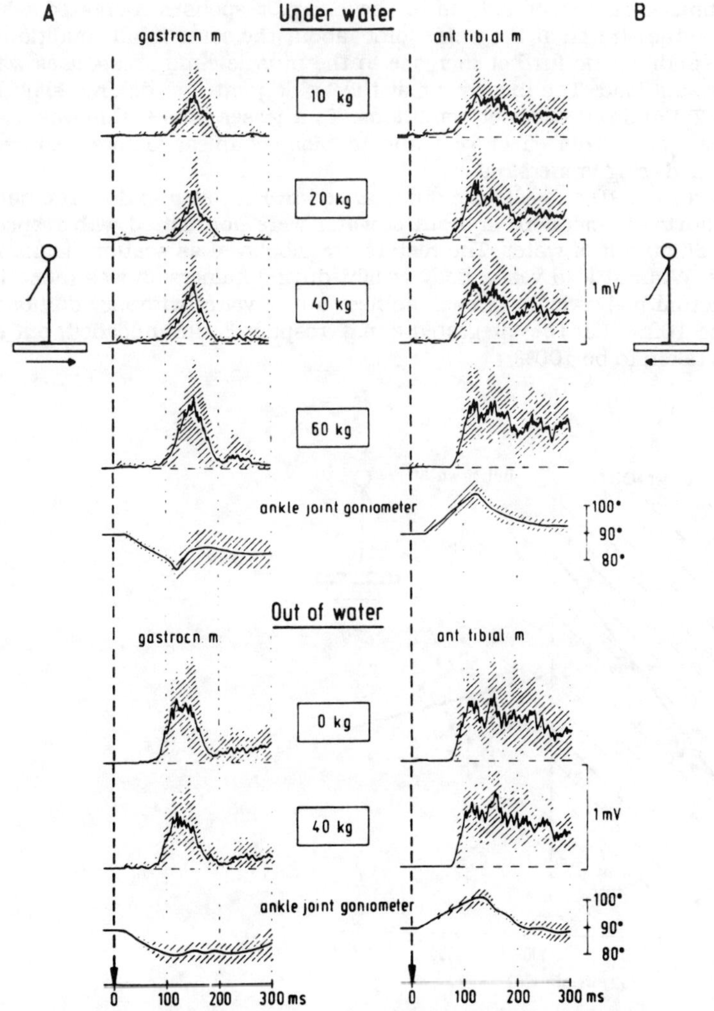

Fig. 4: Grand average of the rectified and integrated e.m.g. responses obtained from 10 subjects (+/- SD) in gastrocnemius m. (A) following backward acceleration and in tibialis anterior m. (B) following forward acceleration at different load conditions, during immersion and out of water. The results of the right leg are displayed and only one ankle joint goniometer curve (no load dependence observable) for the immersion (40 kg load) and for the out of water (0 kg load) condition is displayed. Stimulus was released at time zero (from Horstmann and Dietz 1990).

of water. During immersion, the amplitude of the e.m.g. responses increased with increasing load, while the stretch at the ankle joint about the same in all conditions. In the out of water condition no further increase in the muscle e.m.g. responses was observed with increasing load. The movement at the ankle joint differed slightly between the immersion and out of water conditions. To a lesser extent, this was also the case at the knee joint. This effect was due to biomechanical factors (reduced ground reaction forces during immersion).

The results obtained from 10 subjects are summarized in Figures 5 and 6. The data for all experiments during immersion and out of water were normalized with respect to the unloaded condition out of water. The results are displayed as scatter diagrams and regression lines. Without lead vests, body weight during immersion was taken to be zero, while the actual body weight of each subject out of water without additional load was taken to be 100%. For the integrated e.m.g. responses the unloaded, out of water condition was taken to be 100%.

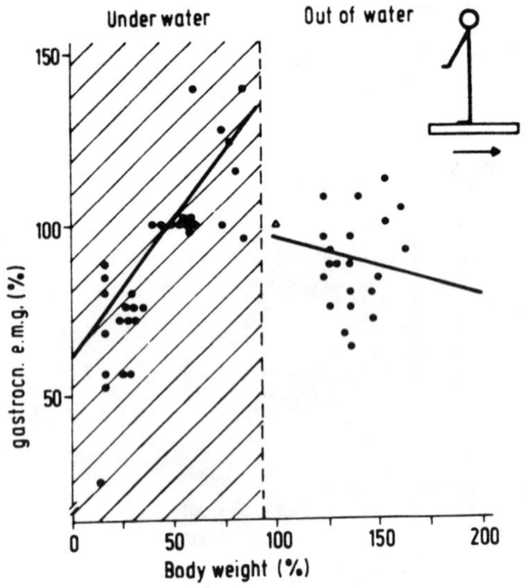

Fig. 5: Scatter diagram of relative gastrocnemius m. e.m.g. response activity versus relative body weight for all impulses during immersion (dashed area) and out of water. The triangle indicates the no load e.m.g. response activity for all subjects, which was taken to be 100 %. The lines indicate the linear correlation function (from Dietz et al. 1989a).

During water immersion a close relationship existed between actual body weight and the magnitude of the e.m.g. responses following backward (fig. 5) and forward (fig. 6) displacements. All correlations were significant ($r = 0.63 - 0.82$, $p < 0.0001$). No significant correlation was found for the late tibialis anterior activation following backward displacement of the platform. In addition, when the correlation between load and the pre-activation e.m.g. level was calculated, no significant correlation was found. Similarly the hip, kee and ankle angles showed no significant correlation with

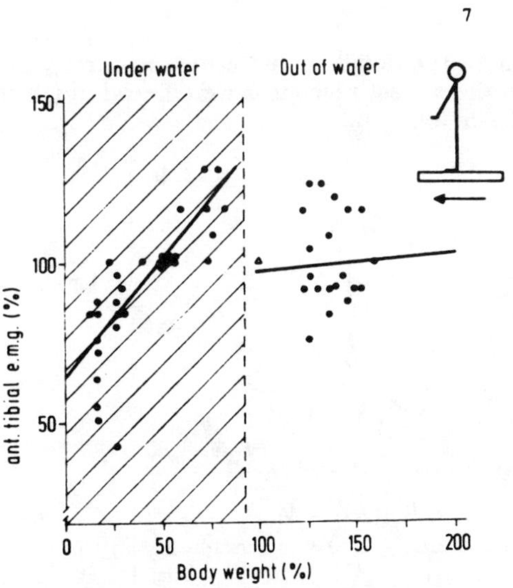

Fig. 6: Scatter diagram of relative tibialis anterior m. e.m.g. activity versus body weight for all impulses during immersion (dashed area) and out of water (see fig. 5) (from Dietz et al. 1989a).

increasing load. The movements at knee and hip joints, however, differed with smaller and slightly slower excursions occurring out of water. This could be the explanation for the smaller e.m.g. amplitudes obtained out of water compared to the largest e.m.g. amplitudes measured during immersion. The obeserved differences were, on the other hand, sufficiently similiar to allow us to assume, that the influence of water viscosity on the e.m.g. responses was negligible. A control experiment during which the subjects voluntarily increased the pre-activation of leg muscles produced no effect on the e.m.g. responses induced following the displacement. No difference was observed in the e.m.g. responses between eyes opened and closed conditions.

In addition, an unloading experiment performed out of water was undertaken in order to evaluate the skeletal/joint contribution to the e.m.g. responses induced by the displacements. The subjects were suspended in a parachute harness with a counterweight of 40 kg attached, via a pulley, above them. E.m.g. and biomechanical parameters were recorded as described above. Three different experimental conditions (unloaded 40 kg plus reloaded 2 x 10 kg at the knees; unloaded 40 kg plus reloaded 2 x 10 kg at the hip, and unloaded 40 kg plus reloaded 2 x 10 kg at the shoulders) were tested for both backward and forward acceleration impulses. This was tested in five subjects. Figure 7 shows the averaged data obtained in one subject during backward (A) and forward (B) displacements. During backward displacement (fig.7, A) the amplitude of the gastrocnemius e.m.g. response was largest when the shoulders were reloaded with 20 kg (a), smallest when the reloading weights were attached over the knees (c), and intermediate if the reloading weights were attached over the the hip (b).

Fig.7 (B) shows the corresponding results for the tibialis anterior m. e.m.g. responses during forward displacement. Once again the largest responses were found in (a), the smallest in (c) and intermediate amplitudes in (b).

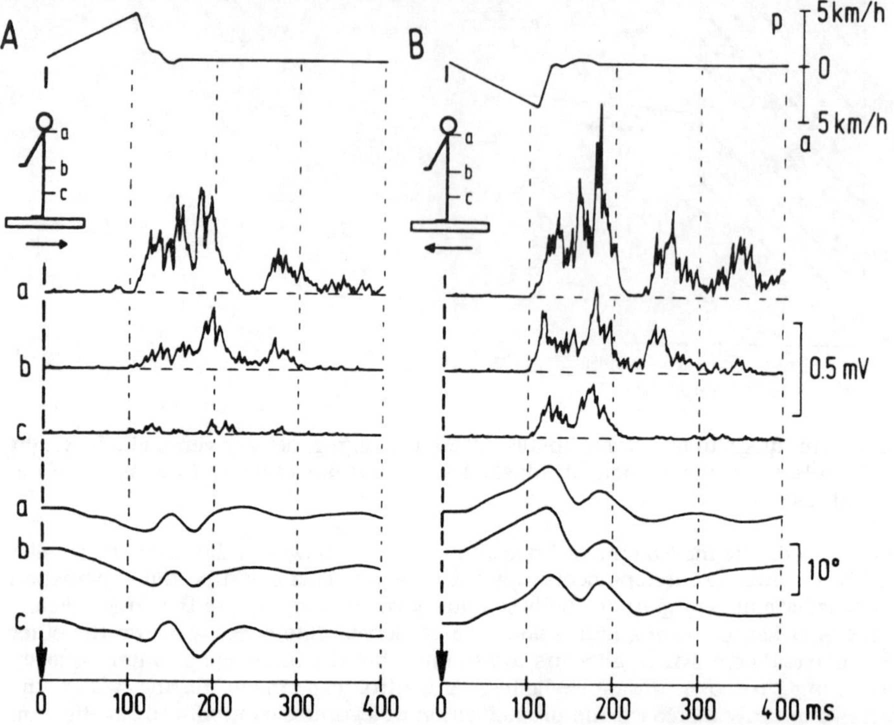

Fig. 7: Average biomechanical and e.m.g. (rectified) data of one subject (22.3 years, 68.5 kg) obtained following ten trials of backward (A) or forward (B) displacements (out of water). In all conditions the subject was unloaded at the shoulders by 40 kg and simultaneously reloaded with 20 kg over the shoulders (a), hip (b) or knees (c) (from Horstmann and Dietz 1990).

3.2 Receptors involved in gain modulation of postural reflexes

For the gain modulation of postural reflexes the following mechanisms may be considered:

1. The strength of background activity in the leg muscles might change with the body weight during stance, leading to a gain modulation of the muscle proprioceptive reflexes during stance. The results of the present study make this rather unlikely, as the quantified leg muscle activity preceding the e.m.g. responses did not systematically change with loading and the level of voluntary leg muscle activity did not visibly affect the compensatory responses.

2. Pressure receptors within the sole of the foot might signal the actual force exerted on the supporting surface. This possibility is also rather unlikely, as the compensatory responses are preserved after ischaemic blockade of the skin afferents of the foot (Berger et al. 1984).

3. Impulses from pressure receptors distributed over the whole body (within joints and the vertebral column) may converge with other reflex pathways on spinal interneuronal circuits, as has been shown to be the case in the cat (Lundberg et al. 1987). The unloading experiment with successive reloading of knees, hip and chest supports this assumption. The successive increase of the e.m.g. responses with increasing height of the attached load, is in line with the idea, that the observed gain modulation of the e.m.g. responses is not primarily influenced by the body weight alone - which was unaltered in the whole experiment - but by the position of the body's centre of mass with respect to the ground - which was progressively elevated during the experiment.

3.3 Possible role of pressure receptors

Proprioceptive reflexes alone, or in combination with vestibulo - spinal reflexes cannot explain the following observations: 1. the change from a rotational towards a translational platform tilt in standing man resulted in a shift from a compensatory tibialis anterior to a gastrocnemius response, even though stretch of the gastrocnemius muscle was roughly similar throughout the experiment (Gollhofer et al. 1989). An experiment in sitting man (Gottlieb and Agarwal, 1979) produced only a monosynaptic stretch reflex response when stretch was applied to the ankle joint by a torque motor, despite a large amount of pre-activity; 2. Dietz et al. (1989b) showed that the strength of stretch reflex activity induced by platform shifts is dependent on the functional implications of the shift with respect to the body's centre of gravity: unilateral displacements evoked a small e.m.g. response in both legs, while bilateral displacements were followed by e.m.g. responses of double amplitude, but were negligible when the legs were displaced in opposite directions.

While the second observation can be explained by an interneuronal network mediating polysynaptic spinal stretch reflexes, the first observation cannot be explained by such a mechanism.

From the observations mentioned in the last section and the results obtained in the under water experiments, there is much evidence that the body's centre of gravity, as well as the actual body weight, plays a major role in the stabilization of human posture. The function of known reflexes involved in the stabilization of human posture (e.g. muscle proprioceptive and vestibulo - spinal reflexes) may depend on the activity of receptors within the body which indicate the body weight or the deviation of the centre of gravity from a certain neutral condition. Slowly adapting receptors within the knee joint, sensitive to externally applied pressure, were found by Clark (1975) in the cat. Burgess et al. (1982) proposed that the knee articular receptors in man do not have an important role in signaling joint position, but do contribute to deep pressure sensation. They may also be responsible for signaling changes in the actual body weight and, by interneuronal calculation, of changes in the center of gravity.

4. Conclusion

It has been demonstrated, that the corrective leg muscle e.m.g. responses during stance perturbations, which were attributed to proprioceptive spinal reflexes, are modulated by changes in body weight, i.e. by contact forces. The gain modulation of these proprioceptive reflexes is dependent on the activity of force dependent receptors within the vertical axis of the body. According to the work of Lundberg et al. (1987) the force dependent receptors which are most likely pressure receptors within the joints and the vertebral column, may converge with other afferent inputs on common spinal interneurons. The latter are assumed to generate an appropriate response in a close interaction with the spinal locomotor centres.

The research was supported by the Deutsche Forschungsgemeinschaft, SFB 325.

References

Allum, J.H.J. and Keshner, E. A.
Vestibular and proprioceptive control of sway stabilization. In: Bles, W. and Brandt, T. (Eds.), Disorders of Posture and Gait, Elsevier, Amsterdam, 1986, 19 - 40.

Allum, J.H.J. and Pfaltz, C.R.
Visual and vestibular contributions to pitch sway stabilization in the ankle muscle of normals and patients with bilateral peripheral vestibular deficits. Exp. Brain. Res. 1985, 58: 82 - 94.

Belyaeva, M.G., Burlachkova, N.I. and Kozlovaskaya, I.B.
The effect of weightlessness on spinal reflex mechanisms. VI. Int. Symp. Motor Control, 1989, Albena, Bulgaria.

Berger, W., Dietz, V. and Quintern, J.
Corrective reactions to stumbling in man: Neuronal co-ordination of bilateral leg muscle activity during gait. J. Physiol., 1984, 357: 109 - 125.

Berger, W., Dietz, V. and Horstmann, G.
Interlimb coordination of posture in man. J. Physiol., 1987, 390: 135P.

Berthoz, A., Lacour, M., Soechting, J.F. and Vidal, P.P.
The role of vision in the control of posture during linear motion. In: Granit, T. and Pompeiano, O. (Eds.), Reflex Control of Posture and Movement, Progrress in Brain Res., Vol. 50, Elsevier, Amsterdam, 1979, 197 - 209.

Burgess, P. R., Wei, J. Y, Clark, F. J. and Simon, J.
Signaling of kinesthetic information by peripheral sensory receptors. Ann. Rev. Neurosci., 1982, 5: 171 - 187.

Clark, F. J.
Information signaled by sensory fibers in medial articular nerve. J. Neurophysiol., 1975, 38: 1464 - 1472.

Clément, G., Gurfinkel, V.S., Lestienne, F., Lipshits, M.I. and Popov, K.E.
Adaptation of postural control in weightlessness. Exp. Brain Res. 1984, 57: 61 - 72.

Christova, L., Gydikov, A., Koslova, V. and Kozlovskaya, I.
Electromyographic investigations of the changes in the antigravity muscles caused by three day immersion. VI. Int. Symp. Motor Control, 1989, Albena, Bulgaria.

Diener, H.C., Dichgans, J., Guschlbauer, B. and Mau, H.
The significance of proprioception on postural stabilization as assessed by ischemia. Brain Res., 1984, 196: 103 - 109.

Dietz, V., Horstmann, G.A., Trippel, M. and Gollhofer, A.
Human postural reflexes and gravity - an under water simulation. Neurosci. Lett., 1989a, 106: 350 - 355.

Dietz, V., Quintern, J. and Berger, W.
Corrective reactions to stumbling in man: functional significance of spinal and transcortical reflexes. Neurosci. Lett., 1984, 44: 131 - 135.

Dietz, V., Quintern, J. and Sillem, M.
Stumbling reactions in man: significance of proprioceptive and pre-programmed mechanisms. J. Physiol., 1987, 386: 149 - 163.

Dietz, V., Horstmann, G. and Berger, W.
Involvement of different receptors in the regulation of human posture. Neurosci. Lett., 1988, 94: 82 - 87.

Dietz, V., Horstmann, G.A. and Berger, W.
Interlimb co-ordination of leg muscle activation during perturbations of stance in humans. J. Neurophysiol., 1989b, 62: 680 - 693.

Forssberg, H. and Nashner, L.M.
Ontogenetic development of postural control in man: adaptation to altered support and visual conditions during stance. J. Neurosci., 1982, 2: 545 - 553.

Gollhofer, A., Horstmann, G.A., Berger, W. and Dietz, V.
Compensation of translational and rotational perturbations in human posture: stabilization of the centre of gravity. Neurosci. Lett., 1989, 105: 73 - 78.

Gottlieb, G.L. and Agarwal, G.C.
Response to sudden torques about ankle in man: myotatic reflex. J. Neurophysiol., 1979, 42: 91 - 106.

Horstmann, G.A. and Dietz, V.
The contribution of vestibular input to the stabilization of human posture: a new experimental approach. Neurosci. Lett., 1988, 95: 179 - 184.

Horstmann, G.A. and Dietz, V.
A basic posture control mechanism: the stabilization of the centre of gravity. Electroencephal. Clin. Neurophysiol., 1990 (in press).

Lundberg, A., Malmgren, K. and Schomburg E.D.
Reflex pathways from group II muscle afferents. Exp. Brain. Res., 1987, 65: 294 - 306.

Mauritz, K.H. and Dietz, V.
Characteristics of postural instability induced by blocking of leg afferents by ischaemia. Exp. Brain. Res., 1980, 38: 117 - 119.

Nashner, L.M.
Adapting reflexes controlling the human posture. Exp. Brain. Res., 1976, 26: 59 - 72.

Nashner, L.M. and Berthoz, A.
Visual contribution to rapid motor responses during postural control. Brain Res., 1978, 150: 403 - 407.

Nashner, L.M., Woollacott, M. and Tuma G.
Organization of rapid response to postural and locomotor-like perturbations of standing man. Exp. Brain Res., 1979, 36: 463 - 476.

Quintern, J., Berger, W. and Dietz, V.
Compensatory reactions to gait perturbations: short and long term effects of neuronal adaptation. Neurosci. Lett., 1985, 62: 371 - 376.

THE DISSOCIATION OF MOTOR SEQUENCES IN CONTROLLING LANDING FROM A JUMP

Patricia McKinley
McGill University
School of Physical and Occupational Therapy
3654 Drummond Avenue
Montreal, Quebec, CANADA H3G 1Y5

INTRODUCTION

Over the past several years in our laboratory at McGill, and in Milan at the Bioengineering center with A. Pedotti, we have been studying the act of landing from a jump-down using various paradigms including: the role of height and vision when planning landing in adults (Thompson and McKinley, 1988), the role of vision in landing from a jump in children (Pelland and McKinley, 1988, Pelland et al., 1990), the effect of similarity or difference in compliance of the takeoff and landing surfaces on jump execution (Bastien and McKinley, 1988) and the roll of skill in landing from a jump onto surfaces of differing compliances (McKinley and Pedotti, 1990).

The basic task across all of these paradigms was to takeoff from a support stage and land within a targeted area with both feet, maintain balance and reach stability quickly. To do so required successful completion of two goals. At takeoff, the body must be projected into space so that the center of gravity is projected onto the landing area. For landing, the individual must anticipate landing forces to control impact and allow for maintenance of stability. In order to evaluate performance of each of these goals, the jump was divided into two phases: takeoff and landing. The takeoff phase includes stance preparation before initiation of the jump, the takeoff motion from the support stage and the ascending part of flight (from toeoff to the time at which the iliac crest reaches peak trajectory and vertical velocity is zero). The landing phase is comprised of the descending limb of flight, impact, yield, and stabilization post landing.

Because these goals are interrelated with respect to successful task performance, we were interested to see whether or not the takeoff and landing phases were programmed independently or as a unified motor plan, and secondly, if the phases are independently planned, was the landing component planned before movement initiation or during execution of the jump. To address the first question 2 sets of experiments are compared where landing parameters are changed by either altering the height of the jump (which examines the effect of changing momentum on landing preparation) or by altering the compliance of the landing surface (which alters the initial impact experienced at landing). To address the second question,

J. Requin and G. E. Stelmach (eds.), Tutorials in Motor Neuroscience, 223–229.

the difference between skilled and unskilled subjects in planning for landing on surfaces of differing compliance will be examined as well as the effect that altering the takeoff surface compliance has on planning the landing phase.

METHODS

To analyze the movement, we collected EMG from 6 leg muscles (Vastus Lateralis, VL; Rectus Femoris, RF; Biceps Femoris, BF; Lateral Gastrocnemious, LG; Soleus, SOL; and Tibialis Anterior,TA), and kinematics for analysis of joint motion at the head, trunk, hip, knee and ankle. In the paradigm involving skilled performers, a force plate (Kistler 9261A) was used at landing to determine a performance index based on center of pressure information and time to stability at landing (McKinley and Pedotti, 1990) and kinematics were collected and analyzed using the ELITE motion analysis system hard and software (50Hz). In the other paradigms, footswitches were used to demarcate timing of takeoff and landing, and kinematics were collected with a Panasonic high speed shutter video camera (WD5000) at 60Hz and analyzed using Peak Performance Technologies software.

RESULTS AND DISCUSSION

Independence of the two phases.

Takeoff Phase. Kinematic and electromyographical assessment across the jumping paradigms revealed two styles of takeoff that were not related to skill, height of the jump, or surface compliance. Instead, they were subject specific. That is, a subject preferentially utilized one of the two styles at least 80% of the time. They were described as the *"Roll off"* and *"Push off"* styles. The *"Push Off"* style was the most common and was characterized by an active yield and propulsion so that the lower limb segments were extended at takeoff from the support stage and included a swing ahead phase during the ascending flight period which was characterized by slight flexion of the joints (Fig 1a). In contrast, the *"Roll off"* style had no well-defined propulsion component. Instead, the lower limb appeared to flex continuously prior to takeoff and the subject rolled off the support stage with the knee, and often the ankle, at a point of maximum flexion. The ascending flight period was characterized by continuous extension of the limb (Fig 1b). EMG during this phase was style dependent. When executing *"roll off"*, activity was found only in the Tibialis Anterior prior to and during takeoff. In contrast, when using *"Push off"*, there was alternate activity in the hip musculature and cocontraction of the ankle muscles.

Landing Phase. The overall joint motion and patterns of EMG were similar in the landing phase, regardless of takeoff style chosen (Fig 1). Typically, the joints reached maximum extension prior to landing so that the hip, knee and ankle were already partially flexed at impact. Prelanding EMG was initiated first in the ankle musculature and then at the knee and hip. However, adjustments in amount of joint flexion and timing and amplitude of prelanding EMG were made that were dependent on whether there were changes in compliance of the landing surface or in the height of the jump (see below).

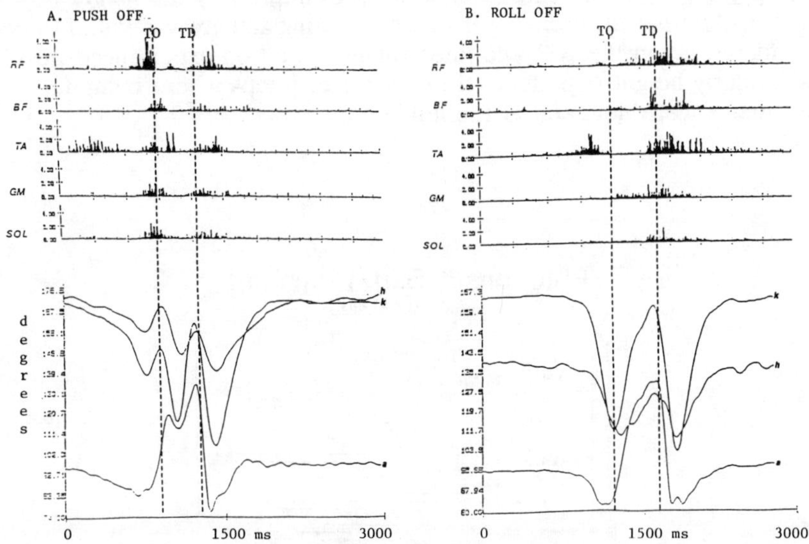

Figure 1. Takeoff and landing kinematics of the ankle (A), knee (K) and hip (H) and accompanying EMG for a) push off and b) roll off styles. TO, toeoff; TD, touchdown.

The effect of surface compliance. The effect of landing on a stiff, metal surface versus landing on a compliant, 3" thick piece of low memory foam was studied in skilled and non-skilled jumpers (McKinley and Pedotti, 1990). We found that the skilled jumpers made adjustments in ankle position so that the ankle was significantly more plantarflexed when landing on a stiff surface than when landing on a compliant one. This had the effect of increasing range of motion during the post-landing flexion when landing on the stiff surface (50 $\pm$3) as compared the to the compliant surface (38 $\pm$2), and placing the body in a more stable position when landing on foam, as the weight was shifted posteriorly by the more dorsiflexed position.

Parallel adjustments were also made in the timing of the onsets and amount of activity in prelanding EMG. When landing on a stiff surface, prelanding activity began first in the LG and SOl, followed by the TA, BF and subsequently the RF and Vl. The onset of activity, particularly in the ankle plantarflexors, began closer to landing when jumping onto the compliant surface in the skilled subjects. This had the effect of activating both the dorsiflexor and the 2 plantarflexors simultaneously (Fig 2a). Activity in the LG and SOL was also significantly damped, while activity in the TA was stable.

Adjustments for landing from different heights. In another set of experiments jump height was varied in three 10cm increments and ranged from 35-55 cm. When well practised subjects jumped from randomly varying heights, we found that the onset of the prelanding EMG bursts was not altered by the height of the jump (Fig 2b). Another difference was that prelanding activity was found to be height dependent only in the RF, and knee range of motion experienced post-landing was similarly height dependent. In contrast to jumps where compliance varied, adjustments were not seen at the ankle.

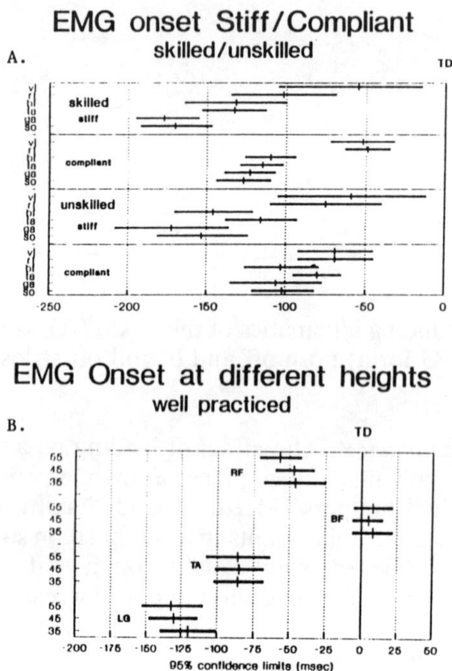

Figure 2. EMG onsets and their 95% confidence intervals for prelanding activity. a) skilled and unskilled subjects when landing on a stiff or a compliant foam surface. b) well-practised subjects when landing on a stiff surface from jumpdowns of three heights: 35, 45 and 55cm.

Independence of the two landing programs. The comparison of these different jump paradigms yielded two points that support the notion that the landing and takeoff phases are planned independently. One is the observation that the takeoff style chosen was independent of skill, compliance, or height of the jump.

The second is the observation that the landing program, while maintaining an overall style, is adjusted to fit the needs of the task. We have postulated that these adjustments are specifically related to the forces experienced at landing. For example, when landing surface compliance is changed, initial impact forces are different, while momentum is similar. In contrast, when landing from jumps of different heights, initial impact forces are similar, while momentum is different. Thus the adjustments observed at the ankle for surfaces of differing compliance (changes in range of motion, EMG onset and activity) as compared to adjustments observed at the knee when landing from different heights (changes in range of motion and RF activity but no change in EMG onset) may reflect this difference.

A third point, not discussed above, relates to our studies of jumpdowns in children. In contrast to adults, two landing styles are often observed when landing on a stiff surface from a constant height with normal vision. One style is seen more frequently in motorically less developed children and is one where joint action is serial or '*dissociated*'. That is, knee flexion or extension during flight and landing precedes that at the hip (Fig 3a). The other is the normal adult pattern where knee and hip flexion or extension occurs in concert, that is, '*coordinated*'(Fig 3b). In the older child (11-13 yrs), these two landing styles are often used interchangeably but are preceded by a consistent, subject-selected takeoff style. Thus it appears that either takeoff or landing styles can be chosen independently of one another, and further, that modifications in the basic landing style can be made to adjust to varying external conditions.

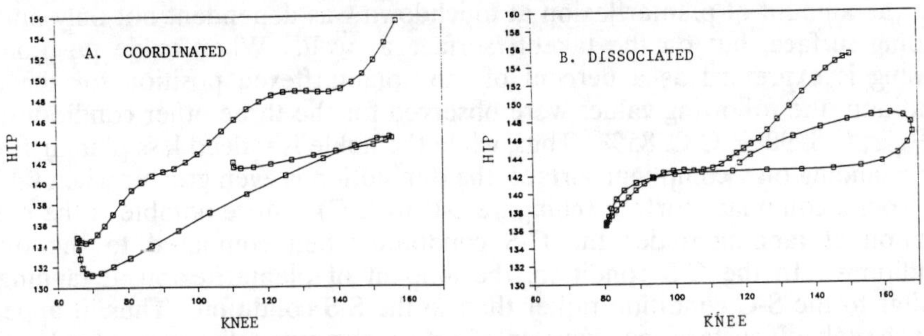

Figure 3. Angle/angle diagrams of the hip/knee for two different landing styles in a 13yr old. a) adult-like, *coordinated*; b) non adult-like, *dissociated*.

Planning of the landing component.

Comparison of skilled and unskilled subjects. As described above, skilled jumpers made adjustments in the amount of plantarflexion at the ankle, so at touchdown, the ankle was more plantarflexed (PF) when landing on a stiff (27° ± 3 PF) than on a compliant (17° ±3 PF) surface. In contrast, the unskilled jumpers did not alter the ankle angle or range of motion for changes in compliance, but adopted an ankle angle for touchdown that was intermediate between the two angles used by the skilled subjects being 23° $\pm$ 3 and 21° $\pm$3 for stiff and compliant surfaces respectively. Thus unskilled subjects appeared to be unable to adapt the basic landing style to adjust for minor differences in landing requirements, but instead used a 'default' angle for both surfaces. And, although the onset of muscle activity when landing on a compliant surface was also delayed in the unskilled subjects, the onset of all three ankle muscles was affected, rather than the shift seen only in the plantarflexors of the skilled subjects (Fig 2a). Thus there was no adaptation in onset sequence, as TA onset occurred after plantarflexor onset regardless of landing surface in the unskilled jumpers. The use of a 'default' program has been described by Ghez and colleagues when faced with a choice reaction time task. They have suggested that the selection of a default program occurs when a motor program is incompletely specified prior to movement initiation (Hening et al., 1988a,b) and that the ability to turn a default response into an accurately matched response is accomplished by mechanisms collectively called 'response specification' which rely on appropriate sensory informational update. These compensatory modifications appear to occur during response execution (Ghez and Gordon, 1987; Hening et al., 1988a,b).

Effects of takeoff surface compliance on landing adjustments. In another set of experiments, both landing and takeoff surfaces were altered so that when landing on a stiff or compliant surface, the takeoff surface was either stiff (S) or compliant (C). Thus, subjects experienced takeoff and landing surfaces that were either similar (C-C, S-S) or different (S-C, C-S). In this paradigm, when ankle adjustments at landing were studied in well practised subjects, it was observed that the amount of plantarflexion at touchdown was dependent not only on the landing surface, but on the takeoff surface as well. When ankle position at landing is expressed as a percent of the plantarflexed position for the S-S condition the following values were observed for the three other conditions: S-C, 92%; C-S, 90%; C-C, 85%. Thus, while the ankle is indeed less plantarflexed when landing on a compliant surface, the diminution is even greater when taking off from a compliant surface (compare S-C to C-C). More notable is the ankle position at landing under the C-S condition when compared to the other conditions. In the C-S condition, the amount of plantarflexion at landing is similar to the S-C condition rather than to the S-S condition. Thus, it appears that the takeoff surface may have an effect on planning adjustments for landing surface. This carry-over effect of the last surface experienced has also been reported in the planning of grasp for objects of varying roughness (Westling and Johannsen, 1984).

The difference between the 'default' program seen in unskilled subjects and the adjustments for landing surfaces seen in skilled subjects suggests that skilled subjects may be able to complete specifications for the landing program during movement execution while unskilled subjects do not have this ability. However, it is not clear when, during the jump, specifications for the landing component are completed. As adjustments for landing surface are influenced by the takeoff surface it may be that specifications are completed during the execution of the takeoff phase.

CONCLUSIONS

In summary, it is suggested that the takeoff and landing components of the jumpdown task are independent programs that are assembled to meet varying requirements of the environment such as surface compliance or height. It is further suggested that the landing component is incompletely specified prior to movement initiation, but may be completed during the takeoff phase of the jump.

This work has been funded in part by an NSERC grant (Canada) and in part by the Fondazione Don Gnocchi pro Juventute (Italy).

REFERENCES

Bastien M., and P.A. McKinley (1988) Effect of compliance on ankle, knee and hip motion in landing from a jump in humans. Soc Neurosci Absts 14:66

Ghez, C., and Gordon, J (1987) Trajectory control in targeted force impulses. I. Roles of opposing muscles. Exp Brain Res 67:225-240

Hening, W., D. Vicario, and C. Ghez (1988a) Trajectory control in targeted force impulses IV. Influences of choice, prior experience and urgency. Exp Brain Res 71:103-115

Hening, W., M. Favilla, and C. Ghez (1988b) Trajectory control in targeted force impulses V. Gradual specification of response amplitude. Exp Brain Res 71: 116-128

McKinley, P.A. and A. Pedotti (1990) Landing from a jump-down in humans. I. Role of skill in adjusting to landing surfaces of differing compliances. Exp Brain Res (in review)

Pelland, L., and P.A. McKinley (1988) Age dependent differences in programming landing from a jump in humans. Soc Neurosci Absts 14:65

Pelland, L., P.A. McKinley and A.Beuter (1990) Age dependent visual control of landing from a jump. in *Disorders of Posture and Gait.* (Eds T Brandt, W Paulus, W. Bles, M. Dieterich, S. Krafczyk, A. Straube), Georg Thieme Verlag, New York. pp 249-252

Thompson, H.W. and P.A. McKinley (1988) Effect of visual perturbations in programming landing from a jump in humans. Soc Neurosci Absts 14:66

Westling, G., and R.S. Johannsen (1984) Factors influencing the force control during precision grip. Exp Brain Res 53:277

SECTION 5

MODELLING APPROACHES TO MOVEMENT CONTROL

ANALOGIC AND SYMBOLIC ASPECTS IN DISTRIBUTED MOTOR CONTROL

PIETRO MORASSO, VITTORIO SANGUINETI, and MASSIMO SOLARI
Department of Computer Science
University of Genoa
Via Opera Pia 11A
I-16145 Genoa
Italy

ABSTRACT. The basic idea of distributed processing is that any given function is performed by a large ensemble of processing elements and, at the same time, any processing element contributes to many different functions. In general, this has the advantage of robustness, because no processing element is crucial, although a distributed system is more difficult to study experimentally and to understand. In the case of motor control, distributed architectures represent efficiently and flexibly constraints and mappings that characterize the co-ordination of multiple degrees of freedom, i.e. synergy formation. This paper reviews some of the trends in the field by comparing four types of neural models which are representative of many of the relevant issues: multi-layer networks for inverse motor mappings (Jordan, 1988, 1989), VITE (Bullock and Grossber (1988, 1989), and two types of composite networks (Kawato et al, 1989, Morasso et al, 1989, 1990).

1. Introduction

As motor control is a difficult problem, investigators tend to be influenced by metaphors which hopefully can help their understanding. Terms like motor program, computational model, etc. clearly allude to the computer metaphor and the computer, for a few decades, has designated a conceptual framework which is characterized by sequentiality, symbolic and localized representations, dualism of software and hardware, etc. Now things are changing, in cognitive science and neuroscience as well as in artificial intelligence; people begin to realize that the established computational paradigm is not the only possible one and the good old-fashioned cybernetic concept of analogic computation is re-gaining its former respectability. This means, among other things, parallelism, distributed representations, unification of software and hardware in the notion of netware, systems dynamics and adaptive control, i.e. learning.

In the motor control area, in particular, this conceptual shift is a good motivation to re-evaluate the role of biomechanical constraints that can be seen as allies, not enemies, of the motor controller, a sort of mechanical analogic machinery to be exploited by a related neural analogic machinery in the brain. At the same time, the return to analogic computation is pushing backward the perceived application domain of the symbolic

J. Requin and G. E. Stelmach (eds.), Tutorials in Motor Neuroscience, 233–252.

approach, contrasting the tendency of AI advocates to express everything in terms of rules and symbolic problem-solving. Nonetheless, the symbolic to analogic interface is a crucial one that still needs a substantial contribution from the scientific community.

The purpose of this paper is to review some of the trends, with particular attention to the problem of co-ordination of multiple degrees of freedom, i.e. synergy formation. Four types of neural models are analysed, which are representative of many of the relevant issues: multi-layer networks for inverse motor mappings (Jordan, 1988, 1989), VITE (Bullock and Grossber (1988, 1989), and two types of composite networks (Kawato et al, 1989, Morasso et al, 1989, 1990).

2. Motor Spaces

If we consider motor control problems that involve a large number of degrees of freedom, we must realize that different motor patterns co-vary in distinct descriptive frameworks, the we way call *motor spaces*: Muscle space, Joint space, and Task space. In each space, we can single out two main types of representation: patterns of *generalized motions* and patterns of *generalized forces* (figure 1). For example, generalized motions are angular rotation patterns in the joint space and muscular shortening/lengthening patterns in the muscles space; the task patterns, on the other hand, depend on the choice of an *end-effector* and the specification of its motion in 3D space.

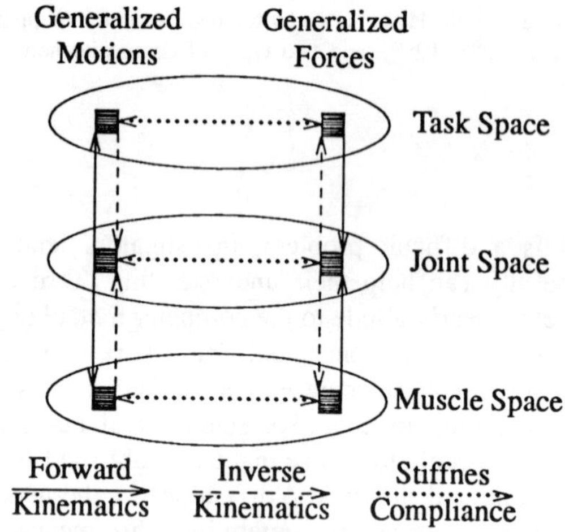

Figure 1. Motor Spaces. Continuous lines: forward kinematic mappings; dashed lines: inverse kinematic mappings; dotted lines: stiffness/compliance.

These patterns must satisfy kinematic constraints that correspond to the musculo-skeletric structure of the body and that determine well-posed mappings (forward kinematics) as well as ill-posed mappings (inverse kinematics). The latter, in particular, are due to the *redundancy* of the motor system, as regards task vs joints and joints vs muscles. Moreover, it is important to understand that if the controller focuses on motion representations it loses control of forces and vice versa. Keeping the middle way, between motion and force representations, means to focus on the stiffness/compliance of muscles that cross-connect the two kinds of representations.

Direct and inverse kinematic mappings are important modules for the theoretical study of motor controller architectures and in the following sections we shall examine some significant examples of neural models. However, it is important to realize that the same mappings are actually implemented by the musculo-skeletal system as real analogic computations: During active movements, input patterns in the muscle space are transformed into output patterns in the task space and during passive movements the opposite mappings are actually computed, in vivo. In other words, the musculo-skeletal system can be viewed by itself as a *kinematic computational engine* and we shall see in the following how it is possible to exploit this concept.

In summary, the consideration of motor spaces brings to our attention the fact that the motor controller must posses a *kinematic competence* (the ability to relate different motor spaces) as well as a *muscular competence* (the ability to relate position and force). To this we must add a *timing competence*, i.e. the capacity to constrain the motor mappings in the same time framework.

3. Multi-Layer Network Models

Let us summarize the equations that characterize a multi-layer network (MLN, figure 2):

- Each layer L_i is an array of units and it has an input vector $\mathbf{I}_i$, an output vector $\mathbf{O}_i$, and a bias vector $\mathbf{B}_i$.
- There is no connection among units of the same layer and the input-output transformation is a smooth sigmoidal non-linearity, applied separately to all vector components:
  $\mathbf{O}_i = f(\mathbf{I}_i + \mathbf{B}_i)$.
- The transformation from the output vector $\mathbf{O}_i$ of one layer and to the input vector of the next layer $\mathbf{I}_{i+1}$ is a linear combination with the weight matrix $\mathbf{W}_i$:
  $\mathbf{I}_{i+1} = \mathbf{W}_i\,\mathbf{O}_i$
- For a given topology and a given set of weight matrices and bias vectors, a MLN implements an input-output mapping that is obtained by means of a forward propagation of an input pattern, through the intermediate (hidden) layers, to an output pattern: $\mathbf{O} = MLN(\mathbf{I})$.

A MLN can be trained by a supervisor on the basis of a teaching set $((\mathbf{I}^*, \mathbf{O}^*), * = a, b, \ldots)$ and a learning technique, known as back propagation (LeCun, 1985, Rumelhart et al, 1986), that consists of three logical steps, repeated a large number of times:

[1] Forward propagation of the input pattern $\mathbf{I}^*$ into an output pattern $\hat{\mathbf{O}}$.

[2] Computation of the output error vector $\delta^* = \mathbf{O}^* - \hat{\mathbf{O}}$
and back propagation of the error to the input of the different layers

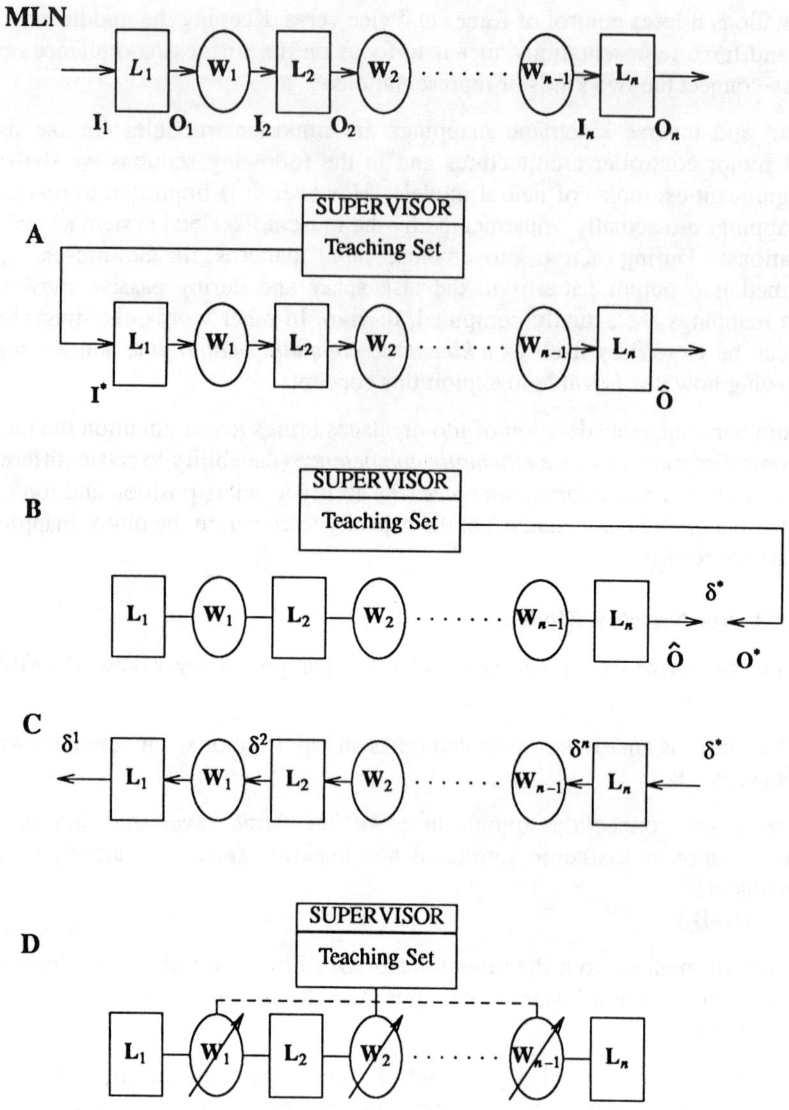

Figure 2. Multi-Layer Network (MLN). A: Forward propagation of teaching input. B: Computation of output error. C: Back-propagation of output error. D: Learning (adaptation of weights). $L_1, L_2, ..., L_n$ are the subsequent layers of units and $W_1, W_2, ..., W_{n-1}$ are the corresponding inter-layer connection weights.

$(\delta^* \rightarrow \delta^n \rightarrow \delta^{n-1} .... \rightarrow \delta^1)$.

[3] Adaptation of the weights, with the purpose of minimizing the output error via gradient descent; this can be carried out with a local computation that only requires the output vector from the previous layer and the error vector at the current layer.

It is also possible to induce the network to learn other constraints in addition to the requirement of reproducing as well as possible the associations of the training set: it is indeed sufficient to add to the output error vector δ^* a term $\gamma\mathbf{C}$, where **C** is the additional criterion and γ is a scalar coefficient that is decreased smoothly to 0 during the training.

3.1 MLN's for inverse kinematics

MLN's have been used for learning direct/inverse mappings between motor spaces. For example, learning direct kinematics is a rather straightforward application where the teaching input is a set of angular patterns and the teaching outputs are the corresponding position/orientation vectors of the end-effector. More interesting is learning an inverse mapping in the case of motor redundancy, because infinite output vectors can be associated with the same input pattern (Eckmiller, 1990). In this case, indeed, the supervisor has the big problem of providing a consistent training set. A technique which has been proposed (Jordan, 1988, 1989, Nguyen and Widrow, 1990) is to use a forward model as a *distant teacher* (figure 3). This technique uses the combination of two MLN's, a forward model (DK) and an inverse model (IK):

- DK is initially trained in order to learn the direct kinematic mapping from joint angular rotations Θ to end-effector motions **X**.
- IK is then linked with DK, with the purpose of learning the global chain of transformations from desired end-effector motion $\mathbf{X}^*$ to Θ and then to the predicted end-effector motion $\hat{\mathbf{X}}$.

The teaching of IK is performed as follows:

[1] $\mathbf{X}^*$ is propagated through the entire network and the output error vector $\delta_x = \mathbf{X}^* - \hat{\mathbf{X}}$ is computed.

[2] This vector is back-propagated through DK without changing the weights, with the only purpose of estimating the error vector δ_θ at the joints, (the error at the output layer of the IK) which is then combined with an additional criterion, as mentioned above, yielding a composite error.

[3] The composite error is back-propagated through IK with adaptation of weights.

After training, the network can perform inverse kinematics in an optimal way, i.e. it is able to co-ordinate patterns of joint rotations that match the desired end-point positions and, at the same time, minimize a desired criterion such as smoothness.

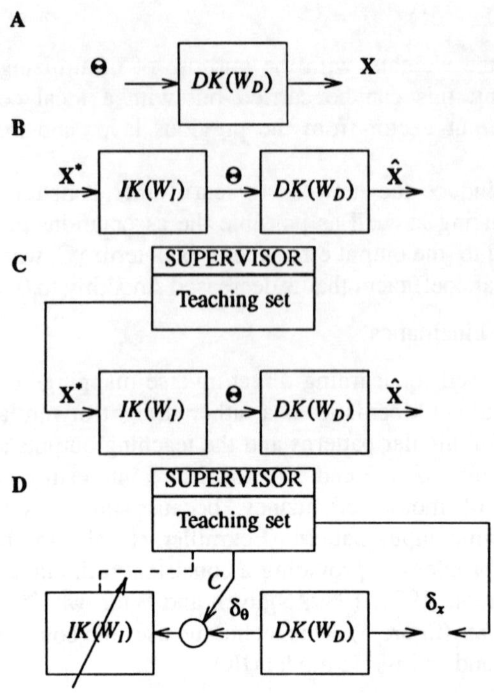

Figure 3. Learning Inverse Kinematics via a Forward Model. A: Trained Direct Kinematics MLN (DK, with weight matrix W_D). B: Linking DK with IK (Inverse Kinematics MLN with weight matrix W_I). C: Forward propagation of the desired motion ($\hat{X}$: desired end-point motion, X: predicted end-point motion, Θ: joint motion). D: Back-propagation of the output error, formation of the composite error and learning (δ_x: end-point error, δ_θ: joint error, C: optimization criterion).

3.2 Recurrent MLN's

Standard MLN's can learn non-linear mappings but are unable to generate time varying motor patterns, for example a trajectory as a response to a target stimulus. However, it is possible to overcome this limitation by means of recurrent MLN's, originally proposed by Jordan (1988, 1989). In the simplest case (shown in figure 4) the input layer, which is meant to represent the plan or target of the movement, is augmented with another set of units that are fed recurrently from the output layer. For example, Massone & Bizzi (1989) implemented a recurrent MLN that can generate patterns of muscle activity for the arm muscles during reaching movements. The teaching set, in this case, consisted of a set of associated pairs: an input stimulus in a 2D domain and the corresponding complete trajectoy.

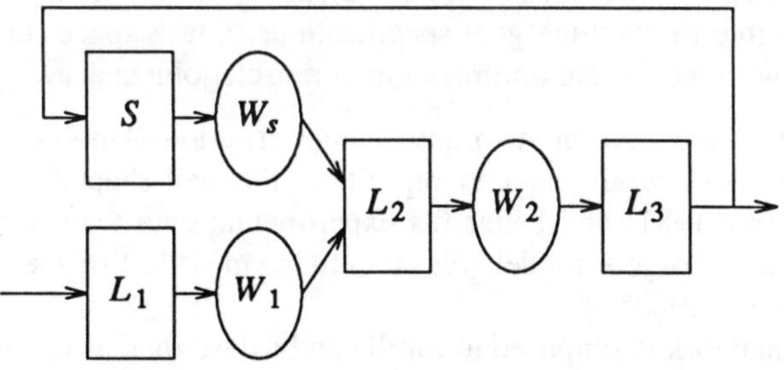

Figure 4. Recurrent multi-layer network. The network consists of a layer of input units (L_1), hidden units (L_2), output units (L_3), and state units (S). W_s, W_1, and W_2 are inter-layer connection weights.

3.3 Pros and Cons of MLN's

MLN's have been been very successful because they are simple mechanisms for representing a wide variety of problems of association, independently of the application domain. However, the price of this generality is that the knowledge that MLN's are able to encapsulate is totally dependent on the teaching set and therefore their generalization capability is intra-task but not inter-task. For example, direct and inverse kinematic mappings between task space and joint space are task-dependent, in the sense that the selection of the end-point is arbitrary and the kinematic mapping learned with one configuration is not easily generalizable to a different one. Therefore, the motor planner would need a very large number of MLN's and teach all of them separately. As regards timing structure, recurrent MLN's are very good at learning them. The point, however, is that they are too good because, again, the learned structure is totally embedded in the training data.

Summing up, although MLN's are not credible models for representing the mappings between motor spaces, they are still good candidates at a more abstract level, for representing the general motor ability of learning specific motor skills in a very focused way.

4. Cascaded Neural Networks with Minimum Torque-Change Criterion

A cascaded neural network with minimum torque-change criterion (CN-MT) is a composite network architecture (Kawato et al., 1990) that combines the paradigm of association (a set of MLN's for representing direct kinematics/dynamics) with the paradigm of relaxation, for representing motor optimization. This network is the result of three main motivations:

- A global view of ill-posed motor problems that implies the requirement for the neural network architecture to represent all of them at the same time.

- An integrated approach to motor spaces, i.e. the concept of designing an architecture in which it is possible to combine goal specifications in task space (target point, via points, obstacles) with movement optimization in muscle/joint space.
- The identification of the minimum torque-change criterion (Uno et al., 1989) as a global *smoothness constraint* that can reproduce the bell-shaped profiles of arm trajectories (Morasso, 1981) and better fits experimental data than purely kinematic criteria like the minimum jerk model (Flash and Hogan, 1985) or the 2/3 power law (Viviani and Terzuolo, 1982).

The function that the network is supposed to fulfill can be described as follows:

[1] For a pre-defined duration T_f of the movement, an outside planner provides, as input to the network, the target (i.e. the end-point position to be reached at the final time) as well as intermediate via points and obstacles, if necessary.

[2] A pattern of torques is produced (for all the joints and the total time span) that is compatible with the task-constraints and, at the same time, minimizes the integral of the squared torque derivatives, computed for the total movement duration and summed over all the joints.

In order to implement the stated function, the model needs to be unfolded in time, i.e. it is necessary to setup a spatial representation of time were system variables and system transformations are replicated for each time instant. The inner skeleton of the network (figure 5, top) is a cascaded network where FD and FK are two MLN's that were previously trained to implement forward dynamic and forward kinematic transformations, respectively. FD, in particular, receives in input the current torque vector and the current state vector (joint angles and joint rotation speeds) and produces in output the next state vector which is the input both to FK, for predicting the end-point movement, and to another copy of FD, in order to propagate the computation to the next time step.

All together, the network maps the global spatio-temporal torque pattern into the predicted trajectory of the end-effector. In actual operation, when the spatio-temporal torque pattern must be produced, the torque signals are generated by a relaxation network which has one unit for each element of the torque vector (figure 5, bottom). The torque units are cross-connected by means of symmetric synapses of constant weight that link each unit (identified by a joint and a time instant) with the two units that correspond to the same joint, one time step before and one step after. These cross-connections induce a relaxation dynamics, (Hopfield, 1982, Cohen and Grossberg, 1983, Hopfield, 1984), that automatically seeks maximally smooth patterns.

The optimal torque patterns are constrained by the task errors that are back propagated through the cascaded MLN's, similarly to the forward model technique described in the previous section. In other words, the errors detected at some of the units of the output layer are back propagated in order to estimate the corresponding torque input errors and these errors, in turn, are spread throughout the torque units via relaxation. The resulting torque variations are then propagated forward for updating the samples of the end-point trajectory and the three processes (forward and back propagation as well as

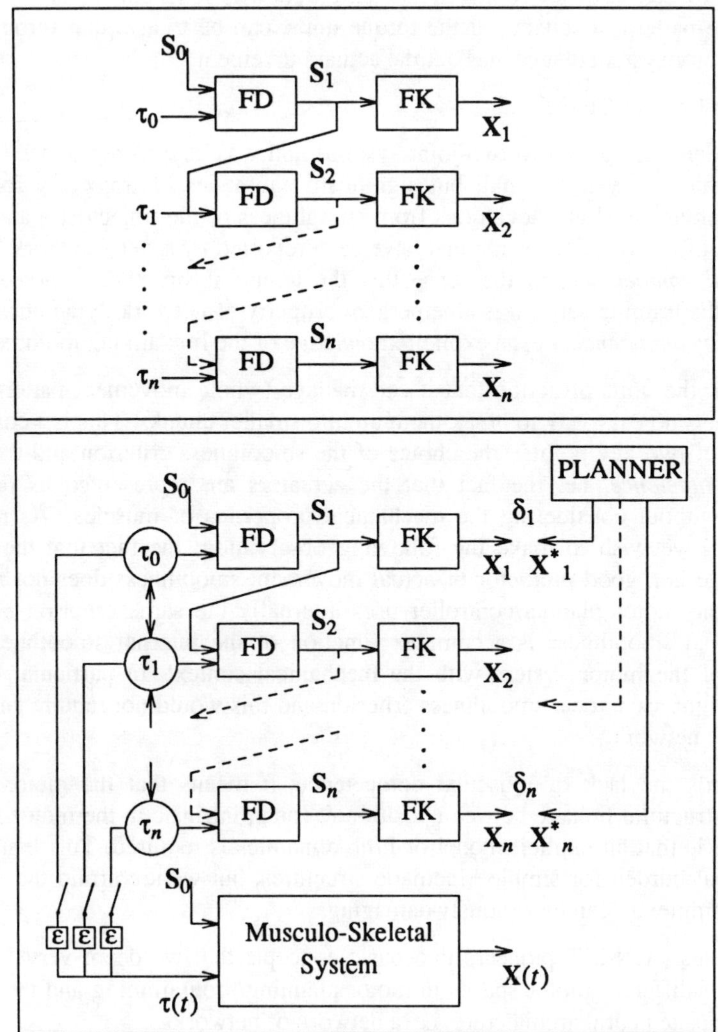

Figure 5. Cascaded neural network with minimum torque-change criterion. Top box: Structure of the network (FD: forward dynamics MLN; FK: forward kinematics MLN; τ_i: joint torque vector at time-step i; $\mathbf{S}_i$: state vector; $\mathbf{X}_i$: position of the end-effector). Bottom box: torque optimization via relaxation of the torque pattern (ε: delay elements).

relaxation) proceed until an equilibrium configuration is reached: in that situation, the whole torque pattern, localized on the torque units, can be transmitted through a delay line to the motor system that carries out the actual movement.

4.1 Pros and Cons of CN-MT

This model was applied to two-joint systems and was able to reproduce bell-shaped velocity profiles as well as small but significant anomalies of trajectory formation in reaching, such as the slight deviations from straightness of the trajectories and the small asymmetries of the velocity profile that have been reported by several authors. The model has *temporal competence*, in the sense that the temporal structure is not exogeneous (implicit in the training set) but is an emergent property of network dynamics, and it also has *kinematic competence*, i.e. an explicit knowledge of the link among motor spaces.

However, the units of action that it can manage (whole movement patterns) are too big and there is no easy way to break them up into smaller chunks. This is a consequence, we believe, of two key points: the choice of the smoothness criterion and the lack of a *muscular competence*, i.e. the fact that the actuators are represented as pure torque generators, without considering the mechanical properties of muscles. As regards the former point, we wish to make the following observation: the fact that the minimum torque-change is a good predictor of actual movement smoothness does not necessarily imply that the motor planner/controller uses internally the same criterion because the global external smoothness is a complex function of the internal smoothness and the interaction of the motor system with the mechanical context. In particular, the motor controller might use a local smoothness criterion and this would not require an unfolding in time of the network.

As regards the lack of muscular competence, it means that the motor controller ignores the structural linkage between position & force, intrinsic in the motor spaces; the consequence is that an explicit model of limb dynamics is required. This is not a heavy computational burden for simple kinematic structures, but we are afraid that scaling up kinematic complexity can be seriously damaging.

Nevertheless, CN-MT pioneers two basic concepts that we deem very fruitful: the integration of different motor spaces in motor planning/programming and the attempt to build a composite neural architecture, i.e. a network of networks.

5. VITE Model

The VITE model (Vector Integration to Endpoint) is motivated by its proponents (Bullock and Grossberg, 1988, 1989) on the basis of the concept that during the development of skills there appears to be a gradual strategic shift from low-compliance/low-speed to high-compliance/high-speed. Therefore, it is important that the neuro-motor system is capable to factor out trajectory and speed, on one hand, and posture and stiffness, on the other. We shall focus on the former factorization because it is closer to the topic of this article. The basic idea is to use a multiplicative gating signal that controls the rate of change of all the elements of the motor array and is instrumental

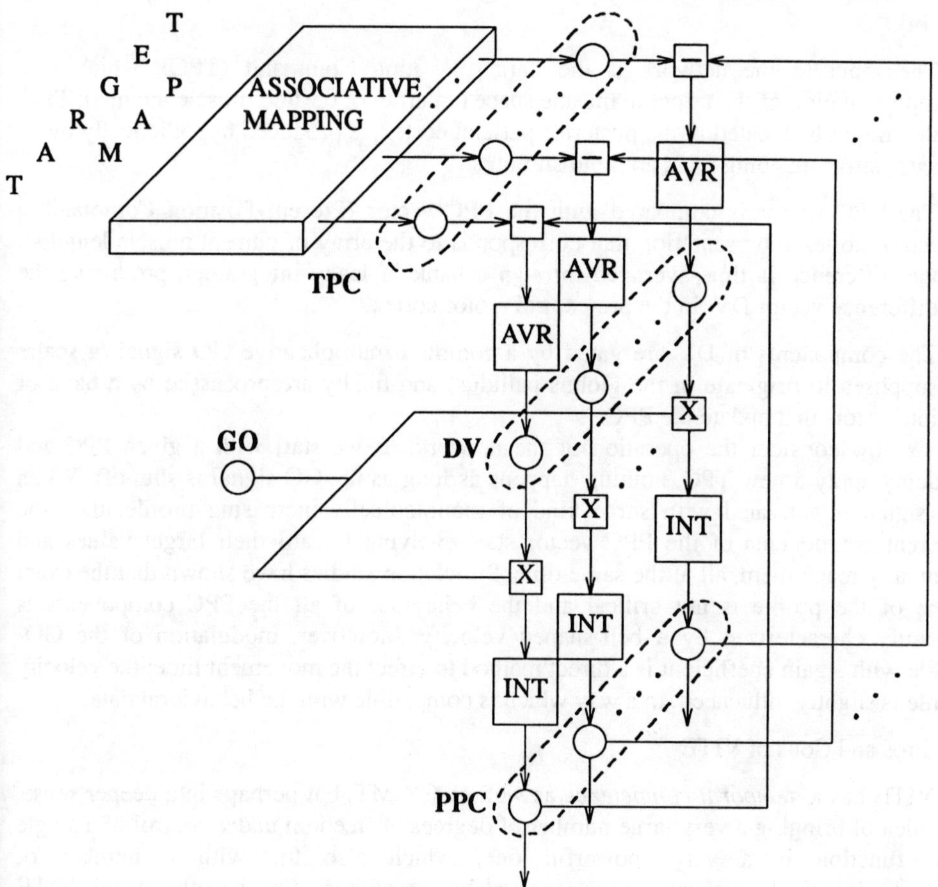

Figure 6. VITE model. An associative mapping transforms a target map into the TPC (Target Position Command) which is compared with PPC (Present Position Command) and the time averaged difference (AVR: bank of leaky integrators) produces DV (Difference Vector). This is gated by GO and integrated over time (INT: bank of integrators), thereby updating the PPC.

for factoring out trajectory and speed. Figure 6 shows a schematic diagram of the network:

- The input to the network is the Target Position Command (TPC), which is a representation of the target in muscle space (an array of desired muscle lengths). TPC, that might be located in the posterior parietal cortex, is produced hypothetically by an associative mapping from an input stimulus.
- The TPC vector is compared with the PPC vector (Present Position Command, a motor cortex representation that corresponds to the array of current muscle lengths); the difference is time-averaged through a bank of leaky integrators, producing the difference vector DV, in the pre-central motor cortex.
- The components of DV are gated by a common multiplicative GO signal (a scalar supposed to originate in the globus pallidus) and finally are processed by a bank of integrators that update the PPC.

Let us now consider the operation of the network. If we start with a given PPC and suddenly apply a new TPC, nothing happens as long as the GO signal is shut off. When this signal is activated with some kind of monotonically increasing profile, then the different components of the PPC vector start evolving toward their target values and eventually reach them, all at the same time. Simulation studies have shown that the exact shape of the profile is not critical and the behaviour of all the PPC components is generally characterized by a bell-shaped velocity. Moreover, modulation of the GO-profile with a gain coefficient is a direct method to affect the movement time; the velocity profile is slightly influenced, in a way which is compatible with the behavioral data.

5.1 Pros and Cons of VITE

VITE has a *temporal competence* as well as CN-MT, but perhaps in a deeper sense. The idea of bringing a very large number of degrees of freedom under control of a single time function is a very powerful one, which also fits with a number of neurophysiological observations, as reported by the authors. On the other hand, VITE does not have a *muscular competence*, because a positional code is assumed for the central motor commands; this is a dual assumption with respect to Kawato, but in both cases the structural linkage between position & force is ignored.

Perhaps more questionable is the choice of the muscle space for the VITE mechanism. In fact, this neural circuit can be considered as a sort of vectorial interpolator which is supposed to operate in muscle space. The point, however, is that muscle lengths are not independent variables and thence, even if PPC (at t=0) and TPC were valid representations, the intervening arrays might be kinematically wrong. Moreover, it is well known that interpolation in joint space generally produces implausible end-point trajectories (Morasso, 1981).

However, the attractiveness of the non specific, scalable, speed control signal (GO) can well survive the problems of muscle space interpolation. One might indeed think of having GO operate in task space rather than in muscle space. In this case, PPC can be thought of as the end-effector, to be compared with the task-defined target: the output of the GO-gating module would then have the meaning of a *moving target* in task space which is the input to some kind of motor programming network (This network carries out the function of the bank of integrators of the VITE model.) A significant advantage of bringing VITE to the task space is that multiple VITE-circuits can be instantiated at the same time, in accordance with the fact that, in many motor skills, multiple end-effectors are concurrently active. In other words, this implies the solution of another, more abstract, factorization problem: the factorization of task dynamics (motor planning) and neuromotor dynamics (motor programming) inside the same global synergy.

6. M-Nets/P-nets: a muscle-oriented synergy formation model

M-Nets (Motor relaxation Networks) attempt to bring *muscular competence* to the synergy formation process, in accordance with the concept that the elastic properties of muscles do not represent only a significant low-level feature of the motor system, but can also provide an organizing principle for the global computational architecture (Mussa Ivaldi et al, 1988, Morasso et al, 1989, 1990, Morasso, 1990).

The main concept is that muscle elasticity allows one to define a global elastic potential function, tuned by the patterns of neuromuscular activity, whose equilibrium configurations determine posture; at the same time, these properties facilitate a unified treatment of posture and movement because a centrally induced modification of the shape of the potential field in the direction of the intended movement is a simple and yet powerful mechanism of trajectory formation. The origin of this concept can be traced back to the pioneering work of Astratyan and Feldman (Astratyan and Feldman, 1965, Feldman, 1966) and to the following development of so called *equilibrium-point models* (Bizzi et al., 1976, Bizzi et al., 1984, Hogan, 1984, Mussa Ivaldi et al., 1985, Feldman, 1986). However, these studies did not provide a computational model of the neural mechanisms that underly the selection and the continuous modification of equilibrium points, compatibly with task constraints. In order to obtain this goal, a principle was formulated (Mussa Ivaldi et al, 1988), called *Passive Motion Paradigm* (PMP), according to which the continuous modification of equilibrium points is produced via a central neural mechanism that simulates *passive movements*, i.e. movements driven by *virtual force impulses* that correspond to the intended movement direction.
The rationale of PMP is based on the following points:

- Passive motion occurs quite naturally in the interaction between the musculo-skeletal system and the environment;
- Passive motion has the remarkable property of selecting patterns of joint-rotation and muscle lengthening/shortening that solve the inverse kinematic problem, independently of the degree of redundancy and the kinematic degeneracy which occurs at the limits of the workspace;
- A central neural mechanism that is capable to simulate passive motion is the most natural way to incorporate musculo-skeletal competence in the central neural controller;

- This mechanism can be used directly for synergy formation in the sense that the muscular and articular variations that result from the simulation are an effective representation of the pattern of motor commands.

M-Nets are parallel and distributed architectures that have been proposed as an implementation of the simulation mechanism discussed above (Morasso et al, 1989, Morasso, 1990, Morasso and Sanguineti, 1990). It is an analogic device in several senses: (i) it operates as a dynamic system (an analogic computer) and not as a symbolic system; (ii) it has a structural similarity with the neuromuscular system, i.e. it is somatotopically organized; (iii) its dynamics is a *neural relaxation* that strictly mirrors the corresponding mechanical relaxation of passive movements. M-Nets are driven by a computational energy analogous to the elastic potential energy of the muscles. The model is constructed as a network of units which correspond to the different constituent parts of the musculo-skeletal system: S-units (skeletal segments), M-units (mono- and poly-articular muscles), and L-units (mono- and poly-articular ligaments).

These units are defined by the local computations that they perform: M-units and S-units behave as impedances, i.e. they receive positional information and react feeding back force information; S-units, on the contrary, behave as admittances, i.e. they receive force information and react, modifying positional parameters. S-units model the different skeletal body segments, considered as rigid bodies to which complex sets of forces are applied: *internal forces*, applied by M&L units for taking into account the biomechanical constraints, and *external forces*, applied as sequences of impulses from the outside of the M-net for expressing intended directions of movement. The activation function of S-units consists of the simulation of the passive motion of the body segment, thereby changing the current position vectors of the insertion points of all the forces impinging on the unit. M-units and L-units model the elastic elements of the system as cables passing through pulleys and fixed at their extremities onto two different bodies. The functions that they compute reflect the length-tension curves that characterize the corresponding real devices. These functions are obviously invariant for L-units but tunable for M-units.

With these units it is possible to build networks that can relax in a similar way to the underlying musculo-skeletal system. In this sense, an M-Net is a computational *body schema* that implements the pattern generator required by the equilibrium-point hypothesis. The inputs to an M-Net are sequences of virtual force impulses that express the intended direction of motion: we may call them *synergy vectors.* These vectors displace the equilibrium state of the network and the ensuing dynamics provides two streams of output data at the same time: a stream of *muscle activations* and a stream of *kinematic expectations.* From the network modelling point of view, M-Nets are dynamic systems with a well defined Liapunov function and therefore are similar to the content addressable memories (CAM) initially investigated by Hopfield (Hopfield, 1982, 1984) and by Cohen and Grossberg (1983):

- The role of neurons in Hopfield nets is played by S-units; the difference is that inputs and outputs in M-Nets are vectors, not scalars, and the activation function is more complex.

- The role of connections (which are linear functions in Hopfield nets) is played by M-units and L-units, which are non linear functions.
- The role of input signals is played by the synergy vectors.

In spite of significant differences, in both cases we have a similar relaxation behaviour which is driven by a potential energy function: in the Hopfield continuos model, this function is the total electrostatic energy stored in the membrane capacitances; in the M-Net model, it is the total mechanical elastic energy, stored in M-units and L-units. However, the purpose of the relaxation is quite different:

- A CAM can store multiple patterns as equilibrium configurations (points of minimum in the energy landscape) and the purpose of the relaxation is to recover a pattern from an initial representation which may be partial and/or corrupted.
- In an M-Net, there is only one significant equilibrium configuration (the current posture) and this is changed during the simulation process in accordance with the passive motion principle.

Therefore, the pseudo connection-weights represented by M-units must be adaptively changed during relaxation and this can be obtained by interleaving a *passive phase* and an *active phase*. In the passive phase, an M-Net reacts to the application of a synergy vector by relaxing to a new equilibrium state in an iso-electric way, i.e. without changing the control variables of M-units; at equilibrium, these variables are changed in order to shift the point of minimum of the energy landscape onto the current configuration: this is the active phase. In other words, the neural relaxation process operates in such a way that the point of minimum in the potential field *tracks* the sequence of equilibrium configurations determined by the sequence of synergy vectors. Symmetrically, the mechanical relaxation induced by the neural relaxation evolves the other way around: the potential field *leads* the current posture and attracts it.

The units in an M-Net must be conceived as cell assemblies (e.g. cortical columns) not as single neurons and we might represent them as small MLN's replicated in a great number of exemplars: their function is to compute the linear and non linear vector operations described above and this can be learned through training.

In the proposed synergy formation architecture (Morasso, 1990, Morasso et al, 1990), an M-Net is the task-independent part that incorporates the basic biomechanical constraints, but it has no competence as regards goal selection and timing. This requires another neural mechanism which is the second half of the synergy formation system and logically operates at a more abstract level: a planning network (P-Net). The nature of M-Nets allows P-Nets to operate on representations that are directly linked to the task, in accordance with the behavior of some populations of cortical neurons in reaching experiments (Georgopulos et al, 1986,Georgopulos, 1988, Caminiti et al, 1990). The P-net is supposed to select one or more end-effectors and then, for each of them, it generates a sequence of synergy vectors which are transmitted to the M-net. The generation function can be performed by means of different types of neural mechanisms, for example a version of the VITE circuit applicable to task-level representations, that operates via integration, or a relaxation circuit, similar to M-Nets (Morasso and

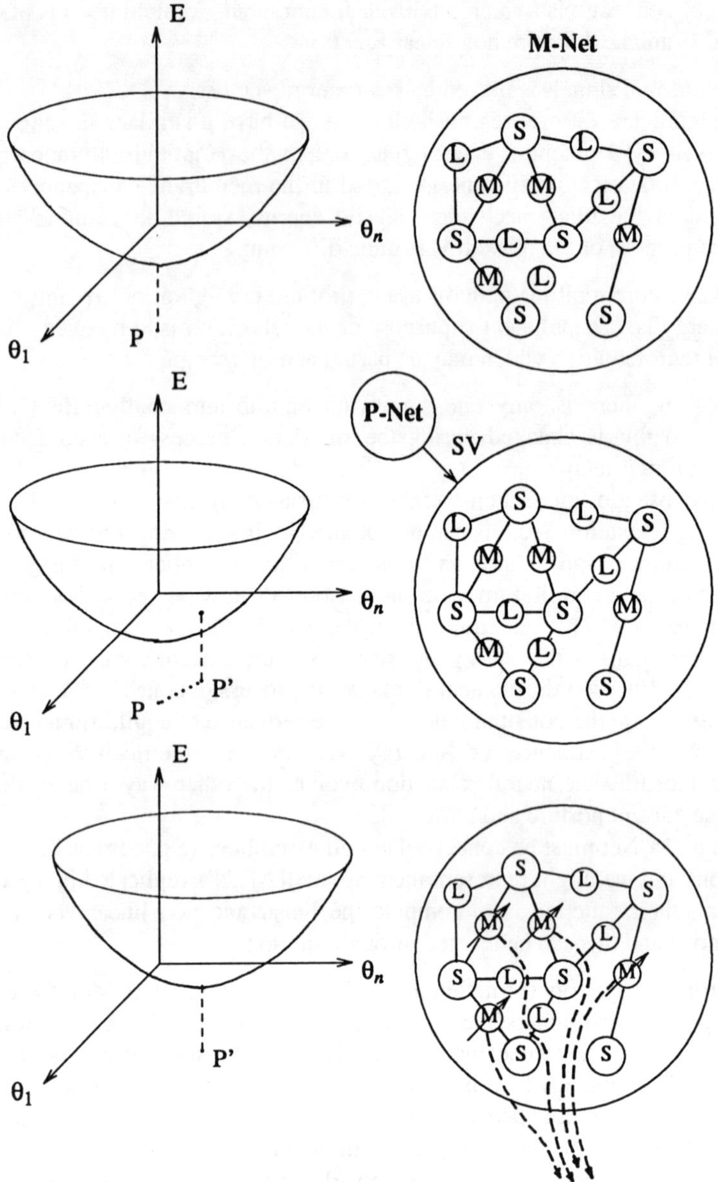

Figure 7. M-Net/P-Net Model. Top part: The M-Net is initially in equilibrium (point P in the potential energy landscape E). Middle part: The P-Net fires a synergy vector (SV) that displaces the equilibrium point of the M-Net from P to P'. Bottom part: The M-units of the M-net are updated in such a way to re-gain equilibrium (the updating pattern is the outflow synergy).

Sanguineti, 1990). Summing up, the synergy formation architecture is composed as follows (figure 7):

- A planning system that has an analogic and a symbolic component: the former one consists of a set of neural networks that behave as synergy vector generators and the latter is a set of rules for setting up the generators and connecting them with the underlying M-Net. The symbolic component has the semantic competence, whereas the analogic components has the timing and targeting competence.
- A motor programming system, an M-Net, which is a somatotopically organized relaxation network of local MLN's. It has the biomechanical competence: it is able to absorb the different sequences of virtual force impulses produced by the planning system and to integrate them in the overall motor pattern.

6.1 Pros and Cons of the P-Net/M-net model

The P-net/M-net model is an attempt to integrate in a composite architecture the different types of motor competence that we discussed above. There are points of contact with CN-MT as regards the general framework and the relaxation dynamics and with VITE as regards the mechanism of timing, whereas the stress on muscular competence is the main element of difference with both models. The model still needs significant developments in many points, such as the interaction among different synergy vector generators in the P-net and the training paradigm of the local MLN's in the M-net.

Finally, we wish to remark that the unfocused nature of M-nets and their relaxation dynamics make them general but slow mechanisms of synergy formation that are not ideal for perfectioning specific motor skills. It is then reasonable to hypothesize some kind of supra-structure that exploits the P-Net/M-net model as a bootstrap of motor learning and later relinquishes control to some form of MLN for specific skills.

7. Conclusions

In the previous sections, we showed that analogic mechanisms, arranged in a variety of architectures, can carry out a great part of the computational burden of the motor planner/programmer, particularly as regards synergy formation. There is still work to be done in the integration of the different modelling frameworks because a single type of network is very unlikely to satisfy all the requirements.

The details of the symbolic to analogic interface need to be investigated much more deeply, but the general framework can already be understood because the availability of quick and robust mechanisms for expanding in space and time very concise task specifications greatly compresses the information exchange and simplifies the complexity of symbolic computation.

Acknowledgments. This work was supported by the Esprit Projects FIRST and ROBIS, by a National Programme on Robotics of the Italian Research Council, and by a National Programme on Bioengineering of the Italian Ministry of University&Research.

References

- Astratyan, D.G., and A.G. Feldman (1966). Functional tuning of the nervous system with control of movement or maintenance of a steady posture. I. Mechanographic analysis of the work of the joint on execution of a postural task. Biophysics, 10, 925-935.
- Bizzi, E., Accornero, N., Chapple, W., and N. Hogan (1984). Posture control and trajectory formation during arm movements. J. Neuroscience, 4, 2738-2744.
- Bullock D. and S. Grossberg (1988) Neural dynamics of planned arm movements: emergent invariants and speed-accuracy properties during trajectory formation. Psychological Review 95:49-90.
- Bullock D. and S. Grossberg (1989) VITE and FLETE: Neural modules for trajectory formation and postural control. In "Volitional Action" (W.A. Hershberger, Editor), North-Holland/Elsevier:Amsterdam, 253-297.
- Caminiti, R., Johnson, P.B., and A. Urbano (1990) Making arm movements within different parts of space. Dynamic aspects in the primate motor cortex. Journal of Neuroscience.
- Cohen, M.A., and S. Grossberg (1983) Absolute stability of global pattern formation and parallel memory storage by competitive neural networks. IEEE Trans. Systems, Man, and Cybernetics, SMC-13, 815-826.
- Eckmiller, R. (1990) Neural computers for motor control. In "Advanced Neural Computers" (R. Eckmiller, Editor), North-Holland/Elsevier, Amsterdam, 357-364.
- Feldman, A.G. (1966). Functional tuning of the nervous system with control of movement or maintenance of a steady posture. II. Controllable parameters of the muscle. Biophysics, 11, 565-578.
- Feldman, A.G. (1986). Once more on the equilibrium-point hypothesis (Lambda Model) for motor control. J. Motor Behavior, 18,17-54.
- Flash, T. and N. Hogan (1985) The coordination of arm movement: an experimentally confirmed mathematical model. J. Neuroscience 5:1688-1703.
- Georgopulos, A.P., Schwartz, A.B., and R.E. Kettner (1986) Neuronal population coding of a movement direction. Science 233:1416-1419.
- Georgopulos, A.P. (1988) Neural integration of movement: role of motor cortex in reaching. Faseb Journal, 2, 2849-2857.
- Hogan, N. (1984). An organizing principle for a class of voluntary movements. J. Neuroscience, 4, 2745-2754.
- Hopfield, J.J. (1982). Neural networks and physical systems with emergent collective computational abilities. Proc. Nat. Acad. Sci. USA, 79, 2554-2558.

- Hopfield, J.J. (1984). Neurons with graded response have collective computational properties like those of two state neurons. Proc. Nat. Acad. Sci. USA, 81, 3088-3092.
- Jordan, M. (1988) Supervised learning and systems with excess degrees of freedom. COINS 88-27, M.I.T.
- Jordan, M.I. (1989) Indeterminate motor skill learning problems. In Attention and Performance XIII (M. Jeannerod, Editor), MIT Press:Cambridge Mass.
- Jordan, M.I. (1989) Serial order: A parallel, distributed processing approach. In "Advances in Connectionist Theory: Speech" (J.L. Elman and D.E. Rumelhart, Editors), Lawrence Erlbaum Ass.:Hillsdale, NJ, 44-93.
- Kawato, M., Maeda, Y., Uno, Y., and R. Suzuki (1990) Trajectory formation of arm movement by cascade neural network model based on minimum torque-change criterion. Biological Cybernetics 62:275-288.
- LeCun, Y. (1985) A learning scheme for asymmetric threshold networks. Proceedings of Cognitiva 85, Paris, France.
- Massone, L. and E. Bizzi (1989) A neural network model for limb trajectory formation. Biological Cybernetics 61, 417-425.
- Morasso, P. (1981) Spatial control of arm movements. Exp. Brain Res. 42:223-227.
- Morasso, P., Mussa Ivaldi, F.A., Vercelli, G., and R. Zaccaria (1989) A connectionist formulation of motor planning. In "Connectionism in Perspective" (Pfeifer, Schreter, Fogelman-Soulie, and Steels, Editors), Elsevier Science Publishers:Amsterdam, 413-420.
- Morasso, P. and V. Sanguineti (1990) Neurocomputing concepts in motor control. In "Brain and Space" (J. Paillard, Editor), Oxford University Press, Oxford, Chapter 21.
- Morasso, P. (1990) Neural representation of motor synergies. In "Advanced Neural Computers" (R. Eckmiller, Editor), North-Holland:Amsterdam, 51-59.
- Mussa Ivaldi, F.A., Hogan, N., and E. Bizzi (1985). Neural, mechanical, and geometric factors subserving arm postures in humans. J. Neuroscience, 5, 2732-2743.
- Mussa Ivaldi, F.A., Morasso, P., and R. Zaccaria (1988). Kinematic networks - A distributed model for representing and regularizing motor redundancy. Biological Cybernetics, 60, 1-16.
- Nguyen, D. and B. Widrow (1990) The truck backer-upper. In "Advanced Neural Computers" (R. Eckmiller, Editor), North-Holland:Amsterdam, 11-17.
- Rumelhart, D.E., Hinton, G.E, and R.J. Williams (1986) Learning internal representations by error propagation. In "Parallel Distributed Processing. Vol. 1" (D.E. Rumelhart and J.L. McClelland, Editors), MIT Press:Cambridge Mass, 318-363.

- Uno, Y., Kawato, M., and R. Suzuki (1989) Formation and control of optimal trajectory in human multijoint arm movement. Minimum torque-change model. Biological Cybernetics 61:89-101.
- Viviani, P. and C. Terzuolo (1982) Trajectory determines movement dynamics. Neuroscience 7:431-437.

OUTLINE FOR A THEORY OF MOTOR LEARNING

JAMES C. HOUK
Department of Physiology
Northwestern University Medical Center
Ward Building 5-319
303 E. Chicago Avenue
Chicago, Illinois 60611

ABSTRACT This paper outlines a new theory of motor learning. The theory is constrained by the anatomy and physiology of sensorimotor pathways through the cerebellum, motor cortex, brainstem and spinal cord. It also draws upon recent knowledge regarding cellular and molecular properties and receives insight from the results of learning experiments with artificial neural networks. The zonal organization that aligns climbing fibers, Purkinje cells and nuclear cells in the cerebellum provides an ideal neuronal architecture for efficient learning. Climbing fibers are postulated to acquire properties that improve their ability to train sets of Purkinje cells. The Purkinje cells then control movements by selectively inhibiting premotor networks. Repetition of this control establishes motor habits that are then carried out automatically by the premotor networks. Long-term repetition is postulated to align the topography of the overall network utilizing trophic mechanisms. These processes in combination are considered capable of explaining the salient features of motor learning.

Introduction

The long-range goal in this work is to develop a comprehensive theory of motor learning. We seek to understand how motor programs are represented in the nervous system, how these programs are adaptively controlled to accommodate novel and complex environments, and how the brain perfects its capabilities for adaptive control. While motor programs are executed on a time scale of milliseconds, adaptation occurs on a time scale of seconds to days, and the ability to adapt is perfected over a developmental time course of years. We seek a theory capable of relating to this spectrum of time scales.

The theory should also be comprehensive in the levels of knowledge that it addresses. The anatomy of the brain has been studied for over a century, and well-established neuroanatomical findings provide invaluable constraints for a theory of motor learning. Similarly, we know much about the physiology of the pathways, and we wish to consult this store of knowledge both for constraints and for inspiration. Recently there has been a major expansion of our knowledge about cellular and molecular properties of neurons, and some of this information would be valuable to include. We should likewise heed findings from biomechanics and

J. Requin and G. E. Stelmach (eds.), Tutorials in Motor Neuroscience, 253–268.

motion studies. Work with artificial neural networks has advanced rapidly in recent years, and insights gained from these experiences would also be valuable in constructing a theory of motor learning. While we cannot possibly deal with each and every detail of this vast, multi-level body of knowledge, neither should we ignore any particular area. Rather, the theory should draw upon key findings in each area. These elements should then be integrated into a model that can be tested and explored.

The goal of this particular paper is to present an outline for a comprehensive theory of motor learning. While we know much about subsystems in motor learning, we have a poor grasp of the total problem. What seems required to foster progress is that we grapple with all aspects of the problem. Although this effort is bound to be speculative, it can be informed. The hope here is to highlight some of the essential elements that need to be included in a comprehensive theory. Considerable additional work will be required to develop a mechanistic model that is actually capable of explaining motor learning.

Anatomical Setting of the Theory

Fig. 1 identifies the main anatomical structures to which this theory relates and shows how we envision information to flow along pathways connecting these structures. This schema is not unique to the present theory; rather it should be looked upon as an outline of fundamental anatomical and physiological knowledge that serves to provide constraints on plausible theories of motor learning. Most of this section reviews the basic operations and interactions envisioned to occur along these pathways. At the end of the section a brief synopsis of the proposed learning mechanisms will be introduced within the framework of this organizational plan.

The bottom of Fig. 1 emphasizes somatomotor interactions with the environment. Motor units produce actions on the environment in response to motor programs generated by premotor networks. Somatosensory receptors sense the environmental responses to these actions. The theory places special emphasis on these tactile and proprioceptive sensors, whereas vision and other special senses are considered to provide higher-level mechanisms that become superimposed on basic somatomotor mechanisms.

The premotor networks in Fig. 1 range in complexity from simple spinal reflexes to motor cortical circuits controlling voluntary movement. Correspondingly, the events that trigger the premotor networks to produce motor programs range from simple somatosensory stimuli to multifarious events such as visual stimuli or internal drives. The motor programs produced by the premotor networks are regulated by output from the cerebellar cortex. These regulatory actions are especially important in controlling voluntary movements but are also involved in modulating basic reflexes. Premotor networks can function without regulation to produce basic reflexes or to perform well-rehearsed, habitual actions.

The cerebellar cortex is needed to control complex, precise and/or novel actions. It accomplishes these functions by regulating a relatively stereotyped set of habitual reactions and basic reflexes. The intricate regulatory actions postulated to occur in the cerebellar cortex require the assimilation of a large amount of information about the external environment and the internal state of the body. The unique neuronal architecture of the cerebellar cortex is considered ideal for this purpose. The output neurons, the Purkinje cells, are exposed to a large state vector and also receive training signals from the inferior olive.

Motor learning is postulated to occur in all three CNS structures shown in Fig. 1. The most rapid and versatile adaptive process is placed in the cerebellar cortex where longitudinal bands of Purkinje cells receive private training signals from small clusters of neurons in the inferior olive. The Purkinje cells learn how to utilize the diverse information supplied by the state vector to regulate premotor circuits in an intelligent fashion. A slower adaptive process is placed in the premotor networks. We assume that this process is use-dependent, allowing the network to learn new habits simply as a consequence of persistent regulatory actions sent from the cerebellar cortex. This frees up Purkinje cells for coordinating more complex and novel actions. A third adaptive process is placed in the inferior olive. According to our theory adaptive mechanisms adjust the response properties of olivary neurons to make them provide more informative training signals. In this manner a climbing fiber progressively learns how to train Purkinje cells more effectively. Finally, we postulate that trophic influences, operating particularly during development but also during adulthood, may function to improve the functional alignment between training signals from the inferior olive and premotor circuits controlling elemental motor programs. These mechanisms in combination are considered capable of accounting for most of the organism's capacity for motor learning.

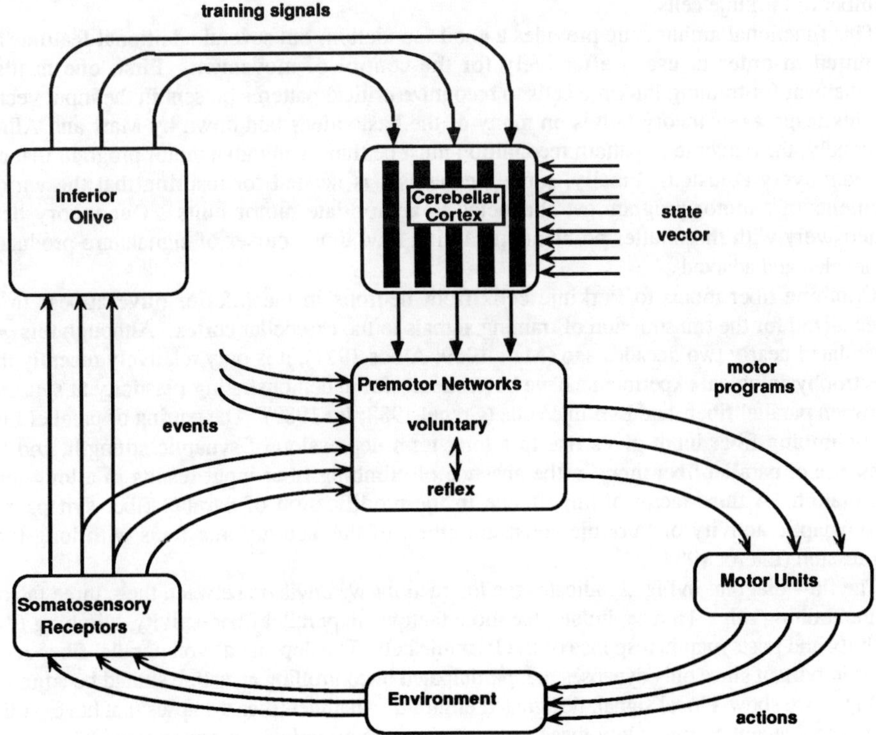

Figure 1: Organizational framework for the proposed theory of motor learning

Adjustable Pattern Generator Model of the Cerebellum

The present theory builds on a recent model that relates the anatomy and physiology of the cerebellum, red nucleus and motor cortex to the concept of motor programs for the control of limb movement (Houk 1989; Houk et al 1990; Barto et al 1990). In this section we review these ideas concerning how motor programs might be stored, recalled and executed using adjustable pattern generator (APG) modules in the cerebellum.

The cerebellar cortex has long been celebrated as a neuronal architecture ideally suited for learning to recognize complex patterns (Marr 1969; Albus 1971). One favorable feature is the enormous divergence in the mossy fiber/parallel fiber input to the cerebellum. This arrangement is ideally suited for distributing a given input to a maximal number of Purkinje cells across large areas of cerebellar cortex. A second favorable feature concerns the tremendous convergence of parallel fibers onto individual Purkinje cells. This arrangement exposes each Purkinje cell to approximately 200,000 potential inputs. The resultant of these divergent and convergent features is that a high dimensional state vector is distributed to a great number of Purkinje cells.

This functional architecture provides a good foundation, but several additional features are required in order to use it effectively for the control of movement. First, one needs a mechanism for training Purkinje cells to recognize critical patterns present in the input vector. In this respect our theory builds on many of the basic ideas laid down by Marr and Albus. Secondly, the outcome of pattern recognition must be translated into a motor program that can be adaptively adjusted. Finally, some mechanism is needed for insuring that the various elements of a motor program get channeled to appropriate motor units. Our theory deals extensively with these latter problems regarding how time courses of signals are produced, channeled and adapted.

Climbing fiber inputs to Purkinje cells from neurons in the inferior olive appear to be specialized for the transmission of training signals to the cerebellar cortex. Although this was postulated nearly two decades ago (Marr 1969; Albus 1971), it is only relatively recently that electrophysiological experiments have been successful in demonstrating plasticity at synapses between parallel fibers and Purkinje cells (Crepel 1988; Ito 1989). The pairing of parallel fiber and climbing fiber input gives rise to a long-term depression of synaptic strength, and the presence of parallel fiber input in the absence of climbing fiber input results in a long-term facilitation. A third factor of importance in the modification of parallel fiber synapses is postsynaptic activity of Purkinje cells; inhibition of this activity interferes with long-term depression (Ekerot 1984).

The flow diagram in Fig. 2 indicates the interactions we envision between these three factors in the training rule. To recapitulate, the three factors are parallel fiber activity, climbing fiber activity and postsynaptic response of the Purkinje cell. The dependence on parallel fiber activity is important since only synapses that participated in controlling an action should be adjusted. In Fig. 2 we show a local signal, designated eligibility, that identifies synapses that have participated in a recent action. Only these synapses become eligible for modification. The dependence on climbing fiber activity is important to include since it communicates an evaluation of how successful the interaction was with the environment. In actuality, climbing fibers fire under unsuccessful conditions, to indicate when synaptic strength should be decreased. The dependence on postsynaptic response is important to include since it indicates whether or not the cell actually participated in controlling the action that was evaluated. Since Purkinje cells function by inhibiting action, it is appropriate that weight decreases depend on the cell responding.

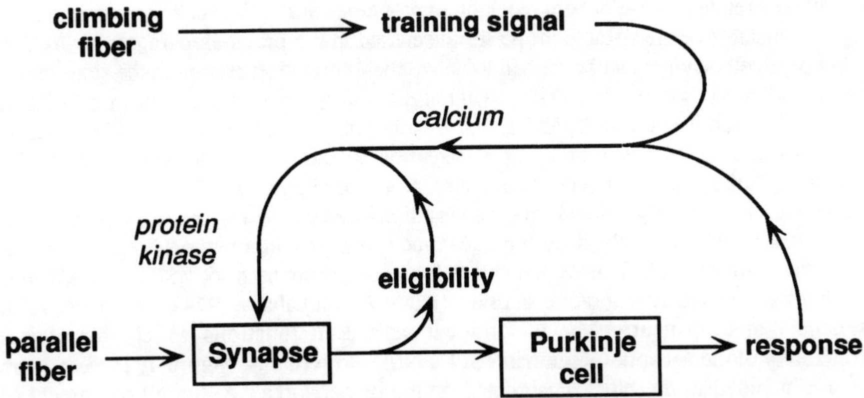

Figure 2: Model of synaptic modification in cerebellar Purkinje cells

The dependence of weight modification on two of these factors, synaptic eligibility and postsynaptic response, is analogous to the Hebbian-like rules discovered for CA1 hippocampal neurons (Brown et al 1989; Stanton and Sejnowski 1989) and thought to apply also to motor cortical neurons (Sakamoto et al 1987). This type of local learning rule works well for the self-organization of sensory channels (Willshaw and Marlsburg 1976; Linsker 1988), and it may also be useful for adapting the simple motor functions that occur in premotor networks as discussed later. However, the learning of complex and novel motor responses can be vastly improved by providing an additional training signal that is based on an evaluation of the organism's performance, particularly if that evaluation is informed. The ability of neurons in the inferior olive to perform informed evaluation is discussed in a later section.

Successful training would result in different Purkinje cells being trained to recognize different critical patterns of activity in the state vector. The more appropriate the pattern, the greater would be the activation state of a Purkinje cell. Translation of this activation state into a motor program is not trivial. The present theory of motor learning builds on our previous work on this problem (Houk 1989; Houk et al 1990; Sinkjaer et al 1990; Barto et al 1990) where we proposed that naturally occurring cerebellar modules function as adjustable pattern generators (APGs). An important operational property of an APG module is its capability for generating temporal patterns of output independently of the time course of the input. This capability, originally motivated by physiological studies of single unit discharge in behaving animals, was accounted for by including two nonlinear dynamical processes in our model.

Abstracting from microelectrode recordings from dendrites (Ekerot 1984; Llinás & Sugimori 1980), Purkinje cells were treated as multistable elements with hysteresis. We assumed that synaptic excitation has a small effect on membrane potential until it achieves a critical value designated the on-threshold; at this point the Purkinje dendrite flips to a depolarized state. It then remains in this depolarized state until synaptic inhibition exceeds an off-threshold, at which point the dendrite flips back to a hyperpolarized state. Individual dendrites were thus modeled as bistable devices with hysteresis. This gives rise to multistability at the level of the cell body where the effects of events in several dendrites are combined. The simplified case of a cell with just one dendrite (bistability instead of multistability) was assumed in the simulation study of our model (Houk et al 1990).

We can now reinterpret the pattern recognition properties of Purkinje cells described earlier in light of the bistable or multistable properties discussed in the previous paragraph. In effect, each Purkinje cell dendrite can be trained to utilize the information present in the state vector to determine when to switch state. This switching is analogous to a decision in a finite state automata. It comprises the recognition that a particular state of the automata has just been achieved and marks this by producing a transition in the state of a Purkinje cell. State transitions in Purkinje cells are then used to control a second process.

The second nonlinear dynamical process was motivated by the reciprocal neuroanatomical pathways that exist between the deep cerebellar nuclei and premotor neurons in red nucleus and motor cortex (Houk 1989). These connections form a recurrent network with positive feedback loops that can sustain regenerative activity (Allen & Tsukahara 1974). We have further postulated that loop neurons have nonlinear activation functions which improves the controllability of the network (Eisenman et al 1990). In terms of Figure 1, these recurrent loops are included in the block designated premotor networks. Although our model of a recurrent network was advanced on the basis of cortico- and rubro-cerebellar circuits, similar concepts probably apply to other premotor networks.

We have postulated that a recurrent network displays stable states of activity and of quiescence. In the quiescent state, most loop neurons are below threshold, and network activity is relatively insensitive to the inhibitory input to it from Purkinje cells. While the reciprocal network is quiescent, Purkinje cells can be turned on and off to prepare for anticipated movements. This is like recalling a motor program from memory, the memory being the synaptic weights of parallel fiber synapses. Once a motor program is recalled from memory, it can then be executed by a transition to the active state.

A transition to the active state of the recurrent network is triggered by a transient input to loop neurons, as might be produced by the event inputs to the premotor networks in Figure 1. Once the network is triggered to its on state, the intensity and spatial distribution of activity tends to build up in a regenerative fashion, due to positive feedback. (We have suggested that reaction times are caused by this progressive spread of recurrent activity.) The final equilibrium state of the network is regulated by the set of inhibitory inputs from Purkinje cells to the recurrent network. In this manner, positive feedback serves as the driving force for generating a motor program, while states of Purkinje cells guide the regenerative process and control the resultant spatiotemporal pattern of activity in the recurrent network. This spatiotemporal pattern defines the motor program produced by a premotor network.

A composite motor program can be thought of as a vector comprised of an array of elemental motor programs. It is natural to associate the elemental programs with the outputs from individual APG modules, and the composite program with an array of modules. The temporal aspect of an elemental program is specified by the firing pattern in the axon of a premotor neuron, and the spatial aspect is specified by the fiber's branching matrix to motor units. The next problem we wish to consider is how the various elements of a composite motor program get channeled to appropriate motor units.

Zonal Organization of the Cerebellar Cortex

While many areas of the brain are organized in a topographic fashion, one of the most striking examples occurs in the cerebellum. Anatomical (Jansen & Brodal 1942; Voogd and Bigaré 1980), physiological (Oscarsson 1980; Andersson et al 1987) and immunohistochemical (Scott

1964; Doré et al 1990; Gravel et al 1987) studies have each demonstrated a highly organized system of longitudinal, parasagittal zones that extend, albeit with some distortion, throughout the entire cerebellar cortex. While the ontogenetic origin of these zones remains poorly understood, the fact that several different classes of intrinsic molecules are organized in longitudinal zones at an early stage of development suggests that some aspects of the zonal organizational plan are genetically prescribed, whereas others are probably shaped by the processes of developmental plasticity.

The zonal organization in the cerebellar cortex relates to the APG modules discussed in the previous section. In Fig. 1, longitudinal zones of Purkinje cells were symbolized by vertical bands oriented orthogonal to the state vector transmitted by parallel fibers, and we implied that the inputs and outputs of these zones are also topographically organized. This input/output organization is more fully illustrated in Fig. 3 where the horizontal rectangles are meant to define longitudinal zones of Purkinje cells. Each longitudinal zone is shown to receive a focused climbing fiber input from a small cluster (symbolized by a circle) of inferior olivary neurons. The axons of the Purkinje cells in the zone project a focus of inhibition to a small cluster of cerebellar nuclear cells. There is considerable specificity in these projections from the inferior olive through the cerebellar cortex and out to the cerebellar nuclei.

There is less specificity in the projection from the cerebellar nuclei through premotor networks to motor units. Fig. 3 was constructed to relate specifically to the projections via red nucleus and motor cortex. The diagram portrays a reciprocity between clusters of cerebellar nuclear neurons and clusters of premotor neurons in red nucleus and/or motor cortex. It also shows divergence and convergence between adjacent recurrent loops. Intra-axonal staining has demonstrated a modest degree of topographic specificity in this system (Shinoda et al 1988). The fibers that descend to the spinal cord to innervate motor units also show divergent branching (Shinoda et al 1986).

One of the consequences of these branching patterns is a loss of specific alignment between single zones of Purkinje cells and individual premotor neurons. This means that APG modules cannot be conceptualized as discrete entities since they will have overlapping boundaries. Overlap is not necessarily an undesirable feature since experience with layered neural networks has indicated that this type of course coding actually improves a network's learning performance and its capacity for generalization. However, overlap does make it more difficult to define an elemental motor program. In the previous section we defined an elemental program as the signal present in an individual premotor neuron. Alternatively we could define it in terms of the output from single zones of Purkinje cells.

The degree of topographic specificity along the sensory pathway to the inferior olive is also less striking than the precision of the zonal organization in cerebellar cortex. The clearest olivary organization occurs in the rostral division of the dorsal accessory olive (rDAO) where there is a complete cutaneous map of the contralateral body surface (Gellman et al 1983). Tracing these pathways backward to sensory relays in dorsal column nuclei, the somatotopic specificity becomes much more difficult to recognize (McCurdy 1988). These and other findings indicate that there is considerable convergence along the input pathways to the inferior olive (Oscarsson 1980; Andersson et al 1987). This convergence may in fact provide the basis for some of the unique physiological properties of olivary neurons described in the next section.

In summary, there appears to be an intriguing gradient in topographic specificity to and from the cerebellum. Topographic specificity increases along the input pathway from somatosensory receptors through the inferior olive to the cerebellar cortex, but then decreases along the output

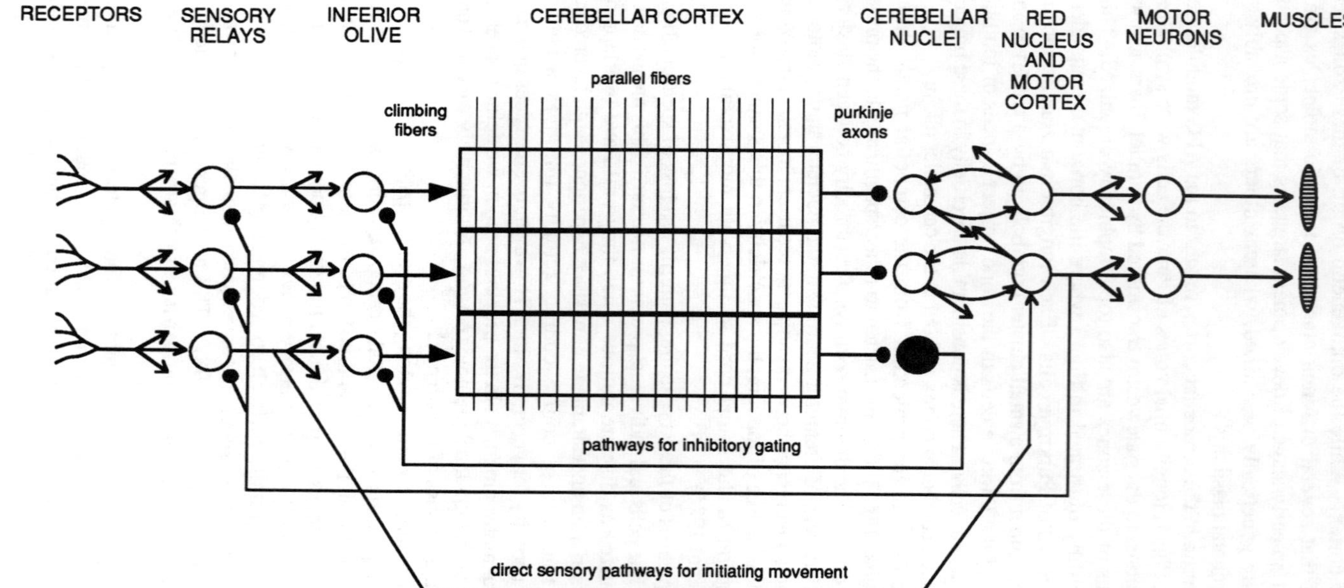

Figure 3: Zonal organization in the cerebellar cortex and its alignment with inputs and outputs

pathway from the cerebellar cortex through premotor networks to motor units. The greatest precision of alignment occurs in the center of the network, in the cerebellar cortex, where climbing fibers, Purkinje cells and Purkinje axons appear to be precisely aligned by molecular markers that are expressed at different stages of development.

Our interpretation of zonal organization in the cerebellar cortex is that it may have evolved to alleviate the "credit assignment" problem. This problem arises when a network is large, since it is then difficult to determine which neural elements and which of their synaptic weights should be given credit for good or for bad network performance. The problem of credit assignment severely interferes with efficient learning when large networks are massively interconnected. The precise alignment of climbing fiber input should help solve this problem by providing private training signals to organized zones of Purkinje cells. For this to work, however, requires special mechanisms for insuring that any given training signal is appropriately tuned for training the particular APG module that it innervates. In the next section we discuss aspects of inferior olive physiology that may contribute to this tuning.

Training Signals from the Inferior Olive

The inferior olive is, with minor exception, the origin of all of the climbing fiber input to APG modules in the cerebellum. The physiology of olivary neurons is unique in several ways, and in this section we explore how these characteristic properties may relate to the postulated role of the olive in motor learning.

First of all, the electrical activity of olivary cells differs strikingly from that of most neurons (Eccles et al 1966). Their action potentials consist of an initial sodium spike followed by a calcium plateau and then a long refractory period (Llinás and Yarom 1981). As a consequence, olivary neurons fire at very low frequencies and usually respond to sensory stimuli with just a single action potential transmitted along their climbing fiber axons. While this binary responsiveness is appropriate for signalling the occurrences of sensory events, it interferes with the transmission of information about the intensity and duration of the stimulus. From the standpoint of postsynaptic effect, olivary refractoriness may be important, since it limits the amount of excitation delivered to Purkinje cells. As a consequence, a climbing fiber can transmit a training signal that is dissociated from the postsynaptic response. This is the justification for assuming that climbing fibers serve as (relatively) pure training signals in the synaptic modification scheme proposed earlier (Fig. 2).

The response properties of olivary neurons are also very intriguing. In studying receptive fields, one finds a strong emphasis on somatosensory modalities, and the cells appear to be specialized for detecting different types of somatic event (Gellman et al 1985). Many are sensitive to contact with a patch of skin or with hairs, and another large group is responsive to small displacements of a limb in a particular direction. Yet others are sensitive to slip, vibration or to tugs on muscles. Even the group of cells receiving visual input seems more specialized for detecting motion of the body, a somatic event detected by retinal slip, rather than being sensitive to the motion or location of a target.

Somatic events might seem to be unusual signals for adaptively controlling movements, since in robotics one typically relies heavily on vision to compute error signals. However, a visual computation of error followed by coordinate transformation into the motor space regulated by the cerebellum represents an elaborate computation. Furthermore, this scheme would not explain infant development where visual skills appear to lag behind abilities to control limb movements (White 1971). It is more plausible that somatosensory training signals are used to calibrate vision than the other way around.

How could responsiveness to low threshold somatic stimuli be useful in training the cerebellum to control movement? We postulate that the utility of light touch and proprioception as training signals derives from their capacity to predict more primal training signals. This would be analogous to the role of an "adaptive critic" in training reinforcement learning controllers in artificial neural networks (Barto et al 1983; Barto 1989). Adaptive critics are particularly efficient as trainers since they learn to predict long-term consequences of controller behavior. A capacity of olivary neurons to predict is probably not originally present but instead is acquired through experience. Before considering how experience might train olivary neurons to become better trainers, let us first consider primal training signals and how they might operate.

Painful events detected by nociceptors located in skin, muscle and joints might serve as primal training signals. When the moving limb bumps hard against an external object, skin nociceptors with receptive fields at the site of contact are activated. These signals could be used to decrease the amount of agonist muscle activation or to increase the activation of antagonists. This would be adaptive since it would decrease the probability of a similar collision in the future. Nociceptors and other high-threshold receptors located in muscle and joint are activated when other types of uncoordinated movements are produced, clumsy movements that unduly stress the ligaments and muscles. If these signals were connected to appropriate pattern generators in the cerebellum, they too could produce adaptive adjustments in motor programs.

Motor learning would be most unpleasant if it had to rely solely on pain as a training signal. However, if olivary neurons acquired an ability to judge when the cerebellum produces movements that come close to causing pain, learning could occur without the occurrence of a painful outcome. Support for this concept comes from studies that compared nociceptive and light cutaneous receptive fields of individual olivary neurons (Ekerot et al 1987). The location of the receptive field and the distribution of sensitivity was essentially identical for noxious and tactile modalities of stimuli. Given these properties, a light cutaneous response becomes an excellent indicator that an agonist muscle is being too strongly activated. A slightly stronger activation in a subsequent occasion could result in a painful collision.

It is not difficult to conceive of a plausible mechanism for training olivary cells to respond to light cutaneous stimulation of the patch of skin overlying the nociceptive field of the neuron. A simple associative mechanism would suffice, since any situation in which the cell's nociceptive input is activated by a strong cutaneous stimulation would be accompanied or preceded by an activation of overlying low-threshold cutaneous receptors. A low-threshold input that initially produces a synaptic potential too small to activate the cell would be strengthened until it becomes capable of producing a response in the absence of the nociceptive input. The cellular mechanism might involve coactivation of a protein kinase by calcium and a local eligibility signal (similar to the situation depicted in Fig. 2 minus the training signal) since nociceptive stimulation strongly activates these cells (the response component in Fig. 2) and would be expected to produce a large influx of calcium by itself without requiring an additional training input.

The generation of appropriate training signals may be additionally facilitated by anatomical pathways that modulate sensory input to the olive (Fig. 3). One such pathway involves collaterals of the fibers that transmit motor programs from the red nucleus and motor cortex to the spinal cord. Another involves special inhibitory neurons that are located in the cerebellar nuclei and project to the inferior olive (Andersson et al 1988; Nelson and Mugnaini 1989).

Electrophysiological studies of the former pathway have revealed an inhibitory gating mechanism that modulates tactile input to the olive (Weiss et al 1990). The pattern of modulation is novel and appears to represent a simple, though elegant, mechanism for detecting when a motor program commands too large a movement. To explain this, we will describe an hypothetical example that extrapolates from the physiological studies (Gellman et al 1985; Weiss et al 1990).

Consider an olivary neuron that is sensitive to light cutaneous stimulation of the palm of the hand. The subject now generates a motor program commanding the arm to reach for a door knob. Transmission of this program to motor units in the spinal cord produces corollary discharge to the sensory relays to the olive (Fig. 3). Extrapolating from the delayed time course of inhibition found in the physiological studies, this corollary mechanism would not inhibit sensory input during the movement, but rather would produce a rebound inhibition just as the motor program is terminated. Because of this timing, the olivary neuron would not fire if the planned movement were of a correct amplitude. This is because a motor program generally ends just before a movement ends, and inertia carries the movement to completion. Thus, if the brain generates a correct command, inhibitory gating would prevent the olivary neuron from responding. However, if the brain commands too large a movement, the hand would contact the door knob before the motor program ends, and the olivary neuron would fire. This response would signify that the motor program was too large. One of the neat things about this mechanism is that it will work, without further modification, for reaches to objects anywhere in the workspace.

The previous example illustrates how a relatively simple mechanism for inhibitory gating can be combined with an elementary tactile receptive field to produce an ingenious mechanism for detecting when a motor program is too large. In effect, these features open a spatiotemporal window for sampling a simple event (touch) by inhibiting responsiveness for times outside of this window. If touch occurs within this spatiotemporal window (when the olive is not inhibited), it signifies the likelihood that a specific type of error has occurred. If the climbing fiber is connected to an appropriate APG, it should be capable of efficiently training that APG to assist in the production of an accurate movement. This theory also predicts the type of connections that would be appropriate. Since climbing fibers depress the responses of Purkinje cells to excitatory input from parallel fibers, and since the Purkinje cells in turn inhibit movement commands, the adaptive loop requires another sign inversion to exhibit stable learning behavior. This would occur if the climbing fiber connected to an APG controlling muscles that oppose the movement that stimulates the climbing fiber.

The previous discussion is directed toward olivary neurons with cutaneous receptive fields that are inhibited at the end of a movement. However, analogous principles may apply to other categories of olivary neurons. Some cells, particularly those with proprioceptive fields, are inhibited during movement (Gellman et al 1985). They appear to resume their sensitivity just after the end of a normal movement, since they respond to perturbations that occur in that time window (Andersson and Armstrong 1987). Elsewhere we have interpreted these data as revealing a mechanism for detecting when a motor program commands too small of a movement (Houk et al 1990). In analogy with the cutaneous neurons discussed in the previous paragraph, the detection mechanism appears to function by opening a spatiotemporal window for sampling a simple somatosensory event. In this case the window occurs at the end of a movement, rather than during it, and leads to a mechanism for detecting when commands are too small, as opposed to too large. This type of climbing fiber is suitable for controlling APGs that innervate agonist muscles.

The detection mechanisms described in the previous paragraphs do not require a very sophisticated inhibitory gating process and are probably explicable in terms of the corollary pathway in Fig. 3 that inhibits sensory relays to the olive. This raises the question as to the function of the special pathway from GABAergic cerebellar nuclear cells to the inferior olive. This pathway is directly controlled by zonal output from the cerebellar cortex, which raises the possibility of it being adaptively adjusted. This might permit an adaptive tuning of the temporal window during which a somatic event is detected and might lead to quite sophisticated training signals.

In summary, there seem to be several ingenious mechanisms for improving the capacity of the inferior olive for training the cerebellum how to generate better motor programs. Intuitively, this cascade of learning processes seems like an efficient way to regulate motor learning, and support for this conclusion comes from neural modeling studies in which an adaptive critic operates on an adaptive controller to solve difficult control problems (Barto et al 1983).

Credit Assignment in Motor Learning

An important theme of the present work is credit assignment in adaptive neural networks controlling movement. Efficient learning in large networks requires getting appropriate training information back to where it is needed. This is the problem of credit assignment. In an earlier section it was suggested that the zonal organization of the cerebellar cortex facilitates the resolution of credit assignment problems by aligning small clusters of olivary neurons with narrow strips of Purkinje cells that in turn control small clusters of cerebellar nuclear neurons. This neuronal architecture connects semi-private training signals to individual APG modules. It provides a good foundation for motor learning, but other problems need to be resolved to efficiently utilize this architecture.

In the previous section it was suggested that olivary training signals are initially specified by the pattern of nociceptive input to the inferior olive, and that these relatively crude punishment signals are progressively refined by two tuning processes. One refinement is provided by an associative mechanism that seeks out low-threshold somatosensory signals in a combination capable of predicting when movements come close to causing pain. This can be thought of as opening a spatial window (the evolving receptive field) for evaluating performance. A second refinement is provided by an inhibitory gating process that superimposes a temporal window of evaluation. The two processes in combination provide a spatiotemporal window that ingeniously bases evaluation on the detection of simple somatic events, such as touch, that occur within this window.

This type of informed evaluation by private training signals could create a very effective learning situation provided each climbing fiber innervated an appropriate APG. An appropriate APG is one that, if activated more strongly, would maximally diminish the firing probability of the climbing fiber innervating the APG. Higher goals in action might then come about through an associative search process (Barto et al 1981) operating in the cerebellar cortex. Starting with the simple view that the training signal transmitted by a climbing fiber is essentially a punishment signal, effective learning would involve modifying action so as to reduce the probability of future punishment. This could involve discovering higher goals as well as simply making accurate movements.

One way to resolve the final issue in the credit assignment problem, i.e., connecting to an appropriate APG, is to postulate an adaptive mechanism for selecting the muscles controlled by a given APG module. In effect, this would translocate the optimal connection problem (part of the spatial credit assignment problem) out of the cerebellum into the premotor network. Rather than reconnecting training signals to APG modules, one simply changes the muscles to which an APG module connects. As indicated earlier, the present theory postulates that a use-dependent Hebbian process occurs at the level of premotor networks. This learning mechanism is thought to explain the associative mechanism in classical conditioning (Houk 1989; 1990) and long-term adaptive changes in the vestibuloocular reflex (Peterson et al, in press). The rapid reorganization in motor cortical representations following peripheral nerve transections speaks to the efficacy of such mechanisms (Sanes et al 1988). The same learning mechanisms could guide an alignment process that would complete the resolution of the credit assignment problem.

The APG model of the cerebellum includes a preselection process during which APG modules, and the Purkinje cells that contribute to them, are selected by causing transitions to the off states of these neurons. The learning rule postulated to operate in the cerebellar cortex reinforces the selection of cells and modules that minimize the probability of receiving future punishment signals from their climbing fibers. By this process, APG modules that connect with muscles that minimize punishment would be extensively used, and those that connect with muscles that increase punishment would suffer from disuse. In this manner simple, use-dependent processes for modifying synaptic weights in the premotor network would act to improve the alignment between training signals and the elemental motor programs that they control.

The scheme outlined in the previous paragraph would further provide an efficient mechanism for guiding longer-term trophic processes that are postulated to regulate the development and maintenance of topographic maps in the nervous system (Purves 1988). According to the trophic theory, nerve ending are continually eliminating old connections and sprouting new ones. This process is thought to be guided by a two-way interaction between arrays of nerve endings and arrays of target cells. The target cells are thought to release trophic substances that promote sprouting to attract new inputs, particularly when they are not adequately activated by their existing inputs. This process essentially extends to a longer time scale the use-dependent modification of input provided by a Hebbian synapse. The main difference is that the trophic mechanism stabilizes new anatomical connections whenever they are heavily used, whereas a Hebbian mechanism merely strengthens existing synapses when they are heavily used. Both processes would be guided by Purkinje cells whose learning is in turn guided by climbing fibers. This total scheme thus comprises a comprehensive theory of motor learning.

References

Albus, J.S. (1971) A theory of cerebellar function. Math Biosci 10: 25-61

Allen, G.I., Tsukahara, N. (1974) Cerebrocerebellar communication systems. Physiol. Rev 54: 957-1006

Andersson, G., Armstrong, D.M. (1987) Complex spikes in Purkinje cells in the lateral vermis (b zone) of the cat cerebellum during locomotion. J. Physiol. Lond. 385: 107-134

Andersson, G., Ekerot, C.-R., Oscarsson, O., Schouenborg, J. (1987) Convergence of afferent paths to olivo-cerebellar complexes. In: Glickstein, M., Yeo, C., Stein, J. (eds) Cerebellum and Neuronal Plasticity. London, Plenum Press, pp 165-173

Andersson, G., Garwicz, M., Hesslow, G. (1988) Evidence for a GABA-mediated cerebellar inhibition of the inferior olive in the cat. Exp. Brain Res 72: 450-456

Barto, A.G. (1989) Connectionist Learning for Control: An Overview. COINS Technical Report, Vol 89-89. Amherst, Massachusetts , Univ. of Massachusetts

Barto, A.G., Berthier, N. , Singh, S., Houk, J.C. (1990) Network model of the cerebellum and motor cortex that learns to control planar limb movement. Soc. Neurosci. abstr

Barto, A.G., Sutton, R.S., Anderson, C.W. (1983) Neuronlike elements that can solve difficult learning control problems. IEEE Transactions on Systems, Man, and Cybernetics 13: 835-846

Barto, A.G., Sutton, R.S., Brouwer, P.S. (1981) Associative search network: reinforcement learning associative memory. IEEE Transactions on Systems, Man, and Cybernetics 40: 201-211

Brown, T.H., Ganong, A.H., Kairiss, E.W., Keenan, C.L., Kelso, S.R. (1989) Long-term potentiation in two synaptic systems of the hippocampal brain slice. In: Byrne, J.H., Berry, W.O. (eds) Neural Models of Plasticity, Ch. 14. San Diego, Academic Press, pp 266-306

Crepel, F.C., Krupa, M. (1988) Activation of protein kinase C induces a long-term depression of glutamate sensitivity of cerebellar Purkinje cells. An in vitro study. Brain Res 458: 397-401

Doré, L., Jacobson, C.D., Hawkes, R. (1990) Organization and postnatal development of Zebrin II antigenic compartmentation in the cerebellar vermis of the grey opossum, *Monodelphis domestica.* J. Comp. Neurol 291: 431-449

Eccles, J.C., Llinás, R., Sasaki, K. (1966) The excitatory synaptic action of climbing fibers on the Purkinje cells of the cerebellum. J. Physiol. London 182: 268-296

Eisenman, L.N., Keifer, J., Houk, J.C. (1990) Computer studies of the role of NMDA receptors and positive feedback loops in the generation of descending motor commands. Soc. Neurosci. abstr

Ekerot, C.-F., Oscarsson, O., Schouenborg, J. (1987) Stimulation of cat cutaneous nociceptive C fibres causing tonic and synchronous activity in climbing fibres. J. Physiol 386: 539-546

Ekerot, C.-F. (1984) Climbing fibre actions of Purkinje cells -- plateau potentials and long-lasting depression of parallel fibre responses. In: Bloedel, J., et al. (eds) Cerebellar Functions. Berlin, Springer-Verlag, pp 268-274

Gellman, R., Gibson, A.R., Houk, J.C. (1985) Inferior olivary neurons in the awake cat: Detection of contact and passive body displacement. J. Neurophysiol 54: 40-60

Gellman, R., Houk, J.C., Gibson, A.R. (1983) Somatosensory properties of the inferior olive of the cat. J. Comp. Neurol 215: 228-243

Gravel, C., Eisenman, L.M., Sasseville, R., Hawkes, R. (1987) Parasagittal organization of the rat cerebellar cortex: Direct correlation between antigenic Purkinje cell bands revealed by mabQ113 and the organization of the olivocerebellar projection. J. Comp. Neurol 265: 294-310

Houk, J.C. (1989) Cooperative control of limb movements by the motor cortex, brainstem and cerebellum. In: Cotterill, R.M.J. (ed) Models of Brain Function. Cambridge, Cambridge Univ Press, pp 309-325

Houk, J.C. (1990) Role of the cerebellum in classical conditioning. Soc. Neurosci. abstr

Houk, J.C., Singh, S.P., Fisher, C., Barto, A.G. (1990) An adaptive sensorimotor network inspired by the anatomy and physiology of the cerebellum. In: Miller, W.T., Sutton, R.S., Werbos, P.J. (eds) Neural Networks for Control. Cambridge, Mass., MIT Press

Ito, M. (1989) Long-term depression. Ann. Rev. Neurosci 12: 85-102

Jansen, J., Brodal, A. (1942) Experimental studies on the intrinsic fibers of the cerebellum. The corticonuclear projection in the rabbit and in the monkey (Macacus rhesus.). Norske Vid. Akad., Oslo, Avh. I. Mat. Natury, K1 3: 1-50

Linsker, R. (1988) Self-organization in a perceptual network. IEEE Computer 105-117

Llinás, R., Sugimori, M. (1980) Electrophysiological properties of *in vitro* Purkinje cell dendrites in mammalian cerebellar slices. J. Physiol. London 305: 197-213

Llinás, R., Yarom, Y. (1981) Electrophysiology of mammalian inferior olivary neurons *in vitro*. Different types of voltage-dependent ionic conductances. J. Physiol. London 315: 549-567

Marr, D. (1969) A theory of cerebellar cortex. J. Physiol 202: 437-470

McCurdy, M. (1988) Sensory input to the forelimb inferior olive and its relationships to motor pathways. Ph.D. Thesis, Northwestern University

Nelson, B.J., Mugnaini, E. (1989) Origins of GABAergic inputs to the inferior olive. Exp. Brain Res, Series 17, pp 86-107

Oscarsson, O. (1980) Functional Organization of Olivary Projection to the Cerebellar Anterior Lobe. In: J. Courville et al. (ed) The Inferior Olivary Nucleus: Anatomy and Physiology. New York, New York, Raven Press, pp 279-289

Peterson, B.W., Baker, J.F., Houk, J.C. (in press) A model of adaptive control of vestibuloocular reflex based on properties of cross-axis adaptation. Ann. N.Y. Acad. Sci.

Purves, D. (1988) Body and Brain: a Trophic Theory of Neural Connections. Cambridge, Mass., Harvard Univ. Press

Sakamoto, T., Porter, L.L., Asanuma, H. (1987) Long-lasting potentiation of synaptic potentials in the motor cortex produced by stimulation of the sensory cortex in the cat: a basis of motor learning. Brain Res 413: 360-364

Sanes, J.N., Suner, S., Lando, J.F., Donoghue, J.P. (1988) Rapid reorganization of adult rat motor cortex somatic representation patterns after motor nerve injury. Proc. Natl. Acad. Sci. USA 85: 2003-2007

Scott, T.G. (1964) A unique pattern of localization within the cerebellum of the mouse. J. Comp. Neurol. 122: 1-8

Shinoda, Y., Futami, T., Mitoma, H., Yokota, J. (1988) Morphology of single neurones in the cerebello-rubrospinal system. Beh. Brain Res 28: 59-64

Shinoda, Y., Yamaguchi, T., Futami, T. (1986) Multiple axon collaterals of single corticospinal axons in the cat spinal cord. J. Neurophysiol 55: 425-448

Sinkjaer, T., Wu, C.H., Barto, A., Houk, J.C. (1990) Cerebellum control of endpoint position—a simulation model. Proc. of Intern'l Joint Conference on Neural Networks, San Diego

Stanton, P.K., Sejnowski, T.J. (1989) Associative long-term depression in the hippocampus induced by hebbian covariance. Nature 339: 215-218

Voogd, J., Bigaré, F. (1980) Topographical distribution of olivary and corticonuclear fibers in the cerebellum. A review. In: Courville, J., Montigny, C. de, Lamarre, Y. (eds) The Inferior Olivary Nucleus. New York, Raven Press, pp 207-234

Weiss C., Houk J.C., Gibson A.R. (1990) Inhibition of sensory responses of cat inferior olive neurons produced by stimulation of red nucleus. *J Neurophysiol* 64: In Press

White, Burton (1971) Human Infants, Experience and Psychological Development. Englewood Cliffs, N.J., Prentice-Hall, Inc.

Willshaw, D.J. and Malsburg von der, C. (1976) How patterned neural connections can be set up by self-organization. Proc. R. Soc. Lond. B. 194: 431-445

Acknowledgement

The author thanks Dr. Andy Barto for many valuable suggestions in the course of this work, Wendy Fraser for editorial assistance, and Damon LaPorte for help with the illustrations. Supported by NIH grant R01-NS21015 and by ONR contract N00014-88-K-0339.

A QUANTITATIVE MODEL OF GRAPHIC PRODUCTION [1)]

ARNOLD J.W.M. THOMASSEN and HEIN J.C.M. TIBOSCH

Nijmegen Institute for Cognition Research and Information Technology (NICI)
University of Nijmegen, PO Box 9104, 6500 HE Nijmegen, The Netherlands.

ABSTRACT. When producing a drawing and when copying a geometrical pattern, the subject organizes the sequence of his or her graphic movements (strokes) in a way which is by no means ad hoc or arbitrary. Specifically, the manual execution of a geometrical pattern appears to be governed by a set of biases determining the starting points, stroke directions and stroke order of the graphic-production sequence. These constraining principles are often called 'rules', although they have a probabilistic rather than a deterministic status. The 'rules' also appear to have different strength. The present study attempts to assess the strength values of each of eight such 'rules' given a specific set of patterns. The experiment collects stroke-sequence data of fifteen adult subjects copying a well-defined set of 149 geometrical patterns twice. Subsequently, their behavior is simulated by a computer algorithm in which the weights of eight 'rules' are optimized, minimizing the deviation between the frequency distributions of the simulated and the original stroke sequences. The model operates in a probabilistic fashion according to several stages dealing with each subsequent stroke successively. The weights obtained allow the model to produce 88 percent of the patterns in the same way as the human subjects did. It is shown that further refinements should improve on this performance. A brief discussion of behavioral data such as reaction times and kinematic features illuminates the facilitatory function of several of the graphic production rules.

1. Introduction

In the planning of complex movement sequences, subjects not only have to prepare each movement segment separately, but they must organize the entire sequence by deciding on the action's appropriate segmentation and by selecting an adequate ordering of the segments. The need for a more systamatic study of such sequencing aspects in complex motor behavior has recently been recognized, e.g., by Rosenbaum (Rosenbaum et al., 1990; Rosenbaum & Jorgensen, 1991).

---

1) The reported research was in part performed by the second author as one of his MA assignments at the Department of Experimental Psychology, University of Nijmegen. It was conducted in the framework of NWO Project 560-259-035 'Graphic production'. Further support was received from ESPRIT Project 419 'Image and movement understanding'. The present article was written while the first author was a Fellow at the Netherlands Institute for Advanced Study (NIAS). Technical assistance by Dr. Frans Maarse and mathematical advice by Professor Eddy Roskam are gratefully acknowledged.

J. Requin and G. E. Stelmach (eds.), Tutorials in Motor Neuroscience, 269–281.

The organization and performance of movement sequencing can be studied fruitfully in the domain of graphic behavior, especially in copying geometrical patterns. A subject who is presented such a pattern visually and instantaneously with the instruction to copy it, is required to 'linearize' the graphic movements leading to its reproduction. This involves the selection of starting points, stroke directions, pen lifts and stroke order. In principle, each pattern can be drawn in a large number of ways, as is shown by the formula

$$S = (N!) * 2 (\exp N),$$

where S is the number of possible sequences and N the number of segments in the pattern. Thus, a three-segment pattern has 48 possible stroke sequences; a six-segment pattern has as many as 46,080 solutions. Obviously, constraints to guide a draughtsman can be extremely useful when he is faced with the task of copying multi-segment patterns.

Like in all motor behavior, the subject will bring his or her implicit knowledge of the perceptual-motor system, including its mechanical characteristics, to the copying task. But also prior learning and educational (especially handwriting) and further cultural factors will impose constraints on the selection procedure by the subject. In general terms, this procedure is guided by economy: Spending a minimal amount of effort to achieve satisfactory performance is also a principle of graphic behavior. Research on this topic over the past two decades has demonstrated the existence of a number of principles, or 'rules', which in Bruner's terminology might together be considered to make up a grammar of (graphic) action. These studies have revealed developmental trends in the obeyance of such rules (Goodnow & Levine, 1973; Lehman & Goodnow, 1975; Nihei, 1983; Ninio & Lieblich, 1976; Thomassen & Teulings, 1979). Also their relation to educational and cultural handwriting biases has been investigated (Goodnow et al., 1973; Nachson, 1983; Nachson & Alek, 1981; Simner, 1981, 1984). Visual-control aspects were recently studied by Smyth (1989) and a more general, cognitive framework was proposed in recent years by Van Sommers (1984; 1989).

What then are the graphic production 'rules' that have been demonstrated in the studies quoted above? If we limit ourselves to righthanded adults educated in our alphabetic (left-to-right) writing system, the following sets of 'rules' have been reported in the literature. (a) starting points of a pattern are preferably chosen at the top or at the left-hand extreme of the pattern, and there is a tendency to start with vertical segments; (b) stroke directions are usually rightwards or downwards; (c) stroke sequences often involve a minimum number of pen lifts (this tencency we will call 'threading' in the rest of this article); (d) following a pen lift, the start of the next stroke is often made from a point Bon a stroke drawn earlier (this tendency we will call 'anchoring'); and (e) identical and parallel lines are preferably drawn in immediate succession (to be called 'parallel performance'). A summary of these 'rules' is presented in Table 1.

It will be clear that the term 'rules' may not be implied here to have the status of rules of grammar, or production rules (following which if a condition is present the action is always performed). Instead, the 'rules' we are discussing are biases, preferences, or constraining principles of a certain strength. Some biases are stronger than - and can dominate over - others; some govern the production of all graphic patterns, others apply only to specific patterns. In the remainder of this article, we will refer to them as graphic-production rules, or simply as rules without quotation marks.

Seen from an economic viewpoint, the following global justifications for graphic production rules may be illuminating. Drawing in a preferred movement direction (rightwards, downwards) can reduce jerk, due to the biomechanics either of the effector or of the drawing materials, or both. Threading avoids the demanding visuo-motor control required to reposision the stylus onto the writing surface. Anchoring facilitates the determination of the locus from where to start the next stroke once the stylus has been lifted. Drawing parallel and identical segments in immediate succession allows the repeated use of the same motor program. In the present context, however, we do not intend to clarify the origin and nature of the rules. This is attempted in a current project at the authors' department (see the references to Thomassen et al. below, and Meulenbroek & Thomassen, 1990; in preparation).

Table 1. Starting an progression rules in graphic production as mentioned in the literature.

a	Start at the top of the pattern
	Start at the left-hand extreme of the pattern
	Start at a vertical segment
b	Draw strokes downwards
	Draw strokes rightwards
c	Thread: continue pen-down
d	Anchor: connect to earlier strokes
e	Draw parallels in immediate succession

In an earlier study, we have shown that the strength of the rules determining the selection of stroke sequences is under the influence of a higher-order control mechanism. For the same patterns, different stroke sequences were adopted when the context was varied between non-linguistic and linguistic (Thomassen, Tibosch & Maarse, 1989). In a very recent study, moreover, we demonstrated that the economy of parallel performance is often anticipated and may influence the organization of early parts of the sequence even if the advantage of using the same motor program more than once is effectuated only in the terminal stage of executing the pattern (Thomassen, Meulenbroek & Hoofs, 1991). We will refer to this study in the Discussion section. In another experiment, to which we will also return, we showed that the operation of graphic production rules is reflected by the latencies and kinematics of the performed pattern strokes (Thomassen, Meulenbroek & Tibosch, 1990).

The present study was undertaken to test the feasibility of a simulation model implementing a set of graphic-production rules with appropriate weights to be determined empirically. The general design of this study comprises two phases, viz., an experiment and a simulation study. The experiment is concerned with collecting the human data. These were simulated in the next phase by a computer model based on a minimum number of assumptions which was designed to assign the optimal weight to each of the implemented rules. Below we will report the experiment and the simulation separately. Major features of the study are thus (i) the collection of empirical data; (ii) the design of a stroke-selection algorithm; (iii) the determination of the weights applying to a given set of patterns; (iv) the assessment of the resulting hierarchical or dominance relationships among the rules; moreover (v) an attempt is made to decrease the number of rules to a minimum.

Indeed, it may be clear from Table 1 that some of the rules overlap with others (Rules 5 and 6; Rules 2 and 8), whereas others - if applied to the same segment - are mutually exclusive (Rules 1 and 3), and others may or may not result in conflict (Rules 2 and 5), depending on the pattern's configuration. Finally, some rules may simply not apply to a specific pattern or to a class of patterns.

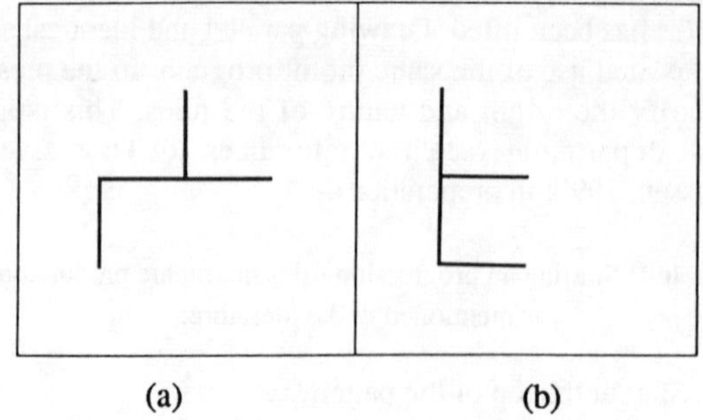

(a) (b)

Figure 1. Two stimulus patterns whose copying may involve low (a) and high (b) compatibility among graphic production rules.

Two examples may illustrate this (see Figure 1). In the pattern of Figure 1a, Rules 1, 2 and 4 apply to the two bottom segments, but their obeyance implies violation of either Rule 6 or Rule 8. Moreover, with respect to the top segment, Rule 3 is incompatible with Rule 6. Rules 4, 5 and 6 may be applied, but they are incompatible with Rules 2 and 3. If Rule 7 is applied, this should lead to the violation of either Rule 1 or Rule 3. In contrast, in the pattern of Figure 1b, Rules 1 through 8 can all be obeyed without conflict if first the long L shape is drawn, followed by the horizontal cross bar. The question now is what are the hierarchical or dominance relationships among the rules? Obviously, any class of patterns may be expected to enhance or decrease the strength of certain rules (e.g., the presence of many parallel lines may boost the strength of Rule 7; the presence of an apex at the top may resolve any conflicts between Rules 2 and 5). It therefore appears necessary to perform the intended analysis on a well-defined, fairly large class of geometrical patterns. We have opted for the class of one, two, and three-segment patterns which can be drawn in an orthogonal matrix of 3 * 3 dots (see Figure 2a).

2. Experiment

2.1. METHOD

2.1.1. *Subjects.* Fifteen students of the Department of Experimental Psychology (2 females, 13 males) took part in the experiment as unpaid volunteers. They were aged between 25 and 30 years and righthanded as determined by their preferred hand for writing and drawing. The subjects were unfamiliar with the topic of investigation and unaware of the purpose of the experiment.

2.1.2. *Materials.* The stimulus patterns were all the possible different patterns composed of one, two, or three connected straight line segments that can be drawn in a virtual orthogonal 3 * 3 dot matrix, excluding oblique segments (see Figure 2a). Every pattern with its unique structural properties was adopted in all its different orientations and rotations. The crossing of line segments could of course occur only in the center of the virtual matrix. In this way, 4 one-segment patterns, 25 two-segment patterns, and 120 three-segment patterns were adopted, totalling 149 stimulus patterns.

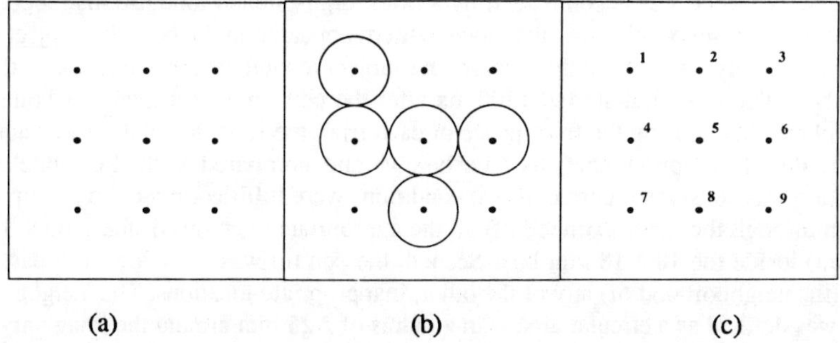

(a) (b) (c)

Figure 2. The virtual 3 * 3 matrix underlying the stimulus patterns (a) , an illustration of the neighborhood of a number of dots (b), and the digits used for coding the trajectories (c).

2.1.3. *Apparatus*. The experiment was controlled by an Olivetti M280 personal computer. The pen-tip displacements in the drawing plane (i.e., on paper) and above this plane were recorded by means of a digitizer (XY tablet; Calcomp 23180). The special laboratory-made pen and the recording and signal-analysis techniques have been described elsewhere (Maarse, 1987; Teulings & Maarse, 1984; Teulings and Thomassen, 1979). Pen-pressure criteria were used to differentiate 'pen-down' movements on paper from 'pen-up' movements above the writing surface. The X and Y coordinates and axial pen pressure were sampled at a rate of 100 Hz. A black-and-white EGA monitor, located about 50 cm before the seated subject, was used for the presentation of the stimulus patterns. Each pattern was presented in a box of 24 * 24 mm in the center of the monitor display. The size of the patterns within this box was limited to 12 * 12 mm, so that the 6-mm outer edge within the box was always empty. The background of the monitor screen was white, the contour of the box was grey, and the pattern segments were black.

The patterns were copied on normal white A4 paper sheets taped onto the digitizer. In the top half of each sheet, a rectangle of 90 mm width and 36 mm height, divided into two rows of five empty boxes of 18 * 18 mm, was printed in thin, grey lines to match the grey lines of the boxes appearing on the display. Only this limited region of the response sheet, allowing no more than ten responses to be entered in the boxes, was used to avoid large differences in arm and hand position as well as inconvenient postures due to a lack of resting space. Before the start of the session, the subject was free to rotate the digitizer, and to place it at a comfortable distance. The resulting orientation always had the horizontal sides of the response boxes at a slight positive angle with respect to the table edge. The ten boxes on each response sheet were used from left to right, the top row preceding the bottom row.

2.1.4. *Procedure*. The 149 patterns to be copied were presented twice in the same random order. There were two such orders, one for eight subjects, the other for the remaining seven subjects. Ten patterns were selected to serve as preliminary practice materials; these patterns occurred also among the experimental stimuli. The subject was required to hold the pen tip a few millimeters above the center of the next box of the response sheet until the next model pattern was displayed, and then to copy it in that box immediately in approximately the same relation to the box. Response sheets were changed after every ten trials.

A trial consisted of three phases. First, and empty box was presented in the center of the screen; its appearance was accompanied by a brief, high-pitched tone (50 ms, 2000 Hz). Then, 500 ms after the onset of this tone, the model pattern appeared in the box; the subject copied this pattern immediately. During the third phase, the subject moved the pen tip to above the center of the next box. The next trial started 1500 ms after the pen tip approached this point above the drawing plane. Also during the third phase of each trial, the recorded pattern was automatically coded for the stroke-order analysis (see below) and compared with the model pattern. A reproduction was considered correct if two conditions were fulfilled. First, the pen tip had to pass pen-down through the (neighborhood of) all the appropriate locations of the virtual 3 * 3 matrix (9 * 9 mm) inside the 18 * 18 mm box. Second, the pen tip was not allowed to pass pen-down through (the neighborhood of) any of the other, inappropriate locations. The 'neighborhood' of a location was defined as a circular area with a radius of 2.25 mm around the imaginary matrix dot (see Figure 2b).

Following each response, the subject was informed as to whether or not his or her graphic production was correct in the above sense; the criteria themselves, however, were unknown to the subject. Non-correct attempts were followed, not by the normal tone, but by a longer-lasting, lower-pitched tone (300 ms, 250 Hz) announcing the next trial. The subjects were told not to let their spontaneity drop by such negative feedback, but just to try and draw more accurately in the next trial. Rejected trials were repeated automatically at the end of the session. A session, including the 10 practice trials, the 298 experimental trials and the repeated trials, lasted 30 to 45 minutes.

2.1.5. *Data analysis.* As stated above, the patterns fitted in a virtual 3 * 3 dot matrix with dots separated 4.5 mm horizontally and vertically. The nine dots were numbered for the purpose of the stroke-sequence analysis (see Figure 2c). Each accepted production could now be coded by the sequence of numbers passed along the drawing track, and by a comma for each pen lift. The patterns depicted in Figures 1a and 1b could, for example, have been produced as 456,47,52 and 1478,45 respectively. The 15 * 2 = 30 productions of each of the 149 patterns performed by the 15 subjects were classified according to these codes. The results take the form of a list of such codes per pattern, together with the frequency of its occurrence (see Table 2 for some examples).

2.2. EXPERIMENTAL RESULTS

As just mentioned, for each of the 149 patterns, the results of the experimental phase consisted of a list of sequence codes with a frequency attached to each code. Usually, only a very limited number of different sequences appeared to be adopted over the 30 productions per pattern. Across all 149 patterns, the mean number of possible sequences was 40.1. The observed mean number of different sequences was only 4.2. This evidently reflects the strongly constraining effect of graphic-production rules. Moreover, there was a high level of conformity among the subjects. The highest frequencies across the 149 lists had a mean of 19.8. In other words, on average, 66 percent of the subjects adopted one and the same stroke sequence for a pattern, for which - on average - there were just over 40 solutions.

There were considerable differences, however, indicating that patterns involving conflicting rules (see Figure 1a) led to a larger variability in adopted stroke sequences than non-conflicting patterns (see Figure 1b). Thus, for each of the 149 patterns, the results of the experimental phase consisted of a list of sequence codes with a frequency attached to each code. A simpler illustration, involving two-segment L shapes, is provided by Table 2. Note that even in the case of conflicting pattern No. 14, only four different sequences were adopted, whereas according to the above formula, for N = 2 segments, there are S = 8 possible sequences.

Table 2. Examples of the stroke-sequence codes and their observed frequencies over 30 productions of two different two-segment patterns

Pattern	Sequence code	Frequency	Total
No. 10	14789	30	30
No. 14	78963	12	
	36987	10	
	789,369	4	
	369,789	4	30

3. Simulation

3.1. GENERAL OUTLINE OF THE MODEL.

The strength of each of the graphic-production rules was estimated by means of an iterative probabilistic procedure aimed at the gradual reduction of the discrepancies between the simulated and the observed stroke-sequence frequencies. The model, whose architecture is depicted in Figure 3, selects and applies rules on the basis of their applicability and effectiveness (usefulness) and rejects rules in cases of conflict. The model is characterized by probabilistic selection procedures. Moreover, for each segment to be copied an identical set of alternatives is considered. In a number of stages, the model arrives at the starting point and direction of the next stroke in the task of copying the pattern currently dealt with. For each segment of the stimulus pattern, every rule is in principle considered. If a rule does not apply to a pattern in its current stage of being copied, that rule is rejected. The probability of a rule to be selected next for consideration is different for each rule. If more than one rule is selected at any stage, and if they result in 'conflict', rules with lower selection probabilities are subsequently eliminated with higher probabilities.

3.2. OPERATION OF THE MODEL

The model first determines the number of segments to be copied. A segment is defined here as a straight line segment in the pattern. In the present study, this number N varies between 1 and 3. The pattern's copy is ready when N segments have been specified ('copied'). As indicated in Figure 3, the first rule is now retrieved at random from the pool of eight rules presented in Table 1. Its retrieval probability is a function of its relative strength as a proportion of the sum of the strengths of the entire pool. If the rule is both applicable and effective, i.e., if it can be applied and if it constrains the number of starting locations, it is put onto an initially empty stack. If the rule is not applicable, or if an applicable rule is not effective (not 'useful'), it is discarded until the next segment is dealt with.

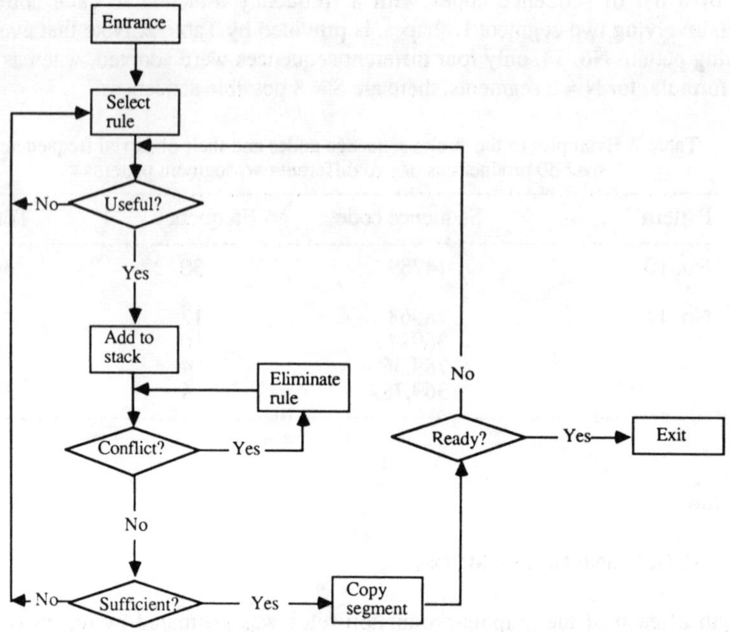

Figure 3. Architecture of the proposed model for simulating stroke sequences.

The stack is now checked as to the presence of conflict and sufficiency. First, the stack of effective rules is stripped of any conflicting rules. Two or more rules are considered conflicting if together they constrain the number of starting rules to zero. At least one of these rules has to be removed from the stack. The criterion for a rule to be maintained is again its relative strength. Weaker rules have a proportionally higher probability of being eliminated from the stack at this stage.

Subsequently, a test is performed as to whether the remaining non- conflicting rules in the stack prescribe with sufficient accuracy the starting point and the stroke direction of the next stroke. If more than one possibility remains, a further rule is retrieved, and if it is effective, it is added to the stack. This may have to be repeated more often until the first stroke is finally defined unambiguously. At this stage, the segment is specified or 'copied' and the number of remaining segments is reduced by unity. The whole procedure is now repeated for the next segment until the number of remaining segments is zero, and the pattern copy is 'ready'.

3.3. WEIGHT DETERMINATION

The simulation study involved a series of runs of simulations. Each simulation consisted of 60 trials, following the above procedure, on each of the 149 patterns. Such a simulation thus yielded 60 stroke sequences distributed over a limited set of different stroke sequences (different 'solutions'). A simulated stroke-sequence frequency distribution for a given pattern was regarded 'correct' if it did not differ significantly from the distribution observed for the 30 human productions in the Experiment. An example of a frequency distribution obtained by the Simulation procedure as compared to one observed in the Experiment, is presented in the top part of Figure 4 below. The test statistic used was Chi-square, with alpha = .01. Yates' correction for continuity was applied. The procedure, furthermore, aimed at maximizing the number of correct stroke- sequence frequency distributions of a simulation (60 sequences per simulation per pattern), using the same set of rule weights over all 149 patterns.

In the first simulation, the eight rules all received the same initial weight value. In the second simulation, the first rule's weight was incremented to twice its initial value; and in the third simulation, the first rule's weight was three times its initial value. All the time, the weights of the other rules were held constant at their original values. This incremental procedure was repeated ten times. After every simulation, the number of correct distributions was noted and the best value for the first rule was retained. The same procedure was now applied successively to each of the seven remaining rules. This constitutes a 'run'. The following run of simulations started with the set of eight 'best' values (rule weights) thus obtained, and the procedure was reversed: the weights of each of the rules were now successively decremented in steps of one-tenth of the latter best' values. The next run again incremented the values, and the subsequent run, in turn, decremented them, always starting with the 'best' values obtained in the preceding simulation run. This procedure was repeated until no further significant changes were observed in the number of correct distributions.

3.4. SIMULATION RESULTS

The optimal values for the weights of the eight rules obtained following the above algorithm are presented in Table 3 (left-hand weights column). The performance of the model equipped with these weights is a proportion of .88 correct distributions. Twelve percent of the distributions deviated significantly ($p <= .01$) from the ones obtained from the human subjects in the Experiment.

Table 3. Graphic production rules and their weights as determined for one pattern set.

Number	Rule	Weight	
1	Threading	23.0	27.0
2	Starting at the left	21.0	24.5
3	Anchoring	16.0	18.5
4	Starting at a vertical	13.0	15.0
5	Starting at the top	12.0	14.0
6	Drawing downwards	6.0	--
7	Parallel performance	6.0	--
8	Drawing rightward	2.0	--

It was, furthermore, tested whether some rules could be abandoned altogether in view of the inherent redundancy in the set, as indicated in the Introduction section. It appeared that when Rules 6, 7, and 8 were effectively discarded by setting their weights at a constant zero value, the number of correct distributions did not decline significantly. The resulting values obtained with the five remaining rules in this reduction study are also given in Table 3 (right-hand weights column).

4. Discussion and conclusion

The huge constraining power of graphic-production rules has again been demonstrated in the Experiment. On average only one-tenth of the possible stroke sequences are adopted by human copiers. The results of the Simulation study show that the model is reasonably successful in simulating quantitatively not only the frequency of the most preferred sequence of a pattern, but also the distribution of the alternative stroke sequences which were adopted at lower frequencies by our subjects. In view of the limited set of assumptions, this is a notable achievement. A number of comments must, however be made with respect to both the model and its results.

Eight possible stroke sequences

Model	11	10	5	4	0	0	0	0
Mean observed	13.3	8.8	5.0	3.0	0	0	0	0
	12	10	4	4	0	0	0	0
	14	11	3	2	0	0	0	0
	11	10	3	6	0	0	0	0
	16	4	10	0	0	0	0	0

Figure 4. Effect of segment size on stroke sequence. The four different patterns (lefthand column) were copied in four different ways (stroke-sequence row at the top, where a dot indicates the start of the sequence); four other solutions were not adopted. In cases of unequal segments (two bottom rows), the stroke sequence depends on the relative positions of the segments. The model, treating these patterns as identical, was reasonably successful in predicting the means observed. It could be adapted to take also segment size into account.

Starting with the results, it can be shown that, by applying additional parameters, the model's performance could be improved considerably. Let us take as an example the length of the segments, which was neglected in the present study. Our model treats, e.g., every inverted L shape as equivalent, irrespective of the position of its long and short segments. Figure 4, however, shows that our subjects did adopt different stroke sequences: They appear to apply the strong threading rule more frequently, for example, if the longest can be copied in a preferred direction (rightwards). It thus appears that subjects select their stroke sequences not only on the basis of the applicability, effectiveness and weights of rules per se, but that the weights of these rules are moderated by the costs and benefits involved in the lengths of the trajectories involved. Such a moderation seems relatively easy to implement in our model.

Much more difficult to handle is a further, more fundamental limitation of the present model. It operates in an ad-hoc, opportunistic mode, not looking further ahead than just the next stroke. A more sophisticated model should take (significant parts of) the whole pattern into account. In doing so, it could anticipate the applicability of one or more stronger rules later in the sequence perhaps at the cost of producing an earlier stroke which violates a weaker rule. In a recent study, specifically devoted to the topic of anticipation (Thomassen, Meulenbroek & Hoofs, 1991), we were able to show that human subjects do anticipate such later advantage. In this case it concerned the benefit of parallel performance, which appeared only as a weak rule in the present study. An entire four-stroke sequence was often organized such that by the end of the movement sequence the repetition profit coud be gained. It seems impossible to implement this kind of anticipation in a model of the type presented above. A connectionist model should, however, be able to handle the remote influence implied in anticipation in graphic production.

With respect to Rule 7 (Parallel performance), it was a little surprising to observe a non-significant contribution. Among the set of 149, however, the number of patterns containing parallel segments of equal size (see Figure 1b for an example), was relatively small. In a stimulus set in which such parallels are systematically present, as in the Thomassen et al. (1991) study just quoted, different results may be expected. This brings us to the issue of generalizability across pattern sets. The present study was designed to test the feasibility of constructing a quantitative model of geometrical-pattern copying. We claim to have been relatively successful in our attempt. This leaves undiscussed, however, the extent to which specific features of a set of stimulus patterns, rather than tendencies in the subjects, determine the sought-after weights of the rules. We believe that this is largely an empirical question. For the present moment, it seems that the stronger rules are more likely to be 'universal' than the weaker ones. We have, for example, learned to regard the Threading rule as a very dominant one in a number of studies. We have also observed across different experiments that in righthanders the left-right tendency is slightly but consistently stronger that the top-down tendency.

Less surprising was our finding of little contribution by Rules 6 and 8. It was argued by Van Sommers (1984) that starting rules (our Rules 5 and 2) are fundamentally different from progression rules (our Rules 6 and 8). Indeed, this author demonstrated that lefthanders and righthanders have the same preference for starting locations, whereas their preferences for stroke directions differ. He also showed that preferences for starting position and stroke direction develop independently in children. In view of our experimental subject population (adult righthanders), however, we did not really expect that the Rules 5 and 2 would contribute much in the context of Rules 6 and 8, respectively.

A final comment is concerned with the concrete behavioral correlates of applying the graphic-production rules, which in a sense may be regarded as abstract principles. Recently, we were able to bridge the gap between these two domains. We demonstrated that the obeyance of abstract graphic-production rules is reflected both by the reaction times and by the kinematic features of the trajectories, on paper as well as above paper. These data indicate that rules generally facilitate the planning and execution stages of the movement sequence. The Anchoring rule, which appears to be associated with a precision strategy, forms an exception. It tends te be accompanied by longer latencies - evidencing higher-order planning - and is characterized by lower pen-down velocities - reflecting considerable on-line guidance and control of the movements.

We conclude that we have been able to provide an adaquate first-order approximation of a formal, quantitative account of the operation of graphic-production rules. We have indicated ways to improve the model's performance, but we have also pointed out the weakness of the 'short-sighted' approach inherent in the present model. Like most forms of complex action, planning a graphic production sequence involves the anticipation of movement segments further ahead. In the case of copying geometrical patterns, visual imagery is most likely a major factor in such anticipation.

REFERENCES

Goodnow, J.J. & Levine, R.A. (1973). "The grammar of action": Sequence and syntax in children's copying. *Cognitive Psychology, 4,* 82-98.

Goodnow, J.J., Friedman, S.L., Bernbaum, M. & Lehman, E.B. (1973). Direction and sequence in copying: The effect of learning to write in English and Hebrew. *Journal of Cross-Cultural Psychology, 4,* 263-282.

Lehman, E.B. & Goodnow, J.J. (1975). Directionality in copying: Memory, handedness, and alignment effects. *Perceptual and Motor Skills, 41,* 863-872.

Maarse, F.J. (1987). *The study of handwriting movement: Peripheral models and signal processing techniques.* Lisse: Swets & Zeitlinger Publishers.

Meulenbroek, R.G.J. & Thomassen, A.J.W.M. (1990). Stroke-direction preferences in drawing and handwriting. *Human Movement Science,* (in press).

Meulenbroek, R.G.J. & Thomassen, A.J.W.M. (in preparation). Stroke-direction preferences as a function of arm position, handedness, and hand posture.

Nachshon, I. (1983). Directional preferences of bilingual children. *Perceptual and Motor Skills, 56,* 747-750.

Nachshon, I. & Alek, M. (1981). The development of directional preferences: Cross-cultural differences. *Psychologia, 24,* 86-96.

Nihei, Y. (1983). Developmental change in drawing and handwriting. *Acta Psychologica, 54,* 221-232.

Ninio, A. & Lieblich, A. (1976). The grammar of action: Phrase structure in children's copying. *Child Development, 47,* 846-849.

Rosenbaum, D.A., and Jorgensen, M.J. (1991). Planning macroscopic aspects of manual control. *Human Movement Science,* (in press).

Rosenbaum, D.A., Vaughan, J., Barnes, H.J., Marchak, F., and Slotta, J. (1990). Constraints on action selection: Overhand versus underhand grips. In M. Jeannerod (Ed.), *Attention and performance XIII* (pp. 321-342). Hillsdale, NJ: Erlbaum.

Simner, M.L. (1981). The grammar of action in children's printing. *Developmental Psychology, 17,* 866-871.

Simner, M.L. (1984). The grammar of action and reversal errors in children's printing. *Developmental Psychology, 20,* 136-142.

Smyth, M.M. (1989). Visual control of movement patterns and the grammar of action. *Acta Psychologica, 70,* 253-265.

Teulings, H.-L. & Maarse, F.J. (1984). Digital recording and processing of handwriting movements. *Human Movement Science, 3,* 193-217.

Teulings, H.-L. & Thomassen, A.J.W.M. (1979). Computer-aided analysis of handwriting movements. *Visible L¬nguage, 13,* 218-231.

Thomassen, A.J.W.M., Meulenbroek, R.G.J. & Hoofs, M.P.E. (1991). Economy and anticipation in graphic stroke sequences. *Human Movement Science,* (in press).

Thomassen, A.J.W.M., Meulenbroek, R.G.J., and Tibosch, H.J.C.M. (1990). Latencies and kinematics reflect graphic production rules. *Human Movement Science,* (in press).

Thomassen, A.J.W.M. & Teulings, H.-L. (1979). The development of diretcional preference in writing movements. *Visible Language, 13,* 299-313.

Thomassen, A.J.W.M., Tibosch, H.J.C.M., and Maarse, F.J. (1989). The effect of context on stroke direction and stroke order in handwriting. In R. PLamondon, C.Y. Suen, and M.L. Simner, (Eds.), *Computer recognition and human production of handwriting.* Singapore: World Scientific.

Van Sommers, P. (1984). *Drawing and cognition.* Cambridge: Cambridge University Press.

Van Sommers, P. (1989). A system for drawing and drawing-related neuropsychology. *Cognitive Neuropsychology, 6,* 117-164.

ON THE ORIGIN OF ASYMMETRIC BELL-SHAPED VELOCITY PROFILES IN RAPID-AIMED MOVEMENTS

RÉJEAN PLAMONDON
Laboratoire Scribens
Département de Génie Électrique et Informatique
Ecole Polytechnique de Montréal
P.O. Box 6079, Station "A"
Montreal QC, CANADA, H3C 3A7

ABSTRACT. This paper proposes a stochastical model for the origin and the invariance of the bell-shaped velocity profiles, as frequently reported in studies dealing with rapid aimed movements. The model describes these movements as originating from the sequential action of a set of velocity generators working in cascade. Applying the central limit theorem to describe the converging behavior of such a system, it is shown that velocity profiles can be described by log-normal curves. Results of an analysis by synthesis experiment are reported and some consequences of this approach to the study of human movement are discussed. A unified view of the speed-accuracy trade-offs is proposed as a direct consequence of this approach.

1. The Invariance of Velocity Profiles

Many investigators have reported that the velocity profiles of rapid-aimed movements are approximately bell-shaped (Beggs and Howard (1972), Georgopoulos et al. (1981), Morasso (1981), Soechting and Laquantini (1981), Abend et al. (1982), Atkeson and Hollerbach (1985)). Moreover the shape of the bell, after appropriate rescaling, is approximately superimposable that is, the shape is almost preserved for movements that vary in duration, distance or peak velocity (Atkeson and Hollerbach (1985), Bullock and Grossberg (1987)). These velocity profiles have also been observed in handwriting, for curvilinear velocity (Morasso and Mussa Ivaldi (1982), Plamondon et al. (1990a)) and for angular velocity as well, (Plamondon (1989), Plamondon and Yergeau (1990)).

This invariance in the shape of the velocity profiles suggests that velocity might play a key role in movement control, and that the CNS might take this information into account for movement planning.

Several indirect evidences also support this hypothesis. Houk and Gibson (1987), Gibson et al. (1985)) have shown that short high frequency bursts transmitted from the red nucleus to the spinal cord in the rubrospinal track, as recorded from monkey subjects that were trained to perform a visual tracking task, code movement velocity in terms of burst frequency of the firing cells. Moreover, it was shown that burst duration correlates closely with movement duration and the number of spikes in a burst correlates with

J. Requin and G. E. Stelmach (eds.), Tutorials in Motor Neuroscience, 283–295.

movement amplitude. Based on these observations, it was suggested that red nucleus signals may serve as a velocity command (Houk (1989)).

Other arguments favorizing a velocity control come from the study of handwriting generation. This phenomenon may be seen as a motor task producing a certain spatial output within relatively stringent time limits. Controlling velocity seems to be the simplest way to perfrom such a task. Handwriting can then be divided into strokes with relatively less activity at the beginning and at the end. These segmentation points can be seen as spatial targets and it is possible to develop a motor program based on such targets in a training or learning phase. Moreover, the velocity vector is the sole dynamic information which is uniquely related to the pentip trajectory: it is always tangent to the trajectory and can thus be recovered, at least partially, from visual inspection.

Finally, in a comparative simulation of 14 simplified models, it has been shown (Plamondon and Maarse (1989)), that velocity controlled models yield the best reconstruction output in an analysis by synthesis by experiment. Similarly, in a study of comparative performance of position, velocity and acceleration signals for automatic signature verification, it has been suggested that velocity domain is one of the best representation space for a 2D signature verification system (Plamondon and Parizeau (1988)).

2. Previous Models of Velocity Profiles

Many studies have been run to describe the velocity profiles mathematically. Although these models were not all developped in the same context nor with a common goal, most of them has been dealing with the generation of handwriting.

In the sixties, Eden (1962) proposed a velocity model where sinusoids were used as basic functions. With the appropriate difference in oscillation frequency of the orthogonal X and Y axes, plus a proper phase shift, handwriting movements could be generated, provided also that a linear trend was added to the horizontal displacement. Mermelstein (1963) modified this representation to incorporate a different velocity amplitude for the rise and fall of input excitation. He found that in this case handwriting could be better simulated, at the expense of using a larger set of fitting parameters: X and Y velocity amplitude, frequency and phase shift between X and Y velocities, and timing of velocity changes. Hollerbach (1981) carried out an extensive study of the practical interest of the sinusoidal model in controlling the shape, height, and slant of handwriting. He analyzed the basis of such a representation in the context of a muscle-spring oscillatory model.

More recently, Morasso and Mussa Ivaldi (1982) used cubic spline functions to simulate the bell-shaped velocity profiles. The generated profiles resulted from a weighted sum of B-splines, which were overlapped in time. Each basic strokes were then described by a set of five parameters: length, tilt, curvature, time delay and duration.

Applying the transfer function of the dc motor used in the Vredenbregt and Koster's (1971) handwriting simulator, Plamondon and Lamarche (1986) described the handwriting processes with a system based on a velocity generator. Curvilinear movements were analyzed with a simplified second order version of this model. It was demonstrated that at least a third-order linear system was necessary to simulate human handwriting performance. This approach was later generalized to the study of 2D pentip trajectory (Plamondon (1989)).

Maarse (1987), among other things, compared different types of velocity functions (step, trapezoidal, exponential, sinusoidal, triangular and bell-shaped from superimposition of sinusoids), fed into an integrator to generate handwriting trajectory. Sinusoidal functions were found to lead to better performance in this experiment.

Bullock and Grossberg (1987) used neural networks to generate velocity. Bell-shaped profiles were produced from that system, provided that a proper GO signal was used as a command. Finally, Plamondon et al. (1990a) have used a piece-wise concatenation of half-gaussians to simulate velocity profiles and generate curvilinear trajectories. This approach was found to give the better results than a second-order system, in terms of mean square error on the velocity and the displacement.

Only a few of these models have been successful in reproducing the real shape of specific velocity profiles. Moreover, most of them remains mainly descriptive, except perhap, for the Bullock and Grossberg (1987) model that provides some internal insight on the trajectory formation, in terms of simple neural networks and also the gaussian model proposed by Plamondon et al. (1990a). This latter model was studied extensively in the context of a psychophysical experiment (Plamondon et al. (1990b), Plamondon and Clément (1990)). The model parameters were interpreted in terms of their central or peripheral levels of influence in movement planning and analysed statistically as a function of the direction of movement and the presence or absence of precue information about these directions. This model was also generalized to the simulation of angular velocity as well (Plamondon (1989), Plamondon and Yergeau (1990)).

What are the causes of these asymetric bell-shaped velocity profiles? Where does their invariance come from? These are the key questions that have to be answered to select a proper set of functions to simulate and to study movement velocity and its invariance. This paper provides an answer to such questions by describing a genesis approach to velocity modelling. Experimental results are also presented to support the model and its interest for the study of human movement is discussed at the end.

3. A Genesis Approach

A general way to look at the origin of the velocity profile is depicted in Figure 1 (Plamondon (1990a,b)). The overall sets of neural and muscle networks involved in the generation of a single aimed-movement is assumed to result from the sequential actions of a set of n stochastic processes.

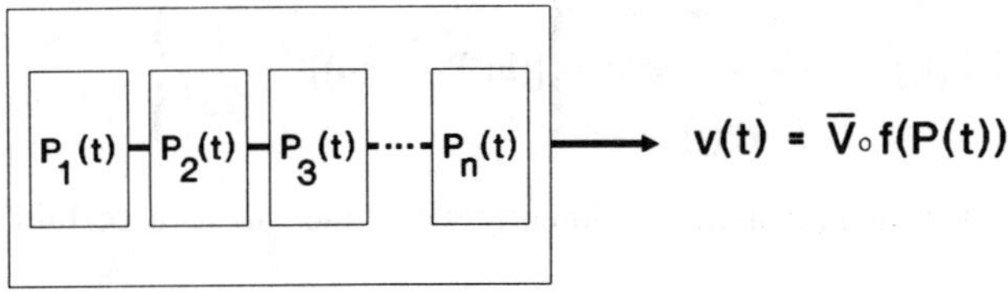

Figure 1 - A general model of rapid-aimed movement.

The internal description of these processing units is irrelevant to the model only their global statistical behavior is of interest. Some of them might be composed of neural networks, others of muscle fiber networks, some might incorporate internal feedback, etc... As long as they are in sufficient number, the theory holds. Assuming the ideal output for such a system to be a constant mean velocity $\bar{V}$ that would result in a displacement D when applied for a time interval $(t_1 - t_0)$:

$$D = \bar{V}\ (t_1 - t_0) \tag{1}$$

one can then consider the real output v(t) to reflect the probability P(t) of producing such an ideal output in a given time interval. In this context, one can write:

$$v(t) = \bar{V}\ f(P(t)) \tag{2}$$

where f(P(t)) is the probability density function of P(t).

If one assumes that all these processes are independent, in a sense that the jth subsystems does not load down the jth-1, the probability P(t) of producing a specific velocity at a given time will be the product of the probabilities $p_i(t)$ that each subsystem produces a response during that specific time interval:

$$P(t) = \prod_{i=1}^{n} p_i(t) \tag{3}$$

By taking the natural logarithm of this product,

$$\ln P(t) = \sum_{i=1}^{n} \ln p_i(t) \tag{4}$$

and assuming that the density function of each $p_i(t)$ satisfies the weak condition of having a finite variance, then, if n is large, the central limit theorem applies, Liapounoff (1900), (1901). The probability density function of ln P(t) is thus normally distributed around its mean μ with a variance σ^2 and P(t) is distributed according to a log-normal density function (McAlister (1879), Kapteyn (1903), Kapteyn and van Euven (1916), Aitchison and Brown (1966)):

$$f(P(t)) = \frac{1}{\sigma\sqrt{2\pi}P(t)} \exp - \left\{ [(\ln P(t) - \mu)]^2 \cdot \frac{1}{2\sigma^2} \right\} \tag{5}$$

If one assumes that P(t) increases linearly with time, up to a certain limit t_1:

$$P(t) = \frac{t - t_0}{t_1 - t_0}\ ,\ \text{for } t_0 \le t \le t_1 \tag{6}$$

where t_0 = time when the probability of response of the system starts increasing from 0 (reaction time of the system)

t_1 = time when the probability of response of the system is equal to 1, that is: t_1-t_0 = movement time.

then by combining the equation (5), (6) and (2), one obtains a mathematical description for the velocity profiles:

$$v(t) = \frac{\bar{V}\,(t_1-t_0)}{\sigma\sqrt{2\pi}(t-t_0)}\ \exp - \left\{\left[\ln\left(\frac{t-t_0}{t_1-t_0}\right) - \mu\right]^2 \cdot \frac{1}{2\sigma^2}\right\} \qquad (7)$$

4. Model Simulation and Validation

Equation (7) constitutes a mathematical description of an asymmetric bell-shaped velocity profile where the shape of the profile can be theoretically affected by five parameters: $\bar{V}$, t_0, t_1, μ and σ. The next figures shows the effect of each of these parameters on the profile. Figure 2 shows the amplitude scaling effect of $\bar{V}$ while Figure 3 depicts the time delay and the time scaling effect introduced by a change in t_0 and Figure 4, the time scaling effect of t_1. Figures 5 and 6 show the effects of a change in μ and σ respectively. As one can anticipate from these figures, these parameter effects can be easily combined to rescale the velocity profiles, as it is observed under different changes of experimental conditions, keeping the general aspect of the profile invariant under these conditions.

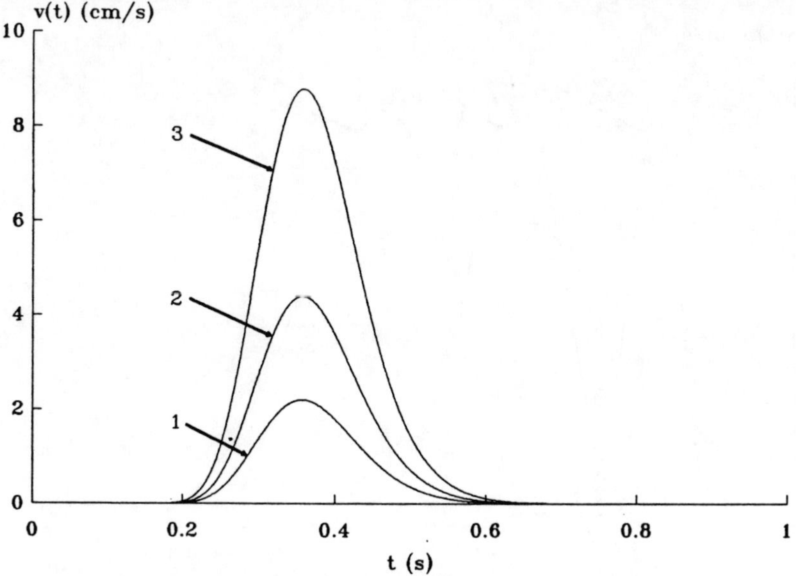

Figure 2 - Effect of the parameter $\bar{V}$ on the velocity profile.
The following parameter values were used in the simulation:
$\bar{V}_{1,2,3}$ = 0.5, 1.0, 2.0 cm/s; t_0 = 0.0 s, t_1 = 1.0 s, μ = -1.0, σ = 0.25.

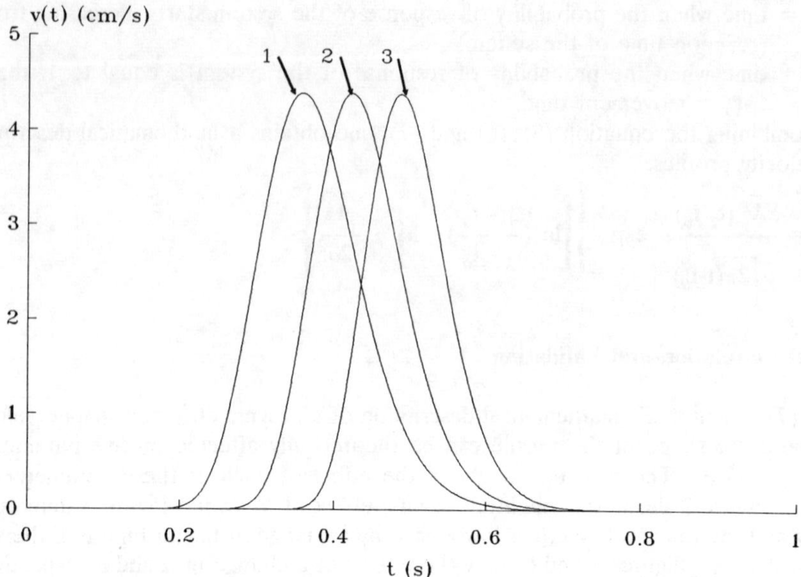

Figure 3 - Effect of the parameter t_0 on the velocity profile. The following parameter values were used in the simulation: $\bar{V}$ = 1.0 cm/s; $t_{0\ 1,2,3}$ = 0.0, 0.1, 0.2 s, t_1 = 1.0 s, μ = -1.0, σ = 0.25.

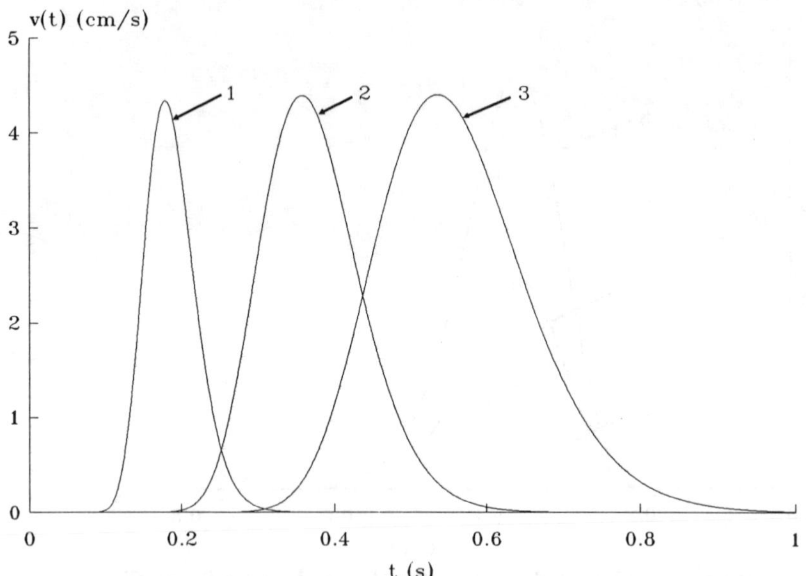

Figure 4 - Effect of the parameter t_1 on the velocity profile. The following parameter values were used in the simulation: $\bar{V}$ = 1.0 cm/s; t_0 = 0.0 s, $t_{1\ 1,2,3}$ = 0.5, 1.0, 1.5 s, μ = -1.0, σ = 0.25.

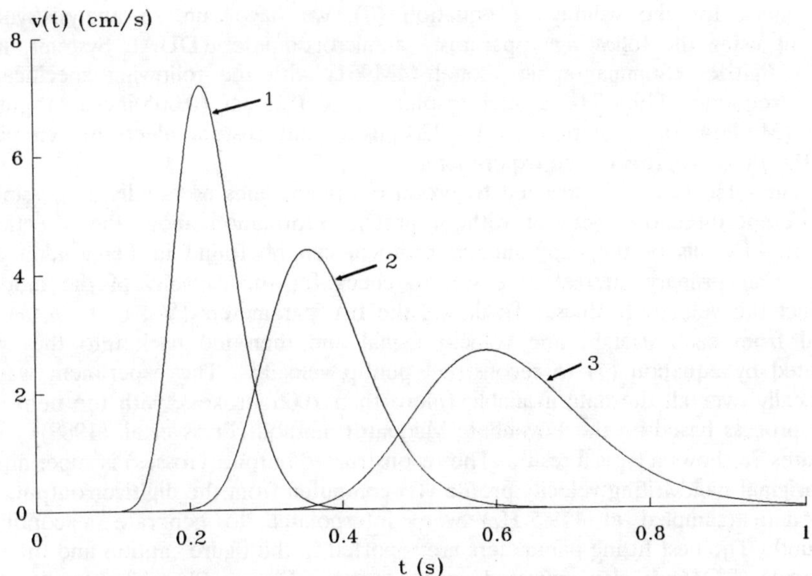

Figure 5 - Effect of the parameter μ on the velocity profile. The following parameter values were used in the simulation: $\bar{V}$ = 1.0 cm/s; t_0 = 0.0 s, t_1 = 1.0 s, $\mu_{1,2,3}$ = -1.5, -1.0, -0.5, σ = 0.25.

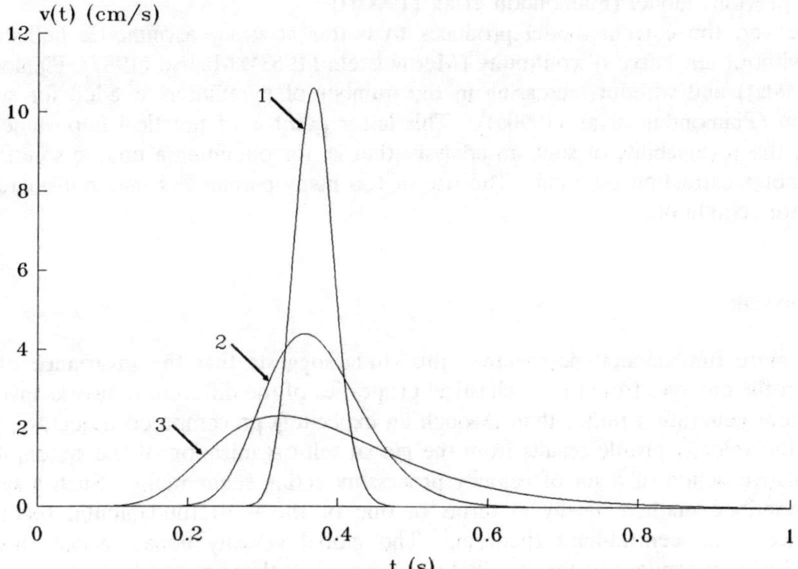

Figure 6 - Effect of the parameter σ on the velocity profile. The following parameter values were used in the simulation: $\bar{V}$ = 1.0 cm/s; t_0 = 0.0 s, t_1 = 1.0 s, μ = -1.0, $\sigma_{1,2,3}$ = 0.1, 0.25, 0.5.

To check for the validity of equation (7), we have run an analysis/synthesis experiment using the following apparatus: a microcomputer (DUAL Systems, model 83/20); a digitizer (Summagraphic, model MM961) with the following specifications: sampling frequency 119.5 Hz, spatial resolution: 0.0125 cm (0.005 inch); a graphics terminal (Modgraph), resolution 1024 x 728 pixels; and custom electronic circuits to control the time progress of the experiment.

Human subjects were instructed to produce straight lines as rapidly as possible, in eight different directions, with or without precue information about the direction of movement. Details of the experimental protocol can be found in Plamondon et al. (1990a). Our primary interest here was to check for the capacity of the model to reconstruct the velocity profiles. To do so, the five parameters ($\bar{V}$, t_1, t_0, μ, σ) were extracted from each straight line velocity signal and then fed back into the system represented by equation (7) to reconstruct pentip velocity. The experiment was run automatically over all the data available (more than 1000 strokes), with the help of an iterative process based on the Levenberg-Maquardt method (Press et al. (1988)).

Figures 7a shows a typical result. The reconstructed output (crosses) is superimposed on the original handwriting velocity profile v(t) computed from the digitizer output. The original data (sampled at 119.5 Hz) were interpolated to generate a continuous background. The best fitting parameters are reported in the figure caption and the mean square error (EQM) is also indicated on the curve. Figures 7b, gives details of the absolute error as a function of time.

Globally in this experiment, it was observed that 81% of the profiles analysed could be reconstructed with EQM < 0.35 cm^2/sec^2. This constitutes a significant improvement over our previous model (Plamondon et al. (1990a)).

Moreover, the current model produces to better accuracy asymmetric bell-shaped profiles without any curve discontinuity (Mermelstein (1963), Maarse (1987), Plamondon et al. (1990a)) and without increasing in the number of parameters needed for system description (Plamondon et al. (1990a)). This latter point is of practical importance for validating the reversability of such an analysis, that is, for obtaining a unique solution to the parameter extraction problem. The use of too many parameters may not guarantee this essential condition.

5. Discussion

From a more fundamental perspective, this study suggests that the invariance of the velocity profile emerges from the stochastical properties of the different networks involved in movement generation rather than through an explicitely precomputed trajectory. The shape of the velocity profile results from the global self-organization of the system, from the cumulative action of a set of velocity processors acting sequentially. Such a system can be described mathematically in terms of one of the most fundamental theory of convergence: the central-limit theorem. The global velocity behavior can thus be considered as independent of the detailed structure of all these networks in cascade and even of the type of input function used to command this system. This latter point is a major improvement as compared to all the previous models proposed in the field.

In a sense, the present study can be seen as a generalization of the neural network approach, showing what is happening when the generation of rapid movements is studied

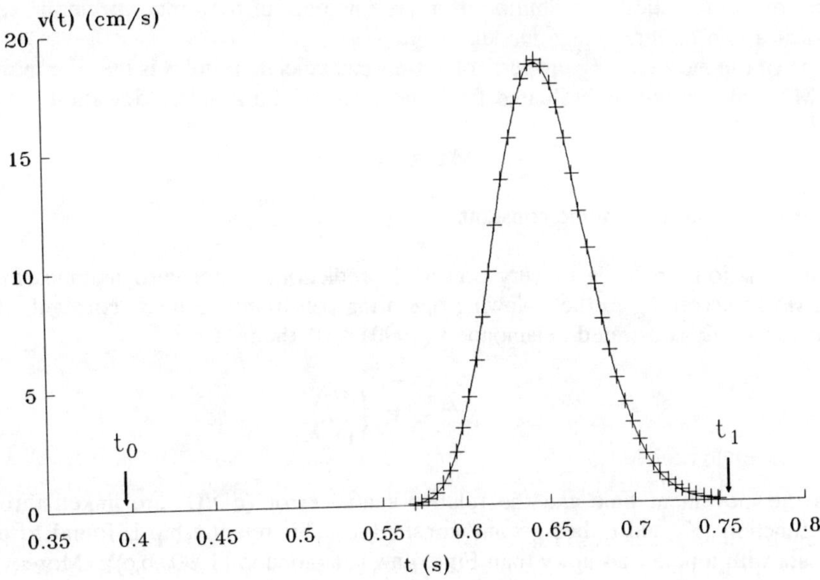

Eqm = 0.0014 cm2/s2

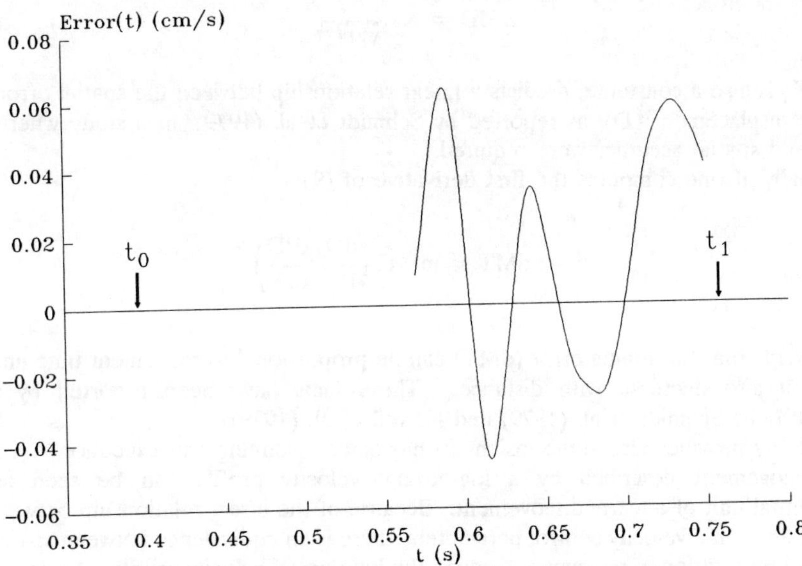

Figure 7 - Typical results of a reconstruction experiment.

(a) Reconstructed output (crosses) superimposed on the original velocity profile. The optimal reconstruction parameters were: $\bar{V} = 3.55$ cm/s; $t_0 = 0.398$ s, $t_1 = 0.753$ s, $\mu = -0.355$, $\sigma = 0.109$.

(b) Absolute reconstruction error, as a function of time.

in terms of the statistical contribution of a large number of networks working in cascade, instead of a two or three stage model.

One of the most striking property of log-normal velocity profiles is that the movement time (MT) can be directly evaluated from the value of the standard deviation (σ) of the profile:

$$MT = K_1 \sigma^n \tag{8}$$

where K_1 and n are some constants.

This relationship leads to very powerful predictions if one also assumes that the system works according to the following operating constraint: $\sigma \; d\sigma$ = constant. In this case, it can be demonstrated (Plamondon (1990a,b,c)) that:

$$MT = K_2 \left(\frac{dD}{D}\right)^{-m} \tag{9}$$

that is the movement time and the relative spatial error (dD/D) are linked through a power function (K_2 and m being some constants). This relationship is found to predict Fitt's data with a better accuracy than Fitts' Law (Plamondon (1990a,b,c)). Moreover the same equation, once properly rewritten:

$$dD = K_3 \frac{D}{MT^{1/m}} \tag{10}$$

where K_3 is also a constante, predicts a linear relationship between the spatial error (dD) and the displacement (D), as reported by Schmidt et al. (1979), in a study where both timing and spatial accuracy were required.

Finally, if one computes the first derivative of (9):

$$dMT = m \; MT \left(\frac{dD}{D} - \frac{d^2D}{dD}\right) \tag{11}$$

one predicts that the timing error (dMT) can be proportional to movement time and that dMT will also decrease with distance. These facts have been reported by many, particularly by Schmidt et al. (1979) and Newell et al. (1979).

This study provides also some insight to movement planning and execution. A rapid-aimed movement described by a log-normal velocity profile can be seen as the fundamental unit of a learned movement. Because of the direct relationship between the distance and some velocity control parameters, there is an equivalence between spatial and timing representation of movement through the log-normal velocity profile. Assuming that this knowledge is available to the CNS, more complex movement can be planned and generated by summing up these bell-shaped velocity profiles. This immediately suggests two types of execution processes: an anticipation process, where complex velocity profiles can be described in terms of superimposed log-normal curves and a correction process were velocity profiles can be reproduced by concatenating log-normal curves without superimposition.

6. Conclusion

The theory presented here is quite general. It shows that the asymmetric bell-shaped velocity profiles results from the global stochastical behavior of the large number of processes involved in the velocity control, in other words the invariance of these profiles can be interpreted as resulting from the global self-organization of the system. In the same context, it is shown that the previously reported relationship between movement time and spatial or timing accuracy are no conflicting observations of different independent phenomena but results from the intrinsic properties of the log-normal velocity profiles.

7. Acknowledgements

Part of this work was done when Réjean Plamondon was a Fellow of the Netherlands Institute for Advanced Study. It was also sponsored by CRSNG grant 0915 from Gouvernement du Canada and FCAR grant CRP2665 from Gouvernement du Québec. The author wants to thank M. Pierre Yergeau, engineer at Laboratoire Scribens, for his kind assistance in computer simulations, model verification and his critical support throughout this work and Hélène Dallaire for the typesetting of this paper.

8. References

Abend, W., Bizzi, E. and Morasso, P. (1982) "Human arm trajectory formation", Brain 105, 331-348.

Aitchison, J. and Brown, J.A.C. (1966) "The lognormal distribution", Cambridge University Press, 176 p.

Atkeson, C.G. and Hollerbach, J.M. (1985) "Kinematic features of unrestrainted vertical arm movements", Journal of Neuroscience 5/9, 2318-2330.

Beggs, W.D.A. and Howarth, C.I. (1972) "The movement of the hand toward a target" Quaterly Journal of Experimental Psychology 24, 448-453.

Bullock, D. and Grossberg, S. (1987) "Neural dynamics of planned arm movements: emergent invariants and speed-accuracy properties during trajectory formation", in S. Grossberg (ed.), Neural Networks and Natural Intelligence, MIT Press, 553 622.

Eden, M. (1962) "Handwriting and pattern recognition", IRE Trans. on Information Theory, vol. IT-8, 160-166.

Fitts, P.M. (1954) "The information capacity of the human motor system in controlling the amplitude of movement", Journal of Experimental Psychology 47/6, 381-391.

Georgopoulos, A.P., Kalaska, J.F. and Massey, J.T. (1981) "Spatial trajectories and reaction time of aimed movements: effects of practice, uncertainty, and change in target location", Journal of Neurophysiology 46/4, 725-743.

Gibson, A.R., Houk, J.C. and Kohlerman, N.J. (1985) "Relation between red nucleus discharge and movement parameters in trained macaque monkeys", Journal of Physiology 358 (London), 551-570.

Houk, J.C. and Gibson, A.R. (1987) "Sensorimotor processing through the cerebellum", in New Concepts in Cerebellar Neurobiology, Alan R. Liss Inc., 387-416.

Houk, J.C. (1989) "Burst of discharge recorded from the red nucleus may provide real measures of Gottlieb's excitation pulses", Behavioral and Brain Sciences 12/2, 224-225.

Kapteyn, J.C. (1903) "Skew frequency curves in biology and statistics", Astronomical Laboratory, Groningen: Noordhoff.

Kapteyn, J.C. and van Uven (1916) "Skew frequency curves in biology and statistics", Groningen, Hoitsema Bros., 75 p.

Liapounoff, A. (1900) "Sur une proposition de la théorie des probabilités", Bulletin de l'Académie des Sciences, St-Petersbourg 13, 359.

Liapounoff, A. (1901) "Nouvelle forme du théorème sur la limite de la probabilité", Mémoire de l'Académie des Sciences, St-Petersbourg 12/5.

Maarse, F.J. (1987) "The study of handwriting movement: peripheral models and signal processing techniques", Swets and Zerthinger, Lisse, The Netherlands, 160 p.

McAlister, D. (1879) "The law of geometric mean", Proceedings of the Royal Society of London 29, 367-376.

Mermelstein, P. (1963) "Study of the handwriting movement", Quaterly Progress Report n 69, MIT Research Laboratory of Electronics, 229-232.

Morasso, P. (1981) "Spatial control of arm movements", Experimental Brain Research 42, 223-227.

Morasso, P. and Mussa Ivaldi, F.A. (1982) "Trajectory formation and handwriting: a computational model", Biological Cybernetics 45, 131-142.

Newell, K.M. (1980) "The speed-accuracy paradox in movement control: errors of time and space", in G.E. Stelmach, J. Requin (eds.), Tutorials in Motor Behavior, North-Holland, 501-510.

Newell, K.M., Hoshizaki, L.E.F., Carlton, M.J. (1979) "Movement time and velocity as determinants of movement timing accuracy", J. Motor Behavior 1, vol. 11, 49-58.

Plamondon, R. (1990a) "A fundamental law of human movment", NIAS Conference on Sequencing and Timing of Human Movement, 19-20.

Plamondon, R. (1990b) "A unified approach to the study of target directed movements", Nato Advanced Study Institute, Tutorials in Motor Neuroscience Corsica, sept. 15-24.

Plamondon, R. (1990c) "A general framework for the understanding of rapid aimed movement", accepted, Human Movement Science, Special Issue on Sequencing and Timing, 22 p.

Plamondon, R., Yu, L.D., Stelmach, G.E. and Clément, B. (1990a) "On the automatic extraction of biomechanical information from handwriting signals", in press, IEEE Trans. on Systems, Man and Cybernetics.

Plamondon, R., Stelmach, G.E. and Teasdale, N. (1990b) «Motor program coding representation from handwriting generators models: the production of lines responses», in press, Biol. Cybernetics.

Plamondon, R. and Yergeau, P. (1990) "A system for the analysis and synthesis of handwriting", Proc. Int. Workshop on Frontier in Handwriting Recognition, 105-111.

Plamondon, R. and Clément, B. (1990) "Dependence of peripheral and central parameters describing handwriting generation on movement direction", in press, Human and Movement Science, Thematic Issue on Handwriting.

Plamondon, R. (1989) "Handwriting control: a functional model", in R. Cotterill (ed.), Models of Brain Functions, Cambridge University Press, 563-574.

Plamondon, R. and Maarse, F.J. (1989) "An evaluation of motor models of handwriting", IEEE Trans. on System, Man and Cybernetics, 1060-1072.

Plamondon, R. and Parizeau, M. (1988) "Signature verification from position velocity and acceleration signals: a comparative study", Proc. 8th Int. Conf. on Pattern Recognition, Rome, 260-265.

Plamondon, R. and Lamarche, F. (1986) "Modelization of handwriting a system approach", in H.S.R. Kao, G.P. van Galen, R. Hoosain (eds.), Graphonomics: Contemporary Research in Handwriting, Elsevier Science Publishers, B.V., 169-183.

Press, W.H., Flannery, B.P., Teukolsky, S.A. and Vetterling, W.T. (1988) "Numerical Recipies in C", Cambridge University Press, Cambridge, 540-547.

Schmidt, R.A., Zelaznik, H.N., Hawkins, B., Frank, J.S. and Quinn, J.T. (1979) "Motor output variability: a theory for the accuracy of rapid motor acts", Psychological Review 86, 415-451.

Soechting, J.F. and Laquantini, F. (1981) "Invariant characteristics of a pointing movement in man", Journal of Neuroscience 1/7, 710-720.

Vredenbregt, J. and Koster, W.G. (1971) "Analysis and synthesis of handwriting", Philips Tech. Rev., vol. 32, 73-78.

SEVERAL EFFECTORS FOR A SINGLE ACT: COORDINATION AND COOPERATION

Y. GUIARD
Laboratory of Functional Neurosciences
Unit of Cognitive Neurosciences
31, Chemin J. Aiguier
13402 Marseille Cedex 9
France

ABSTRACT. This chapter addresses the issue of motor coordination (Bernstein's degrees-of-freedom problem) in the context of cooperativity, in which several motor elements concur to the achievement of a single goal. The obvious reason that biological motion, at every scale, is composite (thereby demanding coordination) is that most motor problems encountered by organisms (e. g., locomotion) are insolvable at the level of a single degree of freedom: Partition of labour, the decomposition of the task into several roles that can be mapped onto a corresponding set of bodily degrees of freedom (actually an inverse degrees-of-freedom problem) is the primitive problem. This problem is posed both at the time scale of phylogenesis (gradual shaping of highly composite musculo-skelettal morphologies), and at the time scale of motor behaviour.

The concept of a motor, kinematically defined as a device that moves some mobile relative to some local reference frame, is introduced. A distinction is proposed between two classes of motors, ordinary motors, which must be assembled (and for this reason are assembled rigidly), and effectors, for which assembly is optional. A further distinction is proposed between a parallel and a serial logic of partition of labour among motor elements, whether ordinary motors or effectors. It is suggested that the fixed partition of labour that is evident among the elemental anatomical degrees of freedom of the body may model the flexible partition of labour that takes place in the diversity of human multi-effector, and notably bimanual, activities.

1. Introduction: The Ubiquitous Necessity of Coordination in Biological Movement

Why is the issue of coordination fascinating ? One reason that has become particularly apparent over the last decade, with the blossoming of the natural-physical approach to biological motion (see Kugler and Turvey, 1987; Zanone, this volume; Swinnen, this volume), is that the study of motor coordination offers researchers privileged opportunities to experience and demonstrate the continuity of psychology with physics. Not only basic activities like walking (Bernstein, 1967), but also quite elaborate, specifically human activities like the production of polyrhythms (Van Wieringen, this volume) or juggling (Beek, 1990) have been shown to be amenable to a few standard principles of nonlinear dynamics.

But besides the fact that coordination has become a particularly attractive topic for scientists, there is another, more profound reason: Coordination is important intrinsically (i. e., for organisms) because it is an absolute necessity for every single instance of movement: Whatever the nature of movement and the scale chosen for its analysis, one faces a number of anatomically distinct motor elements that work together. The most rudimentary prehension gesture involves the coordination of two component acts, a transport of the hand and a grasp. The transport involves two or three joints (the shoulder and the elbow, as well as the wrist if the global orientation of the hand needs to be adjusted to the object), each of which has several degrees of freedom; likewise, the grasp involves at least two opposable fingers, each made up of at least two segments; now, every elemental geometrical degree of freedom at anyone of the joints contributing to the prehension act involves at least two antagonist muscles or muscle groups; and so

J. Requin and G. E. Stelmach (eds.), Tutorials in Motor Neuroscience, 297–304.

forth down the scale. Quite clearly, it is because the movement of living beings is always compound that coordination is an ubiquitous necessity.

2. The Elements of Coordination: Motors and Effectors

To address the coordination issue from a general point of view, one needs some abstract term to denote the elements of motor coordination across scales. To take just one example, the atomism concept (Kugler and Turvey, 1987) may refer in principle to any anatomical entity that participates in movement: An individual fiber involved in the contraction of a muscle, a muscle involved in the control of some skelettal degree(s) of freedom, a finger involved in the activity of the hand, or even an entire upper limb involved in some multi-limb activity.

In recent publications, I have argued for a purely kinematic characterization of motor elements, and put forward the general notion of a *motor* (Guiard, 1987, 1988). Any device, whether natural or artificial, that serves or contributes to move some identifiable mobile relative to some identifiable local reference frame can be called a motor. This definition is kinematic in the sense that it considers displacements, based on the physical dimensions of length and time, but not mass, and therefore ignores the complex biophysical causality of movement. By this definition, it is easy to see that the biceps is a motor (which flexes the forearm, its mobile, relative to the upper arm, its reference frame). But, more interestingly, a handwriter's preferred upper limb, taken as a whole, can just as well be viewed as a motor which moves the pentip relative to the page. In no way is the latter instance more problematic than the former, provided that of course no confusion occurs between levels of description.

There is, however, one important difference between the two instances just evoked. Whereas a biceps cannot work efficiently on its own, an arm can. For the biceps to function in a significant fashion for the organism, obviously its activity must be coupled with the activity of other muscles. At the next higher level, we find the biceps-triceps coupling, which gives rise to a higher-order motor, but at this level action is not yet allowed to take place - except perhaps in some severely impoverished laboratory tasks. In fact one must proceed several steps up this hierarchy of analysis to recognise the possibility of action. The first motor that has the capability of acting is the whole upper limb: With an arm, one can grasp and throw an object, manipulate a racket, switch on the light or wave to someone; that is, on can display proper motor behaviour.

Certainly this latter subset of motors must be distinguished from the rest, and the term of *effector* appears to be quite appropriate. An effector may be defined as a motor (without any reservation with respect to the above definition) that serves some function of relevance to the organism considered as a whole. The specificity of an effector is that its functioning is describable not only in kinematic terms (physical *motion*) but also in behavioural terms (biologically significant *movement*). Obviously, effectors emerge above some critical level on the scale hierarchy and in some sense effectors are to movement science like cells to physiology, or molecules to chemistry.

If an effector is the smallest isolable entity that can act on its own, then it is the smallest isolable entity for which assembly is optional. Ordinary motors, which cannot act in isolation, will naturally tend to be rigidly assembled, as is the case for example with the various rotary motors that form a limb. For effectors, in contrast, the assembly will be temporary and revisable. A musician who is now playing the piano, with his or her two hands involved in the same keyboard activity, can in the next moment radically modify the effectors arrangement and play, for example, the pipe and tabor, with now the left hand working in association with the mouth, and the right hand taking charge of the drum. In sum, above some critical level on the scale hierarchy, we eventually find motors that need not be associated for significant action to emerge. What is most interesting, however, is that they still can.

3. The Source of the Degree-of-Freedom Problem: The Partition of the Motor Problem

So far I have emphasized that in biological movement the necessity of coordination arises from the simple fact that, at every conceivable scale, the movement of living organisms is composite. Put differently, one always finds a many-to-one mapping between biological motors and movement, with the consequent problem of understanding how assembly takes place: This, in essence, constitutes Bernstein's (1967) degrees-of-freedom problem.

It is basically the approach to this problem that seems to be at issue in the so-called motor/action controversy (see, for example, Meijer and Roth, 1988). Schematically, whereas the motor program approach is an attempt to explain the many-to-one mapping of motors onto movement in terms of hierarchically-organized sets of instructions generated by the central nervous system (see Shaffer, this volume), the action approach aims at providing an alternative explanation that only resorts to general principles of nature, and notably to laws of self-organization borrowed from nonlinear dynamics.

We may leave this debate aside and seriously consider the following question, which may look rather naive but is certainly not trivial: Why this pervasive degrees-of-freedom problem, that is, why is biological motion always composite? Here Bernstein's doctrine, which puts a strong emphasis on the primacy of the motor problem, may be worth recalling.

> "If movements are classified from the point of view of their biological significance to the organism making them, it is clear that on the first level of significance we have acts that solve one or another *motor problem* that the organism encounters. (...) *Meaningful* problems that can be solved by motor action arise, as a rule, out of the external environment".
>
> N. Bernstein (1967, p 115; italics in the original)

These lines of Bernstein's read like an ecological manifesto. Much like Gibson's (1966, p. 7) statement that a theory of perception should first stipulate the environment ("the "what" of perception"), Bernstein's suggestion is that a theory of motor action should start with the "what" of action, the motor problem. In some way, Gibson's (1966, p. 7) question "what is there to be perceived?" translates with Bernstein into the question: What is there to be done?

As soon as the primacy of the motor problem is recognized, it becomes apparent that the degrees-of-freedom (or coordination) problem is just one half of the story: The reason that biological motion is composite is that most motor problems are definitely insolvable at the level of a single degree of freedom. For example, locomotion, a displacement problem that arises at the scale of the whole body relative to its surrounding, implies the coordinated mobilization of several body parts, each with several degrees of freedom. Thus, it seems important to recognise that the many-to-one mapping problem has its source in a more primitive *one-to-many* mapping problem, namely the necessity for organisms of partitioning the treatment of the motor problem among several motors.

If we start from this inverse degrees-of-freedom problem, our view of coordination changes. Coordination, indeed, remains a necessary condition for successful action, but an understanding of coordination can no longer be likened to an understanding of motor behaviour. The reason that one cannot be satisfied with an account of how coordination takes place, whether by internal planning or by virtue of self-organization, is simply that this account does not tell whether motor action solves any motor problem, obviously the critical issue from the point of view of animal adaptation. A laboratory coordinative structure may have beauty for the scientist, but it has no ecological relevance whatsoever unless the coordinated elements concur to the solution of a given motor problem. Ultimately, the ecological relevance of a coordinative structure depends on whether the structure constitutes a *cooperative structure*.

4. Partition of Labour and Cooperation

In the coordination relation, the elements of motor action, be they simple motors or true effectors, pre-exist the coordinative structure. That is, their identity *qua* elements is independent of whether coordination among them occurs. We may say that coordination, the process whereby elements assemble into a higher order functional unit, has its locus at the element level - and this is true regardless of whether this process is construed as the product of central planning or emergent order.

In this respect, the relation of coordination is in sharp contrast to that of cooperation. Since to cooperate is to concur to the solution of one and the same problem, elements cannot be said to cooperate until the problem in question has been identified. In the cooperation relation, it is the problem that preexists the elements. Whereas the process critical to coordination is the assembling of elements, the process critical to cooperation is the partitioning of a whole. Now, just as the partitioning of a cinema script yields roles, but certainly not actors, it must be realized that the partitioning of a motor problem yields sub-problems (or task components), but not motors. In other words, the product of task partitioning should be distinguished from the ingredients of motor coordination: Rather than an identity relationship, at this level we have a one-to-one mapping relationship which associates a set of component motor problems and a set of motors.

5. The Taxonomy of Cooperation: The Parallel and the Serial Logics of Partition of Labour

We now turn to the problem of classifying cooperative structures in motor action, which we must examine in terms of partition of labour, as just suggested. The question is, How many ways of decomposing a motor problem can we distinguish ? At the level of effectors, this question seems rather hard to handle, for lack of any strict characterization of motor action. If, however, we consent to retreat to the domain of motion and ordinary motors, unequivocally definable in kinematic terms, it becomes apparent that the question has an answer: As shown in the block diagram of Figure 1, a motion problem can be partitioned in only two ways, in parallel and in series[1].

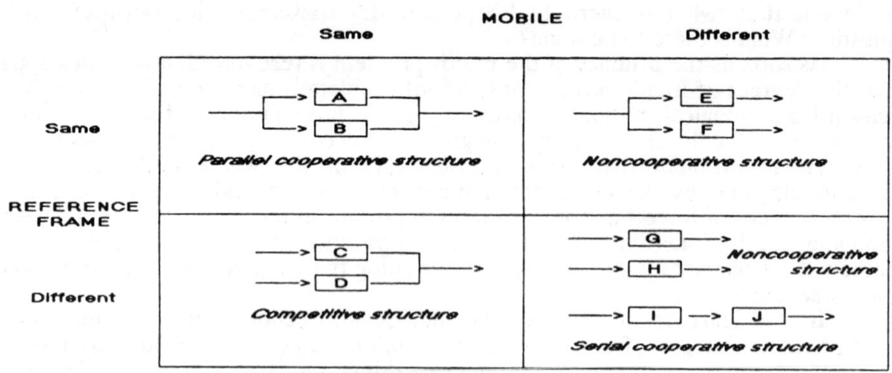

Figure 1. A simple taxonomy of motor assemblies. In all cases but one, at least one arrow is common to the two motors.

The construction of the block diagrams of Figure 1 obeys the constraint, imposed by our definition of motors, that locally each component motor must have its definite mobile and its definite reference-frame. We provide each block with an input arrow (standing for the reference frame) and an output arrow (standing for the mobile), each of which must correspond to some rigid body. In this analysis, we associate two motors in all possible ways and see whether the obtained structures form motors in their own right: For this condition to be met, obviously, there must be a single input to and a single output from the global structure. If this is the case, we have a cooperative structure.

Figure 1 illustrates the four possible ways of mounting two motors together, in the sense of sharing the reference frame and/or the mobile between them, with at least one arrow common to the two motors.

1) *Parallel cooperative structure*: The two motors (A, B) have the same mobile and the same reference frame.

2) *Competitive structure*: The motors (C, D) control the displacement of the same mobile relative to different reference frames.

3) *Noncooperative (and noncompetitive) structure*: The motors control the displacement of different mobiles relative to the same reference frame (E, F), or to different reference frames (G, H; note that in this latter case, mentioned just for the sake of clarity with respect to the logic of the table, the two motors are *not* mounted together).

4) *Serial cooperative structure*: The motors (I, J) have both different reference frames and different mobiles, but they do have an arrow in common, as the output from the proximal motor is the input to the distal motor. In other words, one motor (identifiable as distal) is mounted on top of the other (identifiable as proximal).

In all four cases of this list, the components we deal with are well defined motors. In cases 2 and 3, however, it is clear that the cooperativity criterion is not met: A competitive structure cannot be considered a motor, because it is equivocal with respect to the reference frame; neither can a noncooperative structure, because it has two mobiles (it should be viewed just as a set of two motors).

6. The Locomotion Problem: Parallel and Serial morphogenetic Differentiation

Thus, kinematic cooperativity among motors can be achieved in two fashions, by partitioning the motion problem in parallel and in series. As an illustration of this statement, we may take the example of locomotion, considered from the point of view of evolutionary morphogenesis. Although, at the time scale of behaviour relevant to psychology, the skelettal morphology of an animal constitutes an invariant, of course it actually is the present state of the adaptation of the species to its locomotion problem. We can view the differentiation that has taken place as a partition process, with more and more degrees of freedom involved in the solution of the problem.

If we examine the overall morphology of an animal equipped with a number of locomotor appendages, we can see quite clearly that evolution has partitioned the locomotion problem among dozens of skelettal segments both in parallel and in series.

As illustrated in Figure 2 (which takes the example of the human biped), we first find a parallel logic of partition of labour at a macro-level, between the legs: For the two legs, what is controlled is the displacement of the same mobile (the trunk) relative to the same reference frame (the ground). Note that this remains true regardless of the locomotion regime, e. g., walking vs. running (the latter variable has to do with coordination, with no change in the logic of cooperation; simply, a given cooperative structure may function in several coordination regimes).

At a micro-level, in contrast, the logic of partition of labour is serial, with each leg forming a multi-level kinematic chain. By construction, the function of the musculature working at the ankle is to rotate the foot relative to the lower leg, which in turn can be set into rotation relative to the upper leg by the musculature of the knee, which in turn can be rotated relative to the trunk by the musculature of the hip. As already emphasized in previous articles (Guiard, 1987, 1988), seriality of partition of labour in a kinematic chain entails a hierarchical organization of control: The more distal a motor in the chain, the more finely grained its contribution to overal motion[2].

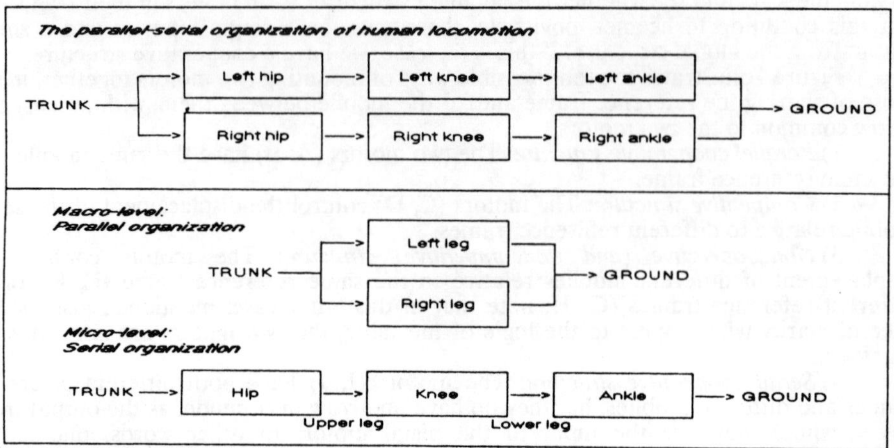

Figure 2. A two-level description of the fixed organization of labour for locomotion in the human biped: Parallel partition of labour at a macro level (between the legs), serial partition of labour at a micro level (within the legs).

7. Two hands for a single act: A Top-Down Approach to Bimanual Gestures

It is an interesting issue whether the rather simple parallel/serial taxonomy that applies to rigidly assembled motors, as just shown for locomotion, also applies to activities involving the revisable assembly of effectors. The domain of bimanual activities seems, for several reasons, of privileged interest in this regard. In bimanual activities (as opposed to multi-effector activities such as speaking) not only are there just two effectors on the stage, but at every instant these two effectors remain distinct anatomical entities. Furthermore, most human manual activities rest, not simply on the special skills of the preferred hand, but on a species-specific ability to have the two hands cooperate in a strongly-differentiated fashion - a fact traditionally overlooked by students of handedness (e. g., Annett, 1985), which have strongly emphasized hand preference and accordingly focused more or less arbitrarily on unimanual activities (see Guiard, 1987, 1990).

To demonstrate the emergence of coordinative structures, that is, of unicity of action out of the multiplicity of effectors, many investigators in practise have used bimanual-task paradigms. Two well-known examples are the double-handed version of Fitts's reciprocal aiming paradigm designed by the Kelso group (Kelso, Southard, and Goodman, 1979; Kelso, Putnam, and Goodman, 1983) and the double-pendulum swinging paradigm of Kugler and Turvey (1987).

Most of this research on bimanual coordination has been done in a bottom-up perspective. In the alternative, top-down perspective, one starts from the motor problem, first asking questions about the task. As an illustration, let us consider a few instances of the cooperative structures represented in Figure 1, chosen in the domain of bimanual activities.

The two hands of a person skipping rope cooperate in parallel, as the person's two hands move the same object (the rope) relative to the same reference frame (the body). Likewise, we find a parallel partition of labour in climbing rope (out-of-phase coordination), and in the so-called Etch-A-Sketch game (in which the two hands separately take charge of the two spatial dimensions of the motion). In the parallel cooperation logic the two roles are functionally equivalent, with a symmetric partition of the task between the left and the right hands. Therefore parallel cooperation paradigms are suitable for studying manual specialization: If some performance asymmetry is observed, it must be attributed to the different capabilities of the two effectors (see, for example, Fagard, 1987).

As an example of serial cooperation, we may take handwriting, in which the nonpreferred hand usually takes charge of the motion of the sheet of paper relative to the writing table (two degrees of freedom for positioning plus one degree of freedom for orienting) whereas the preferred hand takes charge of the motion of the pentip relative to the sheet of paper. In sharp contrast to parallel cooperation, serial cooperation implies a hierarchical differentiation of the component roles. Thus, in serial cooperative structures the left-right partition of labour is asymmetric.

It must be remarked that in the double-handed paradigms of Kelso et al. (1979, 1983) and Kugler and Turvey (1987), subjects are actually presented with *dual* tasks. In instances such as these, it is clear that there is neither cooperation nor competition between the two effectors, because each of them has to manipulate its own mobile (move a stylus in Kelso et al., swing a pendulum in Kugler and Turvey) relative to the general reference frame of the task (respectively, the working table on which the targets are marked and the armchair on which the forearms are resting). Therefore, according to the taxonomy of Figure 1, the bimanual structures built up in such instances must be considered noncooperative.

Finally, one is of course unlikely to find instances of competitive structures in the realm of normal bimanual activities. In the neuropsychological literature, however, there are reports of split-brain patients occasionally displaying symptoms of intermanual competition (diagonistic apraxia), with one hand (usually represented by the patient as a "stranger") tending to undo what the other has undertaken to do (e. g., Poncet, 1990).

Footnotes

(1). Rather than a two-class taxonomy of cooperative structures, in previous articles (Guiard, 1987, 1988) I proposed a three-class taxonomy including what I called the orthogonal case, in addition to the parallel and the serial cases. The reason was that there exists a variety of tasks in which two effectors must independently take charge of the two spatial dimensions x and y of a given motion problem (think, for example, of the so called Etch-a-sketch game, or a milling machine). Obviously this kind of between-hand cooperation must be in some way distinguished from the simple parallel cooperation involved in tasks like, say, skipping rope or rotating a steering wheel with the two hands - in particular because the difficulties of coordination strongly differ. This three-class taxonomy must be amended, however: By the definitions adopted, the orthogonal case is in strict parlance nothing but a sub-class of the parallel case. Here too we have a single mobile and a single reference frame for the two motors. In fact, orthogonality takes place at another, more analytic level of description of the cooperation relationship. At this lower level, we will have to introduce some other important distinctions, such as synergic cooperation vs. antagonistic cooperation (as is the case, respectively, between the two legs in hopping and between the biceps and the triceps of an arm).

(2). Notice that the lower-limb kinematic chain might be oriented the other way round, with the trunk being considered as the mobile and the ground as the frame of reference: Such an exocentrical representation would certainly make sense from a mechanical point of view. It would raise, however, the difficulty that for lower limbs we would have a more finely grained contribution to overall motion from more proximal motors, in contradiction to what happens with an arm. The alternative, egocentrical representation adopted in the text and in Figure 2 solves this contradiction.

References

Annett, M. (1985). *Left, Right, Hand and Brain: The Right Shift Theory*. London (Hillsdale, New Jersey): Lawrence Erlbaum Assoc.

Beek, P. J. (1989). *Juggling Dynamics*. Amsterdam: Free University Press.

Bernstein, N. A. (1967). *The Control and Regulation of Movements*. London: Pergamon Press.

Fagard, J. (1987). Bimanual stereotypes: Bimanual coordination in children as a function of movements and relative velocity. *Journal of Motor Behavior 19*, 355-366.

Gibson, J. J. (1966). *The Senses Considered as Perceptual Systems*. Boston: Houghton Mifflin.

Guiard, Y. (1987). Asymmetric division of labor in human skilled bimanual action: The kinematic chain as a model. *Journal of Motor Behavior 19*, 486-517.

Guiard, Y. (1988). The kinematic chain as a model for human asymmetrical bimanual cooperation. In A. Colley and J. Beech (Eds.), *Cognition and Action in Skilled Behaviour*. Amsterdam: North-Holland, Pp. 205-228.

Guiard, Y. (1990). Lateral asymmetry in the context of two-handed action. *Perceiving/Acting Workshop* 4, 10-13.

Kelso, J. A. S., Putnam, C. A., & Goodman, D. (1983). On the space-time structure of human interlimb coordination. *Quarterly Journal of Experimantal Psychology 35A*, 347-375.

Kelso, J. A. S., Southard, D. L., & Goodman, D. (1979). On the coordination of two-handed movements. *Journal of Experimental Psychology: Human Perception and Performance 5*, 229-238.

Kugler, P. N. & Turvey, M. T. (1987). *Information, Natural Law, and the Self-Assembly of Rhythmic Movement*. Hillsdale: Lawrence Erlbaum.

Meijer, O. G. & Roth, K. (1988). *Complex Movement Behavior: "The" Motor/Action Controversy*. Amsterdam: North-Holland.

Poncet, M. (1990). Apraxie di-agonistique et main étrangère. Oral paper presented at the *Second Workshop of the ARN (Association des Rééducateurs en Neuropsychologie)*. Lyon, May 1990.

An Intermittency Mechanism for Coherent and Flexible Brain and Behavioral Function

J.A.S. Kelso & G.C. DeGuzman
Program in Complex Systems and Brain Sciences
Center for Complex Systems
Florida Atlantic University
Boca Raton, FL 33431

ABSTRACT. Complex biological systems must balance a universal tendency to phase- and frequency synchronize with the requirement to flexibly explore other coordinative states. We identify intermittency, a generic property of dynamical systems near tangent bifurcations, as a unifying mechanism that provides this vital mix of coherent yet flexible function.

Remarkable parallels exist between patterns of coordination in mammalian visual cortex (Gray et al. 1989; Eckhorn et al. 1988), invertebrate and vertebrate neural circuitry for central pattern generation (Cohen et al. 1988; Selverston, 1988) and work on human sensorimotor coordination (e.g. Haken et al. 1985; Kelso, 1984; Kelso, Schöner, Scholz & Haken, 1987; Schöner & Kelso, 1988). In all three fields it is important to note: a) that under well-defined experimental conditions the patterns take the form of phase- and frequency-synchronization; b) that very different kinds of signals are coordinated, e.g. single units and local field potentials (vision), kinematic and EMG (motor control); and, c) that the components synchronize only at certain, readily identifiable points, such as onsets, peaks or valleys. In between, such points signals may vary spatially (e.g. characteristic amplitudes) and temporally (e.g. characteristic period/frequency content).

The main, almost entire focus of theoretical work has been on modeling phase and frequency synchronization in terms of asymptotic solutions of coupled nonlinear differential equations (basically, but nontrivially, the mathematical theory of coupled nonlinear oscillators). This is fine for stability but not much good for flexibility. It may be far more important for minds and brains to have a mechanism for going in and out of stable coherent states, as well as for exploring the range of available possibilities. For flexible biological function, we may presume it is not so good for the system to get stuck in a (mode-locked) attractor.

In Figure 1 we show some experimental records from our recent studies of multifrequency coordination in which absolute phase and frequency synchronization are neither essential nor attainable (DeGuzman & Kelso, in press; Kelso & DeGuzman, 1988; Kelso, DeGuzman & Holroyd, 1991a,b). We believe this to represent the more general case in biological coordination (see also Peper, Beek & Van Wieringen, this volume). Here the frequency and amplitude of one component (the Left Hand; middle plots) are determined passively by a torque motor while the other component (the Right Hand; both plots) actively maintains a base rhythm. The frequency ratio is scanned in ascending and descending order from just below 1:1 to just above 2:1 and *vice-versa* in small steps of .1 Hz. A close look at the trajectories reveals 'slips' and sometimes even 'jumps' in cycles of the LH driven component. For example, when a peak of the LH is just at the edge of a flattened peak of the RH, either the LH jumps over a cycle or shortens the next one. An analogous situation happens in the example of a father walking along with his small child. Since their intrinsic frequencies

J. Requin and G. E. Stelmach (eds.), Tutorials in Motor Neuroscience, 305–310.

are different, synchronization is difficult unless one or both adjust their periods. Like our data, the child sometimes skips a step or the father slows down and we observe *phase slippage*, followed by more or less regular cycles again. These modulations appear rarely and randomly near 1:1 but far more frequently near 2:1, and are due to the cross-coupling between the components. The reader will recall that von Holst (1939/73) coined the terms Magnet or M-effect, and relative coordination to describe such tendencies.

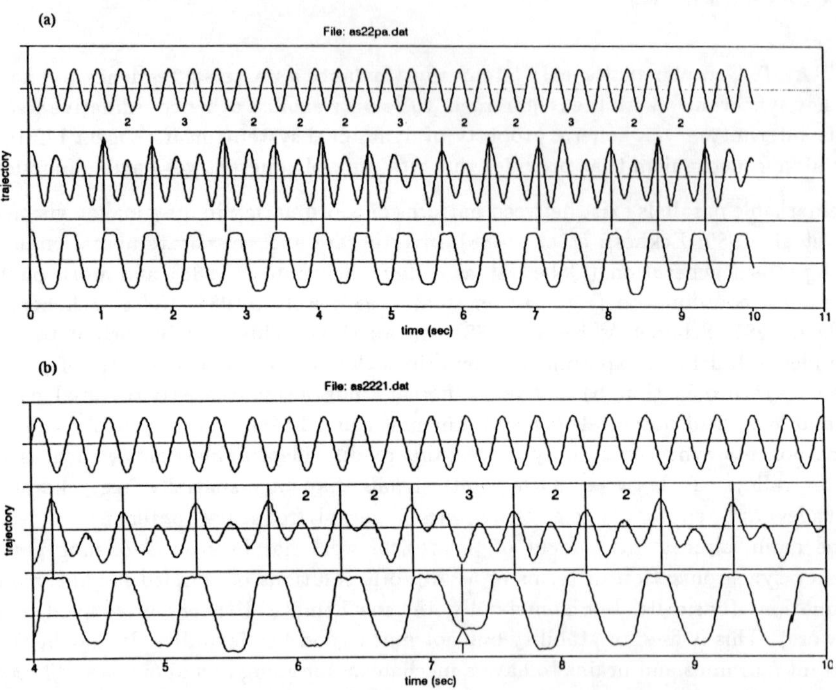

Figure 1. (a) Trajectories of the input signal used to drive the torque motor (top), the actual motion produced by the driven left hand (middle) and the "free" hand (bottom) when the required frequency condition is near 2:1. Enhancement of the peaks near inphase regions points to the discrete nature of the coordination, i.e. intermediate portions of the left hand trajectories are adjusted to sustain the natural tendency to be inphase with the right hand (indicated by the vertical lines). Occasional slips occur when the position of the enhanced peak extends beyond the broadened peak of the right hand (3 steps instead of 2). (b) Similar plot near 2:1 from another time interval showing the slippage effect on the right hand.

To understand this flexible form of coordination in multifrequency situations in which the components synchronize only at certain event times and may not have a well-defined frequency relationship we examine the evolution of the relative phase, ϕ_n calculated experimentally as

$$\phi_n = 2\pi \frac{\tau_n}{T_n} \tag{1}$$

where T_n is the peak-to-peak period of component 1 beginning at time, t_o and τ_n is the interval between that event and the peak onset of component 2. The utility of (1) for expressing coordination in terms of the evolution of ϕ_n at discrete times τ_n is based on earlier phase transition experiments (e.g. Kelso, 1984) and consequent theoretical modeling showing that the relative phase dynamics can be derived from nonlinearly coupled autonomous nonlinear oscillators (Haken, Kelso & Bunz, 1985). We examine the phase at time $n+1$ as a function of the phase at time n, i.e. as a return map. For example, if the system was 1:1 phase and frequency-locked, all consecutive phases would lie on a single fixed point. The map has the form

$$\begin{aligned} \phi_n + 1 &= \Omega + f(\phi_n) \\ &= \phi_n + \Omega + g(2\pi\phi_n) \mod 1 \end{aligned} \tag{2}$$

where Ω is an added constant term corresponding to the initial frequency ratio and g is a smooth periodic function of period 1 in ϕ (for complete details, see DeGuzman & Kelso, 1989/in press).

In Figure 2 we show the map and the resulting evolution of the phase in time. The stability of the fixed points of the map are given by the slope of the function which can be compared to the diagonal line whose slope is unity. In Figure 2(a) a pair of fixed points is shown, one stable (slope $<$ 1) the other unstable. All initial conditions converge on the stable fixed point, i.e. there is frequency locking. The appearance of phase slippage (Fig. 2(b)) means that between two channel crossings one of the oscillators gains a period with respect to the other: exactly what is seen experimentally. Note that close to the tangent bifurcation where the function hits the line, the phase concentrates (*phase attraction persists*) and slows. The reason, of course, is because the iterations are compressed in the channel. In Figure 2(c) through (d) plots of the relative phase in time (x-axis) are shown as the system approaches a fixed point (1:1 frequency and phase-locked state). As the system approaches the tangent it spends longer and longer near a given frequency and phase-locked state (the flat portion of the curves) before abruptly slipping away. We note that the appearance of phase gathering near mode-locked solutions has an *anticipatory* quality about it. A good way to see this is through the corresponding bifurcation diagram over the non-linearity parameter K in the circle map shown in Fig. 2(f). In Fig. 2(f) the fuzzy area reflects quasiperiodic motion because the asymptotic frequency ratio (winding number) between the components is irrational. Notice, however, that the progressive darkening "anticipates" the upcoming stable solution (the single line) which indicates the system is trapped or mode-locked 1:1.

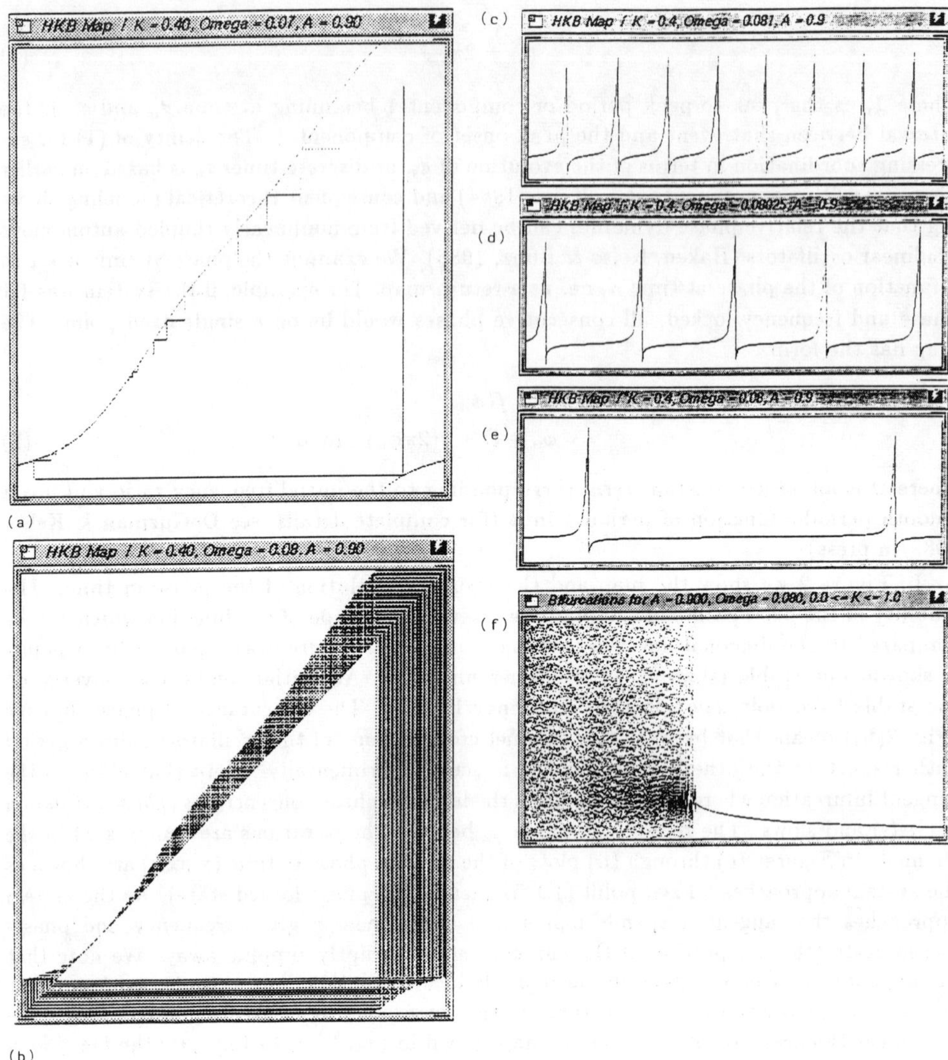

Figure 2. Return maps (Eq. 2) of ϕ_{n+1} (y-axis) against ϕ_n (x-axis), (a) Mode locked behavior at 1:1 showing a pair of fixed points, stable and unstable, close to the origin. (b) intermittency occurs when the graph is lifted slightly above the diagonal (see text for details). (c)-(e) Plots of the time series ϕ_n for decreasing values of Ω, showing the slowing down of the bursts. For Fixed map parameters, the frequency of the burst depends on how far Ω is from the fixed point. (f) Bifurcation diagram showing the phase behavior (y-axis) when crossing the Arnol'd tongues at 1:1. The straight line is the 1:1 frequency and phase-locked solution. In the fuzzy area the frequency-ratio (winding number) is irrational.

Significance

This less rigid, more flexible form of coordination corresponds to *intermittency* which is a generic feature of dynamical systems near tangent bifurcations. Intermittency refers to a region around a destabilized saddle node which contains both a stable (attractive) and an unstable (repelling) direction. Typically, the system spends some time near the attractive fixed point (a frequency and phase synchronized state) and then escapes through the unstable direction before returning via a reinjection process (Berge, et al. 1984). Intermittency provides a mechanism for entering and exiting phase and frequency synchronized states, thus preventing the biological system from getting stuck. It may be important for vision not to get trapped in resonant, mode-locked states, an argument that applies equally well to neuronal circuits for flexible central pattern generation, coordinated action and even cognition itself (cf. Mandell, 1983). Our studies indicate that intermittency is crucial, on the one hand affording the possibility to phase and frequency synchronize but also providing the necessary mechanism for going in and out of coherent states as well as for exploring the full state space. The mechanism we have identified allows biological systems – on several levels – a vital mix of coherent yet flexible function. An important technical point for understanding complex systems like the brain is that this intermittency mechanism is a low dimensional and deterministic dynamical process.

Acknowledgments

The research discussed herein is supported by NIMH (Neuroscience Research Branch) Grant MH42900, ONR contract N00014-88-J1191 and BRS Grant RR07258. We thank Arnold Mandell and Hermann Haken for discussion and encouragement.

References

1. Bergé, P., Pomeau, Y. & Vidal. C. (1984). *Order within chaos: Towards a deterministic approach to turbulence.* John Wilery, New York.

2. Cohen, A.V., S. Rossignol, & S Grillner (Eds.). (1988). *Neural control of rhythmic movements in vertebrates.* New York: John Wiley.

3. DeGuzman, G.C. & Kelso, J.A.S. (1989/in press). Multifrequency behavioral patterns and the phase attractive circle map. FAU Center for Complex Systems preprint, July 1989; To appear in *Biological Cybernetics.*

4. Eckhorn, R., Bauer, R., Jordan, W., Brosch, M., Kruse, W., Monk, M. & Reitboeck, H.J. (1988). Coherent Oscillations: A mechanism of feature linking in the visual cortex? *Biological Cybernetics, 60*, 121-130.

5. Gray, C.M., Konig, P., Engel, A.K. & Singer, W. (1989). Oscillatory responses in cat visual cortex exhibit inter-columnar synchronization which reflects global stimulus properties. *Nature, 338*, 334-337.

6. Haken, H., Kelso, J.A.S. & Bunz, H. (1985). A theoretical model of phase transitions in human hand movements. *Biological Cybernetics, 39*, 139-156.

7. Kelso, J.A.S. (1984). Phase transitions and critical behavior in human bimanual coordination. *American Journal of Physiology: Regulatory, Integrative and Comparative Physiology, 15*, R1000-R10004.

8. Kelso, J.A.S. & DeGuzman, G. (1988). Order in time: How cooperation between the hands informs the design of the brain. In H. Haken (ed.), *Neural and synergetic computers.* Berlin: Springer-Verlag, pp. 180-196.

9. Kelso, J.A.S., DeGuzman, G.C. & Holroyd, T. (1991a). The self-organized phase attractive dynamics of coordination. In A. Babloyantz (Ed.) *Self Organization, Emerging Properties and Learning*, Plenum Press, NY.

10. Kelso, J.A.S., DeGuzman, G.C.& Holroyd, T. (1991b). Synergetic dynamics of biological coordination with special reference to phase attraction and intermittency. In H. Haken & H.P. Koepchen (Eds.) *Synergetics of Rhythms*, Springer, Berlin.

11. Mandell, A.J. (1983). From intermittency to transitivity in neuropsychobiological flows. *American J. Physiol., 245*, R484-R494.

12. Schöner, G. & Kelso, J.A.S. (1988a). Dynamic pattern generation in behavioral and neural systems. *Science, 239*, 1513-1520. Reprinted in K.L. Kelner & D.E. Koshland (Eds.), *From Molecules to Models: Advances in the Neuroscience*, AAAS, Washington, DC.

13. Selverston, A.I. (1988). Switching among functional states by means of neuromodulators in the lobster stomatogastric ganglion. *Experientia, 44*, 376-383.

14. von Holst, E. (1939/1973). Relative coordination as a phenomenon and as a method of analysis of central nervous function. Reprinted in: *The collected papers of Erich von Holst.* Coral Gables, FL: University of Miami Press.

SECTION 6

PROGRAMMING OF MOVEMENT PARAMETERS

INITIATION AND EXECUTION OF MOVEMENT: A UNIFIED APPROACH

MARTIN R. SHERIDAN
Human Performance Laboratory
Department of Psychology
University of Hull
Hull HU6 7RX, UK

ABSTRACT. Motor control theorists have postulated a series of processes involved in converting an initial idea for movement into the movement itself, with motor programming as one component. With the exact nature of the motor output unfolding as the result of the interplay of many factors in the heterarchical system, it is possible for the motor program to have different conceptions at different levels in the distributed system. Many tasks demand some degree of accuracy and the performer has to attune the accuracy of performance to the task requirements; in so doing, the kinematics of the movement are delimited. It is argued that it is most productive to consider the programming of movement in the context of its execution, and that the processes of program formation and movement execution are interrelated. The time to initiate a movement is found to covary with the variability with which the movement is executed. Thus, the lower the endpoint variability (tolerance) allowable in a movement, the longer RT will be, and that this is dependent on the amplitude of movement. Velocity is seen as the intervening factor, with less variable movements requiring a lower velocity, and lower velocity movements having longer RTs. With increasing evidence that there may be a direct relationship between the kinematic parameters of movement and initiation time, it seems prudent not to build conceptually too much into the motor program. By analysing the relationship between endpoint variability and velocity in terms of amplitude, an approach is derived which unifies the previously conflicting results from spatially and temporally constrained aiming tasks.

Motor control theorists have postulated a series of processes involved in converting an initial idea for movement into the movement itself, with motor programming as one component. The nature, or even existence, of motor programs has been the subject of considerable debate (Keele, 1981; Pew, 1984; Reed, 1982, 1984; Requin, Semjen & Bonnet, 1984; Sheridan, 1988). It has been argued that the attack made on a motor program conceived of as a central rigid set of commands is inappropriate because this conception of a motor program is far too limited (Requin et al., 1984; Sheridan, 1988). Early work, together with concepts such as "kinetic formula" (Liepmann, 1900), "motor scheme" (Head, 1920), "schema" (Bartlett, 1932) and "action plan" (Miller, Galanter & Pribram, 1960), are all variants of the general idea that a central representation of action contributes to the structuring of the outputs responsible for movement, but this is quite different from arguing that the final form of the movement is specified centrally. The "central"/"peripheral" dichotomy claimed by some (e.g. Reed, 1982) needs to be rejected. There is no clear border between the perceptual domain and the motor domain in our representation of the outside world, and in action. For years we have talked about perceptual-motor skills meaning just that. One can consider so-called "mixed" theories of motor control in which the central and peripheral origin of motor commands are closely integrated. It is known that control processes occur during responses (Ostry, 1980) and that feedforward control, based on perceptual knowledge, is used in

J. Requin and G. E. Stelmach (eds.), Tutorials in Motor Neuroscience, 313–332.

addition to sensory feedback (Arbib, 1981, 1984). Thus, there are conceptions in which feedback and feedforward mechanisms are the source of the adaptive flexibility of control exerted by different levels in the hierarchy (Brooks, 1979). The program concept has, moreover, been extended to include postural adjustments accompanying and/or anticipating movements (Gahery & Massion, 1981). Such mixed approaches to motor control have been widely advocated (e.g. Bizzi, 1980; Kelso, 1978).

There is increasing evidence that the more viable explanations advanced for the production of movement come from theories that emphasise the cooperation of the various parts of the central nervous system. The kinematic shape of a movement unfolds as the result of the interplay of many factors in the heterarchical system, with feedback and feedforward processes operating to achieve the performer's goal. Moreover, it emerges that even elementary motor processes are "cognitively penetrable" (Pylyshyn, 1980). The context in which the movement occurs, the task demands, the desired outcome and the expectations of the performer, all affect the final detailed specification of the response. It is, therefore, possible for the motor program to have different conceptions at different levels in the distributed system. With the exact nature of the motor output thus determined, it seems reasonable not to build conceptually too much into the motor program, or into reaction time (RT) measures used to index motor programming considered in isolation. Indeed, it has been argued (e.g. Phillips & Hughes, 1988) that circularity of explanation can be a problem, with an increase in RT explained in terms of additional processing operations, but with any additional processing operations being inferred from increases in RT.

Classically, the selection of a response has been considered independent of the execution of the response. Thus, the factors affecting RT have not been thought to be the same as those affecting movement time (MT). In this chapter it is argued, however, that it makes sense to consider the programming of movement in the context of its execution, and that the processes of program formation and movement execution are interrelated. The characteristic of movement execution focused on here is the variability of movement, that is, the within-subject scatter of shots about the target. The reason for concentrating on accuracy of performance is that, what is generally more important to people is achieving a successful goal, rather than the way the goal is achieved. This is, of course, no more than to reiterate the idea of functional equivalence. Many tasks demand some degree of accuracy and the performer has to attune accuracy of performance to the task requirements; in so doing, the kinematics of the movement are delimited. Marteniuk (e.g. Marteniuk, MacKenzie & Leavitt, 1988) and others have argued that motor control is dependent on a knowledge base of integrated sensorimotor information that is acquired and represents knowledge about past movements and their consequences in specific situations. Thus, for example, one reaches for fragile objects more carefully (with a higher degree of accuracy) than for more solid objects. In the present chapter, although concentrating on variability of performance, movement accuracy is discussed in terms of both constant error (CE) and variable error (VE). CE is the within-subject mean of the distribution of movement endpoints, and is a measure of individual bias. VE is the within-subject standard deviation (SD) of the movement endpoints, and reflects an individual's consistency of performance (scatter of shot about the target). Accuracy data are considered in absolute terms rather than as percentages of desired movement length. The reason is that absolute magnitude of error is what has to be compensated for in the 'real world' in order to achieve a successful outcome, and thus is argued to be more ecologically valid from the performer's perspective.

There is evidence to suggest that there is a relationship between RT and characteristics of the impending movement (Falkenberg & Newell, 1980; Sheridan, 1981, 1984). The relationship has been noted between the average velocity of a movement and the time to initiate it. Simply stated, decreasing the velocity of movement increases RT, with for low velocity movements, a distinct negative exponential function. The relationship holds both for total RT and for the fractionated premotor component. Recently, Carlton, Carlton and Newell (1987) reported a similar relationship between RT and the average rate of force production. Since premotor RT typically follows the same basic pattern as total RT (e.g. Carlton, Robertson, Carlton & Newell, 1985; Fischman, 1984; Sheridan, 1984; Siegel, 1988) only total RT will be discussed in the present chapter.

In addition to the suggested relationship with RT, velocity is of course, clearly linked with parameters of the type of movement one is going to execute, the tolerance and the amplitude. Since velocity appears to relate to both movement initiation and the parameters of the movement one is executing, using this kinematic variable it will be argued that the initiation and execution of movement link together, or at least, covary. First, one needs to consider the relationships that have been suggested to exist between velocity and movement variability. The best known speed-accuracy tradeoff is Fitts' law (Fitts, 1954). Fitts proposed a logarithmic speed-accuracy tradeoff, however, despite wide general support in the literature, not everyone has accepted Fitts' law as the best possible account. Some investigators have proposed various modified logarithmic speed-accuracy tradeoff functions (e.g. Crossman & Goodeve, 1963; Sheridan, 1979; Welford, 1968), while others have proposed linear or power functions (e.g. Howarth, Beggs & Bowden, 1971; Kvalseth, 1980; Meyer, Smith & Wright, 1982; Schmidt, Zelaznik & Frank, 1978). Wright and Meyer (1983) have suggested that the exact form of the tradeoff function could depend on the task demands, and they distinguish "spatially constrained" and "temporally constrained" tasks. In spatially constrained tasks subjects attempt to stop within a specified target region while minimising MT (Fitts type aiming task). In temporally constrained tasks, subjects must produce movements of a specified duration while trying to come as close as possible to a target point, rather than a region. It has been argued that the logarithmic tradeoff function accounts better for spatially constrained tasks, while the linear tradeoff function correlates more highly with data from temporally constrained tasks (e.g. Schmidt et al., 1978; Wright & Meyer, 1983). In the present chapter, evidence is presented that spatially and temporally constrained movements can both be described by the *same* relationship, but not the logarithmic function proposed by Fitts. Before reporting the present study, data from well known experiments using temporally and spatially constrained tasks are considered.

Schmidt and colleagues Using a temporally constrained task Schmidt and his colleagues (Schmidt, Zelaznik & Frank, 1978; Schmidt, Zelaznik, Hawkins, Frank & Quinn, 1979) reported a linear relationship between movement variability and average velocity. Figure 1 displays the data from Schmidt et al. (1978) redrawn according to the amplitude of movement.

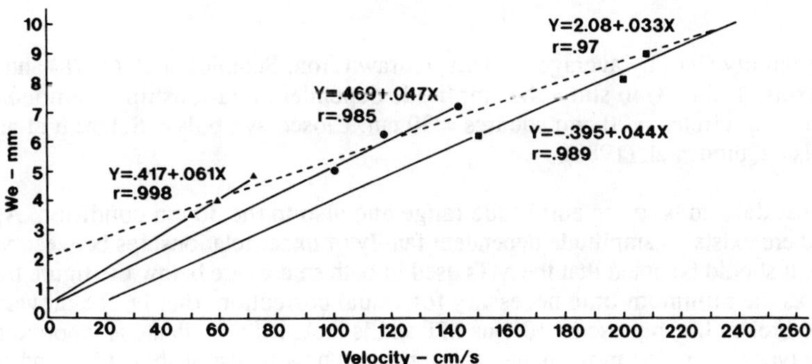

Figure 1. Variability (We) by average velocity, redrawn from Schmidt et al. (1978) to show the amplitude dependent relationship. Triangles = 10 cm; circles = 20 cm; squares = 30 cm. The broken line is the linear fit provided by Schmidt et al..

The general relationship is, as can be seen from Figure 1, the greater the velocity, the greater the variability of movement. Variability in this study was measured by the effective target width (We), and is the within-subject SD of movement endpoints from the centre of the target. The relationship for VE parallels that for We, and is identical when CE is zero. They used a temporally constrained task, and subjects aimed with a stylus at a dot target in a specified MT. The data for Schmidt et al. (1978) were produced for three movement amplitudes (10, 20 & 30 cm), and three movement times (140, 170 & 200 ms). As can be seen from the broken line of Figure 1, the linear relationship suggested by Schmidt accounts for the data reasonably well. However, it is suggested that there is another way of looking at the data, that is, as a family of linear functions relating the variability of movement produced to the average velocity, but dependent upon the amplitude of the movement. Such a limited analysis is not by itself convincing, therefore, further data from Schmidt's group has been added. Figure 2 combines the data from Schmidt et al. (1978) with the data from Quinn, Schmidt, Zelaznik, Hawkins & McFarquhar (1980). They also used a temporally constrained task, but with subjects aiming from different distances at strip targets of different sizes. Fortunately, they also recorded the actual variability of movement endpoints and this is what has been included here.

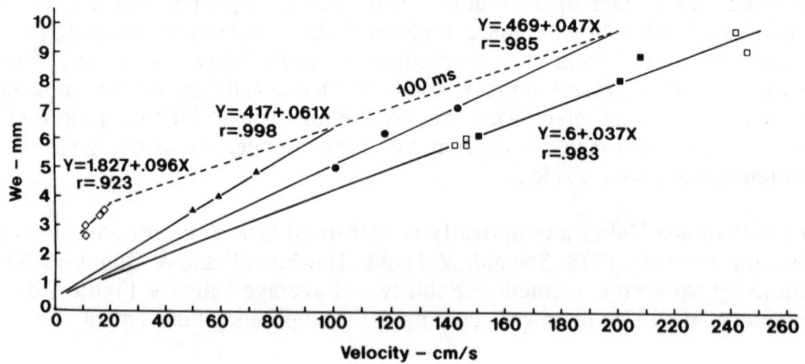

Figure 2. Variability (We) by average velocity, redrawn from Schmidt et al. (1978) and Quinn et al. (1980, Expts. 1, 2 & 3) to show the amplitude dependent relationship. Diamonds = 2 cm; triangles = 10 cm; circles = 20 cm; squares = 30 cm. Closed symbols = Schmidt et al. (1978); open symbols = Quinn et al. (1980).

The additional data adds to the amplitude range and also to the 30 cm condition. Again it is argued that there exists an amplitude dependent family of linear relationships between variability and velocity. It should be noted that the MTs used in both studies are below the figure traditionally accepted as the minimum time necessary for visual correction, that is, the movements are ballistic. The broken line represents 100 ms MT and is included to indicate an approximate ceiling for these types of aiming movements. It can be seen how by using short MTs and approaching this ceiling on the different amplitude functions, a single linear fit of the type proposed by Schmidt only *appears* to describe the data quite well. Proceeding along this idea of a family of variability by velocity functions, Figure 3 illustrates the relationship that exists between the slopes of the variability by velocity functions at the various amplitudes. Using the logarithm of the amplitude by the slope a good linear fit is provided.

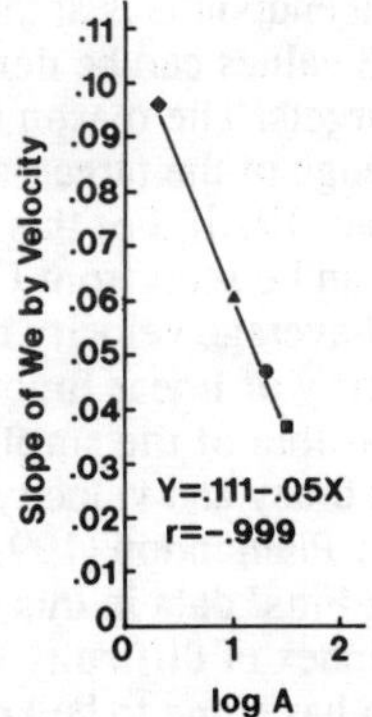

Figure 3. Slopes of the variability (We) by velocity functions plotted against the logarithm of the amplitude. Data from Schmidt et al. (1978) and Quinn et al. (1980).

Fitts (1954). The task is to relate the present analysis of Schmidt's data to Fitts' approach. Schmidt's task is temporally constrained while Fitts' now classic aiming task is spatially constrained. Moreover, movements in the Schmidt studies were of sufficiently short duration that traditionally they would not be thought to involve much, if any, visual correction, while Fitts' data come predominantly from tasks with MTs sufficiently long to involve visual correction. A third difference is that Schmidt's data derive from a discrete aiming task, while Fitts used a continuous aiming task. One appears, therefore, to be considering two completely different sets of data, however, as can be seen from Figure 4, a similar pattern exists in Fitts' data.

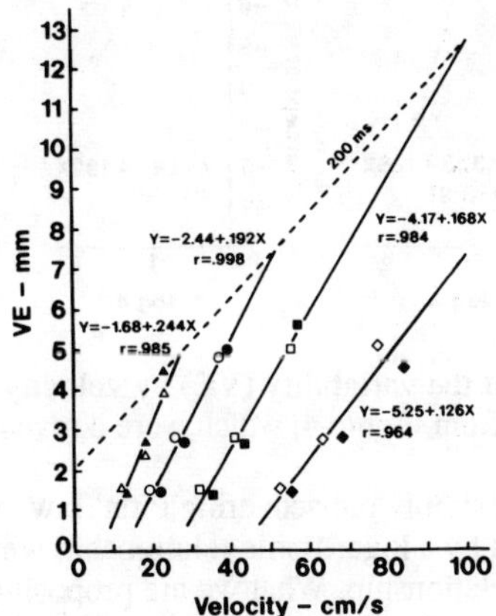

Figure 4. Fitts' (1954) stylus tapping data (Table 1) expressed in terms of variable error by velocity. For amplitude of movement, triangles = 5.08 cm; circles = 10.16 cm; squares = 20.32 cm; diamonds = 40.64 cm. Closed symbols = 1 oz stylus; open symbols = 1 lb stylus.

Although originally presented by Fitts in terms of error scores and target tolerances, his data have been calculated in terms of VE to facilitate comparison with the other studies. By making the assumption that the mean of the movement endpoints is at the centre of the target (i.e. CE is zero) and using the normal distribution, the VE values can be derived. This assumption is reasonable for small targets, but not for the larger targets. The reason is that for larger targets the mean of the distribution can drift toward the near edge of the target, also less than the full target tolerance is used (Beggs & Howarth, 1988; Sheridan, 1976). For this reason the larger target tolerance (2 in) was excluded from the analysis. As can be seen from Figure 4, again there appears to be a linear relationship between variability and average velocity that is dependent on the amplitude of movement. The amplitude dependent family of linear functions is confirmed by again plotting the slopes and intercepts against the logarithm of the amplitude (see Figure 5). The amplitude dependent relationship between spatial accuracy and velocity illustrated here can be derived from and accords with the analysis presented by Plamondon (1991), although he approaches the problem from a different direction. Presenting Fitts' data in this way (see Figure 4), it becomes clear why Fitts found a range of MTs for each index of difficulty (ID), making it easy to fit the quasi-linear function between MT and ID which has come to be known as Fitts' law. Fitts' formula for ID, weights the tolerance and amplitude of movement equally. Thus, halving the tolerance is equivalent to doubling the amplitude. Sheridan (1979) pointed out that, in fact, in Fitts' data tolerance has a greater effect on MT than amplitude. Figure 4 shows why this is the case, with different VE by velocity functions for each amplitude. In Figure 4, for example, 3 mm VE at 5.08 cm, 6 mm VE at 10.16 cm, and 12 mm VE at 20.32 cm all have the *same* ID according to Fitts' formula and so should have the same MT, however, as can be calculated from Figure 4 the MTs for these three combinations are 265 ms, 231 ms, and 211 ms, respectively.

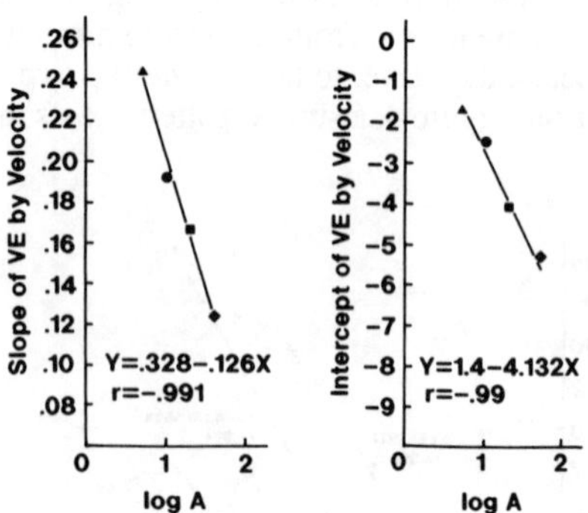

Figure 5. Slopes and intercepts of the variability (VE) by velocity functions plotted against the logarithm of the amplitude. Data from Figure 4, which were derived from Fitts (1954, Table 1).

It should be noted that we are not simply rediscovering Fitts' law, because in the literature, Fitts' data are supposed to be described by a logarithmic relationship while Schmidt's data are argued to be best described by a linear relationship. What we are proposing here is one relationship that describes *both* spatially and temporally constrained tasks. Moreover, the same relationship applies to both tasks involving visual correction (Fitts' data), and those in which movements are too fast for visual correction to take place (Schmidt's data). The relationship proposed here is not the same as that proposed by Fitts. Unlike Fitts' assumption of equal weighting of amplitude and tolerance, tolerance (variability constraint) affects MT more than amplitude. Thus, for the same ID, there is a range of MTs, and because of this range it has been relatively easy for researchers to produce a quasi-linear fit and *appear* to confirm Fitts' law (Sheridan, 1979). A similar argument can be advanced to account for the linear relationship apparently detected in temporally constrained tasks.

Present Study

Following from the arguments outlined earlier, the present study was designed to combine investigation of variability of movement in relation to velocity, with investigation of RT in relation to velocity. One aim of the study is to see if the programming of movement (as measured by RT) can be linked to characteristics of its execution. For example, is the "programming" of movement affected by the defined range of movement endpoint variability? In seeking to investigate further the proposed amplitude dependent relationship between movement variability and velocity, a task compatible with the previous experiments (Fitts, 1954; Schmidt et al., 1978) was chosen. This was a stylus aiming task, but with a wide range of MTs in one experiment. The MTs ranged between those providing ample time for visual correction, to those where traditionally visual correction would not be considered available.

METHOD

Subjects. Eighteen right-handed volunteer undergraduates at Hull University participated. Their ages ranged from 19 to 23 years. The subjects were unpaid, and had normal or corrected vision and were without known motor defects.

Apparatus. The subject was seated in front of a Summagraphics digitizer with an active tablet area of 76 X 102 cm. On the horizontally oriented digitizer was fixed a matt black acrylic sheet with a single target line (20 cm in length) and 3 starting locations marked in yellow. The starting positions were indented to facilitate location of the stylus. The seating arrangement was such that the target line coincided with the sagittal plane of the subject's body and was at right angles to the plane of movement (see Figure 6).

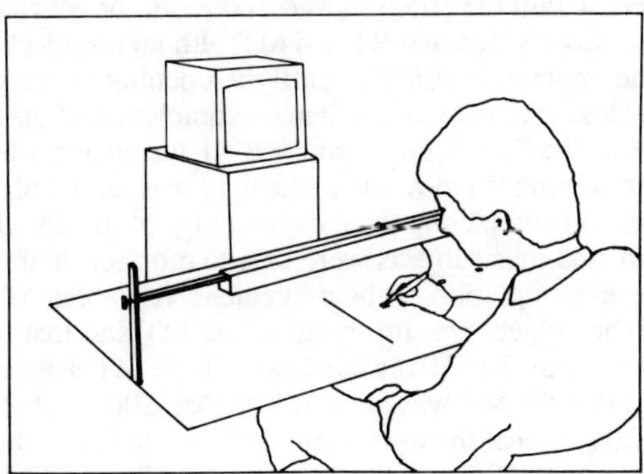

Figure 6. Arrangement of the apparatus and subject.

Subjects received information concerning parameters of the impending movement on a monitor screen placed at eye level 100 cm in front of them. The monitor was also used to provide knowledge of results (KR) about movement time after completion of the movement. With their right hands, subjects held the digitizer stylus (which was 15 cm in length) and used this to make the movements. The imperative stimulus for movement was a 1000 Hz tone. Accuracy of movement was recorded to within 0.1 mm by the digitizer. The only light source for the experiment was a fast decay fluorescent tube placed parallel to and forward of the plane of movement (see Figure 1). When turned off, the light source decayed to 10% luminescence within 10 ms. Data collection and control of the experiment proceeded automatically under computer guidance.

Procedure. The present study was designed to measure accuracy of arm movements on a task which required aimed movements at a target. The basic task required right-handed subjects to move a stylus leftwards from a given peripheral starting position, to a target line located directly in front of them. Subjects were instructed to hit, or get as close as possible to, the target line, within a criterion time. A target line rather than a zone target was used because interest was in variability of stopping (which involves variability in the forces of movement) rather than in errors concerned with directional biases from the starting position.

In a balanced design, 18 subjects made movements over three movement times (150, 400 and 1200 ms), of three different amplitudes (5, 15, and 30 cm), with or without visual feedback; a total of 18 conditions for each subject. The nesting of factors in the balanced design was such that movement duration was the slowest cycling factor (i.e., an entire MT condition was completed before moving on to the next), then movement amplitude, and finally presence or absence of visual feedback was the fastest cycling factor. In each of the 18 blocked conditions subjects made 5 practice movements followed by 10 test movements. Before each movement subjects received on the monitor screen information concerning the parameters of the impending movement; whether it was to be short (150 ms), medium (400 ms) or long (1200 ms) duration; short (5 cm), medium (15 cm) or long (30 cm) amplitude; and whether visual feedback was available or not. Thus for a sequence of 5 practice and 10 test movements the same information was repeated. The experimenter also monitored the information and provided appropriate verbal prompts to the subject. The instructions appeared for 5 s, after which, in their own time, subjects grounded the stylus on the digitizer at the appropriate starting location. This closed a microswitch in the tip of the stylus, and after a variable foreperiod of 1.5 - 2.5 s, the imperative stimulus (1000 Hz tone) sounded. Subjects were instructed to start the movement on detecting the imperative stimulus and to complete the movement within the criterion MT, and to hit, or get as close as possible to, the target line. Response time was divided into RT and MT, although subjects were not aware of this. RT was the time from the imperative stimulus until the opening of the stylus microswitch, MT was from this point until closure of the microswitch on completion of the movement.

In the feedback condition the light source remained on during and after the movement. In the no-feedback condition, on starting the movement, the light was turned off and subjects completed the movement in the dark. On completing the movement by contacting the digitizer surface, the light was turned on again, and thus subjects were able to monitor their terminal accuracy, as in the feedback condition. On completion of the movement, KR about MT was displayed on the monitor screen for 5 s. The subject was informed of the MT and that the movement was "too fast", "okay", or "too slow". For the 150 ms condition, if the MT was between 100 and 200 ms the subject was informed that the MT was okay, if less than 100 ms that the movement was too fast, and if greater than 200 ms that the movement was too slow. For the 400 ms condition, the MT okay message was displayed if the time was between 267 and 533 ms, and for the 1200 ms condition, the movement okay message was displayed if the MT was between 800 and 1600 ms, with for each conditions appropriate messages being displayed for being too fast or too slow. Where necessary the experimenter provided appropriate verbal prompts. The cycle described was repeated for each movement.

Statistical treatment. Reaction time, movement time and movement accuracy data were analysed using analysis of variance and linear regression comparisons.

RESULTS

Reaction time. The pattern of the present RT data conform to that described by Sheridan (1984) and Siegel (1988). RT was less for shorter movement duration conditions than longer duration conditions, with for short, medium and long durations, RTs of 222, 278 and 400 ms, respectively ($F(2,34) = 25.71$, $P < 0.001$). There was a difference in RT between the visual feedback (288 ms) and no visual feedback (313 ms) conditions ($F(1,17) = 8.12$, $P < 0.025$). There was also a difference in RT between the short (319 ms), medium (298 ms) and long (284 ms) amplitude conditions ($F(2,34) = 4.37$, $P < 0.025$). The only other significant effect was an interaction between movement duration and the presence or absence of visual feedback ($F(2,34) = 4.04$, $P < 0.025$); with increasing movement duration, RT in the no visual feedback condition increased more than in the visual feedback condition.

Movement time. As MT was one of the independent variables manipulated in the experiment, the data reported in this section provide an assessment of whether or not subjects complied with experimental requirements concerning movement duration. The main MT effect was that, not surprisingly, short duration (150 ms), medium duration (400 ms) and long duration (1200 ms) movements took differing amounts of time ($F(2,34) = 1703.47$, $P < 0.001$), thus confirming the effectiveness of the manipulation of the independent variable of movement duration. The actual mean times taken for the short, medium and long duration conditions were, 184, 359 and 1098 ms, respectively. The other significant effects were that, MT took longer in the visual feedback condition (560 ms) than in the no visual feedback condition (534 ms) ($F(1,17) = 8.91$, $P < 0.01$), and subjects were unable to maintain uniform MT across different movement amplitudes ($F(2,34) = 16.46$, $P < 0.001$). Shorter movement amplitudes were completed more quickly than longer movement amplitudes (534, 530 and 578 ms, respectively). No other factors were significant.

Movement accuracy. To provide maximum information concerning subjects' performance, movement accuracy is reported in terms of both constant error and variable error, as discussed previously.

Constant error. In terms of CE, there was a difference dependent on the visual feedback condition ($F(1,17) = 26.67$, $P < 0.001$), with subjects in the visual feedback condition overshooting by 0.9 mm, while in the no visual feedback condition undershooting by 1.2 mm. This difference between feedback conditions was accentuated by increasing amplitude as evidenced by the interaction between these two factors ($F(2,34) = 22.43$, $P < 0.001$). For the short amplitude condition both visual feedback and no visual feedback conditions were more of less on target (+0.4 and +0.1 mm, respectively), however, for the long amplitude condition, in the visual feedback condition there was an overshoot of 1.3 mm, while in the no visual feedback condition an undershoot of 2.9 mm. No other CE factors were significant.

Variable error. For the VE data, in terms of main effects, short duration movements were more variable than long duration movements ($F(2,34) = 90.99$, $P < 0.001$) (9.4, 5.2 and 3.1 mm, respectively); movements with visual feedback were less variable (4.7 mm) than movements without visual feedback (7.2 mm) ($F(1,17) = 88.66$, $P < 0.001$); and short amplitude movements were less variable than longer amplitude movements (3.9, 6.1 and 7.8 mm, respectively) ($F(2,34) = 59.11$, $P < 0.001$). In addition, there were a number of interaction effects. Movement duration interacted significantly with the availability of visual feedback ($F(2,34) = 26.37$, $P < 0.001$), with VE decreasing more with increasing movement duration in the visual feedback condition than in the no vision condition. Movement duration also interacted with amplitude ($F(4,68) = 6.38$, $P < 0.001$), with VE increasing more for greater amplitudes the less the movement duration. The only other significant effect was that movement duration interacted with visual feedback and amplitude ($F(4,68) = 3.12$, $P < 0.025$), with the difference in VE between visual feedback and no visual feedback being more pronounced for longer movement durations across greater amplitudes.

DISCUSSION

Consideration will now be given to the variability data presented in the context of velocity, so that comparison can be made with the earlier analysis of Schmidt's and Fitts' data.

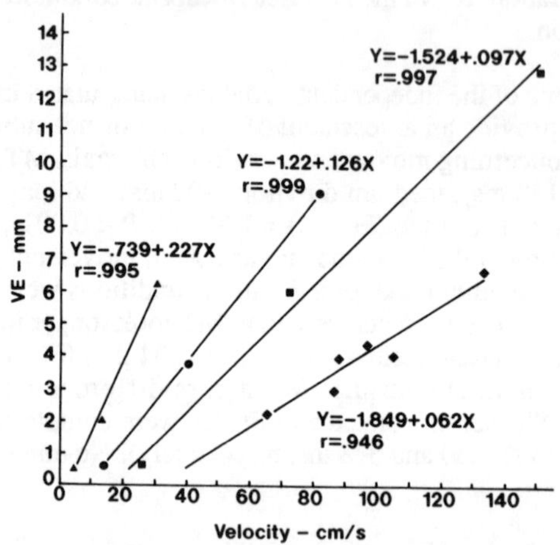

Figure 7. Variability (VE) by average velocity, drawn to show the amplitude dependent relationship for the visual feedback conditions. Triangles, circles and squares = present experiment; diamonds = Beggs and Howarth (1988). Triangles = 5 cm; circles = 15 cm; squares = 30 cm; diamonds = 50 cm.

Movement variability and velocity. In the present study, as with the previous analysis of Schmidt et al.'s (1978) data and Fitts' (1954) data, again a family of variability by velocity functions is detectable. Considering for the moment just the 5, 15 and 30 cm conditions, plotting both the intercept and slope by the logarithm of the amplitude a relationship can be detected ($r = -.999$ and $r = -.984$, respectively). Thus, the same relationship as identified in previous data appears again here. In the present experiment the movement durations ranged from those where there was ample time for visual correction to be effected, to MTs where little or no visual feedback could be used, however, the *same* function fits the range of data. Since there is often doubt about evidence provided by one study using a particular set of task conditions, data from a study by Beggs and Howarth (1988) has been included in Figure 7. The Beggs and Howarth task is completely different from the task used in the present study. Beggs and Howarth had subjects aim with a stylus at vertical strip targets of varying tolerances, using an action similar to throwing a dart. The amplitude of movement was 50 cm and they recorded actual VE in the different conditions, and these data are displayed in Figure 7. As can be seen, the Beggs and Howarth data fall into the same general relationship as data from the present study, with a reduced slope function being associated with increasing amplitude. This family of variability by velocity functions is supported by the analysis presented in Figure 8.

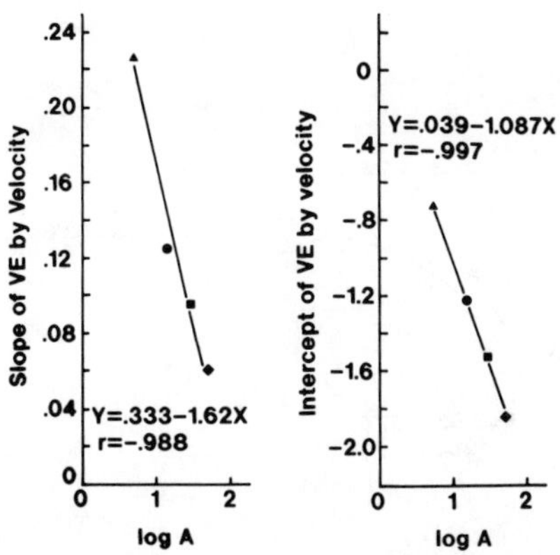

Figure 8. Slopes and intercepts of the variability (VE) by velocity functions plotted against the logarithm of the amplitude. Data from Figure 7 - visual feedback condition.

The amplitude dependent relationship between variability and average velocity is again confirmed by plotting the slopes and intercepts against the logarithm of the amplitude, even though the data come from quite different tasks. In Figure 8 a better fit is actually provided for the slopes by plotting the logarithm of the slope against the logarithm of the amplitude ($r = -.991$).

Previous studies analysed in this chapter have only had visual feedback conditions (Beggs & Howarth, 1988; Fitts, 1954; Schmidt et al., 1979), however, the present study also includes no visual feedback conditions, and these will now be considered. As can be seen from Figure 9, the no visual feedback situation presents a completely different set of functions from the visual feedback situation. Again, the variability by velocity data are presented in terms of the amplitude of movement, and fitted by linear regression equations. It is clear that the 5 cm condition is curvilinear, but is presented linearly here for the purposes of comparison.

Comparing Figures 7 and 9 it can be seen that, as indicated by the ANOVA results, the difference between feedback and no visual feedback is greater for the longer durations and greater amplitudes. It is worth noting that there is a suggestion in the data that for situations involving "high load" (high velocity - 140+ cm/s, large amplitude, and insufficient time to use visual feedback), having vision available leads to more variable performance than in the no vision condition. At the moment, this is no more than a tentative suggestion, although supported by another, as yet, unpublished study. If confirmed, this effect would suggest that performing a task with vision present is, even where insufficient time is available to actually use visual correction, quite different from performing it without vision. It is suggested that in high load conditions, having vision available and attempting to use it but not having time, may actually disrupt performance to a greater extent in terms of variability than not having vision available. Although confirmation of this awaits further work, Beggs and Howarth (1988) have, in passing, noted that the estimate of tc (visual corrective reaction time) seems to be dependent on the velocity of movement (p. 53), with visual guidance being more "difficult" for fast limb movements, thus lengthening tc.

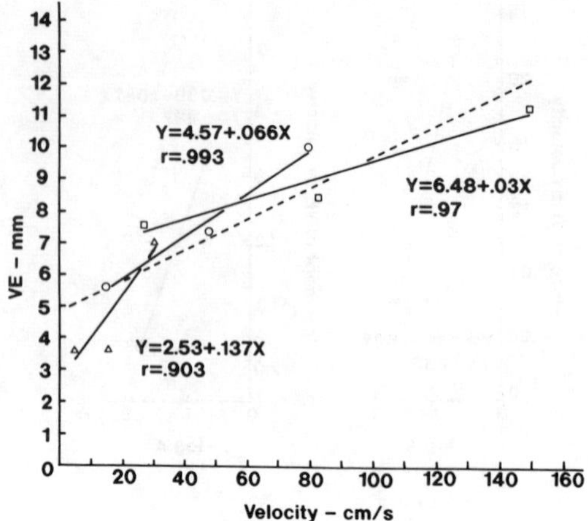

Figure 9. Variability (VE) by average velocity, expressed in terms of amplitude for the no visual feedback conditions. Triangles = 5 cm; circles = 15 cm; squares = 30 cm. Broken line is the fit of all the data points.

Although the no visual feedback situation produces data different in form from the visual feedback situation, as can be seen from Figure 10, a similar relationship still exists for the slopes and intercepts plotted against the logarithm of the amplitude. While more data needs to be collected over a wider range of conditions, nevertheless, an interesting pattern has emerged.

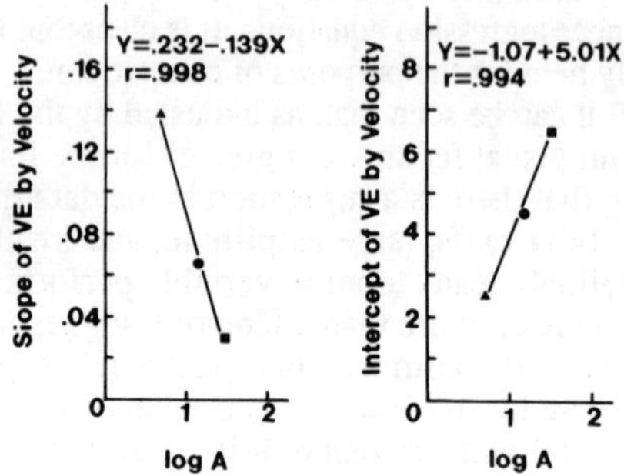

Figure 10. Slopes and intercepts of the variability (VE) by velocity functions plotted against the logarithm of the amplitude. Data from Figure 9 - no visual feedback condition.

Reaction time and velocity. Having considered the relationship between variability of movement endpoints and average velocity, attention will now turn to the relationship between RT and average velocity. Figure 11 displays the relationship for the visual feedback conditions, expressed in terms of the amplitude of movement. As can be seen from Figure 11, again the 5 cm condition is noticeably curvilinear, however, a linear fit is provided here for the purposes of comparison. This negative exponential function for low velocities has been noted by Newell and others (e.g. Falkenberg & Newell, 1980; Sheridan, 1984). As with the variability by velocity data, the slopes and intercepts of the RT by velocity functions were related to the logarithm of the amplitude (slope - r = .968; log slope - r = .999; intercept - r = .995).

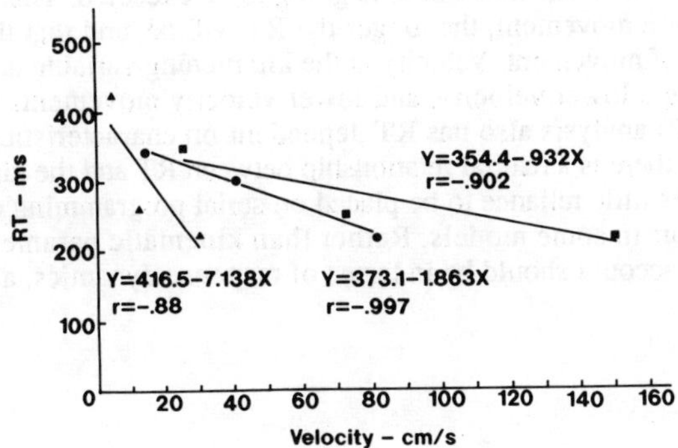

Figure 11. Reaction time by average velocity, expressed in terms of amplitude for the visual feedback conditions. Triangles = 5 cm; circles = 15 cm; squares = 30 cm.

The relationship between RT and velocity for the no visual feedback conditions is expressed in Figure 12. Although the slopes of the functions are greater for no vision condition compared to the vision condition (see Figure 11), and RT is significantly longer, overall the general form of the data is similar, with the relationship being influenced by amplitude. As with the RT by velocity data in the vision condition, the slopes and intercepts for the no vision condition were related to the logarithm of the amplitude (slope - r = .979; log slope - r = .999; intercept - r = .999).

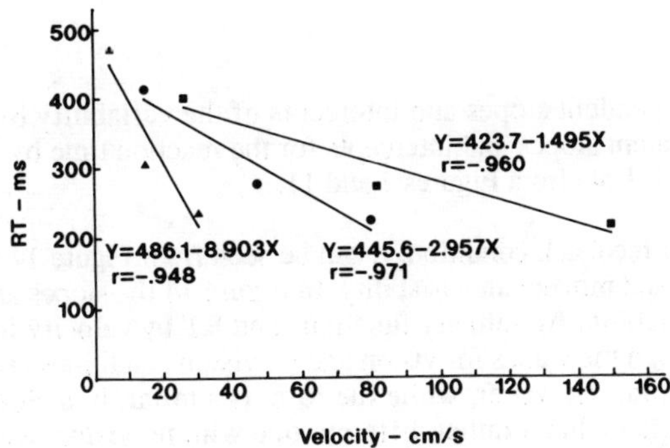

Figure 12. Reaction time by average velocity, expressed in terms of amplitude for the no visual feedback conditions. Triangles = 5 cm; circles = 15 cm; squares = 30 cm.

Reaction time and movement variability. Since there are a set of functions that relate variability to velocity, and a set of functions relating RT to velocity, the question arises if the two are related. In Figure 13 (visual feedback condition), the slopes and intercepts for the amplitude dependent variability by velocity functions and RT by velocity functions are plotted against each other. As can be seen from Figure 13, a strong relationship exits between the two, RT and movement variability appear to interrelate. It seems as if RT is partly determined by, or at least covaries with, the variability with which the movement is going to be executed. Thus, the less the tolerance that is allowable for a movement, the longer the RT will be, and that this is dependent on the particular amplitude of movement. Velocity is the intervening variable here, with less variable movements requiring a lower velocity, and lower velocity movements having lengthened RTs. Plamondon's (1990) analysis also has RT dependent on characteristics of the movement execution. It may be that there is a natural relationship between RT and the kinematic parameters of movement that requires little reliance to be placed on serial programming of specific parameters, as is argued to occur in some models. Rather than kinematic parameters, Carlton et al. (1987) consider that the account should be in terms of response dynamics, although, of course, the two are related.

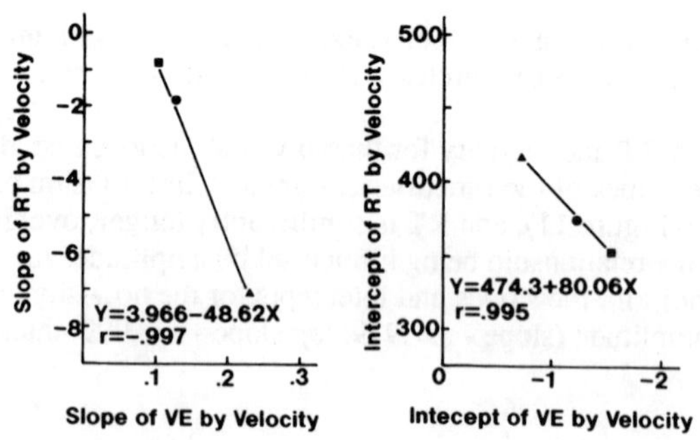

Figure 13. Amplitude dependent slopes and intercepts of the variability by velocity functions plotted against the equivalent slopes and intercepts for the reaction time by velocity functions - visual feedback conditions. Data from Figures 7 and 11.

Considering the no visual feedback condition, it can be seen from Figure 14 that again there is a relationship between RT and movement variability. In Figure 14 the slopes and intercepts for the amplitude dependent variability by velocity functions and RT by velocity functions are plotted against each other. Although the values for vision and no vision conditions are different, the form of the relationship is similar. However, while the form is similar, it is clear that one function operates with vision available but a rather different one with no vision available. With vision available the same function fits the full range of data, from movement durations where there is ample time for visual correction to be used, to MTs where little or no visual correction could be used. The presence or absence of vision changes the way the task is performed, even though insufficient time may be available for visual feedback to be utilised.

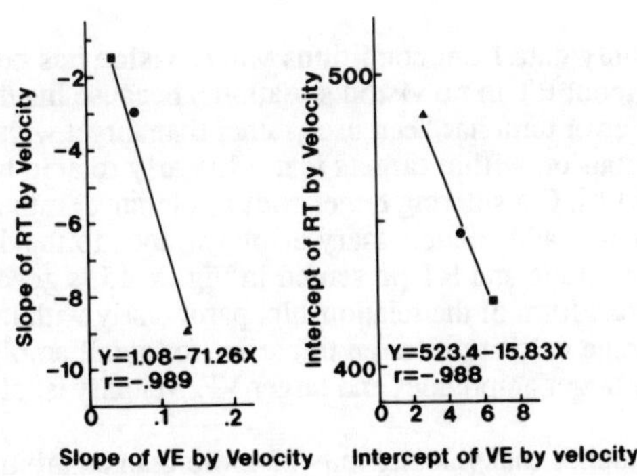

Figure 14. Amplitude dependent slopes and intercepts of the variability by velocity functions plotted against the equivalent slopes and intercepts for the reaction time by velocity functions - no visual feedback conditions. Data from Figures 9 and 12.

As was argued earlier, velocity can be seen as the factor intervening between the parameters of the impending movement and the time required to initiate it. Thus, the lower the endpoint variability allowable in a movement, the lower the required, amplitude dependent, movement velocity, and the longer RT will be. In the literature, RT has a number of times been reported in terms of the tolerance and amplitude of the impending movement. Figure 15 presents a synthesis of the results from a survey of the literature (e.g. Klapp, 1975; Klapp & Greim, 1979; Sheridan, 1981, 1984; Siegel, 1977).

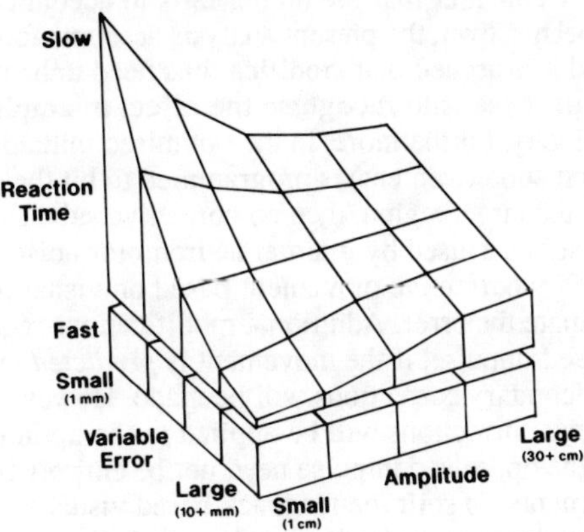

Figure 15. Generalised relationship between variable error, amplitude and reaction time - visual feedback conditions.

Figure 15 summarises only data from conditions where vision has been available. No general comment can be made about RT in no vision situations, because insufficient studies have been conducted. The variable error term has been used rather than target width (tolerance), because the width of the error distribution within targets is not linearly related to target width across the entire range of target widths. Considering target width (tolerance) rather than the actual distribution of movement endpoints adds unnecessary error variance to the data. The relationship between variable error, amplitude and RT presented in Figure 15 is generalised, and some uncertainty exists about the exact form of the relationship, particularly with large VE. The curvilinearity between RT and average velocity referred to earlier for small amplitude movements, can be seen expressed here. For larger amplitudes and larger VE, velocity is, of course, higher.

Conclusion. There is evidence that velocity may be more than an arbitrary kinematic variable in the present analysis. In Plamondon's (1990) model, the bell-shaped velocity profiles of rapid aiming movements are seen as originating from the sequential action of a set of velocity generators working in cascade. Moreover, there is in the model the mathematical prediction that initiation time is related to velocity, and the empirical data here accord with this analysis. There is also evidence that velocity may be coded directly in the CNS. For example, Houk (1989) has suggested that signals from the red nucleus may function as a velocity command. In addition to velocity being coded in terms of the burst frequency of these cells, movement duration was seen to correlate with burst duration, and amplitude of movement with number of spikes in a burst (Gibson, Houk & Kohlerman, 1985; Houk, 1989). Moreover, in relation to the preparation of movement, MacKay and Riehle (1990) report that in parietal area 7a there are task-related cortical neurons differentially sensitive to either small or large amplitude movements, and some neurons related to both extent and direction.

The present variability by velocity results may be accounted for either by Plamondon's (1990) model, or by a significantly modified version of the "optimised initial-impulse model" (Abrams, Kornblum, Meyer & Wright, 1983). The Abrams et al. (1983) model accounts for Fitts' Law through a force-pulse generator that optimally programs the initial ballistic impulse of aimed movements. However, they consider that the model fails to account for data from temporally constrained tasks. As has been shown, the present analysis accounts for both spatially and temporally constrained tasks, and it is argued that modifications need to be made to the optimised initial-impulse model. One of these is to recognise the effect of amplitude on the relationship between variability and velocity. Furthermore, in the optimised initial impulse model it is argued that the initial impulse (first submovement) is programmed to hit the centre of the target. If the initial impulse ends within the target region, then no corrective submovement follows it. However, if as a result of perturbations caused by internal neuromotor noise and impulse variability, a miss occurs, then a secondary corrective movement based on visual feedback is executed after the initial impulse to eliminate the error. Additional modifications required, are that, during the performance of the optimised impulse, if the movement is *predicted* to end within the allowable tolerance range, then no secondary corrections will be made. However, if an inappropriate endpoint is *predicted*, additional corrections will be applied to the optimised impulse, if sufficient time is available. That is, the optimised impulse need not be entirely ballistic. The present view is, for predictable environments, to shift the feedback based visual corrections of the optimised initial-impulse model, to predominantly feedforward control. Here, visual target position information and proprioceptive information about the hand, can be used to generate an intersensory error signal. In this explanation, sight of the hand is used to keep visual and proprioceptive spatial information in synchrony (Beggs & Howarth, 1988). A similar view has been expressed by Jeannerod (1986, 1988), "Visual and proprioceptive maps jointly exert a steering influence on the motor program. Provided the two maps remain interconnected, they contain all the information needed to accurately direct the hand at the target location. This is the basis for a visually calibrated position sense, which allows feedforward control of movements" (1988, p.243). The emphasis on feedforward control does not, of course, exclude the role of feedback in controlling movement.

In the optimised initial-impulse model, variability in movement endpoints is assumed to increase proportionally with the average velocity generated by the initial impulse. Furthermore, the initial impulses are assumed to have an ideal average velocity (dependent on the task parameters) that minimises total MT by making an optimal compromise (tradeoff) between the mean duration of the initial impulses and the mean duration of the secondary corrective movements. Aspects of this model are, in fact, quite similar to the position advanced by Beggs and Howarth (1988). Such theorising is not antithetical to Plamondon's (1990) position. In his model, since there is a direct relationship between distance and some velocity control parameters, there is an equivalence between spatial and timing representation of movement, and such knowledge he considers is available to the CNS. Thus, for the performer, selecting a particular target region for spatial accuracy (spatially constrained task) dictates a certain average velocity for movement, while selecting a particular average velocity (temporally constrained task) dictates a certain spatial variability. The process is bidirectional and the resultant values are equivalent.

Finally, the relationship observed between RT and velocity is suggested by Plamondon's (1990) mathematical modelling procedure. In this model only the global statistical behaviour of the processing units is of interest not the units themselves. There is some suggestion, however, that one processing unit may be the inhibition of muscle fibres, and that this is linked with RT. Basmajian (1977) suggested that movements are controlled through the inhibition of muscle fibres not involved in movements. Thus, low velocity movements (small amplitude movements and/or movements demanding low endpoint variability) may require greater inhibition of motor units, and this could conceivably take more time to achieve (program). RT would then increase with decreasing velocity, as is observed. Houk and Barto (1990) have proposed a more recent model involving inhibitory mechanisms. They suggest that arrays of Purkinje cells in the cerebellar cortex can specify direction, intensity and duration of commanded movements by inhibiting cerebellar nuclear cells. Again, greater inhibition would be required for low velocity movements (small amplitude and/or low endpoint variability), and this may require more time with a consequent increase in RT. At another level in the system, there is increasing evidence that the triphasic EMG pattern of simple voluntary movements is central in origin (Jeannerod, 1988). The first agonist burst (AG1) that precedes movement is clearly of central origin, and the amplitude of AG1 reflects both the extent and the force of the movement (Gordon & Ghez, 1984). Increased burst amplitude and duration of AG1 thus reflects greater movement amplitude and velocity (Lestienne, 1979). Again, greater inhibition may be required for low velocity / small amplitude movements with an attendant increase in RT. In addition, the amplitude and timing of the antagonist burst (ANT) are both influenced by the extent and velocity of movements (Marsden, Obeso & Rothwell, 1983). The ANT burst occurs earlier in small fast movements than in large slow ones, and this again may require differential inhibition. Inhibitory mechanisms may prove to be an important component in the process of motor programming.

With the exact nature of the motor output unfolding as the result of the interplay of many factors in the heterarchical system, it is possible for the motor program to have different conceptions at different levels in the distributed system. Many tasks demand some degree of accuracy and the performer has to attune accuracy to task requirements; in so doing, the kinematics of the movement are delimited. It is argued that it is most productive to consider the programming of movement in the context of its execution, and that the processes of program formation and movement execution are interrelated. However, with increasing evidence that there may be a direct relationship between initiation time and the kinematic parameters of movement, it seems prudent not to build conceptually too much into the motor program.

References

Abrams, R.A., Kornblum, S., Meyer, D.E. & Wright, C.E. (1983). Fitts' Law: Optimization of initial ballistic impulses for aimed movements. *Bulletin of the Psychonomic Society, 22*, 335.

Arbib, M.A. (1981). Perceptual structures and distributed motor control. In V.B. Brooks (Ed.), *Handbook of physiology, Section 1, The nervous system, Vol.II, Part 2, Motor control.* Bethesda, Md.: American Physiological Society.

Arbib, M.A. (1984). From synergies and embryos to motor schemas. In H.T.A. Whiting (Ed.), *Human motor actions: Bernstein reassessed.* Amsterdam: North-Holland.

Bartlett, F.C. (1932). *Remembering*. Cambridge: Cambridge University Press.

Basmajian, J.V. (1977). Motor learning and control: A working hypothesis. *Archives of Physical Medicine and Rehabilitation, 58*, 38-40.

Beggs, W.D.A. & Howarth, C.I. (1988). Unpaced aiming - A control theory explanation. In A.M. Colley & J.R. Beech (Eds.), *Cognition and action in skilled behaviour.* Amsterdam: North-Holland.

Bizzi, E. (1980). Central and peripheral mechanisms in motor control. In G.E. Stelmach & J. Requin (Eds.), *Tutorials in motor behaviour.* Amsterdam: North-Holland.

Brooks, V.B. (1979). Motor programs revisited. In R.E. Talbott & D.R. Humphrey (Eds.), *Posture and movement.* New York: Raven Press.

Carlton, L.G., Carlton, M.J. & Newell, K.M. (1987). Response time and response dynamics. *Quarterly Journal of Experimental Psychology, 39A*, 337-360.

Carlton, M.J., Robertson,, R.N., Carlton, L.G. & Newell, K.M. (1985). Response timing variability: Coherence of kinematic and EMG parameters. *Journal of Motor Behavior, 17*, 301-319.

Crossman, E.R.F.W. & Goodeve, P.J. (1963). Feedback control of hand-movement and Fitts' Law. Published in *Quarterly Journal of Experimental Psychology*, 1983, *35A*, 251-278.

Falkenberg, L.E. & Newell, K.M. (1980). Relative contribution of movement time, amplitude, and velocity to response initiation. *Journal of Experimental Psychology: Human Perception and Performance, 6*, 760-768.

Fischman, M.G. (1984). Programming time as a function of number of movement parts and changes in movement direction. *Journal of Motor Behavior, 16*, 405-423.

Fitts, P.M. (1954). The information capacity of the human motor system in controlling the amplitude of movement. *Journal of Experimental Psychology, 47*, 381-391.

Gahery, Y. & Massion, J. (1981). Co-ordination between posture and movement. *Trends in Neurosciences, 4*, 199-202.

Gibson, A.R., Houk, J.C. & Kohlerman, N.J. (1985). Relation between red nucleus discharge and movement parameters in trained macaque monkeys. *Journal of Physiology, 358*, 551-570.

Gordon, J. & Ghez, C. (1984). EMG patterns in agonist muscles during isometric contraction in man: relations to response dynamics. *Experimental Brain Research, 55*, 167-171.

Head, H. (1920). *Studies in neurology* 2. London: Frowde, Hodder & Stroughton.

Houk, J.C. (1989). Burst of discharge recorded from the red nucleus may provide real measures of Gottlieb's excitation pulses. *Behavioral and Brain Sciences, 12*, 224-225.

Houk, J.C. & Barto, A.G. (1990). A theory of distributed learning in sensorimotor networks. NATO Advanced Study Institute, *Tutorials in Motor Neuroscience,* Corsica, September 15-24.

Howarth, C.I., Beggs, W.D.A. & Bowden, J.M. (1971). The relationship between speed and accuracy of movement aimed at a target. *Acta Psychologica, 35*, 207-218.

Jeannerod, M. (1986). Mechanisms of visuomotor co-ordination: A study in normal and brain-damaged subjects. *Neuropsychologia, 24*, 41-78.

Jeannerod, M. (1988). *The neural and behavioural organization of goal-directed movements.* Oxford: Clarendon Press.

Keele, S.W. (1981). Behavioral analysis of movement. In V.B. Brooks (Ed.), *Handbook of physiology, Section 1, The nervous system, Vol.II, Part 2, Motor control.* Bethesda: American Physiological Society.

Kelso, J.A.S. (1978). Joint receptors do not provide a satisfactory basis for motor timing and positioning. *Psychological Review, 85,* 474-481.

Klapp, S.T. (1975). Feedback versus motor programming in the control of aimed movements. *Journal of Experimental Psychology: Human Perception and Performance, 104,* 147-153.

Klapp, S.T. & Greim, D.M. (1979). Programmed control of aimed movements revisited: The role of target visibility and symmetry. *Journal of Experimental Psychology: Human Perception and Performance, 5,* 509-521.

Kvalseth, T.O. (1980). An alternative to Fitts' Law. *Bulletin of the Psychonomic Society, 16,* 371-373.

Lestienne, F. (1979). Effects of inertial load and velocity on the braking process of voluntary limb movements. *Experimental Brain Research, 35,* 407-418.

Liepmann, V. (1900). Das krankheitsbild der apraxia ("motorischen asymbolie") auf grund eines falles von einseitiger apraxie. *Monatschrift fur Psychiatrie und Neurologie, 8,* 15-44, 102-132, 188-197.

MacKay, W.A. & Riehle, A. (1990). Correlates of preparation of arm reach parameters in parietal area 7a. NATO Advanced Study Institute, *Tutorials in Motor Neuroscience,* Corsica, September 15-24.

Marsden, C.D., Obeso, J.A. & Rothwell, J.C. (1983). The functions of the antagonist muscle during fast limb movements in man. *Journal of Physiology, 335,* 1-13.

Marteniuk, R.G., Mackenzie, C.L., & Leavitt, J.L. (1988). Representational and physical accounts of motor control and learning: Can they account for the data? In A.M. Colley & J.R. Beech (Eds.), *Cognition and action in skilled behavior.* Amsterdam: North-Holland.

Meyer, D.E., Smith, J.E.K. & Wright, C.E. (1982). Models for the speed and accuracy of aimed limb movements. *Psychological Review, 89,* 449-482.

Miller, G.A., Galanter, E., & Pribram, K.H. (1960). *Plans and the structure of behavior.* New York: Holt, Rinehart & Winston.

Ostry, D.J. (1980). Execution-time movement control. In G.E. Stelmach & J. Requin (Eds.), *Tutorials in motor behavior.* Amsterdam: North-Holland.

Pew, R.W. (1984). A distributed processing view of human motor control. In W. Prinz & A.F. Sanders (Eds.) *Cognition and motor processes.* Berlin: Springer-Verlag.

Phillips, J.G. & Hughes, B.G. (1988). Internal consistency of the concept of automaticity. In A.M. Colley & J.R. Beech (Eds.), *Cognition and action in skilled behaviour.* Amsterdam: North-Holland.

Plamondon, R. (1990). A unified approach to the study of target directed movements. NATO Advanced Study Institute, *Tutorials in Motor Neuroscience,* Corsica, September 15-24.

Plamondon, R. (1991). On the origin of asymmetric bell-shaped velocity profiles in rapid-aimed movements. In J. Requin & G.E. Stelmach (Eds.), *Tutorials in Motor Neuroscience.* Dordrecht: Kluwer.

Pylyshyn, Z. (1980). Computational models and empirical constraints. *Behavioral and Brain Sciences, 1,* 93-128.

Quinn, J.T. Jr., Schmidt, R.A., Zelaznik, H.N., Hawkins, B. & McFarquhar, R. (1980). Target-size influences on reaction time with movement time controlled. *Journal of Motor Behavior, 12,* 239-261.

Reed, E.S. (1982). An outline of a theory of action systems. *Journal of Motor Behavior, 14,* 98-134.

Reed, E.S. (1984). From action Gestalts to direct action. In H.T.A. Whiting (Ed.), *Human motor actions: Bernstein reassessed.* Amsterdam: North-Holland.

Requin, J., Semjen, A., & Bonnet, M. (1984). Bernstein's purposeful brain. In H.T.A. Whiting (Ed.), *Human motor actions: Bernstein reassessed.* Amsterdam: North-Holland.

Schmidt, R.A., Zelaznik, H.N. & Frank, J.S. (1978). Sources of inaccuracy in rapid movements. In G.E. Stelmach (Ed.), *Information processing in motor control and learning*. New York: Academic Press.

Schmidt, R.A., Zelaznik, H.N., Hawkins, B., Frank, J.S. & Quinn, J.T. Jr. (1979). Motor output variability: A theory for the accuracy of rapid motor acts. *Psychological Review, 86,* 415-451.

Sheridan, M.R. (1976). *Studies on the control of voluntary movement.* Unpublished Ph.D. thesis, University of Hull.

Sheridan, M.R. (1979). A reappraisal of Fitts' Law. *Journal of Motor Behavior, 11,* 179-188.

Sheridan, M.R. (1981). Response programming and reaction time. *Journal of Motor Behavior, 13,* 161-176.

Sheridan, M.R. (1984). Response programming, response production, and fractionated reaction time. *Psychological Research, 46,* 33-47.

Sheridan, M.R. (1988). Movement metaphors. In A.M. Colley & J.R. Beech (Eds.), *Cognition and action in skilled behaviour.* Amsterdam: North-Holland.

Siegel, D.S. (1977). The effect of movement amplitude and target diameter on reaction time. *Journal of Motor Behavior, 9,* 257-265.

Siegel, D. (1988). Fractionated reaction time and the rate of force development. *Quarterly Journal of Experimental Psychology, 40A,* 545-560.

Welford, A.T. (1968). *Fundamentals of skill.* London: Methuen.

Wright, C.E. & Meyer, D.E. (1983). Sources of the linear speed-accuracy trade-off in aimed movements. *Quarterly Journal of Experimental Psychology, 35A,* 279-296.

NEURAL BASIS OF MOVEMENT REPRESENTATIONS (1)

J. REQUIN
Cognitive Neuroscience Unit
Laboratory of Functional Neuroscience
C.N.R.S.
31, Chemin Joseph Aiguier
13402 Marseille Cedex 9, France

ABSTRACT. In the framework of the cognitive neuroscience research strategy, data are accumulating which provide strong support for the central assumption of motor control studies : the organization of motor behavior would be based upon the utilization by the motor system of information stored in memory in the form of multiple, more or less abstract, hierarchically-organised representations of motor actions. Especially, studies conducted with single-neuron recording techniques in monkeys trained in tasks derived from those used in cognitive psychology, have demonstrated brain mechanisms which can be associated with three levels of representational processes. At the highest level, which may be called "semantic", the action goal would be represented in a non-motoric, holistic, context-independent and symbolic mode, from which the "response" to be made is determined. At the middle level, which may be called "syntactic", the motor features would be represented in a non-motoric, parametric, context-dependent and subsymbolic mode, from which the subroutines of the motor program would be specified and then assembled. At the lowest level, motor commands would be represented in a motoric, anatomical, biomechanically-constrained and neuromuscular mode which, when activated, results in a specific motor output. However, neurophysiological data appear to be increasingly incompatible with the traditional view in which, according to the hypothesis of a one-to-one mapping between functional processes and neural structures, these three representational levels would be implemented in association, premotor and motor cortical areas, respectively. The functional heterogeneity of cortical areas, which differ quantitatively more than qualitatively, as well as the continuum of function for individual neurons, between which differences are also more quantitative than qualitative, suggest another organization of the neocortex : each "behavioral" function would be implemented in a widely distributed neuronal network, which explains that the three different representational functions can be found closely intermixed in the same cortical region. A key point for future research is, thus, to understand how cognitive processes, as representations of action, are implemented in the microstructure of the cortical tissue.

(1) *This work was supported by ONR grant N00014 89 J1557*

J. Requin and G. E. Stelmach (eds.), Tutorials in Motor Neuroscience, 333–345.

1. Introduction

In the framework of the quickly developing cognitive neuroscience approach, based on the research strategy in which concepts and methods of cognitive psychology and neuroscience are combined, data have been accumulated which provide strong support for the central assumption of the current conception of motor control. Disregarding a number of variations, the organization of motor activity is supposed to be based upon the utilization by the motor system of information stored in memory in the form of multiple, more or less abstract, hierarchically-organized "representations" of motor actions. These representations, which are elaborated and permanently updated during ontogenesis and learning, may be viewed as the "language" of the motor system. Such a metaphoric reference to linguistics is recurrent in the domain of motor control, from Lashley (1951), when calling for a "syntax" of movement units (or words) determining the ordering of the movement sequence (or sentence), to, for example, Shaffer (1982), when describing the structure of the motor program as a "set of grammatical representations of intended actions". Note that "multiple", "abstract" and "hierarchical", which define the main features of this representational conception of motor control, were not introduced by the cognitive revolution. They were already the key-words used, for example, by Von Monakow (1914) at the beginning of the century, for describing the brain mechanisms responsible for movement control (cf. Wiesendanger, 1990).

2. Neural mechanisms of action goal representation

At the highest level, which may be called "semantic", motor actions would be represented in a non-motoric, holistic and symbolic mode and could give rise to a conscious experience. Non-motoric means that this level of representation does not contain the spatio-temporal structure of the forthcoming motor action, i.e. the ordering and parameters of the elementary movement units. Holistic means that action is represented by a unique feature which conceptually defines the behavioral goal to be reached (i.e. "the change in the surrounding world that will result from acting", to quote W. James). Symbolic means that this representation of the action goal is stored in an abstract, or conceptual, form: by analogy to concept formation in linguistics, a symbol may be viewed here as some single code which results from extracting a common feature of several more elementary codes. The process by which this context-independent representation of action goal is retrieved from memory is what is called "response" selection or determination in the framework of information processing stage models of cognitive psychology.

First empirical evidence for this highest level of action representation came from early neuropsychological studies identifying, among the different apraxic syndromes, disorders which seemed to specifically result from the inability either to intentionally initiate a well-anticipated action (ideomotor apraxia), which could, however, be automatically performed, or, even, to evoke the image of the forthcoming action (ideatory apraxia). The possible implication of the association cortex, especially the posterior parietal and prefrontal areas, in the setting up of action goal representation and its transmission downstream to the motor system was hypothetized early on the basis of lesions studies (cf. Paillard, 1982). However, it is only recently that studies conducted in the cognitive neuroscience approach with single-neuron recording techniques have provided direct evidence for the neuronal substrate of action goal representation.

The starting point was undoubtedly the influencial work of Mountcastle and his colleagues (1975) when discovering in the posterior parietal areas 5 and 7 changes in neuronal activity which were triggered by an object, provided that this object was the target for limb reaching and/or grasping movements. The hypothesis that this brain area generated a "command" function for initiating motor activity, on the basis of a synthesis of multiple-sensory information, was subsequently challenged (for review, see Lynch, 1980). The debate focussed on whether this increase in neuronal activity after the stimulus and before the movement was due to an attentional facilitation of sensory processes or a preparatory facilitation of motor processes. The impossibility to demonstrate unequivocally the "sensory" vs "motor" function of this "enhancement" phenomenon stressed out the difficulty in delimiting the boundary between perception and action, which is exactly as one would expect for an interfacing neural system responsible for making connections between perceptual and action representations.

Studies conducted during the last 10 years with single neuron techniques have provided strong support for such a view. For instance, John Seal (cf. this volume) has found in area 5 "sensorimotor" neurons which modified their activity in relation to both the stimulus presentation and motor response execution. It must be underlined that changes in activity of these two-component neurons, whose features did not depend upon the stimulus and movement physical parameters have been observed only after monkeys were trained in a task associating quite arbitrarily stimuli and responses. One may thus infer that these neuronal changes appeared only when the stimulus has acquired a behavioral meaning, i.e. became able to evoke a representation of the forthcoming action (Seal, 1989; Seal and Requin, 1987). Very similar neurons have been found in area 7 by Andersen et al. (1987; see also Andersen, 1989), who stressed their possible role in sensorimotor integration and, more precisely, in the implementation of the process responsible for "the formulation of motor behaviors".

Note, however, that recent data suggest that the neural mechanisms associated with action goal representation would be not restricted to the posterior parietal cortex but would extend to the frontal cortex. Rizzolatti and his colleagues (1988) have shown in the inferior part of the premotor area 6, neurons which were specifically activated by the "holistic" meaning of action sequences in term of their behavioral goal, but not in terms of sequence elements or movements features. For instance, some neurons were activated before and during the action of grasping food with the hand, - but not when the hand, although shaped similarly as for grasping, performed a pointing-pushing movement - or before and during the action of grasping with the mouth - but not during mouth movements when feeding the monkey These neuronal activities thus seem to express the basic elements of some kind of "lexicon" of motor actions semantically defined. In such a perspective, the changes in activity of neurons associated with the action of grasping food, whenever the grasp was performed with either the hand or mouth, are even more suggestive. They may be viewed as participating in the process by which symbolic representations - for example that of the "grasping action" - are built by extracting the common concept for a class of similarly goal-directed motor activities.

In some studies, the effects of manipulating the experimental factors known as acting specifically on the response determination stage, i.e. response probability and S-R compatibility (cf. Kornblum, this volume), were examined by combining behavioral and physiological approaches. For example, in experiments designed in the framework of the Eriksen's noise-compatibility paradigm (cf. Eriksen and Eriksen, 1974; see also Eriksen et al., 1985), in which conflict between stimulus features results in a competition between responses, Coles and his colleagues (1988; this volume) have demonstrated that the lateralized readiness potential (LRP) recorded over the motor cortical areas in human subjects in a good index of subthreshold activation of inappropriate responses. The effects of manipulating response probability were analyzed by Requin et al. (1990; cf. also Lecas et al., 1986) in monkeys trained to perform a visuo-manual pointing-task in a between-hands choice RT procedure, i.e; when the biomechanical features of the movement to be performed by either hand did not change. It was shown that changes in preparation-related neuronal activity were closely related to the changes induced experimentally with probabilities for either hand to perform the movement and, thus, to changes in RT. Although they were found more often in the posterior parietal association cortex than in the premotor cortex and more often in the latter than in the motor cortex, these probability-related neurons cannot be considered, however, as characterizing specifically one cortical region.

For ten years, the development of methods making possible to analyze neuronal activity in behaving monkeys has led to discover an increasing number of brain processes which occur during the execution of sensorimotor tasks, but whose the timing and features, being unrelated to any partition of the sensory inputs and motor outputs, are hardly explainable without refering to the notion of conceptual representation of action goal. In the same time, the traditional conception in which association cortical areas are mainly, even exclusively responsible for implementing this highest representational function is increasingly challenged by data which strongly suggest that such a function could be widely distributed in a neuronal network throughout a large set of neocortical structures.

3. Neural mechanisms of movement feature representation

At the intermediary level of representation, which may be called "syntactic", motor actions would be represented in a non-motoric, parametric and subsymbolic mode, and would not give rise to conscious experience. Non-motoric means that the content of this representation, although structured in terms of the features specifying classes of motor activities rather than movements themselves, are not yet ready to be directly used by the neuromuscular system. Parametric means that these features have, however, some homomorphic relationships with either the physical and/or biomechanical parameters with which movements can be described. Subsymbolic is used - by analogy to the meaning of this word in the connectionist approach (cf. Smolensky, 1988) - to describe the form in which movement features are stored in memory: a subsymbolic mode of representation (or a subconcept) is intermediary between the symbolic mode of representation of cognitive architectures - in which the elementary codes from which the symbol is extracted are no longer retrievable - and an elementary code, which results from a simple translation, as, for example, the translation of one stimulus physical parameter into frequency of neural impulses at the level of peripheral sensory coding. Briefly, the main interesting property of a subsymbolic mode of representation would be to have a certain level of abstraction, although maintaining some relationships with the more elementary and specific codes from which it derives. This could provide the theoretical ground for solving the difficult problem of the translation of movement features, represented in some abstract, non motoric, language, into the language of the neuromuscular system, a criticism which was rightfully and recurrently addressed against the motor program concept. The processes by which these representations of motor features are selected in memory and, then, completed, specified and continuously updated according to the detailed requirements of the forthcoming movement are that is currently called "motor programming" in the information processing models. Note that, in most conceptions of motor programming, the motor program, as a context-dependent representation of the future movement is the result of this programing process and not the subset of motor features selected in memory (cf. for example, Schmidt, 1982).

Before the beginning of the last decade, most of the physiological evidence for a stored representation of motor features from which the subroutines of the motor program are selected was mainly indirect (cf. Brooks, 1979). The notion that movement performance was controlled by a prestructured pattern of neural activities which prescribed movement parameters resulted indeed from an inferential reasoning based upon the analysis of the physiological mechanisms observed during the performance of the movement itself, but not of those which were supposed to be associated with the earlier motor programming process. Such indirect evidence was drawn from three kinds of experimental data. The oldest were probably those showing that electrical stimulations specifically localized in the midpart and lower part of the brain could evoke behaviorally significant sequences of movements. A second set of data were provided by studies showing that well-structured spatio-temporal patterns of electromyographic, spinal or more centrally recorded activities (cf. Grillner, 1975) were observed during the performance of rythmic behaviors, as locomotion, even when motor outputs were artificially blocked at the peripheral level. The last, and well-known, line of indirect evidence was drawn from studies demonstrating that skilled and well-organized movements or, even, movement sequences, can be performed when sensory feedbacks are disrupted (cf. Bizzi, 1980; Evarts et al., 1971).

Although such kind of data may be viewed as strongly suggesting a central pattern generator or motor program interpretation, the lack of any direct experimental support for the neural mechanisms of the motor programming process itself left the door open for any likely alternative explanation of movement production. Such an alternative explanation was proposed by proponents of action view when they claimed that the motor program remained an assumption and that order, regularity and invariant features of motor outputs are shaped through the synergy of biological and environmental constraints in a continuous dynamical perspective.

This is the main reason for considering experimental findings collected by recording the activity of neural structures during tasks designed in the framework of the preparation paradigm as directly relevant for the motor programming concept. Here, the neural processes underlying the preparatory phase and the execution phase of movement performance can be dissociated in time and separately analyzed. Provided, first, that the neural processes occurring during the planning period can be modulated by experimentally manipulating prior information about the features of the forthcoming movement, and, second, that these modulations are found to correlate with movement performance speed or accuracy, the rationale of the preparation paradigm necessarily implies the notion of representation. The role of the first, preparatory event is to select in, and to extract from memory the representation of one or several movement features, which will be utilized, after the remaining movement features are selected when the second, imperative event occurs, to assembly the motor program. The greater the number of movement features so preselected and the most accurate this presection process, the faster the program assembling process and the shorter the time to initiate movement performance.

In the framework of this preparation paradigm, the study of neuronal activity related to the programming of movement parameters was initiated by the work of Evarts and his colleagues (for reviews see Evarts, 1984 and Evarts et al., 1984). They developed a prototypic behavioral situation, similar to the RT protocols used with humans, in which monkeys were trained to push or pull a handle by either flexing or extending the forearm. The preparatory stimulus indicated to the animal whether a pull or a push was required after the imperative stimulus. During the preparatory period a large number of pyramidal as well as non-pyramidal tract neurons of the motor cortex (area 4) exhibited changes in their resting discharge frequency, which depended upon advance information provided to the animal about the direction, i.e. flexion vs extension, of the forthcoming movement. The neurons that controlled the muscles to be activated exhibited an increase in their activity, while a decrease in activity was found for neurons that controlled the muscles to be relaxed during movement performance. Similar data were subsequently found, by using not only monoarticular but also polyarticular movements, in the premotor cortex, i.e. the lateral part of area 6 (Godschalk and Lemon, 1983; Wise, 1985; Wise and Mauritz, 1985; Wise et al., 1986) as well as in the supplementary motor area, i.e. the medial part of area 6 (Tanji et al., 1980; Kurata and Tanji, 1985; Tanji and Kurata, 1982).

Such an implication of motor and premotor cortical areas in the programming of movement direction has been confirmed indirectly by using transcortical reflex techniques in human subjects. Changes in the amplitude of the late components of the EMG response to muscle stretch were analyzed during the preparatory period in a procedure requiring the performance of a wrist movement. It was shown that the late components of the long-loop reflex were differently modulated when movement direction was precued, being consistently larger when the stretched muscle was precued as an agonist than as an antagonist in performing the forthcoming movement (Bonnet, 1983; Bonnet et al., in press). Note that when movement extent was precued, no such a differential effect was found.

While evidencing the possibility to evoke in a large set of cortical structures some representation of the movement parameters which could be then incorporated into the motor program, this ensemble of data suffered of various weakness with regard to the constraints of thepreparation paradigm. First, although changes in neuronal activity were observed before movement execution and were found to be sensitive to one parameter of the forthcoming movement, there was no direct evidence for their predictive value for performance speed and/or accuracy. Second, advance information about only movement direction, i.e. a kinematic parameter, was shown to result in changes in neuronal activity, and moreover, most often when movement direction was a dimension confounded with the reciprocal activation of antagonistic muscles.

By adapting to monkeys the movement dimension precuing technique used in human subjects, preparation-related changes in the neuronal activity of the motor and premotor cortex have been examined in detail by Riehle and Requin (1989). While the precuing of movement direction resulted in a large decrease in RT associated with significant changes in neuronal activity during the preparatory period, the precuing of movement extent did not reduce RT and, accordingly, no preparation-related changes in neuronal activity were found. However, when only direction was precued, RT was longer than when both parameters were precued, a result which is compatible with a serial, hierarchical model of movement programming in which movement extent could not be specified before movement direction. Accordingly, some neurons were found, whose preparation-related changes in activity when both parameters were precued were of greater amplitude than when only movement direction was precued. Furthermore, these changes were greatly reduced when movement extent was precued, as well as when no dimensional information was provided in advance.

These data suggest that the representation of a dynamic movement parameter may be evoked by an adequate manipulation of prior information and would be used to build the motor program. They are in agreement with the data collected by Kutas and Donchin ten years ago (1980) who showed in human subjects that the amplitude of the readiness potential recorded before the execution of a limb movement is related to the force to be exerted. Such an effect of the anticipated force to perform the movement on slow brain potentials was recently confirmed by Bonnet and MacKay (1989). Note, along the same lines, that it has also be found that precuing the duration of isometric movements was associated with specific changes in the late component of the CNV (cf. Vidal et al., in press).

Moreover, by using trial-by-trial correlation analyses, changes in neuronal activity during the preparatory period were found to be highly predictive for RT, thus suggesting that these changes are closely associated with the process which participates in movement planning. Lastly, although a number of neurons showing preparation-related changes in activity also showed subsequent execution-related changes in activity, thus indicating that they were involved in controlling movement performance, some of the neurons which exhibited anticipatory effects were not involved at all in the movement execution process.

By showing that it is possible to evoke and to experimentally manipulate patterns of neuronal activities which have homomorphic relationships with some of the dimensions with which motor outputs can be describes, recent data collected with various techniques in the frame of the preparation paradigm therefore provide strong support for the motor programming concept. Especially these data suggest that not only the kinematic but also the dynamic parameters of a movement would be represented in the motor program. This answers one of the current criticisms against the motor control view, i.e. that only thc brain processes associated with the planning of the movement dimensions which would be centrally coded in terms of the topological distribution of the peripheral effectors, could be evidenced.

Finally, recent neurophysiological studies suggest, once again, that the representational function implied by the motor program concept is not implemented in a precisely localized anatomical region of the brain, for example the premotor cortex, as currently proposed. With Alexa Riehle (Riehle and Requin, 1989), we have recently compared the roles played in movement preparation by the motor and premotor cortex respectively and concluded that there are quantitative but not qualitative differences between the two structures. Movement-related neurons, preparation-related neurons, as well as neurons sharing both these properties were found to be closely intermixed in both areas, with the proportion of the first class decreasing and the proportion of the second class increasing from the central sulcus to the arcuate sulcus. Moreover, when the

predictive value for RT of the preparation-related changes in neuronal activity induced by precuing movement parameters was considered, the number, distribution and strength of correlations were found to only slightly differ between the two cortical regions. MacKay and Riehle (this volume) have recently shown that such a conclusion must probably be extended to the posterior parietal areas, thus suggesting that the intermediary level of movement feature representation has also to be viewed as distributed in a widely extending neuronal network.

4. Neural mechanisms of movement command representation

At the lower level of representation, which may be called "phonemic", would be represented the basic movement units, or movement "commands" - in the same sense as that used when saying cortico-spinal commands - which, after being selected acording to program instructions, result in the pattern of neuromuscular activations. This process underlies what is called movement execution or motor output. Although the precise relationships of these basic movement units with peripheral effectors, as controlling muscles, muscle synergies, coordinative structures or, even, pre-structured elementary movements, was and remains controversial, some of their features are logically necessary. Motoric means that the activation of movement commands has mechanical consequences at the periphery. Anatomical means that there is some homomorphic relationships between the distribution of movement units in the central nervous system and the spatial organization of the muscular effectors. Biomechanically-constrained means that the activation of some of the possible patterns of movement units is unlikely because the activation of the corresponding patterns of muscular effectors is excluded by, at least, the biomechanical limitations of the musculo-articulatory system. Neuromuscular refers to the type of language in which movement commands are represented and operate. In contrast to the problems raised by the language in which representations at the higher levels are stored and used, the coding processes by which movement units are activated and communicate with the neuromuscular system, uses obviously the current language of the nervous system, i.e. frequency of neuronal impulses.

A number of chapters in this volume - especially those by Roger Lemon, Marie-Claude Hepp-Reymond and Roberto Caminiti - describe in detail the neurophysiological mechanisms by which movement features are controlled during execution. They emphasize what is probably one of the major findings of the last decade, i.e. that the control of movement parameters during execution would result from the cooperative action of large populations of neurons, rather than from the collection of the independent activations of individual neurons, each of them being connected to peripheral effectors in a one-to-one mapping mode. The work of Georgopoulos (1990) and his group, which was summarized recently by Roger Lemon (1989), is especially relevant in this way.

Neuronal activity of the motor cortex was recorded when monkeys had to move a handle from a central position to one of different targets located around at the same distance. When the activity of individual neurons is considered, each neuron was activated for the differently oriented reaching movements, but showed a maximum of activity for a prefered direction. This first finding suggested that one neuron is not fully specialized in controlling movements performed in a particular direction and, conversely, that one neuron could partly be involved in the control of differently oriented movements. Now, if one considers a population of neurons, in which the amount of activation during the execution of a movement in a particular direction differs from one neuron to another, the sum of the contributions of this neuronal population may be represented by a population vector which precisely points in the direction of the movement. These data suggest that movement trajectory is controlled thanks to the cooperation of a large number of neurons, including neurons whose prefered direction is not that of the movement actually performed.

Although initially demonstrated in the motor cortex, this mode of control of movement trajectory cannot be considered, however, as characterizing specifically the motor cortex only. Very similar data were found in the premotor cortex (cf. Caminiti, this volume), as well as in the posterior parietal area 5 by Kalaska (1988). Together, these data suggest, once again, that even the representation of movement commands is not the the privileged function of a highly specialized cortical region - as initially proposed for the motor cortex - but widely extends over the neocortex.

5. Distributed networks for representational functions

Not only data provided by studies of the neuronal correlates of action and movement representations support a hierarchical conception of the cognitive processes involved in motor control but, more importantly, they add to the current reconsideration of the classical views on the anatomical separation of these cognitive functions, at the macroanatomical level of cortical structures and microanatomical level of neuronal units. Neither the concept of functional homogeneity of cortical areas, as defined by the cytoarchitecture, nor the concept of functional specificity of neurons are still tenable. For some years, revised views of the correlation between the structural and functional aspects of the brain have been stimulated mainly by the influential conception of a modular organization of the neocortex (Mountcastle, 1978). In the modular concept, all cortical areas are formed by aggregates of similar units or modules with the same neuronal circuitry and performing the same basic operation. Networks of modules, which are delimited by the extent of the extramodular input and output connections, are distributed throughout the cortex, or at least a large part of it, and the functions - where function is used in the usual sense of the term - implemented in different networks depend upon the relative weighting of these extramodular connections. Of course, this does not imply that the whole cerebral cortex is an entirely homogenous structure, nor that each neuron is totally unspecific in its functional involvement. Quantitatively different functions between neurons, resulting in variations in the statistical parameters of neuronal populations (cf. Requin et al., 1988, in press), make the existence of topographical differences in function between cortical regions compatible with the distributive parallel network concept (cf. Goldman-Rakic, 1988).

Such a view of cortical organization appears to be particularly appropriate for integrating the two main aspects of neurophysiological data on movement representations. First, the continuum of function for individual neurons implies that each representational function is necessarily implemented in a large neuronal population. Second, the distribution of each representational function over different cortical areas implies that different representational functions overlap in the same cortical area.

In order to develop this conception further, we need to understand how the cognitive processes underlying the planning of motor actions are brought into play, either continuously or in a hierarchical sequence - as suggested above - by the network organization of the neocortex. One may suggest that, now, the key to take a step towards this understanding is to be looked for in the microstructural organization of the cortical tissue - i.e. the functional cooperation between neurons at the modular level - rather than in the interrelationships between macroanatomically defined cortical regions .

REFERENCES

Andersen, R.A. (1989) Visual and eye movement functions of the posterior parietal cortex, Annual Review of Neuroscience, 12, 377-403.

Andersen, R.A., Essick, G.K., and Siegel, R.M. (1987) Neurons of area 7 activated by both visual stimuli and oculomotor behavior, Experimental Brain Research, 67, 316-322.

Bizzi, E. (1980) Central and peripheral mechanisms in motor control, in G.E. Stelmach and J. Requin (eds.), Tutorials in Motor Behavior, North-Holland, Amsterdam, pp. 131-143.

Bonnet, M. (1983) Anticipatory changes of long latency stretch responses during preparation for directional hand movements, Brain Research 280, 51-62.

Bonnet, M., and MacKay, W.A. (1989) Changes in CNV and reaction time related to precueing of direction and force of a forearm movement, Brain, Behavior and Evolution 33, 147-152.

Bonnet, M., Requin, J., and Stelmach, G.E. (in press) Changes in electromyographic responses to muscle stretch, related to the programming of movement spatial parameters, EEG and Clinical Neurophysiology.

Brooks, V.B. (1979) Motor programs revisited, in R.E. Talbott and D.R. Humphrey (eds.), Posture and Movement, Raven Press, New York, pp. 13-49.

Coles, M.G.H., Gratton, C., and Donchin, E. (1988) Detecting early communication: using measures of movement-related potentials to illuminate human information processing, Biological Psychology 26, 69-89.

Eriksen, B.A., and Eriksen, C.W. (1974) Effects of noise letters upon the identification of target letter in visual search, Perception and Psychophysics 16, 143-149.

Eriksen, C.W., Coles, M.G.H., Morris, L.R., and O'Hara, W.P. (1985) An electromyographic examination of response competition, Bulletin of the Psychonomic Society 23, 165-168.

Evarts, E.V. (1984) Neurophysiological approaches to brain mechanisms for preparatory set, in S. Kornblum and J. Requin (eds.), Preparatory States and Processes, Lawrence Erlbaum, Hillsdale, pp. 137-153.

Evarts, E.V., Bizzi, E., Burke, R.E., Delong, M., and Thach, W.T. (1971) Central control of movement, Neurosciences Research Program Bulletin 9, n° 1.

Evarts, E.V., Shinoda, Y., and Wise, S.P. (1984) Neurophysiological approaches to higher brain functions, Wiley and Sons, New York.

Georgopoulos, A.P. (1990) Neurophysiology of reaching, in M. Jeannerod (ed.), Attention and Performance XIII, Lawrence Erlbaum, Hillsdale, pp. 227-263.

Godschalk, M., and Lemon, R.N. (1983) Involvement of monkey premotor cortex in the preparation of arm movements, Experimental Brain Research , suppl. 7, 114-119.

Goldman-Rakic, P.S. (1988) Topography of cognition: parallel distributed networks in primate associative cortex, Annual Review of Neuroscience, 11, 137-156.

Grillner, S. (1975) Locomotion in vertebrates: central mechanisms and reflex interactions, Physiological Reviews 55, 247-304.

Kalaska, J.F. (1988) The representation of arm movements in postcentral and parietal cortex, Canadian Journal of Physiological Pharmacology 66, 455-463.

Kurata, K., and Tanji, J. (1985) Contrasting neuronal activity in supplementary and precentral motor cortex of monkeys. II. Responses to movement triggering vs nontriggering sensory signals, Journal of Neurophysiology, 53, 142-152.

Kutas, M., and Donchin, E. (1980) Preparation to respond as manifested by movement-related brain potentials, Brain Research 202, 95-115.

Lashley, K.S. (1951) The problem of serial order in behavior, in L.A. Jeffress (ed.), Central mechanisms in behavior, Wiley, New York, pp.

Lecas, J.C., Requin, J., Anger, C., and Vitton, N. (1986) Changes in neuronal activity of the monkey precentral cortex during preparation for movement, Journal of Neurophysiology, 56, 1680-1702.

Lemon, R. (1989) Cognitive control of movement, Nature 337, 410-411.

Lynch, J.C. (1980) The functional organization of posterior parietal association cortex, Behavioral and Brain Sciences, 3, 485-534.

Mountcastle, V.B. (1978) An organizing principle for cerebral function: the unit module and the distributed system, in F.O. Schmitt and F.G. Worden (eds.), The Neurosciences. Fourth Study Program, MIT Press, Cambridge, pp. 21-42.

Mountcastle, V.B., Lynch, J.C., Georgopoulos, A., Sakata, H., and Acuna, C. (1975) Posterior parietal association cortex of the monkey: command functions for operations within extrapersonal space, Journal of Neurophysiology 38, 871-908.

Paillard, J. (1982) Apraxia and neurophysiology of motor control, Philosophical Transactions, Royal Society of London. B 298, 111-134.

Requin, J., Lecas, J.C., Vitton, N. (1990) A comparison of preparation-related neuronal activity changes in the prefrontal, premotor, primary motor and posterior parietal areas of the monkey cortex: preliminary results, Neuroscience Letters 111, 151-156.

Requin, J., Riehle, A., and Seal, J. (1988) Neuronal activity and information processing in motor control: from stages to continuous flow, Biological Psychology, 26, 179-198.

Requin, J., Riehle, A., Seal, J. (in press) Neuronal networks for movement preparation, in D.E. Meyer and S. Kornblum (eds.), Attention and Performace XIV. Lawrence Erlbaum, Hillsdale.

Riehle, A., and Requin, J. (1989) Monkey primary motor and premotor cortex: single-cell activity related to prior information about direction and extent of an intended movement, Journal of Neurophysiology, 61 (3), 534-549.

Rizzolatti, G., Camarda, R., Fogassi, L., Gentilucci, M., Luppino, G and Matelli, M. (1988) Functional organization of inferior area 6 in the macaque monkey, Experimental Brain Research 71, 491-507.

Schmidt, R.A. (1982) Motor control and learning. A behavioral emphasis, Human Kinetics, Champaign.

Seal, J. (1989) Sensory and motor functions of the superior parietal cortex of the monkey as revealed by single neuron recordings, Brain, Behavior and Evolution, 33, 113-117.

Seal, J., and Requin, J. (1987) Sensory to motor transformation within area 5 of the posterior parietal cortex in the monkey, Society for Neuroscience Abstracts 13, part 1, p. 673.

Shaffer, L.H. (1982) Rhythm and timing in skill, Psychological Review, 89, 102-122.

Smolensky, P. (1988) On the proper treatment of connectionism, Behavioral and Brain Sciences 11, 1-74.

Tanji, J., and Kurata, K. (1982) Comparison of movement-related activity in two cortical motor areas of primates, Journal of Neurophysiology, 48, 633-653.

Tanji, J., Taniguchi, J., and Saga, T. (1980) Supplementary motor area: neuronal response to motor instructions, Journal of Neurophysiology 44, 60-68.

Vidal, F., Bonnet, M. and Macar, F. (in press) Programming response duration in a precueing reaction time paradigm, Journal of Motor Behavior.

Von Monakow, C. (1914) Die Lokalisation im Grosshirn und der Abbau der Funktion durch kortikale Herde, Bergman, Wiesbaden.

Wiesendanger, M. (1990) The motor cortical areas and the problem of hierarchies, in M. Jeannerod (ed.), Attention and Performance XIII, Lawrence Erlbaum, Hillsdale, pp. 59-75.

Wise, S.P. (1985) The primate premotor cortex: past, present, and preparatory, Annual Review of Neuroscience, 8, 1-19.

Wise, S.P., and Mauritz, K.H. (1985) Set-related neuronal activity in the premotor cortex of rhesus monkey:effects of changes in motor set, Proceedings of the Royal Society of London, B 223, 331-354.

Wise, S.P., Weinrich, M., and Mauritz, K.H. (1986) Movement-related activity in the premotor cortex of rhesus macaques, in H.J. Freund, U. Buttner, B. Cohen and J. Noth (eds.), Progress in Brain Research 4. Elsevier, Amsterdam, pp. 117-131.

CORRELATES OF PREPARATION OF ARM REACH PARAMETERS IN PARIETAL AREA 7A OF THE CEREBRAL CORTEX

William A. MACKAY
Department of Physiology
University of Toronto
Toronto, M5S 1A8 Canada

Alexa RIEHLE
Unité de Neurosciences Cognitives
C.N.R.S.-LNF1
31, Chemin Joseph-Aiguier
13402 Marseille Cédex 9
France

ABSTRACT. A monkey was trained in a reaching task with a delay period: the arm to be used, reach direction and extent could all be varied. Microelectrode recordings were made in area 7a of the parietal lobe in order to determine if the activity of single cortical neurons coded any of these parameters as a precued reach was prepared. Parameter preparation was manifested either as a differential response to the signals providing laterality and target location information, or as differential rates of discharge during the last .4 s of the delay period. Reach direction was the parameter most commonly coded (16%). When extent (small or large) was discriminated (9%), it was usually dependent on the directional parameter. It was, however, independent of spatial position. Laterality preparatory coding was also rare (9%), and furthermore was never observed in differential discharge rates at the end of the delay period. Area 7a appears, therefore, to be a cortical locus of direction preparation.

1. Introduction

The posterior parietal lobe of the cerebral cortex is a key site for the visual and somatosensory integration required to guide limb movements in extrapersonal space (Andersen 1987; Hyvärinen 1982). Lesions in this region, both in monkeys and humans result in the systematic misdirection of reaches. It is, therefore, likely that area 7a is at least partly responsible for constructing an internal model of the spatial relationships of external objects relative to the body (Stein 1989). Parameters which may well be abstracted in such a model could include the direction of a target relative to the hand, and the movement extent, small or large. Moreover, area 7a could possibly determine which hand to use.

These ideas were tested in a monkey ("Prince Rubah") trained to perform a reaching task with either arm. Precued tasks with a delay period were designed so that neuronal responses to laterality and direction/extent information could be individually examined independently of motor execution (cf. Requin et al. 1988).

J. Requin and G. E. Stelmach (eds.), Tutorials in Motor Neuroscience, 347–356.

2. Materials and Methods

Rubah (*Macaca fascicularis*) weighed 5 kg and was carefully selected for his tameness and learning ability (tested by making him remove a lifesaver from a bent wire). For experimental sessions he was seated in a mobile primate chair with both arms completely unrestrained. On the right and left sides of the monkey, rest-plate switches were positioned at waist level. These had to be lightly depressed with the forearms to start a trial. Within reach in front of him was a 20 inch color videomonitor. Rubah was not deprived of food or water, but received no chow in the morning until after the recording session. Rewards for successful trials were nuts, raisins or fruit juice: when Rubah tired of one, another was given until he was satiated and stopped working.

2.1 PROTOCOL 1: BILATERAL REACH

The first task required reaching with either arm to a visual target displayed on the videomonitor. When both arms were positioned on the rest-plates the monitor illuminated with a colored background to indicate the arm to be used (HAND cue), blue for the left arm and violet for the right. After .6 s the target position was marked by an open square at a random location on the screen (TARG cue). Rubah could not move, however, until the GO signal which occured 1 s later. The GO signal was a solid square replacing the open one simultaneously with the sounding of a computer-generated tone. The monkey was then free to lift the appropriate arm and touch the target to get his reward (Fig. 1). If the wrong arm was lifted, or both or neither within 1.5 s, then the trial was aborted and another initiated. The first touch had to be accurate: subsequent corrections were ignored. If the target was attained, the videomonitor flashed a bright green and a burst of random tones in a 1 s sequence was generated for Rubah's amusement.

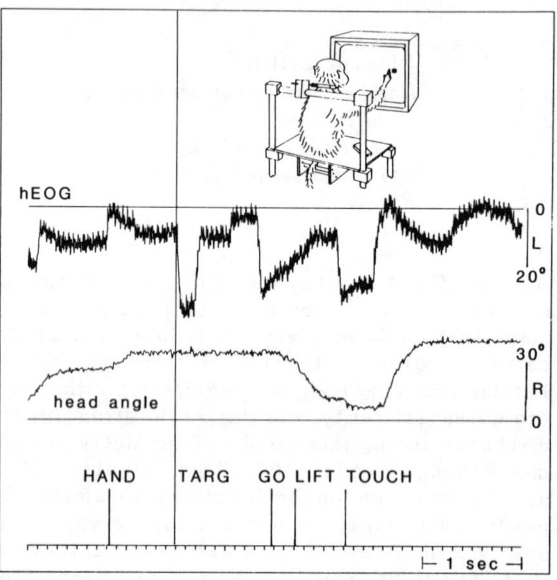

Fig. 1. Protocol 1 sequence. A sample record is given of the horizontal EOG and head rotation which accompanied performance of the task. Target was on the right side of the screen. R, rightward; L, leftward.

2.2 PROTOCOL 2: SEQUENTIAL REACH

The second task involved use of the right arm only. Trials started as before with both arms on the rest-plates. The violet HAND cue appeared on the monitor, then the first target (TARG1) .6 s later. TARG1 was a solid square (Fig. 2) indicating that it could be touched immediately without waiting. When TARG1 was correctly touched, TARG2 appeared .4 s later as an open square (Fig. 2). As in protocol 1, Rubah had to wait for 1 s before the GO signal. This meant maintaining contact with TARG1 for a total time of 1.4 s before reaching to TARG2. The targets were selected from 4 loci spaced 11 cm apart along a horizontal axis. The second target was always 11 cm (small extent) or 22 cm (large extent) away from the first, and either to the left or to the right. Eight different trajectories were selected in random order for Rubah to perform: 2 small rightward, 2 large rightward, 2 small leftward and 2 large leftward. Each member of a pair covered a different screen region. In this way it could be determined if neuronal discriminations of small/large extent (or right/left) were independent of screen position. Small reaches would have subtended an angle at the shoulder of about 10° and large reaches 20°. Again, if either target was incorrectly touched or the wrong arm was lifted, the trial was aborted and another started.

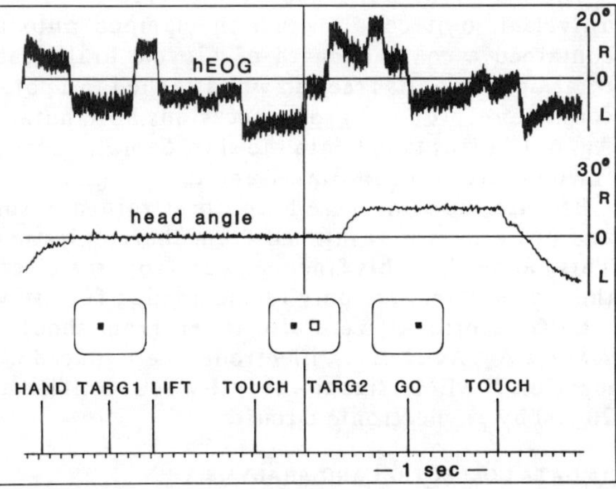

Fig. 2. Protocol 2 sequence (2 targets). The relative position of the 2 targets on the screen is schematically illustrated (small extent, rightward movement).

2.3 SURGICAL IMPLANTATION

After Rubah was trained, a recording chamber (22 mm i.d.) was implanted over the left parietal lobe under general inhalation anesthesia (halothane). Rubah's head was stabilized in a stereotaxic frame and a hole trephined in the skull, the same diameter as the chamber. Additional burr holes were drilled for 10 self-tapping stainless steel screws. The screws were inserted snugly, the chamber mounted in position and a brass post with a flanged end against the skull was held in a position close to the midline. Then dental acrylic was poured over the exposed skull, the screws, the base of the recording chamber and the flanges of the post. As the acrylic solidified, it was shaped to overhang the adjacent skin so that Rubah could not get his fingers at the wound edge. Rubah was kept on antibiotic for a week after the surgery. Recordings started after 2 weeks.

2.4 RECORDING PROCEDURES

Glass-coated platinum-iridium (70/30) microelectrodes were used for recording single unit activity in area 7a. The recording chamber was cleaned out and filled with a low melting point (39°C) paraffin wax to stabilize brain pulsations. Then a hydraulic microdrive (Kopf) was mounted on the chamber and the electrode lowered through the dura mater into the cerebral cortex. Units were pulse height discriminated.

The time and location of screen contacts during reach were monitored by an infra-red grid and microprocessor ("Smart Frame", Carroll Touch) mounted on the front of the videomonitor.

The head was not fixed: it was free to rotate to the right or left. A rod, with a universal joint coupling which clamped onto the brass post of the skull implant, contained a coaxial length of bicycle brake cable. The cable turned with the head rotations and at its free end was attached to a potentiometer so that the head movements could be recorded. On some occasions horizontal EOG was also recorded using surface Ag/AgCl disks taped onto the skin. Sample records of hEOG and head angle are shown for both protocols in Figs. 1 and 2.

Because the arms were both unrestrained, a sombrero with broad brims flaring out and upwards was fashioned from thermoplastic material and fitted to Rubah's head. This barrier kept his fingers away from the electrodes and ground wire. Nevertheless, these procedures are only recommended for use with placid, well-behaved monkeys.

EMG recordings were also taken from shoulder muscles on a few occasions using surface Ag/AgCl disks. Electrodes were placed over the anterior deltoid and trapezius muscles. The EMG signal was full-wave rectified and partially integrated (time constant 20 ms) by an electronic circuit.

2.5 DATA COLLECTION AND ANALYSIS

All of the signals, the discriminated unit activity, switch voltage, task video cues, screen contacts, head angle, hEOG/EMG were collected via a computer interface (CED 1401) and stored on a 386 microcomputer using commercial software (Spike2, Cambridge Electronic Design). All analog signals were sampled at 100 Hz, except for hEOG which was sampled at 200 Hz. Rubah's task was controlled by a separate microcomputer running lab-written programs.

Initial data processing into peri-event time histograms was done within Spike2. Further analysis was done with lab-written software or SigmaPlot (Jandel). Assessment of parameter discrimination involved trial-by-trial spike counts over a fixed interval adjusted to the response pattern (e.g. 0-.3 s after TARG cue or the last .4 s of PP), then a t-test between the two relevant populations (left or right arm, leftward or rightward direction, small or large extent), $p<.05$.

2.6 ELECTRODE LOCALIZATION

An initial estimate of the placement of area 7a within the chamber was obtained from a plasticine impression of the inner surface of the trephined skull disk. The impression clearly demarcated the cortical sulcal pattern. Subsequent electrode tracks confirmed the location of the intraparietal sulcus. Area 7a was readily distinguished from neighboring area 7b by the prominence of visual responses in the former, and of somatosensory responses (especially from the hand) in the latter. Area 7b is heavily innervated from other somatosensory areas but area 7a is not (Cavada & Goldman-Rakic 1989a).

3. Results

In order that the preparatory period (PP) truly reflect an interval of preparation, it was imperative that Rubah not start moving prior to the GO signal. By and large this was the case. Any release of the switch before the GO signal ended the trial and the data was ignored. Rubah learned not to anticipate. EMG recordings from the shoulder muscles indicated that the PP was an interval of stationary activity. In protocol 1 (Fig. 3A), anterior deltoid activity started rising only very slightly in the last 50 ms of the PP. The main burst always occured after the GO signal: mean reaction time was 160 (s.d.

±70) ms. If the contralateral hand was used for the reach, muscle activity did not increase until after the target was touched (Fig. 3B).

Again in protocol 2 there was no net change in shoulder muscle activity during the PP until after the GO signal (Fig. 4). In protocol 2 Rubah had to maintain contact with the first target throughout the PP.

3.1 LATERALITY PREPARATION

A total of 212 neurons was recorded in area 7a as Rubah performed in protocol 1. Of these, 185 were task-related. Table 1 gives a breakdown of the number of neurons which responded either to the HAND cue or showed changes in discharge rate during the PP. "Early" cue responses had latencies <150 ms, "late" >150 ms. Very few cells discriminated significantly between the two hands, and when they did the differences observed were relatively small. As shown by the example in Fig. 5, preparatory activity prior to the GO signal never distinguished laterality.

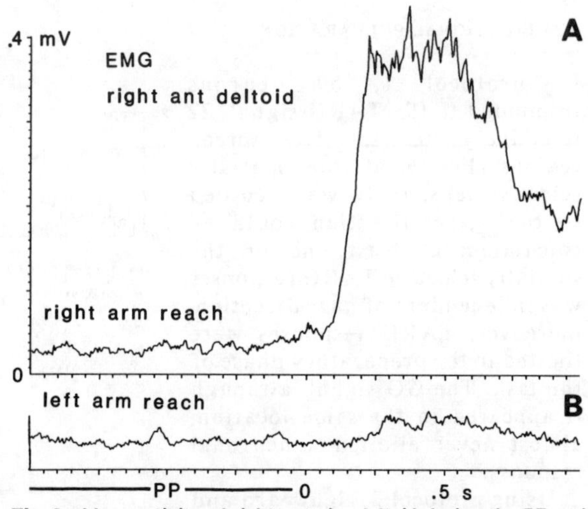

Fig. 3. Mean activity of right anterior deltoid during the PP and subsequent reach (protocol 1). GO signal occured at 0. A. reach with the right arm (n=49). B. reach with the left arm (n=56).

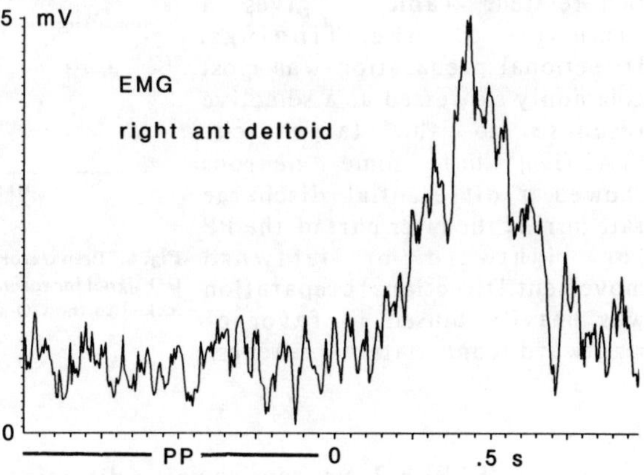

Fig. 4. Mean activity of right anterior deltoid during the PP (holding TARG2) and subsequent reach in protocol 2. GO signal occured at 0 (n=26).

TABLE 1. Neurons showing laterality preparation for protocol 1

Response type	Left hand	Right hand	Both hands
Early hand cue	4	2	56
Late hand cue	7	6	31
PP	0	0	76

3.2 DIRECTIONAL PREPARATION

In protocol 1, 59 neurons responded to the TARG signal, 22 selectively to specific screen regions. For 6 of the spatially selective cells, hEOG was recorded so that gaze direction could be calculated: at least one of the spatially selective TARG responses was independent of gaze direction. Moreover, TARG responses were limited to the preparatory phase of the task. The GO signal, although it appeared in the same location, almost never elicited a neuronal response.

Using protocol 2, rightward and leftward directional preparation was looked for in a total of 160 cells, of which 136 turned out to be task-related. Table 2 gives a summary of the findings. Directional preparation was most commonly expressed as a selective response to the target cue (TARG2), but some neurons showed a differential discharge rate during the later part of the PP for rightward or leftward movement. Directional preparation was heavily biased in favor of rightward (contralateral) reaches.

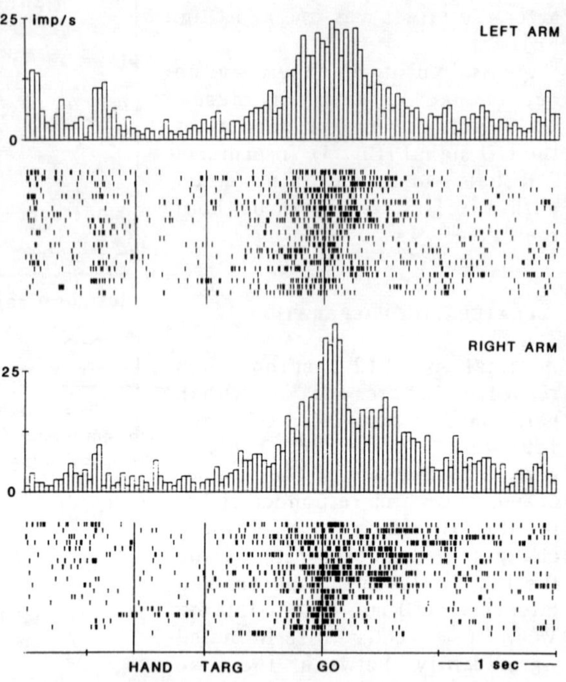

Fig. 5. Preparatory activity of an area 7a neuron in anticipation of GO signal for reach with either arm. Note paucity of responses time-locked to the GO signal in the rasters.

TABLE 2. Neurons showing direction and/or extent preparation with protocol 2

Response type	No.	Direction		Extent			
		Left	Right	Bidirectional		Unidirectional	
				Small	Large	Small	Large
TARG cue	54	5	23	4	0	5	1
PP	22	0	4	1	0	2	2
TARG & PP	26	1	6	0	0	0	0
Total	102	6	33	5	0	7	3

3.3 EXTENT PREPARATION

Also with protocol 2, a small number of neurons were encountered in area 7a which showed some degree of discrimination of reach amplitude (see Table 2). Most of these neurons were additionally influenced by the intended direction of the movement, such that the discrimination of movement extent was only observable for the preferred direction (invariably rightward). In the case illustrated by Fig. 6, the preference for small extent movements was maintained in both directions, and for different screen positions, but impulse rates were greatly increased for the rightward direction. (N.B. The "bidirectional" data in Table 2 lists cells with significant extent discrimination for both directions of movement, but the intended direction could modulate the discharge rates.)

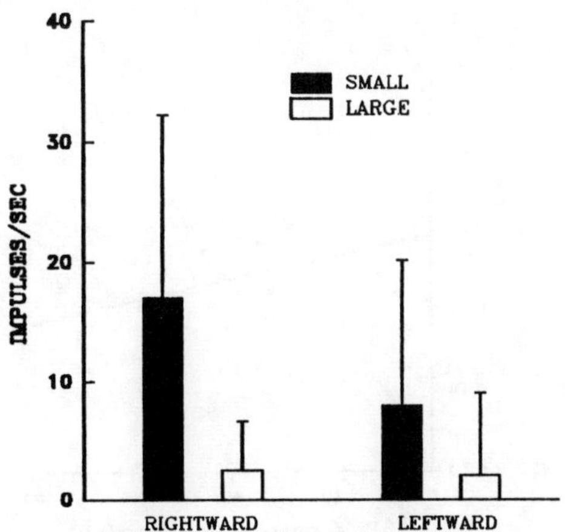

Fig. 6. Bar graph of mean discharge rate (± s.d., n≥14) in response to TARG2 signal (0 - 0.25s) for a single neuron which distinguished between small and large extent movements.

As with direction, extent preparation was manifested either by selective responses to TARG2, or by differential rates of firing during the last half of the PP. In one neuron, the PP activity integrated intended direction and extent into a continuum such that large rightward movements elicited the greatest PP activity, and large leftward the least (Fig. 7). This neuron received weak shoulder somatosensory (deep) input and discharged during rightward reach.

3.4 REACTION TIME CORRELATIONS

Ten cells with significant buildup of discharge during the PP in protocol 1 were analyzed for a relationship with reaction time (RT). The latter was computed as the interval from the onset of the GO signal to the time of switch release. Neuronal spikes were summed over the last .4 s of the PP and the first .1 s after the GO signal, i.e. the period immediately preceding the reach reaction.

For 6 of the cells, the relationship between late PP discharge rate and RT was very scattered. An example is shown in Fig. 8A. A negative trend was usually evident but correlation coefficients were poor (.2 - .4). In the remaining cases a correlation appeared, but always in the general form of a rectangular hyperbola (Fig. 8B). There was no difference for reaches with either arm.

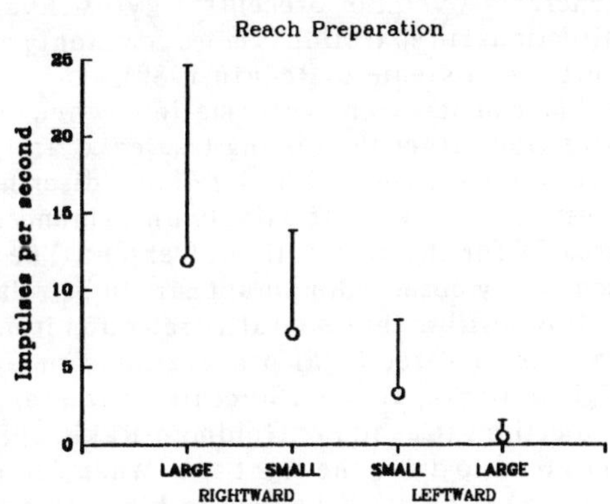

Fig. 7. Mean activity of an area 7a neuron during the 0.4s prior to the GO signal. Bars indicate s.d. (n≥10). For each point, data for 2 screen positions are combined.

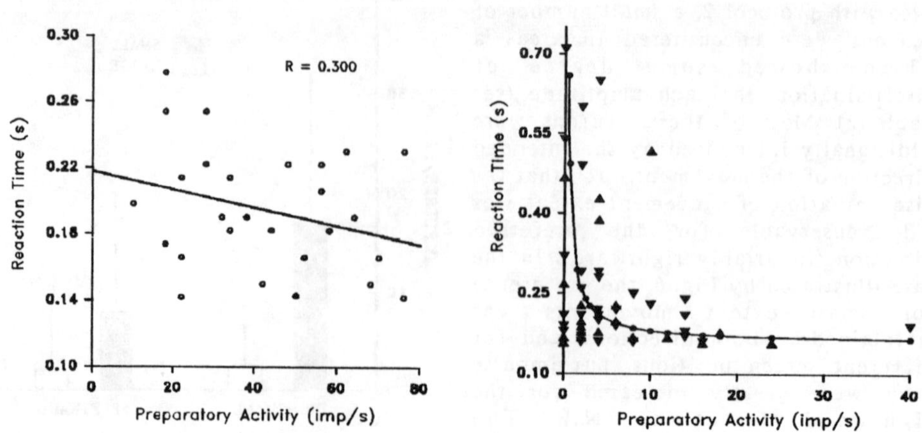

Fig. 8. Relationship between reaction time and neuronal discharge rate at the end of the preparatory period. **A:** neuron with a very weak correlation (both arms). **B:** neuron with a hyperbolic correlation. Symbols: ▲ right arm, ▼ left arm.

4. Discussion

With regard to motor planning, our findings indicate that area 7a is concerned with movement direction. This result was not unexpected. In an EEG study of elbow movement preparation, MacKay and Bonnet (1990) found that the CNV generated over the human parietal lobe was more sensitive to directional preparation than that generated over the precentral gyrus. Even in motor and premotor cortex, however, directional preparation is more commonly observed than extent preparation at the single unit level (Riehle & Requin 1989).

Movement extent information, when coded at all, was integrated together with direction rather than being treated as an independent parameter. Nevertheless, one or two cells were found which reliably discriminated movement extent in different screen positions, and were relatively uninfluenced by direction. Thus there may be a basis in area 7a for the abstraction of extent. The preparation of arm laterality was the least commonly observed, and appears to be relatively unimportant in area 7a.

It is possible that laterality selection is normally manifested by the interhemispheric balance of directional preparation. For example, left area 7 is activated mainly by rightward preparation. Since its ipsilateral projections are more massive than its crossed projections (Cavada & Goldman-Rakic 1989b), it would activate left frontal lobe and thus tend to drive the right arm. When forced to use the left arm for a rightward reach, there must be a mechanism to block this built-in tendency. Since use of either hand usually gave the same results in area 7a, such a mechanism is not likely to reside in the parietal lobe. Perhaps it is in the frontal cortex.

Early hand cue responses could potentially be explained as color discriminations, having nothing to do with laterality preparation for the reach task. Since no evidence for color-selective responsiveness has been found in area 7a (Andersen 1987; Robinson et al. 1978), such an explanation seems unlikely. Early responses would, however, require preset gating of color-selective visual inputs to area 7a.

Most of the extent discriminations and many of the directional discriminations were coded by the magnitude of the visual response to the TARG cue. If gaze were fixed to a standard point at this instant in each trial then the differential responses could be manifestations of retinal receptive fields. Since the eyes and head were free to wander at will, gaze position was far from standardized. In those sessions when hEOG was measured, and gaze position at the time of the TARG signal was calculated, it became clear that extent detection was not dependent on gaze position. As mentioned previously, at least one spatially-selective TARG response was similarly independent of gaze. In addition, a receptive field explanation of extent discrimination fails to explain the strong bias toward small movements. Many area 7a visual receptive fields are peripheral (Motter & Mountcastle 1981).

Neuronal activity related to directional preparation is commonly observed in area 5, anterior to 7a (Crammond & Kalaska 1989). In area 5, activity during the PP seems to anticipate the intended arm position. The neurons usually also receive somatosensory input related to arm position. In area 7a neurons, however, somatosensory input was very weak at best, and discharge rates during the PP, even when they reliably indicated the intended movement direction, bore no consistent relationship to the discharge rates during the actual reach. An exception was the neuron summarized in Fig. 7, which also received weak somatosensory input and thus, in many respects, resembled area 5 cells.

The data emphatically points to a preparatory emphasis of area 7a activity rather than to a motor execution linkage. Responses to the information cues were commonplace, but virtually no neuronal responses were elicited by the GO signal. Furthermore, correlations between preparatory discharge and reaction time indicated a relationship which was highly nonlinear and indirect. Long RTs were often associated with a lack of PP activity, but short RTs could be coupled with a huge range of unit firing rates. The anatomical projection of area 7a to prefrontal cortex, not to motor cortex (Cavada & Goldman-Rakic 1989b) reinforces the notion that it may influence motor planning but not the execution stage. The fact that many area 7a neurons are also active during reach performance or during hand manipulation (Mountcastle et al. 1975) is insufficient evidence for a direct motor link. An interpretation more consistent with our observations using delay task protocols is that area 7a neuronal activity during specific phases of motor performance is a behavioral discrimination which serves to prepare a subsequent action conditional upon it.

5. Acknowledgements

This study was immeasurably facilitated by the expert technical assistance of Antonio Mendonça and his care of the Prince. Many thanks are also due to Alex Palimaka for computer programming, and Mohammed Dhattu and Shardul Mehta for data analysis. The project was funded by MRC of Canada: A.R. was supported by ONR grant N00014-89-J1557 .

6. References

Andersen, R.A. (1987) The role of the inferior parietal lobule in spatial perception and visual-motor integration. In: *The Handbook of Physiology. Section 1: The Nervous System. Volume IV. Higher Functions of the Brain. Part 2*, edited by Plum, F., Mountcastle, V.B. and Geiger, S.T. Bethesda, MD: American Physiological Society, pp. 483-518.

Andersen, R.A. (1989) Visual and eye movement functions of the posterior parietal cortex. *Ann.Rev.Neurosci.* 12: 377-403.

Cavada, C. and Goldman-Rakic, P.S. (1989a) Posterior parietal cortex in rhesus monkey: I. Parcellation of areas based on distinctive limbic and sensory corticocortical connections. *J.Comp.Neurol.* 287: 393-421.

Cavada, C. and Goldman-Rakic, P.S. (1989b) Posterior parietal cortex in rhesus monkeys: II. Evidence for segregated corticocortical networks linking sensory and limbic areas with the frontal lobe. *J.Comp.Neurol.* 287: 422-445.

Crammond, D.J. and Kalaska, J.F. (1989) Neuronal activity in primate parietal cortex area 5 varies with intended movement direction during an instructed-delay period. *Exp.Brain Res.* 76: 458-462.

Hyvärinen, J. (1982) *The parietal cortex of monkey and man,* Berlin:Springer-Verlag, pp. 1-202.

MacKay, W.A. and Bonnet, M. (1990) CNV, stretch reflex and reaction time correlates of preparation for movement direction and force. *Electroenceph.clin.Neurophysiol.* 6: 47-62.

Motter, B.C. and Mountcastle, V.B. (1981) The functional properties of the light-sensitive neurons of the posterior parietal cortex studied in waking monkeys: foveal sparing and opponent vector organization. *J.Neurosci.* 1: 3-26.

Mountcastle, V.B., Lynch, J.C., Georgopoulos, A., Sakata, H. and Acuña, C. (1975) Posterior parietal association cortex of the monkey: command functions for operations within extrapersonal space. *J.Neurophysiol.* 38: 871-908.

Requin, J., Riehle, A. and Seal, J. (1988) Neuronal activity and information processing in motor control: from stages to continuous flow. *Biol.Psychol.* 26: 179-198.

Riehle, A. and Requin, J. (1989) Monkey primary motor and premotor cortex: single-cell activity related to prior information about direction and extent of an intended movement. *J.Neurophysiol.* 61: 534-549.

Robinson, D.L., Goldberg, M.E. and Stanton, G.B. (1978) Parietal association cortex in the primate: sensory mechanisms and behavioral modulations. *J.Neurophysiol.* 41: 910-932.

Stein, J.F. (1989) Representation of egocentric space in the posterior parietal cortex.*Q.J.Exp.Physiol.* 74: 583-606.

EQUILIBRIUM CONTROL VECTORS SUBSERVING RAPID GOAL-DIRECTED ARM MOVEMENTS

J.R. FLANAGAN[1], A.G. FELDMAN[2], and D.J. OSTRY[1]
[1]*McGill University, Montreal, Canada*
[2]*Institute for Information Transmission Problems, Moscow, U.S.S.R.*

ABSTRACT. The composition of central commands underlying rapid goal-directed arm movements to visual targets was explored within the framework of the equilibrium point hypothesis. This hypothesis suggests that movements arise from shifts in equilibrium associated with the dynamic interaction of central commands, reflex mechanisms, muscle properties, and loads. Central commands control this process by regulating muscle threshold lengths (λs) for motoneuron recruitment. Subjects performed rapid arm movements to fixed and displaced targets (LEDs) located in a horizontal plane. The position of the first target and the onset and the position of the second target were varied. Experimental trajectories of the movement endpoint (e.g., the hand or wrist) were compared with simulated trajectories, based on the λ model, generated with theoretical central commands. The findings support the hypothesis that multi-joint arm motions are planned in equilibrium coordinates corresponding to the position of the endpoint. Moreover, the results suggest that, in the absence of overriding constraints, the equilibrium position of the hand is shifted at a constant velocity towards the target. Finally, for rapid movements the rate of shift appears to be the same for movements of different amplitude.

1. Introduction

In this paper we describe experiments in which subjects made goal-directed arm movements to both single-step and double-step targets (LEDS) located in a horizontal plane. Subjects were asked to move rapidly to the presently illuminated target. For the double-step targets different inter-stimulus intervals (ISIs) were used ranging from 50 ms to 700 ms. Thus in some cases the target was shifted prior to the onset of movement whereas in other cases the second target was presented after the movement to the initial target was completed. We also present simulation data generated from a model of two-joint planar arm motion which is based on the equilibrium point (EP) hypothesis. A summary review of the EP hypothesis (λ model) - at the level of a single muscle and at the level of a single joint - is provided.

Figure 1 illustrates several aspects of the λ model at the level of a single muscle acting against a load. Five limb configurations are depicted in Figure 1a. The corresponding levels of depolarization of the motoneurons (MNs) innervating the muscle are shown in Figure 1b. The dashed horizontal line is the depolarization threshold (T_1) at which the first MN is recruited. As the level of depolarization increases beyond this threshold, the number of recruited MNs as well as their firing rates also increase. The level of depolarization of MNs is assumed to depend on the summation of central facilitation (open box) and afferent facilitation (filled box).

In A the system is in equilibrium with muscle length x_a. The level of MN depolarization (T_{L1}) exceeds threshold - with some specific combination of central and afferent facilitation - such that

J. Requin and G. E. Stelmach (eds.), Tutorials in Motor Neuroscience, 357–367.

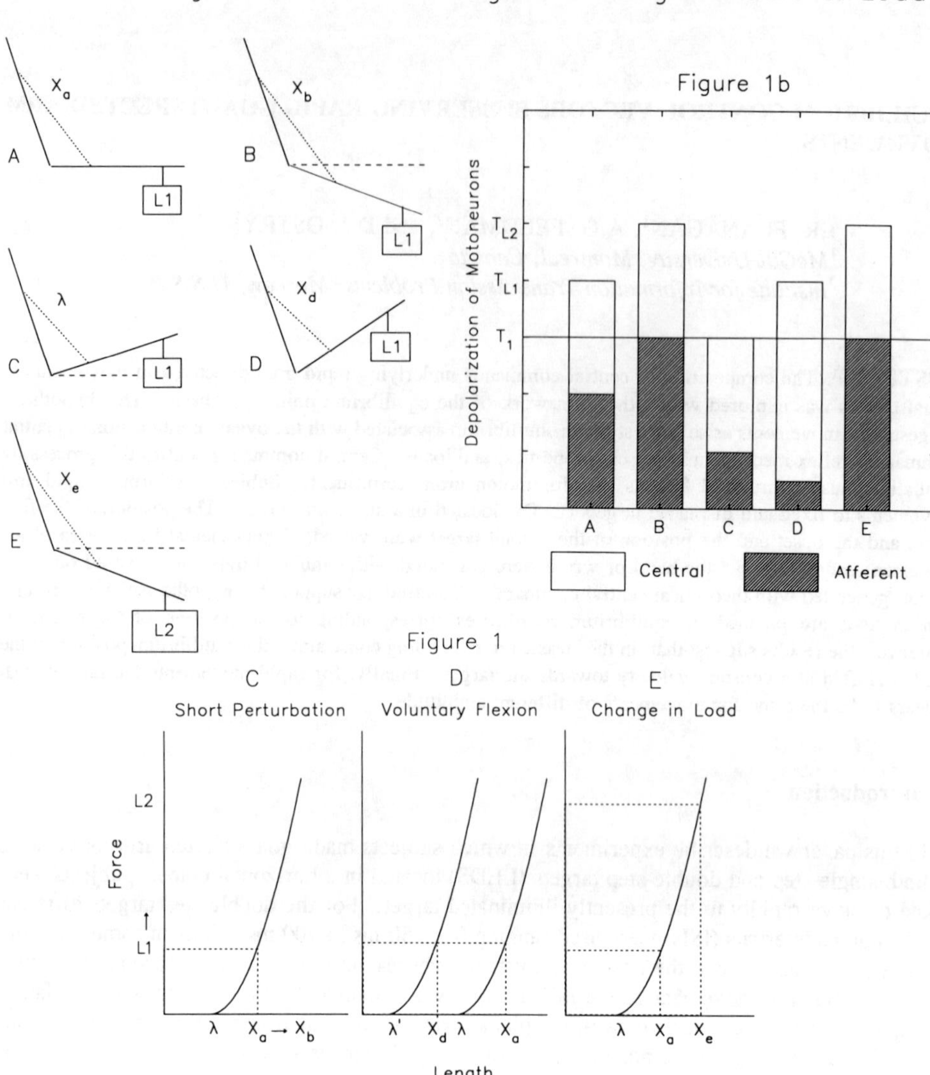
Figure 1a
Figure 1: Single Muscle & Load
Figure 1b
Depolarization of Motoneurons
Central
Afferent
Figure 1
C
Short Perturbation
D
Voluntary Flexion
E
Change in Load
Force
Length

MNs are recruited and muscle force balances the load L1. Consider, in B, the response of the system to a short perturbation which stretches the muscle to length x_b. Due to muscle stretch the level of length dependent afferent facilitation increases, more MNs are recruited, and a flexion torque results which restores the system's equilibrium position to A. Now consider, in C, the response of the system to a brief perturbation which shortens the muscle. Afferent facilitation decreases, the level of depolarization is reduced, MNs are de-recruited, and the load acts to restore the system's equilibrium position. In this particular example, the muscle is shortened to the *recruitment threshold length* - denoted λ - at which all MNs are de-recruited. A biomechanical account of A-C is presented in Figure 1c which shows the muscle force-length curve. Initially the muscle is at length x_a corresponding to force L1. The muscle is then stretched to length x_b and the resulting force exceeds the load. Finally, in C, the muscle is shortened to the threshold length λ and zero force is produced. Figure 1c shows that all possible equilibrium states of the system associated with a constant level of central facilitation can be characterized by an invariant force-length curve which is defined by λ. Thus, the λ parameter is a position or load independent measure of the level of central facilitation.

Now consider, in D, how the nervous system produces a voluntary flexion motion (from A to D). The level of central facilitation is increased so that depolarization increases and more MNs are recruited. The muscle actively shortens to length x_d and length dependent afferent facilitation diminishes. For simplicity, we assume that the load is independent of limb position and that the same level of muscle activation is required to balance the load regardless of muscle length. Thus, when the sum of central and afferent facilitation is restored to the same level as in A, a new equilibrium position is established in which the load is once again balanced. The force-length curves corresponding to A and D are shown in Figure 1d. The effect of increasing central facilitation is to shorten the recruitment threshold length of the muscle from λ to λ'.

In A and D the same load is supported in two different positions with equivalent levels of MN activity. The sole difference between A and D is the relative amounts of central and afferent facilitation. Clearly, models which suggest that MN recruitment (i.e., muscle activity) is directly controlled by the nervous system fail to explain how the system differentiates between postures A to D.

Finally, in E, we may consider how the system in A reacts to a larger load L2. The muscle is stretched to length x_e, the level of afferent facilitation increases, and more MNs are recruited. This effect is illustrated in Figure 1e. Note that the load is balanced without changing the level of central facilitation (i.e., λ). Figure 1E illustrates how, for a given level of central facilitation, afferent feedback establishes a specific mapping between external space (i.e., muscle length) and MN activation. In the deafferented system, this elegant correspondence is lost. In this case, MN activity must be directly (centrally) controlled in order to generate movement. For example, in E, the level of central facilitation would be higher than in A in the deafferented system unlike the normal system. Under normal conditions, however, the model attributes an essential role to afferent regulation of MN activity in the control of posture and movement.

A simple single-joint system with a flexor and an extensor muscle is shown in Figure 2a. The joint angle Θ is defined such that it increases with joint extension. The corresponding torque-angle relationships for both the flexor and extensor muscles are presented in Figure 2b. The system is in static equilibrium; the torque produced by the flexor is equal and opposite to the torque produced by the extensor. Thus, the equilibrium joint angle (R) is specified by the combined actions of both muscles and corresponds to the angle at which the net joint torque is zero. The location of the flexor and extensor torque-angle curves is determined by the threshold lengths for the flexor (λ_f) and extensor (λ_e). Consider now the effects of briefly perturbing the

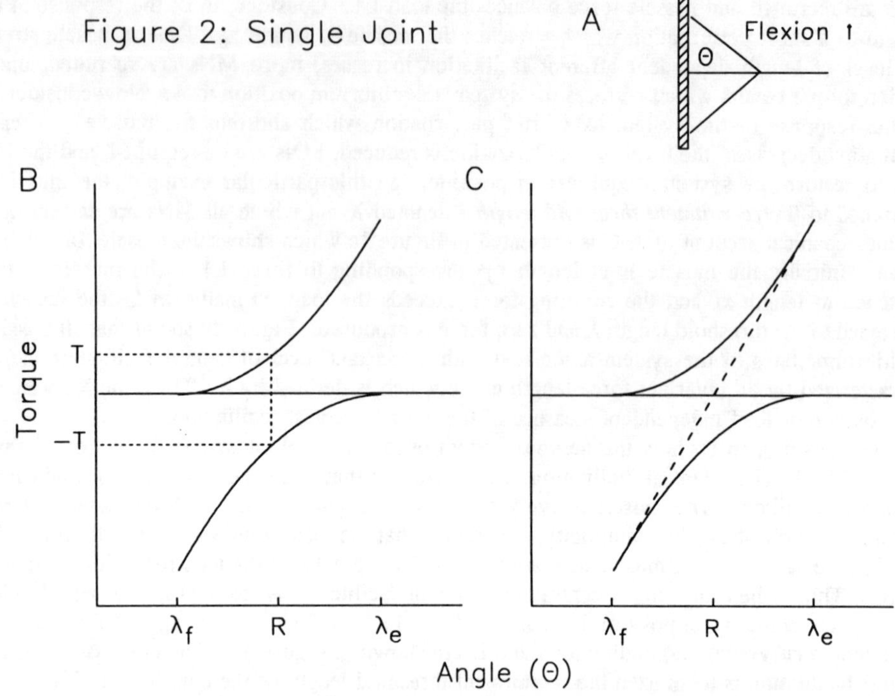

system. If the limb is perturbed into extension (increasing Θ) then the torques produced by the flexor and extensor muscles will increase and decrease respectively and a net flexion torque will restore the system towards equilibrium. Conversely, if the system is perturbed in flexion (decreasing Θ) then the extensor torque will increase while the flexor torque decreases and a net extensor torque will result.

We may make use of Figure 2b to illustrate two orthogonal ways in which the nervous system can centrally control the system. First, by shifting λ_f and λ_e and the corresponding torque-angle curves in opposite directions the nervous system can alter the level muscle of co-activation without changing the equilibrium angle. Second, by shifting λ_f and λ_e in the same direction, the nervous system can change the equilibrium angle without altering the level of muscle co-activation. Thus, in principle - and surely in practice, central signals to the motoneuron pools of both muscles can independently control the equilibrium joint angle as well as the level of muscle co-activation or co-contraction.

Figure 2c illustrates the static relationship between the level of muscle co-activation and joint stiffness. By summing the two muscle torque-angle curves we obtain the total torque-angle relationship for the joint as a whole. It can be shown that when the level of co-activation is increased the slope of the torque-angle relationship for the joint increases. The slope of this relationship represents joint stiffness. Thus, joint stiffness can be controlled independently of joint angle.

In the two-joint planar arm model we have included three pairs of antagonist or opposing muscles. These include a pair of single-joint antagonists at both the shoulder and elbow and a pair of double-joint antagonists spanning both joints. In addition, the two-joint model includes a full geometric and mechanical description of the planar arm; position and velocity (i.e., damping) sensitive afferent feedback; and muscle properties (e.g., relaxation). Central signals specify the equilibrium posture of the arm in terms of the position of the hand in extrinsic coordinates. The corresponding equilibrium joint angles are determined from limb geometry. Central control signals also specify the levels of co-activation for the three antagonist muscle pairs. For a complete description of the two-joint planar model see Feldman, Adamovich, Ostry & Flanagan (1990) and Flanagan, Ostry & Feldman (1990). For a general account of the mechanisms underlying the λ model see Feldman (1986).

2. Method

Subjects were asked to make rapid arm movements to illuminated targets (LEDs) inlaid in a horizontal surface at a level just below the shoulder. They were instructed not to make any corrective movements. Both single-step targets and double-step targets with inter-stimulus intervals (ISIs) of 50, 75, 100, 125, 175, 250, 350 and 700 ms were presented. Single- and double-step trials were randomly mixed. The position of the single-step targets and the position of the second target in the double-step trials were randomly varied.

The three-dimensional position of infrared emitting diodes (IREDS) attached to the hand, wrist, forearm and upper arm were recorded using the Watsmart system at 400 Hz. In addition, electromyographic (EMG) activity was recorded at 1200 Hz for the biceps, triceps, pectoralis, and posterior deltoid muscles. However, in this paper only the kinematics of the hand will be reported. The experiments took place in a darkened room to limit visual feedback of the position of the limb. The target surface was painted matt black and black infrared absorbing form was used to reduce infrared reflections. Position data were numerically filtered using a Butterworth low-pass filter with a cutoff frequency of 20 Hz. The filtered data were then numerically differentiated with a second-order least squares method.

3. Results and Discussion

3.1. MOVEMENTS TO CO-ALIGNED CENTRAL TARGETS

This section of the results deals with single- and double-step movements involving a proximal and a distal target located centrally with respect to the subject. The start position was also located centrally 5 cm in front of the body. Thus, the two targets and the start position were co-aligned in the mid-sagittal plane. The proximal target was 30 cm from the start position and the distal target was 45 cm from the start position.

The Figure 3a shows position and velocity records for a double-step movement with a 700 ms ISI directed initially to the proximal target and then to the distal target. Two distinct movements can be observed. In Figure 3b, simulated position and velocity records corresponding to the

Figure 3: Double-Step Motion

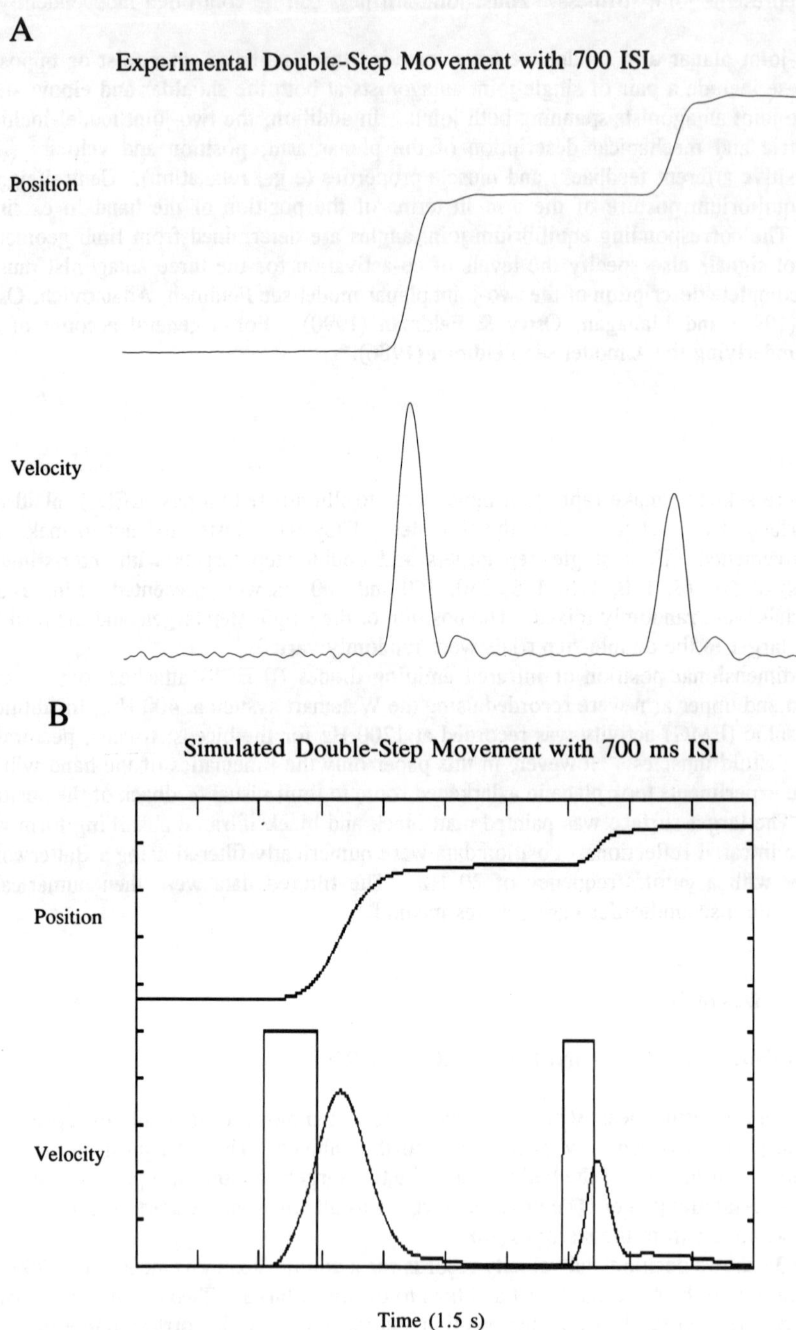

experimental records shown Figure 3a are presented. The model is able to produce smooth bell-shaped velocity profiles consistent with the experimental data.

Figure 4 outlines the predictions of the model for rapid movements of two different amplitudes. Figure 4a shows the equilibrium position of the hand (solid traces) and the predicted actual position of the hand (dotted traces) for a small and a large amplitude movement. The corresponding equilibrium hand velocity profiles are presented in Figure 4b. We assume that the equilibrium position of the hand is shifted at a constant rate in a straight line towards the illuminated target. Moreover, in the case of rapid movements, we assume that the rate of shift is independent of movement amplitude; thus movement amplitude is determined by the duration of the shift. The predicted actual hand positions for the small and large amplitude movements will coincide over the initial phase of the movement (up until the equilibrium position of the hand reaches the proximal target). This prediction was tested experimentally. An advantage of constant velocity shifts in the equilibrium position of the hand is that - in contrast to the minimum jerk model (Flash, 1987) - amplitude need not be specified prior to the initiation of equilibrium shift.

Experimental hand speed profiles (i.e., tangential velocity profiles) for two subjects are presented in Figure 5. The curves have been aligned so that they start at the same time. In all cases, movement onset was taken at the point when the hand speed reached an absolute threshold of .2 m/s. Hand speed profiles for single-step movements to the proximal (thick traces) and distal (light traces) targets are shown in 5a. Consistent with the model prediction, these two sets of profiles are similar over the initial phase of the movement and then diverge. Thus, the rate of shift of the equilibrium position of the hand appears to be unaffected by movement amplitude.

We assume that the equilibrium position of the hand is shifted at a constant rate (in a straight line) towards the presently illuminated target. When the target shifts distally early in the trial, the equilibrium position of the hand simply continues shifting at the same rate towards the second (more distal) target. Therefore the model predicts that double-step movements with small ISIs between the first (proximal) and second (distal) targets should be equivalent to single-step movements to the distal target. This prediction was tested by comparing these double-step movements with single-step movements to both the proximal and the distal targets.

Figure 5b shows hand speed profiles for single-step movements to the proximal target (thick traces) and double-step movements to the proximal and then the distal target with an ISI of 70 ms (thin traces). As expected, these two sets of profiles are similar over the first part of the movement and then diverge. The pattern is similar to that shown in Figure 5a. In Figure 5c, the hand speed profiles for single-step movements to the distal target (thick traces) and the same double-step movements (thin traces) are presented. With a few exceptions, the two sets of profiles are similar as predicted by the model. This suggests that, in the case of the double-step movements, the shift in the equilibrium position of hand is simply continued, at the same rate, towards the second target.

3.2. MOVEMENTS INVOLVING CHANGES IN TARGET DIRECTION

The second part of the results deals with double-step movements in which the position of the second target requires the subject to change movement direction. The same start position and initial target as above were used. Thus, the first target was positioned 30 cm in front of the subject. The second target was positioned either 30 cm to the left or 40 cm to the right of the first target with respect to the subject.

Flash (1990) has modelled double-step movement trajectories as the superposition of two

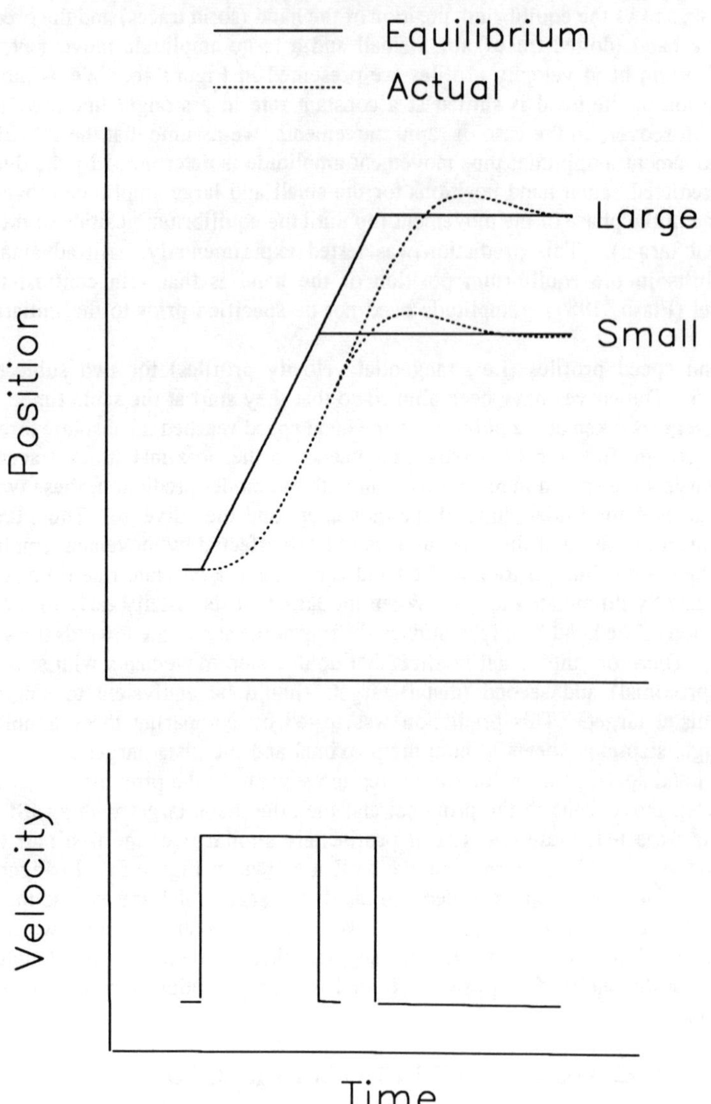
Figure 4: Model Predictions
Equilibrium
Actual
Large
Small
Position
Velocity
Time

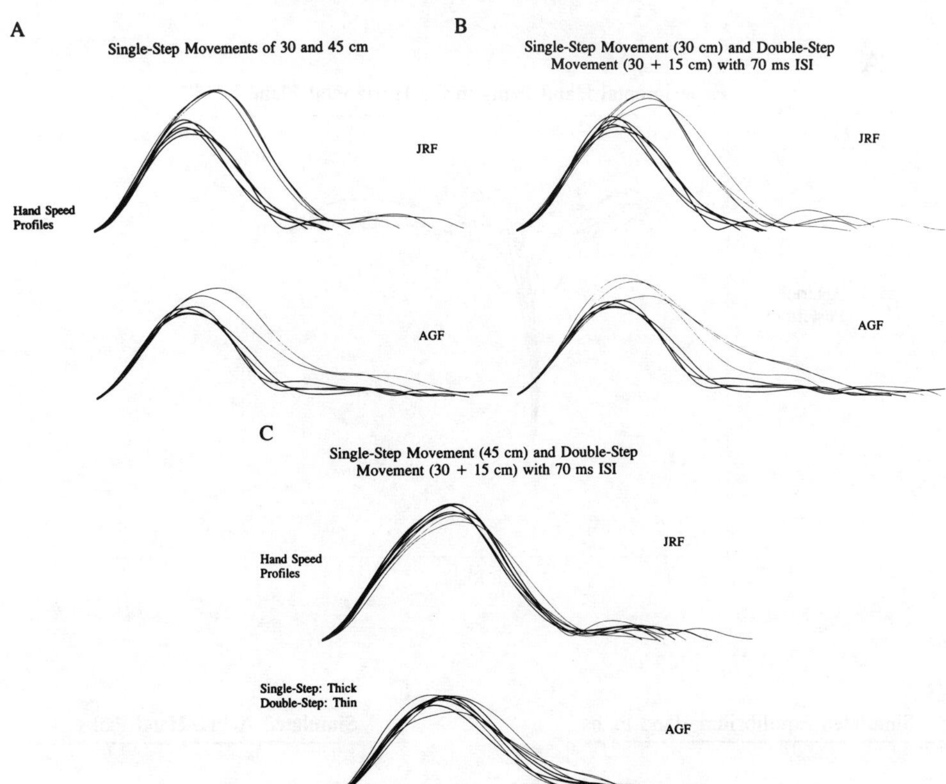

single-step trajectories. The first is from the start position to the initial target and the second is from the initial target to the final target. Both of the superimposed trajectories are assumed to have straight line hand paths and smooth bell-shaped hand speed profiles. The time at which the second trajectory is superimposed on the first will depend on the ISI. This kinematic model can account well for the curved hand paths and smooth bi-modal hand speed profiles observed in double-step movements with sufficiently large ISIs. However, we suggest that curved hand paths and smooth hand speed profiles result from movement dynamics as the equilibrium position of the hand is shifted at a constant rate in the direction of the first target and then (at the same rate) in the direction of the second target. Thus, the equilibrium shift to the first target is terminated as the shift to the second target is initiated.

Figure 6a shows experimental hand paths in the horizontal plane for double-step movements involving both final targets with a range of ISIs. With larger ISIs the hand achieves the first

Figure 6: Motions with Changing Target Direction

A

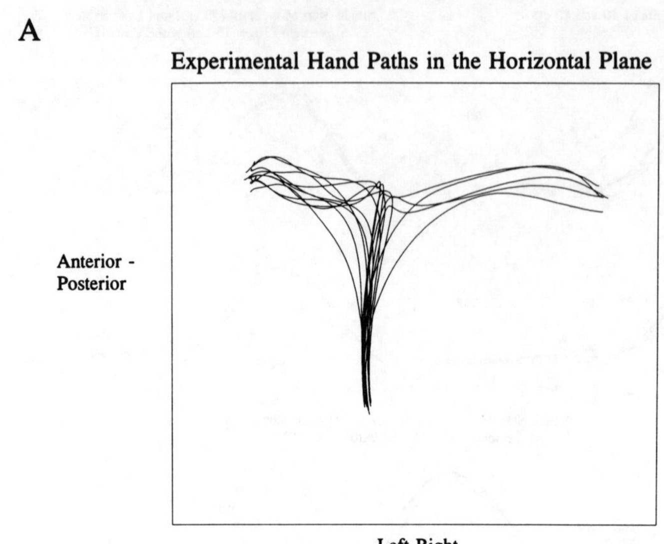

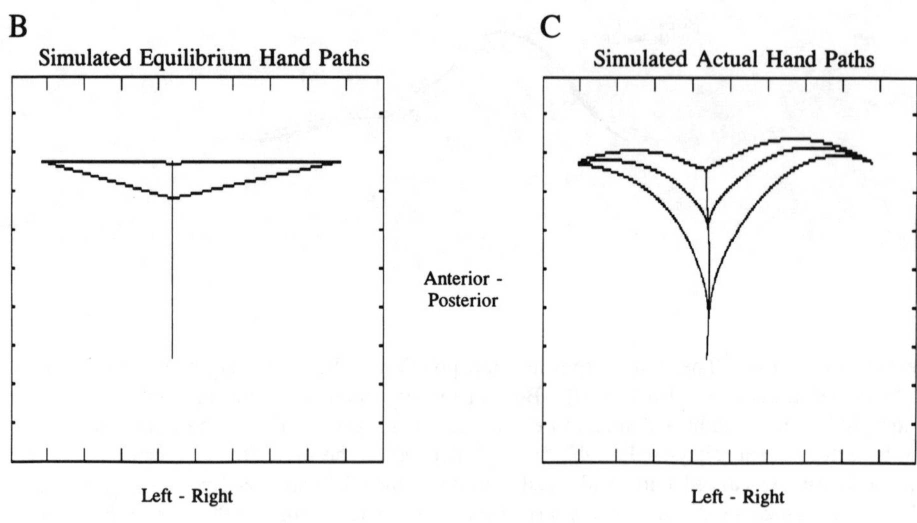

target before moving to the second target whereas with small ISIs the hand begins to curve towards the second target before reaching the first target. Simulated equilibrium and actual hand paths corresponding to these experimental paths are shown in Figure 6b and Figure 6c respectively for different ISIs. With the shortest ISIs, the equilibrium position of the hand shifts in the direction of the second target before it reaches the first target. With intermediate ISIs, the equilibrium position of the hand achieves the first target but is shifted towards the second target before the actual hand position reaches the first target. Finally, with large ISIs, both the equilibrium hand position and the actual hand position reach the first target. Figure 6c illustrates that constant rate shifts in the equilibrium position of the hand directed in a straight line towards the presently illuminated target can account - qualitatively - for the smoothly curved hand paths that are observed experimentally.

4. Conclusions

We suggest that the two-joint horizontal arm movements are planned in coordinates corresponding to the equilibrium position of the hand. In the case of rapid goal-directed movements, it is argued that the equilibrium position of the hand is shifted at a constant rate in a straight line directed towards the presently illuminated target. We have demonstrated that these constant rate shifts in the equilibrium position of the hand can produce predicted actual hand speed profiles that are smooth and bell-shaped. Thus, we do not believe that smoothness is explicitly planned by the central nervous system. Rather, smooth motions arise as a consequence of the dynamic interaction of central commands, afferent reflex mechanisms, muscle properties and limb mechanics. Finally, it is proposed that for rapid goal-directed arm movements, the rate of shift in the equilibrium position of the hand is independent of movement amplitude.

5. References

Feldman, A.G. (1986). Once more on the equilibrium-point hypothesis (λ model) for motor control. *Journal of Motor Behavior*, 18, 17-54.

Feldman, A.G., Adamovich, S.V., Ostry, D.J., & Flanagan, J.R. (1990). The origin of electromyograms - explanations based on the equilibrium point hypothesis. In J. Winters & S. Woo (Eds.), *Multiple muscle systems: biomechanics of movement organization*. Springer-Verlag.

Flanagan, J.R., Feldman, A.G., & Ostry, D.J. (1990). Control of human jaw and multi-joint arm movements. In G. Hammond G (Ed.), *Cerebral control of speech and limb movements*. Springer-Verlag.

Flash, T. (1987). The control of hand equilibrium trajectories in multi-joint arm movements. *Biological Cybernetics*, 57, 57-74.

Flash, T. (1990). The organization of human arm trajectory control. In J. Winters & S. Woo (Eds.), *Multiple muscle systems: biomechanics of movement organization*. Springer-Verlag.

SECTION 7

PLANNING OF MOVEMENT SEQUENCES

COGNITION AND MOTOR PROGRAMMING

L H SHAFFER
Department of Psychology
University of Exeter
Exeter, Devon, EX4 4QG, UK

ABSTRACT. In the recent theory of motor skill a clear distinction has emerged between a cognitive system that plans and organizes a goal structure of movement, and a motor system that organizes movement given information about the goal structure. The representation of goal structure is here called a motor program, and the total activity of planning and control is called motor programming. The explanatory value of this theoretical framework is discussed in the context of performance in skilled touch typing.

1. MOTOR PROGRAMMING

In the past decade an abstract theory of motor control has grown, moving away from details of neuromuscular structures to a principled mathematical account of motor systems. It has become clear on empirical and theoretical grounds how in principle a motor system can act autonomously in organizing movement. Motor systems can be viewed as special cases of self-organizing systems, and this view has been arrived at from a number of theoretical directions (Hinton, 1984; Morasso and Tagliasco, 1986; Saltzman and Kelso, 1987).

A motor system may take in sensory feedback enabling it to test certain conditions of movement completion or target attainment. Thus, in grasping an object, finger closure ceases when a haptic condition for contact is met; in walking, the forward swing of one leg can be initiated only when the support condition for the other leg has been met. The system may also take in sensory information relevant to the task so as to set the parameters of a movement. Thus, in catching a ball, the timing and direction of the catching movement can be tailored given optic information of the trajectory of the ball; the final strides in the run-up for a long-jump can be tailored given optic information of approach to the edge of the jumping board (Lee, Lishman and Thomson, 1982). What is important in the motor theory is that the motor system is open to such information, but is solely responsible for organizing the movement.

It is sometimes supposed that such a theory of motor control makes a concept of motor programming redundant (eg Michaels and Carello, 1981).

J. Requin and G. E. Stelmach (eds.), Tutorials in Motor Neuroscience, 371–383.

On this view the environment directly provides the relevant information as it is needed. There is some merit in attempting to limit the scope of a concept of motor programming, but to dismiss it misses the essential point that most skilled action is directed. It is almost never fully constrained by the environment: on the contrary, motor intentions are often formed under conditions of almost limitless choice. If the formation of a motor intention is an internal event, it follows that the goals of movement are internally represented. Rather than engage in a polemic arising from badly focused attacks on some inept versions of motor programming theory, it seems preferable in the space available here to establish a sound basis for the theory.

In the task Pick up X, X can be anything from a glass of wine to a packed suitcase. For many objects X it is important to know certain properties relevant to the action of picking it up that are not made explicit in the optic array. This can be generic knowledge and therefore abstract. The object need not even be visually present in forming the intention to pick it up. For a variety of reasons, then, the mental planning of an action must often be made in an abstract, symbolic space. In this sense the basis of skilled action is cognitive.

For a given X, the conditions of the task Pick up X can vary widely and skilled action should be adaptive to these contingencies. Either cognitive planning must specify the goal differently for every contingency or it can specify an abstract, canonic goal which can be realised by a flexible motor system. The division of labour presented by the latter option seems optimal for an organism that has to be able to switch rapidly from one task to another, and we have seen that such a division is in principle feasible. Thus we can state a fundamental assumption of skilled action:

> Skilled action is arranged at two levels in the brain, (a) in a cognitive system which plans and represents (symbolically) a goal structure of action, and (b) in a motor system which organizes movements appropriate to the goal.

The first step towards understanding a particular skill involves analysing the separate contributions of cognitive and motor processes.

Ancillary to the assumption is the following definition:

> A motor program is a cognitive representation of the goal structure of an intended action.

Note that this definition comes down on one side of an ambiguity which has pursued the discussion of motor programming over the years. The term motor program has been used in two distinct senses. In one of these a motor program, like a theatre program, is a declaration of intent; in its other sense it is like a computer program, and contains procedures for realizing an intention (eg Rumelhart and Norman, 1982). The latter sense becomes inappropriate given the distinction made here between cognitive and motor processes. It could be retained in a restricted version by considering that in some skills a motor program may have to unfold through different levels of abstraction, for example the mental process of getting from meaning to speech must involve this, in which case the process of unfolding might be thought intrinsic to the motor program. However, it can be theoretically convenient to consider the abstract representations of motor output without being committed to speculation on the processes that construct them, though a complete theory of motor programming has to take account of both representation (a program) and process (its construction).

Given the distinction between cognitive and motor processes, we can discriminate between those applications of a motor program concept to fixed action sequences, such as the songs of certain birds, and those to improvised sequences, such as speech. Fixed action sequences can in principle be handled by an associative process within a motor system. Because of its supposed invariance the sequence need not be represented symbolically, and so the above definition of a motor program need not apply to such actions. Therefore the concept of motor programming seems best restricted to improvised action, and it is in this context that it has been most successfully used, in particular in the context of the skills of language and music. In these skills, at least, there is little choice but to suppose a symbolic planning of action.

Another consideration in the past usage of the concept of motor program is in its application to simple action, having a single goal, like catching a ball, as well as to serial action, having a sequence of goals, like speech. It can be argued that the primary role of the concept is in accounting for the patterning of action, in which case its application to simple action is unclear, and in fact much of the criticism of the concept is in this context. It is within the capabilities of a motor system to organize movement to a single goal, and so indeed the concept of a motor program seems redundant. Yet frequently movement is intentionally given a gestural shape beyond its functional requirement, as in mime or dance, or to create a graphic form, as in drawing and writing, or to create a certain sound as in speech and playing music. In all these examples we have to suppose that the shape is part of the motor goal and hence is programmed. A motor control theory based on assumptions of motor dynamics can only account for a default, functionally optimal movement. This leads us to the assumption:

> Motor programming allows movements and motor sequences to take designed shapes.

Skilled movement is fluent, in the sense that movements can flow one into another without abrupt changes in acceleration and direction (Shaffer, 1982). Viewing a complex movement as a composition of simpler, ballistic movements, Morasso (1986) has shown that fluency depends upon (a) the advance preparation, or programming, of a motor sequence, and (b) the temporal overlap of control of these movements. Since fluency is implicit in the idea of shaping or patterning of movement, we can state as a corollary of the above assumption:

> Motor fluency is achieved only if movement is programmed sufficiently ahead of its execution to allow the motor system time to organize the

motor sequence and arrange to overcome inertial constraints on movement.

Consideration of the designed quality of movements and motor sequences puts motor skill beyond the scope of purely motor dynamic accounts of motor control. Evidence for such design of movement is abundant in speech (Levelt, 1989), musical performance (Shaffer, 1981) and dance (Morasso and Tagliasco, 1986). It is instructive, however, to discuss and demonstrate motor programming in the context of the rather simpler skill of copy typing. The principles established here can be shown to apply to the more complex skills.

In the following discussion the emphasis will be not on exploring motor programs as an end in itself, which seems a wasteful exercise, but in developing the concept as a basis for explaining phenomena of motor skill.

2. MOTOR PROGRAMMING IN TOUCH TYPING

2.1 The Input-Output Span

An important phenomenon of motor skill, studied before the turn of the century, is that as skill is acquired in a transcription task there tends to be an increasing lag between current motor output and the intake of new information from the text. This is known generally as the input-output span, and it was observed by Bryan and Harter (1899) studying trainee telegraphers learning to receive morse code and transmit it into type. The span for copy touch-typing was measured by Butsch (1932), who recorded the timing both of symbols being typed and of fixations in the text. He showed that eye-finger span fluctuates continually around an average that increases with skill to about 7-8 symbols. One can conjecture that the intervening material between fixation and output, held in buffer memory, in some way serves to prepare the motor output. This has been confirmed by creating a window on the text so as to limit the amount of its preview, ie how far ahead of the symbol currently being typed the text is visible (Hershman and Hillix, 1965; Shaffer, 1973).

The result is clearcut. Fast typists can maintain normal speed as long as they can see 8-10 symbols ahead of their type. If preview falls below this then progressively there is a loss in typing speed and an increase in errors. In the extreme, when only the current symbol can be seen, a normal typing speed of 10 symbols per second is reduced to less than 2 symbols per second. Motor programming appears to be necessary, but it has still to be explained how it makes fluency possible. Some insight into this is obtained from another result in the two studies.

In both studies, the text itself as well as preview was manipulated. It could be prose, or prose degraded to randomized words, or to randomized syllables, or to randomized letters. The difference in typing performance between prose and random-word texts was negligible, but speed and errors deteriorated progressively with the more degraded forms of text. In the extreme case of random-letter text, typing speed was reduced to about 4 symbols per second. Fast copy touch-typing thus appears to depend on the existing skills of fluent reading, in particular on using word syntax - spelling and syllable structure - to assist recognition.

Combining the results of controlling preview and text, we are led to the following model of touch typing:

(i) Text is read as a series of linguistic units;

(ii) Each unit makes available an ordered set of symbol codes, represented in a motor program as an ordered set of targets in keyboard space;

(iii) Target codes are accessed in parallel from a linguistic unit, the time required for access being independent of the unit size; and

(iv) The motor system, containing a knowledge of keyboard topography and a standard assignment of fingers to keys, computes the movements of hands and fingers to achieve the succession of motor targets.

Assumption (iii) partly explains the speeds made possible by motor programming, and it generalizes to all transcription skills. It explains how the hierarchic structure in a text can be exploited by motor programming to make information rapidly available to the motor system. It will be shown that this assumption also provides a way of explaining order errors as a side-effect of motor programming.

The method of controlling text preview confounds two factors, the amount of text needed to program efficiently, and the time required to achieve this programming. An alternative method, which separates these factors, allows unrestricted preview of a computer-displayed text, but occasionally and unpredictably changes the text a predetermined distance ahead of the current point of type. The typist is instructed always to type the altered version, and the question is how close to the current point of type the change can occur without disrupting motor fluency. The input-output span estimated in this way, in terms of time, is about 7-8ths of that estimated by controlling preview (Shaffer, 1988).

The distribution of keypress latencies, ie the lapsed time from one keypress to the next, shows that there may be few or no pauses in an extended typing sequence (Shaffer, 1973). The motor system does not need to wait for renewal of a motor program, and this implies that the input process of reading text and the output process of producing type operate not sequentially but in temporal overlap. The motor program currently held in buffer memory serves to decouple input and output processes, allowing them a temporal independence. This is another factor explaining speed in transcription skill. It also explains the fluctuation of input-output span in fluent typing: the input can be read in words or phrases, while output is constrained to a letter by letter sequence.

It follows from this description of motor programming that a motor program is not a fixed but a continually renewing entity. It is constructed dynamically by translating text into ordered sets of motor targets. Reference here to ordered sets rather than sequences permits a distinction between extrinsic and intrinsic ordering of elements. Ordering is extrinsic if it relies upon an association of elements with ordinal labels, and this idea is useful in accounting for order errors.

2.2 Order Errors

The order errors of fluent typing are less exotic than those of speech (Fromkin, 1973). As one might expect from the account of motor programming in typing, order errors tend to occur within a word or a short phrase. Yet they have a systematic structure, and a similar principle is involved in both skills. These errors can be shown to arise as a corruption of ordinal labelling created by contextual factors in a motor program.

In its simplest form an order error in typing is a transposition of two adjacent elements, usually both letters but sometimes a space and the preceding letter. Thus we observe:

ascetci - ascetic
ignroe - ignore
twneyt - twenty
forkc - frock
jounrye - journey
wne todnw - went down

The latter examples show compound transpositions, the last one containing no less than four, one of which involves a space. The examples are taken from a corpus of over 100 errors and they reveal a common pattern. If the letters in the examples are coded L or R, according to the hand typing them, then all the transpositions are seen to be across hands. When the hand sequence is made explicit, there is a statistically significant tendency for the error to anticipate or prolong an LR pattern (Shaffer, 1975). Thus:

ascetic - LLLLLRL	ascetci - LLLLLLR
ignore - RLRRLL	ignroe - RLRLRL
twenty - LLLRLR	twneyt - LLRLRL
frock - LLRLR	forkc - LRLRL
journey - RRRLRLR	jounrye - RRRRLRL
went down - LLRL(R)LRLR	wne todnw - LRL(R)LRLRL

The bracketed R in the last example indicates that the typist used her right thumb to type the space. If we think of letters in a word as being coded simultaneously by order and by hand, eg

L	w	e		t		d		w	
R			n		(sp)		o		n

then the transpositions can be seen as resulting from a phase shift between vectors in a two-dimensional space, with the corruption of hand sequence overriding the spelling order.

Sometimes the letters do not permute into adjacent positions, but they tend to fall within a syllable, and the displaced letter tends to be on the opposite hand to the others. Thus:

sa tur day - LL LRL LLR   sa tur yda - LL LRL RLL
con tri bute - LRR LLR LRLL   con itr bute - LRR RLL LRLL
him self - RRR LLRL   him lsef - RRR RLLL

These errors raise the possibility that syllable also forms a dimension of coding, and that letters are ordered in terms of syllable position. This would account for errors like

dic to tar - dic ta tor
de mo ni ma tion - de no mi na tion
pre cen dent - pre ce dent
ou weight - out weigh
ob ver ved - ob ser ved

in which letters exchange or move to comparable positions in adjacent syllables. Note how the initial switching of m and n in denomination has created a further ambiguity in the identity of the next consonant. This is reminiscent of the error thses (these) observed by Lashley (1951), which led him to postulate motor programming. These syllable errors are akin to the spoonerisms of speech, eg Yew Nork (New York) (Fromkin, 1973).

In spelling a word like "took" we tend to say t-double-o-k, and it appears that in a motor program for typing it is often represented tDok, where D is a command to double the following letter. If the symbol D can also take part in a transposition then tDok may be corrupted to Dtok, toDk, or tDko, leading to the output errors ttok, tokk and tkko respectively. All these possibilities are exemplified in the following actual errors:

diseect - dissect
spluuter - splutter
weel - well
sttod - stood
neceaasrily - necessarily

There is an interesting variant of this kind of error in the following:

Buggourhs - Burroughs
colappse - collapse
fiddicult - difficult

In these the D symbol seems to have jumped a number of intervening symbols, but adjacency is restored in a two-dimensional array. Thus:

```
   L  B   D r      g   s        c        a   s e       d    D f   c     t
   R    u     o u   h             o D l    p             i      i   u l
```

Transpositions within a hand vector will also account for errors like

aosl - also

Whether or not a description in terms of multivariate labelling turns out to be correct, there can be little doubt that order errors occur in symbolic representations of motor output. Furthermore the form they take suggest that there are two levels of representation, one of which is correct and the other incorrect. Both levels may appear in the motor program, or the lower (incorrect) level may be represented in the motor system itself. The latter possibility becomes interesting if we consider a motor system that can compute a complex movement trajectory over a target sequence. Symbols in the motor system would presumably represent movement in a very direct way.

2.3 Motor Fluency

Skilled movement is described by Morasso (1986) as a composition of simple, ballistic trajectories. Continuous movement is achieved by a temporal overlap of production of successive trajectories. This concept of overlapping production is formalized in the mathematical theory of spline functions. A motor system can achieve such overlap if it has access to information specifying a motor sequence and can coordinate this information in arranging movement. Morasso mainly had in mind movements that serially construct complex shapes, as in handwriting and dance, but the concept extends readily to the patterning of concurrent movements in different limbs and articulators. It applies, for instance, to coarticulatory movements in speech (Kent and Minifie, 1977) and typing (Gentner, 1983).

The conventional typewriter keyboard constrains keypresses to be consecutive rather than simultaneous, but filming the hands of skilled typists in action reveals that hand and finger movements overlap in time, such that as one finger presses its key other fingers may be moving in anticipation towards their respective keys. Rumelhart and Norman (1982) have used a computer simulation model to show how a computational motor system can produce such overlapping movements, and have explored the consequences of these in the timing of keypresses.

One of the features of skilled serial action is that its patterning over time is very consistent from one occasion to another, which suggests that it is constructed by a fairly definite algorithm in the motor system. This feature is illustrated in Fig. 1, which shows the latency pattern in typing the sentence "the architect left his card on the table", by a typist who has a speed of about 100 words per minute. This was one of 40 sentences typed in sequence, and the sequence was typed on 8 occasions distributed over about a fortnight. The latency of each keypress, ie the time between that and the previous keypress, is shown on the vertical scale. The lower part of the figure shows the superimposed graphs obtained from the 8 occasions of typing the sentence. The constancy of patterning is easily seen. The composite graph obtained by averaging, shown in the upper part of the figure, is fairly representative of the individual graphs, and the standard deviation at each letter position, shown below this graph, is seen to be uniformly small. The question arises: what is the algorithm that generates these latency patterns?

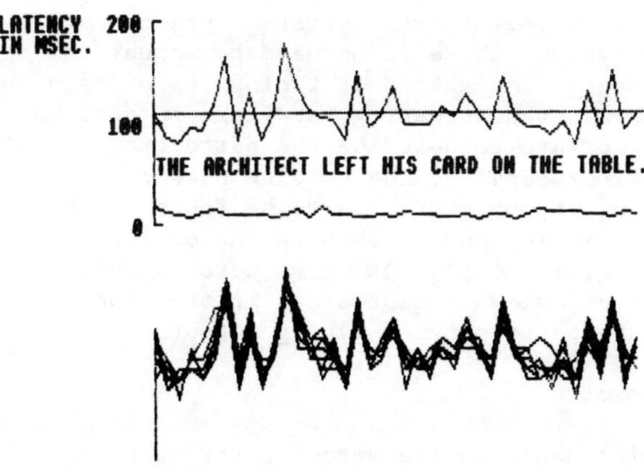

Figure 1. The latency pattern obtained in typing the given sentence. The lower graphs show the superimposed patterns obtained on 8 occasions of typing the sentence; the upper graphs show the average latency pattern, the overall average latency and, below these, the standard deviation of latency at each letter position.

Rumelhart and Norman were able to show that much of the variation in the latency pattern can be accounted for in terms of biomechanical constraint on movement overlap between hands and fingers. Constraint is minimal between hands and maximal in re-using the same finger to type successive letters. Central to the typing model are the assumptions that (a) motor output is programmed, (b) movements are computed in parallel and interactively over a motor sequence, (c) movements have a fairly narrow velocity bandwidth, and (d) successive movements overlap in time depending on mechanical constraint. These assumptions together provide a timing algorithm, though time itself is not a primary variable.

It can be shown that the model leaves a significant part of the timing variation in a latency pattern unexplained, and that the residual variation cannot be readily explained by appeal to other biomechanical factors. It seems to depend rather on idiosyncratic factors of letter grouping which resist a general characterization. For example, in Fig. 1, the f in left is consistently fast, though one should expect the preceding e to be fast because of the cross-hand transition. It is as though the fast f is anticipating the following t, necessarily slow because it involves re-use of the index finger.

Such a shaping of movement over letter groups can still be achieved within the motor system: there is no need to appeal to cognitive factors shaping the movements. The action of typing is not required to be expressive in the way that speaking, drawing and dancing are expressive, and so it is appropriate to describe the patterning over typing sequences as quasi-expressive. The genuine uses of expression, involving a cognitive shaping of movement, are to be found in the prosodies of speech, in playing music, and in shaping the calligraphic line in writing and drawing. The important point here is that the capacity of the motor system to implement expressive information is already demonstrated in the patterning of typing output.

2.4 Feedback Control

Motor programming has been presented so far as a flow of information from a cognitive system to a motor system. We now have to consider a possible flow in the other direction.

When skilled touch typists make errors they often know this at once without consulting their copy, and can stop and initiate correction. Even when they are instructed not to attempt to correct errors but to carry on typing, a tendency to pause seems irresistible (Shaffer, 1975). In the absence of visual confirmation, the motor system must be able to compare directly what it has done with what it should have done. Kinesthetic feedback must be compared with a reference representation, in the motor system itself or in the motor program. If the motor system contains a correct representation then the error can have arisen from noise in the activation levels for different movements, affecting their sequence. Rumelhart and Norman (1982) attempted to explain transposition errors in this way, but it is difficult to see how it could apply to reversals of distant elements. Such errors would suggest that if they arise in the motor system their detection is made with reference to the motor program. This in turn requires that information in the program is not immediately erased following its use, but survives long enough to permit such reference.

The following amusing example demonstrates some aspects of control in such backward reference. In Fig. 2 there are four graphs showing letter latencies for 4 successive occasions on which the same typist copied the sentence "the wedding cake collapsed under its own weight". On the first occasion she typed callapsed, and paused before typing the second a. It is likely that the motor program had become corrupted to callopsed and the pause allowed the o to be corrected to a. On the second occasion she typed callapsed again, but this time paused immediately after the wrong a. On the third occasion she paused before the o and typed the word correctly. On the fourth occasion she typed colappsed, getting the o fluently correct but the subsequent letter doubling wrong, in the way analysed earlier, and paused after the second, incorrect p.

This comedy of errors shows that the control element making use of backward reference can become sensitised to the potential for error at a particular point in the text. Though the nature of the errors and of their detection and (possible) correction indicates that reference is made to an abstract symbolic representation, viz the motor program,

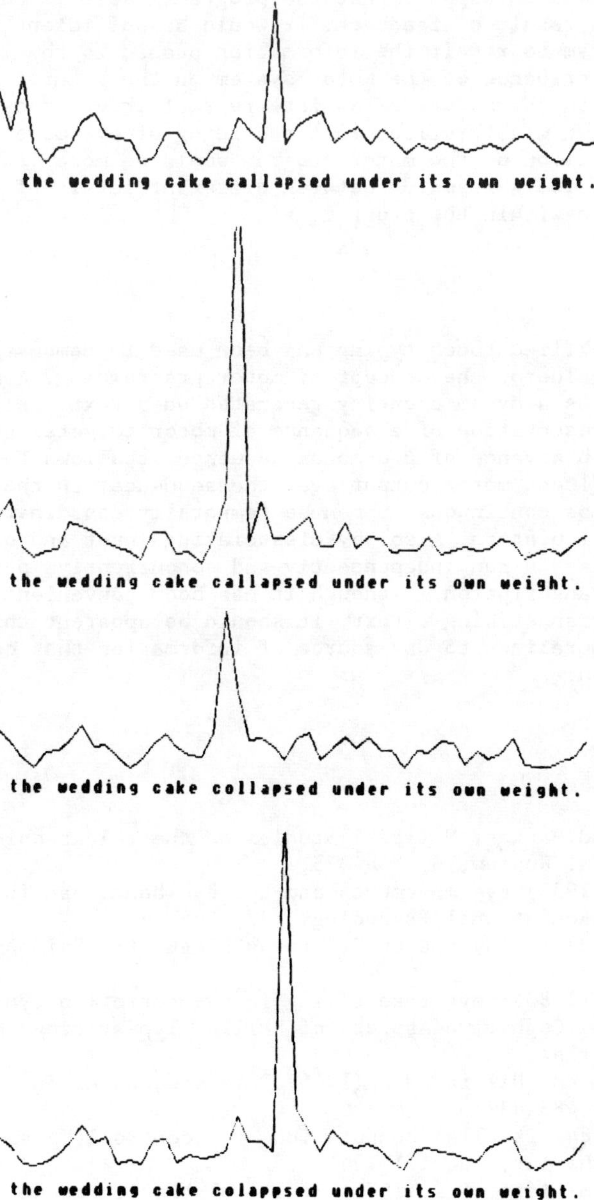

Figure 2. The latency patterns obtained on 4 occasions of typing the same sentence, showing the errors made and the pauses in response to these errors.

there is no need to suppose that the program itself is corrected or modified as a result of feedback. It would be sufficient to use the correct program to repair the information passed to the motor system. If there is an influence of the motor system on the cognitive system, it is likely to be in the nature of an interrupt of input processing, something which might reflect in a recording of eye movements during typing. A revision of the motor program would be more likely to occur if feedback revealed a mismatch between different levels of abstract representation within the program.

3. CONCLUSION

Data from skilled touch typing has been used to demonstrate the explanatory value of the concept of motor programming. A motor program is described as a dynamic entity generated on a text, in the form of an abstract representation of a sequence of motor targets. Being constructed in advance of the motor sequence it allows the motor system to organize fluent motor output over the sequence, in the sense that movement can be continuous or can be temporally coordinated in different limbs and articulators. Also, by dissociating input and output processes it allows these to run independently and concurrently, permitting continuous transcription. Although it has been convenient to describe the task of transcribing a text, it should be apparent that the concept of a text generalizes to any source of information that can be used to plan an action.

4. REFERENCES

Bryan, W L and Harter, N (1899) Studies on the telegraphic language. Psychological Review, 6, 345-375.

Butsch, R L (1932) Eye movements and the eye-hand span in typewriting. Journal of Educational Psychology, 23, 104-121.

Fromkin, V (ed) (1973) Speech Errors as Linguistic Evidence. The Hague: Mouton.

Gentner, D R (1983) Keystroke timing in transcription typing. In W E Cooper (ed), Cognitive Aspects of Skilled Typewriting. New York: Springer-Verlag.

Hershman, R L and Hillix, W A (1965) Data processing in typing. Human Factors, 7, 483-492.

Hinton, G (1984) Parallel computations for controlling an arm. Journal of Motor Behavior, 16, 171-194.

Kent R D and Minifie, F D (1977) Coarticulation in recent speech production models. Journal of Phonetics, 5, 115-133.

Lashley, K S (1951) The problem of serial order in behavior. In L A Jeffress (ed), Cerebral Mechanisms in Behavior. New York: Wiley.

Lee, D N, Lishman, J R and Thomson J A (1982) Regulation of gait in long jumping. Journal of Experimental Psychology: Human Perception and Performance, 8, 448-459.

Levelt, W J M (1989) Speaking. Cambridge Mass.: MIT Press.
Michaels, C F and Carello, C (1981) Direct Perception. Englewood Cliffs, NJ: Prentice-Hall.
Morasso, P (1986) Trajectory formation. In P Morasso and V Tagliasco (eds), Human Movement Understanding. Amsterdam: North-Holland.
Morasso, P and Tagliasco, V (eds) (1986) Human Movement Understanding. Amsterdam: North-Holland.
Rumelhart, D E and Norman, D A (1982) Simulating a skilled typist: A study of skilled cognitive-motor performance. Cognitive Science, 6, 1-36.
Saltzman, E and Kelso, J A S (1987) Skilled actions: A task-dynamic approach. Psychological Review, 94, 84-106.
Shaffer, L H (1973) Latency mechanisms in transcription. In S Kornblum (ed), Attention and Performance IV. New York: Academic Press.
Shaffer, L H (1975) Control processes in typing. Quarterly Journal of Experimental Psychology, 27, 419-432.
Shaffer, L H (1981) Performances of Chopin, Bach and Bartok: Studies in motor programming. Cognitive Psychology, 13, 327-376.
Shaffer, L H (1982) Rhythm and timing in skill. Psychological Review, 89, 109-122.
Shaffer, L H (1988) Forced revision in fast typing. Quarterly Journal of Experimental Psychology, 40A, 581-589.

Levels of Organisation and the Planning of Movement Sequences: Sequential Evidence.

Diana Kornbrot
Psychology Division, Hatfield Polytechnic
College Lane
Hatfield, Hertfordshire AL10 9AB
United Kingdom

ABSTRACT. The structural organisation of motor plans into high and low level units was investigated using a discrete keying task. Motor patterns, in terms of ease of organisation and number of key strokes in a sequence, were experimentally manipulated. The pattern of correlation between interresponse times, both within a single trial sequence and from trial to trial was analysed using factor analytic techniques and compared with theoretical predictions. Within trial analyses identified three factors corresponding to: low level interresponse times; high level interresponse times; and latencies. Between trial analyses, which included lag 1 times for each trial, always identified a single factor for lag 0 and lag 1 low level interresponse times, but separate factors for lag 0 and lag 1 latencies. These results are interpreted in terms of execution processes which persevere from trial to trial and initiation processes which are independent across trials.

1. Introduction

Skilled discrete motor behaviours, such as typing, may be loosely described in terms of loading organised collections of units into a motor buffer; searching through the motor buffer for the next unit to execute, and then executing the found unit. The collections of units loaded into the buffer could be words in typing or bars in music. Or, in more arbitrary experimental tasks, they may be short, easily organised, sequences such as trills. Psychological modelling of such processes usually considers the individual keystroke as the lowest level of output. Loading of one collection of units in to the buffer can and does proceed in parallel with search/execution of other units out of the buffer.

The loose framework described above is a long way from a detailed account of the psychological processes involved. In particular, a major theoretical question concerns the nature of the organisation of the keystroke units in the motor buffer into larger groups. The highly consistent pattern of slow and fast interresponse times found in real world typing tasks and in laboratory tasks has been used as a major source of evidence in

J. Requin and G. E. Stelmach (eds.), Tutorials in Motor Neuroscience, 385–396.

this debate (Shaffer, 1978; Long et al., 1983). Gentner, (1982, 1983) argues that the patterns can be completely accounted for by low level processes of activation and inhibition in muscle groups controlling the hands and fingers. Conversely, Povel and Collard (1982) and Rosenbaum (1983, 1987) argue that the effects are due to number of nodes traversed in searching a motor buffer which is organised hierarchically as a binary tree. Kornbrot (1989) also argues for hierarchical organisation of a motor buffer, but not necessarily as a binary tree. Further important theoretical questions concern the relation of the loading phase to the search/execution phase. In particular, is the organisation of the motor buffer related to the organisation and sequence of the loading of its content?

In this investigation these issues are addressed by investigating the correlations among latencies and interresponse times within and across trials, where each trial comprises the cued performance of a sequence of keystrokes. Conceptually, three kinds of processes can be identified. Firstly, processes concerned with loading into the motor buffer. The first response of a trial sequence giving rise to the initiation time or latency is always determined by the loading process. In addition, if the trial sequence is organised into high level groupings; then the first response of such a group, termed a high level interresponse time might also be effected by the loading process. The second kind of process involves determining which of the units already in the buffer to execute next; either by searching through the buffer, or by building up activation. Finally, there is the process of actually executing the currently active unit.

1.1. RESEARCH PARADIGM

The empirical study used a visually cued keying task, which is formally identical to that of Sternberg et. al. (1978). A sequence of digits appeared on a screen, followed sequentially by the words: MARK; SET; GO. Subjects were required to key the sequence as quickly as possible after receiving the GO signal. A specially designed keyboard was used with eight keys directly under the fingers of the two hands, the keys were labelled "1" through "8" from the left little finger to the right little finger. The time from presentation of the GO signal to the first key press was noted and is described as the initiation time or latency. The times of each subsequent keystroke are also recorded so that interkey response times (IRTs) can be measured. The IRTs are taken to provide measures relevant to the search and execution times of units from the motor buffer.

This paradigm permits one to independently manipulate the complexity of the task in terms of (a) motor pattern which influences how people organise the task into high level groupings and (b) total number of keys to be struck, which influences either the number of groups or the number of units per group, depending on the motor pattern. Results from 4 key, 6 key and 8 key conditions crossed with two distinct motor patterns are reported here. The RUN motor pattern comprised a sequence which , ran from left to right with an interhand transition in the middle. Thus all RUN sequences can be regarded as comprising 2 higher order groups: left hand run followed by right hand run. If this organisation is present, a 6 key run would be comprised of two 3 key units and an 8 key run would be comprised of two 4 key units. The FINGER FOLLOWS (FF) motor pattern has the keystrokes alternating between hands with a given right finger always following its left homologue. This pattern can be regarded as having high level groups comprising pairs of keystrokes with the number of units being half the number of keys.

As well as permitting the manipulation of theoretically significant variables, this paradigm has the advantages that (a) perceptual features are held constant since there is always plenty of time to read the sequence to be keyed; (b) low level motor features are held constant as subjects do not have to move their fingers from one key to another; and (c) information is obtained as to the time to load the motor buffer, i.e. the latency for the first key stroke, as well as the execution times, i.e. the interresponse times (IRTs). The disadvantage, of course, is that the strategies used may differ substantially from those in truly continuous tasks such as typing.

1.2. PREDICTED PATTERNS OF CORRELATIONS

In general latency patterns arise from loading mechanisms and low level IRTs arise from searching.executing mechanisms; while high level IRTs could arise from either kind of mechanism depending on whether they are more closely correlated with latencies or with low level IRTs. Thus, granted that high and low level responses have already been identified via mean IRTs, the following patterns for within trial correlations are of theoretical interest.

A Correlations between latencies (initiation times) and other response times can provide evidence for a separate loading phase

A1 latencies and all IRTs are correlated
  * supports a single mechanism for loading to and executing/searching from the buffer such that if a trial is fast, all units in that trial are fast

A2 latency correlated with high level IRTs, but high & low level responses uncorrelated
  * supports a separate mechanism for loading and searching/executing from buffer

A3 latency not correlated with any interresponse time
  * supports a separate mechanism for loading and searching/executing from buffer, with loading of first unit independent of loading of subsequent units

B Correlations between low and high level IRTs within and across high level groups provide evidence concerning whether searching (decoding) is occurring in the buffer and whether there is a single execution process.

B1 all IRTs correlated
* supports a single mechanism for searching and executing from the buffer

B2 all IRTs (low & high) correlated within, but not between, high level groupings
* supports tree traversal or other within buffer decoding models since all responses within a group share common processes

B3 all low level IRTs correlated within a group, but not between groups or high level
* supports loading as a group, and then decoding of groups in the buffer, with separate mechanism for loading and searching/executing the buffer as in A2

B4 all low level IRTs correlated within & across groups, but not with high level IRTs
* supports single executing mechanism or activation mechanism at lowest level

Once the within trial correlation patterns have been identified it is important to know if these patterns extend across trials. In particular, one may ask separately for loading, searching and executing if being fast on one trial implies that one will be fast on the immediately following trial.

2. Method

This is a brief summary of the method, further details may be found in Kornbrot (1989).

2.1. TASK AND APPARATUS

The subjects' task was to key in a sequence of digits which appeared on a display screen. The keyboard had 10 keys arranged in a shape corresponding to the two hands side by side. In this experiment the thumbs were not used and the other keys were labelled 1 - 8 running from left little finger to right little finger. Subjects were seated with the keyboard under their fingers. The program ARTIST (Kornbrot, 1981) presented number sequences and messages on the screen, and recorded the timing of the key presses.

2.2. SUBJECTS AND DESIGN

The subjects were 10 young adults who performed the experiment as part of the laboratory component of a B.Sc. Psychology degree course.

The number of keys in the sequence was manipulated as a repeated measure blocked within subjects. There were 4 levels: 2 key; 4 key; 6 key; and 8 key. Motor pattern was manipulated as a between subjects factor with 2 levels: RUN; and FINGER FOLLOWS (FF) as described above.

2.3. PROCEDURE

Subjects performed the task in blocks of 120 trials. The same keying sequence was used throughout the block. In every 20 trials 2 trials were randomly assigned to be catch trials.

For each block of trials the subject was seated in an experimental booth where the oscilloscope screen was the only light source. The word START then appeared on the screen for 9 secs followed by 120 trials. On each trial a sequence of digits appeared centred on the screen for 1 sec followed by MARK for 0.7 sec followed by SET for 0.7 sec followed either by GO until the subject completed keying the digit sequence, or by a blank screen for 1 sec (catch trial). After the subject's response, or the 1 sec blank screen, a feedback message was presented for one second. There were 5 possible messages: EARLY ER appeared if the subject responded before the GO signal;

CORRECT appeared if the subject keyed the sequence correctly; ORDER ER appeared if the sequence was out of order or not completed within 6 sec of the GO signal; BLANK CORRECT appeared if the subject (correctly) did not respond on a catch trial; and BLANK ER appeared if the subject did respond when the GO signal failed to appear. The next trial followed immediately after the feedback message. At the end of the block the message THANKS appeared on the screen and the subject took a 2-5 minute rest while the next block of trials was set up.

3. Results

The analyses reported here were performed on data from 5 subjects in the RUN group and 5 subjects in the FINGER FOLLOW group from an experiment reported by Kornbrot (1989). Each of the selected subjects showed the predicted pattern of mean IRTs for the presented motor pattern in both the 6 and the 8 key condition. Thus for these subjects, high and low level IRTs had been identified by by the criterion of relatively high or low mean IRTs before any correlation analysis was performed.

3.1. WITHIN TRIAL ANALYSES

The following analyses were performed separately for each subject in each condition. Firstly a correlation matrix was formed from the latency and the n - 1 IRTs. Then a principle components factor analysis with varimax rotation of the components, constrained to extract 3 factors was performed. (Unconstrained factor analyses extracted 3 factors for most matrices using any of the standard algorithms, so it was decided to analyse the matrices under the assumption that 3 factors were indeed present.). The loadings of each of the variables on the 3 factors were then averaged over the 5 subjects who experienced a given motor pattern, separately for 4, 6 and 8 key conditions. The results are shown in Figure 1 together with the percentage of variance accounted for by each factor.

For both motor patterns and all conditions, a factor emerges which loads only on latency. Not surprisingly, this factor accounts for diminishing proportion of variance as number of keys increases. Latency is 25%, 16.7% and 12.5% of the variables in 4, 6 and 8 key conditions respectively. This corresponds to the A3 pattern. Latency is uncorrelated with any interresponse time.

For the RUN motor pattern separate factors emerge for the high level central interresponse time and all the other IRTs which are low level for this motor pattern. In Figure 1 the RUN IRTs are labelled relative to the central interresponse time i0, with the interresponse time immediately preceding the central one labelled i-1, and that immediately following it i1, i2 etc. The low level factor accounts for about 38% of variance in all conditions, while the high level factor, like the latency factor, accounts for diminishing variance with increasing number of keys. This corresponds to pattern B4 and suggests a single execution mechanism.

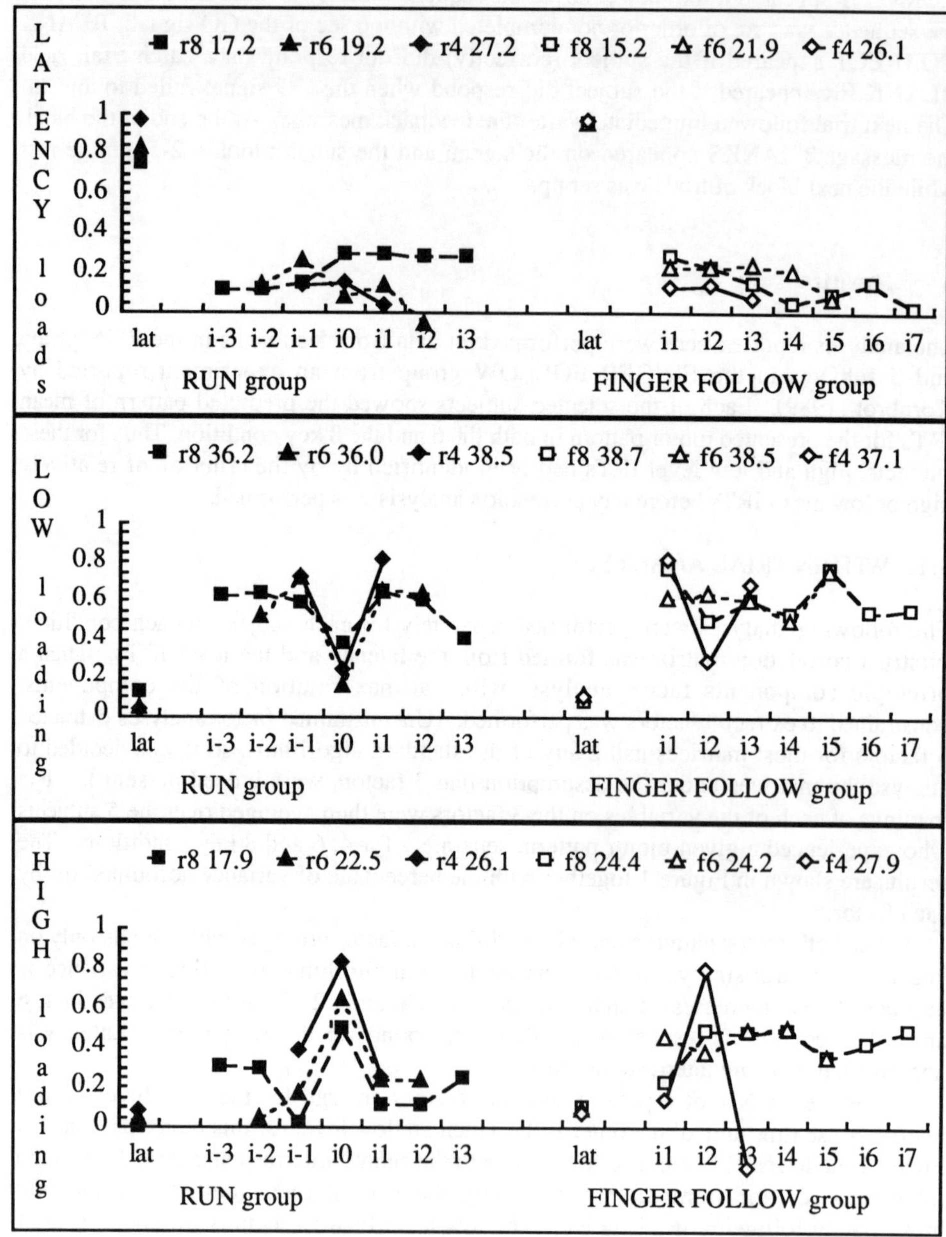

Figure 1. Average loadings of latency and IRT variables on latency (upper panel); low level (middle panel) & high level (bottom panel) factors. RUN, left; FF, right.

For the FINGER FOLLOWS motor pattern the picture is less clear cut. As with the RUN pattern a factor accounting for 38% of variance in all conditions emerges, and this is tentatively labelled "low level", a further factor corresponding about 25% of the variance is tentatively labelled "high level".

3.2. ACROSS TRIALS ANALYSES

Since a particularly clear pattern of factors emerges from within trial analyses for the subjects in the RUN group, it is worth investigating if these factors persevere across trials. So the following across trial analyses were performed for the 8 key RUN condition. Again, factor analyses techniques were used with 4 and 5 factor solutions for subjects KEF and YTB shown in Tables 1 and 2 respectively.

Table 1. Loading on low level; high level and latency factors at lag 0 and lag 1 for Subject KEF in RUN condition: 4 factor and 5 factor analyses

variable		source factor					total variance
		5 factor solution					
response	lag	low	high 0	high 1	latency 0	latency 1	
latency	0	-.06	-.01	-.01	**1.00**	.05	
high IRT	0	-.05	**.93**	.33	-.02	.00	
low IRT	0	**.93**	.14	-.20	.02	-.10	
latency	1	-.06	.00	-.02	.05	**1.00**	
high	1	-.05	.33	**.93**	-.01	-.02	
low	1	**.93**	-.21	.13	-.10	.02	
% variance		28.8	17.3	17.3	16.8	16.8	96.9
		4 factor solution					
response	lag	low	high		latency 0	latency 1	
latency	0	- .06	- .02		**1.00**	-.05	
high IRT	0	- .05	**.89**		.01	.01	
low IRT	0	**.93**	- .04		.03	.11	
latency	1	- .06	- .01		.05	**-1.00**	
high	1	- .05	**.89**		-.02	.01	
low	1	**.93**	- .06		-.11	-.03	
% variance		28.8	26.6		16.8	16.8	89.0

Table 2. Loading on low level; high level and latency factors at lag 0 and lag 1 for Subject YTB in RUN condition: 4 factor and 5 factor analyses.

variable		source factor					total variance
		5 factor solution					
response	lag	low	high 0	high 1	latency 0	latency 1	
latency	0	.07	-.02	-.04	**.99**	.09	
high IRT	0	-.02	**1.00**	.04	-.02	.04	
low IRT	0	**.95**	.06	.05	.02	.07	
latency	1	.11	.04	-.01	.09	**.99**	
high IRT	1	.10	.05	**.99**	-.04	-.01	
low IRT	1	**.94**	-.09	.09	.08	.07	
% variance		30.0	16.8	16.7	16.7	16.6	96.8
		4 factor solution					
response	lag	low	high 0	high 1	latency		
latency	0	.01	-.22	.02	**.84**		
high IRT	0	-.04	**.93**	.07	-.01		
low IRT	0	**.94**	.06	.05	.06		
latency	1	.18	.33	-.09	**.67**		
high IRT	1	.11	.06	**.99**	-.04		
low IRT	1	**.93**	-.09	.09	.11		
% variance		30.1	17.4	16.6	19.6		83.7

Detailed analyses were performed as follows. First a single low level IRT variable was calculated for each trial by averaging IRTs 1, 2, 3, 5, 6 & 7, so that each trial could be characterised by the 6 variables: latency 0; high 0 (corresponding to IRT4); low 0; latency 1; high 1 and low 1. The index 1 indicates that the variable occurred on the previous trial, corresponding to a lag of one trial; while the index 0 indicates that the variable occurred on the current trial corresponding to zero lag. A correlation matrix was then constructed for the 6 variables and a principle components factor analysis performed with varimax rotation. Only correct responses on non-catch trials where a signal was present were included in the analysis. Thus in some cases lag '1' will have had an intervening catch or error trial. This occurred on a minimum of 10% (the catch trials) and a maximum of 20% of trials for the data reported here.

For subject KEF the four factor solution shown in Table 1, accounts for 89.0 % variance, and generates one factor loading highly on low 0 and low 1 variables and minimally on other variables; one factor loading highly on high 0 and high 1 variables and minimally on other variables; and separate factors for latency 0 and latency 1. The additional variance accounted for by a fifth factor is not significant using a chi-square test. Indeed, the five factor solution is rather worse in that the factor designated high 0 loads a substantial .33 on the high 1 variable and similarly, the factor designated high 1 loads .33 on variable high 0. In the four factor solution all loadings are either $\leq$ 0.11 or $\geq$ 0.89 indicating a good identification of variables with factors. Thus for KEF the four factor solution appears most satisfactory and suggests that both high level and low level effects persevere across trials, but latency effects do not.

For subject YTB by contrast the five factor solution seems most satisfactory, accounting for 96.8% of variance, this is 13.1% more than the four factor solution and does give rise to a significant chi-square. All variable loadings are either $\leq$ 0.11 or $\geq$ 0.94, again indicating good identification of factors with variables. The five factor solution identifies one factor for low 0 and low 1; and one each for high 0, high 1, latency 0 and latency 1. The four factor solution also identifies a single low level factor; and a further factor loading on latency 0 (.84) and relatively weakly on latency 1 (.67). Separate factors for high 0 and high 1 are identified, but the high 0 factor also loads on latency 1.

Similar analyses on subjects in other groups and/or conditions also exhibit the general feature that low level effects persevere across trials, while latencies are not correlated with other measures, either within or across trials, is present. High level effects are more ambiguous. In the FF group, although a clear separation into high (slow) and low level (fast) is apparent from mean IRTs, it is not always so clear in each condition from the correlation matrices even within trial let alone across trials (see figure 1). For the RUN group even though high and low levels are well separated from within trial correlation matrices, the pattern across trials is sometimes less clear cut than that shown in Tables 1 and 2.

4. Discussion

4.1. WITHIN TRIAL MECHANISMS

Several inferences can be made from the within trial analyses. Not surprisingly, the loading phase is distinct from the search/execution phase. Fast search/execution as evidenced by fast IRTs is unrelated to fast initiation as evidenced by latencies. Thus there is no 'general' speed factor which influences all aspects of a trial. This is true even though features like complexity of motor pattern, and number of keys do effect both latency and IRT (Sternberg et al., 1978, Kornbrot, 1989).

More surprisingly, low level IRTs are just as highly correlated across high level groups as within high level groups. The major evidence coming from the RUN group which conform to correlation pattern B4. That is, loadings of the IRT variables i-3, i-2, i-1 in the first left hand high level grouping have equivalent loadings on the hypothesised low level factor as the variables i1, i2, i3 in the second right hand high level grouping. This supports a single executing or activating mechanism at the lowest level. The pattern from the Finger Follow group where all IRT variables load at least moderately on both the hypothesised low and high level factors is also consistent with a single execution mechanism.

At the same time, in the RUN group at least, there is no commonality between the single high level IRT corresponding to the inter-hand transition and the other low level IRTs in that group (i.e. on that hand). This corresponds to correlation pattern B3 as opposed to B2 and presents problems for models which suggest any kind of hierarchical search through the buffer, or decoding within the buffer, or parallel activation of high and low level units within a group; since all such models would predict that low and high level IRT in the same group share a common mechanism and hence are correlated.

4.2. ACROSS TRIAL MECHANISMS

The identification of a factor which loads strongly on low level IRT variables for both lag 0 and lag 1 suggests that the mechanism which causes all the low level IRTs within a trial to be correlated perseveres across trials. The execution mechanism which actually performs the currently most active item in the motor buffer is the most likely mechanism for this role. It is as if the last step is to place the active item on a conveyor belt; if that conveyor belt is moving fast then low level IRTs are fast. Furthermore, the speed of the conveyor belt on any given trial is likely to be close to the speed on the previous trial. A fast conveyor belt does not necessarily speed up higher level IRTs since their speed can be limited by the mechanism which places them on the belt.

By contrast, latency on any trial is totally uncorrelated with latency on the previous trial as evidenced by the identification of separate factors for latencies at lag 0 and lag 1. This is different from what is found in choice reaction time experiments where small but significant lag 1 correlations of about 0.15 are common (Laming, 1979). Magnitudes of raw lag 1 correlations in this experiment were between 0.09 and -0.09 and not significant. Raw correlations of low 0 and low 1 were around 0.6, which is really quite a substantial effect.

The results for the high level IRTs are not clear cut. For KEF, there was substantial carryover from trial to trial on the high level IRT which was independent of the carryover for low level IRTs. This highly consistent pattern should be explainable by a good theory. Of course, the RUN pattern only has one high level IRT, so one cannot ask whether there is a high level feed conveyor to the lower level execution conveyor. Results for YTB are easier to explain, as the single high level response acts like a separate latency. If latencies are not correlated across trials then one would not predict correlations between latency and high level IRTs within trial or between high level IRTs across trials. More data is needed from conditions like FF where there is more than one potential high level IRT and subjects have had much more extensive practice on this more difficult motor pattern.

4.3. SUMMARY

Timing correlations in a discrete motor task, both within and across trials, have been studied using factor analytic techniques. The following theoretically important facts emerge from the analyses.

* Separate mechanisms are required to account for the generation of latencies and low level IRTs
* The mechanism generating the latencies may be thought of as loading material into a motor buffer. This mechanism does not persevere across trials.
* The mechanism generating the low level IRTs may be thought of as an execution mechanism. This mechanism does persevere strongly across trials.
* Correlation analyses only unambiguously identify a mechanism generating high level IRTs in the simpler RUN motor pattern. For some subjects the mechanism identified could be the same as that generating latencies. For other subjects, the mechanism generating high level IRTs perseveres across trial.

Current models of motor programming will need to take account of these findings. In particular, the correlation of low level responses across high level groupings, in conjunction with the lack of correlation of low level responses with high level responses in the same group, argues strongly for the importance of organisational structure in the loading process. Furthermore, the strong perseveration of the low level mechanism across trials points to a mechanism for learning consistent with current neural network theories.

5. Acknowledgments

This work is based on research funded by the Economic and Social Research Council (ESRC) reference number H00 23 2044.

6. REFERENCES

Gentner, D. R. (1982) 'Evidence Against a Central Control Model of Timing in Typing', Journal of Experimental Psychology: Human Perception and Performance 6, 793-810.

Gentner, D. R. (1983) 'Keystroke timing in transcription typing', in W.E. Cooper (ed.), Cognitive Aspects of Skilled Typewriting, Springer-Verlag, New York, pp. 95-120.

Kornbrot, D. E. (1981) 'ARTIST: A Computer System for Creating and Running

Psychology Experiments', Behaviour Research Methods and Instrumentation, 13, 351-359.

Kornbrot, D. E. (1989) 'Organisation of Keying Skills', Acta Psychologica,70, 19-41.

Laming, D. R. J. (1979) 'Autocorrelation of Choice-Reaction Times', Acta Psychologica, 43, 381-412.

Long, J., I. Nimmo-Smith, and A. Whitfield, 1983. Skilled typing- a characterisation based on the distribution of times between responses. W.E. Cooper (Ed.), Cognitive aspects of skilled typewriting, Springer-Verlag, New York.

Povel, D. J. and Collard, R. (1982) 'Structural Factors in Patterned Finger-Tapping', Acta Psychologica, 52, 107-123.

Rosenbaum, D. A., Kenny, S. and Derr, M. A. (1983) 'Hierarchical Control of Rapid Movement Sequences', Journal of Experimental Psychology: Human perception and Performance, 9, 86-102.

Rosenbaum, D. A., Van Hindorff, G., and Munro, E. M. (1987) 'Scheduling and Programming of Rapid Movement Sequences', Journal of Experimental Psychology: Human Perception and Performance, 13, 193-203.

Shaffer, L.H., 1978. Timing in the Motor Programming of Typing. Quarterly Journal of Experimental Psychology, 30, 333-345.

Sternberg, S., Monsell, S., Knoll, R.L.and Wright, C.E. (1978) 'The Latency and Duration of Rapid Movement Sequences', in G. E. Stelmach (ed.), Information Processing in Motor Control & Learning, Academic Press, New York, pp. 117-152.

STRUCTURAL FACILITATION OF MOVEMENT SEQUENCE PLANNING

Andras SEMJEN and Robert GOTTSDANKER
C.N.R.S., L.N.F.1., 31 chemin J.Aiguier
13402 Marseille Cedex 9, France, & University of
California, Psychology Department, Santa Barbara,
CA 93106. USA.

ABSTRACT. Previous work has shown that a sequential plan for movements can be selected prior to, and independently of, the selection of its particular effector realization. In the present study, we addressed the question of whether selection of a sequential plan for movements may also be "content free", i.e., based upon relational (structural) properties, at least at some stage of the planning process. If this is the case, selection of a movement sequence should be more rapid if the alternatives represent different realizations of the same serial structure than if not. Moreover, execution of a movement sequence should facilitate subsequent programming of another movement sequence, if they share the same serial structure.

Subjects learned to associate each of four visual symbols with a different finger-tapping sequence. The tapping sequences consisted of strong and weak taps which were tapped at a high rate with the same finger. They differed in serial structure (e.g., OXOXOOOX or OOOXOXOX) and stress (O=strong and X=weak tap, or X=strong and O=weak tap). In one procedure, subjects chose between tapping sequences that either were identical in serial structure (but opposite in stress assignment) or were different in serial structure. The shorter reaction times for the former shows the advantage of relational identity. In another procedure, the subjects were required to read aloud, at the beginning of each trial, a sequence of visually presented CV syllables (e.g.,Ba and Bi) and, shortly after completion of the vocal sequence, to produce the appropriate tapping sequences in response to one of the visual symbols presented as a choice reaction stimulus. The priming of the tapping sequence with the previously pronounced vocal sequence was "Congruent" if both sequences had the same serial structure. Congruent priming generally resulted in shorter reaction time than non-congruent priming, supporting the view that programs for rapid serial movements are generated, at least under certain conditions, from "content-free", relational representations of the required serial order.

1. Introduction

The notion of motor programming refers to the widely held assumption that rapid, skilled, serial actions are organized in advance of their execution. By "organized" it is meant that, before acting, a person generates a representation not only of the ultimate goal of his (her) action, but also of the way in which the action will unfold. However, the thinking about which aspects of the action are represented in the motor program has undergone some rather profound changes over the past 15 years. This evolution can be characterized as an increasing shift from a "muscle-specific command assumption" towards an "abstract-program assumption" (MacKay, 1982). A motor program is no longer conceived of as a set of structured muscle commands (Keele, 1968) but, rather, as "a set of abstract

J. Requin and G. E. Stelmach (eds.), Tutorials in Motor Neuroscience, 397–412.

representations of the intended motor output" (Shaffer, 1984). "Abstractness" refers, in part, to the view that motor programs must define at some point relationships between events, rather than the events themselves. Note that such a conception was advocated as early as 1951 by Lashley (Lashley, 1951). Effector independence is a case in point. It implies that sequence knowledge (e.g., writing one's name) is independent of the effector system (e.g., the particular articulators such as the hand, wrist, forearm, elbow, arm or shoulder) that realizes it.

The idea that the organizational aspects of serial skills are represented in an effector-independent manner largely stemmed from experimental and theoretical work on skill learning and transfer (e.g., Pew, 1974; Schmidt, 1975; Cohen, Ivry and Keele, 1990). Effector-independence appeared, prima facie, characteristic of long-term memory codes. There are, however, a small number of experimental studies showing that, in the course of short-term program construction for an upcoming action, information about sequential (temporal) organization is processed independently of information about effector implementation. In such studies, choice reaction time (CRT) has been used as index of program construction, with subjects either receiving or not receiving information about sequential structure in advance of the response stimulus (RS). In an experiment by Klapp, advance information about the required duration of a single key-press eliminated the RT-difference between short- and long-duration key-presses, usually found when such information is delivered by the RS. The finding was taken as evidence that response duration - a temporal parameter - may be selected in advance of the particular finger that must execute the response (Klapp, 1977). Selection of a sequential plan for more extended finger-tapping sequences has also been shown to occur prior to the selection of a particular effector. An example of effector-free plan-selection comes from a study by Rosenbaum, Inhoff and Gordon (1984). In one of their experiments, subjects chose between finger sequences that either mirrored each other (for example, index-index-middle of the left hand versus the right hand), or did not mirror each other (for example, index-index-middle of the left hand versus index-middle-middle of the right hand). Consistent with the notion that a common sequential plan could be selected in advance in the former, but not in the latter condition, CRT was shorter for mirror-image sequences than for other sequences.

The question arises, then, as to whether plan preselection is content-bound, i.e., limited to the particular description that can be given of the alternative sequences, or may occur in even more abstract, more generally relational terms. Taking the Rosenbaum et al.'s experiment as an example, "index-index-middle" is a description of moderate abstractness, given that it refers to finger-sequences of either the left or right hand. In contrast, "doublet and singleton" or "repeat and shift" are more abstract, rule-like descriptions of the serial order implied in the "index-index-middle" sequence, and such descriptions may allow for generating a far broader class of movement sequences than just the finger sequences mentioned above. The purpose of the present series of experiments was to examine whether such content-free, "generative" representations may intervene in program construction for rapid serial movements. The emphasis is on rapid responding for it has long been known that serial patterns can be encoded and remembered in terms of generative rules (Restle, 1970), or figure-ground relations (Garner, 1962), that are fairly free from concrete details of pattern presentation (for a review see Jones, 1974). Experiment 1 will show that choosing between sequences made up of weak and strong finger-taps takes less time if the alternative sequences derive from the same serial pattern (by interchanging strong and weak taps), than if they derive from different serial patterns. Experiments 2 and 3 will show that the time to choose between sequences of weak and strong finger-taps may be reduced by previously uttering a sequence of binary syllables, if the syllable sequence and the required tapping sequence share the same serial pattern. The conclusion will be advanced that content-free, generative representations may intervene, at least under certain conditions, in program construction for rapid serial movements.

2. Experiment 1

2.1. METHOD

2.1.1. *Task*. Subjects learned to associate each of four visual symbols with a different finger-tapping sequence and to produce them according to the requirements of a CRT paradigm. The eight-tap sequences consisted of combinations of strong and weak taps made with the same finger (index or middle) of the preferred hand. Table 1 shows that the sequences were derived from two basic serial patterns (A and B) by assigning to the O-s and X-s, respectively, either weak (W) and strong (S), or strong and weak taps. These assignments created two "Stress conditions", one of which was called "Normal" (with fewer strong than weak taps), and the other, "Inverted" (with fewer weak than strong taps).

TABLE 1. Sequences used in Experiment 1

PATTERN	STRESS	SEQUENCES
A: O X O X O O O X	Normal	W S W S W W W S (1)
	Inverted	S W S W S S S W (2)
B: O O O X O X O X	Normal	W W W S W S W S (3)
	Inverted	S S S W S W S W (4)

Subjects produced the sequences under two- and four-choice conditions in which the response alternatives had the same relative probability (0.50 or 0.25). There were three types of 2-choice condition: first, between sequences of identical pattern but different stress (i.e., stress-choice, S); second, between sequences of identical stress but different pattern (i.e., pattern-choice, P); third, between sequences of different pattern and different stress (i.e., mixed choice, M).

Each type of choice was represented by two pairings between sequences: 1-2 and 3-4 for S-choice, 1-3 and 2-4 for P-choice, and 1-4 and 2-3 for M-choice. The main hypothesis tested in this experiment was that programming a tapping sequence takes less time (yields a shorter CRT) if the response alternatives derive from the same serial pattern, i.e., in the stress-choice case.

2.1.2.Subjects. Subjects were 6 members of the laboratory staff, 3 men and 3 women. They were paid for their services. Some of them had previously served as subject in other experiments involving production of rapid serial finger-taps.

2.1.3.Response device and stimulus display. The subject was seated facing the response device, with the forearm resting on a molded support, and the fingers held just above the response plate. Given this position of the forearm and hand, finger movements and wrist flexions and extensions contributed to the realization of the tapping sequences. The energy of the impact of the taps was converted into a proportional electric signal by a piezoelectric shock sensor located beneath the response plate. Any contact between the finger and a response plate triggered an electronic circuit whose output indicated the onset of a tap. For each response sequence we recorded: (a) the RT, that is the delay between the RS onset and the onset of the first tap; (b) the intervals between the successive taps, measured from tap-onset to tap-onset; and (c) the force of each tap as the peak output voltage of the piezoelectric sensor. Time was measured to the nearest millisecond, and force in arbitrary units. No particular force values were required; the measures were used in a relative way to ensure that the location of the strong and weak taps was correct.

A slightly inclined rectangular panel was placed in front of the subject on the same plane as the response device. Mounted midline on the panel was a digital display, and close to the left border of the display, a light emitting diode (LED). The visual codes serving as RS consisted of two horizontal bars for sequence 1, and two vertical bars for sequence 3. These codes were supplemented by concomitant illumination of the LED to call for production of these sequences with inverted stress (i.e., sequences 2 and 4).

2.1.4.Procedure and Design. The experimental events and measurements were controlled by an interfaced OLIVETTI M24 PC computer. A trial began with the presentation of a "temporal model" delivered to the subject via headphones. It consisted of 8 clicks of equal intensity separated by 160-ms intervals, and served to impose a high speed of execution. One sec after the last click the RS was presented and remained on until the subject has performed the sequence. A trial was aborted if any interval between successive taps was over 320 msec. The inter-trial interval was 5 sec.

Subjects received two to four practice sessions, depending on their prior experience on comparable finger-tapping tasks. During these sessions, they produced the tapping sequences under simple, 2-choice and 4-choice RT conditions. The data to be reported were recorded after the end of the practice period, in a single session during which the subjects performed 8 blocks of 32 trials. There were six blocks recorded in 2-choice conditions (one for each pairing between sequences), and two blocks in 4-choice condition. Subjects received the trial blocks in a variable order.

2.2. RESULTS

2.2.1.Reaction times. The means of individual median reaction times, based on correct sequences only, are shown in Table 2. The 4-choice RT data are those collected in the second block of trials only.

TABLE 2. Means of individual median RTs (msec)

CHOICE CONDITION	STRESS	
2-CHOICE	NORMAL	INVERTED
S	436	456
P	452	502
M	476	492
4-CHOICE	584	673

The individual median RTs were first subjected to a 2 x 2 x 4 Analysis of Variance (ANOVA) that tested the effects of Pattern (A vs B), Stress (Normal vs Inverted) and Choice conditions (S, P, M, 4-choice). Significant effects were found for Choice condition, $F(3,15)=22.0$, $p<.001$, and Choice condition x Stress interaction, $F(3,15)= 4.04$, $p<.05$. As Table 2 shows, 4-choice RTs were longer than 2-choice RTs (a significant difference in pairwise comparison ($F(1,5)=30.9$, $p<.01$), and the Inverted-Normal difference was larger in the 4-choice condition than in the 2-choice conditions.

Our chief interest was in finding whether the 2-choice conditions referred to as S, P, and M would be different from each other. Table 2 shows that choosing between sequences of different stress resulted in shorter RT than other types of choice. Yet a second ANOVA including only the 2-choice (S, P, and M) conditions indicated that the effect of Choice condition did not reach the significance level, $F(2,10)=2.91$. It must be noted, however, that the 2-choice conditions were different in regard to the subject's uncertainty about the first element of the sequence. In P-choice, the first tap of the sequence was predictable, for it was a weak tap when both sequences had normal stress, and a strong tap when both sequences had inverted stress. In contrast, initial stress was uncertain ($p= .50$) in the S- and M-choice conditions. The RT-difference between these two conditions proved to be significant, $F(1,5)=7.33$, $p<.05$.

Table 3 presents the individual RTs for the S-, P-, and M- choice conditions. The table shows that except for S5, no subject produced a shorter RT on either the M or P condition than on the S condition.

TABLE 3. Individual reaction times in 2-choice conditions (msec)

SUBJECTS	CONDITIONS		
	S	P	M
1.	326	358	378
6.	357	358	408
2.	449	515	501
4.	452	482	487
3.	527	637	597
5.	564	513	535

2.2.2.Errors. A sequence was classified as erroneous if the number of taps was other than eight, or if any of the taps required to be "weak" was as strong as, or stronger than, any of the taps required to be "strong". The proportion of errors was higher for inverted stress than normal stress (18.4% vs. 8.6%). It did not vary significantly across choice conditions (the percent errors for the S-, P-, M-, and 4-choice conditions was, respectively, 14.9%, 13.3%, 11.5%, and 15.7%, $\chi^2=2.71$, d.f.=3, NS).

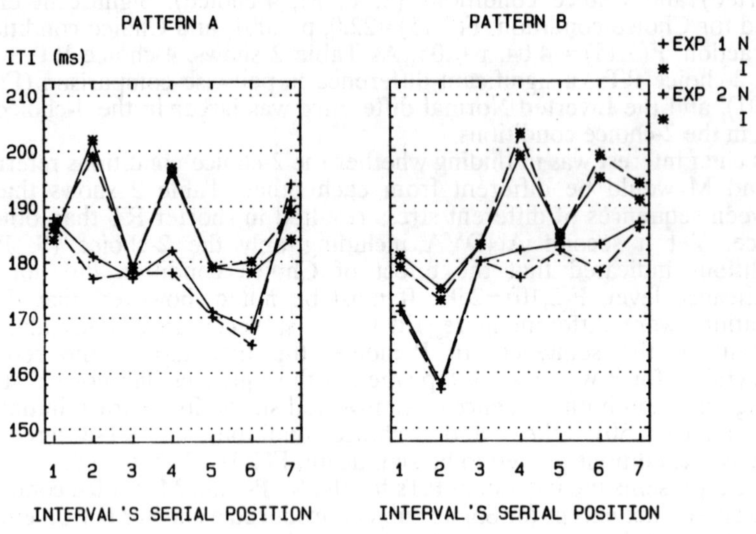

Figure 1. Intertap intervals in sequences with normal and inverted stress. Left: pattern A. Right: pattern B. Data were collapsed over choice conditions (Experiment 1) and sessions (Experiment 2).

2.2.3.Sequence timing. The over-all mean inter-tap interval was 178 msec in sequences with normal stress, and 186 msec in sequences with inverted stress. The individual mean inter-tap intervals were subjected to a Interval's serial position (7) x Stress type (2) x Choice condition (4) ANOVA. Intervals stemming from sequences of pattern A and those stemming from sequences of pattern B were treated separately. Only Interval's serial position yielded a significant main effect ($p < .01$ in both analyses) showing that the inter-tap intervals displayed systematic variations as execution of the sequence unfolded (Figure 1).

2.3. DISCUSSION OF EXPERIMENT 1

The pattern of RTs observed in the 2-choice condition appears consistent with the prediction that programming of a sequence of movements takes less time if the alternative sequences derive from the same serial pattern than if they derive from different patterns. This expectation was based on the hypothesis that motor programs may be generated from abstract plans for sequential organization. A plan of this kind could be selected in advance of receiving the choice stimulus only if the alternative sequences have a common underlying structure. The prediction was most clearly supported by the shorter RTs found in the stress-choice condition as compared to the mixed-choice conditions. There was a suggestion of a similar difference between the stress-choice and pattern-choice conditions. However, the difference might have been reduced because the initial stress was known in advance in the pattern-choice condition. Previous studies have indicated indeed that the time to choose between alternative sequences decreases as the first non-predictable element recedes from the beginning of the sequence towards its end (Rosenbaum, Hindorff and Munro,1987; Garcia-Colera & Semjen, 1988).

The advantage of choosing between sequences that share a common serial structure has been already demonstrated by Rosenbaum et al. (1984). However, the present results extend their observation in that structural identity between sequences was defined, in this experiment, not in terms of "surface descriptions" (such as symmetrical identity of the fingers to be used either by the left or the right hand), but in terms of relations (such as repetitions of, or transitions between, dichotomous events).

Although "relational" identity between tapping sequences with normal and inverted stress appears to facilitate the final motor program construction, generating a program for inverted stress may, in fact, be more demanding that generating a program for normal stress. An observation by Fraisse and Oleron (1954) suggests not only structural equivalence between sequences in which the strong and weak taps have been interchanged, but also greater difficulty in generating inverted stress. These authors had subjects produce repeatedly sequences of finger taps in which there were more strong than weak elements. In the course of sequence production, some subjects spontaneously substituted weak taps for strong taps, and vice versa, giving the sequence a more usual figure-ground relationship. In addition, Semjen, Garcia-Colera and Requin (1984) reported that the RT to choose between finger-tapping sequences was shorter under conditions of initial uncertainty regarding the single tap to be emphasized (normal stress), than the single tap to be de-emphasized (inverted stress). It must be noted, however, that in the Semjen and al.'s (1984) study the number of alternative sequences (2 or 4) and the type of stress (normal or inverted) had additive effects on the CRT, whereas in the present experiment, these factors interacted. The reasons for such a discrepancy need further experimental clarification.

3. Experiment 2

Experiment 1 suggested a higher-level code for sequential organization of the motor output. "Higher-level" means that the code is uncommitted with respect to the particular events, such as strong and weak taps, that fill in the serial order. Strong and weak taps constitute thus a specific implementation of the serial pattern. Experiment 2 was aimed at examining whether a higher-level code for sequential organization may subserve different motor domains, i.e., different kinds of implementation of the same serial pattern. Consider, for example, the following situation. A person is asked to pronounce a sequence of syllables such as Ba Bi Bi Ba and to make, shortly after having uttered the sequence, a speeded choice between several possible sequences of strong and weak finger-taps. If the same higher-level code may subserve the vocal and tapping implementation of the OXXO pattern underlying the vocal sequence, and if that code were to remain activated for some time after the vocal sequence has been pronounced, generating a strong-weak-weak-strong (or a weak-strong-strong-weak) tapping program should take less time than generating a program with a different serial structure. The effect we hypothesize here is reminiscent of the repetition effect, well-known from the literature on sequential CRT, that is, a shortening of the CRT on trial n if the stimulus presented and/or the response required on that trial are the same as on the preceding trial n-1 (e.g., Kornblum, 1973; Holender, 1980; Soetens, Boer and Hueting, 1985). However, in the aforegoing example there are two successive tasks rather than a single task, and that which may be repeated in performing these tasks is neither a stimulus, nor a response, but the serial structure of the sequence of responses called for in each of them. In the present experiment, then, the subjects of Experiment 1 produced the previously learned finger-tapping sequences in a 4-choice paradigm. At the beginning of each trial, they read aloud a sequence of syllables that either derived from the same serial pattern as the subsequently required finger-tapping sequence, or from a different pattern. The main hypothesis tested in this experiment was that programming a tapping sequence takes less time (yields a shorter CRT) if the vocal sequence and the subsequent tapping sequence derive from the same pattern.

3.1. METHOD

3.1.1. Task and procedure. Given the hypothesis that uttering a syllable sequence results in a persisting activation of some code for sequential organization, and that such a code contributes to the final programming of the required tapping sequence, the procedure used in this experience will be referred to as "priming".

The finger-tapping sequences and the visual symbols that identified them were the same as in Experiment 1. Three sequences of syllables were used. One of them derived from pattern A, another from pattern B, whereas the third was unpatterned (Table 4). Therefore, there were three priming conditions: Congruent (C, when the vocal sequence and the tapping sequence in a trial derived from the same pattern), Non-congruent (NC, a vocal sequence of pattern

A followed by a tapping sequence of pattern B, or vice versa), and Unpatterned (UP, the unpatterned vocal sequence followed by any of the tapping sequences).

TABLE 4. Vocal sequences used in Experiment 2

PATTERN	SEQUENCES
Unpatterned	Ba Ba Ba Ba Ba Ba Ba Ba
A	Ba Bi Ba Bi Ba Ba Ba Bi
B	Ba Ba Ba Bi Ba Bi Ba Bi

The syllable sequences were printed on a card, one sequence per row. Three LEDs, one for each row, were also mounted on the card at a distance of 1.5 cm to the left of the first syllable. The card was fixed on a slightly inclined rectangular panel which was disposed in such a way as to prolong upwards the plane of the stimulus display.

A trial was organized in the following way: (1) First, one of the LEDs on the syllable panel was illuminated for 1 sec. In response to this indicator, the subject read aloud the sequence of syllables located on the same row as the illuminated LED. (2) Second, the subject was presented with the temporal model, that is, 8 clicks of equal intensity separated by 160-ms intervals. The delay between the presentation of the initial LED and the first click of the temporal model was 3.5 sec. The uttering of the vocal sequence occurred during this period. (3) Finally, the subject was presented with the RS calling for one of the finger-tapping sequences. The delay between the last click of the temporal model and the RS was 1 sec. For (2) and (3), the trial was organized exactly as in Experiment 1.

Subjects were instructed to articulate clearly each syllable and to pronounce the whole sequence fluently. Furthermore, they were instructed to direct their gaze to the digital display as soon as possible in order to be ready to receive the RS. No measurement was taken of performance of the vocal sequences; however, the experimenter listened to them during the experimental session through an interphone.

3.1.2.Design. Subjects performed three trial blocks during each of three experimental sessions. A trial block consisted of 72 trials. In a block of trials each vocal sequence occurred the same number of times, at random, and was followed by each of the four tapping sequences with the same probability. Therefore, the occurrence of a particular tapping sequence could not be predicted from the particular vocal sequence required on a trial.

3.2. RESULTS

3.2.1.Reaction times. Individual data from within-session trial blocks were pooled. Median RTs were calculated and subjected to a 3 x 2 x 3 ANOVA that tested the effects of Priming condition (Congruent, Non-congruent, Unpatterned), Stress (Normal vs Inverted), and Session. Significant effects were found for Priming condition, $F(2,10)=7.06$, $p<.05$, and Session, $F(2,10)=5.15$, $p<.05$. The effect of Stress fell short of the significance level, $F(1,5)=6.23$, $p<.10$. Means of individual median RTs are shown in Table 5. It can be seen that congruent priming resulted in shorter RTs than did non-congruent priming or priming with an "unpatterned" sequence. Pairwise comparisons confirmed that the congruent-noncongruent and congruent-unpatterned differences were statistically significant ($p < .05$). However, the effect of priming tended to decrease over repeated experimental sessions, especially for sequences with inverted stress. In line with this observation, the Priming condition x Session x Stress interaction was fairly close to the .05 significance level, $F(4,20)=2.74$.

TABLE 5. Means of the individual median RTs (msec)

Session		Priming			Amount of priming effect		
		C	NC	UP	NC-C	UP-C	Mean diff.
1	N	592	608	620	16	28	22
	I	633	664	651	31	18	25
2	N	559	575	581	16	22	19
	I	603	609	601	6	-2	2
3	N	554	575	556	21	2	12
	I	620	622	635	2	15	9
Mean	N	568	586	585	18	17	18
	I	618	631	628	13	11	12

Note: C: congruent priming; N: normal stress
NC: non-congruent priming; I: inverted stress
UP: unpatterned priming;

3.2.2.Errors. Errors were scored as in Experiment 1. When normal stress was required, the proportion of errors was 10.6%. In this case, priming condition did not influence the error scores ($\chi^2=1.90$, df.=2, NS). When inverted stress was required, the proportion of errors was higher, 20.2%. In addition, the percent error varied with priming conditions: 15% with unpatterned priming, 20.3% with congruent priming, and 25.4% with non-congruent priming ($\chi^2=21.27$, df.=2, $p<.001$).

3.2.3.Sequence timing. The over-all mean inter-tap interval was 177 msec with normal stress, and 187 msec with inverted stress. These values reproduce almost exactly those observed in Experiment 1. The mean inter-tap interval in sequences of pattern A was 183, 184, and 183 msec for the congruent, non-congruent, and unpatterned priming, respectively. The corresponding values in sequences of pattern B were 181, 183, and 182 msec. Figure 1 shows that the time-profiles of the tapping sequences were the same as in Experiment 1. It appears, thus, that the priming procedure did not influence the temporal aspects of sequence execution.

3.3. DISCUSSION OF EXPERIMENT 2

The priming procedure used in this experiment was shown to be successful in that congruent priming resulted in shorter over-all CRT than non-congruent and unpatterned priming. It thus appears established that providing subjects with a model of sequential organization in one motor domain may benefit subsequent programming of a sequence of movements in a different motor domain. Such a crossover between motor tasks suggests, in accordance with the hypothesis tested in this experiment, that a higher-level code for sequential organization may subserve different motor domains, i.e., different motor implementations of the same serial pattern.

Although statistically reliable, the priming effect was not strong and was relatively unstable over experimental sessions. Several possible reasons may be suggested to account for the attenuation of the effect. Firstly, there was a rather long delay between the reading of the vocal sequence and the production of the tapping sequence. Such a delay may have exceeded, on some occasion, the period during which the sequential code could remain activated. Secondly, there was an asymmetry between the vocal and tapping implementations of patterns A and B. Whereas the tapping sequences were produced with either normal or inverted stress, the vocal sequences had a single form. Such an asymmetry could have introduced some bias in the amplitude of the priming effect on normal and inverted stress-sequences. Finally, patterns A and B were related in the sense that they could be considered as derived from each other by permutation of their first and second halves: oxox/ooox (A) giving ooox/oxox.(B). Recently, Gordon and Meyer (1987) showed that such permutations are relatively easy in terms of cognitive reorganization. Therefore, the difference between non-congruent and congruent priming may have been attenuated.

Because some or all of the above factors may have contributed to weaken the priming effect over successive sessions, it was decided to run a third experiment with the same subjects in order to find whether the priming effect may be restored by giving the aforementioned factors more appropriate values.

4. Experiment 3

4.1. METHOD AND DESIGN

The subjects were instructed to produce, in response to the previously learned visual symbols, only the first half of the corresponding tapping sequences. As Table 6 shows, pattern A reduced to OXOX, and pattern B to OOOX. Each pattern was produced with normal and inverted stress. Another modification concerned the sequences of syllables. No unpatterned vocal sequence was used. Instead, each of the basic patterns was implemented in two different ways by interchanging the syllables Bi and Ba (see Table 6). As in Experiment 2, the syllable sequences were printed on a card and, in each trial, illumination of a LED identified the sequence that was to be pronounced. Finally, the timing of events within a trial was also modified. The delay between onset of the LED that indicated the vocal sequence to be pronounced and the first click of the temporal model was reduced to 1.2 sec. The temporal model consisted of four clicks of equal intensity, paced at 160 msec. Subjects were urged to utter the sequences quickly so as to finish speaking about the beginning of the temporal model. The RS (one of the visual symbols learned in Experiment 1) was presented 700 msec after the last click of the model.

TABLE 6. Vocal and tapping sequences used in Experiment 3

PATTERN	Vocal sequences	Tapping sequences
A:	Ba Bi Ba Bi	W S W S (N)
0 X 0 X	Bi Ba Bi Ba	S W S W (I)
B:	Ba Ba Ba Bi	W W W S (N)
0 0 0 X	Bi Bi Bi Ba	S S S W (I)

Note: W:weak tap N: normal stress
S:strong tap I: inverted stress

Subjects performed 3 blocks of 64 trials. The vocal sequences were presented as equally likely alternatives in ech trial. Each vocal sequence was followed by one of the tapping sequences with the same probability. Therefore, there were as many congruent as non-congruent priming trials.

4.2. RESULTS AND DISCUSSION

4.2.1.Errors and sequence timing. Errors were defined as in Experiment 1. Sequences with normal stress displayed fewer errors when priming was congruent rather than non-congruent (5.2% vs. 10.4%). No such difference appeared for sequences with inverted stress (11.4% vs. 10.7%). The mean inter-tap interval was 173 msec in normal sequences, and 183 msec in inverted sequences, with both congruent and non-congruent priming.

TABLE 7. Means of the individual median RTs (msec)

	Priming		Amount of priming effect
	C	NC	NC-C
N	582	610	28
I	630	641	11

4.2.2.Reaction times. Individual data were pooled over trial blocks, median RTs were calculated and subjected to a 2 x 2 ANOVA that tested the effects of Priming condition (Congruent vs. Non-congruent) and Stress (Normal vs. Inverted). Only the effect of Priming condition was significant, $F(1,5)=24.86$, $p<.01$, reflecting shorter RT with congruent than with non-congruent priming (see Table 7). These results give new empirical support to our claim that performing a serial order in one motor task may help programming the same serial order in a different motor task.

The hypothesis that motor programs for serial movements may be generated from a "content-free" representation of the sequential order requires, in our experiments, that the positive priming effect not be limited to either the normal or the inverted version of the tapping sequence. The results from Experiment 2 and 3 showed, on the average, concomitant positive priming effects for N and I sequences. Subject-by-subject examination of the individual median RTs from Experiment 2 (three sessions) and from Experiment 3 showed that congruent priming shortened the RT for both stress versions of the sequences in most of the cases (60%). However, there were instances where shortening of the RT for the normal stress version was associated with lengthening of the RT for the inverted stress version (19%), or vice versa (17%). Finally, there were a few cases of lengthened RT for both stress versions (4%). The safest conclusion we may offer at this point is that faster responding to one stress version of the sequence may, but not necessarily, have a cost in responding to the other stress version of the sequence. (Note 1)

5. General Discussion

The purpose of the present set of experiments was to test whether content-free representation of a sequential order may intervene in programming rapid serial movements. It was reasoned that if such were the case, the sequential plan could be activated in advance of knowing the nature of the motor events that must implement the plan, and that final programming would thereby facilitated. The time to choose between alternative movement sequences was taken as an index of such a facilitation. Experiment 1 showed that choosing between sequences made up of weak and strong finger-taps takes less time if the alternative sequences derive from the same serial pattern (by interchanging strong and weak taps) than if they derive from different serial patterns. Experiments 2 and 3 showed that the time to choose between sequences of weak and strong finger-taps may be reduced by previously uttering a sequence of binary syllables which share the same serial pattern. Taken together, these results may be considered evidence for pre-activation of a "content-free" sequential plan that facilitates final motor programming.

In the 2-choice conditions of Experiment 1, subjects received information in advance about which sequences would be paired in a block of trials. Such information may be viewed as a precue allowing the subject to preselect a sequential plan when the alternative sequences derive from the same pattern. In contrast, Experiment 2 and 3 offered no basis for preselecting a sequential plan. Faster reacting under congruent priming may be ascribed to some remnant of the prior action (i.e., pronunciation of a sequence of syllables) facilitating the access to the required sequential plan.

Although the notion of "content-free" representation of serial order appears useful in explaining various empirical findings, it may be hard to tell what such a plan looks like. There are circumstances under which serial patterns are encoded and remembered in terms of generative rules (Restle 1970) or figure-ground relations (Garner, 1962). The organization of the input thus culminates in "serial concepts" (Jones, 1974). It is tempting to suggest that the organization of the output mirrors these processes in that it starts from rule-like representations or, alternatively, from transcodable templates, at least if the internal structure of the intended motor output allows for such representations.

Note 1.

In a follow-up study, Experiment 3 was replicated with the only difference in procedure that the probability of congruent priming was raised from .50 to .66. However, each vocal sequence was followed at equally often by the normal and inverted stress version of each tapping sequence. In accordance with our expectation that the probability bias will strengthen the advantage of congruent priming, normal sequences showed a positive priming effect (i.e., Non-congruent RT minus Congruent RT) of 34 msec, and inverted sequences, an effect of 46 msec. Further studies are necessary to find whether the structural identity between the priming (vocal) and the primed (tapping) sequences is critical in mediating the relative frequency effect.

Acknowledgement

Preparation of this paper was supported in part by a grant for international collaboration in research from the NATO Scientific Affairs Division (reference 0135/87) to Robert Gottsdanker.

References

Cohen A., Ivry, R., & Keele, S.W. (1990). Attention and structure in sequence learning. Journal of Experimental Psychology : Learning, Memory and Cognition, 16, 17-30.

Fraisse, P. & Oléron, G. (1954). La structure intensive des rythmes. L'Année Psychologique, 54, 35-52.

Garcia-Colera, A. & Semjen, A. (1988). Distributed planning of movements sequences. Journal of Motor Behavior, 20, 341-367.

Garner, W.R. (1962). Uncertainty and structure as psychological concepts. New York: Wiley.

Gordon, P.C. & Meyer, D.E. (1987). Control of serial order in rapidly spoken syllable sequences. Journal of Memory and Language, 26, 300-321.

Holender, D. (1980). L'effet de répétition dans les tâches de réaction de choix: préparation volontaire ou activation automatique? In: J.Requin (Ed.), Anticipation et comportement. Paris: Edition du CNRS, pp. 523-542.

Jones, M.R. (1974). Cognitive representations of serial patterns. In: B.H.Kantowitz (Ed.) Human information processing: Tutorials in performance and cognition. Hillsdale, N.J.: Lawrence Erlbaum Associates,pp.187-229.

Keele, S.W. (1968). Movement control in skilled motor performance. Psychological Bulletin, 70, 387-403.

Klapp, S.T. (1977). Response programming, as assessed by reaction time, does not establish commands for particular muscles. Journal of Motor Behavior, 9, 301-312.

Kornblum, S.(1973). Sequential effects in choice reaction time. In S. Kornblum (Ed.): Attention and Performance IV. New York: Academic Press, pp.259-288.

Lashley, K.S. (1951). The problem of serial order in behavior. In L. A. Jeffres (Ed.), Cerebral mechanisms in behavior. New York: Wiley, pp.112-136.

MacKay, D.G. (1982). The problems of flexibility, fluency, and speed-accuracy trade-off in skilled behavior. Psychological Review, 89, 483-506.

Pew, R.W. (1974). Levels of analysis in motor control. Brain Research, 71, 393-400.

Restle, F. (1970). Theory of serial pattern learning: Structural trees. Psychological Review, 77, 481-495.

Rosenbaum, D.A., Inhoff, A.W., & Gordon, A.M. (1984). Choosing between movement sequences: A hierarchical editor model. Journal of Experimental Psychology: General, 113,
372-393.

Rosenbaum, D.A., Hindorff, V, & Munro, E.M. (1987). Scheduling and programming of rapid finger sequences: Tests and elaborations of the Hierarchical Editor Model. Journal of Experimental Psychology: Human Perception and Performance, 13, 193-200.

Schmidt, R.A. (1975). A schema theory of discrete motor skill learning. Psychological Review, 82, 225-260.

Semjen, A., Garcia-Colera, A, & Requin, J. (1984). On controlling force and time in rhythmic movements sequences: the effect of stress location. Annals of the New York Academy of Sciences, 423, pp.168-182.

Shaffer, L.H. (1984). Motor programming in language production. In: H.Bouma and Don G.Bouwhuis (Eds),Attention and Performance X. Control of language processes. London, Lawrence Erlbaum Associates, pp.17-41.

Soetens, E., Boer, L.C., & Hueting, J.E. (1985). Expectancy or automatic facilitation? Separating sequential effects in two-choice reaction time. Journal of Experimental Psychology: Human Perception and Performance, 11, 598-616.

BIFURCATIONS IN POLYRHYTHMIC TAPPING: IN SEARCH OF FAREY PRINCIPLES

C. E. PEPER, P. J. BEEK & P. C. W. VAN WIERINGEN
Department of Psychology
Faculty of Human Movement Sciences
Free University, Amsterdam
The Netherlands

ABSTRACT. Broadly speaking, two approaches can be distinguished in the study of the timing of skilled movement behavior: the representational approach and the dynamical approach. Proponents of the former attempt to account for the timing of movements by invoking symbolic representations (i.e., motor programs with explicit timekeepers). Such models have recently been criticized for being largely data-driven and theoretically underconstrained. In contrast, proponents of the latter approach seek to explain temporal order in movement in terms of physical self-organization (i.e., time as an emergent property) rather than symbolic representations. In the present paper, bifurcations in polyrhythmic tapping are analyzed according to the branching structure of the Farey tree. This tree is a generic mathematical object that summarizes all possible mode locks that complex dynamical systems may attain. Its branching structure coincides with known bifurcation routes in both mathematical and natural systems. Data on tapping 5:2 and 5:3 polyrhythms with increasing cycling frequency reveal that cycling frequency may be conceived as a control parameter. If cycling frequency was increased, bifurcations from mode locks with larger integer ratios to mode locks with smaller integer ratios often occurred in the more skilled subjects. These transitions were generally characterized by a tendency to move to the right (1:1) flank of the Farey tree instead of to its left flank (0:1). Evidence is reported that the stability of tapping polyrhythms is also a function of training.

1. Introduction

Interlimb coordination is a ubiquitous aspect of human and animal movement. In many natural as well as cultural activities, the limbs move rhythmically together in certain phase and frequency relations. Understanding such rhythmic interlimb coordination is a major goal of movement science and a variety of research strategies have been pursued to achieve it. Amongst these research paradigms, the activity of bimanual tapping is rather unique, because it appeals, at the same time, to the presently competing notions of symbolic processes and physical self-organization. A general aim of this paper is to explore how this property of bimanual tapping might be exploited in the construction of a theoretically relevant research program. We will do so by investigating to what extent the temporal order observed in polyrhythmic tapping proves amenable to an analysis in terms of physical self-organization.

J. Requin and G. E. Stelmach (eds.), Tutorials in Motor Neuroscience, 413–431.

2. Representational approaches to bimanual tapping

Usually, in bimanual tapping, each hand taps a cadence at a certain frequency. The temporal structure of the taps of one hand defines the cadence of that hand, while the relation between the one cadence and the other cadence defines the overall rhythm of the activity. Two broad types of rhythms are distinguished: simple rhythms and polyrhythms. In simple rhythms the ratio of the frequencies of the tapping hands can be represented as an integer ratio with 1 as denominator (1:1, 2:1, 3:1, etc.). For polyrhythmic tapping this is impossible because there is no common divisor. Examples of polyrhythmic ratios are 3:2, 4:3, 5:2, 5:3, 5:4, etc. Without exception polyrhythms are more difficult to perform than simple rhythms.

After extensive practice, however, skilled pianists have been shown to perform complex polyrhythms with ratios such as 5:3 and 5:4 very well (Shaffer, 1981, 1982). On the basis of this observation, Shaffer (1982) suggested that enhanced performance in the production of polyrhythms might be due to an increased ability to utilize independent timing structures for each hand, although later studies failed to provide evidence for this. Deutsch (1983), for example, concluded that the performance of her musically trained subjects reflected the development of hierarchical motor programs representing the polyrhythmic patterns as an integrated whole, and that performance was inversely related to the complexity of these programs. Additional evidence for integrated, hierarchical forms of polyrhythms was presented by Jagacinski, Marshburn, Klapp and Jones (1988), who formulated and tested several models of motor organization. They not only distinguished between chain-like and hierarchical organization, but also between integrated and parallel organization. The latter distinction refers to cases in which (i) the cadences of both hands together are controlled as a single temporal pattern (integrated), and (ii) the cadences are controlled separately (parallel). Furthermore, independent and multiplicative hierarchies have been distinguished. Figure 1 depicts the three integrated models schematically. R_1, L_1, etc. represent internal events corresponding to right and left handed tapping responses. Due to motor delays (M_1-M_5), the actually observable intertap intervals are I_1-I_5. The internal timekeeper intervals are represented by A-D. Similar diagrams can be constructed for parallel organizations. In such models, the right and left handed responses are controlled by separate timekeepers, so that the timekeeper intervals are always situated between either two right handed or two left handed responses. Figure 1 illustrates the difference between the two hierarchical types. In an independent hierarchy, as depicted in Figure 1b, B and D do not depend on each other. In Figure 1c, however, PD is a certain proportion of D, and therefore depends on the duration of this interval.

To determine which type of organization applies to bimanual tapping the authors employed Vorberg and Hambuch's (1978) modification of the Wing and Kristofferson (1973) model. The analysis, based on the variance-covariance structures of these models, suggested that, in all subjects, the organization is integrated. Instead of using separate timing mechanisms for each hand, the task is performed as a whole. The multiplicative hierarchical model turned out to be superior. Thus, it was concluded that rhythmic patterns are represented in an integrated hierachical structure or motor program. We may conclude, therefore, that if entirely independent timing of hands during polyrhythmic tapping is at all feasible, this would require very large amounts of practice, as, indeed, is testified to by proficient musicians (Shaffer, 1982; Vorberg & Hambuch, 1984).

It could be argued that the modeling of polyrhythmic tapping by means of motor programs with explicit timekeepers—as in the cognitive approach exemplified above—is, in large part, data-driven, and the resulting models are hardly constrained by principled theoretical considerations. Such programs, so it is assumed, prescribe the timing of the behavior. Thus, there is a fundamental isomorphism between the observed order and the order contained in the program. This isomorphism of programs and behavior has been the target of much criticism. In explaining regularities in movement with the help of characteristics of central devices like motor programs, representational models of movement run a risk of begging the question and ending up in explanatory regress, because the origin of the characteristics of these devices is left unaccounted for.

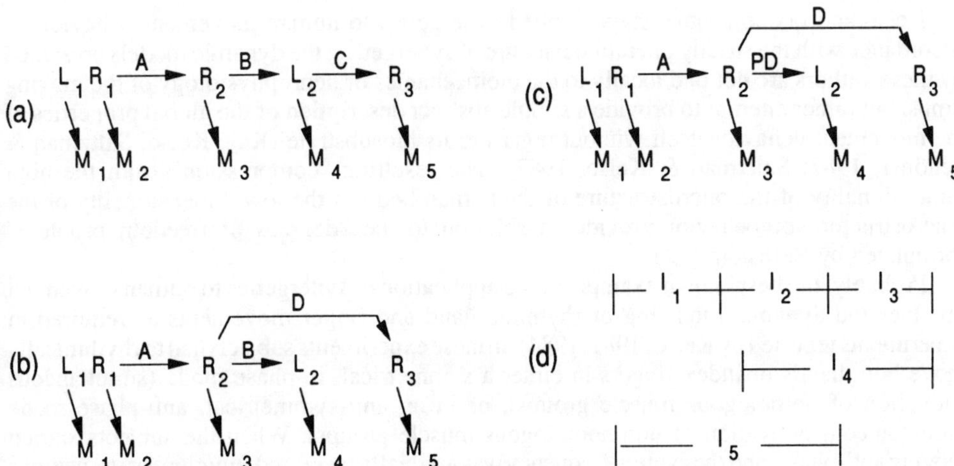

Figure 1. Three models of motor organization. (a) integrated chained; (b) integrated independent hierarchical; (c) integrated multiplicative hierarchical; (d) measured intervals (adpated from Jagacinski, et al., 1988). See text for details.

3. Synergetic approaches to bimanual tapping

During the last decade or so, an alternative approach to movement coordination has been developed which rejects representational accounts of motor behavior, and resorts instead to principles derived from the physics of self-organizing systems. This new approach may be conceived as natural physics (Kugler, 1986; Whiting, Meijer & van Wieringen (Eds.), 1990) or synergetics (Haken, 1977; 1983; Kelso, 1989; Schöner & Kelso, 1988a, 1988b, 1988c). It represents, speaking generally, an interdisciplinary field of study concerned with pattern formation in complex, non-equilibrium systems. In the context of living systems, its significant contribution has been in demonstrating that pattern formation in such open, dissipative systems does not require the existence of independent pattern generators that are ultimately responsible for the observed order, but may be governed by general principles that have no material (symbolic) representation within the system itself. Spatial and temporal patterns may arise spontaneously (i.e., self-organize) when one or a few so-called "control parameters" are changed. These control parameters are unspecific in the sense that they do not prescribe the resulting patterns; they only lead the system through its respective collective states. The macrobehavior of systems with very many degrees of freedom may be completely described by low-dimensional dynamics of a collective variable ("order parameter") characterizing the emergent patterns (Kelso, 1989).

Many examples of self-organization have been found in physical, chemical, and biological systems (Haken, 1983). The analogies between such self-organizing systems are drawn at a high level of abstraction, i.e., at the level of the collective behavior of the system as described by the dynamics of its order parameter(s). When these systems are analyzed in more and more detail down to their subsystems, more and more differences may be revealed between them (Haken, 1983).

Kelso and his colleagues have applied synergetics to human movement behavior. In accordance with the strictly operational nature of synergetics, the dynamic models presented by these authors are not tied locally to the biomechanics or neurophysiology of the moving limbs, but rather attempt to provide a simple abstract description of the global properties of the movement behavior itself without regard to its biosubstrate (Kay, Kelso, Saltzman & Schöner, 1987; Saltzman & Kelso, 1987). The resulting "compression" from the high dimensionality of the microstructure of the human body to the low dimensionality of the macrostructure of behavior provides a solution to the "degrees of freedom problem" formulated by Bernstein (1967).

Probably, the best known example of the application of synergetics to human movement involves the dynamic modeling of rhythmic hand and finger movements as required in experiments reported by Kelso (1981, 1984). In these experiments subjects had to rhythmically move both hands or index fingers in either a symmetrical, in-phase mode (simultaneous activation of homologous muscle groups), or in an anti-symmetrical, anti-phase mode (simultaneous activation of non-homologous muscle groups). When the subjects started moving anti-phase, and the cycling frequency was gradually increased, involuntary transitions to in-phase movements occurred at some critical frequency. Given the characteristics of the transition this change in movement pattern qualified as a spontaneous self-organizing process, as has been described earlier for a number of other open, complex systems by Haken in his pioneering work on synergetics (1977, 1983). After having defined relative phase (the phase difference between the moving limbs) as the order parameter describing the collective behavior of the system and cycling frequency as the control parameter, Haken, Kelso and Bunz (1985) succeeded in modeling the hand and finger movements in terms of a system of non-linearly coupled non-linear oscillators.

Recently, Kelso and his co-workers have also addressed the production of polyrhythms within a synergetic framework. In one of these experiments (deGuzman & Kelso, in press; Kelso & deGuzman, 1988) the (poly)rhythms were brought about in such a way that the driving and the driven oscillating finger were clearly defined. Instead of instructing the subjects to move their index fingers (poly)rhythmically, the researchers invited them to track, with their right index finger, a visual signal moving back and forth on the screen of an oscilloscope. The frequency of the displayed movements was either 1.5 Hz or 2.0 Hz. After about 15 s the signal was turned off and the subjects had to maintain the frequency and amplitude of the finger movements for another 70 s. Meanwhile, the left index finger was passively driven at a certain frequency, forming frequency ratios with the frequency of the right index finger of 4:3, 3:2, 5:3, 2:1, 5:2 or 3:1.

It was found that the average (over 5 subjects) of the produced mean frequency and its variation (SD) were critically affected by the ratio between driving and tracking frequency.

When (in the 1.5 Hz condition) these ratios were 4:3 and 3:2, the actually produced frequencies shifted in the direction of 1:1—a nearby attractor. When the frequency ratio was 5:2, the produced frequency was attracted to 3:1.

The simple frequency ratios (2:1, 3:1, 3:2) resulted in significantly less variable frequencies than did the other ratios. The 5:2 condition proved to be less stable than either the 4:3 or 5:3 condition which were not statistically different.

During the second run of 15 trials learning effects became apparent. For example, with the ratio 4:3 the influence of the neighboring 1:1 state was attenuated relative to the first run, and the frequency distribution was changed from bimodal (maxima at 4:3 and 1:1) to unimodal (maximum at the required 4:3 ratio). Learning, therefore, was interpreted as the creation of an attractor corresponding to the required ratio with an accompanying reduction of the impact of the intrinsic attractor state.

4. Mode-locking structures in dynamic systems

As is illustrated in deGuzman and Kelso's efforts to model the coordination of bimanual polyrhythmic movement, the analysis of dynamic systems, especially systems of coupled oscillators, is often simplified by the use of circle maps. Not only do these maps capture qualitatively the kinds of behavior found in mathematical systems, but, in addition, their scaling behavior quantitatively carries over to real systems. Consequently, circle maps can be used to represent the behavior of complex systems in simpler fashion. It is much easier to identify periodic, quasi-periodic and chaotic solutions by iterating the map (mapping it onto itself) than by numerical integration of the underlying differential equations (Beek, 1989; Jensen, Bak & Bohr, 1984; May, 1976).

A circle map is a first order difference equation. Such equations are discrete, whereas differential equations are continuous. The system's behavior is studied in discrete steps. A general expression of a difference equation is

$$X_{t+1} = F(X_t) \tag{1}$$

Thus, the value of variable X at time t + 1 is a function of its value at time t. Because of this mapping of the variable onto itself, circle maps are called "iterative" or "self-referential" maps.

Mode-locking behavior in systems of coupled oscillators can be studied very well with such circle maps, which, in general, are defined through

$$\theta_{n+1} = f_\Omega(\theta_n) = \theta_n + \Omega + g(\theta_n) \tag{2}$$

where

$$g(\theta_n) = g(\theta_{n+1}) \pmod 1 \tag{3}$$

The variable θ_n represents the phase of the oscillating system measured stroboscopically at periodic time intervals $t_n = 2\pi/\omega_n$, using the frequency of the external oscillator as a clock. A phase shift $\theta_n \rightarrow \theta_{n+1}$ represents a full rotation and therefore functions as the periodic property of g (eq. 3). The term Ω represents the ratio between the period of the external forcing and the eigenperiod of the forced oscillator ($\Omega \equiv T_e/T_o$) in the absence of the nonlinear coupling g.

The iteration of the map is conveniently described by its *winding (or rotation) number* W, defined as $W(K, \Omega) = (\theta_n - \theta_o)/n$ in the limit as the iteration n approaches infinity. The winding number is the time-averaged rate of rotation of the angle θ. Under iteration the variable θ_n may converge to a series which is either periodic, with rational winding number W = N/M (N and M are integers); quasiperiodic with irrational (non-integer) winding number; or chaotic, where the series behave irregularly. The winding number depends on the value of the coupling parameter K in the function $g(\theta_n)$, as in, e.g., $g(\theta_n) = -(K/2\pi)\sin 2\pi\theta_n$, and on the value of Ω. By increasing the control parameter K transitions between winding numbers can be induced and in this way routes to chaos can be studied. The ranges of Ω's within which certain rational winding numbers are found are given as a function of K in a *régime diagram.* The coupled oscillators are mode locked in régimes which are known as Arnold tongues (Beek, 1989; Glass & Mackey, 1988) as a function of the control parameter. Figure 2 represents such stable entrainment or resonance regions. As the coupling strength increases the higher order ratios bifurcate to lower order ratios. If this increase of K continues even very stable ratios such as 1:1 and 2:1 bifurcate to chaos (when all the resonance tongues begin to overlap). Between the entrainment regions quasi-periodic behavior is observed, in which the ratio Ω is an irrational number.

The larger the surface area of the Arnold tongue, the more stable is the ratio. A régime diagram is structured as follows: Between every two entrainment tongues corresponding to the ratios N_1:M_1 and N_2:M_2 another, smaller, tongue is situated: $(N_1 + N_2)$:$(M_1 + M_2)$ (Cvitanovic, Shraiman & Söderberg, 1985; Glass & Mackey, 1988; González & Piro, 1985). This generally implies that higher order integer ratios are less stable than lower order ones. As a function of the control parameter K, ratios can become more or less stable, or the stability can even completely disappear to give way to another, more stable, ratio, depending on the initial conditions and the structure of the régime diagram. Thus, transitions from less stable ratios to more stable ratios can be induced by manipulating the coupling strength.

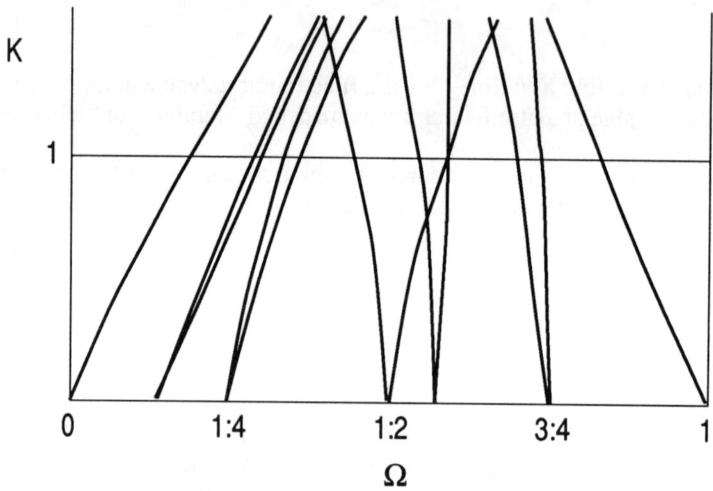

Figure 2. Schematic (fictitious) régime diagram for circle map in (Ω,K) space. Note the Arnold tongues where the winding number W assumes locked rational values.

The ratio $(N_1 + N_2):(M_1 + M_2)$ is the result of the so-called Farey sum of the ratios $N_1:M_1$ and $N_2:M_2$:

$$N_1:M_1 \oplus N_2:M_2 = (N_1 + N_2):(M_1 + M_2) \qquad (4)$$

where the symbol ⊕ denotes the Farey sum (Cvitanovic et al., 1985; Keener & Glass, 1984; González & Piro, 1985). All rational numbers between 0 and 1 (all possible entrainment régimes) can be organized in the so-called *Farey (or Stern Peirce) tree* (Allen, 1983), by application of this summation, starting with the parents 0:0 and 1:1. The zero*th* level is defined as 0:1 ⊕ 1:1 = 1:2. Through Farey addition of these ratios the first level of the tree yields 0:1 ⊕ 1:2 = 1:3 and 1:2 ⊕ 1:1 = 2:3. In this way each subsequent level of the tree can be constructed (see Figure 3). At the *nth* level of the tree 2^n elements are present (Beek, 1989).

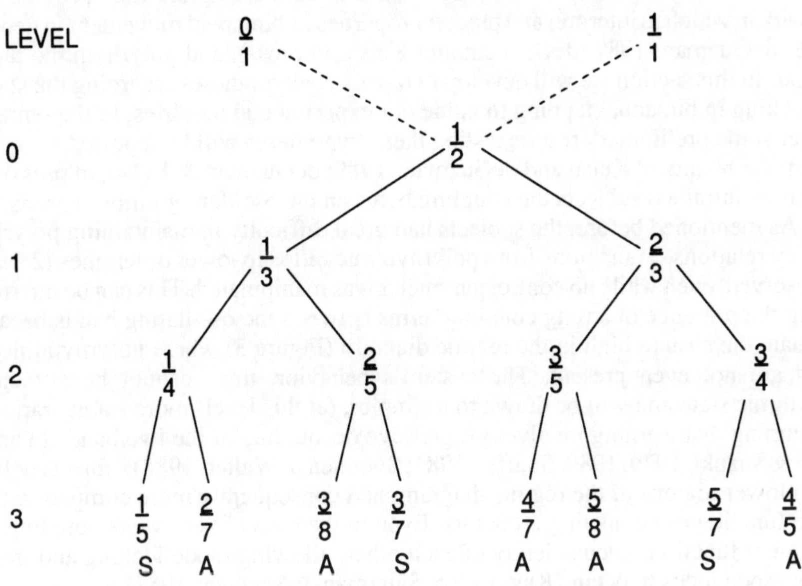

Figure 3. The Farey tree. S = symmetric (sum of numerator and denominator is even), A = asymmetric (sum of numerator and denominator is odd). The series SAASAA re-occurs layer after layer, pointing to the high degree of symmetry in the tree.

The Farey tree is a generic mathematical object that summarizes all the possible mode locks that complex dynamical systems may attain. It derives its significance in the analysis of such systems from the fact that its branching structure coincides with known bifurcation routes in both mathematical and natural systems.

If the coupling strength (control parameter) increases, transitions occur from higher order ratios to one of their (lower order) primitives. Starting with a higher order ratio an increasing coupling strength thus effects bifurcations along the branches of the tree (Allen, 1983). The path 1:1, 1:2, 2:3, 3:5, 5:8 in Figure 3 represents the Fibonacci ratio series, which is formed by dividing each element of the Fibonacci series by its successor. The Fibonacci series starts with 0 and 1. Every successive element equals the sum of the two preceeding elements: 0, 1, 1, 2, 3, 5, 8, 13, etc. (West & Goldberger, 1987). The Fibonacci ratio series represents a frequently occurring stable pattern in nature. Generally, the Fibonacci ratio series, which provides the best rational approximation to the golden mean $(\sqrt{5}-1)/2$, appears to be the most stable path, along which periodic behavior is preserved the longest. Experimentalists who are able to preset winding numbers therefore often induce this path in their experiments (e.g., Feigenbaum, Kadanoff & Shenker, 1982; Fein, Heutmaker & Gollub, 1985; Stavans, Heslot & Libchaber, 1985).

5. Polyrhythmic tapping and the Farey tree: Two hypotheses

The related conceptions of the Farey tree and the régime diagram may provide a useful framework in which to interpret and predict properties of bimanual movement in general (see Kelso & deGuzman, 1989; deGuzman and Kelso, in press), and polyrhythmic tapping in particular. In this section we will develop two working hypotheses regarding the stability of mode locking in bimanual tapping to guide our experimental inquiries. In the remainder of the paper some preliminary results testing these hypotheses will be reported.

From the results of Kelso and deGuzman (1989; deGuzman & Kelso, in press) one can infer that in untrained subjects the coupling between the oscillating limbs is probably quite strong. As mentioned before, the subjects had great difficulty in maintaining polyrhythmic frequency relations. Transitions from polyrhythmic ratios to lower order ones (2:1 and 1:1) were observed, even while no control parameter was manipulated. This can be interpreted as indexing the presence of strong coupling terms between the oscillating hands because this will situate the system high in the régime diagram (Figure 3) where polyrhythmic Arnold tongues are not even present. The system's behavior, thus, cannot be attracted to a polyrhythmic state and will be drawn to a simpler, (at this level) more stable, ratio.

Assuming that learning involves progressive decoupling of the two hands (Yamanishi, Kawato & Suzuki, 1979, 1980; Shaffer, 1981; Swinnen & Walter, 1988) subjects will be able to attain lower regions of the régime diagram, and consequently more complex ratios may begin to function as competing attractors. Even in the case of very weak coupling the two limbs do not function independently of each other, allowing mode locking and transitions between mode locks to occur (Kay, Kelso, Saltzman & Schöner, 1987).

In all earlier studies on bimanual polyrhythmic movements subjects moved their limbs with constant movement frequencies. Although some researchers studied the stability of different mode locks performed with different cycle times, no explicit attempts were made to account for the effect of tapping rate on performance, leaving it unclear what motivated the experimental manipulation. On the basis of the results of Kelso's experiments on phase transitions in rhythmic hand and finger movements, however, it may be argued that movement frequency qualifies as a control parameter in polyrhythmic movements. One could also argue, on the basis of circle map considerations, that the coupling strength is another control parameter for polyrhythmic tapping, but it is not unlikely that coupling strength is, in fact, a function of movement frequency (and perhaps other factors, such as skill). If so, then frequency, being the truly unspecific (indirect) parameter, is the appropriate control parameter to consider (also because, in polyrhythmic tapping, it is impossible to prescribe a particular coupling strength, whereas it is possible to prescribe a frequency).

Stability will be tested by gradually decreasing the cycle time for each ratio and determining at which cycle frequency performance of the required polyrhythm can no longer be sustained. In case of a purely autonomous system it would be expected, without any further knowledge of the system, that 5:3 will be more stable than 5:2, as is usually the case in natural systems obeying the Farey tree. On the other hand, several studies on polyrhythmic tapping (e.g., Deutsch, 1983) demonstrated that 5:2 is more stable than 5:3. It will be investigated if the difference in stability between 5:2 and 5:3 is a function of practice. Because the coupling between the two hands may be assumed to be strong in the early stages of learning, the mode locks will have to be realized by external forcing, which will be easier for the 5:2 than for the 5:3 ratio because 5:2 has a more symmetric distribution of taps than 5:3 (5:2 has a smaller common multiple than 5:3). However, if this coupling is weakened by training, the system will begin to display more autonomous behavior, resulting in more stable performance of the 5:3 ratio than the 5:2 ratio.

These considerations leave us with two tentative hypotheses:

Hypothesis 1: With tapping frequency as the control parameter, bifurcations in polyrhythmic tapping will occur that reflect the branching structure of the Farey tree. The induced transitions will have the following characteristics: (i) they are transitions to lower order integer ratios, located nearby in the Farey tree; (ii) they have a tendency to move more towards the 1:1 flank of the Farey tree than to the 0:1 flank, because the former still reflects coordinated tapping and the latter does not. The ratios to be tested are 5:2 and 5:3. They are positioned in the inner branches of the Farey tree at the same level.

Hypothesis 2: Whereas untrained subjects will maintain the 5:2 polyrhythm up to higher tapping frequencies than the 5:3 polyrhythm, the opposite will hold for well-trained subjects.

6. Experiments

6.1. EXPERIMENT 1

METHOD

Subjects
Four subjects, 3 males and 1 female, participated in the experiment. Their ages ranged between 25 and 30 years. They were all unfamiliar with making music (none of them actively played or had played an instrument), untrained in tapping, and naive with regard to the experiment. They were paid a small fee per hour for their participation.

Apparatus and set-up
Stereo signals were generated by means of a Modula-2 program on an Amiga 500 and presented through a headphone (Sennheiser HD222). The hand movements were registrated with a Selspot-like system (MILCU) with a sampling rate of 122 Hz. A LED was attached to the tip of each middle finger. Data acquisition with MILCU was controlled with a program run on an Olivetti Primavera M24, which was linked to the Amiga to guarantee optimal synchronization between signal presentation and movement registration.

Signals
Three types of signals were presented. Two signal trains with 50 ms sine-wave beeps were generated. The difference in pitch of the two beep trains was small: 440 Hz (A) for the fast

train beeps and 554 Hz (C#) for the slow train beeps. The cadances of the two beep trains related in polyrhythmic fashion, either as 5:2 or 5:3. In all cases the fast train of the ratio started with 500 ms intervals between the middle of two adjacent beeps. The interval time of the slow train was easy to calculate. In one cycle the fast train presented 5 beeps and the slow train either 2 or 3 (see Figure 4).

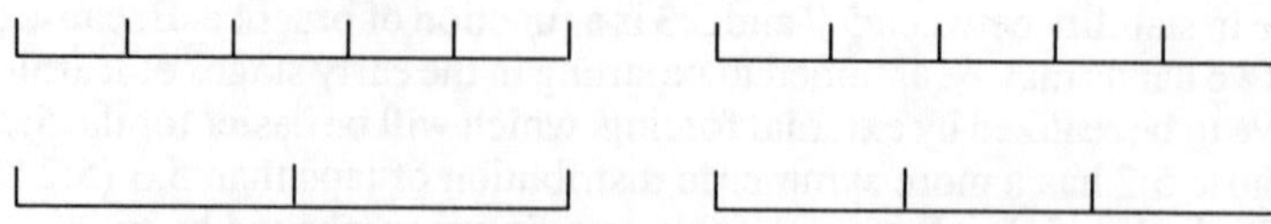

Figure 4. Schematic representation of a 5:2 and a 5:3 cycle. The vertical bars indicate the temporal location of the beeps.

1. Signal type 1 consisted of 25 cycles with constant rate. The signal time was 62.5 s.
2. Signal type 2 started with 10 cycles having a constant rate. Subsequently, the rate was increased by 4%, so that each consecutive cycle was 4% shorter than its predecessor. The signal stopped when the interval time between the beeps of the fast train was shorter than 100 ms. Thus, the total signal duration was 75 s.
3. Signal type 3 was structured in the same way as signal type 2. However, after 8 cycles (when the rate was still constant) only the first beeps of each cycle (when the two trains blip at the same time) were presented. Also in this signal type each new cycle period was decreased by 4%, starting with the eleventh. Of course, the total signal duration was again 75 s.

Procedure

The subject, wearing headphones, sat in an upright position with the underarms resting on a black tabletop surface. He or she was instructed to synchronize tapping with the left hand to the beeps in the left channel and with the right hand to the beeps in the right channel. In case of signal type 3, the subject was instructed to continue tapping the prescribed rhythm even when only the first beeps of each cycle were presented, and to synchronize the first taps of each cycle to these beeps. The subject was encouraged to sustain tapping as long as possible, and to continue even if the required rhythm was lost. The subject was further instructed to tap with the hands and not just with two fingers.

One ratio block consisted of 7 signal presentations. First, signal type 1 was presented in a practice trial. Second, the subject practised tapping with an increasing frequency as specified by signal type 2. Signal type 3 was used for the test trials. The advantage of this signal type over signal type 2 is that if the subject bifurcates to another rhythm the signal beeps cannot interfere with the performed rhythm. The stability of the 5:2 and 5:3 ratios was tested over 5 trials. After each trial the subject was allowed to rest for 60 s. Both ratios were presented in an alternating way, four blocks of each. Halfway through the session a 10 minute break was given. One session lasted about 2.5 hours. Two sessions were conducted. The experimental conditions were counterbalanced.

Analysis of stability

Stability was operationalized in terms of the bifurcation frequency, i.e., the critical cycle frequency at which performance of the required frequency ratio could no longer be maintained. This frequency was further defined as the frequency of the fast moving hand at the moment of occurrence of the second of the last two adjacent correctly performed cycles. In the algorithm used to determine this moment in time the required frequency ratios were considered to be accurately performed as long as the actually realized ratios did not deviate more than 0.10 from the required ratio, i.e., the 5:2 ratio was considered to be present as long as the actual ratio was between 2.40 and 2.60, and 5:3 was considered to be present as long as the actual ratio was between 1.57 and 1.77. When the required polyrhythm was not realized at all, a default value of 2 Hz for the bifurcation frequency was used.

Note that both conditions (5:2 and 5:3) were compared for the hand tapping five times within a period. The differences between these frequencies were tested in a 4 X 2 X 8 X 5 ANOVA with the factors subject, ratio, block, and measurement.

RESULTS

Stability

Figure 5 presents the bifurcation frequencies for the two ratios, averaged over the four subjects, as a function of blocks.

The results of the ANOVA showed significant main effects for the factors ratio ($F(1,3) = 144.70$, $p < .005$) and block ($F(7,21) = 3.92$, $p < .01$), but no significant interaction.

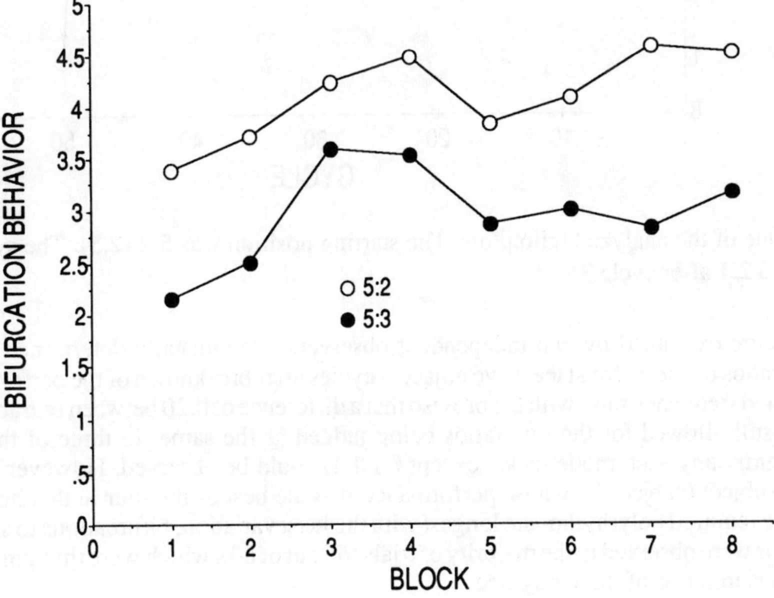

Figure 5. The bifurcation frequencies per block for each ratio condition averaged over subjects. Blocks 1-4 were tested on day 1 and blocks 5-8 on day 2.

It has to be concluded, therefore, that the 5:2 polyrhythm was performed more stably than the 5:3 rhythm, and that this difference did not change as a function of practice in non-musicians. The latter conclusion, however, may be biased due to the use of the default value of 2 Hz in case of an unsuccessful trial, which occurred more often in the 5:3 than in the 5:2 condition.

Analysis of bifurcation routes
For each subject the actual frequency ratios realized over the course of each experimental trial were plotted as a function of the successive cycles in that trial (see Figure 6 for an example).

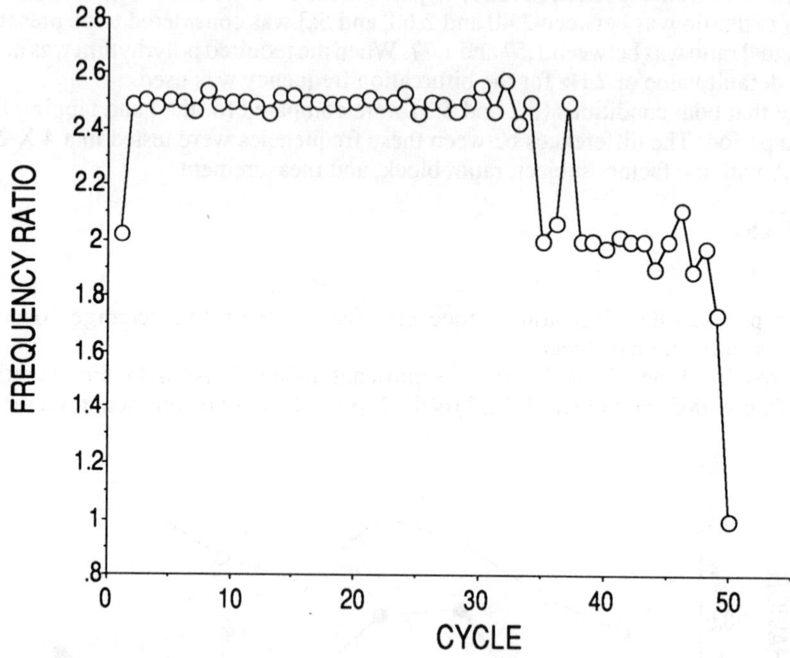

Figure 6. One of the analyzed ratio plots. The starting position was 5:2 (2.5). The system bifurcates to 2:1 after cycle 34.

The plots were examined by two independent observers, who visually determined which frequency ratios occurred for at least two adjacent cycles after breakdown of the performance of the required frequency ratio with the proviso that a difference of 0.20 between two adjacent ratios was still allowed for the two ratios being judged as the same. In three of the four subjects, hardly any such mode locks, except for 1:1, could be observed. However, in the remaining subject (subject 1), whose performance was the best of the four in that he could maintain the required polyrhythm the longest with the least variation, bifurcations to smaller integer ratios were observed in the majority of trials (66 out of 80), which were thus amenable to an analysis in terms of the Farey tree.

Bifurcation routes
Table 1 presents the bifurcation routes identified under the two frequency ratio conditions for subject 1. To get an indication of the inter-observer reliability in visually determining the mode locks, the percentage of mode locks identified as such by the two observers relative to the total number of mode locks identified by either one or both observers was calculated. This resulted in a percentage agreement of 79.8.

Table 1. The frequency of occurrence of bifurcation routes observed in subject 1. The complete Farey route corresponding to the starting ratio is displayed at the top of each series.

Ratio 5:2					Ratio 5:3				
route				frequency	route				frequency
5:2	3:1	2:1	1:1	0	5:3	3:2	2:1	1:1	0
5:2		2:1	1:1	22	5:3	3:2		1:1	3
5:2		2:1		1	5:3		2:1	1:1	4
5:2			1:1	12	5:3	3:2			2
					5:3		2:1		7
					5:3			1:1	14
					5:3	4:3	3:2	1:1	1
			total	35				total	31

In the 5:2 condition all the observed bifurcation routes followed the branches of the Farey tree, although the complete 5:2, 3:1, 2:1, 1:1 route was not found.

In the 5:3 condition all the observed routes but one (5:3, 4:3, 3:2, 1:1) occurred along branches of the Farey tree; again the complete route (5:3, 3:2, 2:1, 1:1) was not among them.

Whereas transitions to 3:2 were observed in 5 out of 31 observed bifurcation routes in the 5:3 condition, no such transitions occurred in any of the 35 observed bifurcation routes in the 5:2 condition. This difference between the two conditions was significant when tested nonparametrically ($\chi^2 = 4.02$, df = 1, $p < .05$). It is this result which provides, admittedly weak and provisional, evidence that bifurcations in bimanual rhythmic movements may indeed be predicted by the structure of the Farey tree.

6.2. EXPERIMENT 2

The absence of an increasing stability of the 5:3 polyrhythm relative to the 5:2 polyrhythm in Experiment 1, and the lack of bifurcations between mode locks in three of the four subjects, might have been due to the fact that the amount of practice of the untrained subjects had been too limited to reveal such effects. In a number of trials, no steady state performance was attained at all. Therefore, it was decided to replicate the experiment with two musically trained subjects, who received (in order to arrive at a fuller picture of the learning process) 15 blocks of trials instead of 8.

METHOD

Subjects were two experienced male drummers (subjects A and B), 26 and 27 years of age. All aspects of the method were identical to those of Experiment 1, except for the number of blocks of trials in each ratio condition (15 as opposed to 8). In each of five consecutive days three blocks of trials with ratio 5:2 and three blocks of trials with ratio 5:3 were presented alternatingly. The fast cadence was always performed with the right hand.

RESULTS

Stability
The bifurcation frequencies of subjects A and B are displayed in Figures 7 and 8.

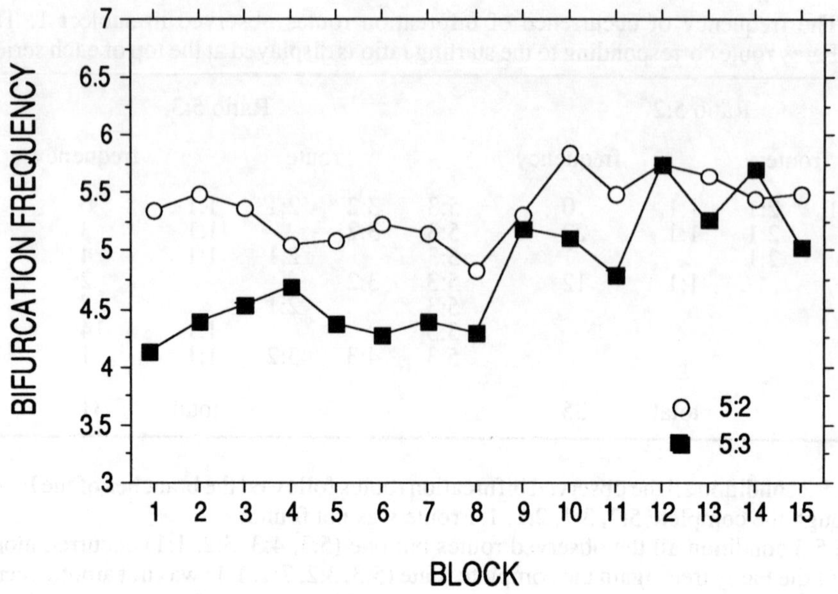

Figure 7. Bifuraction frequencies of subject A as a function of ratio and block.

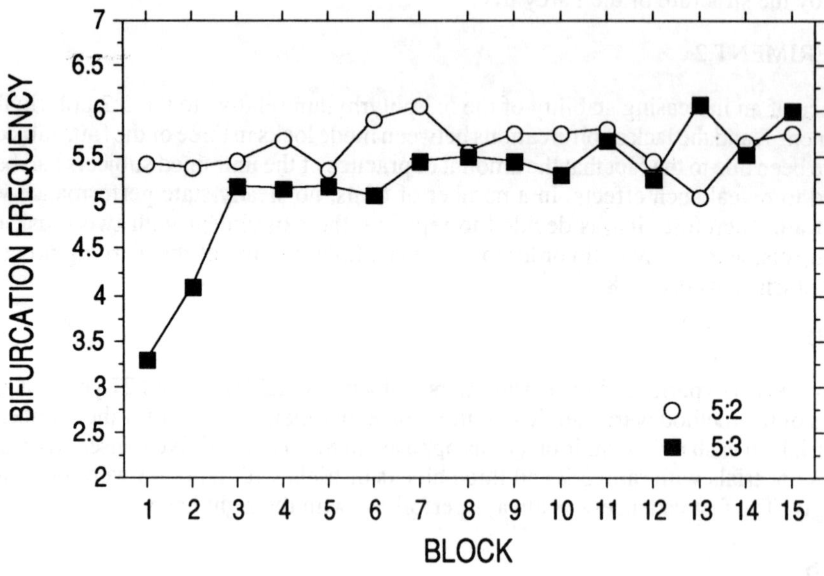

Figure 8. Bifuraction frequencies of subject B as a function of ratio and block.

These bifurcation frequencies were tested, for each subject separately, by means of a two factor ANOVA with factors ratio (5:2 and 5:3) and blocks (1-15), treating each subject as a population from which the bifurcation frequencies were drawn independently from one another. For subject A the ANOVA revealed significant main effects for both factors (ratio: $F(1,120) = 36.04$, $p < .0001$; block: $F(14,120) = 3.73$, $p < .001$). The interaction effect between the two factors, however, was not significant ($F(14,120) = 1.19$, $p = .29$). Also for subject B the ANOVA revealed significant main effects for ratio ($F(1,120) = 24.95$, $p. < .0001$) and block ($F(14,120) = 6.79$; $p < .0001$). In this case also the interaction effect between the two factors was significant ($F(14,120) = 5.40$, $p < .0001$), indicating a differential learning effect for both ratios. Application of multiple independent t-tests revealed that significant differences in favor of the 5:2 ratio on the first two blocks were no longer present on the last five blocks.

Analysis of bifurcation routes

Only the bifurcation routes for subject A have, as yet, been analyzed. The same method of analysis was used as in Experiment 1. The percentage agreement between the two observers with respect to the identified mode locks was 94%.

Bifurcation routes

The identified bifurcation routes for subject A are summarized in Table 2. Without going into too much detail, general trends are clearly observable. The 5:2 ratio bifurcated directly to 2:1 in the vast majority of cases (66 out of 71 times). As in Experiment 1 no direct transitions from 5:2 to 3:2 occurred. Interestingly, however, 7 out of the 66 transitions to 2:1 were followed by transitions to 3:2, i.e., to a higher order instead of a lower order ratio. In the 5:3 condition, 5 out of 72 times the rhythm bifurcated directly to 3:2. In the majority of cases (64 out of 72), however, 64 times the 5:3 bifurcated to 4:3, followed in 11 cases by a further bifurcation to a 3:2 mode lock.

Table 2. The frequency of occurrence of bifurcation routes observed in subject A. The complete Farey route corresponding to the starting ratio is displayed at the top of each series.

Ratio 5:2					Ratio 5:3				
route				frequency	route				frequency
5:2	3:1	2:1	1:1	0	5:3	3:2	2:1	1:1	0
5:2	3:1	2:1		2	5:3	3:2			4
5:2	3:1			1	5:3	3:2	4:3		1
5:2		2:1		54	5:3	2:1	4:3		2
5:2		2:1	1:1	2					
5:2		2:1	3:2	7	5:3	4:3			52
5:2		2:1	4:3	2	5:3	4:3	3:2		11
5:2		2:1	5:3	1	5:3	4:3	2:1		1
5:2	4:1			1					
5:2	4:3			1	5:3	5:4	4:3	3:2	1
			total	71				total	72

7. General discussion

Experiments on polyrhythms were reported in which tapping frequency (cycle time) was manipulated while subjects performed either 5:2 or 5:3 polyrhythms—the assumption being that tapping frequency would serve as a control parameter guiding the system via a series of transitions (bifurcations) from one collective steady state (mode lock) to another. As anticipated, bifurcations to other, smaller integer, mode locks occurred in a reasonably systematic and reliable fashion, provided that the performance of the 5:2 or 5:3 polyrhythm was sufficiently stable (i.e., truly steady state behavior).

In subject 1, who lacked musical training, the 5:2 ratio bifurcated to 2:1, while the 5:3 ratio broke away to either 2:1 or 3:2. The latter ratio never occurred when the subject started in the 5:2 mode. The observed bifurcations are consistent with the bifurcation routes reflected in the branching structure of the Farey tree, identifying the bifurcation behavior of known dynamical systems en route to chaos.

The fact that bifurcations to the 3:2 mode lock only occurred in the 5:3 ratio, but not in the 5:2 ratio, might be difficult to explain in terms of motor programming. For example, according to Deutsch's (1983) model, the hierarchical representation of the 3:2 polyrhythm is less complex than the representation of both the 5:3 and 5:2 ratios. If one wishes to account for the observed bifurcations in terms of changing from a polyrhythm involving a more complex representation to one involving a less complex one, the frequency ratio 3:2 might be expected to occur both after breaking down of the performance in the 5:2 and in the 5:3 modes, unless some additional mechanism were to be introduced.

In subject A, a skilled drummer, the 5:2 ratio nearly always bifurcated to 2:1, consistent with the expected bifurcation route in the Farey tree. However, in this subject 5:3 bifurcated to 4:3 in the vast majority of cases (64 out of 72), a portion of which (11 out of 64) later bifurcated to 3:2. The transition from 5:3 to 4:3, which occurred only once in subject 1, is not immediately transparent in the light of the theory advanced in section 4, although it meets the criterion of moving towards a smaller integer (but not a lower order) frequency ratio. It is also in accordance with the postulated tendency of the system to seek the right flank of the tree, a tendency which is also manifest in the almost complete absence of the 3:1 mode lock in the observed bifurcation routes originating from the 5:2 ratio. A possible explanation for the "attractiveness" of the 4:3 ratio in this subject may be found in his learning history. Being a professional drummer, he had extensive experience with the production of polyrhythms in quadruple time (i.e., 4:2, 4:3 and 4:4), among which the 4:3 rhythm is prevalent. The 7 observed bifurcations from 2:1 to 3:2, i.e., to a larger integer frequency lock, remind us that understanding the fine structure of bifurcation routes in tapping should await a more detailed analysis.

The persistency of the transition from the 5:3 to the 4:3 mode in the experienced drummer may be interpreted as evidence that the intrinsic dynamics of biological systems, or, more specifically, the widths of the Arnold tongues associated with different mode locks, may be changed by learning on the basis of information specifying new behavioral patterns, as, indeed, has been suggested recently by Schöner and Kelso (1988a, 1988b), Kelso and deGuzman (in press) and Schöner (1989). This, in turn, implies that learning itself may be conceptualized as a dynamic process in which the system's intrinsic dynamics—given the intention to perform a to be learned task—are "perturbed" (or "forced") in a task-specific way until the design of the system (i.e., its attractor landscape) is so modified that the performance of the task can emerge autonomously when the appropriate boundary conditions are set. A hallmark property of such learned attractors is that they are global and, therefore, relatively robust to changes in the initial conditions of the system (Koditschek, 1990).

In agreement with earlier findings of, for example, Deutsch (1983), but counter to Kelso and deGuzman (1988), the 5:2 ratio was performed more stably than the 5:3 ratio. Our expectancy that increasing autonomy due to practice of these polyrhythms would result in the opposite, was not borne out in the present experiments. However, in the two musically trained subjects (A and B), the initial difference between the stabilities in favor of 5:2 seemed to disappear, although the interaction effect was only significant for one of them. As indicated before, the learning curves for the 5:3 and 5:2 ratios may have been influenced by the use of a default value of 2 Hz for the bifurcation frequencies in those trials in which the required ratio was not attained at all. Therefore, a definite conclusion regarding differential learning effects for both ratios should await further analysis.

In any event, the issue of learning in polyrhythmic tapping remains crucially important. With many cognitive psychologists we share the conviction that an understanding of the production of polyrhythms must necessarily incorporate an account of skill level and learning. If it would be possible to define a yardstick for skill level, and make this parameter part of the intrinsic tapping dynamics for specific individuals, then it would perhaps also be possible to predict the stability of different mode locks and the bifurcation routes among them during different stages of learning.

Acknowledgements

The authors would like to thank W. J. Beek for his useful comments on an earlier draft of this paper, and B. P. L. M. den Brinker and T. de Haan for their technical assistance in data acquisition and processing.

References

Allen, T. (1983). On the arithmetic of phase-locking: Coupled neurons as a lattice on R^2. *Physica*, **6D**, 305-320.

Beek, P. J. (1989). *Juggling Dynamics.* Ph.D. Thesis. Amsterdam: Free University Press.

Bernstein, N. A. (1967). *The Co-ordination and Regulation of Movements.* London: Pergamon.

Cvitanovic, P., Shraiman, B., & Söderberg, B. (1985). Scaling laws of mode lockings in circle maps. *Physica Scripta*, **32**, 263-270.

deGuzman, C. G., & Kelso, J. A. S. (in press). Multifrequency behavioral patterns and the phase attractive circle maps. *Physica D.*

Deutsch, D. (1983). The generation of two isochronous sequences in parallel. *Perception and Psychophysics*, **34**, 331-337.

Feigenbaum, M. J., Kadanoff, L. P., & Shenker, S. J. (1982). Quasiperiodicity in dissipative systems: A renormalization group analysis. *Physica*, **5D**, 370-386.

Fein, A. P., Heutmaker, M. S., & Gollub, J. P. (1985). Scaling at the transition from quasiperiodicity to chaos in a hydrodynamic system. *Physica Scripta*, **T9**, 79-84.

Glass, L., & Mackey, M. C. (1988). *From Clocks to Chaos: The rhythms of life.* Princeton: University Press.

González, D. L., & Piro, O. (1985). Symmetric kicked self-oscillators: Iterated maps, strange attractors, and symmetry of phase-locking Farey hierarchy. *Physical Review Letters*, **55**, 17-20.

Haken, H. (1977). *Synergetics: An introduction.* Berlin: Springer.

Haken, H. (1983). *Advanced Synergetics.* Berlin: Springer.

Haken, H., Kelso, J.A.S., & Bunz, H. (1985). A theoretical model of phase transitions in human hand movements. *Biological Cybernetics*, **51**, 347-356.

Jagacinski, R. J., Marshburn, E., Klapp, S. T., & Jones, M. R. (1988). Test of parallel versus integrated structure in polyrhythmic tapping. *Journal of Motor Behavior*, **20**, 416-442.

Jensen, M. H., Bak, P., & Bohr, T. (1984). Transitions to chaos by interaction of resonances in dissipative systems: I. Circle maps. *Physical Review A*, **30**, 1960-1969.

Kay, B. A., Kelso, J. A. S., Saltzman, E. S., & Schöner, G. (1987). The space-time behavior of single and bimanual rhythmical movements. *Journal of Experimental Psychology: Human Perception and Performance*, **13**, 564-583.

Keener, J. P., & Glass, L. (1984). Global bifurcations of the periodically forced nonlinear oscillator. *Journal of Mathematical Biology*, **21**, 175-190.

Kelso, J. A. S. (1981). On the oscillatory basis of movement. *Bulletin of the Psychonomic Society*, **18**, 63.

Kelso, J. A. S. (1984). Phase transitions and critical behavior in human bimanual coordination. *American Journal of Physiology: Regulatory, Integration and Comparative Physiology*, **15**, R1000-R1004.

Kelso, J. A. S. (1989). Phase transitions: Foundations of behavior. In H. Haken & M. Stadler (Eds.), *Synergetics of Cognition.* Heidelberg: Springer.

Kelso, J. A. S., & deGuzman, G.C. (1988). Order in time: How the cooperatioon between the hands informs the design of the brain. In H. Haken (Ed.), *Neural and Synergetic Computers.* Berlin: Springer.

Koditschek, D. E. (1990). Globally stable closed loops imply autonomous behavior. Paper presented at the Fifth IEEE International Symposium on Intelligent Control, Philadelphia, PA, U.S.A.

Kugler, P. N. (1986). A morphological perspective on the origin and evolution of movement patterns. In M. G. Wade & H. T. A. Whiting (Eds.), *Motor Development in Children: Aspects of Coordination and Control.* The Hague: Nijhoff.

May, R. M. (1976). Simple mathematical models with very complicated dynamics. *Nature*, **261**, 459-467.

Saltzman, E. L., & Kelso, J. A. S. (1987). Skilled actions: A task-dynamic approach. *Psychological Review*, **7**, 733-740.

Schöner, G. (1990). Learning and recall in a dynamic theory of coordination patterns. *Biological Cybernetics*, **62**, 39-54.

Schöner, G., & Kelso, J. A. S. (1988a). A synergetic theory of environmentally-specified and learned patterns of movement coordination. 1. Relative phase dynamics. *Biological Cybernetics*, **58**, 71-80.

Schöner, G., & Kelso, J. A. S. (1988b). A synergetic theory of environmentally-specified and learned patterns of movement coordination. 2. Component oscillator dynamics. *Biological Cybernetics*, **58**, 81-89.

Schöner, G., & Kelso, J. A. S. (1988c). Dynamic patterns of biological coordination: Theoretical strategy and new results. In J. A. S. Kelso, A. Y. Mandell & M. F. Shlesinger (Eds.), *Dynamic Patterns in Complex Systems.* Singapore: World Scientific.

Shaffer, L. H. (1981). Performances of Chopin, Bach, and Bartok: Studies in motor programming. *Cognitive Psychology*, **13**, 326-369.

Shaffer, L. H. (1982). Rhythm and timing in skill. *Psychological Review*, **89**, 109-122.

Stavans, J., Heslot, F., & Libchaber, A. (1985). Fixed winding number and the quasiperiodic route to chaos in a convective fluid. *Physical Review Letters*, **55**, 596-599.

Swinnen, S. P., & Walter, C. B. (1988). Constraints in coordinating limb movements. In A. M. Colley & J. R. Beech (Eds.), *Cognition and Action in Skilled Behaviour*. Amsterdam: Elsevier.

Vorberg, D., & Hambuch, R. (1978). On the temporal control of rhythmic performance. In J. Requin (Ed.), *Attention and Performance VII*. Hillsdale, NJ: Lawrence Erlbaum Associates.

Vorberg, D., & Hambuch, R. (1984). Timing of two-handed rhythmic performance. In J. Gibbon & L. Allen (Eds.), *Timing and Time Perception*. New York: The New York Academy of Sciences.

Whiting, H. T. A., Meijer, O. G., & van Wieringen, P. C. W. (1990, Eds.). *The Natural-Physical Approach to Movement Control*. Amsterdam: VU University Press.

West, B. J., & Goldberger, A. L. (1987). Physiology in fractal dimensions. *American Scientist*, **75**, 354-365.

Wing, A. M., & Kristofferson, A. B. (1973). Response delays and the timing of discrete motor responses. *Perception and Psychophysics*, **14**, 3-12.

Yamanishi, J., Kawato, M., & Suzuki, R. (1979). Studies on human finger tapping neural networks by phase transition curves. *Biological Cybernetics*, **33**, 199-208.

Yamanishi, J., Kawato, M., & Suzuki, R. (1980). Two coupled oscillators as a model for the coordinated finger tapping by both hands. *Biological Cybernetics*, **37**, 219-225.

Stavans, J., Heslot, F. & Libchaber, A. (1985). Fixed winding number and the quasiperiodic route to chaos in a convective fluid. *Physical Review Letters*, 55, 596-599.
Summers, J. J. & [illegible] (1988). Coordinating [illegible]. In A. M. Colley & J. R. Beech (Eds.) *Cognition and Action in Skilled Behaviour*. Amsterdam: Elsevier.
Vorberg, D. & Hambuch, R. (1978). On the temporal control of rhythmic performance. In J. Requin (Ed.), *Attention and Performance VII*. Hillsdale, NJ: Lawrence Erlbaum Associates.
Vorberg, D. & Hambuch, R. (1984). Timing of two-handed rhythmic performance. In J. Gibbon & L. Allan (Eds.), *Timing and Time Perception*. New York: The New York Academy of Sciences.
Whiting, H. T. A., Meijer, O. G. & van Wieringen, P. C. W. (1990, Eds.), *The Natural-Physical Approach to Movement Control*. Amsterdam: VU University Press.
West, B. J. & Goldberger, A. L. (1987). Physiology in fractal dimensions. *American Scientist*, 75, 354-365.
Wing, A. M. & Kristofferson, A. B. (1973). Response delays and the timing of discrete motor responses. *Perception & Psychophysics*, 14, 5-12.
Yamanishi, J., Kawato, M. & Suzuki, R. (1979). Studies on human finger tapping neural networks by phase transition curves. *Biological Cybernetics*, 33, 199-208.
Yamanishi, J., Kawato, M. & Suzuki, R. (1980). Two coupled oscillators as a model for the coordinated finger tapping by both hands. *Biological Cybernetics*, 37, 219-225.

SECTION 8

CONTROL OF MOVEMENT KINEMATICS

KINEMATIC TRANSFORMATIONS FOR ARM MOVEMENTS IN THREE-DIMENSIONAL SPACE

J. F. SOECHTING and M. FLANDERS
Department of Physiology
6-257 Millard Hall
University of Minnesota
Minneapolis, MN 55455 USA

ABSTRACT. An algorithm is presented whereby information about the location of a target in extrapersonal space can be combined with information about arm posture to determine the kinematics of targeted arm movements. Experimental evidence which led to the model and additional experiments meant to test the model are also presented.

INTRODUCTION

In this paper we will summarize the results of some of our recent psychophysical studies concerning human arm movements to targets located in three-dimensional space. Information about the target's location is normally provided by the visual system and the question we set out to address was: what are the rules which the central nervous system utilizes to generate the appropriate pattern of muscle activity from the information provided by the visual system?

Our investigations were based on several premises. The first premise was that these rules and their implementation by neural circuits involve approximations to the exact solution. If this premise is correct, then there should be consistent errors in the performance of human subjects in pointing to the targets. If subjects do indeed make such errors, then it would be possible to work backwards and to deduce the rules from the errors. (Error-free performance, instead, would be compatible with any number of algorithms.)

The second premise was that feedback is used, when available, to correct for movement errors. For pointing movements to a target, such feedback could be provided by the visual system. Vision of the target, and of the arm as it approaches the target, could be used to correct movement errors no matter

J. Requin and G. E. Stelmach (eds.), Tutorials in Motor Neuroscience, 435–442.

what their source. Thus, to uncover computational errors, it was necessary to eliminate visual feedback.

Our third premise was that the solution to the problem (moving to a visually detected target) is achieved in several steps. Alternatively, one might guess that the problem could be solved directly by assigning a particular pattern of muscle activation to each retinal locus. Such a direct solution will not work since it does not account for the fact that the eye's position in the head and head's orientation relative to the trunk can vary and that, therefore, there is no one-to-one correspondence between retinal loci and points in space. Neural networks have been successfully trained to compute motor signals directly from target maps (Kuperstein 1988). However, we believe that the neural solution involves additional intermediate steps to generate a kinematic specification of the movement and that the patterning of muscle activities is derived from this kinematic specification. (By kinematics we mean a description of movement in terms of spatial distance and direction or changes in joint angles, and the derivatives of these quantities.)

Based on these premises, we were able to develop an algorithmic model for the kinematic transformations involved in moving the arm to a target (Soechting and Flanders, 1990; Flanders, Helms Tillery and Soechting, 1991). In brief, the model predicts that movement kinematics are computed from the arm's position at movement onset (based on kinesthetic information) and from the target's location (based on visual information), and that information in the visual coordinate system is transformed into the kinesthetic coordinate system. The model makes some specific predictions concerning how this transformation would be accomplished. We were able to test some of these predictions based on the following reasoning: subjects should be able to perform with usual accuracy on motor tasks which involve those computations which are normally performed by the central nervous system. However, subjects should make much larger errors when the motor task requires computations which are different from those usually performed.

In the following sections we will describe the model and the experimental results on which it is based and we will also summarize the results of experiments which were designed to test the model's predictions.

Errors in Moving to Remembered Target Locations

We began by investigating the errors subjects made in pointing to the remembered location of a target (Soechting and Flanders, 1989a). To do so, we presented subjects with a target, asked them to remember its location, turned off the lights and asked them to move their right arm so as to place the index finger on the target. We did this for 60 - 100 targets, at random locations within the subjects' reach.

Subjects made consistent errors on this task. For targets that were located distally from the subjects, they tended to point to locations which were short of the target, by as much as 15 cm. It is important to note that these errors were not related to the amplitude of the movement, i.e. the distance from the hand's starting position (at waist level) to the target. Rather, the errors were proportional to the distance of the target from the shoulder (the center of rotation of the pointing movement).

Directional errors in movement were much less, when target direction was measured from the shoulder. We used two parameters to describe direction : azimuth (medial-lateral angle) and elevation (up-down angle). The slopes of the regression between target and finger azimuth and elevation were both close to unity (perfect performance), whereas the slope of the regression for distance was much less (0.65). Furthermore, the subjects' variability in performance was much greater for the distance parameter than for the directional parameters (Soechting and Flanders, 1989a), as measured by the sensorimotor efficiency of the transmission of information (Sakitt, 1980; Georgopoulos and Massey, 1988).

The subjects' performance on this task did not improve when the arm was in sight, i.e. when they moved their arm to a remembered target location with the lights on.

One possible explanation for the errors in distance is that subjects misperceived the targets' locations. We were able to rule out this explanation (Soechting and Flanders, 1989a). When we gave subjects a pointer and asked them to put the tip of the pointer at the remembered location of the target with the lights on, they were able to do so. They no longer exhibited errors in distance as they did in the other situations.

In these pointing tasks, subjects could make effective use of kinesthetic information concerning arm orientation (Soechting and Ross, 1984). We demonstrated this by passively displacing subjects' arms in the dark (Helms Tillery, Flanders and Soechting, 1991) or in the light (Flanders and Soechting, 1989). Subjects were able to reproduce the location of their index finger with little error.

Thus our subjects knew where the target was and they also could sense the orientation of their arm in space and yet they made large and consistent errors in pointing to remembered targets. We have already mentioned the clue which suggested to us the reason for these errors: the errors were proportional to the targets' distance from the shoulder.

This observation led to the hypothesis that at one level of the neural network underlying the computations for goal-directed arm movements, target location is specified in terms of distance and direction (azimuth and elevation) measured from the shoulder. From this specification, arm orientation angles (yaw and elevation of the upper arm and forearm) that would place the finger on the target could be computed by approximation. The approximation in this step would be responsible for the observed errors.

Following this hypothesis, we performed a regression analysis between the angles describing the arm's orientation at the end of the subject's movement to the remembered target location and the parameters specifying the actual location of the target (distance and direction). We found that these orientation angles were linearly related to target location parameters when the subjects made movement errors, as in pointing to the remembered target location in the dark (Soechting and Flanders, 1989b). In particular, the elevation angles of the upper arm and of the forearm depended on a linear combination of target distance and elevation, and did not depend on target azimuth. Upper arm and forearm yaw angles depended linearly on target azimuth, and only to a small and variable extent on target elevation and distance.

By contrast, when subjects made accurate movements (pointing to a physical target in the light), the relations between orientation angles and target parameters were markedly nonlinear. These findings were in agreement with the hypothesis and indicated that the approximation involved a linearization of the exact, nonlinear relations between target parameters and arm angles.

An Algorithm for Coordinate Transformations

When we attempted to put our hypothesis into the context of existing neurophysiological data, we found that there were enough constraints to permit us to develop a relatively comprehensive algorithmic model. This model has been described in detail elsewhere (Soechting and Flanders, 1990; Flanders, Helms Tillery and Soechting, 1991).

According to the model, the task "move to a target" requires that both the position of the target and of the arm be known before movement onset. Arm position is described in terms of yaw and elevation angles (Soechting and Ross, 1984) and is derived from kinesthetic afferents. Target position is initially described in a retinocentric frame of reference. There are two subsequent coordinate transformations:

1) Retinocentric coordinates --------> Head-centered coordinates
2) Head-centered coordinates ------> Shoulder-centered coordinates.

In shoulder-centered coordinates, target location is described by three parameters: 1) distance, 2) elevation and 3) azimuth. From these parameters, final arm orientation is computed, by approximation in two parallel channels:

1) Target distance and elevation -----> arm elevation
2) Target azimuth ------------------------> arm yaw.

This visually-derived estimate of arm orientation angles is compared with the kinesthetically-derived measure of present arm orientation to obtain the amount by which each of the arm angles must change to reach the target. We also suggested, based on electrophysiological evidence (Georgopoulos, Kettner and Schwartz, 1988), that these changes in arm angles are used to compute a spatial description of the movement's kinematics, i.e. the direction and the amplitude of the movement.

Tests of the Algorithm's Predictions

In our subsequent work we attempted to test some of the algorithm's predictions more rigorously. These further experiments were based on the additional premise that if we asked subjects to perform motor tasks that were congruent computationally with the algorithm, the subjects should be able to perform the task with little error. However, if we asked them to perform motor tasks that were not congruent with the algorithm, their performance should deteriorate. In other words, tasks that required transformations which were different from those of the algorithm should prove to be difficult.

The first set of experiments tested the hypothesis that the representation of target location was transformed from retinocentric to shoulder-centered coordinates (Soechting, Tillery and Flanders, 1990). The original evidence in favor of this step was the errors in distance of pointing were proportional to the target's distance from the shoulder, but that they were not proportional to distance measured from other reference points such as the head. We now sought to determine if target direction was also specified using the shoulder as the reference point.

To do so, we asked subjects to move half-way to remembered target locations. (By intentionally increasing errors in distance, we could obtain a more sensitive measure of directional errors.) To make the task more precise, we asked subjects to move either half-way to a point on a line from the head to the target or half-way to a point on a line from the shoulder to the target. We predicted that moving half-way to a point on the line from the head to the target should be easy, whether or not our hypothesized algorithm was correct. In terms of our algorithm, subjects would maintain unaltered their initial retinocentric definition of target direction, and halve the distance.

Moving half-way to a point on the line from shoulder to target would be a difficult problem computationally if the computations were performed in head-centered coordinates, but it would be a simple problem in shoulder-centered coordinates. In fact, subjects performed both tasks (half-way from the head and half-way from the shoulder) with equal accuracy, lending support for the hypothesized coordinate transformation. These experiments also provided additional details. The origin for the specification of target direction did not appear to be shifted all the way to the shoulder (which we defined as the approximate center of rotation of the gleno-humeral joint). Instead, our best estimate for the origin was at about shoulder height, 2/3 of the distance from head to shoulder.

Another prediction of the algorithm is that information about target location in shoulder-centered coordinates is processed in two parallel channels to obtain estimates of arm orientation angles. One channel involves only information about target azimuth, whereas the other channel processes information about target distance and elevation together. If these two channels are separate, then subjects should be able to move to a target's azimuthal coordinate while ignoring distance and elevation. They should also be able to ignore target azimuth and move to its elevation and distance. However, the algorithm predicts that they should not be able to move to the target's elevation, while ignoring azimuth and distance because distance and elevation information are processed in the same channel. These predictions of the algorithm were fulfilled (Flanders and Soechting, 1990).

Finally, according to the algorithm, the spatial locus of the target (in visual coordinates) is transformed into arm orientation angles that would place the finger on the target (in kinesthetic coordinates). In an alternative scheme, one could suppose that the initial arm orientation angles are used to derive an estimate of the initial spatial locus of the hand (in visual or spatial coordinates). The computation of movement kinematics could then take place in this extrinsic coordinate frame. This would imply that subjects should be able to synthesize an estimate of the location of their hand in space based solely on kinesthetic information.

We have already mentioned that subjects were able to accurately reposition their arm to match a passive displacement. While this experiment demonstrates that subjects were able to effectively use kinesthetic information, its interpretation is ambiguous: subjects could have attempted to match either the location of the hand in space or to match the orientation angles of the arm (Helms Tillery, Flanders and Soechting, 1991). To resolve this ambiguity, we performed one additional experiment. We passively displaced subjects' arms in the dark, asked them to remember the location of their index finger, moved the arm back to the side and gave them a pointer to indicate, in the light, the remembered location of the finger. They made large errors on this task. (Recall that subjects could use a pointer successfully to indicate the remembered location of a visual target.) They also made large errors when they were asked to indicate the remembered spatial location of their right index finger with their left index finger (Helms Tillery, Flanders and Soechting, submitted manuscript).

Thus all of the experiments we were able to devise to test the algorithm gave results which were in accordance with the algorithm's predictions. Furthermore, they have also tended to rule out alternate schemes for solving the computational problem. Therefore, we believe that the algorithm can provide a framework for understanding the early stages in the sensorimotor transformation that leads to goal-directed arm movement (Flanders, Helms Tillery and Soechting, 1991).

Subsequent stages to transform movement kinematics to patterned muscular activity are required. While we have begun to address this issue (Flanders and Soechting, 1991), a detailed model for this aspect of the problem appears premature. Where in the nervous system, and how, each of the steps in the algorithm are implemented also remain unanswered questions.

Acknowledgment. This work was supported by USPHS Grants NS-15018 and NS-27484.

REFERENCES

Helms Tillery, S. I., Flanders, M. and Soechting, J. F. (1991) 'A coordinate system for the synthesis of visual and kinesthetic information', *J. Neurosci.* (in press).

Helms Tillery, S. I., Flanders, M. and Soechting, J. F. (1991) 'Errors in the use of kinesthetic information for bimanual pointing', *Exp. Brain Res.* (submitted).

Flanders, M. and Soechting, J. F. (1990) 'Parcellation of sensorimotor transformations for arm movements', *J. Neurosci.* **10**, 2420-2427.

Flanders, M., Helms Tillery, S. I. and Soechting, J. F. (1991) ' Early stages in a sensorimotor transformation', *Behav. Brain Sci.* (in press).

Flanders, M. and Soechting, J. F. (1991) 'Arm muscle activation for static forces in three-dimensional space', *J. Neurophysiol.* (in press).

Georgopoulos, A. P., Kettner, R. E. and Schwartz, A. B. (1988) 'Primate motor cortex and free arm movements to visual targets in three-dimensional space. II. Coding of direction by a neuronal population', *J. Neurosci.* **8**, 2928-2937.

Georgopoulos, A. P. and Massey, J. T. (1988) 'Cognitive spatial-motor processes. 2. Information transmitted by the direction of two-dimensional arm movements and by neuronal populations in primate motor cortex and area 5', *Exp. Brain Res.* **69**, 315-326.

Kuperstein, M. (1988) 'Neural model of adaptive hand-eye coordination for simple postures', *Science* **239**, 1308-1310.

Sakitt, B. (1980) 'Visual-motor efficiency (VME) and the information transmitted in visual-motor tasks' *Bull. Psychonom. Soc.* **16**, 329-332.

Soechting, J. F. and Flanders, M. (1989a) 'Sensorimotor representations for pointing to targets in three-dimensional space', *J. Neurophysiol.* **62**, 582-594.

Soechting, J. F. and Flanders, M. (1989b) 'Errors in pointing are due to approximations in sensorimotor transformations', *J. Neurophysiol.* **62**, 595-608.

Soechting, J. F. and Flanders, M. (1990) 'Deducing central algorithms of arm movement control from kinematics', in D. R. Humphrey and H.-J. Freund (eds.) *Motor Control: Concepts and Issues*, John Wiley and Sons Ltd., NY, (in press).

Soechting, J. F. and Ross, B. (1984) 'Psychophysical determination of coordinate representation of human arm orientation', *Neurosci.* **1 3**, 595-604.

Soechting, J. F., Tillery, S. I. H. and Flanders, M. (1990) 'Transformation from head- to shoulder-centered representation of target direction in arm movements', *J. Cogn. Neurosci.* **2**, 32-43.

THE INTEGRATION OF "NOISE" INTO THE STRUCTURE OF MOVEMENTS

C.J. WORRINGHAM
Department of Movement Science
The University of Michigan
401 Washtenaw Avenue
Ann Arbor, MI 48109-2214
U.S.A.

ABSTRACT. Conventional views of motor control, which state or imply that it is a deterministic system, are being challenged by the recognition that the stochastic properties of motor control may offer clues to some of its operations. This paper considers evidence that there exists a general property of trajectory formation which is a consequence of the inherent variability of movements. Specifically, the amplitude of the initial sub-movement of motions containing two or more such sub-movements is planned so as to take into account its spatial variability. Examples are given of cases where such a principle might be expected to apply, and the usefulness of this phenomenon for examining individual and strategy differences in aiming tasks is discussed.

1. Introduction

Traditional neurophysiological accounts of motor control have tended to emphasize its hicrarchichal nature, inspired by the distinct anatomical levels of those CNS structures which subserve movement, from cortical areas down to individual motor units. Perhaps because of this, such accounts have typically imparted a strong flavor of determinism to motor control. Explicitly or implicitly, the link between the centrally generated command for a movement and the ensuing movement has generally been considered a stable and consistent one, in which exactly the same movement invariably follows a given command. The challenges of early motor neuroscience were, of course, sufficiently great that to describe the properties of the system's *signal* seemed a sufficiently ambitious goal, and little attention was paid to its *noise* characteristics.

Motor output variability or impulse variability theory, a contribution originally made eleven years ago (Schmidt, Sherwood, Zelaznik, Hawkins, Frank, and Quinn, 1979) has brought about a

J. Requin and G. E. Stelmach (eds.), Tutorials in Motor Neuroscience, 443–456.

somewhat different view, in which motor control has come to be seen as a *stochastic* process. In this view, motor commands are viewed as producing inherently variable results: a movement as actually executed will indeed approximate that which was intended, but will differ in certain respects (for example, distance, direction or duration), the size of this disparity being described by some probability function. The source of this variability is unknown, but is widely assumed to reflect the presence of "neuromotor noise" at various levels of the central nervous system (Darling, 1989). In the ensuing years much effort has been expended on debates over the mathematical underpinnings and assumptions of rival models (e.g. Tsiboulevsky, 1981; Meyer, Smith and Wright, 1982; Schmidt, Sherwood, Zelaznik, and Leikind, 1985). A productive recent development of this basic idea is its application to movements which are of relatively long duration (substantially greater than 200 ms) and which are normally considered to come under closed-loop control. Thus Meyer et al. (1988) have proposed that spatial variability is a property of each sub-movement in aiming motions comprising two or more such sub-movements, and that the overall duration of the movement may be minimized by an optimization process. The reasoning is that the time savings offered by a very rapid initial sub-movement may be more than offset by its elevated spatial variability, which may compromise the rapid conduct of the ensuing sub-movement. Conversely, if made too slowly, the low level of variability present in the initial sub-movement may enhance the conduct of later sub-movements but at a prohibitive cost of additional time. In essence, this stochastic sub-movement optimization model describes a scheme by which the motor control system selects an optimal set of sub-movement velocities. This is a significant extension, for although the original authors allowed for the possibility that these principles may also apply to longer movements in which error corrections were present (Schmidt et al., p. 448), their original data emphasized the variability characteristics of discrete, short duration aiming motions. No specific mechanism was offered at that time to incorporate the mechanisms of motor output variability into closed-loop movements, although some evidence was given that longer duration motions, of around 500 ms, exhibited similar variability properties if subjects were prevented from fully attending to these movements (Zelaznik, Shapiro, & McColsky, 1981).

While the view that motor control is a stochastic process is gaining some currency, the full implications of this approach may not yet have been fully realized. Thus far, published work on motor output variability has focussed on one of the following themes: 1) spatial end-point variability in discrete movements (e.g. Schmidt et al., 1979; Meyer et al., 1982); 2) continuous measures of spatial variability throughout a movement, especially with practice (Darling and Cooke, 1987a & b; Marteniuk and Romanow, 1983, Moore and Marteniuk, 1986); 3) spatial variability in multi-sub-movement actions (Meyer et al. 1988); 4) variability in timing tasks and in temporally constrained tasks (Schmidt et al., 1975; Wright and Meyer 1983). The theme of this paper is that motor output variability, far from being moribund and mired in technical criticism, continues to promote new insights into motor behavior phenomena, of which the idea presented below is but one.

2. Spatial Variability and Trajectory Formation

2.1. HOW AND WHY SPATIAL VARIABILITY MAY INFLUENCE UNDERSHOOT

This paper presents evidence for a phenomenon I shall refer to as "variability-related undershoot", which, if it proves to be widespread, may shed light not so much on the internal *temporal* structure of movement (as does the stochastic sub-movement optimization model from which it draws and with which it appears quite consistent (Meyer et al. 1988)), but on lawful features of *trajectory formation*. The idea may best be expressed by an example. In reaching to quickly grasp a small, fragile object, whose size and location ensure that a single sub-movement will not suffice to attain it, motor commands must be generated which bring the limb into the general vicinity of the object, preparatory for the actual grasp motion. The division of aiming and reaching movements into distinct phases has been recognized for a century (Woodworth's (1899) "initial impulse" and "current control" stages), and has been reiterated many times using different terms: Taylor and Birmingham (1948) labelled them "ballistic" and "continuously controlled" phases; Welford (1968) described them as "a faster distance covering phase and a slower phase of homing on to the target". How far will the limb travel in the "initial", "ballistic", "transport" stage? Given that this initial segment is widely thought to be open-loop - evidence for very early corrections (Pélisson, Prablanc, Goodale and Jeannerod, 1986; Gordon and Ghez, 1987) notwithstanding - some active selection of its extent must presumably comprise part of the planning. The current literature gives rather limited guidance. One notion is that the distance covered is some fixed proportion of the total distance to the target (Keele, 1968; Carlton, 1979), figures of 93-96% have been suggested. This may be correct for a particular task, but it is empirically rather than theoretically derived. It also carries the implication of determinacy, as discussed in the opening section.

The answer suggested here is that - at least in some cases - the spatial variability manifest in the the initial sub-movement (this more neutral term will be used in preference to "ballistic" etc.) is associated with and may lead directly to the selection of its mean extent. The principal reasons for this supposition have to do with the spatial and temporal constraints offered by the task. Taking the spatial constraints first, it must first be recognized that uncontrolled impact with objects is rarely a goal of movement. The class of striking motions to impart velocity to a projectile is a specific exception. Picking up a full wine-glass, however, requires that a more measured, low-velocity approach take place. In such a case, the limb could be driven towards the object with an initial sub-movement gauged so as to stop short of the object by an amount related to the inherent spatial variability in that sub-movement. Motions arrested by muscular forces do indeed differ from those stopped by impact, both kinematically and electromyographically (Soechting, 1984; Waters and Strick, 1980). There is a sense in which spatial variability - conventionally and necessarily measured *across* repetitions of the same movement seems to lose meaning when thought of as a property of an *individual* movement. How can a single motion possess any spatial variability? Surely it cannot be measured, since we have no direct information about how much such a single movement departs from that which was intended. The solution to this paradox is to conceive of spatial variability as being the expression across many trials of the unpredictability of a single trial. For the motor system, the existence of some level of unpredictability in the upcoming motion is a characteristic which can be taken into account when planning the movement. It could minimize this unpredictability by reducing the velocity of the movement - indeed this idea lies at the heart of the recently proposed stochastic optimization model (Meyer et al., 1988). There are limits on this, however, since the price of additional time to completion is often not acceptable. A second means of actively planning for spatial variability is to undershoot the object or target on the first sub-movement by some amount which is in proportion to the expected level of spatial variability. This would ensure that there is a low probability of the initial sub-movement striking the target (e.g. knocking over the wine-glass) even if this sub-movement were, on a particular trial, to be longer than intended. If it exists, this process could be seen as a type of optimization since an ultra-conservative strategy also has costs: to undershoot by too great an amount with the first sub-movement may necessitate more sub-movements than are really needed to acquire the target, costing additional time.

There is a second theoretical reason for undershooting a target by an amount related to variability, which concerns the difference between under- and over-shooting. Of course, reaching to physical objects or moving the hand, finger or an implement to a location on a surface perpendicular to the motion may physically restrict the first sub-movement to an undershoot. If a virtual target is used, in which the limb must be brought to a stop in some location without any impact and purely through muscular forces, then overshoots are also possible. In some such cases, initial overshoots may be seen, especially if the movement is under-damped. A subject can simply "shoot" the limb towards the target with no allowance for variability, attempting to end the movement directly on target. Inertia may carry the limb beyond that position, with the elastic properties of the muscle-segment system rather than any "correction" command restoring the limb to the target position. We have observed such cases on occasion in our laboratory, for wrist supinations with no physical stop (unlike the experiments reported below). A characteristic of such motions is a sinusoidal displacement profile in the vicinity of the target, with decaying amplitude ("ringing"). Such actions look much like the "move" and "hold" movements discussed by Brooks (1985) in which terminal co-contractions of agonists and antagonists clamp the limb on target but make the limb's "mass-spring" oscillatory properties manifest. A more common situation, however, may occur with multi-segment movements which are critically damped or over-damped, especially if the limb is unsupported. Here reliance on passive properties to restore the limb's position is excluded. The subject is now confronted with three choices: bias the initial sub-movement towards an overshoot, bias it towards an undershoot, or try to aim exactly for the target with the initial sub-movement. Should the target be of large enough size the last strategy may be adequate, since the attendant terminal spatial variability may be wholly encompassed by the generous size of the target. For smaller targets it may be very costly, since corrective sub-movements following the first will have a direction which cannot be known - or intelligently guessed - in advance, leaving some more or less time-consuming decision process until the limb's location at the end of the first sub-movement is known.

If the target is almost certainly unattainable with a single sub-movement, it would seem preferable to make a small deliberate error of extent. In this way the subject has a high probability of correctly guessing, in advance of its completion, on which side of the target the limb will be. In this way, the direction, if not the magnitude of a subsequent correction will not have to be specified later. If the empirical evidence for the feasibility of partial advance specification of movements (c.f. Rosenbaum, 1980) is correct, and if such specification can occur in multiple sub-movement actions, some time should be saved by this strategy, because only the magnitude of the correction will have to be planned. It is not the place here to consider if such a strategy is made consciously, but it is certainly reminiscent of one conscious process - a navigational trick called "aiming-off" practiced by orienteers. Orienteers are athletes who run through forests to specific check-points using a map and compass, in competitions organized as time-trials. When it is necessary to run using a compass bearing (because of the lack of recognizeable features on the ground or poor visibility), and the next location that must be reached lies on a "linear feature" such as a fence, trail, river or vegetation boundary which can be seen on the map and is perpendicular to the compass bearing, many orienteers add (or subtract) ten or so degrees from the true bearing, deliberately running left or right of the desired position. The slight cost in extra distance (two sides of a triangle) is more than recouped by the knowledge of which way to turn on arriving at the linear feature, e.g. "aim off left, turn right at the fence". It is certainly conceivable that aiming movements are subject to a similar process. If so, the final choice, then, remains that between a slight deliberate undershoot and a slight deliberate overshoot. The former would seem preferable in general, if only because it should be subject to less spatial variability than an overshoot, by virtue of its smaller amplitude and velocity. There are two final aspects of this problem which can be explored, and for which predictions can be made, by pursuing the "aiming-off" analogy. The orienteer uses this tactic because he or she is aware that random directional errors occur in the "planning" and "execution" of a run on a compass bearing. Clearly, though, one would not deliberately err by as much as 50 degrees, for the additional distance run would be extravagant. The ideal performer would have established some prior knowledge of the *size* of his or her typical directional error, and would pare down the size of the aiming-off error accordingly. Were this operation done in a formal mathematical way, knowledge of the distribution of random directional errors would allow the selection of an aiming-off error that would have the correct outcome (i.e. turning in the correct direction on coming to the fence) on some proportion of attempts. This proportion which can be set according to the competitor's own criteria. This introduces the final issue: strategic aspects of task performance which must be superimposed on the properties of the motor system. Were the orienteer to be in a desparate race for the fastest time, the aiming-off error can be reduced to some smaller multiple of random directional error. This is a risky strategy with a high potential pay-off: saving enough seconds to win the competition. Conversely, if the risk of a large error (turning the wrong way) is deemed too costly, a conservative performer will aim off by a generous amount, thereby adding a little time but greatly reducing the likelihood of an even more expensive mistake. In the same way, general strategic factors as well as individual characteristics may show up in the size of undershoots (or overshoots) made in reaching and aiming motions. If the full wine-glass is about to be toppled by a child, we will accept the risk that our own rapid, somewhat unpredictable initial sub-movement which ends near the glass, may itself spill the contents. In other circumstances, the initial sub-movement may not only be slower, it may be planned so as to end somewhat further from the glass. Even given the same implicit and explicit task constraints, different people may lawfully manifest their own characteristics in movement execution. An individual whose movement are "noisier" may find it necessary to undershoot the target by a greater amount, and/or move more slowly than will someone whose movements are more consistent. A "conservative" individual may undershoot with the initial sub-movement by more than an "impulsive" individual.

In the following section, preliminary evidence is provided for a variability-related undershoot representing one aspect of trajectory formation. The features mentioned above will be considered, including indications that strategies and individual differences in aiming and reaching motions may be studied through such an analysis.

2.2. EXPERIMENTAL EVIDENCE

Three experiments were conducted to examine this issue. The first two experiments briefly described below are more fully described in Worringham (1991, in press).

2.2.1. *Reversal Movements*

Subjects performed an aiming task with a direction reversal incorporated. The task required a free hand movement to direct a hand-held stylus to a vertical target. Four different amplitudes, ranging from 6.48 to 30 cm were used. The hand had to be moved towards the body from an initially extended position, past the plane of the target, and then reverse direction to strike the target. The three-dimensional path of the stylus was tracked using an ultrasonic method. As indexed by mean within-subject standard deviations of each of the three coordinates, the longer and faster movements of greater amplitude had more spatially variable "reversal points" (the positions at which movements changed direction along the sagittal axis from approaching to going away from the body). The key result here (see Table 1) was the fact that the

TABLE 1. Overshoot and variability by amplitude

Amplitude Condition	Overshoot			Variability		
	S	L	V	S	L	V
6.48	2.11	.70	.96	.79	.53	.62
10.80	2.23	1.00	.94	.85	.65	.62
18.00	2.77	1.35	1.09	.91	.72	.72
30.00	3.19	1.75	1.30	1.11	.79	.83

Overshoot: distance from mean reversal point to target center. Variability: S.D. of reversal point. S: sagittal, L: lateral, V: vertical. Units: cm position in space at which the reversal took place changed with the amplitude, speed and spatial variability of the phase preceding reversal. The more variable conditions had reversals further from the target, vertically, in the dimension of the movement's long axis, and in the horizontal dimension perpendicular to it (Table 1).

There was no instruction to make the reversal in any particular position. Consequently, this increase in distance to the target from reversal would seem to represent the spontaneous modification of the trajectory in response to experimental conditions. It is consistent with an attempt to take the increased spatial variability into account, as outlined above. A difference in detail from the proposed model, of course, is that here subjects *overshot* the target, as necessitated by the special requirements of the task.

2.2.2. *Wrist Rotation Movements* A second task with one-degree-of-freedom was used in a subsequent experiment. Subjects were required to make rapid forearm supinations using a bearing-mounted handle through angles which varied from 30 to 90 degrees, to direct a pointer projecting from the handle to a target zone four degrees in width. Movements had to be carried out as quickly as possible, subject to the requirement to try to avoid the pointer striking a wooden block at the far end of the target zone. The subjects were therefore forced not to overshoot. Mean within-subject standard deviation of the position at which the initial sub-movement was determined to have ended served as an index of spatial variability. Criteria used for identifying the boundary between initial and second sub-movements were direction reversals, velocity minima, or local minima in acceleration profiles prior to reaching the target. The result of interest here was the tendency for subjects to fall short of the target by approximately two standard deviations, across the five amplitude conditions. Figure 1 depicts this outcome: the five conditions are arrayed vertically. The starting positions are represented by the dots, and the target by the open rectangle. The mean position of the end of the first sub-movement is shown as the dot. Bars are plus or minus two mean within-subject standard deviations. It should be noted that the relationship is not simply linear, it is also very close to proportional.

2.2.3. *Wrist Rotation Movements - Laterality Effects* One of several predictions which can be made concerning this model is that factors which influence "noise" levels in the neuromotor system should also affect the undershoot of targets by initial sub-movements. Several kinds of factors may have this effect, some artificial (such as pharmacological agents which induce tremor) some natural. Amongst the latter is the hand used to perform the task. It has been quite well documented that the trajectory taken by the dominant hand

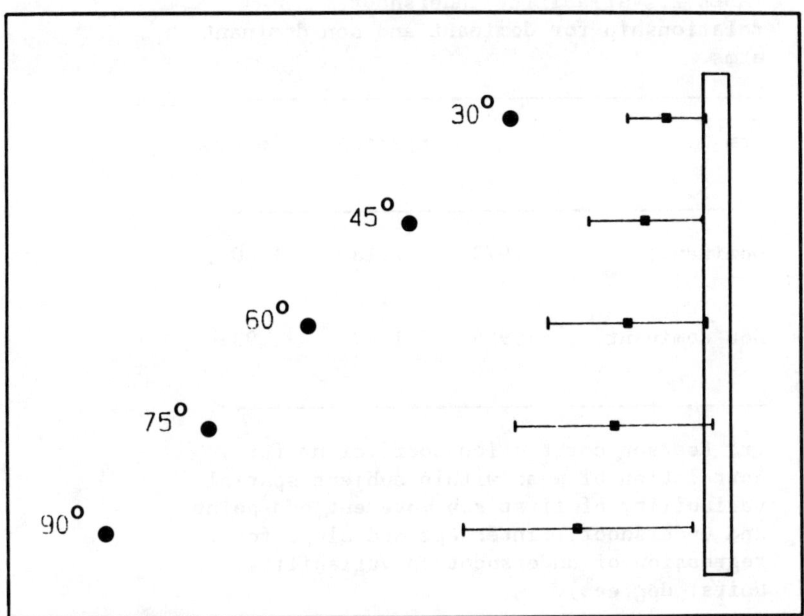

Figure 1. Undershoot of target by initial sub-movement for the five amplitude conditions. See text for explanation. (Figure reproduced from Worringham (1991, in press) Journal of Motor Behavior, by permission of Heldref Publications).

is less variable than the non-dominant in aiming movements (e.g. Todor and Cisneros, 1985; Annett, Annett, Hudson, and Turner, 1979). With this in mind, the dominant limb should approach a target more closely, on average, than the non-dominant limb, according to the current reasoning. This idea was put to the test with a group of subjects varying in hand-dominance from strong left- to strong right-handers (according to scores on a handedness inventory). The task given was essentially the same as that in the previous section. In this initial test the limb used was confounded with anatomical direction of motion (supination for the right hand, pronation for the left hand) because of limitations imposed by the apparatus. The results are equivocal. Foremost is a failure to find a difference in the levels of variability in the terminal position of the first sub-movement for the dominant and non-dominant limbs, despite faster movement times for the dominant limb. Given this unexpected outcome, the absence of a limb difference for the degree of undershoot does not disconfirm but certainly does not support the hypothesized mechanism. On the other hand, the undershoot/variability relationship was present for both dominant and non-dominant limbs and is approximately proportional (Table 2).

TABLE 2. Variability-undershoot relationship for dominant and non-dominant arms

Arm	r	intercept	slope
Dominant:	.973	2.18	1.60
Non-dominant:	.995	-1.09	1.93

(r: Pearson correlation coefficient for correlation of mean within-subject spatial variability of first sub-movement end-point and undershoot. Intercept and slope for regression of undershoot on variability Units: degrees)

2.2.4. *Wrist Rotation Movements - Individual Differences* Individual subject data from the first of the two wrist rotation experiments was inspected for evidence of an association between and degree of spatial variability at the end of the initial sub-movement and the extent of the undershoot. The reasoning outlined earlier was that those whose variability levels are higher should also tend to undershoot the target by a greater amount. After ranking subjects on mean undershoot and mean variability (collapsed across amplitude conditions), a positive correlation emerged (Spearman's rho = .83, $p < .05$). These data were consistent with the hypothesized relationship.

In the second of the two wrist rotation experiments, similar relationships emerged: the individuals with the slowest overall movement times were those with the highest levels of spatial variability, as estimated from either the location of the end of the first sub-movement ($r = .89$) or the location of the peak velocity of that sub-movement ($r = .90$). Those with longer undershoots also tended to have longer mean movement times ($r = .81$). Finally, those subjects with the greatest levels of spatial variability also tended to have longer undershoots ($r = .80$), results which are consistent with the current hypothesis. A curious feature of these individual difference data is that the individual levels of spatial variability in the first sub-movement were not positively correlated with peak velocities ($r = -.70$), as one would expect to find within a single person's data across different movement speeds. This suggests that the measured properties reflect some pattern of individual differences which is not secondary to movement velocities. At least in this data set those with the slowest movement times were more spatially variable, undershot the target by more, and had initial movements with slower peak velocities.

3. Summary and Conclusions

The major idea proposed here is that at least one facet of trajectory formation may be predicted from spatial variability characteristics of tasks, specifically the location relative to the target at which the initial sub-movement ends. This location is, admittedly, but one spatial landmark in what can be quite complex limb trajectories, so even if it can be well predicted, it by no means explains the entire process. A legacy of the traditional reductionist approach in motor behavior has been, however, that many phenomena are kept separate which are in fact closely linked. Thus trajectory formation issues have been the province, primarily, of those whose level of analysis is mechanical (e.g. Hollerbach, 1982), but the last decade's work on variability may clarify some of these issues in a way which no amount of mechanical theorizing can.

The proposed mechanism must, of course, be subjected to additional rigorous testing. One first step will be to determine if the undershoot of targets is truly related to variability, rather than to some other factor highly correlated with variability (e.g. amplitude or velocity). This requires that variability be manipulated other than through varying the amplitude, and consequently, the speed of the subject's movements. A second crucial step will be to establish if spatial variability, as measured in one aiming task, can predict an individual's undershoot of targets in a different task. This should be possible if individuals exhibit reasonably stable levels of variability and of "conservatism" in aiming tasks, and would be a prerequisite to the study of individual differences in movement with this approach. The issue of practice has not been raised here, but this is another obvious test-bed for this mechanism, because ample evidence has accrued that spatial variability is susceptible to reduction through practice (e.g. Moore and Marteniuk, 1986, Darling and Cooke, 1987a and b).

Finally, there is the possibility that other aspects of human motor behavior can be partly understood through examining undershoot-variability relationships. We need look no further than the opposite ends of the life-span. In infancy and childhood reaching and aiming movements are non-optimal, have high variability, and also tend to undershoot the target by a wide margin (Fetters and Todd, 1987). Could early efforts at reaching for objects consist, in part, of crude efforts to adjust the extent of initial movements to take account of the high level of variability? In aging, the general proposition has been made that increased neural noise contributes to altered patterns of motor control (Welford, 1984). If this is true, we might expect to find an age-related increase in mean undershoot with the initial sub-movement.

4. References

Annett, J., Annett, M., Hudson, P.T.W., & Turner, A. (1979). The control of movement in the preferred and non-preferred hands. Quarterly Journal of Experimental Psychology, 31, 641-652.

Brooks, V.B. (1985). How are "move" and "hold" programs matched? In H.J. Dichgans, W.J. Bloedel,, and W. Precht, (Eds.) Proc. life sciences, Berlin: Springer Verlag, pp. 1-23.

Carlton, L.G. (1979). Control processes in the production of discrete aiming responses. Journal of Human Movement Studies, 5, 115-124.

Darling, W.G. & Cooke, J.D. (1987a). Changes in the variability of movement trajectories with practice. Journal of Motor Behavior, 19, 291-309.

Darling, W.G. & Cooke, J.D. (1987b). Movement related EMGs become more variable during learning of fast accurate movements. Journal of Motor Behavior, 19, 311-331.

Darling, W.G. (1989) Neural mechanisms underlying motor output variability. In C.J. Worringham (Ed.) (1989). Spatial, temporal and electromyographical variability in human motor control (Proceedings of a symposium held in Ann Arbor, Michigan, 17-18 February, 1989, pp. 10-13.

Fetters, L. & Todd, J. (1987). Quantitative assessment of infant reaching movements. Journal of Motor Behavior, 19, 147-166.

Gordon, J., & Ghez, C. (1987). Trajectory control in targeted force impulses. III. Compensatory adjustments for initial errors. Experimental Brain Research, 67, 253-269.

Hollerbach, J.M. (1982). Dynamic interactions between limb segments during planar arm movement. Biological Cybernetics, 39, 139-156.

Keele, S.W. (1968). Movement control in skilled motor performance. Psychological Bulletin, 70, 387-403.

Marteniuk, R.G. & Romanow, S.K.E. (1982). Human movement organization and learning as revealed by variability of movement, use of kinematic information and Fourier analysis. In R.A. Magill (Ed.) Memory and Control of Action, Amsterdam: North-Holland.

Meyer, D.E., Smith, J.E.K., & Wright, C.E. (1982). Models for the speed and accuracy of aimed movement. Psychological Review, **89**, 449-482.

Meyer, D.E., Abrams, R.A., Kornblum, S., Wright, C.E. & Smith, J.E.K. (1988). Optimality in human motor performance: Ideal control of rapid aimed movements. Psychological Review, **95**, 340-370.

Meyer, D., Smith, J.E.K., Kornblum, S., Abrams, R.A., & Wright, C.E. (1989). Speed-accuracy tradeoffs in aimed movements: Towards a theory of rapid voluntary action. In M. Jeannerod (Ed.) Attention and Performance XIII, Hillsdale, NJ: Lawrence Erlbaum.

Moore, S. & Marteniuk, R.G. (1986). Kinematic and electromyographic changes that occur as a function of learning a time-constrained aiming task. Journal of Motor Behavior, **18**, 397-426.

Pélisson, D., Prablanc, C., Goodale, M.A., & Jeannerod, M. (1986). Visual control of reaching movements without vision of the limb. II. Evidence of fast unconscious processes correcting the trajectory of the hand to the final position of a double-step stimulus. Experimental Brain Research, **62**, 303-311.

Rosenbaum, D. (1980). Human movement initiation: Specification of arm, direction, and extent. Journal of Experimental Psychology: General, **109**, 444-474.

Schmidt, R.A., Zelaznik, H.N., Hawkins, B., Frank, J.S., & Quinn, J.T. (1979). Motor output variability: A theory for the accuracy of rapid motor acts. Psychological Review, **86**, 415-451.

Schmidt, R.A., Sherwood, D.E., Zelaznik, H.N., & Leikind, B.J. (1985). Speed-accuracy trade-offs in motor behavior: Theories of impulse variability. In H. Heuer, U. Kleinbeck, & K.-H. Schmidt (Eds.), Motor behavior: Programming control, and acquisition (pp. 79-123). Berlin: Springer Verlag.

Soechting, J.F. (1984). Effect of target size on spatial and temporal characteristics of a pointing movement in man. Experimental Brain Research, **54**, 121-132.

Taylor, F.V., & Birmingham, H.P. (1948). Studies of tracking behavior, II: The acceleration pattern of quick manual corrective responses. Journal of Experimental Psychology, **28**, 783-795.

Todor, J.I., & Cisneros, J. (1985). Accomodation to increased accuracy demands by the right and left hands. Journal of Motor Behavior, **17**, 355-372.

Tsiboulevsky, I.E. (1981). Notes on the theory of the accuracy of rapid movements proposed by R. Schmidt and co-authors. (R. Browning, trans.) Voprosy Psikhologii, No. 3, 127-131.

Waters, P., & Strick, P.L. (1980). Influence of "strategy" on muscle activity during ballistic movements. Brain Research, **207**, 189-197.

Welford, A.T. (1968). The fundamentals of skill. London: Methuen.

Welford, A.T. (1984). Between bodily changes and performance: some possible reasons for slowing with age. Experimental Aging Research, **10**, 73-88.

Woodworth, R.S. (1899). The accuracy of voluntary movement. psychological Review, **3**, 1-114.

Worringham, C.J. (Ed.) (1989). Spatial, temporal and electromyographical variability in human motor control (Proceedings of a symposium held in Ann Arbor, Michigan, 17-18 February, 1989.

Worringham, C.J. (1991). Variability effects on the internal structure of rapid aiming movements. Journal of Motor Behavior (in press).

Wright, C.E., & Meyer, D.E. (1983). Conditions for a linear speed-accuracy trade-off in aimed movements. Quarterly Journal of Experimental Psychology: Human Experimental Psychology, **35A**, 279-296.

Zelaznik, H.N., Shapiro, D.C., & McColsky, D. (1981) Effects of a secondary task on the accuracy of single aiming movements. Journal of Experimental Psychology: Human Perception and Performance, **7**, 1007-1018.

REACHING TO VISUAL TARGETS: COORDINATE SYSTEMS REPRESENTATION IN PREMOTOR AND MOTOR CORTICES

R. CAMINITI
P. B. JOHNSON
S. FERRAINA
Y. BURNOD
Institute of Physiology
University "La Sapienza"
P.le A. Moro 5, 00185 Rome
Italy

ABSTRACT. Individual arm-related neurons in both motor (area 4) and premotor (area 6) cortices of the monkey are directionally tuned. We studied these directional neurons while monkeys made arm movements of similar directions within different parts of 3-D space. The behavioral task was aimed at dissociating the direction of movement, which remained similar across the work space, from the pattern of muscular activity and joint rotations underlying these movements. Within a given part of space, motor and premotor cortical cells fired most for a given preferred direction and less for other directions of movement. These preferred directions covered the directional continuum in a uniform fashion across the work space. As movements of similar directions were made within different parts of the work space, the cells' preferred directions in both motor and premotor cortices changed their orientation. Although these changes had different magnitudes for different cells, at the population level, they followed closely the changes in orientation of the arm necessary to move the hand from one part of the work space to another.

In both premotor and motor cortices, neuronal movement population vectors accurately described the direction of movement. In contrast to the individual cells, neuronal movement population vectors did not change their spatial orientation across the work space, suggesting that they remain good predictors of movement direction regardless of the region of space in which movements are made.

Introduction

Reaching to visual targets provides an excellent model for studying the coordinate system used by the cerebral cortex to represent arm movement in 3-D space. In addition, it offers the opportunity to approach the analysis of the visuo-motor transformations required to move the hand toward a desired target which has been located in space by using visual information.

To address these issues we studied the activity of individual neurons in the premotor and motor cortices of monkeys trained to perform arm movements of similar direction within different parts of space. In this way the direction of movement was kept constant across the work space while the underlying patterns of muscle activity and joint excursions changed. In so doing, intrinsic coordinates were dissociated from extrinsic ones.

J. Requin and G. E. Stelmach (eds.), Tutorials in Motor Neuroscience, 457–462.

Materials and Methods

Macaca nemestrina monkeys were trained to perform arm movements at visual targets within 3 (left, center, right) different parts of the space directly in front of them. Within each sub-space (Fig. 1) monkeys made equal-amplitude arm movements of common origin in 8 different directions.

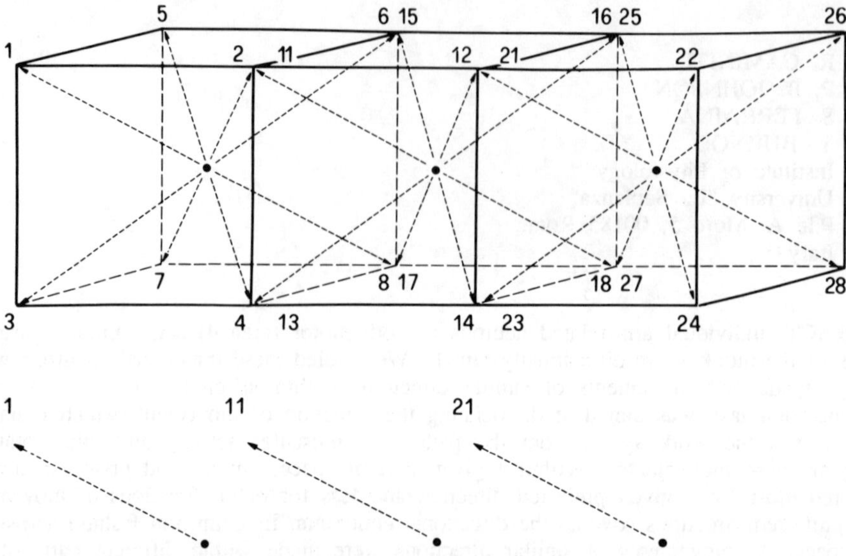

Fig. 1. Apparatus and task. Top, layout of the work space. The animal was seated on a primate chair, 25 cm away from the front lights. The center of the center cube was aligned with the body midline at shoulder height. The animal performed 3 sets of movement directions in the left, center and right parts of the work space. Black dots indicate the 3 movement origins within each part of space where monkeys made equal-amplitude (8.7 cm) movements of common origin in 8 different directions (arrows). Certain push-buttons are labeled by 2 numbers (2, 11; 12, 21; etc.) because they were targets of movements of two different origins. Numbers identify directions of movement. Bottom, triplet (1-11-21) of movements of similar direction performed across the work space.

Across the entire work space they made 8 triplets (1-11-21, 2-12-22, etc.) of movements of similar direction requiring different patterns of muscle activity and shoulder joint rotations. Hand trajectories were recorded by using a 3-D sonic tracking system while the activity of several muscles acting at the shoulder joint was monitored through intramuscular electrodes.

The activity of individual neurons was recorded extracellularly from the hemisphere contralateral to the performing arm and the location of microelectrode penetrations was verified by using conventional histological procedures.

Results

BEHAVIORAL STUDIES

Figure 2 show movement trajectories in directions 4-14-24. These trajectories had very similar spatial orientations but their performance required, as for the other 7 triplets of parallel movements, initial arm placements in 3 different parts of space. Rotations of the shoulder joint of 18° and 20° were necessary to bring the arm from left (trajectory 4) to center (trajectory 14), and center to right (trajectory 24) initial positions respectively.

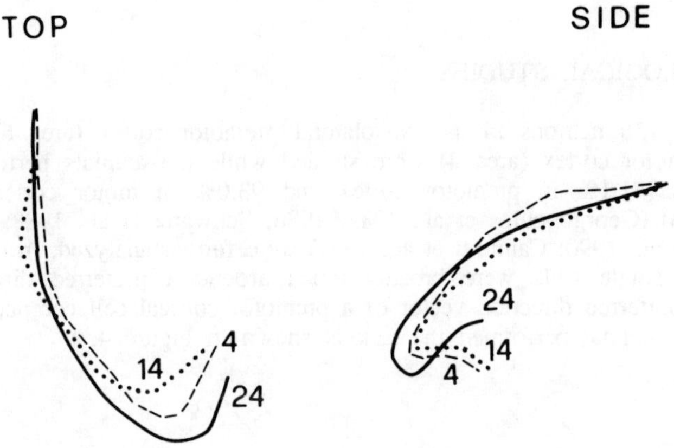

Fig. 2. Two-dimensional plots of top and side views of hand trajectories for movements in directions 4, 14 and, 24. Trajectories were superimposed to the same movement origin to show their degree of similarity.

The pattern of activity of several muscles underlying movement trajectories in direction 4-14-24 is shown in Figure 3.

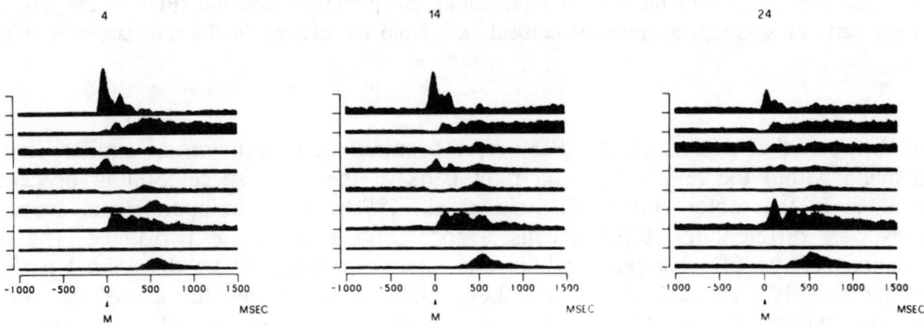

Fig. 3. EMG activity of 9 muscles recorded during movement traveling along trajectories 4-14-24. Muscles, from top to bottom are: caudal trapezius, cranial trapezius, spinal deltoid, clavicular deltoid,

long head of triceps, lateral head of triceps, biceps longus, teres major, and pectoralis. Numbers on the abscissa are in milliseconds relative to the movement onset (0). Records are aligned to the movement onset.

It can be seen that the activity of many muscles changed notably when these movements were performed across the work space. Statistically significant space-direction interactions (ANOVA, F-test, $p < 0.05$) were observed for the triceps, caudal trapezius, spinal deltoid, clavicular deltoid and pectoralis. For all triplets of parallel movement directions the pattern of synergy changed because of significant changes occurring in the activity of at least 2 muscles.

NEUROPHYSIOLOGICAL STUDIES

The activities of 156 neurons in the dorsolateral premotor cortex (area 6) and of 207 neurons in the motor cortex (area 4) were studied while the animals performed the task. Only those cells (71.1% in premotor cortex and 93.0% in motor cortex) which were directionally tuned (Georgopoulos et al., 1982; 1986; Schwartz et al., 1988; Kalaska et al., 1989; Caminiti et al., 1990; Caminiti et al., 1991) were further analyzed. In both these areas the activities of single cells were broadly tuned around a preferred direction (PD) of movement. The preferred direction vector of a premotor cortical cell computed from neural activity while the animal performed the task is shown in Figure 4.

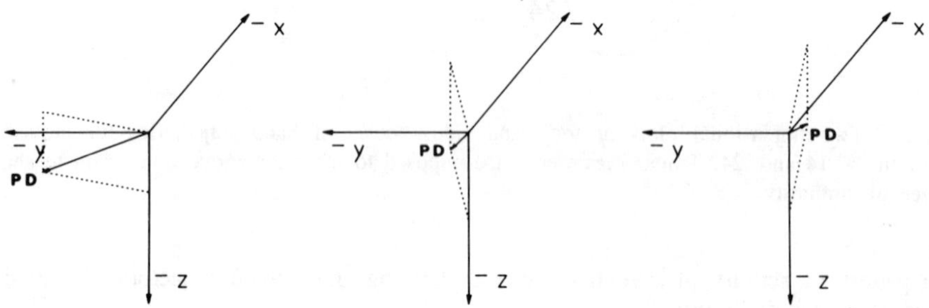

Fig. 4. Spatial orientation of the vectors representing the preferred direction (PD), in the left, center, and right parts of space of a premotor cortical cell. Note the change in the orientation of this cell's PD.

It can be seen that a shift of this PD vector occurred as movements of similar directions were made within the left, center and right parts of the work space. Similar observations were made in the motor cortex (Caminiti et al., 1990). Since in both frontal areas these changes were different for different cells, a population analysis was performed. The spatial distributions of the PDs computed within the 3 parts of space where the animal performed were significantly correlated (Fisher and Lee, 1986). This allowed the use of the spherical regression analysis (Jupp and Mardia, 1980) to detect whether an orderly change of the population of cell PDs had occurred and, if so, along which spatial axis. In area 6, the results showed a rotation of 17.2° between PDs from left and those from the center parts of the work space; a similar rotation of 17.6° was observed between PDs from center and right parts of space while a rotation of 42° separated PDs from left and right parts of the work space. These rotations occurred around the Z axis, i.e., on the horizontal plane. Rotations around the X and Y axes were small and not significantly different from zero. Very similar results were obtained in the motor cortex (Caminiti et al., 1990). Thus, in both frontal areas the rotation of PDs parallels very closely the rotation of the shoulder joint necessary to bring the arm into the 3 parts of the work space were the animal performed the task.

An additional question concerns whether the shift of cell PDs had any effect on the spatial orientation of the neuronal movement population vector (MPV; Georgopoulos et al., 1983, 1986; Kalaska et al., 1989; Caminiti et al., 1990), which is a good predictor of movement direction. A spherical regression analysis performed on the MPVs from the left, center and right parts of the work space showed that they did not rotate with respect to each other.

Discussion

There are two main points to be discussed in this study. The first relates to the shift of cell PDs observed as movements of similar direction were performed within different parts of space. The second refers to the invariance of the spatial orientation of the MPV observed in the same conditions.

In both premotor and motor cortices, the shift of PDs with the initial position of the arm indicates the existence of an invariant relationships between cell PD and arm orientation in space and suggests that these frontal areas use common mechanisms to code arm movement direction. The primate frontal lobe would therefore contain an internal representation of space where coding of arm movement occurs within a coordinate system centered on the shoulder joint. Within this frame of reference the computation of a command appropriate to move the arm toward a visual target would occur through the combination of two different inputs, an intrinsic one concerning the position of the arm relative to the body (angles of the shoulder joint) and an extrinsic one concerning the trajectory in space as defined by the visual system. Within this frame, the vectorial information about the desired trajectory of the hand is "projected" on that concerning the orientation of the arm in space as a way of directly relating the visual input to the motor output. These frontal cortical neurons may therefore be assigned a role in the transformation of visually-derived information into an appropriate motor command. It is interesting that similar conclusion were drawn by Soechting and Flanders (1989 a, 1989 b) on the basis of behavioral studies.

The broad directional tuning of motor (Georgopoulos et al., 1982, 1986; Schwartz et al., 1988; Kalaska et al., 1989; Caminiti et al., 1990) and premotor (Caminiti et al., 1990) cortical cells raises interesting questions about thc possibility that movement direction is a population rather than a single cell code. Broad or coarse tuning is a prerequisite of population codes (Sejnowski, 1986). The actual direction of the movement of the arm is, in fact, predicted by a population of cortical neurons: by summing the vectorial contribution of their preferred directions weighted by their activities, the resulting neuronal movement population vectors will predict well the direction of the forthcoming arm movement (Georgopoulos et al., 1983, 1986; Kalaska et al., 1989; Caminiti et al., 1990). When movement of similar direction were performed within different parts of space, in both premotor (Caminiti et al., 1990) and motor (Caminiti et al., 1990) cortices, neuronal movement population vectors, unlike the individual cells upon which they are based, did not change their spatial orientation, suggesting that they remain excellent representations of movement direction regardless of where in space movements are made and that they relate to movement kinematics more than to movement dynamics.

Acknowledgements: This work was supported by The European Economic Community (grant SCI 0028-c/A), the Consiglio Nazionale delle Ricerche, Rome, Italy, and by USPHS grant NS07166.

References

Caminiti R, Johnson, P. B. and Urbano, A. (1990) "Making arm movements within different parts of space: Dynamic aspects in the primate motor cortex", J. Neurosci. 10, 2939-2058.

Caminiti, R., Johnson, P. B., Burnod, Y., Galli, C., Ferraina, S. and Urbano A. (1990) "Shift of preferred directions of premotor cortical cells with arm movements performed across the work-space", Exp. Brain Res. (in press).

Caminiti, R., Johnson, P. B., Galli, C., Ferraina, S. and Burnod, Y. (1991) "Making arm movements within different parts of space: The premotor and motor cortical representation of a coordinate system for reaching to visual targets", J. Neurosci. (in press).

Fisher, N. I. and Lee, A. J. (1983) "Correlation coefficients for random variables on a unit sphere or hypersphere", Biometrika 73, 159-164.

Georgopoulos, A. P., Kalaska, J. F., Caminiti, R. and Massey, J. T. (1982) "On the relations between two-dimesional arm movements and cell discharge in Primate motor cortex", J. Neurosci. 11, 1527-1537.

Georgopoulos, A. P., Caminiti, R., Kalaska, J. F. and Massey, J. T. (1983) "Spatial coding of movement: A hypothesis concerning the coding of movement direction by motor cortical populations", Exp. Brain Res. (Suppl.) 7, 327-336.

Georgopoulos, A. P., Schwartz, A. B. and Kettner, R. (1986) "Neuronal population coding of movement direction", Science, 233, 1416-1419.

Jupp, P. E. and Mardia, K. V. (1980) "A general correlation coefficient for directional data and related regression problems", Biometrika 67, 163-173.

Kalaska, J. F., Cohen, D. A. D., Hyde, M. L. and Prud'homme, M. (1989) "A comparison of movement direction-related vs. load direction-related activity in primate motor cortex, using a two dimensional reaching task", J. Neurosci. 9, 2080-2102.

Schwartz, A. B., Kettner, R. E. and Georgopoulos, A. P. (1988) "Primate motor cortex and free arm movements to visual targets in three-dimensional space. I. Relations between single cell discharge and direction of movement", J. Neurosci. 8, 2913-2927.

Sejnowski, T. J. (1986) "Open questions about computation in cerebral cortex", in J. A. Feldman, P. J. Hayes and D. E. Rumelhart (eds), Parallel Distributed Processing, Vol. 2, MIT Press, Cambridge, pp. 372-389, Massachusetts.

Soechting, J. F., and Flanders, M. (1989a) "Sensorimotor representations for pointing to targets in three-dimensional space", J. Neurophysiol. 62, 582-594.

Soechting, J. F., and Flanders, M. (1989b) "Errors in pointing are due to approximations in sensorimotor transformations", J. Neurophysiol. 62, 595-608.

VISUAL GUIDANCE OF POINTING MOVEMENTS: KINEMATIC EVIDENCE FOR STATIC AND KINETIC FEEDBACK CHANNELS

N. Teasdale, J. Blouin, C. Bard, & M. Fleury
Université Laval
Laboratoire de Performance Motrice Humaine
Ste-Foy, Québec, G1K 7P4

KEYWORDS / ABSTRACT: movement control / vision / aiming movements / control loops / visual feedback / direction

Based on physiological and psychophysical studies, Paillard (1980) has suggested that two visual feedback channels (static and kinetic) are involved for the guidance of goal directed movements. The present experiment was designed to test the importance of the vision during the initial phase of the movement when this phase is under control of the kinetic channel. Results showed that vision of the initial portion of the trajectory was essential for the control of the directional component of the movements.

Introduction

The role of visual information in the control of movements is a central issue in the study of motor behavior. For aiming-pointing movements, it is generally believed that an initial preprogramed phase brings the limb toward the target, and that a terminal phase, through control loops, is responsible for corrective sub-movements necessary to stop the limb precisely at the target (Meyer, Abrams, Kornblum, Wright, & Smith, 1988; Woodworth, 1899). According to this framework, the contribution of visual feedback in the initial portion of the trajectory is negligible. For instance, Carlton (1981) and Beaubaton and Hay (1986) for pointing movements, have reported that withdrawing visual feedback in the initial phase of the movements had no effect on terminal accuracy, whereas withdrawing visual feedback

J. Requin and G. E. Stelmach (eds.), Tutorials in Motor Neuroscience, 463–475.

during the terminal phase of the movements resulted in decreased accuracy.

Paillard (1980, 1982) proposed that on-line correction of visually guided movement proceeds through two different feedback loops: (a) peripheral vision, mostly sensitive to visual movement cues, provides directional error signals that allow correction of the trajectory in the initial phase of the movements; (b) central vision, highly sensitive to position cues, provides positional error signals that allow the homing in of the hand on the target in the terminal phase of a pointing movement. Paillard and Amblard (1985), based on neurophysiological and psychophysical studies (e.g., Bonnet & Renaud, 1977; Orban et al., 1981), also suggested a neural model for the segregation of the two visual channels. The static channel is subserved by the X fibres originating predominantly from the central retina, whereas the kinetic channel is supported by the Y fibres originating from the peripheral retina. Table 1 presents the respective characteristics of the two channels.

TABLE 1. Respective characteristics for the static and kinetic channels.

	STATIC	KINETIC
Frequency filters:		
spatial	**High-pass**	**Low-pass**
temporal	**Low-pass**	**High-pass**
Retinal Field:	**Central up to 15°**	**Peripheral**
Feature analysis:	**Stationary pattern**	**Movement**
Movement perception:	**Low vel. range**	**High vel. range**
	Up to 15°/s	**From 10 to 200°/s**
Reactivity types:	**X fibres**	**Y fibres**
	Sustained	**Transient**

Adapted from Paillard & Amblard (1985)

The possible contribution of peripheral vision to the correction of pure directional movements (without amplitude requirement) has been advanced in a series of experiments on children (Bard, Hay & Fleury, 1985a, 1990) and adults (Bard et al., 1985b). Recently, Bard, Paillard, Fleury, Hay & Larue (1990) have addressed the validation of the hypothesis in pointing and aiming tasks. Results have shown that amplitude accuracy requires a signal error provided by central vision (up to 10 degrees eccentricity), whereas direction accuracy is significantly improved when vision of the trajectory is restricted to the first part of the trajectory and to peripheral vision.

In the present experiment, a similar experimental paradigm to Bard et al. (1990) was used. The working range of the kinetic and static channels were restrained to the beginning and end portions of the pointing trajectory to evaluate their respective contribution for movement control (i.e., directional control for the kinetic system and homing in positional control for the static system). Aiming and pointing accuracy improvements may obviously have their counterpart in some changes in the kinematic parameters of the trajectory, thereby providing a crucial test for conforming or disproving the hypothesis of a dual channel for the guidance of visually guided movements.

Method

SUBJECTS

Six volunteers, naive to the purpose of the experiment, participated in the experiment.

APPARATUS

Subjects sat on an adjustable seat, with their chest leaning against a vertical restraint. A hand-held pointer extended from the floor between the legs. The pointer could be moved within the transversal plane and was mounted onto a universal joint attached to the floor. A small light emitting diode (LED) was fixed on the tip of the pointer. In the starting position, the tip of the pointer was at eye level, 38 cm from a target-LED. In that position, the target was binocularly focused in central vision, whereas the retinal image of the initial position of the pointer was at 40 degrees eccentricity (nasal). Therefore, the first half of the trajectory swept the peripheral retina toward the fovea (see Figure 1).

The position of the tip of the pointer was obtained through two linear potentiometers fixed on a steel frame facing the subject; small gauge wires were attached from 4 cm below the

tip of the pointer to these potentiometers. The signals from the potentiometers were digitized at 500 Hz and the cartesian position of the pointer was obtained through

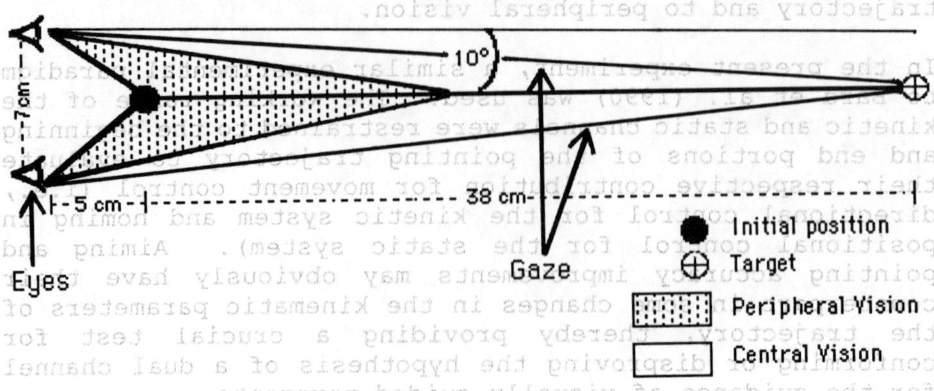

Figure 1. Vision of the subject in the experimental situation.

trigonometric transformations. The displacement-time data were filtered twice with a Butterworth second-order low-pass filter with a cutoff frequency of 8 Hz. This procedure results in a fourth-order, zero-phase shift filter with a cutoff frequency of 6.4 Hz (Oppenheim & Willsky, 1983; Winter, 1979). The displacement signals were then differentiated numerically twice with a central finite difference technique to obtain velocity- and acceleration-time curves. This technique has been shown to provide valid estimates of first and second derivatives (Pezzack, Normand & Winter, 1977; van der Meulen, Gooskens, Denier van der Gon, Gielen & Wilhelm, 1990; Wood, 1982) and is thought to attenuate movement components possibly associated with physiological tremor (Stein & Lee, 1981) or oscillations originating from the mechanical characteristics of the arm (van der Meulen, 1990; Walter, 1985).

EXPERIMENTAL CONDITIONS

The subjects were instructed to move the pointer as fast and as precisely as possible from the initial position to the target under three different constraints. Specifically, subjects had to (a) stop under the target (Amplitude +

Directional requirements), (b) stop under the target, but horizontal tracks were added to remove the directional component of the movement (Amplitude requirement only), and (c) pass under the target and follow through the movement without stopping under the target (Directional requirement only).

For these three constraints, all movements were performed in total darkness and vision of the pointer was varied such that it was (a) not available throughout the trajectory (open loop), (b) not available during the first 1/3 of the trajectory, (c) available during the first 1/3 of the trajectory, and (d) available throughout the trajectory (closed loop).

For all constraints and visual conditions, 40 trials were given; the first 15 trials were considered practice trials. For each task constraint (randomly presented), the closed-loop conditions were always presented first. Other visual conditions were randomly presented.

Results

Data obtained for the different dependent variables were all submitted to a 3 (task constraints) x 4 (visual conditions) analysis of variance with repeated measures on both factors.

DIRECTIONAL ERRORS FOR THE DIRECTION AND AMPLITUDE + DIRECTION TASKS

One purpose of the experiment was to demonstrate the importance of vision in the initial phase of the trajectory when the directional component is under the control of the kinetic system. On average, the duration of the initial part of the trajectory (i.e., first third) lasted 80 ms. Directional errors for the Direction and the Amplitude + Direction tasks for the different visual conditions are presented in Figure 2. Results showed a significant main effect of Vision (F (3,15) = 3.48, $p < .05$), but no effect of Task (F (1,5) = 0.10, $p > .05$). A comparison of means showed that, when vision was available for the initial portion only, the directional errors (-6.5 mm) were smaller than in the open-loop condition (-13.4 mm, F (1,15) = 7.18, $p < .05$), and similar to the closed loop (-5.9 mm, F (1,15) = 8.36, $p < .05$). Therefore, vision in the initial portion of the trajectory alone, contributed to the reduction of the directional error when compared to an open loop condition.

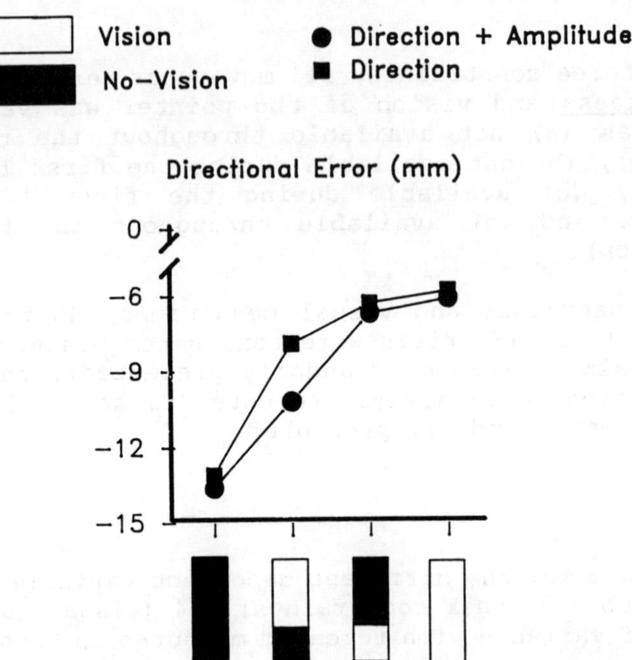

Figure 2. Directional errors for the Direction and Amplitude + Direction tasks. For the visual legends, the starting position of the movement is represented at the bottom, and the target position at the top.

AMPLITUDE ERRORS FOR THE AMPLITUDE AND AMPLITUDE + DIRECTION TASKS

The ANOVA for the errors in amplitude showed no difference across the different tasks and visual conditions (on average, 7.5 mm, $ps > .05$). The static positional system is responsible for the correction of errors in amplitude through a comparison of the retinal coordinates of the pointer and the target. The static system is thought of as a low-pass velocity filter. The task constraints permitted the static positional system to operate only during the last portion of the trajectory (on average, 78 ms for the Amplitude and Amplitude + Direction tasks) and did not allow for rapid on-line corrections.

MOVEMENT TIME (MT)

MTs for the different tasks and visual conditions are presented in Figure 3. The ANOVA showed significant main effects of Vision ($\underline{F}$ (3,15) = 27.83, $\underline{p}$ < .001) and Task ($\underline{F}$ (2,10) = 133.38, $\underline{p}$ < .001). A comparison of means showed that MTs for the three constraints were all different from each other (132, 159, and 232 ms for the Direction, Direction + Amplitude, and Amplitude tasks, respectively, $\underline{p}$s < .001). The interaction of Vision x Task was also

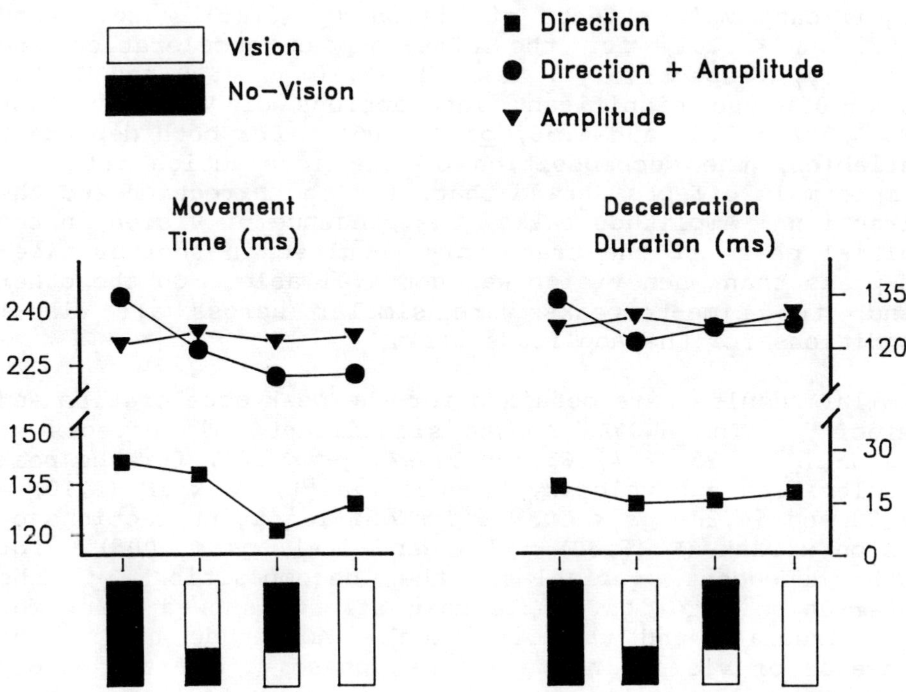

Figure 3. Movement times and duration of the deceleration phases for the different task and visual conditions. For the visual legends, the starting position of the movement is represented at the bottom, and the target position at the top.

statistically significant (F (6,30) = 6.87, p < .001). A decomposition of the interaction into its simple main effects showed that for the Direction and the Amplitude + Direction constraints, the presence of vision in the initial phase of the trajectory yielded shorter MTs than when visionwas not available (F (3,43) = 7.98 and 26.42, ps < .001, respectively). On the other hand, MTs were constant across the different visual conditions for the Amplitude task (F (3,43) = 0.21, p > .05).

TASK CONSTRAINTS, VISUAL CONDITIONS AND MOVEMENT KINEMATICS

The data obtained for the time-to-peak acceleration and velocity and their respective amplitudes are presented in Figure 4. The results obtained from the ANOVAs, for both the time-to-peak acceleration and velocity, showed significant main effects of Vision (F (3,15) = 20.70 and 6.56, ps < .01, for the time-to-peak acceleration and velocity, respectively), Task (F (2,10) = 6.19 and 21.55, ps < .05) and significant interactions of Vision x Task (F (6,30) = 3.50 and 6.56, ps < .005). For both dependent variables, the decomposition of the interaction into its simple main effects showed that, for the Direction and the Direction + Amplitude tasks, the presence of vision in the initial phase of the trajectory resulted in shorter time-to-peaks than when vision was not available. On the other hand, the time-to-peaks were similar across all visual conditions for the Amplitude task.

Similar results were obtained for the peak acceleration and velocity. The ANOVAs showed significant main effects of Vision (F (3,15) = 41.61 and 11.44, ps < .01, for the peak acceleration and velocity, respectively), Task (F (2,10) = 17.33 and 64.24, ps < .05) and significant interactions of Vision x Task (F (6,30) = 9.48 and 3.91, ps < .005). For both dependent variables, the decomposition of the interaction into its simple main effects showed that, for the Direction and the Direction + Amplitude tasks, the presence of vision in the initial phase of the trajectory resulted in larger peaks than when vision was not available. On the other hand, the peaks were similar across all visual conditions for the Amplitude task.

The duration of the deceleration phase (Figure 3) varied as a function of the Task (17, 125, and 128 ms for the Direction, Amplitude + Direction and Amplitude tasks;

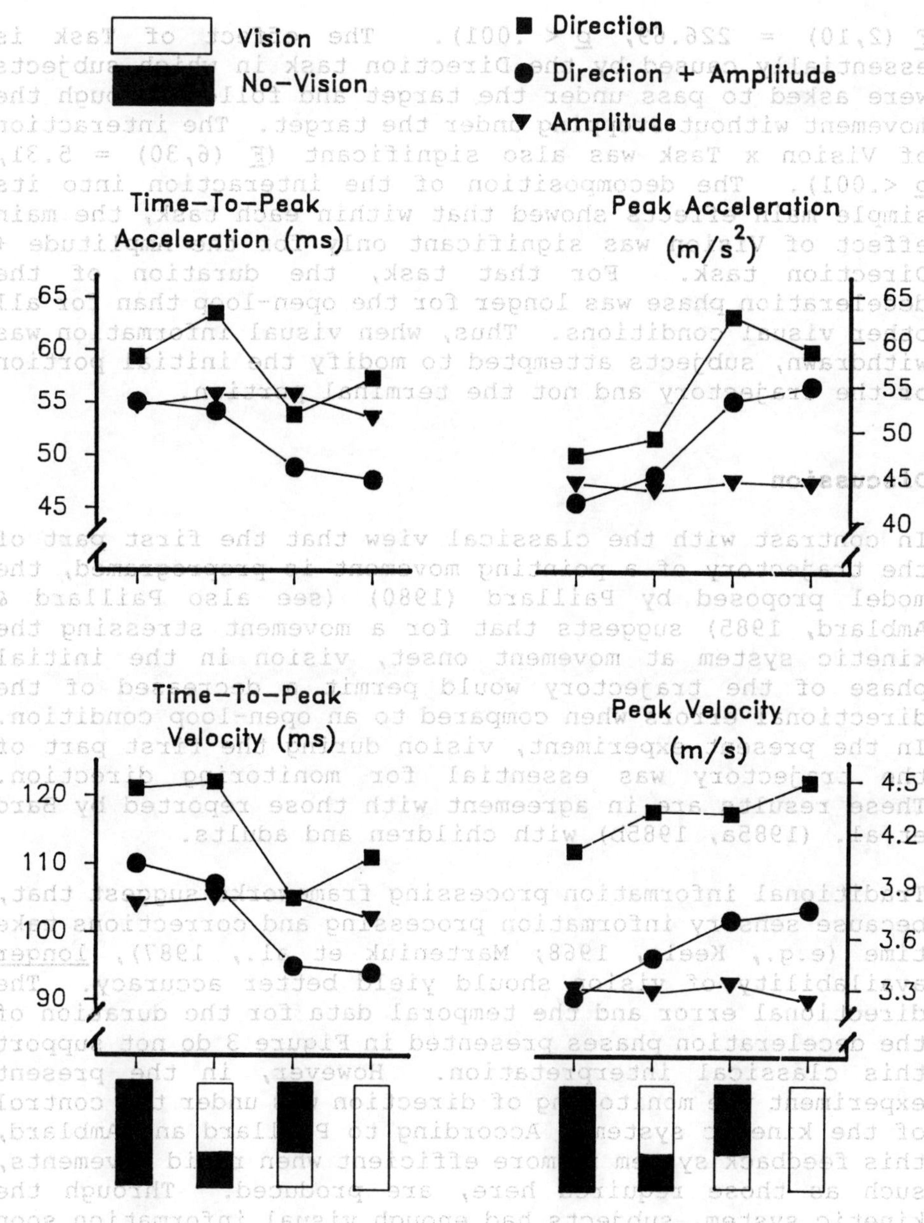

Figure 4. Time-to-peak acceleration and velocity and their respective amplitudes for the different tasks and visual conditions. For the visual legends, the starting position of the movement is represented at the bottom, and the target position at the top.

$\underline{F}$ (2,10) = 226.69, $\underline{p}$ < .001). The effect of Task is essentially caused by the Direction task in which subjects were asked to pass under the target and follow through the movement without stopping under the target. The interaction of Vision x Task was also significant ($\underline{F}$ (6,30) = 5.31, $\underline{p}$ <.001). The decomposition of the interaction into its simple main effects showed that within each task, the main effect of Vision was significant only for the Amplitude + Direction task. For that task, the duration of the deceleration phase was longer for the open-loop than for all other visual conditions. Thus, when visual information was withdrawn, subjects attempted to modify the initial portion of the trajectory and not the terminal portion.

Discussion

In contrast with the classical view that the first part of the trajectory of a pointing movement is preprogramed, the model proposed by Paillard (1980) (see also Paillard & Amblard, 1985) suggests that for a movement stressing the kinetic system at movement onset, vision in the initial phase of the trajectory would permit a decreased of the directional errors when compared to an open-loop condition. In the present experiment, vision during the first part of the trajectory was essential for monitoring direction. These results are in agreement with those reported by Bard et al. (1985a, 1985b) with children and adults.

Traditional information processing frameworks suggest that, because sensory information processing and corrections take time (e.g., Keele, 1968; Marteniuk et al., 1987), longer availability of vision should yield better accuracy. The directional error and the temporal data for the duration of the deceleration phases presented in Figure 3 do not support this classical interpretation. However, in the present experiment the monitoring of direction was under the control of the kinetic system. According to Paillard and Amblard, this feedback system is more efficient when rapid movements, such as those required here, are produced. Through the kinetic system, subjects had enough visual information soon after the initiation of the movement (less than 78 ms) to control the directional component and could also respond to the high speed requirements of the task (trials with MT > 250 ms were rejected).

When vision was not available in the initial phase of movements with a directional component, subjects modified the accelerative (not the decelerative) portion of their movement (except for the open-loop condition in the Direction + Amplitude task). Thus, these results highlight the importance of vision during the initial phase for directional control of goal-directed movements. Results obtained for the Amplitude task (triangles in Figures) suggest that kinematic modifications, observed in the initial phase of the trajectory, were essentially associated with directional control. Indeed, when the movement had no directional components, the role of the kinetic system was negligible. Further, the movement speed did not permit the static system to operate in its optimal working range. For example, the durations of the last third of the trajectory lasted on average 78 ms. Past experiments have shown that, for movements performed in central vision, this duration is too short to permit positional feedback loops to operate at optimal level (Beaubaton & Hay, 1986; Carlton, 1981; Keele, 1968; Zelaznik, Hawkins, & Kisselburgh, 1983). As a result, subjects unconsciously adopted a control strategy that was similar across all visual conditions.

Overall, the data support the existence of two corrective visual channels. Depending upon task requirements and environmental constraints, the visual channels complement, supplement or mutually exclude each other. Clearly, future models of aiming-pointing movements will need to account for the existence of kinetic and static visual channels.

References

Bard, C., Hay, L., & Fleury, M. (1985a). Contribution of vision to the performance and learning of a directional aiming task in children aged 6, 9, and 11. In Y. E. Clark & J. H. Humphrey (Eds.), Motor development current selected research, Vol. 1 (pp. 19-33). Princeton: Princeton Book Company.

Bard, C., Hay, L., & Fleury, M. (1985b). Role of peripheral vision in the directional control of rapid aiming movements. Canadian Journal of Psychology, 39, 151-161.

Bard, C., Hay, L., & Fleury, M. (1990). Timing and accuracy of visually directed movements in children: control of direction and amplitude components. Journal of Experimental Child Psychology, 50, 102-118.

Bard, C., Paillard, J., Fleury, M., Hay, L., & Larue, J. (1990). Positional versus directional control loops in visuomotor pointing. European Bulletin of Cognitive Psychology, 2, 145-156.

Beaubaton, D., & Hay, L. (1986). Contribution of visual information to feedforward and feedback processes in rapid pointing movements. Human Movement Science, 5, 19-34.

Bonnet, C., & Renaud, C. (1977). La détection du mouvement visuel en vision centrale et en vision périphérique. L'Année Psychologique, 77, 113-121.

Carlton, L. G. (1981). Processing visual feedback information for movement control. Journal of Experimental Psychology: Human Perception and Performance, 7, 1019-1030.

Meyer, D. E., Abrams, R. A., Kornblum, S., Wright, C. E., & Smith, J. E. K. (1988). Optimality in human motor performance: Ideal control of rapid aimed movements. Psychological Review, 95, 340-370.

Oppenheim, A. V., & Willsky, A. S. (1983). Signals and systems. New Jersey: Prentice-Hall.

Orban, G. A., & Kennedy, H., & Maes, H. (1981). Response to movement of neurons in areas 17 and 18 of the cat: direction selectivity. Journal of Neurophysiology, 45, 1059-1073.

Paillard, J. (1980). The multichanneling of visual cues and the organization of a visually guided response. In G. E. Stelmach & J. Requin (Eds.), Tutorials in motor behavior (pp. 259-279). Amsterdam: North Holland.

Paillard, J. (1982). The contribution of peripheral and central vision to visually guided reaching. In D. J. Ingle, M. A. Goodale, & R. J. W. Mansfield (Eds.), Analysis of visual behavior (pp. 367-385). Cambridge: The MIT Press.

Paillard, J., & Amblard, B. (1985). Static versus kinetic visual cues for the processing of spatial relationships. In D.J. Ingle, M. Jeannerod, & D. N. Lee (Eds.), Brain mechanisms of spatial vision (pp. 367-385). La Haye: Martinus Nijhoff.

Pezzack, J. C., Normand, R. W., & Winter, D. A. (1977). An assessment of derivative determining techniques used for motion analysis. Journal of Biomechanics, 10, 377-382.

Stein, R. B., & Lee, R. G. (1981). Tremor and clonus. In V. B. Brooks (Ed.), Handbook of physiology: Motor control (pp. 325-344). Bethesda: American Physiological Society.

Van der Meulen, J. H. P., Gooskens, R. H. J. M., Denier van der Gon, J. J., Gielen, C. C. A. M., & Wilhelm, K. (1990). Mechanisms underlying accuracy in fast goal-directed arm movements in man. Journal of Motor Behavior, 22, 67-84.

Winter, D. (1979). Biomechanics of human movement. New York: Academic Press.

Walter, C. B. (1985). Independent control of initial kinematics and terminal oscillations of rapid positioning movements. Experimental Brain Research, 60, 402-406.

Wood, G. A. (1982). Data smoothing and differentiation procedures in biomechanics. Exercise and Sport Sciences Reviews, 10, 308-362.

Woodworth, R. S. (1899). The accuracy of voluntary movements. Psychological Review, 3, 54-59.

Authors' note

This work was supported by NSERC grants to Normand Teasdale, Chantal Bard and Michelle Fleury. Thanks to Dr. Jacques Paillard for his insightful comments on a previous version of this paper and to Benoit Genest and Gilles Bouchard for programming and technical expertise.

THE CORTICO-MOTOR SUBSTRATE FOR SKILLED MOVEMENTS OF THE PRIMATE HAND.

R.N. Lemon, K.M. Bennett and W.Werner
Department of Anatomy
Cambridge University
Cambridge, England

ABSTRACT. This article relates the biomechanical structure of the hand to the neurophysiology and neuroanatomy of the cortico-motoneuronal component of its control system. Biomechanically, the hand coordinates various stabilising and movement configurations in order to perform a variety of functions including prehension and manipulation. Columns and rows of intercalated bony segments are moulded by the interplay of both the passive and active properties of the intrinsic and extrinsic hand musculature. Anatomical and physiological studies indicate that the monosynaptic connections between cortical pyramidal cells and spinal motoneurones provide the substrate for skilled movements of the primate hand. The organisation of the cortico-motoneuronal (CM) input to the hand has been studied using spike-triggered averaging and cross-correlation techniques. Corticospinal neurones with rapidly and slowly conducting axons directly facilitate motoneurones supplying hand muscles. These axons diverge intraspinally and make functional contact with a restricted number of different target muscles, the 'muscle field' of the CM cell. The significance of the pattern of branching, and the relative strength of facilitation of different target muscles is discussed in relation to the fractionation of muscle activity for independent finger control. Finally, the cortical organisation of the CM system is considered in terms of the output map within the motor cortex. The features of this map, and especially the multiple representation of a single muscle, and the overlap of cortical territories influencing different muscles is discussed in terms of the control of skilled hand movements.

1. THE BIOMECHANICAL SUBSTRATE FOR SKILLED HAND MOVEMENTS

Skilled use of the hand as a prehensile and manipulative machine requires independent finger movements. Higher primates achieve opposition by rotating the thumb so that its pulp contacts that of another digit (Napier, 1961). The human, with a greater proportional thumb length and a relatively mobile carpo-metacarpal (CMC) saddle joint, achieves the most efficient contact area between pulps of opposing digits. The efficiency and precision of this movement makes possible a wide range of prehensile and exploratory hand functions. As the evolutionary scale is climbed, the muscle bellies of the flexor digitorum profundus (FDP) have become separated and greater control of digit movement has been assigned to these long muscles (Landsmeer, 1989). Thus, the human hand demonstrates marked fractionation of both thumb and finger movement (Kuypers, 1982). The variety of movement and

J. Requin and G. E. Stelmach (eds.), Tutorials in Motor Neuroscience, 477–495.

gripping functions is the product of an articulated hand skeleton capable of many configurations controlled by the skilled interplay of the extrinsic and intrinsic muscles which move and stabilise it.

Each digit comprises a multi-articulate column of bony segments extending from the proximal carpal row to the distal phalanx. With an isometric opposition grip between the index finger and thumb two such columns must maintain their palmar concavity while exerting force at the finger tip or pulp. Although the gross alignment and force production is achieved by the long flexors (primarily flexor pollicis longus of the thumb and FDP of the index finger), stabilisation of each bone is finely modulated by joint geometry and intrinsic musculature. The flat geometry of the CMC joint of the index finger ensures its stability during the production of force but all other joints need additional support to prevent collapse of the arch (Spoor, 1983). For example, the ulnar directional force produced by the thumb upon the finger must be counteracted by the first dorsal interosseous (1DI). Similarly the inherent mobility at the CMC joint of the thumb is restrained by the concerted activity of all the thenar muscles (Chao et al. 1989). Some muscles appear to have a dual role: the transverse fibres of adductor pollicis can, for example, assist in production of MCP joint flexion while its oblique fibres provide MCP joint stabilisation (Chao et al 1989). In addition to the primary action in generating and maintaining a functional grip each active muscle has additional actions which may need to be countered. FDP, for example, pulls the proximal phalanx of the index finger in an ulnar direction, and this action can be counteracted by the 1DI. The interossei and lumbrical muscles exert axial rotary moments which are subtly balanced by appropriate antagonists.

In an 'auxotonic' or compliant grip, for example holding a pair of forceps, the picture becomes more complex. Landsmeer (1976), in his two-dimensional model of the hand, proposes that simultaneous activity of long finger extensors and flexors leads to a collapse whereby the finger becomes clawed (MCP extension and interphalangeal (IP) flexion). The oblique nature of the interossei, traversing both the palmar aspect of the MCP joint and the dorsal aspect of the proximal IP joint, counteracts this abnormal posture and promotes joint independence whereby the MCP joint angulates reciprocally with respect to the interphalangeal joint complex. EMG studies support this view and demonstrate the importance of the interossei in moulding the fingers to the form of an object (Long et al., 1970).

In the opening phase of the grip the interplay between the extrinsic and intrinsic muscles is again apparent. The extensor digitorum communis (EDC) is active whenever phalangeal extension occurs but also shows some activity during flexion movements (Close & Kidd, 1969; Long et al., 1970). This lengthening, or eccentric, control could reflect the need to maintain slight tension in the extensor hood to counteract the strong flexor pull and to provide a firm insertion point for the interossei and lumbricals (Chao et al. 1989). A synergistic equilibrium exists between EDC and the interossei with interphalangeal extension. With maximal MCP joint flexion and thus a reduced mechanical potential for the lateral expansions of the interossei, the efficiency of the EDC for extension is enhanced. If the MCP joint is extended, EDC is placed more at a mechanical disadvantage while the lateral interossei bands exert a strong extensor pull on the phalanges. This explains the importance of the 1DI in a precision grip posture requiring MCP flexion combined with IP extension.

The lumbricals are active only during IP extension and, as such, establish a link between the flexor and extensor assemblies to affect the balance and rate of distribution of flexor and extensor tension (Landsmeer, 1976). During interphalangeal extension, lumbrical contraction decreases the viscoelastic tension in the distal tendon of FDP (Landsmeer, 1976) while promoting the movement via the extensor apparatus (Kapandji, 1970).

Orthogonal to the proximo-distal columnar arrangement of carpals and metacarpals contributing to the movement of individual digits, is a sequence of bony rows arranged in an radio-ulnar orientation and extending from the proximal carpus to the distal phalangeal row. During most grips they adopt a concavity towards the palm to shape the hand to the object, for instance in gripping a ball. Such configurations can be thought of as postural sets for promotion of precision movements. Carpal stability is enhanced by a continuous bony adaptation of the distal carpal row to the proximal (Landsmeer, 1976) and by the concerted shaping action of the interossei, thenar and hypothenar muscular complexes. This provides a firm origin from which the shortening intrinsic muscles can exert their pulls.

For a given precision grip task, a number of preferable muscle force distribution patterns may exist, but a unique set of fixed muscular contributions across subjects is unlikely. In our studies of the performance of two precision tasks a low coefficient of variation was found when one subject performed a given task. Despite the recruitment of similar muscle groups, intersubject variability was, however, significant for both the degree of the activity of each muscle and the relative contributions of each muscle (Dudwal, Bennett and Lemon, unpublished observations). Each digit appears to be able to produce the same functional contribution through different combinations of muscle activities (Maier, Hepp-Reymond and Meyer, 1990).

All natural movements involving either manipulation or prehension thus require the activity of many different muscles. During manipulatory movements, in which the fingers are moving against negligible loads, the dominant pattern is one of fractionation of muscular activity, with bursts of activity in different muscle groups occurring at different times (Muir, 1985). During the prehensile or gripping phase, with force being developed between the tips of the digits, this changes into a pattern of co-contraction with a precise patterning of the level of activity in groups of different muscles (Smith, 1981; Maier et al., 1990). The number of muscles involved in the co-contractile pattern depends upon the degree of force exerted and clearly serves to stabilise the more proximal parts of the limb during grip. Biomechanical analysis of the hand means discarding traditional concepts of agonists and antagonists seen for muscles acting at the wrist and elbow joints.

2. THE CORTICO-MOTONEURONAL INPUT TO THE HAND

Although the mechanical arrangements for independent finger movements are present in many mammals, the **neural** machinery for selecting and controlling individual digit movements is, with a few exceptions, only developed in primates (Phillips, 1971). A large amount of anatomical, behavioural and electrophysiological evidence suggests that the principal feature of the primate motor system which underpins the performance of relatively independent finger movements (RIFM) is the cortico-motoneuronal synapse (Kuypers, 1981, 1982). In macaque monkeys, the CM system is largely concerned with control of muscles acting on the hand and fingers. The ontogeny of these connections is late and parallels the development of fine finger movements in infant monkeys (Kuypers, 1962; Lawrence and Hopkins, 1976). Flament, Hall, Lemon and Simpson (1990) recently showed that non-invasive electromagnetic stimulation of the cortex in newborn Macaque monkeys fails to elicit any short-latency responses in hand and forearm muscles, but that these responses develop between the fourth and sixth month at a time when these animals first begin to use their hands for precise manipulation and exploration.

Direct connections from the motor cortex to spinal motoneurones innervating hand and finger muscles make their appearance in lower primates and become increasingly prominent in the Old World monkeys, apes and in man. Heffner and Masterton (1975, 1983), found that the number and distribution of CM synapses formed the best neuroanatomical correlate of digital dexterity in different species.

2.1 The Identification of CM neurones.

CM neurones can be identified in conscious monkeys by the spike-triggered averaging (STA) technique (Fetz and Cheney, 1980; Lemon, Mantel and Muir, 1986). The monkeys (adult **M. nemestrina**) were trained over a period of several months to perform a precision grip task. This task (see Fig. 1) required the monkey to use its thumb and index finger to move two small levers into an electronically-defined target zone, usually (3-5mm from the lever rest position). Note that the ulnar fingers are flexed out of the way, while the index finger is extended (Fig.1B & C): i.e. a fractionated pattern of finger movement. Both levers had to be held within their target zones continuously for at least 1s, and if this was achieved the monkey was rewarded with a small piece of fruit. The levers could either be spring-loaded (auxotonic task) or fixed (isometric task). In the latter case strain gauges detected the isometric force exerted by the monkey, which was rewarded for maintaining the force level between 0.4 and 1.0N. These **M. nemestrina** monkeys have larger hands than either Rhesus or cynomolgus monkeys and have no difficulty in exerting precision grip forces within this range.

After training was complete the monkey was prepared, under deep anaesthesia and full aseptic conditions, for chronic single unit recording from the motor cortex contralateral to the trained hand (see Lemon et al., 1986). Two fine tungsten electrodes were permanently

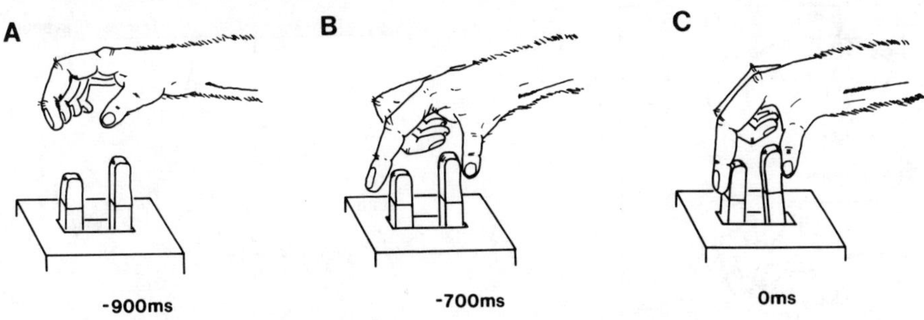

Figure 1. Drawings, taken from video frames, of the hand of **M. nemestrina** at different stages before the onset of lever movement (C).

implanted in the pyramidal tract at the upper medullary level, one electrode being placed approximately 5mm rostral to the other. These electrodes were positioned under stereotaxic guidance and the final positions determined on the basis of the antidromic volleys recorded from the dural surface of the exposed ipsilateral motor cortex. The position of these electrodes within the pyramidal tract was confirmed histologically at the end of each experiment, when the monkey was killed by an overdose of Nembutal and perfused through the heart. Histological reconstructions were made from 40μm frozen sections cut in the stereotaxic plane.

During each experimental session, records were made with glass-insulated platinum-iridium electrodes with tip impedances of 0.8-3MΩ at 1KHz. Most penetrations were directed into the rostral bank of the central sulcus, usually 16-19mm lateral from the midline. After penetrating the dura, the microelectrode was slowly advanced while trains of intracortical microstimulation (ICMS; maximum current, 20μA) were delivered through the electrode tip. The movements evoked were noted at each depth. Recording commenced once a point was reached from which movement of the digits could be evoked with ICMS of less than 10-12μA. As the electrode was advanced further the pyramidal tract was stimulated with single shocks of up to 400μA until an identified pyramidal tract neurone (PTN) was isolated. The pyramidal shock was then turned off, and the task-related nature of the PTN's discharge was assessed by constructing on-line response averages, usually referenced to completion of a successful trial.

On-line STA was made by passing the PTN spikes through a double voltage-time window discriminator to produce a train of TTL trigger pulses corresponding to the spike train of the cell. A digital storage oscilloscope was used to display the full waveform of each spike that had given rise to a trigger pulse to confirm that triggers were derived from one and same cell throughout the averaging period (usually 10-20 min). These trigger pulses were used to generate STA of

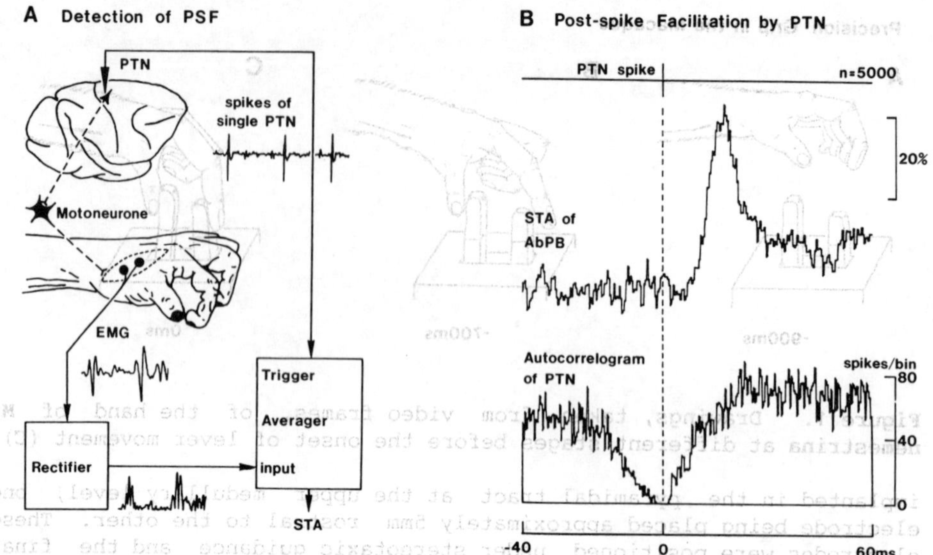

Figure 2. **A.** The use of spike-triggered averaging to detect post-spike facilitation (PSF) of EMG recorded from monkey hand muscles. The putative cortico-motoneuronal connection from a single pyramidal tract neurone (PTN) to the motoneurones of the muscle is indicated by a dashed line. EMG is rectified and averaged with respect to PTN spikes. **B.** Example of PSF produced in a spike-triggered average (STA) of EMG from a thumb muscle (AbPB) by 5000 PTN spikes. Vertical scale indicates 20% modulation of background EMG level. PTN discharge is at time zero. Autocorrelogram of the PTN spikes is shown below.

full-wave rectified EMG recorded from a variety of intrinsic hand and forearm muscles using either surface or intramuscular wire electrodes. Up to ten EMGs were recorded in each session. The arrangement is shown in Fig. 2A. An example of an STA is shown in Fig. 2B, together with the auto-correlogram of the triggering PTN. This PTN produced a large post-spike facilitation (PSF) of the activity recorded from the abductor pollicis brevis muscle (AbPB). The PSF began 10 ms after PTN discharge, reached a peak at 18 ms and the principal peak of the PSF lasted for about 12 ms. The amplitude of such effects was usually expressed as the percentage modulation (height of peak/height of baseline x100) of the background EMG level. The fluctuations due to noise in the STA were subtracted before the percentage modulation was calculated (see Lemon and Mantel, 1989). The percentage modulation of the PSF in Fig. 2B is 40%. The tests used to establish confidence in the PSF as a genuine spike-related event have been described previously (Lemon et al., 1986). The presence of post-spike facilitation in the average is consistent with a direct, monosynaptic connection between the neurone and the motoneurones of the sampled muscles (Mantel & Lemon, 1987). All STAs were repeated off-line from tape-recorded analog data. Most STAs were made with all spikes, i.e. trigger spikes were not selected according to the phase of the precision grip task into which they fell.

In the present study we have sought to exclude any PSF effects which are clearly contaminated by the effects of synchrony (Lemon, Mantel and Muir, 1985; Smith and Fetz, 1989). In general these averages a) show elevations in EMG activity which are much broader than those normally observed (>20ms compared to 10-15ms) b) often appear to begin at unrealistically short times after spike discharge, and indeed sometimes begin **before** such discharge and c) lack any sharp upward inflection in the post-spike peak. In such cases we have found that if cross-correlations are made between the triggering cell and the activity of single motor units in the muscle, correlation peaks with broad half-widths (>5ms) are usually found.

All CM cells facilitating hand muscles in the monkey are particularly active during finger **movement**, with a somewhat lower rate of discharge during the maintenance of precision grip force between the finger tips. CM cells facilitating the intrinsic hand muscles appear to be preferentially active during tasks requiring RIFM (Muir and Lemon, 1983).

2.2. The branching patterns of single CM cells

It was an early finding that CM cells rarely facilitate the EMG activity of only one muscle. Fetz and Cheney (1980) used the term **'muscle field'** to describe the set of muscles facilitated by a single CM cell. An example is shown in Fig. 3A. Buys, Lemon, Mantel & Muir (1986) first analysed the muscle fields of a sample of CM cells active during precision grip. They concluded that most CM cells facilitated a relatively restricted group of muscles. Results obtained from a total of 80 CM cells recorded together with at least five hand and forearm muscles are shown in Fig. 3B. These cells were recorded in five **M. nemestrina** monkeys. For these 80 cells the average number of muscles recorded per cell was 7.2 (SD± 1.8). The majority of the recordings were from the intrinsic hand muscles (see Fig. 3C).

Fig. 3B shows that for this sample, each CM cell facilitated on average 27.5% (SD± 15.4%) of the muscles with which it was sampled. The range was from 10% (1 of 10 muscles sampled) to 67% (6 of 9). Only 9 cells (11%) facilitated more than half the sampled muscles. Nineteen facilitated only one muscle (<15% of the sample) which was, in 16 cases, an intrinsic hand muscle. If only those cells sampled with 8-10 different muscles are considered, 14 CM cells facilitated only one muscle. This was an intrinsic hand muscle for 12 of these cells, and in 6 cases, this was a thumb muscle, suggesting a highly selective control of this important digit. As shown in Fig. 3C, for this population of cells, post-spike facilitation was much more common amongst the intrinsic hand muscles (121 of the 151 PSF effects seen or 80%) and was found more often than would have been expected from the proportion of CM cell-intrinsic muscle pairs in the sample (319 of the 570 pairs investigated, or 56%).

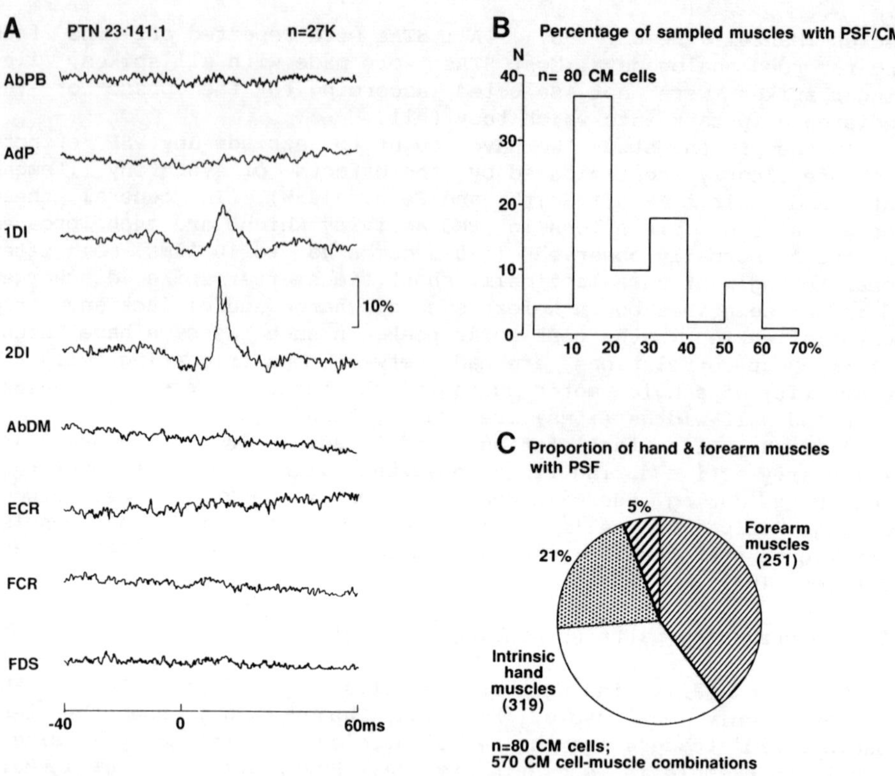

Figure 3. Post-spike effects indicating branching in CM cells. **A.** STAs of EMG from eight hand and forearm muscles averaged with respect to 27000 spikes from a single CM cell (discharge at time zero). Only two of the eight muscles showed PSF (1st and 2nd dorsal interosseous). **B.** Percentage of muscles sampled with a CM cell which showed clear PSF. Number of muscles varied from 5 to 10. For 80 CM cells the mean value was 27.5%. **C.** Proportion of the 570 CM cell-muscle combinations that were investigated by STA which showed PSF. Heavy dots and hatching indicate PSF in intrinsic hand (121 PSFs, 21%) and in forearm muscles (30 PSFs, 5%), respectively.

The amplitude of PSF was larger in intrinsic hand muscles than in forearm muscles. In the former, the mean size of the 121 PSFs was 11.4% (SD$\pm$ 6.6%) compared to 9.1% (SD$\pm$ 4.4%) for 30 forearm PSFs. But the 10 **largest** PSFs found in the intrinsic muscles, which ranged from 21 to 42%, had a mean amplitude (27.6 SD$\pm$ 6.8%), twice as large as that for the 10 largest forearm PSFs (mean 13.5 SD$\pm$ 4.4%, range 9-21%), clearly suggesting a stronger influence of these CM cells over hand muscles.

2.3 The Branching of CM cells within the Motor Nucleus of a Target Muscle

Lemon et al., (1986) originally showed that the PSF of a whole EMG was contributed by the activity of motor units of different sizes that were each correlated with the discharge of the CM cell. Subsequently we developed a methodology which allowed us to make stable recordings from single motor units within a hand muscle (Lemon, Mantel and Rea, 1990), and this has allowed us to explore in some detail the branching of a single CM cell axon within the column of motoneurones supplying a single muscle. The muscles chosen for investigation were the short abductor and adductor of the thumb, and the 1DI, since PSF in these muscles was strongest and most frequent for CM cells in the hand area (Buys et al., 1986). The findings to date suggest a widespread branching of a CM cell amongst the motoneurones of its target muscle (Mantel and Lemon, 1987; Lemon, 1987). Corticospinal collaterals are chiefly orientated in a rostro-caudal direction within the motoneuronal cell groups (Shinoda, Yokota and Futami, 1981; Lawrence, Porter & Redman, 1985). This pattern of organisation would allow a single corticospinal neurone to make contact with motoneurones all supplying a single anatomical muscle, because these motoneurones are organised in long narrow columns spanning two or more spinal segments (Sherrington, 1898; Jenny and Inukai, 1985).

2.4 The Proportion of Motor Cortex Neurones which Produce CM Effects.

In the initial studies this was extremely difficult to estimate because of the selective nature of the neurones sampled. Fetz and Cheney (1980) examined 370 motor cortex neurones discharging during a wrist flexion-extension task; 27% of these cells produced clear PSF in at least one of the wrist flexor or extensor muscles studied. But the yield from neurones identified as PTNs was substantially higher at 55%.

In their study, Lemon et al. (1986) attempted to increase the yield of such neurones by selecting neurones for STA on the basis of four criteria. These were: a) **Location**: within the hand region of area 4, often deep within the rostral bank of the central sulcus, b) **Microstimulation**: neurones were selected if they were found at loci which yielded movements of the digits with low threshold (<10μA) ICMS, c) **Pyramidal tract neurones (PTNs)**: most selected neurones were fast PTNs (antidromic latency from the pyramid 0.7-2.0 ms, conduction velocity estimated at 35-100 m/s) and d) **Task relationship**: neurones were selected if their activity was deeply modulated during the precision grip task which was performed by the monkeys. The yield of neurones for which all criteria held was much higher (44/60 PTNs or 73%) than for unselected neurones (14/45 or 31%). In a more recent study (monkey D24) we made a total of 32 penetrations in the right motor cortex in which 81 PTNs were recorded and used for on-line STA with 6-8 hand and forearm muscles. Of these 81 PTNs, 56 were fast PTNs (antidromic latency <2.0ms; see below) and of these 32 (57%) were positively identified as CM cells (cf Fetz and Cheney, 1980). Some of the slow PTNs also produced PSF (see section d. below).

In conclusion, we know that the proportion of cortical output neurones which are PTNs is small. Powell has estimated that a cylinder of motor cortex 1 sq.mm in area would contain approximately 90 large and 18,000 small pyramidal cells (Landgren et al., 1962). Only about 10-20% of the layer V neurones will be pyramidal tract or corticospinal neurones sending their axons into the spinal cord (Humphrey and Corrie, 1978). However, the PTNs represent the final stage of the cortical processing, and their discharge probably results from activity in many thousands of other neurones, including those located in the cerebellum, basal ganglia and other cortical areas without direct projections to the spinal cord. The significance of these PTNs is therefore probably out of all proportion to their number.

2.5 Do both 'Fast' and 'Slow' PTNs exert CM Effects?

In primates, it is probable that most of the large corticospinal fibres, (axon diameter greater than 5μm, conduction velocity 25-80 m/s), the so-called 'fast PTNs', make direct cortico-motoneuronal connections (Phillips and Porter, 1977). Most of the responses evoked in the EMG recorded from hand muscles by magnetic stimulation of the human motor cortex appear to be contributed by these fast PTNs (Day et al., 1989). Indeed it is the presence of such large fibres in the primate corticospinal tract, as compared to that of sub-primate species such as the cat and rat, that has lead to the suggestion that it is these fibres, and these fibres alone, which contribute CM connections. Both anatomical and electrophysiological studies have provided evidence against the existence of CM connections in these 'lower' species (Kuypers, 1981). The anatomical conclusion is based principally on the paucity or absence of corticospinal terminals amongst the motoneuronal cell groups of lamina IX of Rexed. It should be stressed, however, that the extensive dendritic tree of most motoneurones means that CM connections need not necessarily be restricted to terminals in lamina IX (Lawrence et al., 1985). There is a report of monosynaptic EPSPs in cervical motoneurones in the rat following 'epicortical' stimulation (Elger et al., 1977), but these EPSPs could not have been corticospinal in origin since the authors calculated that the conduction velocity of the descending fibres which generated these EPSPs was around 60 m/s i.e. around three to four times faster than the fastest corticospinal fibres in the rat (Catsman-Berrevoets et al., 1979; see Mediratta and Nicoll, 1983).

In general, the suggestion has been that since slowly conducting fibres in the cat and rat make few, if any, CM contacts, that the same is true in the primate. It is interesting to note that Heffner and Masterton (1975) found a weaker correlation (r=0.26) between digital dexterity and the size of the largest fibres in the corticospinal tract (the seal possesses the largest!) than between digital dexterity and the presence of corticospinal projections to lamina IX (r=0.66). Interestingly, they also found a weak correlation between largest fibre size and the presence of such terminations (r=0.52). This result could be interpreted as indicating a CM contribution from the slower components of the corticospinal tract.

In the primate, electrophysiological evidence for slowly conducting corticospinal fibres giving rise to monosynaptic inputs to motoneurones is difficult to obtain because it is impossible to assess the segmental delay of such effects. In the case of the large fibres, a monosynaptic action of volleys excited in the corticospinal tract by cortical or pyramidal tract stimulation could be determined by measuring the delay between the arrival of the earliest volley at the spinal segment in which the motoneurone is located and the onset of the EPSP, recorded intracellularly from the motoneurone (Phillips and Porter, 1977). This approach is impracticable for the slower fibres for three reasons: 1) the dispersion of their action potentials (due to variation in their conduction velocities) makes it difficult to detect any synchronous volley representative of such a fibre population within the corticospinal tract, 2) the resultant temporal dispersion of postsynaptic action means that no late, compound EPSP can be detected in the motoneuronal record and 3) these late EPSPs may be masked by the earlier EPSPs and IPSPs generated by faster fibres.

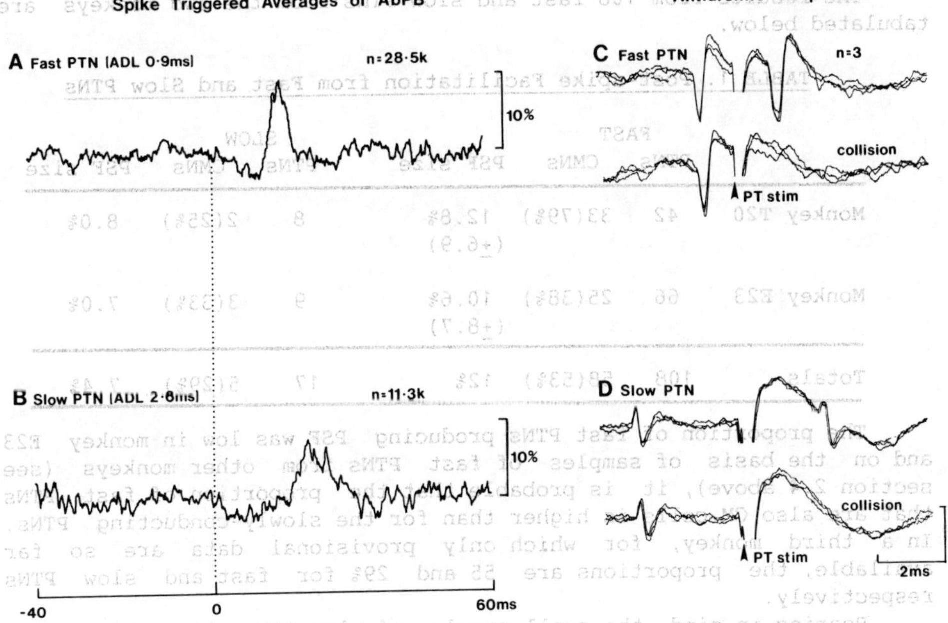

Figure 4. A and B. PSF generated in the same muscle (AbPB) from a fast PTN (A) and from a slow PTN (B) recorded in the same penetration. Note the earlier onset of PSF in A (10.7 ms) compared to B (16.6 ms). C and D. Superimposed antidromic responses of the PTNs. Note longer antidromic latency (ADL) and collision interval of the slow PTN. Voltage calibration, 100μV

One solution to this problem is the application of spike-triggered averaging to the population of 'slow' PTNs. Fetz and Cheney (1980) reported that 4 slow PTNs gave rise to post-spike facilitation, and that, as expected, the onset latency of the PSF was later than that found for fast PTNs. Evidence for PSF from slow PTNs was also found by Lemon et al., (1986). We have recently re-investigated this problem in two monkeys in which we sought both slow and fast PTNs located within the hand region of area 4. We compared the frequency of occurrence of PSF for fast and slow PTNs. These PTNs were categorised as 'fast' or 'slow' on the basis of their antidromic latency from the rostral part of the medullary pyramid (implanted electrodes at A+2 and at P-3mm stereotaxic levels). 'Fast' and 'slow' PTNs had latencies of shorter, or longer than 2.0ms, respectively. This boundary value corresponds approximately to an axonal conduction velocity of 55 m/s (Lemon et al., 1986). PSF produced by a fast and by a slow PTN recorded in the same microelectrode penetration upon the same target muscle (AbPB) is shown in Fig. 4. The fast PTN had an antidromic latency of 0.9ms from the medullary pyramid (Fig. 4C), compared to 2.6ms for the slow PTN (Fig. 4D). The PSF amplitude is the same in each case. Note the later onset and longer duration of the PSF produced by the slow PTN (Fig. 4B).

The results from 108 fast and slow PTNs from the two monkeys are tabulated below.

TABLE 1. Post-spike Facilitation from Fast and Slow PTNs

	FAST			SLOW		
	PTNs	CMNs	PSF size	PTNs	CMNs	PSF size
Monkey T20	42	33(79%)	12.8% (±6.9)	8	2(25%)	8.0%
Monkey E23	66	25(38%)	10.6% (±8.7)	9	3(33%)	7.0%
Totals	108	58(53%)	12%	17	5(29%)	7.4%

The proportion of fast PTNs producing PSF was low in monkey E23 and on the basis of samples of fast PTNs from other monkeys (see section 2.4 above), it is probable that the proportion of fast PTNs that are also CM cells is higher than for the slowly-conducting PTNs. In a third monkey, for which only provisional data are so far available, the proportions are 55 and 29% for fast and slow PTNs respectively.

Bearing in mind the small sample of slow PTNs it would appear that on average, slow PTNs produced smaller PSF than large PTNs. The mean amplitudes given in Table 1 are for the largest PSF effect observed for each PTN in the sample. Despite the number of fast PTNs in the sample, there was a big standard deviation for the amplitude of PSF produced by these PTNs, which results from the large range of effects observed (6 to 42% in T20 and 4 to 35% in E23), whereas only a small range of weak effects was seen with the slow PTNs (4 to 10%). We have only once observed a really strong PSF with a slow PTN, and this effect was clearly contaminated by synchrony (see section 2.1 above). We have found no differences in the proportion of sampled muscles facilitated by the two types of PTN. Shinoda, Yamaguchi and Futami (1986) found no difference in the extent of branching of slow versus fast PTNs in the cat.

Because of their slower conduction velocity, one might predict that the onset of PSF produced in a given muscle by a slow PTN would be delayed. This was the case. In both monkeys, PSF onset latency in the intrinsic hand muscles was 11.6ms (SD± 1.6ms), compared to 15.1ms (SD± 2.1ms) for the slow PTNs.

In conclusion, there can now be doubt that slow PTNs do produce post-spike facilitation, and that some of these neurones should be considered as part of the CM system.

2.6 Which features of the CM system might contribute towards the capacity to execute RIFM?

In the monkey there are 29 muscles acting on the digits. Most of these are active during the precision grip task, and many are recruited in a highly fractionated fashion during the movement phase of the task (Muir, 1985). We have seen that most CM cells facilitate a relatively restricted group of these muscles, and it seems quite possible that this focused pattern of facilitation might contribute to the fractionation of muscle activity during precision finger movements. It is also probable that this distributed output is the physiological manifestation of the intraspinal branching observed with both electrophysiological and intraxonal staining techniques (Shinoda, Zarzecki and Asanumua, 1979; Shinoda et al., 1981; Lawrence et al., 1985).

Because we have not been able to sample all the muscles moving the fingers we can only estimate the real proportion that would be facilitated by a single cortico-motoneuronal cell. In this respect it is interesting that Buys et al. (1986) found that the number of facilitated muscles rose together with the number of sampled muscles (suggesting facilitation of a relatively constant proportion of those sampled) until ten muscles were sampled, when the proportion fell. This effect was due to the increasing proportion of forearm muscles in the sample, in which PSF is a relatively infrequent phenomenon for this population of CM cells (see Fig. 4). This result suggested that sampling a greater number of muscles would not yield more facilitated muscles, and confirmed the relatively focused output of the CM cell population.

The features of the branching patterns of CM cells among the motoneurone pools pose two interesting questions. First, does the group of muscles facilitated by a single CM cell correspond to a 'task-group', and second, is the variation in the strength of facilitation across the muscle field important for setting the balance of activity within it?

In answer to the first point, some CM cells do indeed appear to facilitate functionally related sets of muscles (Buys et al., 1986). Some of these relationships are functional in an anatomical sense. An example here would be those CM cells which facilitated both the 1DI and the EDC: both muscles are involved in extension of the phalangeal joints (see section 1 above). Others had truly task-related synergies: for instance, we found that a proportion of those CM cells (n=19) which facilitated the 1DI or the thumb adductor (AdP), facilitated both muscles (n=7). These muscles could be considered as two 'prime movers' within the intrinsic group for the production of precision grip. Another interesting functional pairing concerns the short abductor and adductor (AbPB and AdP): these muscles cooperate in the sense that when the thumb is abducted by the AbPB, its position is such as to allow the adductor fibres to work at their optimal length, advancing the thumb towards the index finger as the aperture of the grip closes (Jeannerod, 1984).

The pattern of facilitation generated by a given CM cell varies considerably from one cell to another, and may well be a unique feature of each cell. The strength of facilitation is different for each of the muscles within the CM cell's muscle field. To assess the significance of this pattern, we have recently investigated the activity of CM cells during periods in which their target muscles were recruited in two quite different ways (Bennett and Lemon, 1990). In one group of selected periods, the muscle which received strong PSF from the cell was active (the 'dominant' muscle), while a second muscle, also facilitated by the same cell but with weak PSF (the 'non-dominant' muscle), showed little activity. During a second set of periods the situation was reversed. These periods of fractionated muscle activity levels all occurred during the dynamic phase of the precision grip task when the monkey was pressing its digits upon the manipulandum levers. In summary, for 5/6 CM cells analysed in this way, we found that the CM cell discharged at a significantly higher frequency when the 'dominant' muscle was the more active. In addition, there was evidence for a more efficient post-spike facilitation of this muscle during such periods. The specific recruitment of the CM cell during these periods suggests that the pattern of synaptic influence detected by the STA technique is of real significance for determining the relative levels of activity in different intrinsic hand muscles; the mechanisms described above would tend to enhance the fractionation of muscle activity which underlies the performance of relatively independent finger movements.

3. CORTICAL ORGANISATION OF THE CORTICO-MOTONEURONAL SYSTEM.

What features of the cortical organisation of CM cell clusters might subserve relatively independent finger movements? The CM system is only one component of the corticofugal output controlling motoneuronal activity. This output has traditionally been mapped by electrical stimulation and especially in recent years by use of ICMS in conscious, behaving monkeys (see Lemon, 1988, 1990). Although it

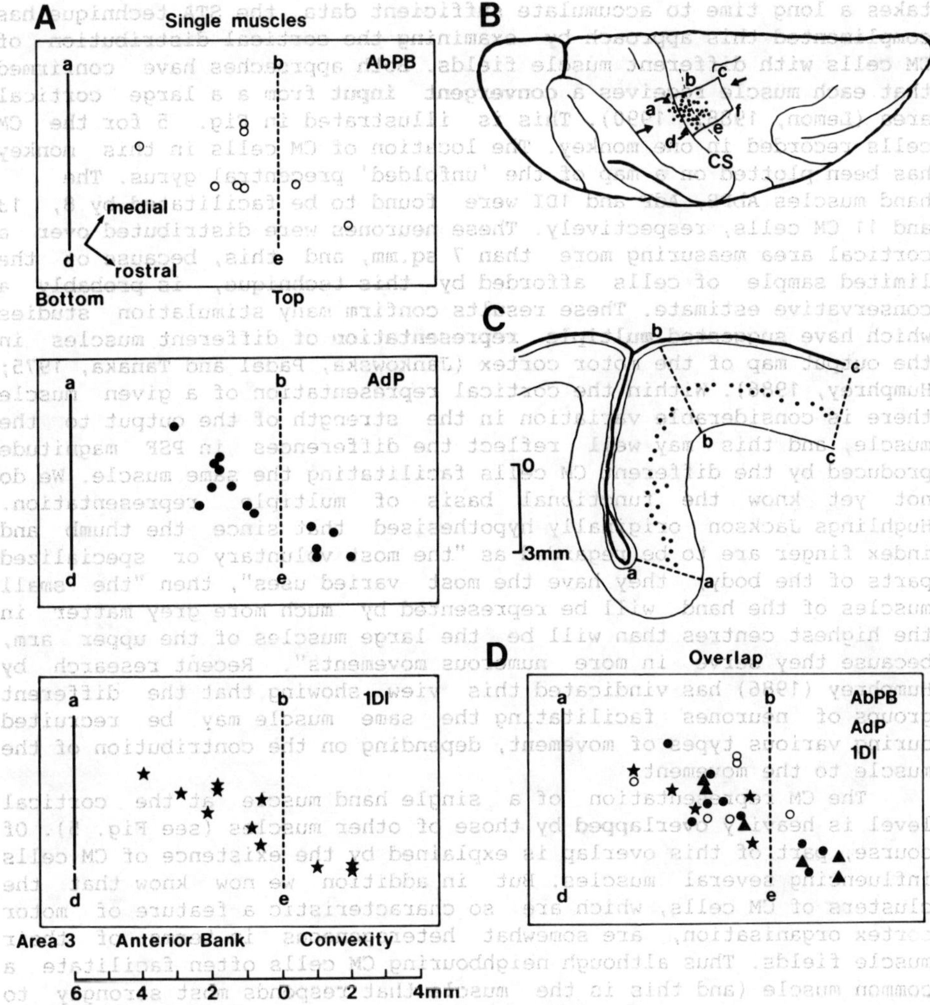

Figure 5. Multiple representation and overlapping representation of intrinsic hand muscles in monkey motor cortex. **A**. Distribution of CM neurones which produced post-spike facilitation in the e.m.g of AbPB, AdP (abductor and adductor pollicis) and 1DI (first dorsal interosseous). **B** shows the surface topography of microelectrode penetrations, marked by dots. The area investigated is indicated by the quadrilateral abcdef. **C** represents a sagittal section of the central sulcus (arrows in B). In **A** and **D** the gyrus was unfolded along lamina V, marked by the presence of Betz cells (dots in C). The lines a-d and b-e represent the bottom and top of the sulcus, respectively. **D** shows the overlap between CM cells projecting to the three muscles. Triangles indicate CM cells facilitating more than one of the muscles.

takes a long time to accumulate sufficient data, the STA technique has complimented this approach by examining the cortical distribution of CM cells with different muscle fields. Both approaches have confirmed that each muscle receives a **convergent input** from a a large cortical area (Lemon, 1988, 1990). This is illustrated in Fig. 5 for the CM cells recorded in one monkey. The location of CM cells in this monkey has been plotted on a map of the 'unfolded' precentral gyrus. The hand muscles AbPB, AdP and 1DI were found to be facilitated by 8, 13 and 11 CM cells, respectively. These neurones were distributed over a cortical area measuring more than 7 sq.mm, and this, because of the limited sample of cells afforded by this technique, is probably a conservative estimate. These results confirm many stimulation studies which have suggested **multiple representation** of different muscles in the output map of the motor cortex (Jankowska, Padel and Tanaka, 1975; Humphrey, 1986). Within the cortical representation of a given muscle there is considerable variation in the strength of the output to the muscle, and this may well reflect the differences in PSF magnitude produced by the different CM cells facilitating the same muscle. We do not yet know the functional basis of multiple representation. Hughlings Jackson originally hypothesised that since the thumb and index finger are to be regarded as "the most voluntary or specialized parts of the body; they have the most varied uses", then "the small muscles of the hand will be represented by much more grey matter in the highest centres than will be the large muscles of the upper arm, because they serve in more numerous movements". Recent research by Humphrey (1986) has vindicated this view, showing that the different groups of neurones facilitating the same muscle may be recruited during various types of movement, depending on the contribution of the muscle to the movement.

The CM representation of a single hand muscle at the cortical level is heavily **overlapped** by those of other muscles (see Fig. 5). Of course, part of this overlap is explained by the existence of CM cells influencing several muscles. But in addition we now know that the clusters of CM cells, which are so characteristic a feature of motor cortex organisation, are somewhat heterogeneous in terms of their muscle fields. Thus although neighbouring CM cells often facilitate a common muscle (and this is the muscle that responds most strongly to ICMS within such a cluster), their muscle fields may differ in composition (Lemon, 1990). This overlap of functional representation may represent a higher level of coordination between the different target muscles than can be achieved by the use of single neurones with branched outputs. Such an organization would allow the motor system to employ the relatively small number of muscles moving the hand and fingers in so rich and varied a manner, allowing a large number of different solutions to a particular motor task.

Acknowledgements. We gratefully acknowledge the expert technical assistance of Lyn Cummings, and the collaboration of Dr. Didier Flament. Supported by the Medical Research Council, the National Fund for Research into Crippling Diseasses (Action Research), The Felice Rosemary Lloyd Trust and the Deutsche Forschungsgemeinschaft.

REFERENCES

Bennett, K.M. and Lemon, R.N. (1990) 'The activity of monkey cortico-motoneuronal (CM) cells is related to their pattern of post-spike facilitation of intrinsic hand muscles', J.Physiol. (in press)

Buys, E.J., Lemon, R.N., Mantel, G.W.H. and Muir, R.B. (1986) 'Selective facilitation of different hand muscles by single corticospinal neurones in the conscious monkey', J. Physiol. **381**, 529-549.

Catsman-Berrevoets, C.E., Lemon, R.N., Verburgh, C.A., Bentivoglio, M. and Kuypers, H.G.J.M. (1979) 'Absence of callosal collaterals derived from rat corticospinal neurones. A study using fluorescent retrograde tracing and electrophysiological techniques', Exp. Brain Research **39**, 433-440.

Chao, E.Y.S., An, K.-N., Cooney, W.P. and Linscheid, R.L. (1989) 'Biomechanics of the Hand', World Scientific Publishing, Singapore. pp 31-72

Close, J.R. and Kidd, C.C. (1969) 'The functions of the muscles of the thumb, the index and long fingers', J. Bone Joint Surgery, **51-A**, 1601-1620.

Day, B.L., Dressler, D., Maertens de Noordhout, A., Marsden, C.D., Nakashima, K., Rothwell, J.C. and Thompson, P.D. (1989) 'Electric and magnetic stimulation of human motor cortex: surface EMG and single motor unit responses', J. Physiol. **412**, 449-473.

Elger, C.E., Speckmann, E.-J., Caspers, H. and Janzen, R.W.C. (1977) 'Corticospinal connections in the rat. I. Monosynaptic and polysynaptic responses of cervical motoneurons to epicortical stimulation', Exp. Brain Research **28**, 385-404.

Fetz, E.E., and Cheney, P.D. (1980). 'Postspike facilitation of forelimb muscle activity by primate corticomotoneuronal cells', J. Neurophysiol. **44**, 751-772.

Flament, D., Hall, E.J., Lemon, R.N. and Simpson, M. (1990) 'The development of cortically evoked responses in infant Macaque monkeys studied with electromagnetic brain stimulation', J.Physiol., **426**, 105P.

Heffner, R. and Masterton, B. (1975) 'Variation in form of the pyramidal tract and its relationship to digital dexterity', Brain Behav. Evolution **12**, 161-200.

Heffner, R. and Masterton, B. (1983) 'The role of the corticospinal tract in the evolution of human digital dexterity', Brain Behav. Evolution **23**, 165-183.

Humphrey, D.R. (1986) 'Representation of movements and muscles within the primate precentral motor cortex: historical and current perspectives', Fed. Proc. **45**, 2687-2699.

Humphrey, D.R. and Corrie, W.S. (1978) 'Properties of pyramidal tract neuron system within a functionally defined subregion of primate motor cortex', J. Neurophysiol. **41**, 216-243.

Jankowska, E., Padel, Y. and Tanaka, R. (1975) 'Projections of pyramidal tract cells to α-motoneurones inervating hindlimb muscles in the monkey', J. Physiol. **249**, 637-667.

Jeannerod, M. (1984) 'The timing of natural prehension movements', J. Motor Behaviour **16**, 235-254.

Jenny, A.B., and Inukai, J. (1985) 'Principles of motor organization of the monkey cervical spinal cord', J. Neuroscience 3, 567-575.

Kapandji, I. A. (1970) The physiology of the joints. Volume 1. Upper limb, Churchill Livingstone, Edinburgh. pp 146-203

Kuypers, H.G.J.M. (1962) 'Corticospinal connections: Postnatal development in the rhesus monkey', Science **138**, 678-680.

Kuypers, H.G.J.M. (1981) 'Anatomy of the descending pathways', in J.M. Brookhart and V.B. Mountcastle (eds.), Handbook of Physiology, American Physiological Society, Bethesda, Maryland, pp.597-666.

Kuypers, H.G.J.M. (1982) 'A new look at the organization of the motor system', Prog. Brain Research **57**, 381-404.

Landgren, S., Phillips, C.G. and Porter, R. (1962) 'Cortical fields of origin of the monosynaptic pyramidal pathways to some alpha motoneurones of the baboon's hand and forearm', J. Physiol. **161**, 112-135.

Landsmeer, J.M.F. (1976) Atlas of Anatomy of the hand, Churchill Livingstone, Edinburgh. pp 315-349.

Landsmeer, J.M.F. (1989) 'A comparison of fingers and hand in varanus, opussum and primates', Acta Morphol., Neerl-Scand **24**, 193-221.

Lawrence, Porter and Redman (1985) 'Corticomotoneuronal synapses in the monkey: Light microscopic localization upon motoneurons of intrinsic muscles of the hand', J. Comp. Neurol. **232**, 499-510.

Lawrence, D.G. and Hopkins, D.A. (1976) 'The development of motor control in the rhesus monkey: evidence concerning the role of corticomotoneuronal connections', Brain **99**, 235-254.

Lemon, R.N. (1987) in 'Motor Areas of the Cerebral Cortex ', CIBA Foundation Symposium No. 132 Wiley Interscience, Chichester p.118-122

Lemon, R.N. (1988) 'The output map of the primate motor cortex', Trends in Neuroscience **11**, 501-506.

Lemon, R.N. (1990) 'Mapping the output functions of the motor cortex', in G.M. Edelman, W.E. Gall and W.M. Cowan (eds.), Signal and Sense, Wiley-Liss, New York, pp. 315-356.

Lemon, R.N., Mantel, G.W.H. and Muir, R.B. (1986) 'Corticospinal facilitation of hand muscles during voluntary movement in the conscious monkey', J. Physiol. **381**, 497-527.

Lemon, R.N. and Mantel, G.W.H. (1989) 'The influence of changes in the discharge frequency of corticospinal neurones on hand muscles in the monkey', J. Physiol. **413**, 351-378.

Lemon, R.N. and Mantel, G.W.H. and Muir, R.B. (1985) 'The consequences of synchronization among cortico-motor neurones in the monkey', J. Physiol. **371**, 46P.

Lemon, R.N., Mantel, G.W.H. and Rea, P.A. (1990) 'Recording and identification of single motor units in the free to move primate hand', Exp. Brain Research **81**, 95-106.

Long, C., Conrad, W., Hall, E.A. and Furler, S.L. (1970) 'Intrinsic-extrinsic muscle control of the hand in power grip and precision handling', J. Bone Joint Surgery **52-A**, 853-867.

Maier M.A., Hepp-Reymond, M.-C. and Meyer, M. (1990) 'EMG coactivation patterns and isometric grip force in human' Eur. J. Neurosci. Suppl. **3**, 1288

Mantel, G.W.H. and Lemon, R.N. (1987) 'Cross-correlation reveals facilitation of single motor units in thenar muscles by single corticospinal neurones in the conscious monkey', Neuroscience Letters **77**, 113-118.

Mediratta, N.K. and Nicoll, J.A.R. (1983) 'Conduction velocities of corticospinal axons in the rat studied by recording cortical antidromic responses', J. Physiol. **336**, 545-561.

Muir, R.B. (1985) 'Small hand muscles in precision grip', in A.W. Goodwin and I. Darian-Smith (eds.), Hand Function and the Neocortex, Springer-Verlag, Berlin, pp. 155-174.

Muir, R.B. and Lemon, R.N. (1983) 'Corticospinal neurons with a special role in precision grip', Brain Research **261**, 312-316.

Napier, J.R. (1961) 'Prehensibility and opposability in the hands of primates', Symposium Zoological Society London **5**, 115-132.

Phillips, C.G. (1971) 'Evolution of the corticospinal tract in primates with special reference to the hand', Proc. 3rd Int. Congress Primates, Zurich 2, 2-23.

Phillips, C.G. and Porter, R. (1977) Corticospinal Neurones, Academic Press, London. pp 139-152

Sherrington, C.S. (1898) 'Experiments in examination of the peripheral distribution of the fibres of the posterior roots of some spinal nerves', Phil. Trans. Roy. Soc.Lond. **190B**, 45-186.

Shinoda, Y., Yokota, J-I. and Futami, T. (1981) 'Divergent projection of individual corticospinal axons to motoneurons of multiple muscles in the monkey', Neuroscience Letters **23**, 7-12.

Shinoda, Y., Yamaguchi, T. and Futami, T. (1986) 'Multiple axon collaterals of single corticospinal axons in the cat spinal cord', J. Neurophysiol. **55**, 425-448.

Shinoda, Y., Zarzecki, P. and Asanuma, H. (1979) 'Spinal branching of pyramidal tract neurons in the monkey', Exp. Brain Research **34**, 59-72.

Smith, A.M. (1981) 'The coactivation of antagonist muscles' Can. J. Physiol. Pharmacol. **59**, 733-747

Smith, W.S. and Fetz, E.E. (1989). 'Effects of synchrony between corticomotoneuronal cells on post-spike facilitation of muscles and motor units', Neuroscience Letters **96**, 76-81.

Spoor, C. (1983) 'Balancing a force on the finger tip of a two-dimensional finger without intrinsic muscles', J. Biomechanics **16**, 497-504.

SECTION 9

CONTROL OF MOVEMENT DYNAMICS

THE DYNAMIC CONTROL OF SINGLE JOINT MOVEMENTS

G. L. GOTTLIEB
Department of Physiology
Rush Medical College
1753 W. Congress Parkway
Chicago, IL 60612, USA

ABSTRACT. Relatively fast, single-joint human movements are performed by stereotyped activation pulses of an agonist muscle and slightly delayed, of its antagonist muscle. It is their combined action that first accelerates the limb and then arrests it at a desired position. The control of such movements requires that the intensity and duration of both pulses and their relative timing be specified either explicitly, or that such parameters emerge from the interaction of descending control signals and afferent feedback from the ongoing movement. This work demonstrates how muscle myoelectrical activity can help us interpret the kinetic actions of the muscles. We will further show that this requires that we take account of two patterns of single-joint control. These patterns emerge from two movement strategies that differ in the way that subjects control the forces to produce movement. For each strategy, different rules exist to control the degree and timing of activation of the agonist and antagonist muscles. The basis on which a strategy is chosen, the rules that characterize each and supporting data are presented.

1. Introduction

The study of how normal human single joint movements are controlled has often been guided by three questions. How do kinematic features of the mover's behavior vary when a single parameter of the movement task is systematically altered? What are the accompanying electromyographic (EMG) changes? What inferences can be made about the motor system's control mechanisms from the relations found in answering the first two questions?

We will develop an hypothesis for the control of single joint movements which is based upon an alternative to the first question. Regularities should be sought not in kinematics but in kinetics. We should not look first at the angular trajectories but at the joint torques. From the torques and the laws of mechanics, kinematics emerge. Further, both muscle tension and myoelectrical activity arise from the same neural input to the muscle and therefore will reveal some common features of that control.

We propose four organizing principles for the control of single joint movements (Gottlieb, Corcos et al. (1989b)).

I. A task requires a "strategy" for performing movements to satisfy constraints that are explicitly or implicitly a part of it. Strategies also help resolve the problem of what to do when there are many possible movements that can accomplish the task.

J. Requin and G. E. Stelmach (eds.), Tutorials in Motor Neuroscience, 499–515.

II. Strategies are operationalized in terms of sets of rules for muscle activation.
III. Rules for muscle activation lead to consistent patterns of both muscle tension and EMG signals. Because of their common drive, many measures of tension and EMG will be highly correlated.
IV. Muscle forces interact with the dynamic properties of the limb and load to produce movement. Because the same forces will produce different movements for different loads, correlations between task variables and EMG and force measures will be more generally descriptive than those between task variables and any kinematic measures.

A corollary to principle III above is that the same relations between EMG and tension should apply to both agonist and antagonist muscles. Differences in EMG patterns therefore should be explainable in terms of the differences in joint torques produced by muscle actions on different sides of the joint and acting under different biomechanical conditions.

In this paper we will describe experimental evidence in support of two general principles stated above. First we will show the correlation between myoelectrical and kinetic measures is the most descriptive association that exists between the two physically different expressions of muscle contraction. We will show this under a variety of single-joint movement tasks for both agonist and antagonist muscles. We will then show how control patterns that are expressed in both mechanical and electrical measurements are separable into two qualitatively different patterns.

2. The Dual-Strategy Hypothesis

2.1. THE EXCITATION PULSE

In order to describe how muscles are controlled for fast movements, many investigators have considered that the large bursts seen in the EMGs of each muscle at or near the onsets of their activity to be the most important feature that needed interpretation. We have carried this one step farther in arguing that the muscles' EMGs can be interpreted, not merely as measures of muscle activity, but as indirect measures of the net excitation of the motoneuron pool. Although the precise relation between excitation taking place in the anterior horn and the EMG measured peripherally is likely to be very complex, our understanding of neuron behavior leads us to suggest that a useful first approximation to this relationship is a linear, first-order differential equation (eq. 1) (Knox (1974)).

$$\frac{de(t)}{dt} + \frac{1}{a} e(t) = x(t) \tag{1}$$

Here $e(t)$ is the muscle EMG signal, $x(t)$ is the motoneuron pool excitation and a is the characteristic time constant. This is the equation of a low-pass filter.

We further suggest that for the control of fast movements the excitation signal, $x(t)$, can be approximated by a rectangular pulse of controllable height (H) and width (W). The assumptions of a rectangular excitation pulse passing through a first-order low-pass filter are not a unique combination for producing signals that have EMG burst-like properties. They are however the simplest set of assumptions that give us the minimum necessary properties. Such highly simplifying assumptions cannot possible capture all the details of muscle excitation but can, we suggest, capture the most important ones. Furthermore, we do not wish to imply that while pulses are the controlling waveform for certain fast movements, they are the only waveforms that the CNS is capable of creating when required to do so by features of the task (Corcos, Agarwal et al. (1990)).

What are the essential features of the EMG bursts that are captured by this model? The transformation of a rectangular pulse *x(t)* through a filter described by equation 1 leads to smoothly rising and falling pulses for *e(t)*. The myoelectrical waveform *e(t)* cannot change instantaneously but only at a finite rate that depends upon the ratio H/a. If the excitation pulse is sufficiently long (eg $W > 3a$), then *e(t)* will approach an equilibrium value before the pulse ends that is proportional to H. However if W is briefer than this, the peak value of *e(t)* will rise monotonically with W, even though H may be constant. The area of *e(t)* will be proportional to the product HW which is the area of the excitation pulse.

2.2. TWO STRATEGIES

We use the term "strategy" to describe the set of rules that define the parameters of the excitation pulses for agonist and antagonist muscles. There are five parameters to completely describe a two pulse, agonist-antagonist burst pattern. These are their heights and widths and also the latency or delay of the antagonist onset relative to that of the agonist. We have found that two different strategies or sets of rules are necessary to describe how the parameters of the excitation pulses for fast movements vary for different distances, with different inertial loads and at different intended speeds. The two strategies that we have described are called speed-sensitive (SS) (Corcos, Gottlieb et al. (1989)) and speed-insensitive (SI) (Gottlieb, Corcos et al. (1989a)). The rules for these are summarized in the Table below:

TABLE. The rules for adjusting the parameters of the excitation pulses for the Speed Sensitive (SS) strategy that is used for intentionally changing movement speed and the Speed-Insensitive (SI) strategy that is used for changing movement distance or adapting to changing loads. H is pulse height, W is pulse width and L is the relative latency of the antagonist burst.

	SS	*SI*
H	*covaries with speed*	*is constant*
W	*is constant*	*covaries with distance or load*
L	*covaries inversely with speed*	*covaries with distance or load*

The notion that there are qualitative differences between classes of movements has been suggested before (Brown and Cooke (1981); Mustard and Lee (1987)). It is implicit in the imposition of kinematic boundary constrictions on descriptive rules for movement (Wallace (1981)). Our definition of strategies provides an *a priori* basis for how movements will be performed without requiring that some behavioral feature (such as movement velocity) be held constant. The SI rules do not imply constant speed (and in fact are never associated with it); only that the subject not be intending to change speed. Movements of constant speed (or movement time) over different distances would be performed by an SS strategy (Mustard and Lee (1987); Wallace and Wright (1982)), as would movements of different distances in the same time (Freund and Budingen (1978); Gielen, van den Oosten et al. (1985); Shapiro and Walter (1986)).

Each strategy implies many kinematic and kinetic correlates that follow from their rules, as do similar myoelectrical correlates. Because both the myoelectrical and mechanical responses are direct consequences of the same excitation pulses, these diverse relations are not independent so that the rules of the strategies also predict specific EMG and mechanical correlates. It is these that we will discuss in the experimental section below.

2.3. DISTINGUISHING BETWEEN STRATEGIES

The defining features of strategies are to be found in the excitation pulses which are not accessible to direct measurement. The variables we can measure are EMG and kinematic. For freely moving limbs we cannot measure muscle forces but can estimate joint torque, at least early in a trajectory, by calculating inertial torque from measurements of the acceleration. Equation 1 requires that the initial rate of change of EMG will be proportional to the height of the excitation pulse. Therefore, the rules for the SS strategy indicate that when subjects intentionally change speed this will be associated with corresponding changes in the rising phases of the EMG bursts. Furthermore, the SI rules indicate that for movements of different distances or of different loads over one distance, the rising phases will be insensitive to distance and load as long as the subject makes no attempt to adjust movement speed. Note that this does not imply that movement speed will remain constant. On the contrary, it is well known that movement speed increases with distance if the target is of constant size (Fitts (1954); Hancock and Newell (1985)) and that it decreases if the load increases (Danoff (1979); Lestienne (1979); Wadman, Denier van der Gon et al. (1979)). However, these changes are not induced by the subjects deliberate intent or by instruction. Because the contractile process is also a form of low-pass filter, the rates of rise of muscle tension will behave in the same manner described for the EMGs. That is, the initial rates should vary when speed is deliberately modified but should be unaffected by distance or load.

3. Methods

3.1. GENERAL FEATURES

We will confine our discussion to flexion of the elbow in the horizontal plane. In all the experiments to be considered here, seated subjects faced a computer monitor placed 1.5 m in front of them. The right shoulder was abducted 90° and the forearm pointed forward. A computer tone signalled the subject to make a flexion movement to a target, defined by a horizontal band on the monitor that also provided an indication of the subjects' joint angle. Subjects made 11 movements under each set of task conditions. The first movement was discarded as were any that did not have well defined EMG burst patterns. Measurements were made from each EMG record of areas and latencies. The records were aligned to the onset of the agonist burst and averaged.

All EMGs were measured with surface electrodes (Boston Myoelectrodes) amplified 1600X and bandpass filtered at 60 and 500 Hz before digitization at 1000/s. We also measured joint angle and joint acceleration. Velocity was computed.

3.2. MOVEMENT CLASSES

Our subjects were asked to perform three categories of movements. They were always asked to move accurately to the target. In one series they were told to move as fast as possible to targets of different distances. In a second, they were asked to move a single distance with different loads.

In a third, they were given instructions that influenced them to adjust their speed to some lower level. Hence, two of the tasks (varying distance and load) involved what we have called an SI strategy while one (varying speed) involved the SS strategy.

3.3. QUANTIFICATION OF VARIABLES

The estimates of the degree to which muscles were activated was done by integrating the area under the full-wave rectified EMG signal. The area under the agonist burst (Q_{ag}) was integrated from the onset of the burst to the reversal of acceleration (equal to the moment of peak velocity). The area under the antagonist burst was rectified from agonist onset to approximately the time the limb returned to zero velocity. To estimate the rate at which the agonist burst rose, we calculated the area under the first 30 ms of the rectified EMG waveform. These methods are described in greater detail in (Gottlieb, Corcos et al. (1989a))

In earlier work cited above, we used peak inertial torque as the mechanical measure of muscle performance. Here we will use mechanical impulse. Mechanical impulse is defined by the following equation.

$$\text{Impulse} = \int_{t_1}^{t_2} \tau \, dt \tag{2}$$

where τ is joint torque. The integration interval is from the movement's onset to peak velocity which is the time when τ reverses sign. This measure is the equivalent of peak momentum. We cannot measure joint torque directly, except for that component accelerating the manipulandum. This inertial component reflects only a fraction of the torque produced by the muscles. We can get a better estimate of inertial torque using the total moment of inertia of the limb and manipulandum. However, this is still an incomplete measure of the muscle's mechanical performance since it omits the dependence of force on velocity (Hill (1940)) and length (Abbott and Aubert (1952); Joyce and Westbury (1969)). We will consider velocity dependent or viscous forces below but will neglect those that depend upon length.

4. Results

4.1. KINEMATIC AND KINETIC SIMILARITY

Similarity between sets of movements may be found in their kinetic and neural pattern or in their kinematic patterns. Figure 1 shows three families of movements, illustrating that by appropriately choosing pairs of movement sets, each of these kinds of similarities can be found. Figure 1 (left) shows movements of a constant inertial load over four different distances while figure 1 (middle) shows kinetically similar movements of four different inertial loads over a single distance. The lower three frames show the inertial torque and EMG envelopes of a single agonist (biceps) and antagonist (triceps) muscle. There are several kinetic features that are noteworthy. These are: 1) rising torques that have similar slopes, independent of distance or load. Smaller distances or loads reach smaller and earlier peaks because of an earlier decrease in their rate of rise. 2)The agonist EMG bursts all rise at similar rates that are independent of distance or load. The areas of these bursts increase with distance or load. Peak EMG often increases with distance or load but an equally common pattern is a flattening of the tops of the bursts at a level that is similar

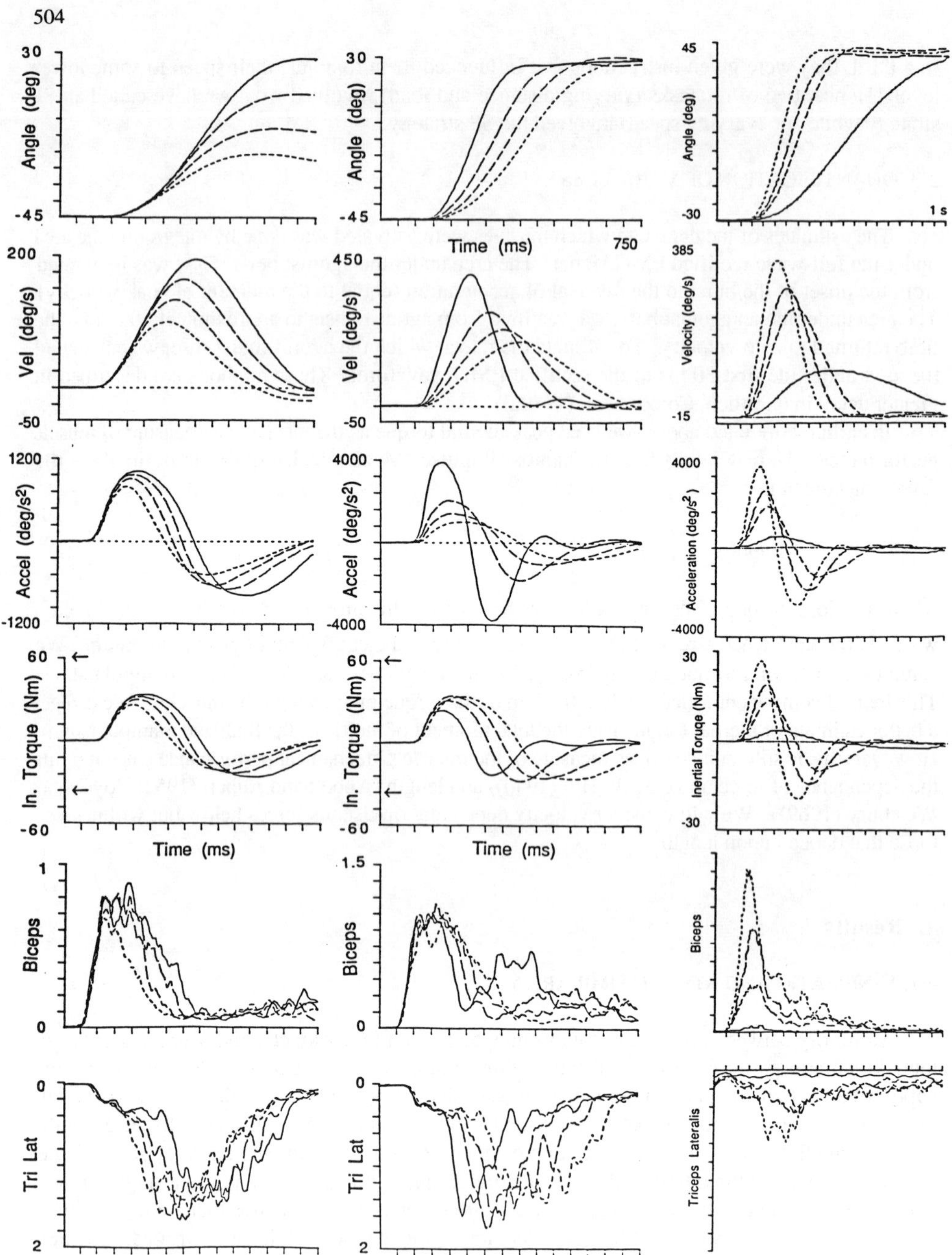

Figure 1. At the left are average movements over four different distances (18°, 36°, 54°, 72°). In the middle are four sets of 72° movements with four different loads. At the right are 72° movements at four different speeds. Plotted are, from top to bottom, joint angle, velocity and acceleration, inertial torque, biceps and triceps EMG.

for different distances and loads. 3) The antagonist EMG shows a small, very early component (latency relative to the onset of the agonist burst is about 25-30 ms), the amplitude of which is independent of distance or load. 4) the principal burst of antagonist activity occurs at a latency that increases with distance or load. Its slope is relatively independent of distance or load. Its area increases with load but behaves inconsistently with distance. On the left, the area of the antagonist burst is smallest for the 72° movement. The most common pattern is for its area to show a decreasing trend with distance although we will show examples of moderate increase as well.

The upper three panels of the figure demonstrate that kinetic similarity does not imply kinematic similarity. The accelerations on the left differ from the inertial torques only by a multiplicative constant moment of inertia. Moving larger distances is associated with increases in peak accelerations and velocities. These larger peaks occur later and overall movement time increases. In the middle however, increases in the inertial load of the manipulandum lead to reductions in the rates and peaks of acceleration and velocity. While movement distance remains constant, the times of the kinematic peaks and the total movement times increase with load. In both cases, average and peak velocities behave similarly.

A different kind of similarity, kinematic similarity, is to be found in a comparison of the middle and right hand parts of figure 1. All the kinematic features of middle set that are described in the preceding paragraph are equally true of the right hand set. In the latter case however, different speeds are a result of deliberate changes by the subject in the way the muscles are activated rather than consequences of externally imposed changes in load. This is readily evident from the lower three panels at the right. The different rates of acceleration arise from correspondingly different rates of rise in inertial torque. This difference must follow from different patterns of muscle activation which is implied by the different rates at which the agonist EMG burst rises. Note too that the durations of the agonist bursts are relatively constant while the latencies of the antagonist bursts increase with movement time.

4.2. KINEMATIC AND MYOELECTRICAL CORRELATES - SPEED INSENSITIVE STRATEGY

In searching for invariant relations between electromyographic and mechanical measures of a movement, the most commonly used measures are the area of the EMG bursts and a kinematic feature. Figure 2 shows the area of the agonist burst plotted against peak movement velocity (part a) and peak acceleration (part b). Figure 3 shows similar relations for the antagonist using the same peak velocity for part a but using peak deceleration for part b.

For a fixed inertial load, the peak movement velocity and acceleration increase with distance and the agonist burst increases with them (regression lines rise upward to the right). This correlation between kinematics and EMG is not independent of the inertial load however. This is indicated by up and leftward pointing arrows that illustrate the rising slopes of the regression curves for increasing inertial loads. In other words, a movement that achieves a given peak velocity (or acceleration) with a heavy load will be associated with a great deal more agonist activity then if the same velocity is achieved with a light load. As a result, regression curves between EMG and peak velocity or acceleration for movements of the same distance but with different inertial loads (not drawn) slope downward to the right.

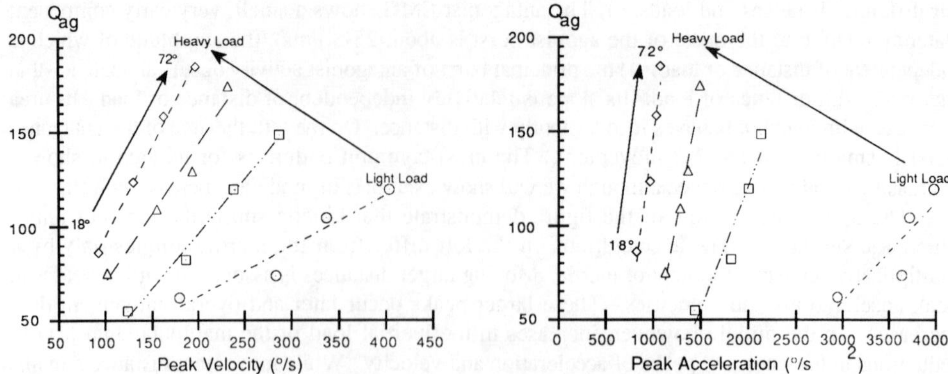

Figure 2. The area of the agonist burst Q_{ag} for movements of different distances and loads (a) Q_{ag} vs Peak Velocity and (b) Q_{ag} vs Peak Acceleration. This and the succeeding figures show data for a single subject. Each symbol represents an average for ten movements.

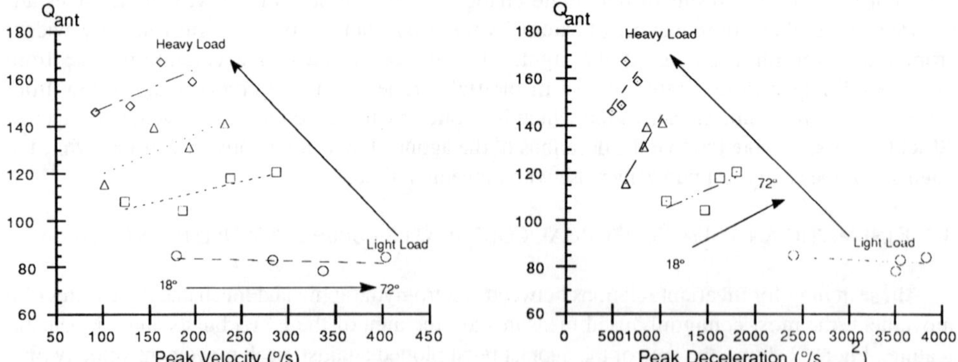

Figure 3. The area of the antagonist burst Q_{ant} for movements of different distances and loads (a) Q_{ant} vs Peak Velocity and (b) Q_{ant} vs Peak Deceleration

The antagonist burst increases with peak velocity or deceleration when those increases occur as a consequence of reduced inertial load for movements of given distance, except at the lightest inertial load. For a fixed inertial load, the relation between peak velocity or deceleration is weak or nonexistent. The lightest load (corresponding to the arm and manipulandum with no added inertia) produces virtually no increase in EMG, in spite of a substantial increase in the kinematic peaks. In other tasks (cf figure 5 below) there is even a decrease in EMG.

4.3 KINEMATIC AND MYOELECTRICAL CORRELATES - SPEED SENSITIVE STRATEGY

The same analysis can be performed on movements of deliberately different speeds such as in Figure 1 (right). Figures 4 and 5 show agonist and antagonist EMG burst areas as they vary with peak movement speed and acceleration. With constant inertial load, increases in speed that are either deliberate or are consequences of increases in distance are clearly associated with increases in the area of the agonist burst. It is significant, as we will discuss below, that to achieve a given peak velocity over a short distance, requires considerably greater activation of the agonist than to reach that same peak velocity for a movement of a longer distance. However, when one looks at peak acceleration, the amount of agonist activity is indifferent to which of the two movement tasks was performed.

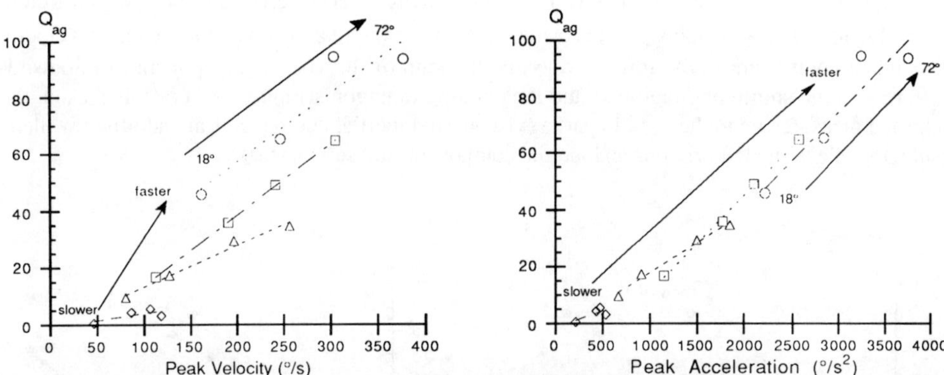

Figure 4. The area of the agonist burst Q_{ag} for movements of different distances with four different instructions affecting speed (a) Q_{ag} vs Peak Velocity and (b) Q_{ag} vs Peak Acceleration

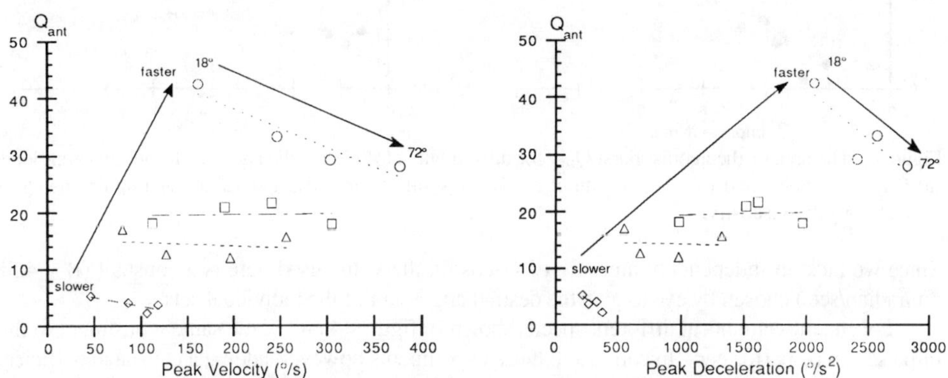

Figure 5. The area of the antagonist burst Q_{ant} for movements of different distances with four different instructions affecting speed (a) Q_{ant} vs Peak Velocity and (b) Q_{ant} vs Peak Deceleration.

Examination of the behavior of the antagonist gives a very different picture. Deliberate increases in peak speed or acceleration result in large increases in the activation of the antagonist. However, in the lightly loaded movements from which the measurements of figures 4 and 5 come, increases in speed that are secondary to increased movement distance are as likely to show a decrease as an increase in the area of the antagonist burst.

The implication of these data is that under three different manipulations of task, the quantity of the EMGs in neither the principal agonist nor the antagonist can be interpreted in uniform or consistent terms of the kinematic variables peak acceleration or peak velocity.

4.4 . KINETIC AND MYOELECTRICAL CORRELATES

The kinetic measure that we have chosen is impulse (eq. 2), calculated both from acceleration alone and with the addition of a velocity dependent component. Figure 6 shows the agonist and antagonist EMGs versus impulse for the same experiments that are shown in figures 2 and 3. In each figure, the open symbols show EMG of movements over different distances with different inertial loads. The • symbols plot the same EMGs with impulse computed to include a velocity dependent component to the force. Note that the sign of the viscous component of impulse is different in the shortening (agonist) and lengthening (antagonist) muscles. For this reason, the adjusted measures lie to the right in part a (viscous and inertial components are additive) while in part b they lie to the left (viscous and inertial components are subtracted).

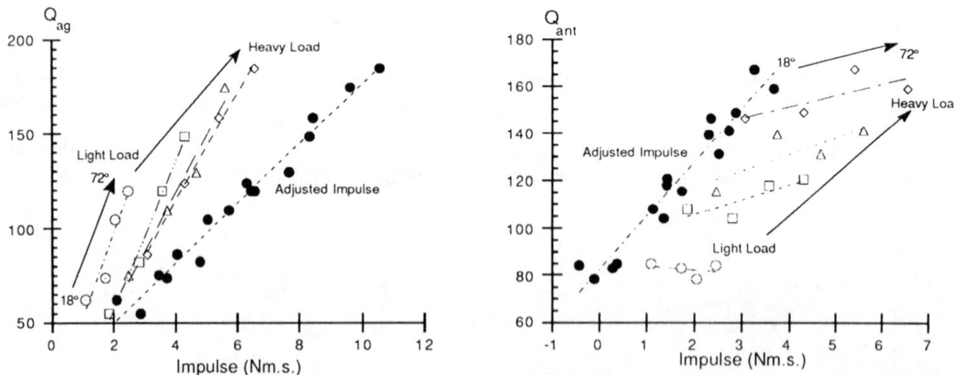

Figure 6. The area of the agonist burst Q_{ag} for movements of different distances with four different loads (a) Q_{ag} vs impulse and (b) Q_{ant} vs impulse. The • symbols are adjusted values of impulse using an estimate of joint viscosity.

Since we lack an independent measure of viscosity, the value used here is a constant (B = 4.58 nm/radian/sec) chosen by eye to give the desired alignment of the individual sets.

For the movements at different speeds shown in figures 4 and 5, the same adjustment to the impulse can be performed. In contrast to those experiments however, adding a constant coefficient viscous component will not improve the correlation of impulse with the agonist or the antagonist burst and in fact, will tend to further disperse the data by introducing a clockwise rotation and stretching of each of the regression curves in figure 7a. Subtracting a constant coefficient viscous component for the antagonist will reduce the dispersion of the regression curves in figure 7b but because the dispersion of the fast movements is much larger than that of the slower ones, the convergence will not be nearly as complete as in figure 6b. We will discuss this below.

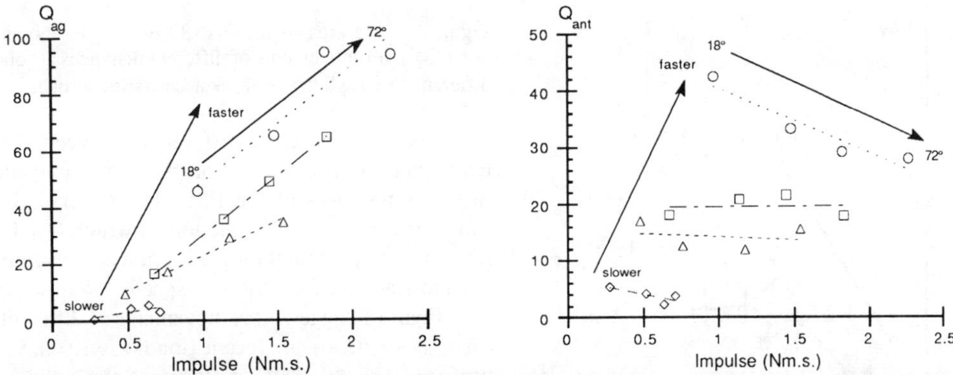

Figure 7. The area of the agonist burst Q_{ag} for movements of different distances with four different instructions affecting speed (a) Q_{ag} vs impulse and (b) Q_{ant} vs impulse.

4.5 ANTAGONIST LATENCY

Accurate movement requires muscle activation that is correct both in degree and in timing. The latency of the antagonist burst will have a very strong effect upon the position at which the limb stops. Figure 8 shows the latency of the antagonist burst for movements of different distances and different loads versus our estimate for adjusted impulse. Note that we have used the calculation of the accelerating impulse as the independent variable rather than that of decelerating impulse.

Figure 8. The latency of the antagonist burst (in ms) is plotted vs. the accelerative impulse. Impulse includes both an inertial and a viscous component to the joint torque.

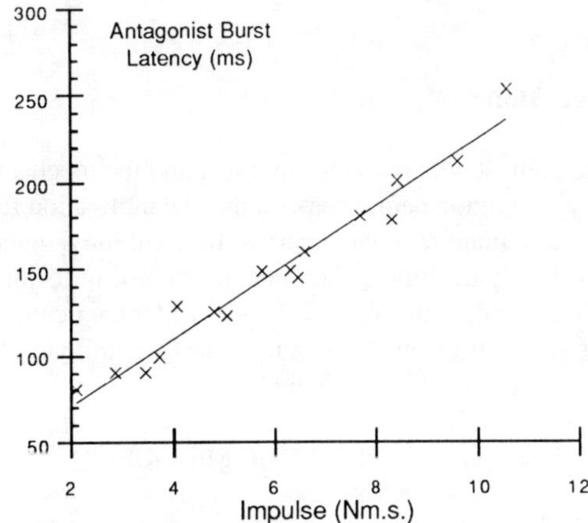

4.6. MYOELECTRICAL DIFFERENCES BETWEEN STRATEGIES

In all of the figures above, there is no clear distinction between movements based on one or the other strategies based upon global myoelectrical and kinetic correlates. There is the suggestion of a difference however, based upon the failure of "adjusted" impulse to improve the kinetic-EMG correlation coefficients under SS conditions (ie fig 7).

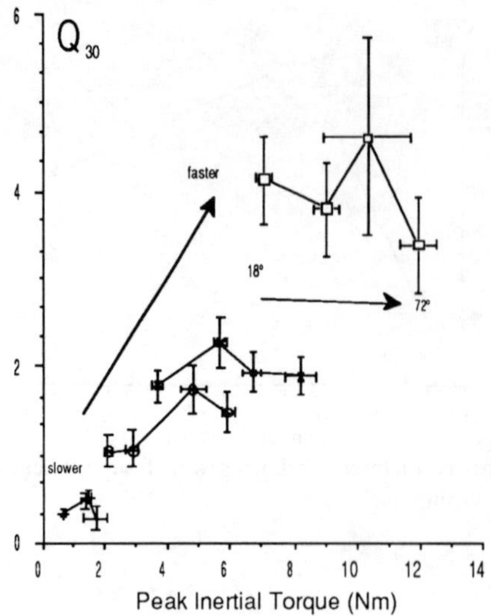

Figure 9. The area of the first 30 ms of the agonist burst Q_{30} for movements of different distances at four different speeds plotted vs. peak accelerating torque.

There is one clear difference between the EMG patterns of the two strategies that is evident in the initial rises of the EMG bursts. This is illustrated by figure 9 at the left in which Q_{30} is plotted vs peak inertial torque for movements four distances and for four different speeds.

Figure 1 suggests that movements of different distances (or of different loads) will have uniformly rising EMG patterns of the agonist bursts. This is demonstrable by integrating the first 30 ms of those bursts and plotting against a suitable kinetic measure as done at the left. For any series of movements in which speed is not intentionally altered, this EMG measure is not dependent upon the distance moved (for three of the four speeds, r^2 is less than 0.15). For any single distance however, the EMG measure is significantly dependent upon the chosen speed of movement.

5. Discussion

The premise that both the electrical and the mechanical behaviors of contracting muscle are driven by a common neural command is the motivation for trying to identify appropriate measures of each that remain correlated across different movement tasks. This has lead us to postulate the utility of first partitioning movements according to the strategies used by the central nervous system to control them. We will first consider the degree to which electromechanical correlations are consistent for different movement tasks employing the same strategy. We will then turn to consistency across different strategies.

5.1 THE NECESSITY FOR KINETIC MEASURES

Mechanical impulse provides a way of accounting for muscle performance that can be applied uniformly to both the agonist and antagonist muscles. This brings consistency between independent measures of the muscles' electrical and mechanical performance.

5.2. THE "PROBLEM" OF THE SPEED SENSITIVE STRATEGY

Although an impulse measure that includes a speculative estimate of viscous muscle forces offers a unifying perspective of movements performed under a Speed Insensitive strategy, it is less successful in improving our description of movements performed under a Speed Sensitive strategy. The argument we will advance is that this failure represents an incorrect estimate of effective viscosity. The principal of electro-mechanical correlation applies across strategies but its quantitative application is difficult.

Figure 10. Relative effects of single manipulations of impulse on the area of the agonist EMG burst.

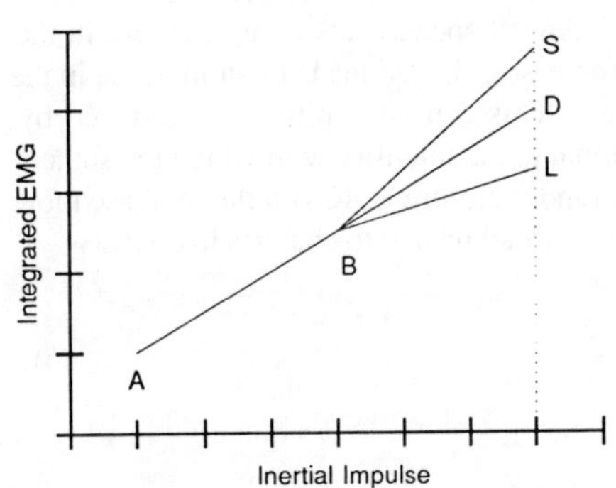

In figure 10 at the left, we have drawn a line from A to B representing the relationship observed between *inertial* impulse and the agonist burst for movements over different distances with a constant inertial load. We know that in going from A to B, subjects perform movements faster. The antagonist burst changes little in area but gets more delayed. Further increases in distance would extend the relationship from point B to point D. If the distance at point B is kept fixed and the subject asked to move a heavier load, the movement will slow but inertial impulse increase. The agonist burst would increase to point L that is below the level of D. This can be deduced from figure 6a since if regression lines were drawn through points of constant distance and different load, the slopes would be less than those drawn in the figure for constant load and different distances. If we specify the same distance and load of point B and ask the subject to move faster, we will again measure an increased inertial impulse. From figure 4a we can deduce that the agonist burst area will increase to a point S that will be larger than that of D. Unlike kinematic measures, impulse increases for each of the three kinds of task manipulations, although not to the same degree. Let us consider next, the reason for the different slopes.

If we let the torque τ of eq. 2 equal a sum of inertial and simple linear viscous components, it is expressed by eq. 3

$$\tau = M \frac{d^2\theta}{dt^2} + B \frac{d\theta}{dt} \tag{3}$$

and the equation for impulse for a symmetrical movement reduces to

$$\text{Impulse} = M \frac{d\theta}{dt}_{max} + B \frac{\theta_{target}}{2} \tag{4}$$

The first term on the right is due to the inertial component and the second to the viscous component. The increase in EMG from point B to D results from the need for the muscle to produce an additional amount of impulse. Changes in distance lead to increases in both the inertial component (peak velocity will increase) and the viscous component (the task definition will increase the far right term of eq. 4) of impulse. A greater load load increases only the inertial component (although peak velocity decreases) of the above relation and therefore does not require as great an increase in impulse or muscle activation.

These movements both use a Speed Insensitive strategy in which the duration of the agonist muscle's excitation pulse lengthens but its intensity is constant. We make a crucial assumption that since excitation intensity determines the frequency of motoneuron firing and the number of active motor units, constant excitation intensity can be associated with a certain constancy of muscle viscosity[1]. It is this relative constancy that allows eq. 4 to align the separate lines in figure 6 into one.

According to the rules of strategies, instructing the subject to change speed leads to a change in the intensity of the excitation pulse. We propose that this has a direct effect on the velocity dependent properties of the muscles. An increase in intended speed leads to an increase in the inertial component of impulse because peak velocity increases. It also leads to an increase in the viscous component because the effective B increases. This can be explicitly introduced by changing eq. 4 to eq. 5 by the term S which is proportional to the intensity with which the subject contracts the limb muscles. This implies that impulse (and therefore EMG) could decrease more steeply for deliberate changes in speed than for changes in speed related to changes in distance.

$$\text{Impulse} = M \frac{d\theta}{dt}_{max} + SB \frac{\theta_{target}}{2} \tag{5}$$

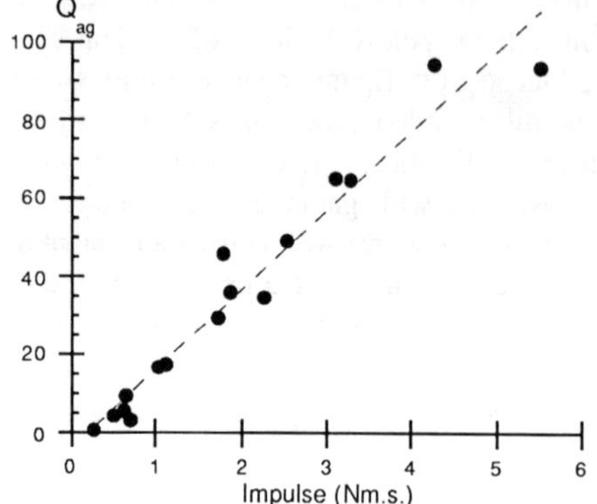

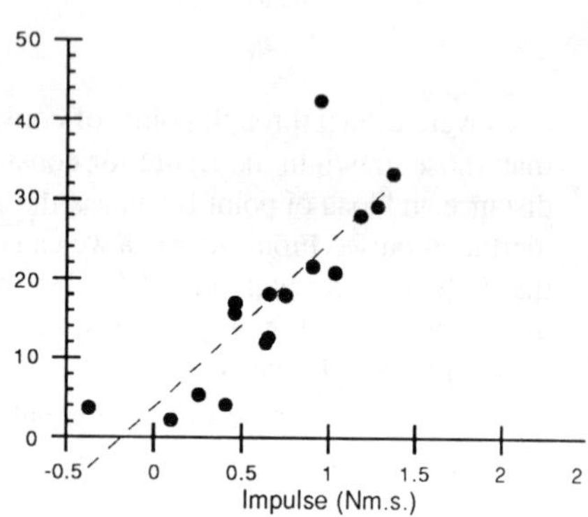

[1]We do not need to assume that the viscosity is constant but rather, that it varies over time in a manner similar to the way that muscle tension varies over time. Uniform intensities of excitation lead to uniformly rising torques that diverge when the excitation pulses diverge. In like manner, we propose that the viscous property of the muscle varies in the same manner and differ at times into the trajectory that depend upon the duration of the excitation pulse.

Figure 11. Agonist (a) and Antagonist (b) EMG for movements of different distances and speeds with impulses computed with an instruction-dependent viscous term. For a, the ratios of S are 0:1:2:4.5; in b they are 1:3:10:15

We can show that this manipulation of the data is capable of aligning the separate regression lines by using instruction dependent values of S. This is done for the agonist in figure 9a and for the antagonist in figure 9b. Comparison of the scales of figures 7 and 9 show that viscous and inertial forces must be of comparable magnitudes if this transformation is reasonably correct. The problem in testing this hypothesis is that we lack an independent measure of the viscous forces.

5.3 CONCLUSIONS

The first conclusion we wish to draw from the above data is that the theoretical premise that muscle force and the electromyogram represent independent observations of the same controlling input. A consequence of this fact is that they must share qualitative and quantitative features of that signal. This has been experimentally supported for both agonist and antagonist muscles and across three different tasks. This is, to our knowledge, the only successful attempt to describe both muscle groups by a uniform set of rules.

The second conclusion is that it is necessary to distinguish at least two strategies for controlling single joint movements. In the face of different kinematic and kinetic demands by different tasks, the CNS selects either a pulse-height or a pulse-width modulation scheme for the motoneuron pools. This distinction must be invoked to account for the way in which movements may share kinematic but not kinetic similarity or alternatively, share kinetic but not kinematic similarity. The quantitative differences between strategies are most clearly evident in the earliest moments of the myoelectrical signals or in the torque trajectories. They may not be apparent from observations confined to the kinematic trajectories.

The suggestion that movements are controlled by the selection of appropriate patterns for muscle activation does not answer the question of how these patterns are able to move the limb to specific targets and sometimes, along specific trajectories. Most human movements are goal directed and have trajectories that appear well chosen to achieve that goal. To move along a specific trajectory requires a well chosen set of net torques to be applied to each joint. To compute those torques from knowledge of the desired trajectory and the load of the mechanical system being moved is called the inverse kinematic problem. Solving this problem is relatively simple for single joints but becomes rapidly more complex as the number of joints increases. Such complexity has inspired many to speculate that the real-time calculation of such solutions, while potentially a practical approach for some robotic systems, is not likely to be the way the CNS controls movement. If this is true, how can the CNS know what forces to apply and therefore, how can it know what parameters to select for the excitation pulses?

Our answer to this must necessarily be highly speculative. We would suggest that to perform accurate, fast movements, the CNS has a very good idea of the forces necessary to perform such movements based upon prior movement experience. This allows it to make one arbitrary selection; excitation pulse width if the task allows an SI strategy or excitation pulse height if an SS strategy is required. The calculation of the other parameters is a matter of linear approximation from prior movements around the work space. This implies that if we lack adequate information to make accurate calculations we will make errors of movement. Alternatively, we will make slower movements to allow visual or proprioceptive corrections to be made as we approach the target. A further implication is that to improve movement accuracy under a variety of conditions and to a variety of endpoints, we must practice. None of these suggestions seem inconsistent with normally observed behavior.

5.4 SELECTING THE PARAMETERS OF THE EXCITATION PULSE

Let us consider one possible way that the CNS might go about "computing" the parameters of agonist and antagonist excitation pulses when asked to move a given distance "as fast and accurately as possible." Fitts' showed how these two criteria of speed and accuracy are competitive. Therefore, the subject will choose an agonist excitation pulse intensity that is large but usually not the largest possible. Having done so, the net agonist excitation will require specification of pulse duration. This is done based upon prior experience of moving the limb over similar distances with similar loads. It is a guess.

Determining the agonist excitation pulse parameters places rigid constraints upon the allowable parameters of the antagonist pulse. The strength of the antagonist's contraction is not uniquely specified however until the latency of that contraction has been set. This is where the observation of Figure 8 is important because it implies that latency of the antagonist's excitation pulse in the SI strategy is completely specified by the duration of the agonist excitation pulse.[2] The simplest rule is to make the antagonist a scale model of the agonist. This would require the specification of only one additional constant which must take into account the characteristics of the load. This too is a guess.

Of the five parameters needed to specify agonist and antagonist excitation pulses, one is chosen based on external criteria and the rest are coupled together such that only two of them, agonist width and antagonist scaling parameters need be "tuned" to the specific features of the movement. Precisely how they are adjusted by trial and error to correct for undershoot and overshoot or misjudgments in the dynamics of the load is an area that requires further investigation.

6. Acknowledgements

This work was supported in part by NIH grant AR 33189 and NS 23593. Appreciation is expressed for the assistance of Drs. Gyan Agarwal, Dan Corcos, Mark Latash and Mr. Om Paul.

7. References

Abbott, B. C. and X. M. Aubert. (1952). "The force exerted by active striated muscle during and after change of length." J. Physiology (London). 117: 77-86.

---

[2] The area of an EMG burst will be proportional to the area of its excitation pulse. In the SI strategy, these areas will be vary with the width of the pulse. Therefore, the covariance of Q_{ag} and antagonist latency implies that a similar covariance exists with the width of the agonist excitation pulse. The same rule cannot apply for the SS strategy. For that strategy the antagonist's latency is inversely proportional to the agonist's intensity. Since each strategy varies only one of the excitation pulse parameters, these two rules are compatible and a simple model of antagonist latency is $L \approx L_0 + W/H$.

Brown, S. H. C. and J. D. Cooke. (1981). "Amplitude- and instruction-dependent modulation of movement-related electromyogram activity in humans." J. Physiol. (Lond.). 316: 97-107.

Corcos, D. M., G. C. Agarwal, B. Flaherty and G. L. Gottlieb. (1990). "Organizing principles for single joint movement: IV - Implications for isometric contractions." J. Neurophysiol. (in press):

Corcos, D. M., G. L. Gottlieb and G. C. Agarwal. (1989). "Organizing principles for single joint movements: II - A speed-sensitive strategy." J. Neurophysiol. 62: 358-368.

Danoff, J. V. (1979). "The integrated electromyogram related to angular velocity." EMG & Clin. Neurophysiol. 19: 165-174.

Fitts, P. M. (1954). "The information capacity of the human motor system in controlling the amplitude of movement." J. Exp. Psy. 47: 381-391.

Freund, H. and H. J. Budingen. (1978). "The relationship between speed and amplitude of the fastest voluntary contractions of human arm muscles." Exp. Brain Res. 31: 1-12.

Gielen, C. C. A. M., K. van den Oosten and F. Pull ter Gunne. (1985). "Relation between EMG activation patterns and kinematic properties of aimed movements." J. Mot. Beh. 17: 421-442.

Gottlieb, G. L., D. M. Corcos and G. C. Agarwal. (1989a). "Organizing principles for single joint movements: I - A Speed-Insensitive strategy." J. Neurophysiol. 62(2): 342-357.

Gottlieb, G. L., D. M. Corcos and G. C. Agarwal. (1989b). "Strategies for the control of single mechanical degree of freedom voluntary movements." Behav. & Brain Sci. 12(2): 189-210.

Hancock, P. A. and K. M. Newell. (1985). The movement speed-accuracy relationship in space-time. in Motor Behavior: Programming, Control and Acquisition, H. Heuer, U. Kleinbeck and K. H. Schmidt (eds), Springer-Verlag, New York.

Hill, A. V. (1940). "The dynamic constants of human muscle." Proc. Royal Society London, Series B. 128: 263-274.

Joyce, G. C., Rack, P. M. H. and D. R. Westbury. (1969). "The mechanical properties of cat soleus muscle during controlled lengthening and shortening movements." J. Physiol. (London). 204: 461-474.

Knox, C. K. (1974). "Cross-correlational functions for a neuronal model." Biophysical J. 14: 567-582.

Lestienne, F. (1979). "Effects of Inertial load and velocity on the braking process of voluntary limb movements." Exp. Brain Res. 35: 407-418.

Mustard, B. E. and R. G. Lee. (1987). "Relationship between EMG patterns and kinematic properties for flexion movements at the human wrist." Exp. Brain Res. 66: 247-2256.

Shapiro, D. C. and C. B. Walter. (1986). "An examination of rapid positioning movements with spatiotemporal constraints." J. Motor Beh. 18: 372-395.

Wadman, W. J., J. J. Denier van der Gon, R. H. Geuze and C. R. Mol. (1979). "Control of fast goal-directed arm movements." J. Human Move. Studies. 5: 3-17.

Wallace, S. A. (1981). "An impulse-timing theory for reciprocal control of muscular activity in rapid, discrete movements." J. of Mot. Beh. 13: 144-1160.

Wallace, S. A. and L. Wright. (1982). "Distance and movement effects on the timing of agonist and antagonist muscles: a test of the impulse-timing theory." J. of Mot. Beh. 14: 341-352.

CENTRAL AND PERIPHERAL CONTROL OF DYNAMICS IN FINGER MOVEMENTS AND PRECISION GRIP

M.-C. HEPP-REYMOND and M.A. MAIER
Brain Research Institute
August-Forel-Str.1
CH-8029 Zurich
Switzerland

ABSTRACT. The control of movement dynamics has been investigated in the precision grip. The task required the generation of fine-graded forces between thumb and index finger. EMG activity, recorded in human subjects and monkeys, have disclosed activation and modulation with force in almost all the finger muscles and shown the existence of muscular synergies between muscle-pairs. In the monkey, force-related neuronal activity was recorded in the motor and somatosensory cortex, thalamus and globus pallidus. These investigations have disclosed similar classes of discharge patterns in all 4 regions, however with some differences in their relative distribution, in their timing with respect to force increase and their relationship with force. Most similar to area 4 was the activity of a class of thalamic neurons, whereas the characteristics of the neurons in the somatosensory cortex seem to indicate that this region is mainly involved in a feedback loop from the periphery.

Introduction

Among the many important problems of motor control, the specification of movement kinematics and dynamics by the activity of single neurons, or population of neurons, deserves a central position. The search for causal relations between spike trains and specific aspects of movement has first been focused on the primary motor cortex because of its direct access to the spinal motor apparatus. Evarts (1968), in his pioneering work raised thc question as to whether pyramidal tract neurons in area 4 were specifying muscle length or muscle tension, i.e. displacement or force. Since then, new insight has been obtained for various types of movements and of movement parameters, under diverse experimental conditions and in several cortical and subcortical regions (Hepp-Reymond, 1988 for review).

Two main requirements have to be satisfied in order to assess neuronal correlates of movement parameters. On the one hand, the experimental paradigm should yield the isolation of one movement parameter or allow the dissociation of several. On the other, the relationship between the discharge rate and the investigated parameters should be quantitatively evaluated and should make explicit both average behavior and variability.

Our investigations have been mainly addressed to the encoding of dynamics by area 4 neurons. For this purpose we have chosen prehension, in particular the precision grip which belongs to the primate repertoire. The finger movements are

J. Requin and G. E. Stelmach (eds.), Tutorials in Motor Neuroscience, 517–527.

under preferential cortical control, since direct connections from some area 4 neurons to the spinal motoneurons innervating hand and finger muscles have been demonstrated both anatomically and electrophysiologically (Phillips and Porter, 1977, Fetz and Cheney, 1980, Kuypers, 1981, Lemon et al., 1986). These cortico-motoneuronal CM projections, which increase in importance in monkey and man, have been related to a primate specialisation, the fractionated finger movements. The precision grip presents another important advantage: the force exerted between thumb and index finger can be studied under nearly isometric conditions, fulfilling one of the conditions for investigating neuronal correlates of movement dynamics. Furthermore, by using a paradigm which requires the control of fine-graded force, the changes in muscular and neuronal activity as a function of force can be quantified trial by trial, thus fulfilling the second condition mentioned above.

In view of the large number of muscles acting on the various finger joints and therefore participating in the grip, the question arises how an efficient force control can be achieved in such a multi-articulate system. Three main questions will be discussed here.

1. Which are the muscles mainly involved in the precision grip and what is their individual contribution to the control of fine-graded forces?

2. How do precentral neurons encode constant force exerted between thumb and index finger in the conscious monkey?

3. Are other cortical and subcortical regions participating to the control of movement dynamics? And how?

Methods

EXPERIMENTAL PARADIGM

The paradigm has been already described in detail previously (Hepp-Reymond et al., 1978, Hepp-Reymond, 1988), as well as the force transducer, its fixation and calibration (Smith et al., 1975). Briefly monkeys (*Macaca fascicularis*) were trained to generate and control force isometrically on a transducer held between thumb and index finger. Signals on a screen indicated the required and exerted force. The task consisted in generating and maintaining a force for 1.5 s at a first low level (indicated by two lines). At the end of this period, the displacement of the two lines instructed the monkey to increase force to a higher level and hold it for one more second. The forces had to be maintained within narrow limits and varied between 0.1 and 0.4 N for the low level and between 0.4 and 1.0 N for the higher one. A correct sequence was rewarded with some drops of fruit juice.

RECORDING PROCEDURES

At the end of the training period a recording chamber was implanted under general anesthesia (pentobarbital sodium after induction with ketamine hydrochloride) contralaterally to the trained hand. Single cells were recorded using varnished tungsten microelectrodes. Neuronal activity showing some relation to the task was digitized on-line and/or recorded on a magnetic tape. Intracortical microstimulation (60 ms train of cathodal pulses at 300 Hz, pulse width 0.2 ms, 5 - 15 occasionally 30 µA) was applied through the microelectrode close to the recorded neurons. Receptive field properties of the neurons were in general assessed by light touch of the skin and manipulation of muscles and joints.

In some monkeys, electromyographic (EMG) activity was recorded with percutaneously implanted bipolar thin wire electrodes, stored on magnetic tape and digitized off-line (Rufener and Hepp-Reymond, 1988). The muscles were identified by the stimulation applied through the electrodes at the end of the recording session.

HISTOLOGY AND RECONSTRUCTION

After perfusion under a lethal dosis of anesthetics, the brains were prepared histologically, cut parasagitally, parallely to the electrodes penetrations and the sections stained generally with Cresyl violet. The reconstruction of the investigated region was made from the histological sections, using landmarks made by small electrolytic lesions, and the experimental protocol. For the cortical recordings, two-dimensional reconstructions were made from the sections. The areas were defined following the description given by Powell and Mountcastle (1958) and by Sessle and Wiesendanger (1982) for the primary motor cortex.

RECORDING OF ELECTROMYOGRAPHIC ACTIVITY IN HUMAN SUBJECTS

Recording of EMG activity was also made in 4 human subjects. The motor task was basically similar and required the generation of fine-graded isometric forces between thumb and index finger in a ramp-and-hold paradigm with three consecutive force steps. The required and actual force (0.5 to 3.0 N) were continuously displayed on a video screen. The force transducer was the same as that used in the monkey experiments but with an enlarged contact surface for the finger tips. The wrist of the experimental subjects was maintained at an angle of about 30° in an individually molded cast. Recordings were made in up to 6 muscles simultaneously with conventional needle electrodes. The EMG activity was stored on a magnetic tape recorder and, after full-wave rectification and averaging, digitized off-line and analysed quantitatively. The muscles were identified functionally.

QUANTITATIVE ANALYSIS OF NEURON AND MUSCLE ACTIVITY

To assess quantitatively the relationship between force and either neuronal activity, or EMG activity, a trial by trial analysis of the digitized data was performed. Time segments of constant force during hold periods were chosen from successful individual trials. For each segment, both force and neuronal firing frequency, or EMG amplitude, were averaged over time, displayed in scatter diagrams, and linear regression coefficients calculated. The slopes of the regression lines were used as index of force-sensitivity for the single neurons recorded in area 4, in somatosensory cortex, or in other subcortical regions.

For the investigation in human subjects, the EMG activity averaged over time for individual force segments was also used to calculate correlation coefficients between the activity of pairs of muscles recorded simultaneously.

Results

MUSCLE ACTIVATION PATTERNS

The analysis of EMG activity of 23 muscles in two monkeys demonstrated that the control of force in the precision grip requires the activity of a large number of muscles, mainly intrinsic and extrinsic finger muscles. Elbow and shoulder muscles were irregularly active, rarely during the force holding itself. In the two force-steps paradigm, several characteristics were disclosed (Rufener and Hepp-Reymond, 1988). First, all the finger muscles, agonists as well as anatomical antagonists (extensor digitorum communis EDC, abductor pollicis brevis AbPB and longus AbPL), modulated their activity with force. Some muscles, like the adductor pollicis AdP, were often recruited at higher force levels. Secondly, the EMG patterns had a certain degree of individuality: some muscles were mainly active during the increase in force from one level to the next in a phasic fashion, others only during the holding of force, and others with both. Thirdly, the computation of correlation coefficients between force and rectified EMG amplitude revealed highly significant values for most intrinsic finger muscles and for some extrinsic ones, with a relatively large scatter of the data points. Muscles like EDC or APL had generally moderate to low correlation coefficients. This indicates that these muscles have most probably a stabilizing function in the precision grip rather than a direct participation in the force generation.

These observations on 2 monkeys have led us to investigate systematically in human subjects the EMG patterns with grip force. A series of experiments were designed to answer questions raised by the monkeys' data, i.e. what are the "prime force generators" in the precision grip if any, and how is the coupling between the active muscles. Recent findings in 4 subjects and for the 7 intrinsic and 8 extrinsic finger muscles acting on thumb and index finger disclosed that all these 15 muscles are coactivated during the precision grip, regardless of their role as functional agonists or antagonists. All the intrinsic finger muscles investigated increased their activity linearly as a function of force. The highest correlations were found for the AdP and the 1st dorsal interosseus 1DI. Some muscles with a presumably more postural function rarely modulated their activity with respect to force in a significant manner (EDC, AbPL). Occasionally, the AbPB decreased in activity with force increase. Despite of the relative high correlations, the function of EMG activity against force showed for all muscles a considerable scatter.

Pair-wise comparison of two muscles recorded simultaneously showed occasional coupling between muscles. This analysis made up to now for 36 pairs revealed mainly coactivation with parallel increase of activity in both muscles (50%). Coactivation was also found for pairs of muscles acting on different fingers, as for the AdP and 1DI, and pairs of flexors and extensors of the thumb. Trade-off synergism with simultaneous reciprocal EMG-changes (Sirin and Patla, 1987) was found in 10% of the cases (e.g. AbPB and AdP). No evident, or variable, coupling was found in 40% of the muscle pairs. Some of these coactivation patterns appear to be reproducible intra- and interindividually.

NEURONAL CORRELATES OF FORCE IN MOTOR CORTEX.

In the finger region of the primary motor cortex, defined by microstimulation, several classes of neurons have been disclosed, whose neuronal activity was clearly related to the precision grip task (Fig.1). These 5 types of neurons with phasic,

phasic-tonic, tonic, decreasing and mixed firing patterns have been found in all the monkeys recorded from and in several investigations using a similar paradigm (Hepp-Reymond et al., 1978, Hepp-Reymond and Diener, 1983, Hepp-Reymond, 1988, Hepp-Reymond et al., 1989).

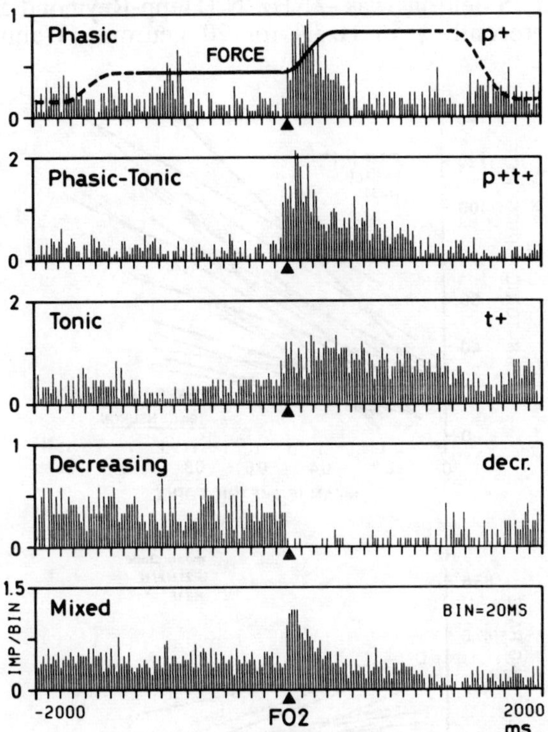

Figure 1. Discharge patterns of neurons in area 4 during force control in the precision grip. Peri-response time histograms of the firing rate of 5 neurons, aligned on the increase of force from the low to the high level (FO2). Display time: 4 s. Top: histogram with superimposed idealized force curve.

Relationship to static force in the hold period were found for the neurons with tonic and phasic-tonic firing patterns. These neurons increased discharge rate with force in a monotonic fashion and, for the majority, the correlation coefficients were highly significant. The linear covariation of the firing rate was limited to a relatively low and narrow range (between 0.1 and 0.8 N) above which saturation or larger data scatter occurred. For neurons with monotonic increase in firing rate with force, the mean force-sensitivity (calculated from the rate-force slopes) was, in a first group of monkeys, 66.5 Hz/N (Hepp-Reymond and Diener, 1983, see Fig.2). In a more recent investigation, a similar value was obtained with a mean rate-force slope of 69 Hz/N for 24 neurons (Hepp-Reymond et al., 1989, Wannier et al., to be published).

The investigations disclosed that other types of neurons also modulated their activity with the force exerted in the grip. These neurons showed a decrement of firing rate with force increase. These decreasing firing patterns can be also classified into phasic, phasic-tonic and tonic. A small group of neurons had a phasic increase followed by a tonic decrease in activity, the so-called mixed firing pattern (Fig.1). For the neurons with tonic decrease of firing rate with static force, significant negative regression coefficients were also disclosed. The mean negative rate-force slope for 15 neurons was -21 Hz/N (Hepp-Reymond and Diener, 1983, Fig.2)) and more recently - 56 Hz/N for 20 neurons (Wannier et al., to be published).

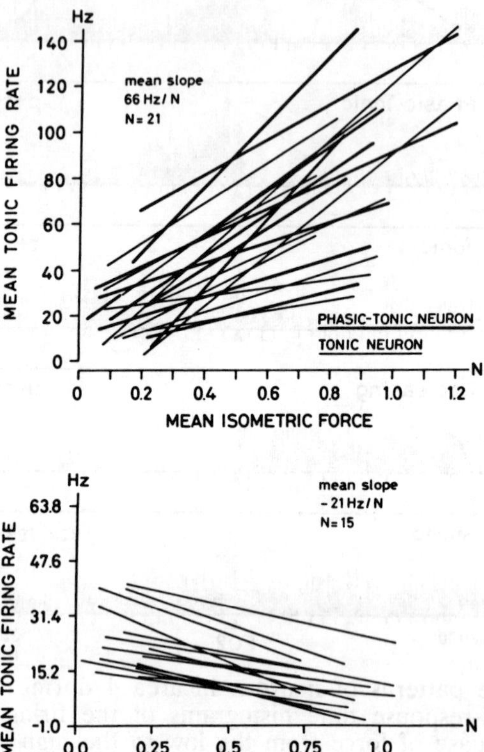

Figure 2. Regression lines calculated for area 4 neurons with statistical significant monotonic changes in firing rate with force. Above: regression lines for 33 tonic and phasic-tonic neurons with positive correlations. Below: negative regression lines for 15 phasic-tonic neurons with decrease in frequency with force increase. The length of the lines corresponds to the force range tested. Slope: mean rate-force slope calculated from the regression lines (from Hepp-Reymond, 1988).

The question was raised as to whether these decreasing neurons could belong to the class of precentral neurons projecting to the spinal motoneurons. In a recent collaborative work with Bennett and Lemon, we have been able to demonstrate negative correlations between mean discharge rate and mean static grip force also for some CM cells with facilitatory projections to 1 or more finger muscles (Maier et al., 1990). From 22 neurons recorded in a one force-step paradigm and identified as CM cells by the spike-triggered-averaging technique (Fetz and

Cheney, 1980, Lemon et al., 1986), three had significant negative correlation between firing rate and force. This new result indicates that some of the decreasing neurons can be connected monosynaptically to motoneurons innervating the distal finger muscles.

A relatively large group of the neurons modulated their activity with force in a non-monotonic fashion over the force range tested, i.e. they had clear activation or deactivation threshold with increasing force. Thus, some neurons with tonic firing patterns were inactive at low grip force but activated only above a certain force level. These neurons had some kind of recruitment threshold, similar to the high threshold motor units in the muscles. In the class of the neurons with decreasing activity, some were clearly tonically active at low force levels but completely silent at higher ones. In a recent investigation, the neurons active only on one force level represented 42 % of the force-related neuron population. The majority belonged to the neurons with total deactivation with force and their function is more difficult to interpret, specifically in view of the fact that 3 of the 22 CM cells analysed also belonged to this neuronal category.

PARTICIPATION OF OTHER REGIONS IN FORCE CONTROL

To investigate this question, the first candidates are cortical and subcortical structures providing important input information to area 4. Neuronal activity in the "motor" nuclei of the thalamus (VLo/VPLo), in the globus pallidus GP and in the cortical somatosensory areas has been recorded under the same conditions as in area 4. In some cases, recordings were made in two regions in the same monkeys. Task-related neurons were found which had discharge patterns similar to those in area 4 and which clearly modulated their firing rate with static grip force in all the three regions investigated. However, some differences could be observed between these regions and motor cortex.

The most striking differences were uncovered for the postcentral neurons (Wannier et al., 1986, Hepp-Reymond et al., 1989). The majority of the neurons located in area 3a, 3b, 1 and 2 were receiving cutaneous afferents (68%), whereas the area 4 neurons were mainly activated by manipulation of joints and muscles (73%). More than 60% of the neurons in somatosensory areas had strong phasic component in their force-related activity in contrast to 27% only in motor cortex. Only 13% of the postcentral neurons showed changes in firing rate before the onset of force increase from one level to the next, in contrast to 56% in area 4 of the same monkeys. Furthermore, the rate-force slopes of 20 neurons covered a wide range, from 16.4 to 212 Hz/N, and had a bimodal distribution, which suggests the existence of two classes of neurons receiving differential input from the periphery or of one class with a single type of input with variations in sensitivity over its receptive field. Neurons with monotonic decrease in firing rate were quite rare in the somatosensory cortex, as compared to their large occurence in area 4 (Wannier et al., to be published).

In contrast, a group of thalamic, and pallidal, neurons had much resemblance with those in the motor cortex (Allum et al., 1983, Anner-Baratti et al., 1986). The results of these investigations have been in large part published and will be only summarized here. These so-called "typical" neurons represented 58% of the VLo/VPLo neurons recorded and 59% of the pallidal ones. They exhibited modulation of their discharge rate in a manner quite similar to those patterns observed in the finger region of area 4. The largest group among these "typical" cells also were the neurons with tonic increase and with tonic decrease in firing rate with force. The distribution of the latencies of activity changes from the low to the high force level also disclosed similarities between thalamic and area 4 neurons. Phasic and phasic-tonic neurons had on average an early onset of activity changes, i.e. before the onset of force increase, and the tonic cells were in the mean late, similar to findings in motor cortex. The pallidal cells in contrast had latencies which in the mean, for the "typical" neurons, were later than the thalamic ones. An important finding was that a group of VLo/VPLo and GP "typical" and "atypical" neurons modulated their firing rate specifically in relation to the isometric force exerted in the grip.

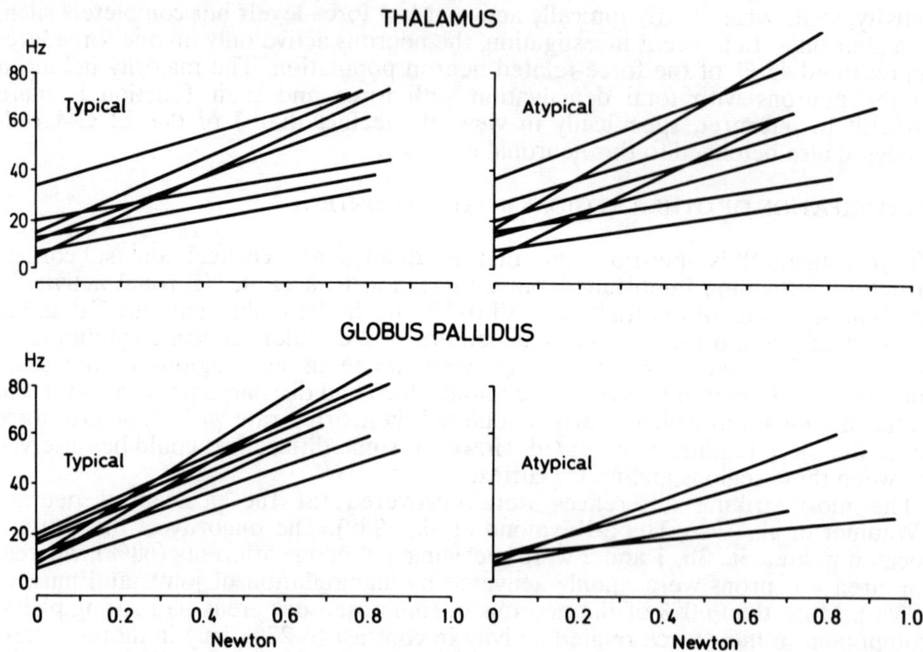

Figure 3. Regression lines calculated for VLo/VPLo and GP neurons with statistically significant monotonic increase in firing rate with force. Left: "typical", right: "atypical" neurons. Length of lines: see legend Fig. 2.

For 15 thalamic and 12 pallidal neurons increasing firing rate with force, the linear regression coefficients calculated for the data gathered on many trials were significant. The mean rate-force slopes were 54.5 Hz/N for the thalamic and 66.7 Hz/N for the pallidal neurons, without differences between "typical" and "atypical" ones (Fig.3). These values do not differ much from that of area 4. Neurons with recruitment threshold were also found in both the VLo/VPLo and the GP, in similar proportions to those in area 4. In the thalamic motor nuclei, only a small percentage were monotonically decreasing in firing rate with force (21%). Thus the proportion of neurons totally deactivated at higher levels was larger than in area 4.

Discussion

The control of force in the precision grip requires the harmonious cooperation of a large number of muscles. For a one force-step task, Smith (1981) reported that almost all intrinsic and extrinsic finger muscles were activated. Our findings in the two force-steps paradigm confirm his observations and bring some new important facts. They demonstrate that fine-graded force exerted in the precision grip is generated by the combined action of many muscles which, in monkeys more than in human subjects, seem to have a certain degree of individual specialization, some being responsible for force holding, others for force development and a few for postural adaptation. The contribution of each muscle to the generated force appears, however, to be relatively inaccurate and variable. This implies that the fine-graded force control is the result of the synergistic action of the muscles, or at least of the most important ones. In fact, our data disclose clear "muscular synergies" between pairs of muscles with a predominance of coactivation synergy. Trade-off synergism (Sirin and Patla, 1987) with reciprocal EMG activity changes in muscle-pairs, was also present but more rare. Some of the coupling patterns, in particular those between muscles acting on the same finger, can be explained by biomechanical properties. Others in contrast, like the coactivation synergies between two muscles, one acting on the thumb and the other on the index finger, are suggesting the existence of centrally organized coupling. It is likely that the CM cells which terminate onto several motoneuronal pools are responsible for these synergistic patterns. (Shinoda et al., 1981, Buys et al., 1986, Lemon, this volume).

The production of force in the grip, as a multi-joint system, requires from the central structures involved some economical control strategy (Bernstein, 1967). The collective action of several neuronal populations located in area 4 should play here a central, if not unique, role. Among these neurons, some are clearly encoding the resultant force exerted in the grip, and have either positive or negative correlations between their firing rate and force. Other investigators have also stressed the relationship between neuronal activity in area 4 and force, or torque, for movements of the wrist (Cheney and Fetz, 1980), the jaw (Hoffman and Luschei, 1980), the forearm (Evarts et al., 1983) and recently for reaching movements in 2D space (Kalaska et al., 1989). In these investigations, mainly in those with reciprocal movements about a joint, only positive relations between neuronal activity and force have been described. Thus, the presence of a large population of neurons decreasing their firing rate with force seems to be unique to a situation requiring muscular cocontraction. The question still remains, whether the neuronal activity recorded in the various investigations is encoding the force exerted about the joints or the tension developed in individual muscles. Our recent findings that some identified CM cells may also decrease their activity with force, despite of the clear activation of their facilitated muscles, give some support to the hypothesis that the collective action of several populations of area 4 neurons specifies the total force required. However, many questions are raised by the existence of the decremental CM cells, their possible function and spinal connectivity.

The control of force in the grip not only depends on the motor commands distributed to the many muscles, but also on the integration of the somatosensory information from the receptors in the skin and in the muscles, joints and tendons. These can be forwarded to the motor cortex either by the intermediate of the somatosensory cortex or of the motor thalamic nuclei (reviewed by Hepp-

Reymond, 1988). Our approach with systematic comparison of neuronal activity under the same experimental conditions in various regions interconnected with area 4 have demonstrated that the encoding of dynamics is distributed among several central nervous structures. Thus, neuronal correlates of force are not only found in area 4, but also belong to the characteristics of neurons in the somatosensory cortex, "motor" thalamus and globus pallidus. However, the differences observed suggest some degree of functional specialization in these regions. The subcortical structures appear to participate to the control of grip force in a way similar to area 4, whereas the role of the somatosensory cortical areas seems to be quite remote. These areas had, in the precision grip task, clear sensory features and could be mainly involved in a feedback loop from the periphery. In contrast, the motor thalamic nuclei may integrate input originating in the periphery and subcortical regions, and forward this information, which may have major importance for the movement execution, to motor and premotor cortex. Recent investigations addressed to the control of movement dynamics and movements kinematics in several cortical and subcortical regions, give support to both possibilities, the distribution of functional characteristics and the regional specialization (Kalaska et al., 1989, Alexander and Crutcher, 1990, Crutcher and Alexander, 1990).

Acknowledgments

This research was supported by the Swiss National Science Foundation Grant 3.549.86, the Hartmann-Muller and the Dr. Slack-Gyr Foundations.

References

Alexander, G.A. and Crutcher, M.D. (1990) Neural representations of the target (goal) of visually guided arm movements in three motor areas of the monkey, J. Neurophysiol. 64, 164-178.

Allum, J.H.J., Anner-Baratti, R., Hepp-Reymond, M.-C. (1983) Activity of neurones in the "motor" thalamus and globus pallidus during the control of isometric finger force in the monkey, Exp.Brain Res. Suppl. 7, 194-203.

Anner-Baratti, R., Allum, J.H.J., Hepp-Reymond, M.-C. (1986) Neural correlates of isometric force in the "motor" thalamus, Exp.Brain Res. 63, 567-580.

Bernstein, N. (1967) The Co-ordination and Regulation of Movements. Pergamon Press, Oxford.

Buys, E.J., Lemon, R.N., Mantel, G.W.H., Muir, R.B. (1986) Selective facilitation of different hand-muscles by single corticospinal neurones in the conscious monkey, J. Physiol. (London) 381, 529-549.

Cheney, P.D., Fetz, E.E. (1980) Functional classes of primate cortico-motoneuronal cells and their relation to active force, J.Neurophysiol. 44, 773-791.

Crutcher, M.D. and Alexander, G.A. (1990) Movement-related neuronal activity selectively coding direction or muscle pattern in three motor areas of the monkey, J. Neurophysiol. 64, 151-163.

Evarts, E.V. (1968) Relation of pyramidal tract activity to force exerted during voluntary movements, J.Neurophysiol. 31, 14-27.

Evarts, E.V., Fromm, C., Kroeller, J., Jennings, V.A. (1983) Motor cortex control of fine-graded forces, J.Neurophysiol. 49, 1199-1215.

Hepp-Reymond, M.-C. (1988) Functional organization of motor cortex and its participation in voluntary movements. In: Comparative Primate Biology, Vol 4 (Steklis,H.D., Erwin,J. eds) Alan Liss, New York, pp. 501-624.

Hepp-Reymond, M.-C., Diener, R. (1983) Neural coding of force and of rate of force change in the precentral finger region of the monkey, Exp.Brain Res. Suppl.7, 315-326.

Hepp-Reymond, M.-C., Wannier, T.M.J., Maier, M.A., Rufener, E.A. (1989) Sensorimotor cortical control of isometric force in the monkey, Prog. Brain Res., 80, 451-463.

Hepp-Reymond, M.-C., Wyss, U.R., Anner, R. (1978) Neuronal coding of static force in the primate motor cortex, J.Physiol.,Paris 74, 287-291.

Hoffman, D.S., Luschei, E.S. (1980) Responses of monkey precentral cortical cells during a controlled jaw bite task, J.Neurophysiol. 44, 333-348.

Kalaska, J.F., Cohen, D.A., Hyde, M.L., Prud'homme M. (1989) A comparison of movement direction-related versus load direction-related activity in primate motor cortex, using a two-dimensional reaching task, J. Neurosci. 9, 2080-2102.

Kuypers, H.G.J.M. (1981) Anatomy of the descending pathways, in Handbook of Physiology, The Nervous System, Part 2, (V.B. Brooks, ed.), American Physiological Society, Bethesda, Maryland, pp. 597-666.

Lemon, R.N., Mantel,G.W.H., Muir, R.B. (1986)Corticospinal facilitation of hand muscles during voluntary mouvement in conscious monkeys, J. Physiol. (London) 381, 497-527.

Maier, M.A., Bennett, K., Hepp-Reymond, M.-C., Lemon, R.N. (1990) Monkey cortico-motoneuronal cells facilitating hand muscles show both positive and negative correlations with grip force, European J. Neurosci. Suppl. 3, 57.

Phillips, C.G. and Porter, R. (1977) Corticospinal neurones. Their role in movement, Academic Press, London.

Powell, T.P.S. and Mountcastle, V.B. (1959) Some aspects of the functional organization of the cortex of the postcentral gyrus of the monkey: a correlation of findings obtained in a single unit analysis with cytoarchitecture, Bull. Johns Hopkins Hosp. 105, 133-162.

Rufener, E.A., Hepp-Reymond, M.-C. (1988) Muscle coactivation patterns in the precision grip, Adv.Biosci. 70, 169-172.

Sessle, B.J. and Wiesendanger M. (1982) Structural and functional definition of the motor cortex in the monkey (Macaca fascicularis), J. Physiol. (London) 323, 245-265.

Shinoda, Y., Yokota, J., Futami, T. (1981) Divergent projections of individual corticospinal axons to motoneurons of multiple muscles in the monkey, Neurosci. Letters 23, 7-12.

Sirin, A.V. and Patla, A.E. (1987) Myoelectric changes in the triceps surae muscles under sustained contractions, Eur. J. Appl. Physiol. 56, 238-244.

Smith, A.M., Hepp-Reymond,M.-C., Wyss,U.R. (1975) Relation of activity in precentral cortical neurons to force and rate of force change during isometric contractions of finger muscles, Exp.Brain Res. 23, 315-332.

Wannier, T.M.J., Töeltl, M., Hepp-Reymond, M.-C. (1986) Neuronal activity in the postcentral cortex related to force regulation during a precision grip task, Brain Res. 382, 427-432.

Wannier, T.M.J., Maier, M.A., Hepp-Reymond, M.-C. (1991) Contrasting properties of monkey somatosensory and motor cortex neurons activated during the control of force in precision grip, J.Neurophysiol. 44, 572-589.

CENTRAL GATING OF MYOTATIC RESPONSES IN ELBOW MUSCLES

F. LACQUANITI, N.A. BORGHESE and M. CARROZZO
Istituto di Fisiologia dei Centri Nervosi
Consiglio Nazionale delle Ricerche
Via Mario Bianco 9
20131 Milano
Italy

Abstract. It is traditionally believed that the behavior of the myotatic reflex is stereotyped, insofar as the direction of the response depends uniquely on the direction of the peripheral stimulus. Contrary to this notion, we here demonstrate a transient reversal of the direction of the myotatic reflex evoked by torque motor perturbations in human elbow muscles during a catching task.

Introduction

Gating of spinal reflexes is well demonstrated for the responses to cutaneous stimuli delivered during different phases of the walking cycle (cf. Grillner, 1981). Typically, a cutaneous stimulus applied to the hindlimb of a cat results in excitation of the extensor muscles in the stance phase and of the flexor muscles in the swing phase (Forssberg et al., 1975; Duysens and Pearson, 1976). Recently, a phase-dependent reversal of a cutaneous reflex has been demonstrated in man: the direction of the middle latency response changes from excitation to inhibition within a single muscle (Yang and Stein, 1990).

By contrast, the behavior of the short-latency myotatic reflex is considered relatively fixed (cf. Houk and Rymer, 1981). In fact, it is generally thought that the direction of myotatic responses depends solely on the direction of the peripheral stimulus. Thus, muscle stretch will result in reflex activation and muscle shortening in reflex inhibition (Matthews, 1972). According to this notion, all flexibility of motor repertoire resides in longer latency responses and in higher levels of sensorimotor integration.

J. Requin and G. E. Stelmach (eds.), Tutorials in Motor Neuroscience, 529–533.

Here we show that also short-latency myotatic responses can be centrally gated in man: The direction of reflex responses evoked by constant torque motor perturbations reverses transiently during the execution of a catching task.

Methods

Subjects were asked to catch a ball falling from 1.6 m. Their forearm was strapped to a goniometer coupled to the shaft of a torque motor. Vision was permitted. EMG activity of elbow flexors (biceps and brachio-radialis) and extensors (triceps) was recorded by means of surface electrodes.

Trains of torque pulses of a pseudo-random binary sequence (PRBS) were delivered continuously by the motor during each trial (Soechting et al., 1981). The ball was dropped 1 sec after the beginning of the trial. The sequence was shifted by one element in each successive trial.

The average time-varying impulse responses of each muscle to a 20-ms torque pulse were obtained by cross-correlating the PRBS perturbations with the EMG activities. These impulse responses were smoothed (45 Hz cutoff).

Results

Figure 1 shows the impulse responses of triceps and biceps EMG activity during the catching task. The oblique axis represents time measured from trial onset. The vertical lines crossing such axis indicate the time of ball release (1 s after trial onset) and the time of ball impact on the hand (1.55 s). Each trace corresponds to the average response to a 20-ms torque pulse tending to flex the elbow that was applied by the motor at that time.

As expected, initial responses (prior to ball release) conformed to the law of reciprocal innervation of antagonist muscles. Thus, the flexor pulse resulted in an increase of EMG activity of the stretched triceps and a decrease of activity of the shortening biceps. Response latency was about 20 ms and peak time about 40 ms.

The changes of triceps responses during the task were modest. The amplitude of the responses increased slightly just prior to impact time and subsequently decreased.

Biceps responses did not change significantly until just prior to ball impact. At that time, the previously negative response became positive. This reversal of the direction of the responses lasted until after impact, when the basal response was observed again.

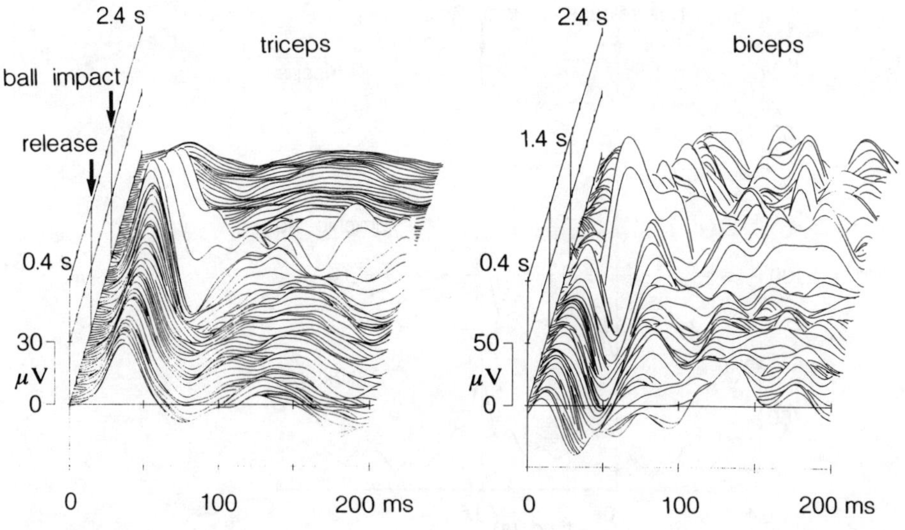

Figure 1. **Impulse responses of triceps and biceps EMG activity from one experiment. Each trace corresponds to the EMG response to a torque pulse tending to flex the elbow. The pulse was applied at the time indicated by the oblique axis, time being measured from the onset of the PRBS perturbations. The vertical lines denote time of release (1 s) and time of impact of the ball on the hand (1.55 s).**

The time course of the changes in the mean amplitude of the reflex responses of elbow muscles is plotted in Fig. 2 (data from an experiment different from that of Fig. 1). The mean amplitude of the responses was calculated over the 20-60 ms interval from pulse onset. The mean response in both elbow flexors investigated here (biceps and brachio-radialis) switched from a negative value to a positive value at about 40 ms prior to ball impact, and returned negative at about 40 ms after impact. As a result of this transient reversal, muscles acting antagonistically around the elbow joint were reflexly coactivated around impact time.

Discussion

The pattern of the short-latency EMG responses evoked by the torque motor perturbations under basal conditions (prior to ball drop) is comparable to that classically described: The muscles that are stretched as a result of the perturbation

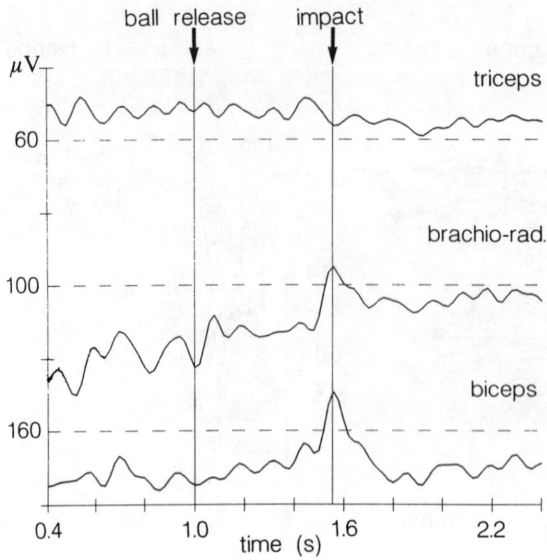

Figure 2. **Time course of the changes in the mean amplitude of the impulse responses of the indicated muscles. The mean amplitude has been computed over the 20-60 ms interval from pulse onset and plotted on the same time frame as the oblique axis in Fig. 1. The zero-lines are dashed.**

are reflexly activated, while the shortening muscles relax (cf. Soechting et al., 1981). However, the pattern of the reflex responses evoked by the same perturbations around impact time is completely different, since both stretched and shortening muscles are coactivated. This coactivation is due to a transient reversal of the direction of the responses of elbow flexors.

After the time of impact, the cutaneous input related to the contact of the hand with the ball adds up to the effects of the torque motor perturbations on the arm. Thus, the peripheral conditions under which such perturbations are applied might differ from the basal conditions. However, it can be safely assumed that the peripheral effects of the perturbations do not change substantially prior to ball impact. Since the reversal of the myotatic responses initiates before impact, one can conclude that it depends on a gating of the reflex by the Central Nervous System.

Reciprocal inhibition is established as the basic pattern of spinal interneuronal connection between the α-motoneurone pools of antagonist muscles (cf. Day et al., 1984). However, stimulation of both Ia and Ib afferents from limb muscles in cat can also evoke widespread coexcitation of antagonist α-motoneurones (Eccles et al, 1957; Willis et al, 1966; Jankowska et al, 1981). This coexcitation is oligosynaptically mediated by lamina V-VI interneurones that are shared by both Ia and Ib afferents from several limb muscles (Jankowska et al., 1981). It is also known that the excitability of these interneurones, similarly to that of Ia inhibitory interneurones, can be extensively modulated by descending tracts.

Thus, one might hypothesize that the reversal of the responses we described depends on a switching from the spinal pathways of reciprocal inhibition to those of coexcitation. In this respect, then, myotatic reflexes can be gated in the same manner as cutaneous reflexes. As for the precise timing of the reflex reversal on impact, it has previously been shown that vision (specifically, optical flow field) can afford an accurate estimate of the time-to-contact during hitting and catching (Lee, 1980; Lacquaniti and Maioli, 1989).

Acknowledgements

We wish to thank Mr L. Chiumiento and F. Neutro for technical help. This work was partially supported by grant n. 3149 ESPRIT II Basic Research awarded by the Commission of the European Communities.

References

Day, B.L., Marsden, C.D., Obeso, J.A. and Rothwell, J.C. (1984) 'Reciprocal inhibition between the muscles of the human forearm', J. Physiol. (Lond.) 349, 519-534.

Duysens, J. and Pearson, K.G. (1976) 'The role of cutaneous afferents from the distal hindlimb in the regulation of the step cycle of thalamic cats', Exp. Brain Res. 24, 245-255.

Eccles, J.C., Eccles, R.M. and Lundberg, A. (1957) 'The convergence of monosynaptic excitatory afferents onto many different species of alpha motoneurones', J. Physiol. (Lond.) 137, 22-50.

Forssberg, H., Grillner, S. and Rossignol, S. (1975) 'Phase dependent reflex reversal during walking in chronic spinal cats', Brain Res. 85, 103-107.

Grillner, S. (1981) 'Control of locomotion in bipeds, tetrapods, and fish', in J.M. Brookhart and V.B. Mountcastle (eds.), *Handbook of Physiology*, Sect. 1, Vol. 2, Part 1, American Physiological Society, Bethesda, pp. 1179-1236.

Houk, J.C. and Rymer, W.Z. (1981) 'Neural control of muscle length and tension', in J.M. Brookhart and V.B. Mountcastle (eds.), *Handbook of Physiology*, Sect. 1, Vol. 1, Part 1, American Physiological Society, Bethesda, pp. 257-324.

Jankowska, E., McCrea, D. and Mackel, R. (1981) 'Oligosynaptic excitation of motoneurones by impulses in group Ia muscle spindle afferents in the cat', J. Physiol. (Lond.) 316, 411-425.

Lacquaniti, F. and Maioli, C. (1989) 'The role of preparation in tuning anticipatory and reflex responses during catching', J. Neurosci. 9, 134-148.

Lee, D. (1980) 'Visuo-motor coordination in space-time', in G.E. Stelmach and J. Requin (eds.), *Tutorials in Motor Behavior*, North-Holland, Amsterdam, pp. 281-295.

Matthews, P.B.C. (1972) *Mammalian Muscle Receptors and their Central Actions*, Arnold, London.

Soechting, J.F., Dufresne, J.R. and Lacquaniti, F. (1981) 'Time-varying properties of myotatic response in man during some simple motor tasks', J. Neurophysiol. 46, 1226-1243.

Willis, W.D., Tate, G.W., Ashworth, R.D. and Willis, J.C. (1966) 'Monosynaptic excitation of motoneurones of individual forelimb muscles', J. Neurophysiol. 29, 410-424.

Yang, J.F. and Stein, R.B. (1990) 'Phase-dependent reflex reversal in human leg muscles during walking', J. Neurophysiol. 63, 1109-1117.

HUMAN JAW MOTION CONTROL IN MASTICATION AND SPEECH

D.J. OSTRY, J.R. FLANAGAN, A.G. FELDMAN and K.G. MUNHALL
McGill University, Montreal, Canada
Institute for Information Transmission Problems, Moscow, USSR
Queen's University, Kingston, Canada

ABSTRACT. Detailed kinematics are presented of two-dimensional jaw motion in mastication and speech. The relationship between jaw translation and jaw rotation is described and experimental records are compared with simulations based on the equilibrium point hypothesis (λ model). In general, in both speech and mastication, jaw rotation and jaw translation were found to start and end simultaneously and their coordination was typically characterized by straight line paths when rotation was plotted against translation. A number of manipulations are described which suggest that jaw rotation and jaw translation can be separately controlled. For example, when jaw movements in speech were examined, the slope of the relationship between rotation and translation varied with the consonant but did not depend on the vowel or speech rate. The kinematic details of jaw motion are well accounted for by the λ model. The model demonstrates that separate central commands can be defined associated with jaw translation, jaw rotation, and co-activation of muscles without motion. Central commands may be superimposed to produce combinations of rotation, translation and muscle coactivation. Empirical patterns can be captured by the model under the assumption of simple constant velocity shifts in equilibrium governed by central commands.

1. Introduction

In this paper, we examine the kinematics of two-dimensional human jaw movement in mastication and speech. We focus on the relationship between jaw translation and jaw rotation and present evidence that these can be independently controlled. A computer model of jaw movement based on the equilibrium point hypothesis (λ model) is briefly described. The model includes separate central control signals for jaw rotation, jaw translation and muscle co-contraction without motion (see Flanagan, Ostry & Feldman, in press, for details). The model shows that simple constant velocity shifts in the threshold lengths of all muscles simultaneously can account for jaw movements in both mastication and speech. The λ model has been previously applied to both human single-joint (Feldman, 1986) and two-joint arm movements (Feldman, Adamovich, Ostry & Flanagan, 1990; Flanagan et al., in press) as well as to human eye movements (Feldman, 1981). The planar jaw system is different than the planar arm system in that the two degrees of freedom are located at the same joint rather than at two different joints.

During jaw opening, the jaw rotates downward and translates forward; the opposite pattern is observed during closing. These motions in the sagittal plane are produced by three groups of muscles. Thus, jaw closers, such as masseter and medial pterygoid, act to both raise and

J. Requin and G. E. Stelmach (eds.), Tutorials in Motor Neuroscience, 535–543.

retract the jaw; jaw openers, such as the anterior belly of the digastric, lower and retract the jaw; the lateral pterygoid produces jaw protrusion. The closers and openers both rotate and translate the jaw whereas the protruders tend to translate the jaw without rotation. Because of this mapping between muscle actions and kinematic degrees of freedom, all three muscle groups must be coordinated to produce rotation, translation or co-contraction without motion. With the model we have shown that it is possible to define central commands which produce pure rotation, pure translation and pure co-contraction independent of the jaw position.

We have recorded the kinematics of human jaw movements using the X-ray microbeam system (Abbs, Nadler & Fujimura, 1988; Westbury, in press). The microbeam is a low-dosage narrow beam X-ray which "tracks" the two-dimensional motions of multiple radiodense markers (typically 2-3 mm spherical gold pellets). The rotation of the condyle and the translation of its axis of rotation along the articular eminence are calculated from the motion of X-ray tracking pellets on the jaw.

2. Methods

Mid-sagittal recordings of tongue and jaw movement were obtained using the X-ray microbeam system. Tracking pellets were attached to the tongue and jaw using a dental cement. Additional pellets were used to correct for planar head motion and to locate the occlusal plane. The projected positions of mandibular and maxillary pellets located off the image plane were corrected using a simple geometric transformation (Westbury, in press). Microbeam mid-sagittal palate tracings were obtained for each subject.

Coordinate systems were defined in both cartesian oral cavity space and mandibular joint coordinates (see Munhall, Ostry & Flanagan, in press for a discussion of orofacial coordinate systems). In the oral cavity coordinate system one axis coincides with the occlusal plane and the other is perpendicular to the occlusal plane and passes through the origin at the tip of the maxillary incisors. In the joint space representation movements are defined in terms of rotation of the jaw about the mandibular condyle and translation of the condyle along the articular eminence.

In speech trials, subjects produced cyclical consonant-vowel combinations at different rates and volumes. Consonants *t* and *k* and vowels *a* and *e* were used. Syllables involving the vowel *a* are associated with large amplitude jaw movements whereas syllables involving the vowel *e* are associated with smaller amplitude movements. In the mastication trials, the subjects chewed unilaterally on rubber tubing of varying compliance. Jaw tracking pellets were sampled at frequencies between 60 and 90 Hz and then low-passed filtered at 8 to 10 Hz. Mandible rotation and translation were computed from the motion of the tracking pellets on the jaw.

3. Jaw Motion Kinematics

In this section jaw movements will be considered in both oral cavity and joint based coordinate frames. Empirical data will be presented which suggest that jaw rotation and jaw translation can be controlled independently by the nervous system.

Figure 1 shows the motion paths of jaw pellets in oral cavity coordinates, projected onto the mid-sagittal plane. The paths for both mastication (solid line) and speech (dotted line) are presented for the full data set. The paths to the right of the figure are for movements of the

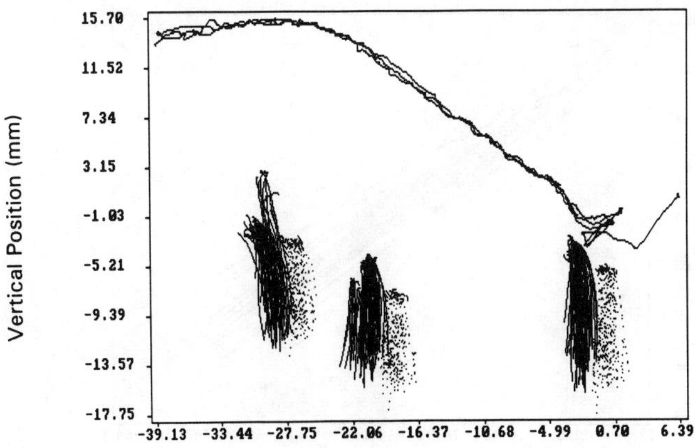

Figure 1. Oral cavity motion paths of jaw pellets during mastication (solid) and speech (dots). The palate tracing at the top of the panel is provided for reference. The positions of all pellets were projected onto the mid-sagittal plane and corrected for off-image plane distance.

mandibular incisors; the paths to the left are for movements of pellets attached to left and right side mandibular molars. A mid-sagittal palate trace is also shown. The amplitude of jaw movements can be seen to be greater in mastication than in speech. For this subject, the jaw appears to be more protruded in the production of speech sounds. However, the tendency for the jaw to be more protruded in mastication has been observed in other subjects.

The motion paths of the jaw were also examined in mandibular joint coordinates. Figure 2 shows the rotation of the jaw plotted as a function of horizontal jaw translation for cyclical jaw movements in both mastication (solid lines) and speech (dotted lines). It can be seen that straight line paths are obtained. This indicates that rotation and translation begin and end at the same time and that the ratio of their amplitudes is preserved over the movement.

Figure 2 also shows that the coordination between translation and rotation is not fixed. Moreover, different subjects reveal different patterns of coordination. In some subjects, we have observed that the slopes of rotation plotted against translation differ for mastication and speech. In other cases, the slopes are similar for speech and mastication but the two kinds of movements are produced in different parts of the workspace. Figure 2 shows a complex situation: the slopes and intercepts for mastication and speech can be seen to differ; the slopes and intercepts for speech alone also differ.

A number of manipulations involving both mastication and speech suggest that jaw rotation and jaw translation can be separately controlled. When jaw movements in speech were examined, the slope of the relationship between rotation and translation, and thus the

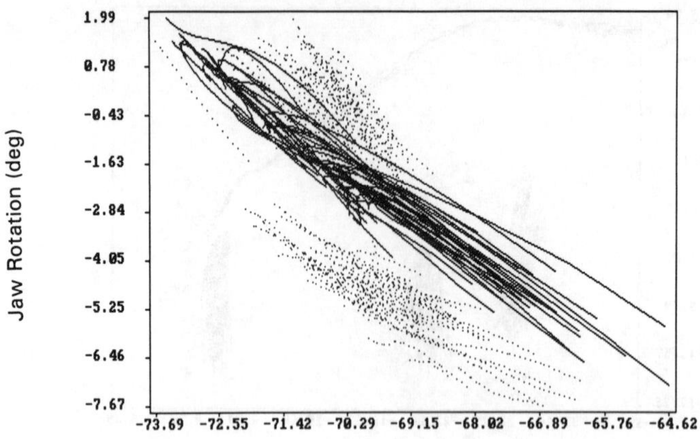

Figure 2. Joint space motion paths for jaw movements in mastication (solid) and speech (dotted). Jaw rotation is plotted as a function of horizontal jaw translation. (The horizontal axis is parallel to the occlusal plane).

coordination between these two degrees of freedom, varied with the consonant but did not depend on the vowel or speech rate. Figure 3 shows jaw motion paths in speech for the same subject as displayed in Figure 2. It can be seen that the slope of the relationship between jaw rotation and jaw translation was steeper for *t* than for *k*. In addition, movements for *k* are produced from a more protruded position.

A number of further manipulations are consistent with the view that jaw rotation and jaw translation can be separately controlled. When loud and fast speech were compared during cyclical repetition of the syllables *sa* and *ka* the slope of the relationship between rotation and translation was affected by the consonant but not by speech volume or rate (Figure 4). This indicates that the coordination between rotation and translation can be altered by the nervous system. Translation may also occur without affecting the balance of rotation and translation. Thus, in repetitions of *sa* the jaw translated forward for loud speech but the slope of path in joint coordinates in repetitions of *sa* was unaffected. The coordination between jaw rotation and translation also varied under different mastication conditions. For example, at fast chewing rates, jaw rotation was in some cases observed without any accompanying translation whereas at slower rates, both rotation and translation were observed.

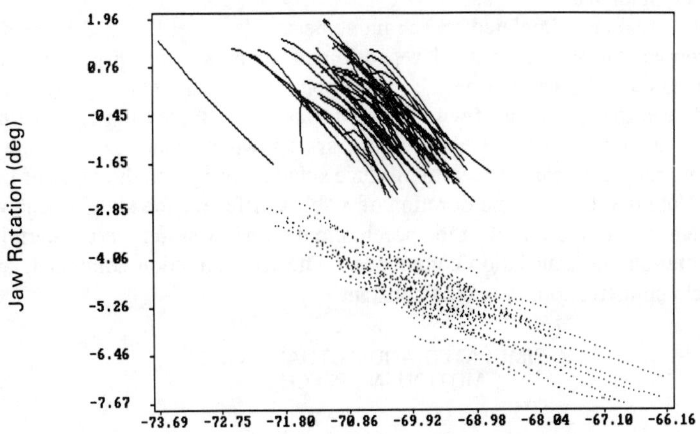

Figure 3. Joint space paths for cyclical speech movements. The solid lines are for the syllables *ta* and *te*; the dotted lines are for *ka* and *ke*.

4. λ Model

The kinematics of jaw movement in mastication and speech are well accounted for by the λ model. Briefly, the λ model suggests that voluntary movements are a consequence of shifts in the equilibrium state of the system. The equilibrium state is determined by the interaction of central neural control signals, segmental reflex mechanisms, muscle properties and load. Central commands control this process by setting, in different combinations, the motoneuron recruitment threshold lengths of multiple jaw muscles (see Flanagan, Ostry & Feldman, in press, for a detailed treatment of the jaw model).

The λ model for the jaw combines three representative muscles (closer, opener, protruder) with central commands, reflex mechanisms and muscle properties including force-length relationships. The origins, insertions, and force generating capabilities of the muscles in the model have been selected to approximate as best possible those of the actual muscles which they represent. Equations of motion for the jaw model were derived by representing the jaw as a moving pendulum which is free to rotate about a suspension point which itself can translate diagonally along the articular eminence. The slope of the eminence was determined from empirical data.

The model posits separate central commands which independently control jaw translation and jaw rotation, and in addition, produce co-activation of muscles without motion. Each of these commands affects the recruitment threshold lengths (λs) of all three muscles. Central commands may be superimposed to produce combinations of rotation, translation and co-activation. Empirical patterns can be captured by the model under the assumption of simple constant velocity shifts in equilibrium associated with simultaneous changes in the threshold lengths of the three modelled muscles.

Figure 4 shows actual and simulated speech movements. In panel B actual jaw rotation and translation are shown during the cyclical repetition of *sa* at a loud speech volume. The simulated movements are shown in panel C. Jaw opening which is indicated by a decreasing value of α can be seen to be associated with forward translation (increasing τ). The individual changes to muscle λs which produce this pattern are shown in panel D. Simple constant velocity changes in muscle λs can be seen to produce smooth changes in the patterns of rotation and translation. Note that the rate and duration of λ shifts differ for the simulated jaw opening and closing movements in speech. In the speech simulations, λ shifts were determined by a combination of rotation and translation commands. The co-contraction command, and hence the level of muscle co-activation, was held constant.

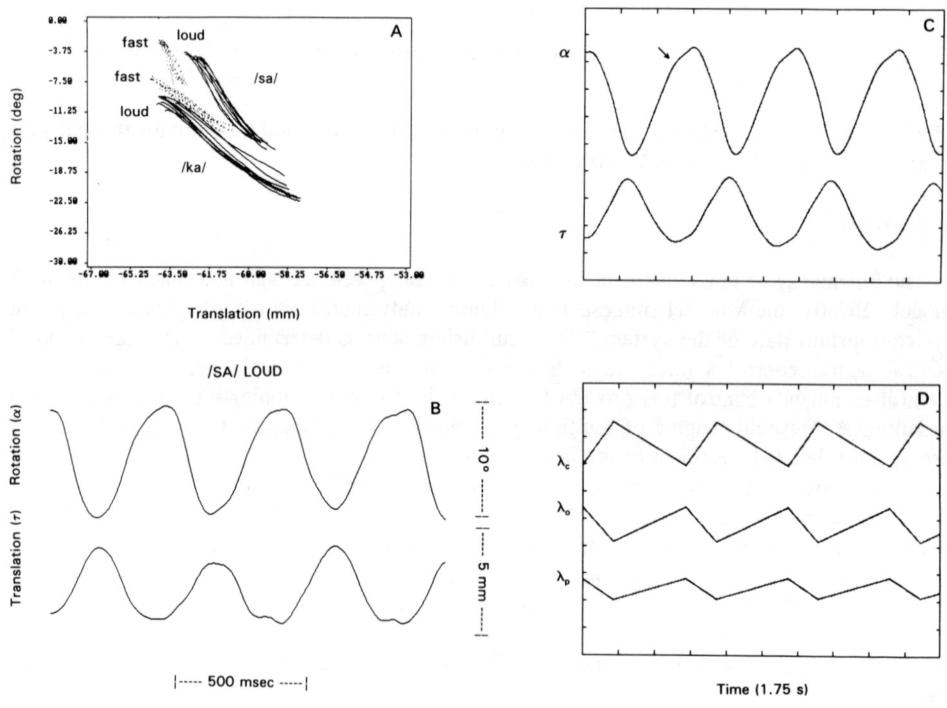

Figure 4. Actual and simulated jaw motion in speech. Panel A shows jaw motion paths during cyclical repetitions of *sa* and *ka* produced at both a fast speech rate and a loud volume. Actual patterns of jaw rotation (α) and translation (τ) are shown in panel B. The patterns are well accommodated by simulations (C) based on constant velocity shifts in the equilibrium angles of closer, opener and protruder muscles (D). (From Flanagan, Ostry & Feldman, in press.)

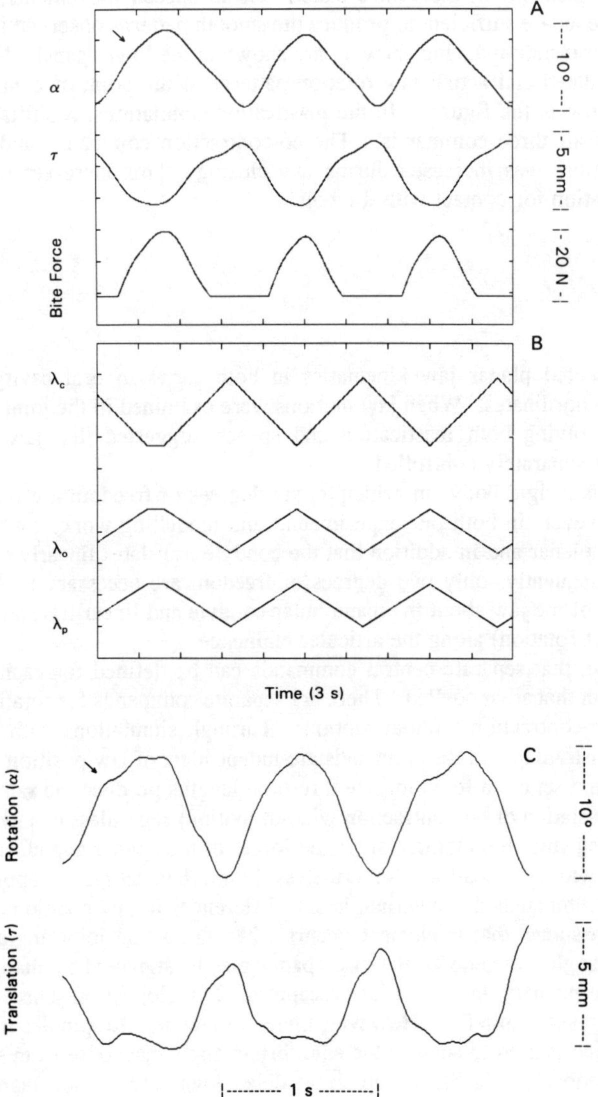

Figure 5. Actual and simulated jaw motion in mastication. Panel A shows simulated rotation (α), translation (τ) and predicted bite force during bolus contact. The equilibrium shifts controlling this movement are shown in panel B. An actual record of jaw rotation and translation during mastication is shown in panel C. (From Flanagan, Ostry & Feldman, in press.)

Actual and simulated patterns in mastication are shown in Figure 5. The upper panel gives simulated patterns of jaw rotation and translation and the predicted bite force developed during cyclical chewing movements. The corresponding changes to muscle λs necessary to produce this movement are shown in the centre panel. As in speech movements, constant velocity changes in muscle λs are sufficient to produce the smooth patterns observed in the data. Actual jaw rotation and translation during chewing are shown in the lower panel. Note, the similarity between the simulated and actual jaw rotation patterns at the point of contact with the bolus (shown as an arrow in the figure). In the mastication simulations, λ shifts were obtained by a combination of all three commands. The co-contraction command, and thus the level of muscle co-activation, was increased during jaw closing. This increases the stiffness of the system in preparation for contact with the bolus.

5. Discussion

We have presented planar jaw kinematics in both cartesian oral cavity coordinates and mandibular joint coordinates. When jaw motions were examined in the joint coordinate frame, manipulations involving both mastication and speech suggested that jaw rotation and jaw translation can be separately controlled.

Since the jaw is a rigid body, in principle, six degrees of freedom are required to describe its position. However, in both our experimental and modelling work, we have assumed that jaw movement is planar and in addition that the condyle translates linearly (along the articular eminence). Consequently, only two degrees of freedom are necessary to describe the jaw's position: rotation of the jaw about the mandibular condyle and linear translation of the condyle (i.e., the centre of rotation) along the articular eminence.

We have shown that separate central commands can be defined for each of the kinematic degrees of freedom that are modelled. There are separate commands for rotation and translation as well as for co-contraction without motion. Through simulations with the model it was possible to demonstrate that these commands are independent of jaw position. That is, for each command, the same set of shifts in muscle threshold lengths produce the same result (i.e. pure rotation, pure translation or co-contraction without motion) regardless of jaw position. This is a surprising finding since the moment arms and forces of the opener and closer muscles change in different ways with jaw location. Nevertheless, biomechanical factors appear to be balanced such that central commands are invariant across differences in jaw position.

The λ model assumes that movement occurs when the actual joint angle differs from the equilibrium joint angle. In general, the discrepancy may be signalled by changes in the afferent activity of spindle primary and secondary receptors. Jaw closing muscles in humans contain large numbers of muscle spindles. However, there are few muscle spindles in the jaw openers. The feedback which is used to achieve the equilibrium angle may arise from sources other than muscle spindle feedback. Mechanoreceptors associated with the temporomandibular joint may provide this information. However, sensory feedback is not strictly required to achieve the equilibrium position. According to the λ model the equilibrium state is determined by a combination of central commands and afferent feedback. Thus, central commands alone may enable the nervous system to specify desired positions without the use of afferent information.

6. References

Abbs, J.H., Nadler, R.D. & Fujimura, O. X-ray microbeams track the shape of speech, SOMA, 29-34, 1988.

Feldman, A.G. Composition of central programs subserving horizontal eye movement in man. Biol.Cybern., 42: 107-116, 1980.

Feldman, A.G. Once more for the equilibrium-point hypothesis (λ model) for motor control. J.Motor Behav. 18: 17-54, 1986.

Feldman, A.G., Adamovich, S.V., Ostry, D.J. & Flanagan, J.R. The origins of electromyograms - explanations based on the equilibrium point hypothesis. In J. Winters and S. Woo (Eds.), *Multiple muscle systems: Biomechanics and movement organization*. London: Springer-Verlag, 1990.

Flanagan, J.R., Ostry, D.J. & Feldman, A.G. Control of human jaw and multi-joint arm movements. In G. Hammond (Ed.), *Cerebral control of speech and limb movements*. London: Springer-Verlag, in press.

Munhall, K.G., Ostry, D.J. & Flanagan, J.R. Coordinate spaces in speech planning. J.Phonetics, in press.

Westbury, J.R. Head position during X-ray microbeam experiments: Its significance and measurement. J.Acoust.Soc.Am., in press.

SECTION 10

SENSORY MOTOR INTEGRATION

CURRENT VIEWS ON THE MECHANISMS OF EYE-HEAD COORDINATION

D. GUITTON
Montreal Neurological Institute and
Department of Neurology and Neurosurgery,
McGill University, 3801 University Street,
Montréal, Québec, Canada, H3A 2B4

ABSTRACT. Considerable evidence has now accumulated suggesting that the eye movement control system, which to date has been almost exclusively studied in animals whose heads are restrained, is but a special case of a more general gaze control system that controls displacements of the visual axis when the head is unrestrained. The gaze control system was viewed, originally, as being oculocentric in nature: i.e., the trajectory of the visual axis in space was thought to be independent of head motion. A mechanism was proposed whereby the head's contribution to a gaze shift could be subtracted out by the action of the vestibulo-ocular reflex. This view has been considerably modified in recent years and the gaze control system has now become an elegant example of how the brain controls and coordinates two independently moving body segments.

1. Introduction

Almost two decades ago, Bizzi and collaborators described how monkeys move their eyes and head simultaneously to acquire a visual target (Bizzi et al., 1971; Dichgans et al., 1973; Morasso et al., 1973). The fundamental feature of this hypothesized motor strategy is that the same saccadic eye movement is programmed to acquire the target irrespective of whether the head moves or not. If the head movement starts after the saccade has terminated, the gaze is maintained on target by the action of the vestibulo-ocular reflex (VOR) which moves the eye, relative to the orbit, with a velocity equal and opposite to that of the head. Normally, in monkey, a saccade starts only just slightly (40ms) before any visible head acceleration and it is hypothesized that any head movement occurring during the saccade attenuates, using the VOR, the saccade amplitude by an amount equal to the head

J. Requin and G. E. Stelmach (eds.), Tutorials in Motor Neuroscience, 547–561.

displacement during the saccade. The influence of the cervico-ocular reflex is negligible (Bizzi, 1981; Dichgans et al., 1973). Therefore the head both adds and linearly subtracts its contribution to the motion of the visual axis and the gaze trajectory (= eye-in-space = eye-in-head + head-in-space) is the same irrespective of whether the head is restrained (head-fixed) or unrestrained (head-free).

This view of eye-head coordination was essentially an "oculocentric" one in that the presence or absence of head motion was considered irrelevant to gaze control. We now know that this view is only partly correct. The following 3 sections explain why.

2. Neural Limits to Ocular Motility

One way of experimentally demonstrating this strategy is to suddenly and unexpectedly brake the head just before it is to move (Bizzi, 1981; Dichgans et al., 1973; Morasso et al., 1973). In this condition, visual feedback is too slow (150-200 ms) to modulate the saccade and neck proprioceptive influences are negligible (Bizzi, 1981; Dichgans et al., 1973). Thus with the head suddenly immobilized, and neither vision, proprioception nor VOR to modulate the saccade, the resulting saccade amplitude, according to the oculocentric theory should be equal to the target offset angle.

To verify this we unexpectedly prevented head motion before gaze shifts of different amplitudes, and found that maximum saccade amplitude was limited neurally, not mechanically, to about 45° in humans and 15° in cats (Guitton et al., 1984; Guitton and Volle, 1987). For targets within these values, a saccadic eye movement alone can attain a target irrespective of whether or not the head moves. An analogous result has been found by Tomlinson and Bahra (1986) in the monkey. For target offsets beyond these values, the saccade would be of insufficient amplitude to reach the target.

3. Interaction Between the Saccade and Oppositely Directed Vestibularly-induced Slow Eye Movement

The manner in which the saccade signal is attenuated by head motion is of considerable interest. One way of experimentally investigating this problem is to brake the head suddenly and unexpectedly **during** the saccade. Again, in this condition, visual feedback is too slow to modulate the saccade and neck proprioceptive influences are negligible. This experimental manipulation should suddenly remove the influence of the VOR on the saccade pulse generator and the saccade, in principle, should accelerate within 10-20 msec.

Both Fuller et al. (1983) and Guitton et al. (1984) used a brake to interrupt the motion of a cat's head for short periods during eye-head orienting gaze shifts. Braking could be timed to occur at different points in the eye trajectory. The former authors braked the head only during those rapid eye movements that occurred in and about the middle portion of large gaze shifts and concluded that no VOR-saccade interaction existed. Visual inspection and comparison of the eye, head, and gaze traces suggested to them that there was interaction during the first saccade. Guitton et al. (1984) braked the head during the first saccade and indeed reported VOR-saccade interaction. A change in eye velocity compensated for the change in head velocity so as to leave the trajectory of the initial gaze saccade unaltered. Blakemore and Donaghy (1980) applied motor-driven, unexpected, head movements during saccades made by cats whose heads were otherwise held fixed. These perturbations caused changes in eye velocity compatible with linear summation between semicircular canal and saccade signals with the consequence that gaze trajectories were essentially unaffected by the head perturbations.

Tomlinson and Bahra (1986) performed similar experiments in monkeys. A monkey's head was perturbed by a torque motor with the advantage that the head could either be decelerated or accelerated during the saccade. They concluded that there was no VOR-saccade interaction when the gaze shift was approximately >40°, but for smaller gaze shifts, particularly <10°, the VOR and saccade signals clearly interacted. There was a transition zone between 40° and 10° within which the VOR operated with a gain less than unity. They also reported that linear summation was absent only in the plane of the gaze shift but that the VOR slow phase remained functional in the orthogonal direction.

Laurutis and Robinson (1986) mechanically perturbed head movements during head-fixed and head-free gaze shifts in two human subjects. For gaze shift amplitudes >40° they concluded that head perturbations did not affect the eye saccade trajectory, but for smaller movements the results were less clear.

Analogous experiments by Guitton and Volle (1987) in

four human subjects yielded results in which VOR-saccade interaction was much more variable. In two subjects a sudden and unexpected head deceleration yielded no concomitant acceleration of the eye for gaze shift amplitudes ranging between 20° and 120° and, as expected, in these two subjects the maximum saccadic eye velocity ($\dot{E}_{max}$) was independent of head velocity. There was evidence in these two subjects that VOR-saccade interaction depended on gaze error--the angle between current gaze and desired gaze positions--since for small gaze errors (about <20°) there was interaction between the VOR and saccade signals. By comparison in a third subject, a head perturbation consistently caused an oppositely directed change in eye velocity during the entire gaze shift, at least for the gaze amplitude range tested (20° to 80°). Responses in a forth subject were variable and appeared to be task dependent: gaze shifts to unpredictable targets showed VOR-saccade interaction whereas gaze shifts to targets of known predictable amplitude showed no interaction.

The results of Guitton and Volle (1987) do not support the strong conclusion of Laurutis and Robinson (1986) that linear summation is absent for human gaze shifts >40°. Their results favour a more temperate view suggesting considerable variability in the interaction between the VOR and saccade signals.

Pélisson et al. 1988 imposed passive head rotations during gaze shifts in 16 human subjects. They concluded that for a 40° gaze shift, canal signal inhibition ranged from 40% to 96%, and that this inhibition increased almost linearly with gaze shift amplitude. (On average there was even a small inhibition for saccades as small as 10°). The results of Tomlinson and Bahra (1986), Guitton and Volle (1987), and Pélisson et al. (1988) concur to show in general that canal signal inhibition is determined by the size of gaze error.

The neurophysiological mechanisms that permit a head velocity signal to affect a saccade signal are unclear. Two important inputs to ocular motoneurons come from vestibuloocular relay neurons that receive head velocity signals from semicircular canals and thus form the middle portion of the three neuron arc of the VOR, and from burst neurons in the reticular formation, whose activity produces the rapid muscle contractions during saccades (for review see Fuchs et al., 1985). The failure of linear summation between saccade and canal signals is compatible with evidence that vestibuloocular relay neurons pause during saccades (e.g. see King et al., 1976;

Pola and Robinson, 1978; Tomlinson and Robinson, 1984). However, other evidence supporting linear summation suggests that the frequency of some burst neurons may be modulated by a head velocity signal, thus allowing canal signals to alter the saccade trajectory (Whittington et al., 1984). To complicate matters, the discharge rate of motoneurons during saccades may also be influenced by certain cells in the vestibular nuclei that burst during all saccades, whereas others burst and pause for ipsilateral and contralateral saccades, respectively (Fuchs and Kimm, 1975; Pola and Robinson, 1978; Tomlinson and Robinson, 1984). Thus the burst signal responsible for driving a rapid eye movement may result from the complex interplay between signals from different sources impinging on ipsilateral and contralateral motoneurons and at this stage in our knowledge it is impossible to clearly link the behavioral observations to the physiology.

4. The Gaze Feedback Hypothesis

The data for humans, monkey, and cat, summarized above, show that the mechanisms responsible for controlling eye-head movements are far more complex than originally proposed by Bizzi and colleagues. The observations vary and the question that arises is whether there are unifying features permitting a generalized scheme of the gaze control system in humans, monkey, and cat.

First there is general agreement that the oculocentric strategy--involving both VOR-saccade interaction and the coding of a saccade whose amplitude equals target offset angle--is valid for gaze shifts of about <10° in humans, monkey, and cat. Second, for larger gaze shifts the strategy appears to change although the precise "transition" point may be highly variable, even between individuals of one species or in one individual from day to day. Third, in those gaze shifts in which VOR-saccade interaction is initially absent there is clearly a time, within the gaze shift, at which the VOR signal again becomes functional in the classic sense and stabilizes the visual axis in space. This reactivation of the VOR occurs when gaze position error--the difference between current and desired gaze positions--approaches or equals zero, i.e., when gaze is on or near the target.

Since the predictions of the oculocentric hypothesis have not been upheld experimentally, it is necessary to develop a substitute scheme. Quite clearly a mechanism that focuses on the control of **both** the eye and head systems is required.

Such an alternative mechanism was first proposed to explain results of collicular stimulation in the cat (Guitton et al., 1980; Roucoux et al., 1980a,b), which revealed orienting mechanisms very similar to those now described in normal humans, monkey, and cat. This hypothesis was based on the principle of gaze feedback: an internal copy of actual gaze position is obtained by summing internal copies of eye and head positions. This sum is then subtracted from desired gaze position to yield a gaze position error signal, which drives the oculomotor and head motor machineries. The principle that gaze itself is the controlled variable has been reiterated by others (Fuller et at., 1983; Guitton et al., 1984; Tomlinson and Bahra, 1986; Pélisson and Prablanc, 1986, 1987; Guitton and Volle, 1987; Pélisson et al., 1988) and in a more quantitative formulation by Laurutis and Robinson (1986) and Guitton et al. (1990).

5. The Eyes and Head Share a Common Motor Drive

In head-restrained humans, monkeys, and cats, the <u>tonic</u> level of EMG activity in many dorsal neck muscles is modulated in relation to the position of the eye in the orbit. For example, as the eye moves progressively more in one direction there can be a concomitant tonic increase in the activity of ipsilateral neck muscles that participate in driving the head in that same direction (Vidal et al., 1982; Roucoux et al., 1982; André-Deshays et al., 1988). In addition to this tonic component there is evidence in both the head-fixed and head-free conditions that phasic bursts of EMG activity accompany saccades (reviewed in Bizzi et al., 1971; Fuller, 1980, Berthoz and Grantyn, 1986; Grantyn and Berthoz, 1987).

Much of this evidence, obtained in head-fixed animals, suggests the existence of common driver signals to the eye and head motor systems. There is considerable anatomical evidence supporting this. Indeed, in cat, the two prime brainstem immediate premotor systems implicated in gaze control - the tecto-reticulo-spinal (TRS) and reticulo-spinal (RS) systems - show profuse collaterali-zation to both oculomotor and head-motor circuits (Grantyn and Grantyn, 1982; Grantyn et al., 1987; Grantyn, 1988).

We have examined coordinated eye-head orienting movements made by cats trained to orient to targets in tasks which affect the metrics of both the eye and head components (Guitton et al., 1990). In the head-fixed monkey it has been shown that saccadic eye movements to visual targets are faster than those made to predictive targets (Hikosaka and Wurtz, 1985). In the head free cat, both the eye and head components were faster when the target was visible compared to when it was predictive. The result is that a visually triggered gaze shift was faster than a predictive gaze shift. Furthermore, despite significant differences between the metrics of movements to predicted and visible targets there was little difference in the relative contributions of the eye movements to the overall gaze displacement in the two conditions. The reason is that the eye and head velocities covaried in each condition such that at any point in the gaze shift the relative contributions of each to gaze amplitude remained about the same.

The eyeball with its connective tissues and muscles, called the eye plant, constitutes a viscoelastic system. It is now well known that the forces controlling ocular rotation during a saccade are dominated by viscous elements (reviewed by Fuchs et al., 1985). By comparison, the important mass of the head suggests that inertial forces dominate during rapid head movements. These arguments suggest that, to a first approximation, a pulse sent to the extraocular muscles should control eye velocity whereas the same pulse sent to the head plant should determine head acceleration. It follows that the head acceleration ($\ddot{H}$) trace should resemble the eye velocity ($\dot{E}$) trace and by extension that the $\dot{H}$ and E traces should be similar. Indeed, there was a remarkable correspondence between the $\ddot{H}$ and $\dot{E}$ and between the $\dot{H}$ and E traces in all gaze shifts studied in cat, over most of the duration of the head movement (Guitton et al., 1990). Particularly noteworthy was the tight correlation between the time of occurrence and magnitude of peaks and valleys in the $\ddot{H}$ and $\dot{E}$, and in the $\dot{H}$ and E traces respectively.

6. A Gaze Feedback Model

In the usual approach to modelling the saccadic system in the head-fixed animal--the so-called local feedback model (Robinson 1975)--the burst generator is driven by a motor error signal obtained by subtracting actual eye position from desired eye position.

An extension of the local feedback model to the head-free condition has been proposed: namely that brain stem eye and head motor circuits are driven by a gaze-motor error signal obtained by subtracting actual gaze position from desired gaze position (See a previous section).

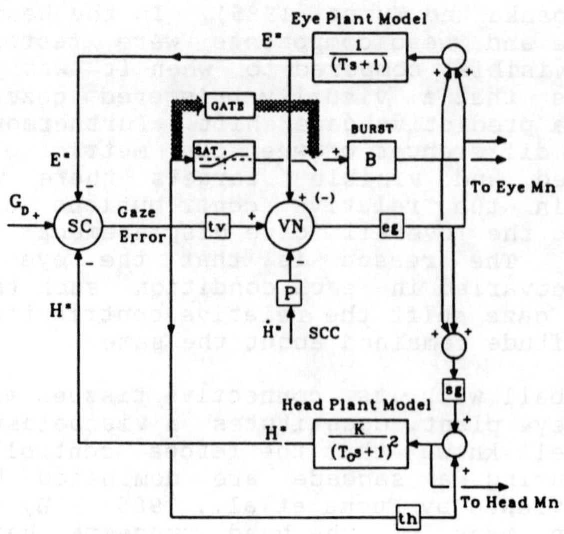

Figure 1. Schematic model of eye-head coordination during head-free gaze shifts in the horizontal plane. Common driver signal to the eye (top half) and head (bottom half) motor systems is gaze error obtained by subtracting the sum of current eye (E*) and head (H*) positions from desired gaze position (G_D). E* and H* are efference copies of eye-in-head and head-in-body, respectively. An important feature of the model is that the head motor system receives a copy of all ocular premotor signals, most notably the burster (B) output. Note that the sign of eye-plant-model input to VN changes between slow-phase (positive) and saccade (negative) modes (see text for more details). SC, superior colliculus; SCC, semicircular canals; VN, vestibular nuclei; B, burst generator; Gate, enables burst generator; SAT, saturation; Mc, moto-neuron; tv, th, eg, sg, K, and P, gains; T_0 and T, time constants.

The schematic in Fig. 1 presents our latest approach (Guitton et al., 1990) to modelling the gaze-control system. It is an extension of the concepts contained in the bilateral model of the VOR originally proposed by Galiana and Outerbridge (1984), here collapsed into its one-sided form. There are fundamental differences between the model of Fig. 1 and those based on the classic local feedback model (Robinson 1975). The first is how the ocular motorneural signal is created. In the Robinson model, the "inverse eye plant" approach is used; premotor circuits compensate for plant dynamics, and motoneurons are driven by the sum of two pathways respectively coding eye velocity and its integral, eye position. In the Galiana approach used here, the motoneurons are driven via feedback circuits through models of the eye and head plants.

A second difference is that the present model assumes that the SC lies within the gaze feedback loop. Although this is a controversial issue, there is mounting evidence to support this proposition (Munoz 1988; Munoz and Guitton 1989a,b, and unpublished observations; Waitzman et al. 1988), and its implementation avoids the widely assumed, though uncomfortable, alternative exemplified in the Scudder (1988) approach: that the SC--functioning open loop--knows a priori how many spikes it should send to the brain stem to drive a given amplitude gaze shift.

On the left of Fig. 1, desired gaze position (G_D or target-position-in-body) is compared with actual gaze position via efference copies of eye-in-head (E*) and head-in-body (H*) to produce an estimate of gaze error. This signal is assumed to project to head and various ocular premotor centres including a burst generator and VN. Note that each motor system to be controlled has an internal representation of its dynamics (eye and head models) that lie in the feedback loops including both the SC and VN summation points. In addition, afferent signals from the semicircular canals project via VN to the eye- and head-plant models.

As in the previously published bilateral model of the VOR (Galiana and Outerbridge 1984), the action of reticular inhibitory burst circuits on premotor networks is presumed to eliminate certain pathways and unmask others thereby leading to different dynamic properties of the oculomotor circuitry in the saccade and slow-phase modes. This important mechanism is expressed in Fig. 1 as a sign change of the efference copy (E*) projection on VN: positive during slow-phase segments when gaze is stable in space and negative during the rapid gaze shifts. An automatic strategy to perform this structural change in vestibular nystagmus is described elsewhere (Galiana 1990).

Note that the SC and VN summation points represent the difference between left and right sides of the brain, and, hence, because of axons crossing the midline they drive the eye and head (and their models) through a plus sign (right-going motion is positive). During a saccade to the right for example, right excitatory burst circuits and left VN cells (such as burst-tonic neurons) drive both the eye and head toward the right.

6.1 OCULOMOTOR SYSTEM

The top one-half of Fig. 1 is concerned with how the oculomotor system is organized and has been considered at length elsewhere (Fuchs et al. 1985; Galiana and Outerbridge 1984; Munoz and Guitton 1989a; Robinson 1975). An internal model of the eye plant provides an efference copy of eye position continuously updated by collaterals from ocular premotor elements or motoneurons. During compensation for head movements, gaze error can interact with the canal signals to produce enhanced or suppressed VOR responses as required to facilitate target acquisition (Guitton and Volle, 1987; Guitton, 1988; Pélisson et al., 1988). In the schematic, a gate [e.g., inhibitory omni-pause neurons (OPNs) in the reticular formation] releases the burst generator when gaze error from a selected target exceeds a threshold (Munoz 1988; Munoz and Guitton 1989a, unpublished observations).

During saccades gaze error is passed on to the burst generator via a saturation (in, say, long-lead burst neurons). This saturation limits not only eye velocity during a saccade but also, during a head-free gaze shift, imposes a neural rather than a mechanical limit to maximum eye deviations in the orbit (see above section). An efference copy of eye position (E*) is used to update both gaze error and the premotor firing rates of VN cells. In a manner analogous to the classic local feedback model (Robinson 1975), the pulse generators are driven until gaze error is within some threshold of zero.

Because Guitton and Volle (1987) found, in human, no interaction between canal and saccade signals during large-amplitude gaze shifts having intentionally slow head velocities, we simply assume that the canal signal is centrally disabled during the saccades.

6.2 HEAD MOTOR SYSTEM

The bottom one-half of Fig. 1 is concerned with the head-motor system. Here inertia plays an important role so that the head-plant model is approximated by a critically damped second-order system. An important new concept is that the head plant is driven not only by gaze error but also by the same signals sent to the eye motoneurons through collateralization (Grantyn and Berthoz 1987; Grantyn and Grantyn 1982; Grantyn et al. 1987, 1988).

This additional pathway accounts for the tight eye-head coupling described in an earlier section.

Simulations using this model compare well with the behavioral data (Guitton et al., 1990). Most importantly they predict new gaze-related properties of "oculomotor" brainstem neurons. For example, according to this model omni-pause neurons should pause for the entire direction of gaze shifts irrespective of the eyes trajectory in the head. This has indeed been shown in cat (Paré and Guitton, in press).

ACKNOWLEDGEMENTS. This work was supported by the Medical Research Council of Canada and le Fonds de la Recherche en Santé du Québec.

André-Deshays, C., Berthoz, A., and Revel, M. (1988). Eye-head coupling in humans. I. Simultaneous recording of isolated motor units in dorsal neck muscles and horizontal eye movements. Exp. Brain Res. 69:399-406.

Berthoz, A., and Grantyn, A. (1976). Neuronal mechanisms underlying eye-head coordination. In: Progress in Brain Res. (H.J. Freund, U. Büttner, B. Cohen and J. Noth, eds.), pp. 325-343, Elsevier, North-Holland, Amsterdam.

Bizzi, E. (1981). Eye-head coordination. In: Handbook of Physiology Vol. II, Section I: the Nervous System (V.B. Brooks, ed.), pp. 1321-1336, American Physiology Society, Bethesda, MD.

Bizzi, E., Kalil, R.E., and Tagliasco, V. (1971). Eye-head coordination in monkeys: evidence for centrally patterned organization. Science 173:452-454.

Blakemore, C., and Donaghy, M. (1980). Coordination of head and eyes in the gaze changing behaviour of cats. J. Physiol. (London) 300:317-335.

Dichgans, J., Bizzi, E., Morasso, P., and Tagliasco, V. (1973). Mechanisms underlying recovery of eye-head coordination following bilateral labyrinthectomy in monkeys. Exp. Brain Res. 18:548-562.

Fuchs, A.F., Kaneko, C.R.S., and Scudder, C.A. (1985). Brainstem control of saccadic eye movements. Annu. Rev. Neurosci. 8:307-337.

Fuchs, A.F., and Kimm, J. (1975). Unit activity in vestibular nucleus of alert monkey during horizontal angular acceleration and eye movement. J. Neurophysiol. 38:1140-1161.

Fuller, J.H. (1980). Linkage of eye and head movements in the alert rabbit. Brain Res. 194:219-222.

Fuller, J.H., Maldonado, H., and Schlag J. (1983). Vestibular-oculomotor interaction in cat eye-head movements. Brain Res. 271:241-250.

Galiana, H.L., and Outerbridge, J.S. (1984). A bilateral model for central neural pathways in the vestibulo-ocular reflex. J. Neurophysiol. 51:210-241.

Grantyn, A., and Berthoz, A. (1987). Reticulo-spinal neurons participating in the control of synergistic eye and head movements during orienting in the cat. I. Behavioral properties. Exp. Brain Res. 66:339-354.

Grantyn, A., and Grantyn, R. (1982). Axonal patterns and sites of termination of cat superior colliculus neurons projecting in the tecto-bulbo-spinal tract. Exp. Brain Res. 46:243-256.

Grantyn, A., Hardy, O., and Berthoz, A. (1988). Activity and ponto-bulbar connections of reticulo-spinal neurons subserving visually triggered orienting eye and head movements. Soc. Neurosci. Abstr. 14:959.

Grantyn, A., Ong-Meang, J.V., and Berthoz, A. (1987). Reticulo-spinal neurons participating in the control of synergic eye and head movements during orienting in the cat. II. Morphological properties as revealed by intra-axonal injections of horseradish peroxidase. Exp. Brain Res. 66:355-377.

Grantyn, R. (1988). Gaze control through superior colliculus: structure and function. In: Neuroanatomy of the Oculomotor System (J. Büttner-Ennever, ed.), pp. 273-333, Elsevier, North-Holland, Amsterdam.

Guitton, D. (1988). Eye-head coordination in gaze control. In: Control of Head Movement (B.W. Peterson and F.J. Richmond, eds.), pp. 196-207, Oxford Univ. Press, Oxford, UK.

Guitton, D., and Volle, M. (1987). Gaze control in humans: eye-head coordination during orienting movements to targets within and beyond the oculomotor range. J. Neurophysiol. 58:427-459.

Guitton, D., Crommelinck, M., and Roucoux, A. (1980). Stimulation of the superior colliculus in the alert cat. I. Eye movements and neck EMG activity evoked when the head is restrained. Exp. Brain Res. 39: 63-73.

Guitton, D., Douglas, R.M., and Volle, M. (1984). Eye-head coordination in cats. J. Neurophysiol. 52:1030-1050.

Guitton, D., Munoz, D.P., and Galiana, H.L. (1990). Gaze control in the cat: Studies and modeling of the coupling between orienting eye and head movements in different behavioral tasks. J. Neurophysiol. 64:509-531.

Hikosaka, O., and Wurtz, R.H. (1985). Modification of saccadic eye movements by GABA-related substances. I. Effect of muscimol and bicuculline in monkey superior colliculus. J. Neurophysiol. 53:266-291.

King, W.M., Lisberger, S.G., and Fuchs, A.F. (1976). Responses of fibres in medial longitudinal fasciculus (MLF) of alert monkeys during horizontal and vertical conjugate eye movements evoked by vestibular or visual stimuli. J. Neurophysiol. 39:1135-1149.

Laurutis, V.P., and Robinson, D.A. (1986). The vestibulo-ocular reflex during human saccadic eye movements. J. Physiol. (London) 373:209-233.

Morasso, P., Bizzi, E., and Dichgans, J. (1973). Adjustment of saccade characteristics during head movements. Exp. Brain Res. 16:492-500.

Munoz, D.P. (1988). On the Role of the Tecto-Reticulo-Spinal System in Gaze Control (Ph.D. thesis). McGill Univ., Montréal, Canada.

Munoz, D.P., and Guitton, D. (1989). Fixation and orientation control by the tecto-reticulo-spinal system in the cat whose head is unrestrained. Rev. Neurol. Paris 145:567-579.

Munoz, D.P., Guitton, D., and Pélisson, D. (1989). Gaze control in the head-free cat. II. Spatio-temporal variations in the discharge of superior colliculus output neurons. Soc. Neurosci. Abstr. 15:807.

Paré, M., and Guitton, D. (1990). Gaze related activity

of brainstem omnipause neurons during combined eye-head gaze shifts in the alert cat. Exp. Brain Res. in press.

Pélisson, D., and Prablanc, C. (1986). Vestibulo-ocular reflex (VOR) induced by passive head rotation and goal directed saccadic eye movements do not simply add in man. Brain Res. 380:397-400.

Pélisson, D., and Prablanc, C. (1987). Gaze control in man: Evidence for vestibulo-ocular reflex inhibition during goal directed saccade eye movements. In Eye Movements: From Physiology to Cognition (J.K. O'Regan and A. Lévy-Schoen, eds.), Elsevier, North-Holland, Amsterdam.

Pélisson, D., Prablanc, C., and Urquizar, C. (1988). Vestibulo-ocular reflex inhibition and gaze saccade control characteristics during eye-head orientation in humans. J. Neurophysiol. 59:997-1013.

Pola, J., and Robinson, D.A. (1978). Oculomotor signals in medial longitudinal fasciculus of the monkey. J. Neurophysiol. 41:245-259.

Robinson, D.A. (1975). Oculomotor control signals. In: Basic Mechanisms of Ocular Motility and Their Clinical Implications (G. Lennerstrand and P. Bach-y-Rita, eds.), pp. 337-374, Pergamon, Oxford, UK.

Roucoux, A., Crommelinck, M., Guerit, J.M., and Meulders, M. (1980a). Two modes of eye-head coordination and the role of the vestibulo-ocular reflex in these two strategies. In Progress in Oculomotor Research (A.F. Fuchs and W. Becker, eds.), pp. 309-315. Elsevier, North-Holland, Amsterdam.

Roucoux, A., Crommelinck, M., and Guitton, D. (1980b). Stimulation of the superior colliculus in the alert cat. II. Eye and head movements evoked when the head is unrestrained. Exp. Brain Res. 39:75-85.

Roucoux, A., Vidal, P.P., Veraart, C., Crommelinck, M., and Berthoz, A. (1982). The relation of neck muscle activity to horizontal eye position in the alert cat. I. Head-fixed. In: Physiological and Pathological Aspects of Eye Movements (A. Roucoux and M. Crommelinck, eds.), pp. 371-378, Junk, The Hague.

Scudder, C. (1988). A new local feedback model of the saccadic burst generator. J. Neurophysiol. 59:1455-1575.

Tomlinson, R.D., and Robinson, D.A. (1984). Signals in vestibular nucleus mediating vertical eye movements in the monkey. J. Neurophysiol. **51**:1121-1136.

Tomlinson, R.D., and Bahra, P.S. (1986). Combined eye-head gaze shifts in the primate. II. Interactions between saccades and the vestibulo-ocular reflex. J. Neurophysiol. **56**:1558-1570.

Vidal, P.P., Roucoux, A., and Berthoz, A. (1982). Horizontal eye position related activity in neck muscles of the alert cat. Exp. Brain Res. **46**:448-453.

Waitzman, D.M., Ma, T.P., Optican, L.M., and Wurtz, R.H. (1988). Superior colliculus neurons provide the saccadic motor error signal. Exp. Brain Res. **72**:649-652.

Whittington, D.A., Lestienne, F., and Bizzi, E. (1984). Behavior of preoculomotor burst neurons during eye-head coordination. Exp. Brain Res. 55:215-222.

PLASTICITY OF METRICAL AND DYNAMICAL ASPECTS OF SACCADIC EYE MOVEMENTS

HEINER DEUBEL
Max-Planck-Institut fuer Verhaltensphysiologie
Abteilung Mittelstaedt
D - 8130 Seewiesen, FRG.

ABSTRACT. Efficacy of saccadic eye movements presumes both high metrical response precision and postsaccadic eye stabilization. Appropriate performance is maintained by adaptive mechanisms based upon the evaluation of visual reafference. Two series of experiments are presented which compare properties of the adaptive control of saccade metrics with the adaptive response to consistent post-saccadic retinal image motion. The findings are discussed in the context of current models of saccade control.

1. INTRODUCTION

The primate retina is highly inhomogeneous with respect to both morphology and performance. Precise, gaze-shifting eye movements are therefore required that bring the central fovea to small objects in the visual field to allow for further scrutinizing. Since the moving eye cannot really see (Westheimer and McKee, 1975), these eye movements have to be as fast as possible. Moreover, since remaining motor errors have to be eliminated by further corrections, high metrical precision of the movement is obligatory. Finally, the eye has to come to a complete rest immediately after the fast gaze shift. Saccades are in perfect agreement with these requirements showing a stereotyped, step-like trajectory with remarkable metrical precision and the high speed characteristic of an open-loop response.

For biological open-loop systems, continuous monitoring and control of system performance by means of adaptive mechanisms is mandatory. Indeed, ample evidence for the ability of the saccadic system to adaptively repair for pathological dysmetria as well as post-saccadic drift (PSD) has accumulated from both clinical observations and experimental lesion studies (e.g. Kommerell et al., 1976; Abel et al., 1978; Optican and Robinson, 1980). Recent findings suggest that two well-separated neural structures, both in the cerebellum, are essential for maintaining metricity and post-saccadic ocular stability, respectively. So, ablation of the cerebellar midline vermis and the fastigial nuclei in monkeys leads to saccadic hypermetria (Optican and Robinson, 1980). Floccular lesions, on the other hand, abolish the adaptive control of post-saccadic drift (Optican et al., 1986).

Adaptive control of both, saccadic metricity and post-saccadic ocular stability relies on the evaluation of visual reafference. For metricity

J. Requin and G. E. Stelmach (eds.), Tutorials in Motor Neuroscience, 563–579.

control this has been demonstrated in a number of behavioral studies in human subjects in which the saccade targets underwent a systematic, step-wise displacement during the saccade (Henson, 1978; Miller et al., 1981; Deubel et al., 1986; Deubel, 1987). Subjects soon begin to compensate and in minutes are acquiring the displaced targets with normal accuracy. It has been shown more recently, moreover, that systematically induced, exponential drift movements of a full-field stimulus after each saccade induce, within days in the monkey and within several hours in humans, an exponential, zero latency post-saccadic ocular drift which also persists for saccades in darkness (Optican and Miles, 1985; Kapoula et al., 1989).

This paper summarizes my work on the adaptive control of saccadic metricity and presents novel findings on the adaptive induction of post-saccadic ocular drift. Experiments which address both mechanisms separately are described which are intended to elucidate similarities and differences. Specific emphasis will be given to the question as to what extent the induced changes are specific to the spatial range where saccadic training occurred. This is of high importance since the capability of the adaptive mechanism to cope with a variety of demands resulting from central and peripheral diseases, fatique, aging and so forth is closely related to the question of how *specifically* the adaptive compensation can react to the specific stimulation properties. Moreover, it will be suggested that analyzing the specificity of the induced changes to magnitude and direction of the saccades may give insight into the topological organization of the respective adaptive mechanism.

2. PLASTICITY OF SACCADE METRICS

2.1. METHODS

2.1.1. Experimental Set-Up. Eye movements in both dimension were recorded from 5 members of the laboratory and two adult rhesus monkeys (Macaca mulatta) with the scleral search coil technique introduced by Robinson (1963). The targets consisting of small bright spots of light projected on a large screen were moved by servo-controlled mirror galvanometers. For the human subjects, vision was binocular and the viewing distance was 110 cm. The rhesus monkeys were implanted with a head holder and a scleral search coil placed in one eye using the method of Judge et al. (1980). The other eye was patched. The animals were well-trained to follow and fixate small spots of light in the dark. They were rewarded for their performance with drops of water, thus earning their daily fluid uptake to satiety. During the experiments, the animals were seated in a plastic chair, with their heads fixed, at a distance of 97 cm from a tangent screen.

In order to avoid irritations of the cornea by the scleral search coil, durations of the sessions with the human subjects were limited to about 30 minutes, yielding 600 to 800 single trials. The monkeys with the chronically implanted coils performed 2000 to 3000 trials in a single session of two to three hours duration. The study as a whole is based on about 15000 single trials with human subjects and 60 000 trials with the two rhesus monkeys.

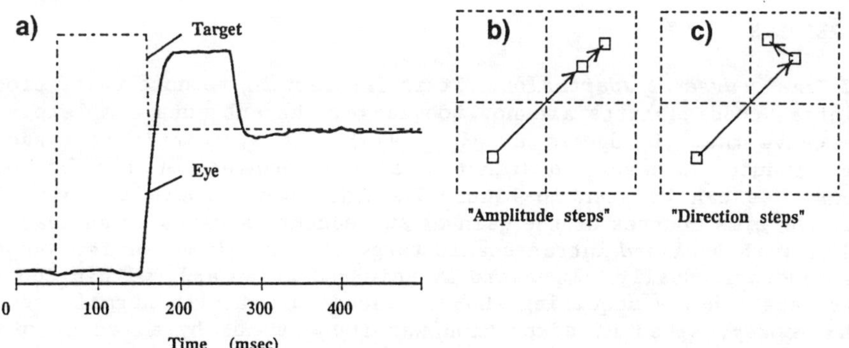

Figure 1: Experimental paradigms for induction of metricity adaptation.

2.1.2. Experimental Procedure

The experimental procedure is described in more detail elsewhere (Deubel, 1987). Briefly, each session consisted of three phases:

- A pre-adaptation test phase, in which the normal parameters of the saccadic response were obtained. In this phase of the experiment, only single target steps were given.
- A long adaptation phase. In order to induce saccadic adaptation, in this phase of the experiments intrasaccadic target steps were elicited. These saccade-triggered, secondary steps occurred only on the specific, pre-selected meridians under adaptation; for saccades into other directions, no secondary target displacements occurred.
- A post-adaptation test phase where the effects of the adaptive changes were determined. For this purpose, either an open-loop condition inducing a 500 msec blanking period after the initial saccade or a condition similar to the adaptation paradigm was applied.

The subjects were instructed to follow the small target "as fast and as precisely as possible". Each trial began with a step-wise target displacement in the range of 6 to 12 deg (Figure 1a). In the adaptation phase, these target steps only occurred on a preselected meridian. When the saccade to acquire this target was on the way, the stimulus was systematically stepped by a certain percentage (in most cases 30%) of the initial step amplitude, and into a consistent direction relative to the initial step. Figure 1a presents an example for a backward target step. The final target position after the second step then served as starting position for the next double-step stimulus.

Depending on the direction of the intrasaccadic step two basic conditions were distinguished:

- "Amplitude steps" occurred in the same (or opposite) direction of the initial step (Figure 1b). Here, the adaptive controller was expected to increase (or decrease, respectively) saccadic gain.
- "Direction steps" occurred orthogonally to the direction of the first target step (Figure 1c). This paradigm was expected to elicit changes of saccadic direction.

2.2. RESULTS

2.2.1 Time course of adaptation. It is interesting to note that, provided the intrasaccadic shifts are not too large, the subjects completely fail to perceive them (Bridgeman et al., 1975). The systematic post-saccadic errors induce, however, distinct adaptive changes in the oculomotor response, as can be seen in Figure 2a. The open symbols in this figure reveal the time courses of the gain of subsequent saccades in an adaptation paradigm with *backward* intrasaccadic target steps, given for two subjects. The subjects gradually compensate by gain reduction and, within 200 - 300 trials, are finally acquiring the displaced target with normal accuracy. In the monkey, speed of adaptation was found to be by a factor of 5-10 slower than in the human subjects.

The speed of the adaptive process is dependent on the kind of adaptation demanded: The fast time constants indicated by the open symbols in Figure 2a only hold for experimental situations where the adaptive system is required to *reduce* saccadic gain. Gain *increases*, on the other hand, occur considerably slower, as can be seen from the closed symbols in Figure 2a representing data from an experimental session where the intrasaccadic target shifts occurred into the *same* direction as the initial steps. Also, adaptation under this condition often tends to be incomplete. This asymmetry seems to indicate that systematic undershoots are largely tolerated by the adaptive controller, whereas the occurrence of saccadic overshoots leads to immediate gain reduction. The crosses in Figure 2a display the time course for the recovery from adaptation (i.e. when normal feedback is provided). The data demonstrate that the time course is similar to adaptation in the gain increase condition. So, in the cases where gain increases are required, recovery from adaptation is slower that learning.

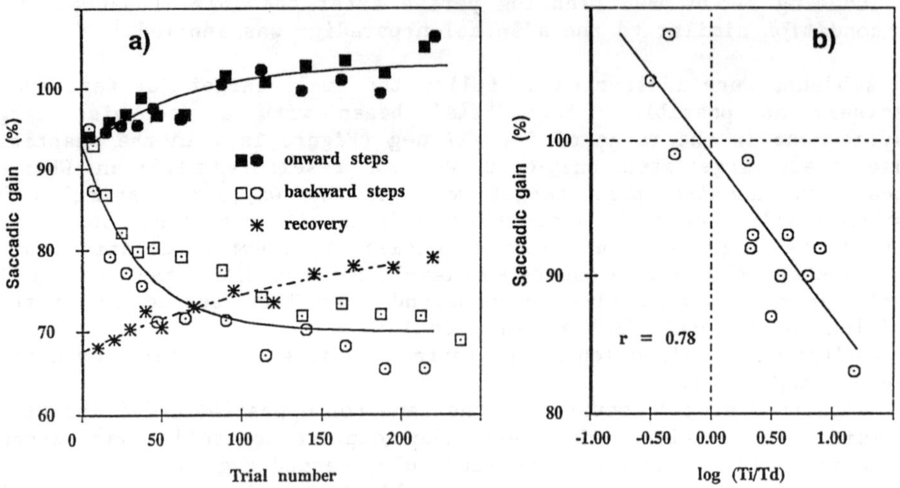

Figure 2: a) Time courses of the gain of saccades as a function of trial number. 2 subjects. b) Mean saccadic gains of unadapted subjects as a function of the ratio of the time constants for gain increases vs. decreases.

It should be further mentioned that the asymmetries are also dependent, to some extent, on the movement directions and on the species; so, in the monkey, for vertical upward saccades, gain increases occur faster than gain decreases.

The time courses reflect the relative ease with which the subject is able to modify saccadic gain in response to the afferent information. The question arises whether the asymmetries in learning speed observed in these experiments relate to the "normal" gain of the unadapted system. This is demonstrated in Figure 2b showing typical saccadic gain values for various saccade directions, taken from one subject and both monkeys, as a function of the ratio of the time needed to increase (T_i) vs. decrease (T_d) saccade gain. It can be seen that both variables correlate significantly. This is of some interest since it suggests that the typical undershooting behavior of the saccade is the consequence of unequal receptiveness of the adaptive controller to react to overshoots vs. undershoots.

For the direction step experiments (c.f. Figure 1c) data not shown here demonstrated that saccadic direction is as easily adaptable as saccade gain, yielding very similar time courses (Deubel, 1987). In general, the time constants for small direction changes were in the range of 30 - 200 trials in the humans, and in the range of 500-1000 in the monkeys.

2.2.2. Generalization of induced metrical changes to untrained saccade amplitudes. The question arises as to what extent metricity adaptation can be specific to amplitude and direction of the saccadic eye movement. Experimental work from our laboratory (Deubel et al., 1986) yielded that induced gain adaptation is *not* specific to the saccadic amplitude that was conditioned in the training phase. This basic finding is demonstrated in Figure 3 showing data from an experiments where subjects were given horizontal target steps ranging in size from 3 up to 20 deg. The open and filled circles in Figure 3 compare pre- and post-adaptive undershoot of the primary saccades with respect to the initial target position as a function of step sizes. Here, for adaptation, a second, intrasaccadic target shift directed opposite to the initial step was again provided, but only for the saccades to 8 deg target steps, while for the other step sizes normal visual feedback (i.e.

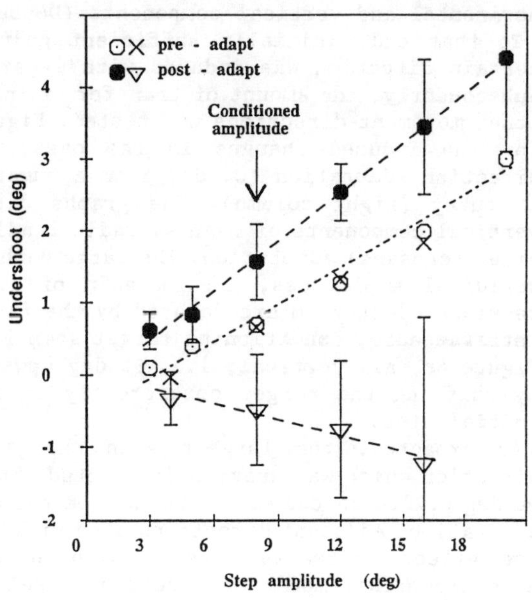

Figure 3: Mean saccadic undershoots as a function of step angle before and after adaptation. The conditioning stimulus was only provided for the 8 deg target steps.

without intrasaccadic displacement) was given. Although the conditioning stimulus was only delivered for the 8 deg saccades, saccades of *all* sizes are equally affected by the adaptation, as becomes obvious from the increased amount of undershoot for all step amplitudes. This means that the gain reduction induced for the 8 deg steps transfers to other amplitudes and indicates that gain control is indeed parametric with a single parameter determining saccadic amplitude in response to all target step sizes. Very similar results are obtained in cases where the intrasaccadic step systematically went into the same direction as the primary saccade, calling for an adaptive gain increase (Figure 3, crosses and triangles, respectively).

It is concluded that, for both, gain decrease and increase, gain changes can *not* be made specific to the amplitude of the movement. Conflicting results indicating some ability for amplitude-specific adaptation (Miller et al., 1981; Semmlow et al., 1989) may be attributed to the fact that in these studies, spatial loci for initial fixation position and final target positions were not sufficiently varied. Therefore, the results from these workers may reflect the effect of position-specific motor learning rather than indicating amplitude specificity of saccadic gain control.

2.2.3 Directional selectivity of induced metrical adaptation. It has long been known for horizontal eye movements that, say, adaptation of the gain of leftward saccades does not transfer to rightward saccades (Abel et al., 1978; Miller et al., 1981). This kind of directional selectivity was recently investigated more closely, now for oblique saccades with both horizontal and vertical components (Deubel, 1987).

To that end, initially sufficient gain adaptation of saccades into a certain direction was induced with a paradigm as described in Figure 1. Subsequently, the amount of transfer of induced adaptation to saccades into other movement directions was tested. Figure 4 shows four typical examples for the induced changes in the cases of gain adaptation (a, b) and direction adaptation (c, d), from a human subject (left column) and from a monkey (right column). The graphs display normalized horizontal and vertical components of mean saccadic landing position before (circles) and after (crosses) adaptation. The large dashed circle in each graph indicates veridical amplitudes, i.e. a gain of 1. The saccadic directions which received adaptation are denoted by the fat lines, and the direction of the intrasaccadic, conditioning target step is indicated by the arrows. So, in Figure 4a, all vertical, i.e. 90 deg upward saccades underwent adaptation by shifting the target consistently into the opposite direction of the initial step.

As expected, the largest gain change is revealed for the movement direction which was previously adapted. Thus, in example a) the gain of the 90 deg upward saccade is reduced from the nominal value of 86% to 56% after adaptation. Adjacent directions also yield reduced saccadic gains. However, the effect decreases steeply with angular distance from the adapted direction. The amount of directional selectivity is shown in more detail in Figure 5 presenting the induced gain change as a function of angular distance from the adapted direction, for several adapted saccade directions. The data can be fit by a Gaussian function with a 'standard deviation' of 26.5 deg. So, evidently, the adaptation is tightly tuned;

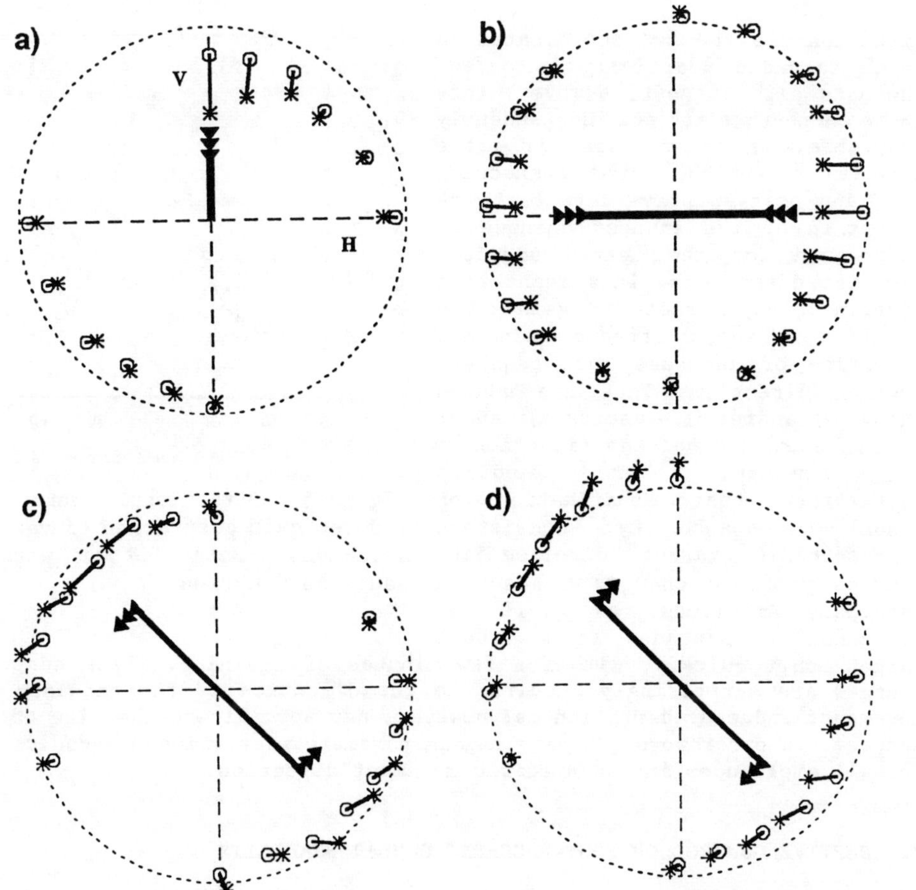

Figure 4: Horizontal and vertical components of mean saccade landing positions before adaptation (circles) and after adaptation (crosses). Left: Human data. Right: Monkey data. Upper graphs: Gain adaptation. Lower graphs: Direction adaptation.

gain can be adjusted independently for saccadic directions with only moderate angular distance. It is concluded that a high directional selectivity of the induced gain modification exists.

As demonstrated in Figure 4 c and d, similar conclusions can be drawn for adaptation of saccadic direction. Obviously, the angular change is highest for the adapted direction but decreases rapidly with angular distance, so that only saccadic directions close to the adapted meridian become affected by the adaptation. The tuning width of the adaptation effect is very similar to the results from the previous gain change experiments. So, evidently, this type of adaptation is also highly directionally selective.

The picture emerging from these data is that magnitude and angular direction are the two coordinates in which saccadic adaptivity is coded. The data also suggest, however, that these components are not independently adaptable. It can be seen from the orientation of the lines connecting pre- and post-adaptive data that the direction of the induced changes is maintained for the tested adjacent saccadic directions. This means that adapting, e.g., saccadic gain on a specific meridian affects gain *and* direction of saccades into adjacent movement directions. Thus, the induced changes transfer in a vectorial manner to the untrained saccade directions.

In summary, these findings demonstrate that modification of visual feedback by consistent intrasaccadic target displacements induces profound changes of saccade metrics. Amplitude as well as direction of saccades to a visual target adapt quickly, within a few minutes of training. These adaptive changes are surprisingly specific to the direction of the saccadic eye movement. Induced adaptation is, however, not specific to the size of the saccade. In other words, a single gain parameter determines saccadic gain for all amplitudes into a specific movement direction.

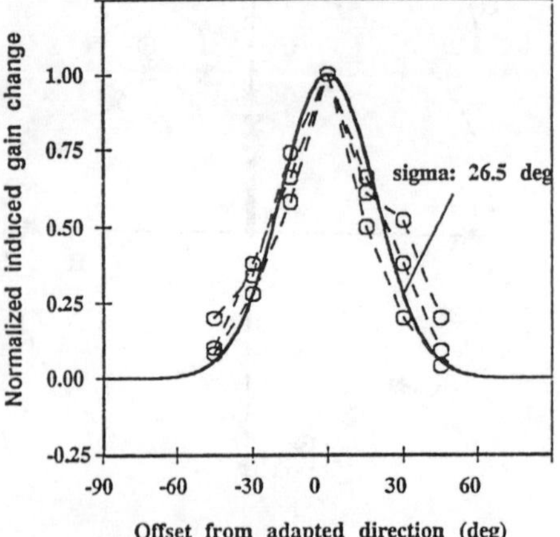

Figure 5: Directional tuning of induced gain adaptation of various meridians. Data fitted with a Gaussian function.

3. ADAPTIVE CONTROL OF POST-SACCADIC OCULAR STABILITY

3.1. METHODS

Four members of the laboratory served as subjects. They were sitting at a distance of 1.5 m from a tangent screen in darkness and were instructed to keep on following a little bright laser spot which was moved by a fast x-y galvanometer scanner. Vision was binocular and eye movements were measured with the scleral search coil technique. An off-line interactive computer program analyzed the traces for the saccade parameters. For the analysis of PSD, the end of the pulse-driven part of the saccade was determined which can be done conveniently by detecting the zero-crossings of the acceleration trace and then adjusting for the inherent filter delays. Then, onset, amplitude and time constant of PSD were estimated considering only eye position data from the first 150 msec after the saccade. In most of the experimental sessions, only on eye was monitored. A few pilot studies with binocular registration however ensured that the induced adaptation effects were truly conjugate (version) movements, and not due to vergence movement components.

A typical experimental session again consisted of a pre-adaptation test phase where the normal parameters of the responses were determined, a long adaptation phase and, finally, a post-adaptation test phase. In all three phases, the target stepped about every 2 sec. Size of these initial target steps was randomly selected between 8, 12 and 16 deg. The target never left the central 20 by 20 deg range. In the pre-adaptation phase, the directions of the steps were taken at random from a few pre-selected meridians going through the center of the screen.

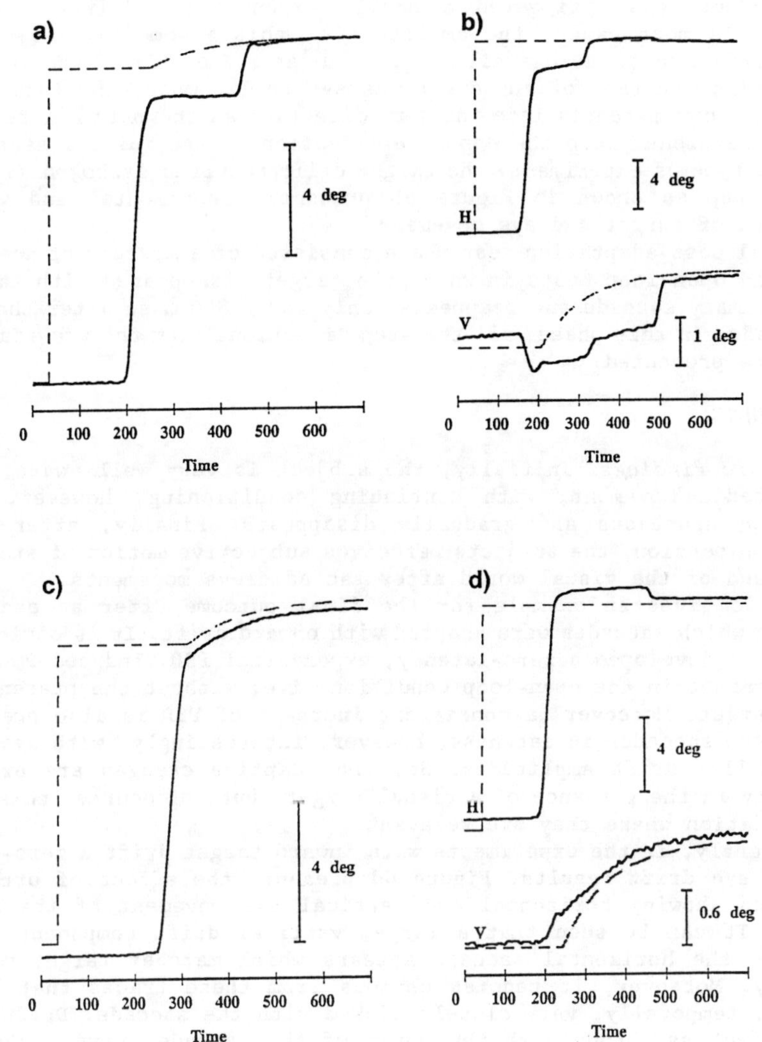

Figure 6: a,b) Exemplary target and eye movements in the experimental paradigms for the induction of postsaccadic drift. a) Onward target drift. b) Orthogonal target drift. c,d) Typical post-adaptive eye movement responses.

The adaptation phase included 600-800 training trials and lasted 20 - 30 minutes. In this phase, again only one saccade direction was trained selectively by providing target movements only on one pre-selected meridian. Figures 6 a and b show the stimuli used in more detail and presents also typical eye movements of the unadapted subject. For the induction of post-saccadic drift, the computer analyzed the eye movements on-line for the end of the primary saccade using a 50 deg/sec velocity criterion and then triggered a small, exponential, drift-like target movement. In most cases, the amplitude of this movement simulating the retinal effect of pathologically large PSD was 10% or 15% of the step size, and the time constant of the drift was set to 120 msec. The target drift could occur consistently into the *same* direction as shown in Figure 6a, or, in other sessions, into the *opposite* direction of the initial saccade. In a further type of experiment, the target drift occurred *orthogonally* to the initial step as shown in Figure 6b presenting horizontal and vertical components of target and eye movements.

The final post-adaptation test phase consisted of a mixture of adaptation trials and open-loop tests in which the target disappeared with the onset of the primary saccade and reappeared only 400 - 500 msec after the end of the saccade. In this phase, all the step directions from the pre-adaptation phase were presented.

3.2. RESULTS

3.2.1 Basic findings. Initially, the subject is very well aware of this post-saccadic movement; with continuing conditioning, however, motion perception decreases and gradually disappears. Finally, after a long adaptation session, the subjects perceives subjective motion of stationary targets and of the visual world after saccadic eye movements.

Figure 6c gives an example for the final outcome after an adaptation period in which saccades were adapted with onward drift. It is obvious that the eye has developed a zero-latency, exponential PSD. Induced PSD occurs in like manner in the open-loop condition, i.e. without the presence of a visual target. Moreover, a consistent increase of PSD is also present in spontaneous saccades in darkness, however, interestingly, with systematically smaller drift amplitudes. So, the adaptive changes are expressed fully only in the presence of a visual target, but, of course, this is the only condition where they are relevant.

Equivalently, in the experiments with inward target drift a zero-latency backward eye drift results. Figure 6d presents the effect of orthogonal adaptation showing horizontal and vertical eye movement of the adapted subject. It can be seen that a large, vertical drift component tightly linked to the horizontal saccade appears which matches target movement perfectly. Moreover, it becomes obvious from these traces that induced drift is, temporally, very closely linked with the saccade. Drift starts with perfect synchrony with the onset of the saccade, never before and never later.

Obviously, this kind of motor learning occurred within 20 to 30 minutes which is considerably shorter than the time courses reported by Optican and Miles(1985) and by Kapoula et al.(1989). This indicates that drifting single targets forms a more adequate stimulus than full-field shifts related with rather spontaneous saccades. However, it should be mentioned that speed of adaptation varies largely between subjects and is direction-dependent; so, some subjects have symmetrical time courses with time constants of 300 - 600 trials, for onward and backward drift adaptation. In another subject, however, who was found to adapt quickly to onward target drift, it was not possible, even with several thousand training epochs, to adaptively induce a backward postsaccadic eye drift.

Moreover, data not shown here suggest that recovery from adaptation is somewhat faster than learning, but nevertheless not immediate. Rather, recovery also takes a few hundreds of trials, implying that it is not possible to voluntarily "switch-off" the induced motor pattern.

3.2.2. Generalization of induced PSD to untrained saccadic amplitudes. The magnitude of PSD has been shown to be linearly related to saccade size (e.g. Optican and Miles, 1985). In a specific experiment I tested if the system has the ability for amplitude-specific compensation for systematic post-saccadic image motion. Figure 7 displays data from an experiment where, in the adaptation phase, saccades to 8 deg target steps received onward target drift, while, at the same time, saccades to 16 deg target steps received normal visual feedback (i.e. no postsaccadic drifts). 12 deg target steps were delivered only in the test phases. Crosses indicate mean pre-adaptive eye drift, circles indicate mean post-adaptive eye drift, as a function of saccade size. The data clearly show that not only the 8 deg saccades were affected, but also the 16 deg saccades were loaded with a (proportional) post-saccadic drift component. So, PSD adaptation is parametric and can not be induced specifically for individual saccade amplitudes.

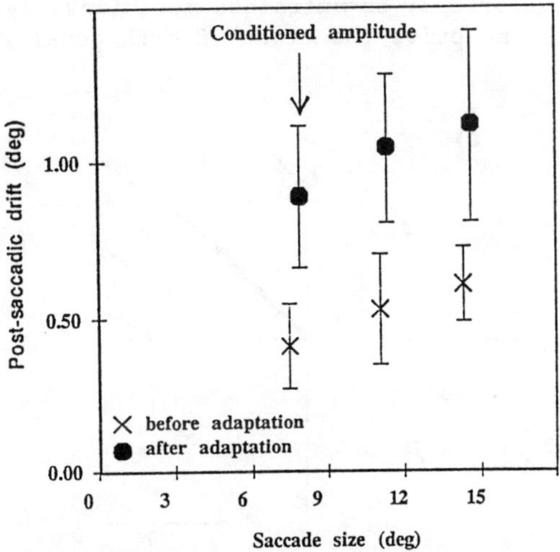

Figure 7: Transfer of PSD induced on saccades conditioned with 8 deg target steps to unconditioned saccadic magnitudes.

3.2.3. Directional specificity of PSD adaptation. Again it is of high interest to investigate the question of how induced changes of post-saccadic eye drift affected other directions which received no stimulation nor training. Typical examples for the consequences of adaptation of specific saccadic directions with consistent post-saccadic target drifts are given in Figure 8. The data points in each graph represent the horizontal and vertical components of the mean, normalized landing positions of saccades in various directions. The fat lines indicate the saccadic meridian which received training; the arrows display the direction of induced target drift. The thin lines attached to the data points represent magnitude and direction of PSD induced by the adaptation. PSD magnitude is given in percent of saccade size. Figure 8a shows data from a session where, for adaptation, oblique target steps were followed by oblique, onward target drift. Figure 8b represents data from a session where rightward horizontal saccades were loaded with onward drift. It can be seen that the induced eye drift follows the direction of the stimulus movement and is highest for saccades in the direction which received training. Adjacent saccade angles are also affected, and they are loaded with drift into the same direction as adapted. All vectorial changes reflect the direction into which the drift had occurred. Thus, as was found for metricity adaptation (Figure 4), the induced effect transfers *vectorially* to untrained meridians. Again, the findings are analogous for the adaptive induction of orthogonal PSD (Figure 8 c and d).

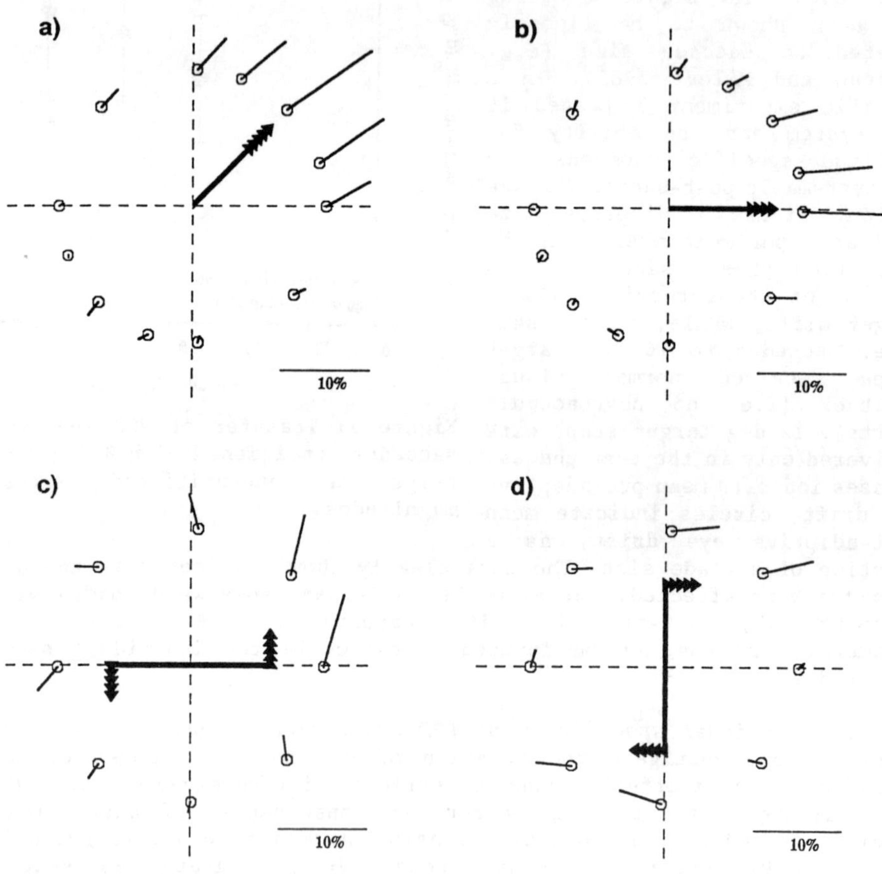

Figure 8: Transfer of induced PSD to untrained directions. Circles denote mean saccade end positions. Attached lines show magnitude and direction of induced eye drift. Fat arrows indicate trained direction and target drift direction.

Figure 9 gives an impression of the angular (directional) tuning of the effect showing induced drift amplitude as a function of the angular distance of the test saccade from the trained meridian. Comparison with Figure 5 reveals that directional tuning is considerably broader than for metricity adaptation, leading to some amount of transfer of induced adaptation even to saccades directed orthogonally to the adapted direction.

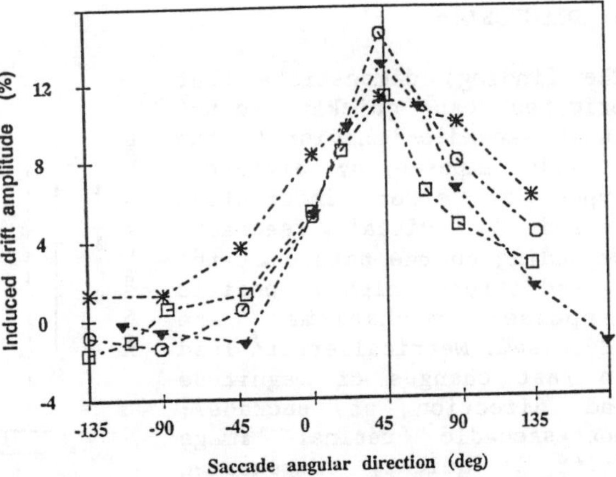

Figure 9: Induced PSD magnitude in percent of saccade size as a function of saccade direction. Adaptation occurred at the 45 deg (oblique) meridian. 4 subjects.

3.2.4 Further observations. The results presented in the previous chapter suggest that, by appropriate stimulation, a slow drift component of any arbitrary direction with respect to the saccade can be appended to the fast part of the trajectory. The question arises if not only direction, but also the waveform of the induced drift is determined by the retinal image movement pattern. This was tested in experiments where, for adaptation, an accelerating type of an exponential, post-saccadic target movement was consistently presented. Target movement and a typical saccadic response after long training are displayed in Figure 10. It is obvious that only the normal, exponential PSD was induced. In confirmation of previous findings (Optican and Miles, 1985), further work not shown here however suggested that the time constant of the exponential movement is, to some extent, dependent on the time constant of the stimulus movement.

Finally, PSD adaptation transfers to the quick phases of optokinetic nystagmus (OKN). While slow post-saccadic drift is normally hidden in the slow phases of OKN, orthogonal drift adaptation allows to elegantly separate both components. So, Figure 11 shows the horizontal and vertical components of OKN elicited by horizontal optokinetic stimulation, after vertical PSD had been induced on horizontal saccades in a long adaptation session. It can be seen that the vertical eye movement component exhibits small but consistent vertical exponential drift movements associated with each horizontal quick phase.

In summary, PSD can be induced quickly by consistent post-saccadic target drifts. As compared to the findings for metricity adaptation, this adaptive process exhibits only moderate directional selectivity. Interestingly, orthogonal drift can be generated as easily, and the induced effects transfer in a vectorial way to saccades into untrained directions.

4. DISCUSSION

The findings demonstrate that primates can quickly adjust their saccadic behavior to the demands imposed by different types of error information sensed by visual feedback. Depending on the nature of the sensed errors, highly specific response mechanisms are addressed. Metrical errors lead to fast changes of magnitude and direction of saccades; post-saccadic retinal image drift is quickly compensated for by the adaptive attachment of a slow, exponential drift component to the fast saccade. The directional tuning properties different for both mechanisms suggest that independent neural mechanisms are involved.

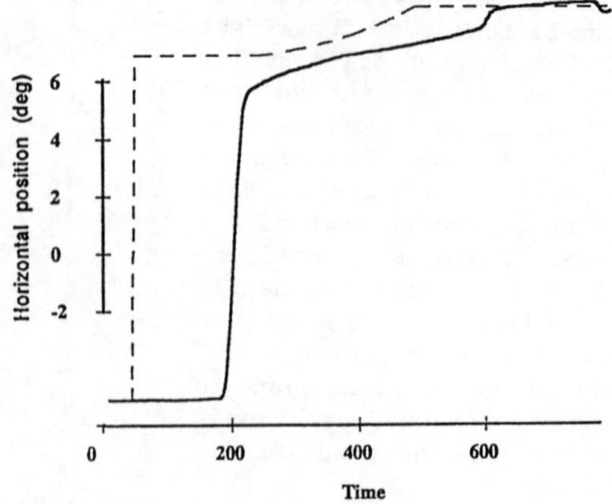

Figure 10: Conditioning stimulus with post-saccadic accelerating target motion and typical eye movement response of an adapted subject.

The experimental data reveal a number of puzzling features. So, speed of adaptation is much faster than one would expect from a mechanism responsible for the repair for disease or aging. Also, it is difficult to understand why a low-level motor adaptation mechanism should account for stimulus modality as is suggested by the reduced adaptation effects found for spontaneous saccades in darkness. These context-dependencies imply some amount of higher-level, supposedly cognitive involvement in the determination of the final oculomotor output.

The main finding of this study is that both mechanisms exhibit directional selectivity, and, for a given direction, adjustments can not be induced specifically to the magnitude of the movement. This suggests that the coding of the adaptive response does not occur in the independent horizontal/vertical components of the pre-motor structures, nor is it associated with the directions of the extraocular muscle forces. Rather, the results are compatible with the hypothesis that, at the level where adaptation occurs, the saccadic system is organized in the components of a polar coordinate frame. It is proposed that at this level of oculomotor control, independent "motor channels" exist, specific for the individual movement directions. Saccadic direction here would be *spatially* coded, i.e. determined by *which* of these elements is active. Saccade magnitude would be already *temporally* coded, namely by the amount of excitation of the active channel. This results in a total saccade vector which is the "population vector", i.e. the vectorial sum of the contributions of the individual direction channels. Within this framework, saccadic metricity is ensured by the appropriate adjustment of the vectorial contributions of each motor channel. This leads to a conceptual model for saccadic metricity

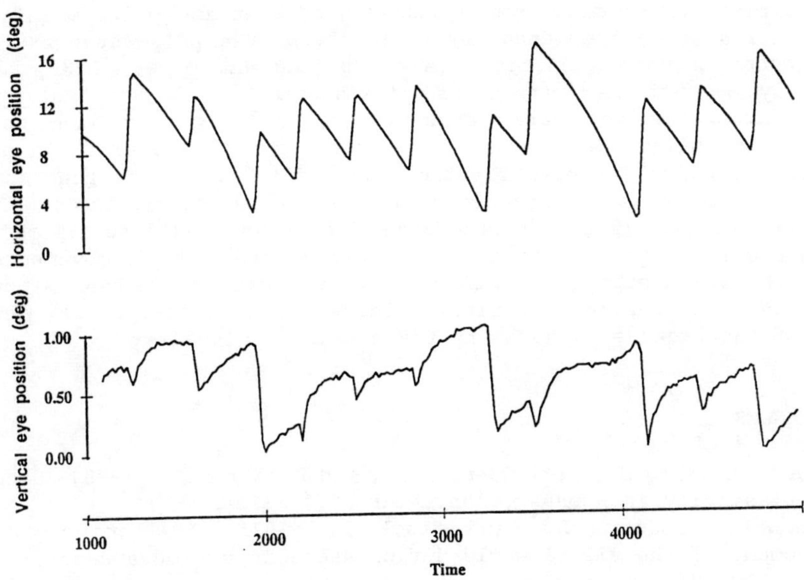

Figure 11: Horizontal and vertical components of horizontal optokinetic nystagmus. Before testing, the subject was adapted with vertical post-saccadic target drifts following horizontal saccades.

adaptation put forward in more detail by Deubel (1987).

The model is close to the "common source" model proposed by van Gisbergen et al. (1985) in which both components of oblique saccades are produced by a *single* pulse generator ("Vectorial burster") which converts the desired vector amplitude into a signal representing the vectorial velocity of the saccade. This vectorial signal would then be decomposed by appropriate weighing to yield the velocity of the (cartesian) horizontal and vertical components.

Further evidence for the existence of directionally selective subsystems in saccade generation comes from neurophysiological studies by Hepp and Henn (1983). These authors demonstrated in single-unit recordings in the monkey paramedian pontine reticular formation the existence of burst neurons whose firing is closely associated with saccades in specific directions. Firing rates of these cells are proportional to the magnitude of the movement.

Moreover, the presented findings are of potential interest for our understanding of PSD. It is widely accepted that the tonic component (the "step") which holds the eye in the orbital position is generated from the velocity signal (the "pulse") in neural structures which are, functionally and neuroanatomically, well-separated for horizontal and vertical eye movements. So, at this level, the saccade is suggested to be organized componentwise in a cartesian coordinate frame. Current conceptions also postulate that PSD results from a mismatch of step and pulse magnitudes. PSD adaptation would then occur by re-adjusting the pulse/step ratios of the brainstem saccade generators (e.g. Optican and Miles, 1985). It is immediately clear that adjustment of pulse/step ratios in a cartesian model cannot account for the data demonstrated. So, e.g., loading oblique saccades with oblique drift would lead to horizontal eye drift in purely horizontal saccades. Moreover, orthogonal drifts cannot be generated by such a model: where there is no vertical pulse, there should be no vertical PSD (Kapoula et al., 1988). As an alternative, I would like to suggest that the data presented on the induction of post-saccadic drift movements may indicate the existence of an active, context-sensitive mechanism outside the saccade generator for the final molding of the rather crude, default version of the saccade produced by the brainstem circuitry.

5. REFERENCES

Abel, L.A., Schmidt, D., Dell'Osso, L.F., and Daroff R.B. (1978) 'Saccadic system plasticity in humans', Ann. Neurol. 3, 313-318.

Bridgeman, B., Hendry, D., and Stark L. (1975) 'Failure to detect displacement of the visual world during saccadic eye movements', Vision Res. 15, 719-722.

Deubel, H. (1987) 'Adaptivity of gain and direction in oblique saccades', in J.K O'Regan and A. Levy-Schoen (eds.), Eye movements: From physiology to cognition, Elsevier North Holland, pp. 181-190.

Deubel, H., Wolf, W., and Hauske, G. (1986) 'Adaptive gain control of saccadic eye movements', Hum. Neurobiol. 5, 245-253.

Gisbergen, J.A.M.v, Opstal, A.J.v., and Schoenmakers, J.J.M. (1985) 'Experimental test of two model for the generation of oblique saccades', Exp. Brain Res. 57, 321-336.

Judge, S.J.., Richmond, B.J., and Chu, F.C. (1980) 'Implantation of magnetic search coils for measurement of eye position: an improved method', Vision Res. 20, 535-538

Kapoula, Z., Optican, L.M., and Robinson, D.A. (1989) 'Visually induced plasticity of postsaccadic ocular drift in normal humans', J. Neurophysiol. 61, 879-891.

Kapoula, Z., Robinson, D.A., and Optican, L.M. (1988) 'Visually induced cross-axis plasticity of post-saccadic drift', Soc. Neurosci. Abstr. 14, 613.

Kommerell, G., Olivier, D., and Theopold, H. (1976) 'Adaptive programming of phasic and tonic components in saccadic eye movements, Invest. Ophthalmol. 15, 657-660.

Miller, J.M., Anstis, T., and Templeton, W.B. (1981) 'Saccadic plasticity: parametric adaptive control by retinal feedback', J. Exp. Psychol.: Hum. Percept. Perform. 7, 356-366.

Optican, L.M. and Robinson, D.A. (1980) 'Cerebellar-dependent adaptive control of primate saccadic system', J. Neurophysiol. 44, 1058-1076.

Optican, L.M., and Miles, F.A. (1985) 'Visually induced adaptive changes in primate oculomotor control signals', J. Neurophysiol. 54, 940-958.

Optican, L.M., Zee,D.S., and Miles, F.A. (1986) 'Floccular lesions abolish adaptive control of post-saccadic ocular drift in primates', Exp. Brain Res. 64, 596-598.

Optican, L.M., Zee, D.S., and Chu, F.C. (1985) 'Adaptive changes due to ocular muscle weakness in human pursuit and saccadic eye movements'. J. Neurophysiol. 54, 110-122.

Robinson, D.A. (1963) 'A method for measuring eye movements using a scleral search coil in a magnetic field', IEEE Trans. Biomed. Eng. 26:, 37-145.

Semmlow, J.L., Gauthier, G.M., and Vercher, J.L. (1989) 'Mechanisms of short-term saccadic adaptation', J. Exp. Psychol.: Hum. Percept. Perform. 15, 249-258.

Westheimer, G., and McKee, S.P. (1975) 'Visual acuity in the presence of retinal-image motion', J. Opt. Soc. Am. 65, 847-850.

TWO-DIMENSIONAL CONTROL OF TRAJECTORIES TOWARDS UNCONSCIOUSLY DETECTED DOUBLE STEP TARGETS.

Olivier MARTIN* and Claude PRABLANC**
* Laboratoire RE.S.ACT UFR-APS Equipe Comportement Moteur
Université Joseph Fourier 38041 Grenoble Cedex (FRANCE).
**Laboratoire Vision et Motricité Unité 94 I.N.S.E.R.M.
16 av. du Doyen Lépine 69500 Bron (FRANCE).

1. Introduction

The understanding of how such a simple pointing movement with the hand towards a target in the prehension space is carried out, has to distinghish two essential sources in an erroneous movement : one on the perceptual side in the mechanisms that allow to build up a more or less accurate representation of the object location with respect to the body and the other one consisting in an encoding error on the motor side.

Much knowledge about the perceptual side, through studies on vision, eye and head motor control, on the vestibulo-ocular, the pursuit and optokinetic reflexes has been accumulated since, and though some uncertainties persist about the role of inflow-versus outflow in the knowledge of eye position signals within the orbit; most is known on the mechanisms by which the gaze captures a target with both eye and head (see for reviews Robinson, 1975, Carpenter, 1988). The spatio-temporal characteristics of those orienting mechanisms are fairly well known and the way perception is built starting from both vision, eye and head motor control systems, begins to be elucidated. When a stimulus appears on the peripheral retina, the acuity of its location is poor, but as soon an orienting saccade towards the stimulus is executed, the location of that stimulus begins to be sharper. Generally the first initial saccade brings the stimulus on a perifoveal region where a first re-updating of the stimulus occurs during the last part of the decelerating saccade (Prablanc et al., 1978). This is generally followed by a second corrective saccade 150 to 200 msec later which brings the stimulus exactly in foveal vision, allowing a perfect localiziation of the target.

On the motor side, the mechanisms involved in the correction of errors in either hand pointing or in gaze saccadic response (either if the goal has been erroneously encoded or if it has changed in between the reaction time) have been extensively analyzed through a large number of experimental paradigms; however few studies have considered the spatio-temporal characteristics of the spatial perceptual system and of the hand motor system, when both are involved in a synergic gaze and hand capture of a peripheral stimulus (Prablanc et al., 1979; Fisk and Goodale, 1985; Mather and Fisk, 1985). Though initial peripheral stimulus is poorly localized on the retina, the motor system can use this limited information to initiate a response and, if the visual information is reupdated through foveation during the last tens of msec after the saccade, it may adequately modulate its amplitude (Prablanc et al., 1986; Pélisson et al., 1986) to correct for the initial programming error.

Among the studies focusing on the hand motor control alone, two basic theoretical frame-

J. Requin and G. E. Stelmach (eds.), Tutorials in Motor Neuroscience, 581–598.

works have guided the investigations, one putting more emphasis on a feedback and the other more on a feedforward mode of control.

The feedback process which corrects for errors departing from a reference, is the most familiar to an engineering approach. It has multimodal aspects : visuo-visual, visual-kinaesthetic, memory-kinaesthetic. The visuo-visual loop, which has been the most extensively analyzed, consists in the simultaneous vision of the target stimulus and of the body limbs participating to the motor response, (also called «visuomotor feedback», «visual closed loop», or «visual reafferences»). The minimum time to amend those closed loop visuomotor errors has been approached first by alternate hitting movements or by single aimed movements. Fitts (1954) brought the most significant contribution by showing that the minimum duration of a successful closed loop movement could be formulated as a mathematical function of the index of difficulty. For instance for hitting a disk, this index of difficulty was expressed as the ratio of the distance to move the hand over the diameter of the disk. «Fitts» law has proved to be very robust and generalizable to many types of movements, even for movements where visual reafferences from the hand were not available. (Prablanc et al., 1979).

The opposite approach to the feedback control is the feedforward control. Basically in order to reach a given point in space, the solution of the required levels of muscle activity at that particular point is supposed to be known; and whatever the initial conditions (i.e. the starting point of the final effector, the configuration of the intermediate effectors), the application of control signals which converge towards this final preknown state of equilibrium will be a solution to the problem. This kind of control does not need any visual feedback from the hand nor any kinaesthetic feedback. It entirely relies on a strict correspondence between any spatial point and a set of muscle activities, which is supposed to have been learnt. This concept has been developed by Feldman (1966) and Bizzi (1975) and is known under the name of «final equilibrium point» or «mass spring model». This model has been shown to work experimentally in deafferented monkeys, but limited only to single joint movements where none of the visual feedback or the kinaesthetic feedback were available. Artificial neural networks have been developped using these notions and the feasibility of such systems has been demonstrated at least for the solution of static correspondences between an endpoint position and the corresponding joint angles (Massone and Bizzi, 1989, Jordan, 1990). However, the execution of complex movements involving multijoint motions becomes practically impossible in deafferented animals; this points out the difficulty of implementing complex feedforward mechanisms.

Another approach to the understanding of visuomotor coordination has been the introduction of changes in the goal either prior, during or after the initiation of the motor response. This kind of double-step paradigm has been initially introduced by Wheeles (1966) in the study of the oculomotor system in order to get some insight about the rules of organization of this simple motor system. The questions basically addressed in this kind of paradigm were the evaluation of a period of refractoriness to any kind of stimulus and of a serial versus parallel processing of incoming information. This paradigm, further developed by Becker and Jurgens (1979) showed up some parallel processing in the oculomotor system. The same paradigm applied to the hand motor control system by Soechting and Lacquaniti (1983) in man and Georgopoulos et al. (1981) in monkey showed even more parallel processing. Those double-step studies were followed by others, all showing the same basic result : that an aimed motor command can be emitted in a nearly continuous mode which can be adjusted to cope with changes in the goal (Gielen et al., 1984, Vicario and Ghez, 1984, van Sonderen et al., 1988, De Jong et al., 1990). The above double step experiments favor parallel processing and can be seen as feedforward actions, as responses involve mainly intentional corrections which in most cases are substitution responses.

This kind of paradigm raises however some basic questions about the meaning of the corrections observed : Do the subjects react «passively» to a double-step stimulus and in that case the behavior may reflect the structural constraints of the motor response system, or does the double-step stimulus induces by learning a set of alternative strategies ? and what is the role of prediction or expectation in the elaboration of reaction time and movement organization ?

In order to avoid those problems, Goodale et al. (1986), Pélisson et al. (1986) imagined a paradigm that we will call the «synchronous double step feedback», where the second step would always be triggered around the onset of the hand movement, and by an artificial technique would not be consciously detected by the subject. One underlying hypothesis was that an unintentional correction of non consciously detected perturbation would rely more on a feedback if any than on a feedforward process. Viewed from the subject, the target was perceived as unique and his strategy was not «point at the last seen stimulus», but point at the stimulus. It was equivalent to a naturally inaccurate motor response performed towards a stationary single step stimulus. The uniqueness of stimulus perception was checked by a perceptual forced choice technique without hand pointing. In Goodale et al.'s experiment the artificial error was aligned on the axis of pointing, in such a way that only the extent of the movement needed a correction throughout the response, but not its general direction, which was preserved.

RATIONALE

This paper intended to generalize the findings from Goodale et al. (1986) and specifically answer .that question : Will the online unconscious corrections of trajectory (path and kinematics) be stillobserved when not only the amplitude of the goal is modified in flight but also its spatial direction ? Are the corrections for changes in direction more time consuming than those for changes in amplitude ? One of the main purposes of this experiment was also to evaluate at what time normal and perturbed trajectories would significantly depart from each other.

2. Methods and Procedure

The horizontal table on which hand movements were performed was, 1.5O depth by 2.00 m wide flat isotropic surface, no detail being perceptible, thus without visual frame of reference. Underneath the table a matrix of red light emitting diodes (l.e.d.) disposed in polar coordinates with a 65cm radius angles could select one of the leds (ranging from 0 to 10, 20, 40, 50 degrees on the right hemispace, see fig. 1), around a circle centered on the subject's head. The initial starting point (resting position RP) for the hand was in the subject's sagittal plane 35 cm ahead of the subject's eyes). A chin and forehead rest restricted subject's position who was instructed not to move his head. Eye movements were monitored with a D.C E.O.G method. The subject's task was to look and point with his fingertip as quickly and accurately as possible onto the target when it jumped from a central to a peripheral position. As soon as the finger left the surface, the whole limb and body parts became unvisible, through the action of an electronic shutter turning the ambiant light off. The saccadic eye movement towards PT produced either a second step a few degrees in (pulse-step) or out (double-step) along the radius of PTs, or no change at all (single step). The PT (single or double step) presentation lasted for 2 seconds, and went back to the central point, followed by the eye while the hand came back to the resting point, 20 cm behind the fixation point.

2.1. RECORDING TECHNIQUES

The experiments were conducted on a PDP11/73. All the data were collected at 333 Hz (every 3msec). The x-y horizontal components of the hand pointing were recorded through a small l.e.d. (emitting in the infra-red spectrum) 3x2mm, stuck on the nail of the fingertip and coupled with a Selspot II system. The eye movements were also simultaneously recorded.

The Selspot II camera, placed vertically two meters above the working surface, recorded the x-y displacements of the fingertip (with an accuracy of +/-2 mm). When the hand began to move, a logical pulse was sent to an electronic shutter which cut-off, within 5 msec. the illumination of the room (open loop) (see fig. 1).

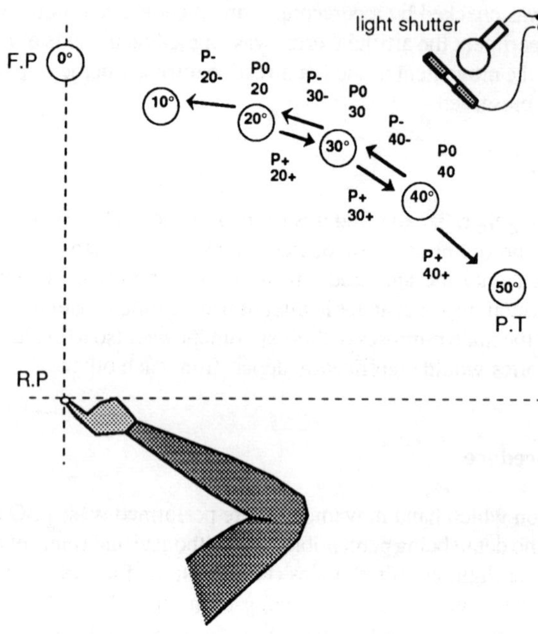

Figure 1. Experimental display. The subject had to point with his forefinger tip on which was stuck on an infra-red light emiting diode (i.r.l.e.d) to visual targets which randomly appeared in his right hemivisual field. Three types of targets were presented: the unperturbed single step targets (P0) at 20, 30, 40 degrees of eccentricity, the perturbed double step targets (P-), targets from 20 to 10 degrees noticed 20-, 30 to 20 degrees noticed 30-, 40 to 30 degrees noticed 40-, and the perturbed double step targets (P+), targets from 20 to 30 degrees : 20+, 30 to 40 degrees : 30+, 40 to 50 degrees : 40+. A fast light electronic shutter (3 to 5 ms response time) was used to cut off all subjects visual reafferences from his limbs and body, while maintaining vision of the target. RP : resting position (initial starting point of the hand); FP : visual fixation point; PT : pointing target.

An experimental session was composed of an equal number of unperturbed targets, perturbed targets to the left side and perturbed targets to the right side (P+),(see fig 2). The 9 different types of stimulations were 20 to 10 degrees, 20 deg. stationary, 20 to 30 deg., 30 to 20 deg.; 30 deg. stationary, 30 to 40 deg., 40 to 30 deg.; 40 deg. stationary, 40 to 50 deg. and were randomized. Each type of stimulation was repeated 5 times every session. One session included thus 9x5=45 trials recorded.

The experiment was carried out in dynamic «open loop», i.e. vision of the moving limbs disappeared at the onset of hand movement (visual feedback beeing prevented throughout all the pointings) and was restored only when the hand moved back to the resting point.

Six healthy naive right-handed subjects, with a normal visual acuity, from 20 to 26 years old, underwent the experiment. Each session was repeated twice for all subjects.

There was thus a number of 2 (session) x 6 (subjects) experimental sessions, each one including 9 (types of stimulation) x 5 (repetitions) i.e. 540 trials. For each subject and condition the two (identical) sessions were grouped for the analysis. The 9 types of stimulations corresponded to 3 target eccentricity x 3 jump direction. For each parameter measured, a 2 way analysis of variance was performed : target eccentricity (20, 30, 40 deg.) x jump direction (perturbation Left (P-), no perturbation (P0), perturbation Right (P+).

Means, standard deviations were computed for all parameters. Position, velocity, acceleration were averaged by synchronizing the traces at the onset of eye movement (for eye) and the onset of hand movement (for hand).

Statistical analyzes were performed to detect the earliest point where the trajectories to perturbed targets began to deviate from the trajectories to stationary targets.

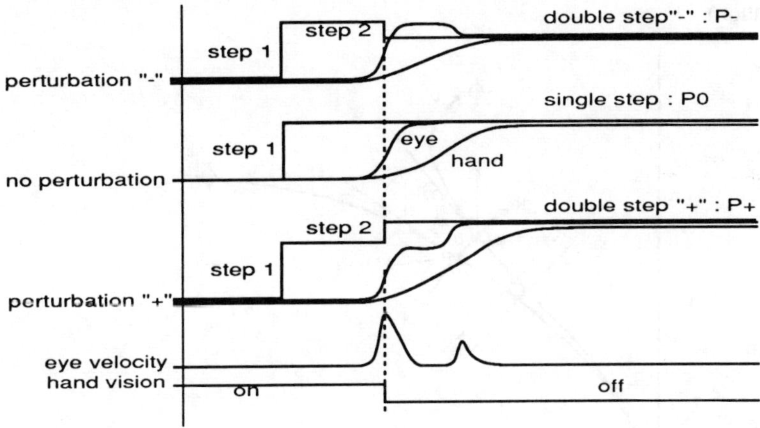

Figure 2. Diagram of spatio-temporal organization of eye and hand movement in the three different task of pointing target. First and third trace : double step stimulation where the target jumped once more after the first step (step1) either to a more eccentric position (double step «+» or «P+») or to a less eccentric position (double step «-» or «P-»). Second trace : single step stimulation (P0) where the target jumped from its central reference position to a randomly determined stationary peripheral position. At the peak eye velocity, the vision of the hand was cut off, and simultaneously the double step stimulation was applied, corresponding also to time when hand had begun to start (see vertical dashed line).

2.2. COMPUTATIONAL METHODS

The signals were numerically filtered with a zero phase Finite Impulse Response (FIR) filter (at 20 Hz for x y of the hand, and at 30 Hz or the eye). The eye, hand velocities and accelerations were computed from the filtered position signals by a least square method. The net acceleration, which was the vectorial sum of the curvilinear acceleration and of the acceleration component orthogonal to the tangential velocity, was computed from the x and y velocity components. The angular orientation in space of the tangential velocity and of the net acceleration were aso computed.

To determine the time at which the perturbed responses differed significantly from the normal responses, we used two different methods according to the type of perturbation : Left or Righ. All curves were synchronized with respect to their mean onset. Then they were all aligned on the family curves of the unperturbed responses.

Perturbations Left : The tests were carried out for each 20, 30 and 40 degrees targets. At every 9 msec time step, we tested according to a t test the significant differences in spatial angles between the acceleration vectors of the normal responses and the acceleration vectors of the perturbed-left responses (fig. 3). This test was run as soon as the modulus of the vectors were emerging from noise level. The same test was applied to the angles of the velocity vectors, and then to the instantaneous distances between the two families of path curves Perturbations Right For this latter case we observed that most of the initial pathes were common to unperturbed and perturbed right responses. The earliest change of the responses was thus calculated from the divergence time of the net acceleration amplitudes; idem for the tangential velocities and for the instantaneous distances between pathes. The reaction time to the perturbation was computed as the difference between the time of divergence minus the time of peak saccadic eye velocity (which was also the instant of perturbation).

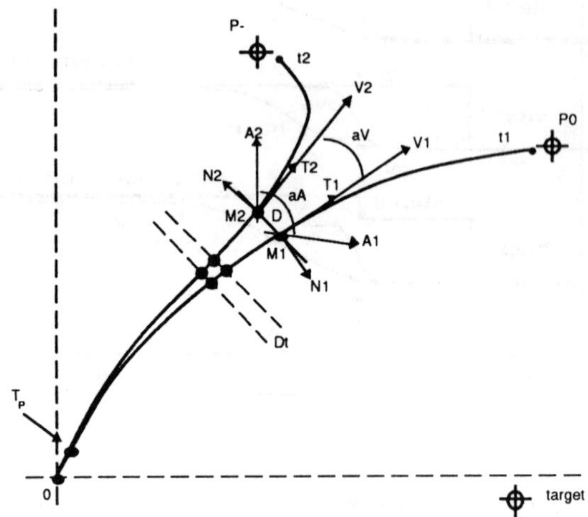

Figure 3. Diagram of comparative kinematic parameters of hand trajectory to perturbed and unperturbed target. 0 : initial hand position; t1 : unperturbed hand trajectory; t2 : perturbed trajectory (P-); Tp : time of perturbation; aA : divergence angle between net acceleration vectors M1A1 and M2A2 of t1 and t2; aV : divergence angle between tangential velocity vector M1V1 and M2V2; M1T1, M2T2 : tangential acceleration of t1 and t2; Dt : time interval between two divergence angle analysis (9 ms); frequency filtering for all hand parameters : 20 Hz.

Real latencies are merely to be lower than the above reaction time. As during the saccade the retinal signals are centrally and peripherally inhibited, reupdating of retinal signals occurs only towards the end of the saccade when the eye velocity decreases under about 100 deg./sec (Prablanc, et al., 1978). If we take that value as the time at which the perturbation becomes centrally available, the reaction time must be decreased of about 2/3 of the deceleration time of the saccade. Thus for a mean saccade amplitude of 30 degrees, the deceleration phase of the saccade beeing 60 msec long, the reaction time must be diminished by about 40 msec.

The measures (based on acceleration, velocity and distance) were averaged over subjects and over the three eccentricities (20, 30, 40 degrees).

3. Results

3.1. QUALITATIVE OBSERVATIONS

At the end of the sessions all subjects were questionned about their sensations. None reported the clear perception of a double step-stimulus, but they all had the impression to be inaccurate, and sometimes for the nearest targets to the midline felt an unintentionnal motor correction. If we look at the profiles of the pathes these cases correspond to the targets appearing initially at 20 degrees and then jumping, during the saccade, to a 10 degrees position. In a few cases the second step stimulus jumped, not at peak velocity of the orienting saccade, but at peak velocity of an involuntary saccade or on a blink eyelids. Even in those cases the subjects did not clearly detected the jump, but had a bizare sensation.

3.2. QUANTITATIVE RESULTS

Table 1 summerizes the means and standard deviations of movement parameters analysed.

3.2.1. *Eye Latency.* As expected, It did not vary with the eccentricity as well for the unperturbed targets (from 249 to 251 msec) as for the perturbed targets (from 250 to to 275 msec).

3.2.2. *Hand Latency.* It did not depend either on the eccentricity of the initial target or on the perturbation (means ranging from 277 to 301 msec). On the average the hand began to move about 20 msec after the eye with a mean standard deviation of 52 msec. Though the two eye and hand motor commands had close timings they were not tightly correlated (with correlation coefficients ranging from 0.33 to 0.54).

3.2.3. *Hand Movement Duration.* It depended heavily on the target eccentricity : for stationnary targets ranging from 20 to 40 degres .it varied from 406 to 430 msec; for perturbations (P+), with initial targets ranging from 20 to 40 d.egrees, the durations varied from 459 to 517 msec; for perturbations (P-), the corresponding durations varied from 501 to 479 msec, i.e. in the reverse order (see fig4).

The other strong factor influencing hand movement duration was the occurence of a perturbation, the mean duration (pooled over eccentricity) for unperturbed responses (P0) was 416 msec while it was 487 msec for (P-) and 489 msec for (P+); thus the perturbation lengthened by about 72 msec the movement duration. There was a strong interaction between target eccentricity and perturbation; perturbation Left (P-) gave a decreasing duration with increasing target eccentricity, while a perturbation Right (P+) gave an increasing duration with increasing target eccentricity. This phenomenon is due to the qualitative structure of the spatial path of (P+) responses which is roughly

TABLE 1. Means (first row) and standart deviations (second row) of movement parameters in open loop condition for perturbed (P-, P+) and unperturbed (P0) responses to different target eccentricity (20, 30, 40); (n=60).

OPEN LOOP POINTING

targetperturbation	P-			P0			P+		
target eccentricity	20-	30-	40-	20	30	40	20+	30+	40+
- TIME TO DOUBLE STEP	275	297	295	-	-	-	300	291	298
(msec)	41	50	46	-	-	-	43	68	48
- HAND LATENCY	278	288	277	301	284	286	285	277	283
(msec)	52	58	54	50	49	48	44	66	52
- HAND DURATION	501	481	479	406	413	430	459	491	517
(msec)	61	54	53	60	30	42	71	69	84
- TIME TO PEAK	58	59	67	57	64	69	57	61	65
ACCELERATION (msec)	8	9	16	5	14	16	6	6	16
- PEAK ACCELERATION	1643	1830	1920	1769	1767	1880	1828	1817	1883
AMPLITUDE (cm/sec2)	401	427	567	291	474	407	660	487	679
- ANGLE OF PEAK ACCEL.	33	40	48	33	41	49	32	40	48
VECTOR (deg.)	8	9	9	8	7	7	8	7	7
- ACCELERATION TIME	161	166	178	176	188	194	195	189	194
(msec)	20	13	28	17	17	20	37	19	2
- TIME TO PEAK	161	166	178	176	188	194	195	189	194
VELOCITY (msec)	20	13	28	17	17	20	37	19	23
- PEAK VELOCITY	143	163	183	161	174	190	168	179	191
AMPLITUDE(cm/sec)	17	17	29	26	20	25	29	27	31
- TIME TO PEAK	251	252	276	291	309	324	327	327	352
DECELERATION (msec)	36	33	45	52	38	45	58	50	60

the same as for (P0) responses, while it differs considerably for (P-) responses, this phenomenon beeing accentuated for (P-) at small eccentricities (large path disimilarities) and for (P+) at large eccentricities (close path similarities).

If we compare the two trajectories ending up at 20 degrees : the unpertubed 20 degrees and the perturbed (P-) 30 deg. their respective duration is 406 and 481 msec i.e. a difference of 75 msec. If we now compare the two trajectories ending up at 40 degrees i.e. the 40 degrees unperturbed and the (P+) 30 degrees, their respective durations are 430 and 491 msec i.e. a difference of 61 msec.

3.2.4. *Peak Hand Net Acceleration.*

Modulus : None of the factors eccentricity or perturbation had any significance on the modulus of the peak hand acceleration.

Angle of the peak net acceleration : The only strong factor influencing the angle of the peak hand acceleration was the eccentricity; the angles varied from 33.4 degrees to 50.2 degrees as target eccentricity varied from 20 to 40 degrees. None of the other factors had any significance.

Time to peak acceleration : This is the image of the rise time of the net force at the end of the tip. None of the factors influenced the time to peak acceleration. Its mean value was 63 msec. To summarize, the spatio-temporal pattern of peak acceleration is uninfluenced by the perturbations; its rise time and modulus are constant and its spatial orientation depends only on the orientation of the initial targets.

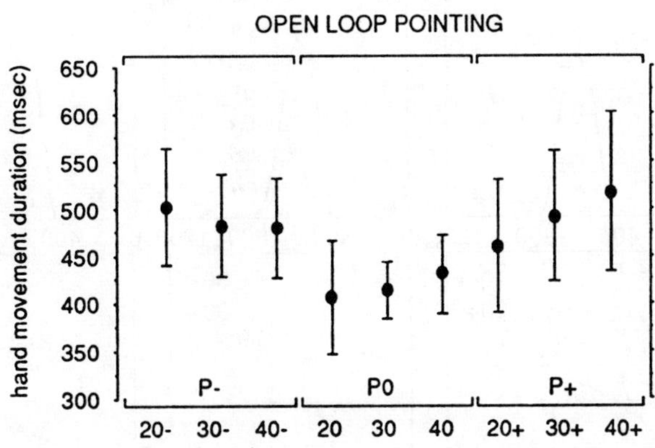

Figure 4. Relationship between mean hand movement durations and eccentricity for the different types of unperturbed and perturbed targets in open-loop condition. Vertical lines indicate standard deviations; (n=60).

3.2.5. *Peak Hand Tangential Velocity*. The eccentricity and perturbation factors were highly significant (p<0,001). With unperturbed responses the peak velocity ranged from 161 deg./sec to 190 deg./sec for target eccentricities from 20 to 40 degrees of visual angle. Perturbations Left (P-) decreased peak velocity (163 deg./sec) with respect to unperturbed targets (P0) (175 deg./sec) while perturbations Right (P+) increased it (179 deg./sec).

There was a significant interaction between eccentricity and perturbation, the tangential velocity of (P-) responses being increased for 40 degrees eccentricity and decreased for 20 degrees eccentricity (see fig.5).

3.2.6. *Time To Peak Hand Tangential Velocity (Or Acceleration Phase)*. Though the time from hand movement onset to peak velocity varied from 176 to 194 msec for unperturbed responses from 20 to 40 degrees targets, it did not dependent significantly on the eccentricity. It was highly dependent on the perturbation factor (p<0,001). For (P0) responses this time was 186 msec, while it was 168 msec for (P-) and 193 msec for (P+) responses, pooled over eccentricity. None of the interactions between factors had any significance.

3.2.7. *Absolute Error*. The absolute error was the distance between the pointing position at the time of impact and the final target position. It was found to depend on the perturbation factor

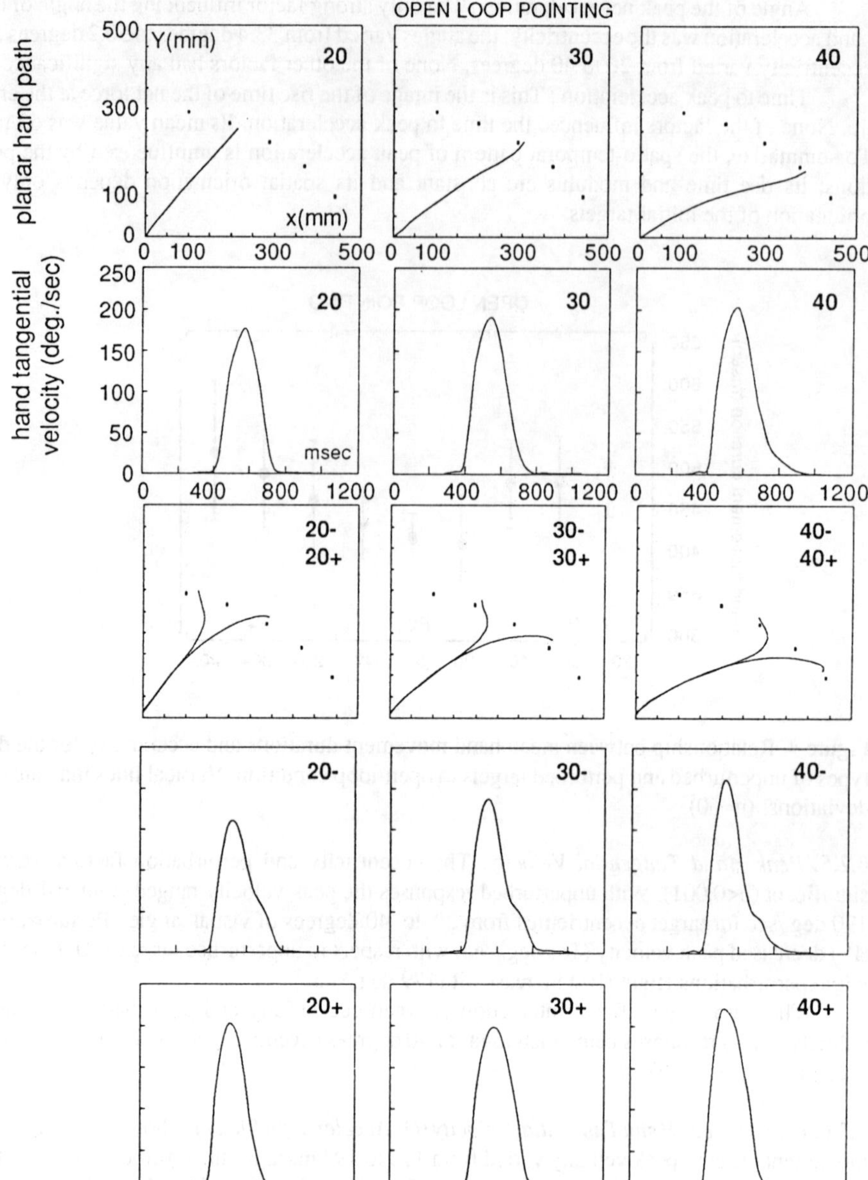

Figure 5. Mean hand trajectories and corresponding normalized tangential velocity to different unperturbed and perturbed targets in open-loop condition. First, second, and third column groups respectively data of 20, 30, 40 degrees centered targets, layed out in row, for trajectory and velocity of unperturbed targets (rows 1,2), trajectories and velocity of perturbed (P-) and perturbed (P+) targets (rows 3,4,5). Subject A.P; (n=10).

(p<0,01),and not on the eccentricity. The perturbations Left (P-) and Right (P+) gave overall errors of 17 and 27 mm respectively, against an error of 19 mm for the unperturbed (P0) responses (see fig.6A).

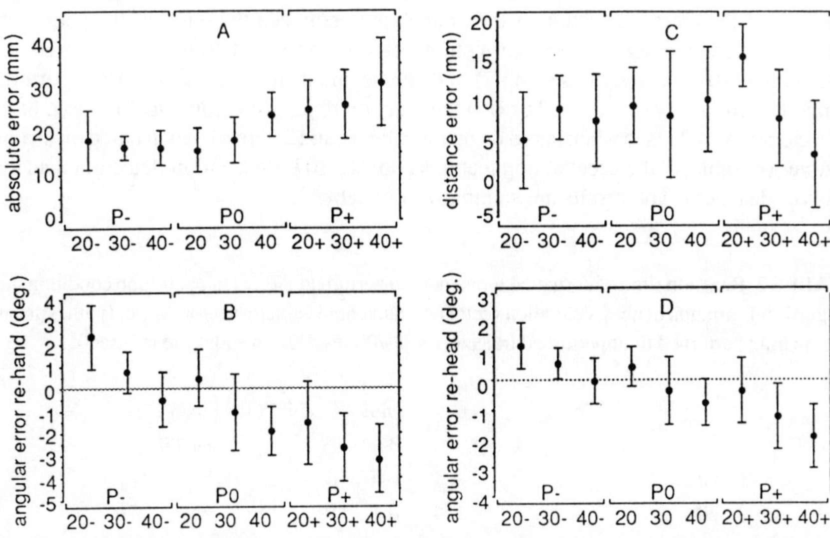

Figure 6 A, B, C, D. Relationship between absolute error means (A) (mm), angular error means (B) (deg), distance error means (C)(mm), absolute error (D)(mm), and the different types of unperturbed and perturbed responses in open-loop condition. Vertical lines indicate standard deviations; (n=60).

3.2.8. *Distance Error*. If RP is the resting starting hand position, PT2 the target final position and HP the position of the hand pointing when it touches the surface, the distance error can be defined as the difference between the length of the segments (RP-HP) and (RP-PT2), positive values beeing noticed as overshoots in depth and negative values as undershoots. None of the main factors influenced the distance error, with means equal to 7 mm, 8 mm and 8 mm for respectively (P-), (P0) and (P+) (see fig.6B).

3.2.9. *Angular Error*. The angular error referenced in a hand coordinate sytem was defined as the angle between the vectors, 1 (RP-HP) and 2 (RP-TP2) Positive errors corresponded to overshoots while negative errors correponded to undershoots. The overall angular errors for perturbation Left (P-), for (P0) and for (P+) were respectively 0.8, -0.8 and -2.4 degree. When considered from a head coordinate system those angular errors became respectively 0.7, -0.2 and -1.2 degrees (see fig.6C).

To summarize, the errors in distance were always overshoots which did not depend upon the perturbation. Conversely the angular errors of the movement in a head coordinate system with respect to the «unperturbed» errors were respectively +0.9 deg. (overshoot) for the perturbations left (P-) and of -1 deg. (undershoot) for the perturbations right (P+) (see fig.6D). As the perturbation in head coordinate system was of 10 degrees the percentage of angular correction was thus of 90% for both types of perturbation.

Considering that the targets lied on a radius of 65 cm centered on the head, an angular error of one degree corresponded to approximately 12 mm. Thus the maximum angular error measured in an equivalent scale of distance error was about 12 mm.

3.2.10. *Reaction Time of the Sensorimotor System To The Perturbation.* This reaction time has been defined in the methods section as the time elapsed between the onset of the perturbation and the earliest detectable signs of the corrections. It was measured from the acceleration vector, the tangential velocity vector (see fig.7) and the segment distance between the perturbed and the unperturbed family of curves. For (P-) the reaction times were equal to 145 msec based on phase acceleration, 182 msec when based on phase velocity and 252 msec when based on distance between curves (Looking at the acceleration features allows a 107 msec earlier detection than looking at the curves distance). The results are summarized on table 2.

TABLE 2. Reaction time of corrected movement to perturbed targets in open-loop condition measured from significant variation of net acceleration vector (A), tangential velocity vector (V), and pathes distance (D) when comparing perturbed to unperturbed trajectories (P-/P0; P+/P0); (n=60; time in msec.).

		OPEN LOOP POINTING				
			P-/P0		P+/P0	
target		A	V	D	V	D
20	means	144	161	195	225	270
	sd	62	54	103	44	65
30	means	126	164	202	327	303
	sd	42	61	116	87	59
40	means	164	220	360	342	352
	sd	45	45	58	112	71
overall means		145	182	252	298	308
overall sd		50	53	92	81	65
		*	*	**	*	**

* : phase detection; ** : amplitude detection.

4. Discussion

The most basic findings of these series of experiment is the generalization of the online mechanisms of control in aimed movements, in a very automatic way, previously described by Goodale et al. (1986), Pélisson et al. (1986). In the Goodale et al. experiment the perturbation involved, at the level of the end point fingertip movement, a change in amplitude of the movement (lengthening or shortening according to the type of perturbation), not of its general direction. In the present experiment, with a similar paradigm, the perturbation involved mainly a modification of the orientation of the movement and to a lesser extent of its amplitude. As other authors have shown some evidence of independant programming of amplitude and direction of a motor response (Vicario and Ghez, 1984; Favilla et al., 1990), the generalization of corrections from amplitude to orientation was an important stage in the demonstration of a non intentionnal general corrective process. Basically our perturbations were at the orientation level in a head centered coordinate

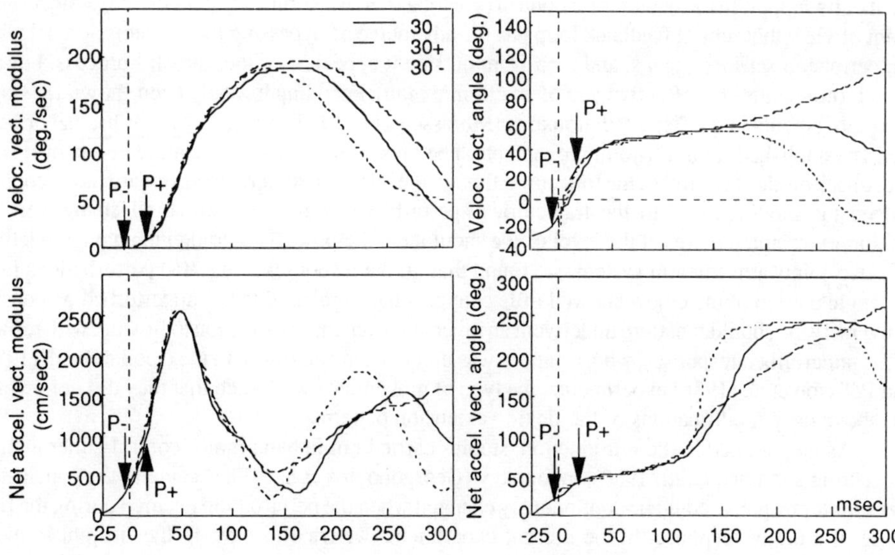

Figure 7. Comparative variation of tangential velocity, and net acceleration (two left frames), and corresponding comparative divergence of tangential velocity and net acceleration vector angles (two right frames) for pointing movement directed to targets at 30, 30-, 30+, in open-loop condition.

system along a constant radius. However the movement of the hand had its origin 30 cm ahead of the head. Thus movements even to 10, 20, 30, 40 and 50 degrees stationary targets were of unequal amplitude.

By applying to our experimental data independent analyses on pointing errors in distance and in orientation we found no cues privileging one type of error over the other; mean errors in distance from the final target were of the same order of magnitude as errors in the orientation. Would orientation and distance be coded by independant channels, the above results indicate that, if the mechanisms of amending trajectories are controlled by different channels, they share at least the same flexibility.

There have been many studies, carried out first on the oculomotor system, to understand how a «programmed» response may be altered by the occurence of a second stimulus considered as the new goal for the motor response (Wheeless et al., 1966; Becker and Jurgens, 1979; Van Gisbergen et al., 1987). All those studies came to the conclusion that some parallel processing occurs and that a decision to move somewhere may be cancelled and replaced by another one, with intermediate solutions embedding the two responses (and defined by a probability transition function). Those first types of studies would fall into the category of either feedforward or central feedback actions, where the processing occurs prior to the first movement. In our experiment such a process occured with the oculomotor system in the few cases where artifacts produced a double step prior to the oculomotor response to the first step; the first saccade was cancelled and replaced latter on by a primary saccade directed towards an imaginary target in between the two steps.

The notion that motor control relies more on central loops where the feedback loop is an efference-copy, rather than the actual sensory information from the moving organ, has received most of its support in the oculomotor control (see for a Review Robinson, 1975). From a functionnal point of view this central feedback loop has the advantage of bypassing the transmission delays of the peripheral sensory organs, and seen from outside it appears as a feedforward process. For the hand motor control the effectiveness of an efferent motor encoding has only been shown for single joint movements in deafferented monkeys (Morasso et al., 1973; Bizzi, 1975) but has never been clearly established for multi-joint movements. The present experiments are more compatible with a feedback mode of control of the trajectory, though they do not exclude a feedforward mode control. Indeed the modifications to the trajectories «perturbed in» involved structural changes in the flexion-extension patterns at the level of the shoulder and elbow. By a crude kinematic modelling of a two-joints arm forearm system, we found that, in the responses to the (P-) perturbations Left, the angle pattern of the elbow showed litlle change when compared to the unperturbed responses; conversely the shoulder pattern underwent an inversion from intial extension to flexion, with respect to the unperturbed responses, whose pattern was uniquely an extension. In the Goodale et al. (1986) and Pélisson et al. (1986) experiments, the type of movements were such that they did not involve the above deep modifications of the flexion extension patterns.

As mentionned in the introduction, studies carried out in hand motor control either in man (Soechting and Lacquaniti, 1983), in monkey (Georgopoulos et al., 1981) came to the conclusion that a motor response could be continuously changed when the perturbation occured during the first phase of a motor response. In the present experiments we tried to have further insight in these corrections by looking at the early cues of change in the kinematic parameters and more specifically to the orientation (or spatial phase) and amplitude of the acceleration and velocity vectors at the same time as the distance from the two pathes. When judged from the phase of the acceleration vectors, they showed up the earliest modifications with a latency of 104 msec. Thus looking at simultaneou-sly the amplitude and the direction of the fingertip acceleration vector allowed us to detect the earliest changes in forces or torques (which are closely time-locked to the nervous command). However the peak acceleration of the hand, characterized by both its modulus, phase angle and time of occurence (63 msec from the movement onset), was totally uninfluenced by the perturbation; the (P-) or (P+) peak vectors always pointed in the same direction as the (P0) unperturbed peak vector. Paulignan et al. (1990), in a reaching and grasping study, with a perturbation of the location of the object at the onset of hand movement, found changes in the peak and time to peak acceleration of the ongoing movement. Our peak and time to peak acceleration was insensitive to the perturbation; but with an average value of 104 msec our latency corresponded to roughly the same value as theirs, though their perturbations were consciously detected, in contrast with ours.

In our (P-) condition the short latencies also looked very similar to the double-step studies with intentionnal corrections to an ongoing response, though our experimental conditions were very different; our subjects reported only a sensation of inaccuracy whereas they actually fully corrected for the displacement.

For the (P+) perturbations occuring in the same general direction as the initial movement, the 284 msec latency was much longer than for the (P-) perturbations, and however produced unexpectedly less increase in duration (61 against 86 msec) with a comparable angular accuracy. At a first glance this result may appear surprising as the late taking into account of the perturbation neither lengthened very much the movement duration nor produced a larger error than the opposite (P-) perturbation.

From other points of view, such as the duration of the movement, our results appear different. Indeed in the classical double-step experiments, the duration from the onset of the correction to the end of the movement appeared nearly equal to the duration of normal response to a single step stimulus, though the authors did not investigate it precisely. Goodale et al. (1986) and Pélisson et al. (1986) using an «unconscious» target jump in amplitude only could not find any difference in movement duration for a unperturbed and a perturbed response of the same final amplitude. They suggested that the modifications of trajectory occured very early as perturbed and unperturbed responses had both a bell-shaped profile, without noticeable changes on the velocity and acceleration traces.

In our experiments the movement duration was dependent upon the type of perturbation. We obtained for the same final target position at 20 degrees, durations of 406 msec for the 20 deg. (P0) target and of 481 msec for the 30 deg. (P-) target, i.e. an increase of 75 msec, whereas the latency to the pertubation was about 86 msec. For the same final position at 40 degrees, durations were 430 msec for the stationnary 40 deg. (P0) target and 491 msec for the 30 deg. (P+) target, i.e. an increase of only 61 msec, while the corresponding latency was of 287 msec. This long processing delay is due to the fact that both 40 deg. (P0) and 30 deg. (P+) trajectories share very close common pathes during the major part of the movement with regards to the accuracy of the pointings.

5.Conclusion

The present experiments have clearly shown that for most of its structure a complex multi-joint movement is under the control of internal loops which guide automatically the hand onto its goal. The latencies to a perturbation (not consciously detected as such) may be short (about 100 msec) or long (about 280 msec) according to the severity of the path inflexion, but in both cases with very small increases in movement duration (from 75 to 61 msec).

Though the short latencies we observed were similar to those of the double-step experiments from Soechting and Lacquaniti (1983), one may think about whether they reflect the same mechanism, as our subjects were never able to report the perturbation nor in most cases the sensation of a motor correction. The conscious perception was unique : i.e. a single step stimulus, the goal beeing also unique : point at the peripheral target; however the nearly perfect corrections were implemented in very automatic way independently from will and with a very small additionnal time. This is reminiscent from Abbs and Gracco experiments (1984) where a perturbation of the mandibula during the pronounciation of a phoneme did not interfere with the goal through the compensatory action of quick internal loops. Though their modalities and ours were very different, in both cases there was an unicity of the goal and a necessity of accuracy.

In our experiment, the corrections, which were not mediated by reafferent visual signals, were implemented like if the goal of the corrections was to bring the end point effector to the target. As the exteroceptive visuo-visual loop was precluded, it favored the role of an inner feedback loop, would it be operative (either an efference copy or a kinaesthetic signal). Such a conception of an internal servo-system has been previously proposed by Abbs and Gracco (1984) in lip control during speech, Laurutis and Robinson (1986), Pélisson et al. (1988) and Guitton et al. (1987) for the eye-head orienting system, and by Goodale et al. (1986), Prablanc and Pelisson (1990) for the multi-joint hand motor control; this view is a top down approach where the controlled variables even at low levels are finalized variables such as gaze position or fingertip position, and where intermediate variables such as joint angles, torques and muscle forces are devoted to meet the goal of the endpoint effector.

There are essentially two ways to implement these corrections : through kinaesthetic feedback loops and/or through feedforward actions, as the visual feedback from the hand was precluded in our experiment.

The respective role of feedback and feedforward in our experiments was not the main goal of this study; however some hypotheses can be suggested by the comparison between Goodale et al. (1986) experiments and ours. In the former case only the extent of the movement was changed by the perturbation, not its orientation. The general pattern of flexion extension did not involve complex reverting pattern in either type of perturbation (just a shortening or lenghthening of the movement). Thus the necessary action could have been limited to an earlier active deceleration on each shoulder and elbow joint in the shortened movement and to a an active reacceleration on both joints. It could as well be easily explained by a kinaesthetic feedback from the hand.

In our experiment perturbations were of two different kinds : the (P+) perturbed responses corresponded to the same curvature of the pathes as for the unperturbed responses and were qualitatively the same as in the Goodale et al. experiments. The (P-) responses instead corresponded to inversions of the initial curvature of the movement, with a kinematic pattern which did not correspond to a simple common scaling on each joint (as revealed by a raw modeling of the two joints (shoulder-elbow) system. Such a complex change of neural commands, though it can not be totally excluded in terms of feedforward action, seems unlikely and the alternative explanation using sensory feedback is much easily conceivable.

However future experiments are needed to test the kinaesthetic hypothesis, with the introduction of perturbations of the sensory message from the arm during the correction to the visual perturbation.

References

Abbs, J.H. and Gracco, V.L. (1984) 'Control of complex gesture : oro-facial muscle responses to load perturbations of lip during speech.', Journal of Neurophysiology, 51, 705-723.

Alstermark, B., Gorska, T., Lundberg, A .and Petterson, L.G. (1990) 'Integration in descending motor pathways controlling the forelimb in cat. 16.Visually guided switching of target reaching', Experimental Brain Research, 80, 1-11.

Becker, W. and Jürgens, R. (1975) 'Saccadic reactions to double step stimuli : evidence for model feedback and continuos information uptake', in G. Lennerstrand and P. Bach-y-Rita (eds), Basic mecanism of ocular mobility and their clinical implications, Pergamon Press, Oxford and New-York, pp. 519-524.

Becker, W. and Jürgens, R. (1979) 'An analysis of the saccadic system by means of double step stimuli', Vision Research, 19, 967-983.

Bizzi, E.(1975) 'Motor coordination : central and peripheral control during eye-head movement', in M.S. Gazzaniga and C. Blakemore (eds), Handbook of Psychology, Academic Press, New York, pp.427-437. .

Carpenter, R.H.S. (1988) 'Toward a synthesis in movements of the eyes', in R.H.S. Carpenter (ed). Pion, London, pp. 343-373.

De Jong, R., , Coles, M.G.H., Logan, G.D. and Gratton, G. (1990) 'In search of the point of no return : The control of response processes', Journal of Experimental Psychology, 16, 164-182.

Favilla, M., Gordon, J., Hening, W. and Ghez, C. (1990) 'Trajectory control in targeted force impulses. VII. Independant setting of amplitude and direction in response prepartion', Experimental Brain Research, 79, 530-538.

Feldman, A.G. (1966) 'Functional tuning of the nervous system during control of movement of a steady posture. II. Controllable parameters of the muscle', Biophysics, 11, 566-578.

Fisk, J.D. and Goodale, M.A. (1985) 'The organization of eye and limb movements during unrestricted reaching to target in contralateral and ipsilataral visual space', Experimental Brain Research, 60, 159-178.

Fitts, P.M. (1954) 'The information capacity of the human motor system in controlling the amplitude of movements', Journal of Experimental Psychology, 47, 381-391.

Georgopoulos, A.P., Kalaska, J.F. and Massey, J.T. (1981) 'Spatial trajectories and reaction times of aimed movements : effects of practice, uncertainty, and change in target location', Journal of Neurophysiology, 46, 725-743.

Gielan, C.C.A.M., van der Heuvel, P.J. and Denier van der Gon, J.J. (1984) 'Control of amplitude of fast goal directed arm movements to double step target displacements', Journal of Motor Behaviour, 16, 2-19..

Gisbergen, J.A.M. van, Opstal, A.J. van and Roebroek, J.G.H. (1987) 'Stimuli-induced midflight modification of saccade trajectories', in J.K. O'Reagan and A. Levy-Schoen (eds), Eye movements : From physiology to cognition, Elsevier North-Holland, Amsterdam.

Goodale, M.A., Pélisson, D. and Prablanc, C. (1986) 'Large adjustements in visually guided reaching do not depend on vision of the hand or perception of target displacement', Nature, 320, 748-750.

Guitton, D. and Volle, M. (1987) 'Gaze control in human : Eye-head coordination during orienting movements to targets within and beyond the oculomotor range', Journal of Neurophysiology, 58, 427-459.

Jordan, M.I. (1990) 'Motor learning and the degree of freedom problem', in M.Jeannerod.(Ed.) Attention and Performance XIII, Motor representation and control, Laurence Erlbaum, Hillsdale, New Jersey, pp. 796-838.

Laurutis, V.P. and Robinson, D.A. (1986) 'The vestibulo-ocular reflex during human saccadic eye movements', Journal of Physiology (London), 373, 209-233.

Massone, L. and Bizzi, E. (1989) 'A neural network model for limb trajectory formation', Biological Cybernetics, 61, 417-425.

Mather, J.A. and Fisk, J.D. (1985) 'Orienting to targets by loocking and pointing : Parallels and interactions in ocular and manual performance', Quaterly Journal of Experimental Psychology, 37A, 315-338.

Morasso, P., Bizzi, E. and Dichgans, J. (1973) 'Ajustment of saccade characteristics during head movements', Experimental Brain Research, 16, 492-500.

Paulignan, Y., Mac Kenzie, C., Marteniuk, R. and Jeannerod, M. (1990) 'The coupling of arm and finger movements during prehension', Experimental Brain Research, 79, 431-435.

Pélisson, D., Prablanc, C., Goodale, M.A. and Jeannerod, M. (1986) 'Visual control of reaching movements without vision of the limb : II Evidence for a fast unconscious process correcting the trajectory of the hand to the final position of a target step', Experimental Brain Research,62, 303-311.

Prablanc, C. and Pélisson, D.(1990) 'Gaze saccade orienting and hand pointing are locked to their goal by quick internal loops', In M. Jeannerod (ed), Attention and Performance XIII, Motor representation and control,. Laurence Erlbaum, Hillsdale, New Jersey, pp. 653-676.

Prablanc, C., Echallier , J.F. and Masse, D. (1978) 'Error correcting mechanisms in large saccades', Vision Research, 18, 557-560.

Prablanc, C., Echallier, J.F.,Komilis, E. and Jeannerod, M. (1979) 'Optimal response of eye and hand motor systems in pointing at a visual target : I Spatio-temporal characteristics of eye and hand movements and their relationships when varying the amount of visual information', Biological Cybernetics, 35, 113-124

Prablanc, C., Pélisson, D. and Goodale, M.A.(1986) 'Visual control of reaching movements without vision of limb. I. Role of retinal feedback of target position in guiding the hand', Experimental Brain Research, 62, 293-302.

Robinson, D.A. (1975) 'Oculo-motor signals', in G. Lennerstrand and P. Bach-y-Rita (Eds) Basic mecanism of ocular mobility and their clinical implications. Pergamon Press, Oxford and New-York, pp. 337-374 .

Soechting, J.F. and Lacquaniti, F. (1983) 'Modification of trajectory of a pointing movement in response to a change in tar-get location', Journal of Neurophysiology, 49, 548-564.

Sonderen, J.F. van, Deniervan der Gon, J.J. and Gielen, C.C.A M (1988) 'Conditions determining early modification of motor programmes in responses to changes in target location', Experimental Brain Research, 71, 320-328.

Vicario, D.S. and Ghez, C. (1984) 'The control of the rapid limb movement in the cat. IV. Updating of ongoing isometric responses', Experimental Brain Research, 55, 134-144.

Wheeless, L., Jr., Boynton, R. and Cohen, G. (1966) 'Eye movement responses to step and pulse step stimuli', Journal of Optical Socity of America, 56, 956-960.

"DELAYED VOR" :
AN ASSESSMENT OF VESTIBULAR MEMORY FOR SELF MOTION

I.Israël[1], S.Rivaud[2], C.Pierrot-Deseilligny[2] & A.Berthoz[1]

1: CNRS - Laboratoire de Physiologie Neurosensorielle
15, rue de l'Ecole de Médecine - 75006 PARIS
2: INSERM - U. 289 - Hôpital de la Salpêtrière
47, Bd. de l'Hôpital - 75013 PARIS

ABSTRACT. This study investigates the ability of humans to use the vestibular system in "path integration" and to memorize the displacement of the head. Human subjects were submitted, in complete darkness, to passive angular whole-body displacements about a vertical axis, inducing a vestibular stimulation. They were presented, before the rotation, a stationary visual target. During the rotation, they had to fixate a target which moved with the head. After a memorization delay of either less than 20 sec, about 1 min, or 5 min, after the end of the rotation, they were required to make an eye saccade toward the "memorized" stationary target. The results show that the amplitude of an angular displacement detected by the vestibular system can be estimated and correctly memorized during 1 min. At 5 min, however, rotation angles below 25 deg are overestimated.

1. Introduction

Recently, quantitative studies about the reproduction, with eye movements, of passive horizontal whole-body displacements in darkness, have been performed with humans by Bloomberg et al. (1988 and in press) for angular displacements, and by Israël & Berthoz (1989) for linear displacements. The results showed that humans can correctly match the amplitude of a preceding memorized head displacement with a voluntary ocular saccade of equal but opposing amplitude. These authors not only showed that the amplitude of a body transport in darkness is correctly estimated by the brain, but also that this information can be stored and adequately retrieved. This performance is possible only if the vestibular system is intact (Israël & Berthoz, 1989).

Besides, it has recently been suggested (Berthoz, 1989; Collewijn, 1989; Bloomberg et al., in press) that the VOR is not a simple vestibular-induced reflex, but proceeds from an internal reconstruction of a "reference" target. These hypotheses are supported by the recent finding by Ventre & Faugier-Grimaud (1988) of cortical afferences on the vestibular nuclei. Segal & Katsarkas (1988) have also suggested that the VOR is goal-directed. A further support for this centrally controlled VOR would be the storage and retrieval of VOR function : the "delayed VOR".

Bloomberg et al. (1988) have invented this paradigm of delayed VOR, but with a short memorization delay (about 2 sec). In the experiments described in the present paper, performed with humans, we have studied in more details the duration of this storage of vestibular information.

J. Requin and G. E. Stelmach (eds.), Tutorials in Motor Neuroscience, 599–607.

2. Methods

The subject sat on a chair which could be manually rotated about the vertical axis, in complete darkness, with the head fixed to the chair. Horizontal eye position was measured by Electro-Oculography (EOG), and head position with a potentiometer. The Earth-Fixed Target (EFT) was a red spot projected on a vertical screen, straight ahead of the subject at the initial position. The Chair-Fixed Target (CFT) was a small faint red LED fixed to a L-shaped bracket attached to the chair. The vertical position of both targets was adjusted at subject's eye level.

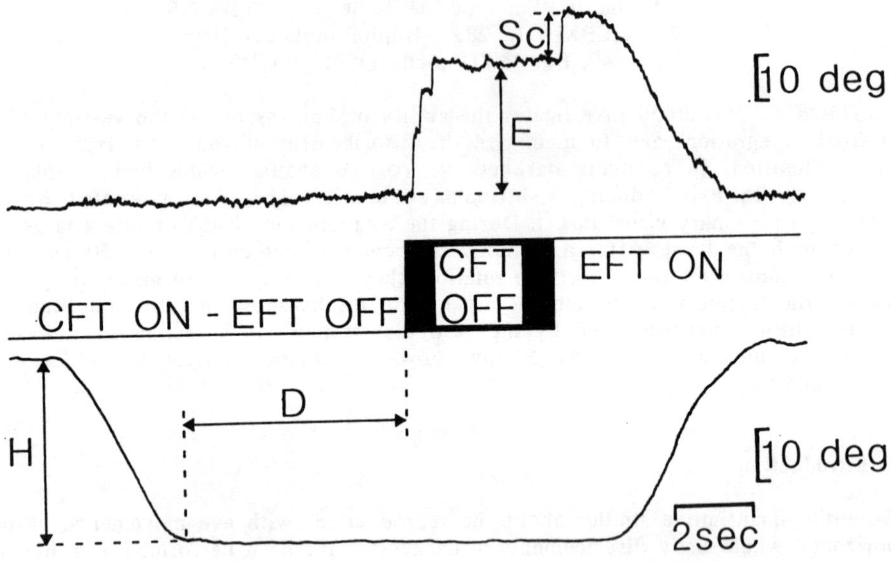

Fig. 1 : A trial of the "delayed VOR" paradigm. Upper trace : horizontal eye position. Middle trace : targets ON/OFF timing signals. Lower trace : head position. H : head displacement amplitude. D : retention duration. E : eye movement from the CFT to the memorized EFT position. Sc : correction saccade from the memorized EFT position to the visual EFT position.

The experimental protocol was as follows (Fig. 1):

(1) The visual EFT was presented straight ahead to the subject during about 5 sec; the subject was required to fixate it.

(2) The EFT was turned off and replaced by the CFT, which was superimposed to the EFT location when the chair was at the initial position. Then, the chair was manually rotated rightwards or leftwards, with a non-predictible angular amplitude H, of 5, 10, 20, 30, 40 or 50 deg. During the rotation, the subject was required to keep his eyes on the CFT (VOR suppression).

(3) After the rotation, when the chair had stopped moving, the subject was asked to keep maintaining his eyes on the CFT during a varying memorization delay. This delay duration D was either shorter than 20 sec, or about 1 min or 5 min.

(4) After this delay, the CFT was turned off. This was the signal for the subject to move his eyes (E) to the memorized EFT position previously shown, and to keep his eyes on that position, in complete darkness.

(5) The EFT was turned on again, about 2 sec after the CFT had been turned off. The subject made, if necessary, a corrective saccade (Sc) to foveate the now visible EFT.

(6) The chair was then rotated back to its initial position, with the subject still foveating the visual EFT.

3. Results

Mean chair rotation duration was 1.97 ± 1.02 sec (n = 150). During this rather short and slow rotation, when subjects fixated the CFT, the eyes stayed in the primary position and the VOR was completely suppressed.

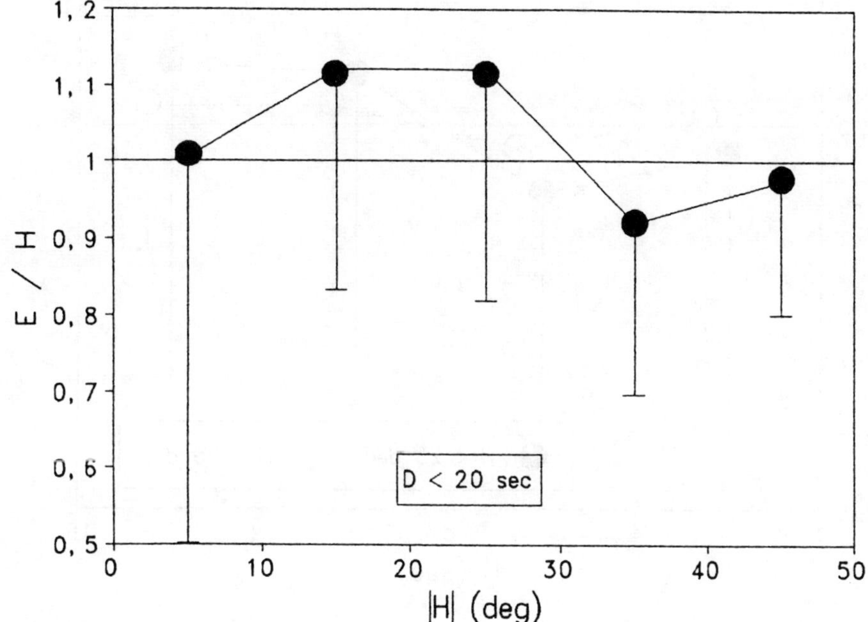

Fig. 2 : Amplitude ratio with respect to head displacement amplitude, for D shorter than 20 sec. Circles show the average of at least 20 data points, and vertical lines are the standard deviation.

All subjects experienced a very strong illusion of persistent target rotation when looking at the CFT, during 1 to 2 sec after chair rotation end. This apparent target motion is probably due to the oculogyral illusion (Graybiel & Hupp, 1946).

After the CFT was switched off, subjects shifted their eyes to the presumed EFT location with one single saccade, or more often with a primary saccade followed by one or two smaller ones in the same direction. They reached their final gaze destination, at E deg from the CFT (Fig. 1), with only one saccade in about 40% of the trials, and made one primary hypermetric saccade, followed by smaller corrective ones in the opposite direction, in less than 10% of the trials. A strong centripetal drift between those saccades was often observed. During the short moment (about 2 sec) when the CFT was off, and the EFT not yet on, the eyes could also drift, mostly in the centripetal direction. When the EFT was switched on again, subjects usually made one corrective saccade (Sc, Fig. 1) toward it.

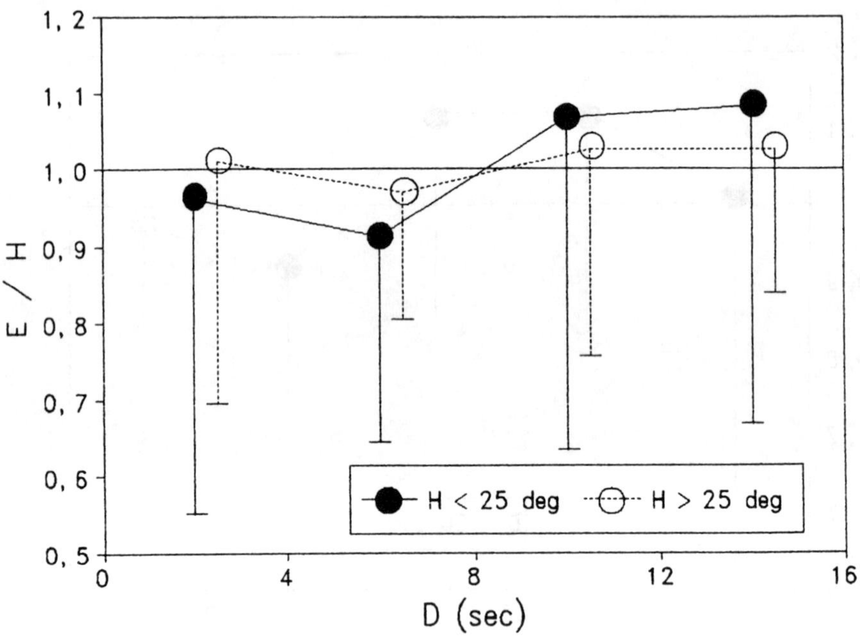

Fig. 3 : Amplitude ratio with respect to the retention duration, for D shorter than 20 sec. Same notation as in Fig. 2, with each circle showing the average of at least 10 data points.

We first investigated subjects' responses with respect to the different stimuli : head rotation amplitude H. For the shorter memorization delays (D shorter than 20 sec), the average amplitude ratio of eye rotation (E) to head rotation (H) was quite close to the ideal unity ratio, throughout the whole range of angular displacements H tested (Fig. 2).

However, subjects slightly overshot the EFT when H was smaller than 25 deg, and undershot it when H was larger than 25 deg, revealing the existence of a "range effect" (Kapoula, 1985) in this task. This can also be seen on Fig. 4, where the regression line for D = 4 sec is below the ideal diagonal for the higher H amplitudes.

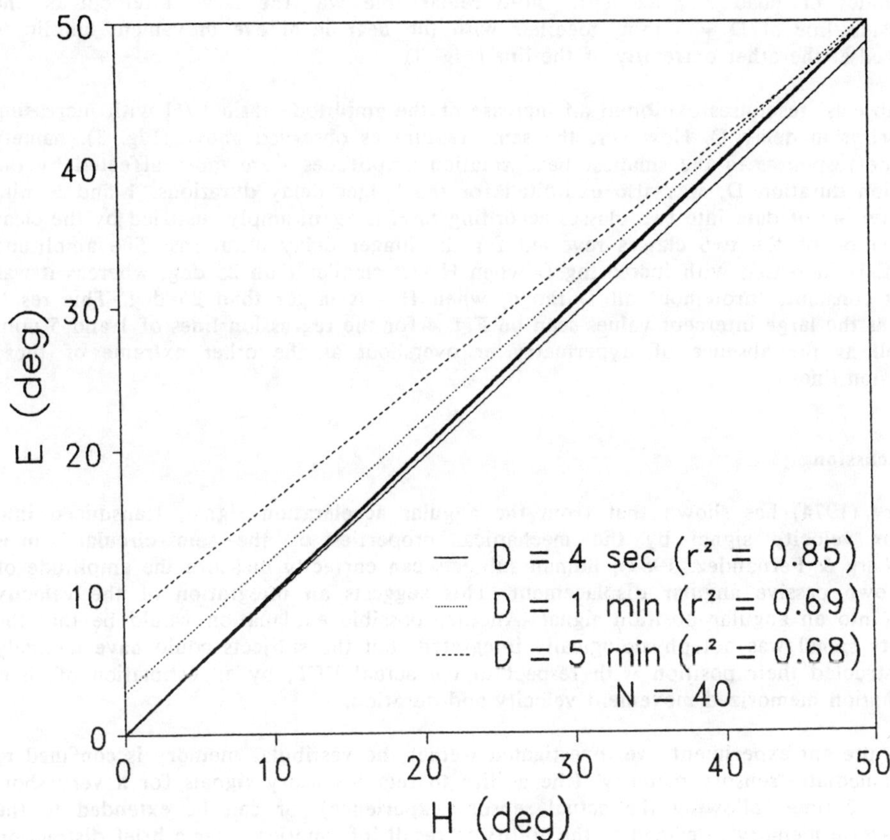

Fig. 4 : regression lines between E (response) and H (stimulus), for the three different D domains used in this experiment. The thick line is the ideal responses line.

We then studied wether there was a change in the performances at retention durations around 8-10 sec, as several experiments have revealed that this duration was a critical delay for human spatial memory (Thomson, 1980; Kasai, 1987; Israël and Berthoz, 1989). Because of the difference observed above (Fig. 2) between subjects' responses to head displacements smaller and larger than about 25 deg, experimental data have been divided into two classes, corresponding to H smaller than 25 deg, and H larger than 25 deg. It can be seen on Fig. 3 that the amplitude ratio E/H slightly increases when D increased from below 8 sec to above 8 sec, for H smaller than 25 deg, whereas it almost does not change when H was larger than 25 deg.

For the longer memorization delays (1 and 5 min), the tendency for overshoot, seen above when H was lower than 25 deg, was increased and extended up to the highest amplitudes of head displacement. Quite remarkable was the large intercept of the regression line at D = 5 min, together with the near ideal eye movement amplitude observed at the other extremity of the line (Fig. 4).

All subjects' responses exhibited an increase of the amplitude ratio E/H with increasing memorization delay D. However, the same feature as observed above (Fig. 3), namely that the responses to the smallest head rotation amplitudes were more affected by the retention duration D, was also exhibited for the longer delay durations, 1 and 5 min. The division of data into two classes according to H is again amply justified by the clear dissociation of the two classes revealed for the longer delay durations. The amplitude ratio E/H increased with increasing D when H was smaller than 25 deg, whereas it was almost constant, throughout all D range, when H was larger than 25 deg. This result explains the large intercept values seen on Fig. 4 for the regression lines of 1 and 5 min, as well as the absence of hypermetry or overshoot at the other extreme of those regression lines.

4. Discussion

Guedry (1974) has shown that from the angular acceleration signal, transduced into angular velocity signal by the mechanical properties of the semi-circular canals (Goldberg & Fernandez, 1984), human subjects can correctly measure the amplitude of their own passive angular displacement. This suggests an integration of the velocity signal into an angular position signal. Another possible explanation would be that the velocity signal was not physiologically integrated, but the subjects could have mentally reconstructed their position with respect to the actual EFT, by an estimation of their self-motion memorized movement velocity and duration.

In the present experiment, we investigated wether the vestibular memory is confined to the immediate "sensory memory" (the ability to retain sensory signals for a very short period of time following the actual sensory experience), or can be extended to the "short-term memory", defined as the ability to recall information after a brief distraction or delay, often several minutes (Kritchevsky, 1988).

Our data, as well as Bloomberg et al.'s (1988), first show that this "delayed VOR", does not use the same neural pathway as the usual VOR in the dark : subjects made saccades to reach the memorized EFT. This cooperative aspect of the saccadic system and the VOR has been previously shown by Segal & Katsarkas (1988). In addition, the role of the saccadic system in the substitution of other deficient gaze control subsystems has been suggested by Berthoz (1985).

Moreover, when considering the goal-directed functional significance of the VOR, Segal & Katsarkas (1988) found a VOR amplitude gain (E/H), when slow phases and saccades were combined, very close to unity : 1.01 ± 0.026. Hence, the performances are as good when the VOR is delayed by less than 20 sec, where the amplitude ratio is 1.03 ± 0.36, as when it is not delayed. This result supports the suggestions of Berthoz (1989), Collewijn (1989), and Bloomberg et al. (in press), namely that the VOR gain is under control of cerebral (cortical) representations of the vestibularly perceived self-rotation and of target spatial localization.

Our data show that vestibular information on body passive displacement can be stored without significant distortion during 1 minute, hence in the short-term memory. This cannot be explained by the time constant of the semi-circular canals, which is about 5 sec (Goldberg & Fernandez, 1984). The storage of the displacement amplitude estimation into a spatial-memory buffer is probably involved. This suggests a fundamental and powerful contribution to short range "navigation" in immediate extrapersonal space and path integration (Beritoff, 1965; Mittelstaedt, 1985). When the memorization time reaches 5 minutes, however, the estimation of small displacements amplitude is biased in the memorization process.

Interestingly, in an experiment of Hansen and Skavenski (1977), the paradigm was similar, but without the CFT fixated by the subject during rotation, i.e. whithout VOR suppression. The initial target to be memorized (EFT) was not straight-ahead before the rotation, but at different positions on the horizontal axis. The aim of the authors was actually to passively displace the eye from its initial position in the orbit. The average error was about 2 deg, and the interpretation offered was that the oculomotor system receives an accurate signal of the position of the eye during slow eye movements; the error was attributed to a weak visuo-spatial memory (with 5 sec memorization delay). The error in our experiment, even when D is shorter than 20 sec, is far larger than 2 deg. It is then quite probable that the movement of the eye was indeed memorized in Hansen and Skavenski's experiment, and added to the pure vestibular memory tested in the present experiment.

We thank M. Ehrette and M. Loiron for their contribution to the installation of the experimental set-up. I.Israël was supported by the "Centre National d'Etudes Spatiales", France.

REFERENCES

Beritoff JS (1965) Spatial Orientation in Man. In: Liberson WT (ed) Neural Mechanisms of higher vertebrate behavior. Little, Brown and Company, Boston, pp 137-157

Berthoz A (1985) Adaptive mechanisms in eye-head coordination. In: Berthoz A, Melvill Jones G (eds) Adaptative Mechanisms in Gaze Control. Elsevier, Amsterdam, pp 177-201

Berthoz A (1989) Coopération et substitution entre le système saccadique et les réflexes d'origine vestibulaires: faut-il réviser la notion de réflexe?. *Rev Neurol Paris* 145: 513-526

Bloomberg J, Melvill Jones G, Segal BN, McFarlane S, Soul J (1988) Vestibular-contingent voluntary saccades based on cognitive estimates of remembered vestibular information. In: Pirodda E, Pompeiano O (eds) Physiology of the vestibular system. Adv. Oto-Rhino-Laryngol. Vol. 40. Karger, Basel, pp 71-75

Bloomberg J, Melvill Jones G, Segal BN (1990) Adaptive modification of vestibularly perceived self-rotation. *Exp Brain Res* (in press)

Collewijn H (1989) The vestibulo-ocular reflex: an outdated concept ?. In: Allum JHJ, Hulliger M (eds) Afferent control of posture and locomotion. Elsevier, Amsterdam, pp 197-209

Goldberg JM, Fernandez C (1984) The vestibular system. In: Darian-Smith I (ed) Handbook of physiology - The nervous system III. American Physiological Society, Bethesda, pp 916-977

Graybiel AM, Hupp DL (1946) The oculogyral illusion: a form of apparent motion which may be observed following stimulation of the semicircular canals. *J Aviat Med* 3: 1-12

Guedry FE (1974) Psychophysics of vestibular sensation. In: Kornhuber HH (ed) Handbook of sensory physiology, Vol. VI/2 :Vestibular system, Part 2 :Psychophysics, applied aspects and general interpretations. Springer, Berlin. Heidelberg. New York, pp 3-154

Hansen RM, Skavenski AA (1977) Accuracy of eye position information for motor control. *Vision Res* 17: 919-926

Israël I, Berthoz A (1989) Contribution of the otoliths to the calculation of linear displacement. *J Neurophysiol* 62(1): 247-263

Kapoula Z (1985) Evidence for a range effect in the saccadic system. *Vision Res* 25(8): 1155-1157

Kasai T (1987) Compensatory eye movement for linear head motion. *Neuroscience (IBRO Abstracts)* 22, suppl.731: (Abstract)

Kritchevsky M (1988) Elementary spatial functions of the brain. In: Stiles-Davis J, Kritchevsky M, Bellugi U (eds) Spatial cognition. Brain bases and development. Lawrence Erlbaum Assoc., Hillsdale, New Jersey, pp 111-140

Mittelstaedt H (1985) Analytical cybernetics in spider navigation. In: Barth FG (ed) Neurobiology of arachnids. Springer Verlag, Berlin, Heidelberg, New York, Tokyo, pp 298-316

Segal BN, Katsarkas A (1988) Goal-directed vestibulo-ocular function in man: gaze stabilization by slow-phase and saccadic eye movements. *Exp Brain Res* 70: 26-32

Thomson JA (1980) How do we use visual information to control locomotion?. *TINS* 280: 247-250

Ventre J, Faugier-Grimaud S (1988) Projections of the temporo-parietal cortex on vestibular complex in the macaque monkey (Macaca fascicularis). *Exp Brain Res* 72: 653-658

Young LR (1984) Perception of the body in space : mechanisms. In: Darian-Smith I (ed) Handbook of Physiology - The nervous system III. American Physiological Society, Bethesda, pp 978-1023

SECTION 11

NEURAL PLASTICITY IN MOTOR SYSTEMS

FURTHER CONSIDERATIONS ON THE CELLULAR MECHANISMS OF NEURONAL PLASTICITY

L. KERKERIAN-LE GOFF, A. DASZUTA and A. NIEOULLON
Neurochemistry Unit
CNRS Laboratory of Functional Neuroscience
31, chemin Joseph Aiguier
13402 Marseille cedex 9
France

ABSTRACT. Recent evidences indicate that functional recovery after brain injury in adults is dependent on extreme adaptive capacities of mature neuronal circuits. Neuroplasticity at the cellular level would involve three sets of mechanisms referring to structural, metabolic and genomic adaptive changes. Most of studies have examined the effects of elimination of afferent pathways to particular brain regions and more recently the effects of intracerebral transplantations. For a long time, neuroplasticity then appeared as a compensatory structural or functional reorganization of neuronal networks triggered by brain damage. However, the concept has recently emerged that the adaptive capacities of the adult brain at cellular level are also expressed in the absence of denervation, and would represent a basic mechanism for learning and memory processes as well as for behavioural adjustements to permanent changes in the external context.

Introduction

Despite the dramatic consequences of brain injury in human neuropathology, data from experimental animal models however have indicated that functional deficits triggered by acute neuronal lesion can gradually rcceed and sometimes disappear. In addition, it has been shown that major cellular loss can be induced in certain brain areas without any apparent deficit occuring in the gross behaviour. This capacity to compensate or recover from brain damage in adults has been initially suggested to be mainly dependent on the development of new behavioural strategies. However, recent advances pointed to the fact that neuroplasticity may also actually contribute to functional recovery.

The adaptive capacities of the adult central nervous system (CNS) have been long underestimated. That may be primarily due to the theorical view that neuroplasticity necessarily needs anatomical supports and that the capacities of anatomical restructuration of mature neuronal circuits are limited (KENNARD, 1940), in view of the high degree of differentiation of mature neurons and of the loss of cell mitosis and axon regeneration properties. Within the last twenty years, however, physiological, biochemical and anatomical evidences were provided that extensive, although local,

J. Requin and G. E. Stelmach (eds.), Tutorials in Motor Neuroscience, 611–624.

anatomical restructurations can occur in the adult brain (for review see FLOHR, 1988). In addition, owing to the development of neurochemical and molecular biology approaches, postlesional adaptive changes in neuronal activity at both metabolic and genomic levels were evidenced, leading to the emergence of the concept of functional plasticity, as opposed to the structural one. In the current view of neuroplasticity in the adult CNS, three sets of mechanisms referring to structural, metabolic and genomic adaptations can be thus considered. They may occur separately or more likely together in response to brain injuries. Evidences were provided, from experimental animal models, that adaptive mechanisms may fully compensate for cellular losses or dysfunctions in the course of certain neurological diseases, so that deficits do not appear gradually but suddenly when these mechanisms are overwhelmed. Then paradoxally, the efficiency of the adaptive processes may have contributed to provide a view of the adult brain as a rigid unadaptable structure.

CNS adaptive capacities are not solely expressed in response to brain lesion but also in normal conditions to allow behavioural adjustments to continuous changes in the individual external environment conditions and internal state. According to recent concepts, the brain organization is viewed as a superimposition of two types of neuronal systems acting in a complementary manner. On the one hand, neuronal networks composed by highly organized systems would process the basic information, as for example the motor responses triggered by primary sensory information. These systems are termed "processing systems". On the other hand, superimposed neuronal systems with more diffuse anatomical organization, termed "modulatory or regulatory systems", would act to adjust the basic information processing as a function of motivational, emotional or more cognitive information they process. These latter systems are mainly represented by the monoaminergic pathways and are known to actually influence the transmission of information in the processing networks by means of presynaptic control mechanisms and of postsynaptic interaction processes at target neurons level. Within this model CNS adaptability is then represented in large part by the dynamic interactions occuring between the two different types of systems at synaptic level. According to this conception it can be assumed that damage to the processing systems would result in marked functional deficits, whereas selective damage to the neuromodulatory systems would alter the fine functional adjustements, then rigidify the behaviours, without suppressing the global function. Finally it is worth noting that plasticity is not synonymous with functional recovery potency. Indeed, if neuronal adaptive responses to brain damage have been evidenced in a few cases to actually result in functional recovery, in other cases they have been shown to reinforce the primary deficit, and even to lead to the appearence of secondary deficits contributing to the syndrom triggered by the lesion.

I - Brain restructuration after lesion.

The most commun view of structural plasticity in the adult CNS is the axonal collateral sprouting and the subsequent reactive synaptogenesis developing in response to selective deafferentation in the target structures (see COTMAN and NADLER, 1978; TSUKAHARA and MURAKAMI, 1983). Sprouting of this kind differs from the axonal sprouting observed after axotomy at peripheral or central levels, since it does not involve a regeneration process of the damaged axons but changes in the terminal field of spared axons. Data from numerous studies have suggested that different types of structural reorganization of neuronal networks can occur depending on the deafferentation extent.

In case of partial lesion of a given pathway, the remaining axons of the same pathway in the neighbourhood of the synaptic sites vacated by the lesioned afferents can develop sprouts, and provide the deafferented target neurons with new functional connections. The density of the neoformed terminals is supposed to be regulated by the target cells so that the reinnervation, from the functional point of view, would be close to normal. From various studies it appeared that this process termed "homotypic axonal collateral sprouting" may primarily concern the monoaminergic and cholinergic pathways and illustrative examples were provided by studies conducted at septal and hippocampal levels (BJORKLUND and STENEVI, 1979; GAGE et al., 1983). However, the homotypic sprouting is not exclusive for the monoaminergic systems. One of the most illustrative example for that was provided by a study in the rat cerebellar cortex on the innervation of Purkinje cells by the climbing fibers, presumably using aspartate as neurotransmitter (ROSSI et al., 1989). After subtotal lesion of the inferior olive at the origin of these fibers, the surviving climbing fibers were shown to emit new collateral branches and form new functional synapses on the deafferented Purkinje cells. Therefore, this reorganization process totally modified the normally occuring "point to point" climbing fiber innervation of the Purkinje cells.

The reactive sprouting and synaptogenesis phenomenon do not limit itself to a reorganization of the systems directly involved in the lesion. "Heterotypic" reaction can occur at two levels in neuronal network. In response to the total lesion of a given input, axon terminals of another pathway may reoccupy the vacated sites and form aberrant, but functional, connections with the deafferented target neurons. Reorganizations of this kind involve neuronal systems the terminal fields of which overlap that of the destroyed afferents, and can consist in an actual sprouting process but also in a rerouting of part of the terminals from their initial position towards the deafferented sites. Moreover, it appeared that the denervated target neurons themselves can exhibit axonal sprouting and develop new connections in the deafferented structure.

An illustration of both type of heterotypic reactive synaptogenesis has been provided by studies performed on the red nucleus. In normal conditions red nucleus neurons are contacted at the level of their soma and proximal dendrites by cerebellar afferents from the contralateral nucleus interpositus, and at the level of their distal dendrites by ipsilateral cortical afferents. After lesion-induced disappearcnce of the interposito-rubral fibers, electrophysiological evidences have been provided that axon terminals of corticorubral fibers have reoccupied the vacated sites on the proximal dendrites and soma of red nucleus neurons (see TSUKAHARA, 1981). This observation has been first interpreted as a sprouting of cortico-rubral axons. However, this hypothesis did not fit with biochemical data obtained in the same model, suggesting that the total number of cortical terminals did not increase in the red nucleus in contrast to what would be expected in case of sprouting (see NIEOULLON et al., 1988). Further morphological studies have shown that reorganization of the cortical input in fact consisted of a shift of part of the corticorubral synapses from their initial distal position towards the deafferented sites at proximal level (MURAKAMI et al., 1982). For the neoformed connections to be functional, one has to assume that, parallely to the presynaptic rearrangement of corticorubral terminals, there may be postsynaptic rearrangements of the membrane receptors present on red nucleus neurons, and particularly of receptors sensitive to the neurotransmitter released by the cortical terminals. Again at postsynatic level, we have provided morphological evidences that GABA-containing interneurons of the red nucleus may display axonal sprouting in response to similar cerebellar deafferentation (see NIEOULLON et al., 1988).

Conversely, the structural reorganization may also concern an input deprived from its natural target. This remodelling may involve retrograde degeneration processes in particular when targets normally provide the presynaptic axons with trophic factors and/or a rerouting of nerve terminals towards new targets. Taken all together these data illustrate the complexity of the anatomical restructurations triggered by selective brain lesions in adults, and further raise the view that functional changes at metabolic and/or genomic level may accompany such structural reorganization.

Interestingly, induction of synaptic vacancy is not necessary for axonal sprouting and subsequent generation of new synapses to occur. In that case, brain restructuration would result from polysynaptic changes in neuronal activity. For example, it has been long conjectured that synaptic remodelling may be a neuronal mechanism underlying basically learning and memory processes (HEBB, 1949). A conclusive demonstration for the occurence of restructurations triggered by changes in neuronal network activity was recently provided by KELLER and coworkers (1990). Unilateral lesion of the deep cerebellar nuclei which do not directly project to the cerebral cortex has been shown to result in a sprouting of corticocortical projections from the somatosensory cortex to the motor cortex, accompanying the disappearence of the motor deficits induced by the cerebellar lesion. Subsequent lesion of the somatosensory cortex in animals with the cerebellar lesion induced the reappearance of the cerebellar syndrom, which provided direct evidence that remodelling of corticocortical connections actually underlied the functional recovery.

Apparent reorganization of neuronal networks might also result from the activation of "silent neurons" the existence of which was postulated by WALL many years ago (1976; MERRILL and WALL, 1978). For example it has been reported that the number of substantia nigra neurons at the origin of crossed projections to the contralateral striatum, detected using horseradish peroxydase retrograde transport, considerably increase after selective lesion of the substantia nigra ipsilateral to the considered striatal side (PRITZEL et al., 1983). Similar increase in the crossed nigrostriatal projections have been evidenced to occur after unilateral sensory deprivation (HUSTON, 1990). The hypothesis that certain "silent" neurons in the substantia nigra have been activated and have newly get the ability to take up and transport retrograde markers injected in the contralateral striatum appears more likely than that of the creation of new connections between the two structures.

II - Adaptive metabolic cellular responses to brain lesions.

From the biochemical point of view, neuronal activity is expressed in terms of neurotransmitter biosynthesis, release and inactivation at presynaptic level and in terms of receptor activation at postsynaptic level. Metabolic adaptations in response to brain lesions are then manifested by alterations in these synaptic mechanisms contributing to functional adjustements of neuronal circuits activity. As reported before concerning the structural reorganization, the adaptive responses to selective deafferentation can be manifested by the remaining intact fibers of the damaged pathway in case of partial lesion, by another input to the deafferented structure or by the deafferented target neurons themselves. This form of plasticity may be underlied by dynamic changes in various mechanisms normally involved in the regulation of neuronal transmission, such as autoregulatory processes, presynaptic interaction processes and postsynaptic receptor-receptor interactions. In view of the multiple pre- and postsynaptic mechanisms linking the activity of different neuronal systems present in a given structure, it appears likely that selective deafferentation may induce a cascade of reactions of the remaining intact systems, including neuronal elements which are not directly under the control of the damaged input in normal conditions. Metabolic adaptations in response to brain damage can also accompany structural reorganizations. For example, the model of cerebellar deafferentation of the red nucleus still provided evidences for such an adaptive response. Indeed, we have shown that an increase in the synaptic transmission of corticorubral glutamatergic fibers is associated to the rearrangement of these fibers terminal field (see NIEOULLON 1984; NIEOULLON et al., 1988).

One of the best illustration of the dynamic adaptive capacities of various neuronal elements in response to deafferentation was provided by studying the effects of lesion of the dopaminergic input to the striatum originating in the substantia nigra. The degeneration of this nigrostriatal dopaminergic pathway is basically at the origin of the parkinsonian syndrom and can be reproduced in animals by the use of 6-hydroxydopamine, a neurotoxin selective for dopamine neurons. Partial lesions of the nigrostriatal dopaminergic system have been shown to primarily induce adaptations in the activity of striatal dopaminergic elements. At presynaptic level, an increase in the synthesis and release of dopamine from the spared nigrostriatal axons was shown to occur (AGID et al., 1973). This increased turnover may result in part from the alteration of inhibitory autoregulatory processes and apparently compensate for the deafferentation until about 20% of the dopaminergic neurons remain present in the substantia nigra. For more extensive lesions, pharmacological studies in animals have shown that in addition to the presynaptic activation of the remaining intact fibers, postsynaptic dopaminergic receptor supersensitivity develops (UNGERSTEDT, 1971). Binding studies have further revealed that this response consists of an increase in the number (CREESE and SNYDER, 1979; SAVASTA et al., 1988) and affinity of the dopaminergic receptors (FEUERSTEIN et al., 1981, FENTON et al., 1985; HERVE et al., 1988) compensating for the decrease in the quantity of dopamine released untill about 10% of the neurotransmitter remains present in the striatum. Then, the behavioural syndrom corresponding to nigrostriatal dopaminergic deafferentation in animals and the clinical expression of Parkinson's disease may appear only when these compensatory processes are overwhelmed, that is for lesions involving about 90% of the dopaminergic neuronal population (ZIGMOND et al., 1990).

Extensive dopaminergic deafferentation has been reported to also result in adaptive changes in the activity of striatal intrinsic neurons. In particular, cholinergic interneurons were shown for a long time to present an hyperactivity response (see SCATTON, 1982) which may contribute to the expression of Parkinson's disease syndrom since anticholinergic compounds have potent antiparkinsonian properties. In addition, it has been reported that dopaminergic deafferentation can result in a presynaptic activation of one of the other main inputs to the striatum, that is the corticostriatal glutamatergic pathway (KERKERIAN et al., 1987; LINDEFORS and UNGERSTEDT, 1990). We have obtained evidence that this heterotypic hyperactivity response may be underlied by alterations in the reciprocal presynaptic interactions normally linking nigrostriatal dopaminergic and corticostriatal glutamatergic transmissions (see NIEOULLON et al., 1983). Since the corticostriatal pathway is activatory, we have hypothetized that, besides the loss of dopamine inhibitory influence, the hyperactivity of the corticostriatal activatory input on striatal neurons may contribute to the expression of Parkinson's disease. Finally, postlesional adaptive changes at target neurons level can also involve alterations in receptor-receptor interactions, contributing in normal conditions to closely associate neighbouring synapse activities. Data have been obtained indicating that activation of glutamate receptors of the NMDA subtype results in a decrease in dopamine postsynaptic action through D1 receptor by favouring the dephosphorylation of the DARPP-32 protein. This effect is suppressed by NMDA receptor antagonist. Therefore it has been suggested that pharmacological blockade of the NMDA receptors would induce a facilitation of dopamine action in the striatum (see GIRAULT et al., 1990). Conversely, it has been reported that dopamine receptor activation may decrease the synaptic efficiency of glutamate. In this respect, we provided evidence that nigrostriatal dopaminergic deafferentation results in an increased number of NMDA receptors (SAMUEL et al., 1990). It can be then assumed that reduced dopamine receptor activation subsequent to dopaminergic deafferentation may increase the synaptic action of glutamate, which in turn would reinforce the decreased dopaminergic transmission. Taken altogether these data indicate that dopaminergic deafferentation induces an increase in glutamatergic transmission at both pre- and post-synaptic levels which rises the view that antiglutamatergic compounds may be potent antiparkinsonian agents.

These already complex adaptive responses to brain damage are probably complicated by the frequent coexistence of different neuroactive substances in the same neuronal system. In most of the cases of neurotransmitter colocalisation, neurons present an association between a classical neurotransmitter (acetylcholine, monoamines, excitatory aminoacids and GABA) with a neuroactive peptide. Synaptic transmission in these neurons is complicated by the fact that the two neurotransmitters can interact with each other in different ways : the one being able to facilitate or inhibit the release of the other one, or being able to reduce or potentiate the postsynaptic action of the other neurotransmitter. Considering for example the case of the septal cholinergic innervation of the hippocampus, it has been shown that acetylcholine is colocalized with the neuropeptide galanin. From the functional point of view acetylcholine is released in basal conditions and acts as the main neurotransmitter of septohippocampal fibers, whereas galanin is released only in case of hyperactivity of septohippocampal fibers and acts to decrease acetylcholine release (FISONE et al., 1987). From these data it has been suggested that, in case of partial lesion of this pathway, a reactive hyperactivity of the remaining intact fibers (similar to that reported concerning the nigrostriatal dopaminergic afferents) would lead to galanin release, which could totally block acetylcholine release and then reinforce the decrease in cholinergic function at hippocampal level. This functional hypothesis has been developed in an attempt to explain the high vulnerability of this cholinergic input with regards to other cholinergic systems in experimentally induced or pathological neurodegenerative processes. Interestingly, comparison of data on nigrostriatal dopaminergic and septohippocampal cholinergic transmission indicate that the same adaptive process in one case would contribute to compensate for the deficit and in the other case would reinforce the primary deficit.

III - Genomic adaptive responses to brain damage.

The observations of receptor supersensitivity response and of the formation of new functional synapses from heterotypic axonal sprouting raised the idea that adaptation of receptor gene expression may be triggered by brain damage. Further observations have indicated that adaptations are not restricted to receptor gene expression but extend to classical neurotransmitter synthesizing enzymes, such as tyrosine hydroxylase (see BLACK et al., 1987), neuropeptide precursors (see LINDEFORS et al., 1990) and other kinds of proteins. For example, reactive synaptogenesis can be said to reflect an alteration of gene expression of proteins constituting the cytoskeleton. Adaptations at genomic level do not only consist of modifications (increases or decreases) of normally expressed gene activities, but can consist in activation of normally non expressed gene, or in suppression of a given protein gene expression. With regards to the mechanisms presumably underlying these adaptations, it was shown that intracellular events triggered by membranar receptor activation, such as specific phosphorylation processes, may influence protein gene expression. Although the mechanisms involved in this control remains to be clarified it thus appeared that postlesional adaptive changes in gene expression may be triggered by modifications in membrane receptor activity.

Considering again a model of selective deafferentation at least three types of adaptive changes in neurotransmitter gene expression (that is changes in classical neurotransmitter synthesizing enzyme or neuropeptide precursor gene expression) can be considered. First, increases or decreases in the gene expression of the neurotransmitters used by the target neurons are commonly observed. Second, actual change in the target neurons phenotype can occur, leading to the activation of normally non expressed neurotransmitter gene or to the suppression of a normally expressed neurotransmitter gene activity in these neurons. Finally deafferentation can result in changes in the phenotype of another input to the denervated structure. The most illustrative example of such changes in afferent input phenotype was provided by immunocytochemical studies performed at the level of the iris (BJORKLUND et al., 1985). The iris receives both sympathetic afferents using noradrenaline and neuropeptide Y (NPY) as cotransmitters and parasympathetic cholinergic afferents. Lesion of the superior cervical ganglion at the origin of the sympathetic input has been shown to initially result in the disappearence of the fibers containing tyrosine hydroxylase (TH), the catecholamine synthesizing enzyme, and NPY in the iris. Later on, the number of TH/NPY-immunoreactive fibers gradually increased overtime and at one month post-lesion a dense plexus of these terminals reappeared. Further removal of the parasympathetic ciliary ganglion at the origin of the iris cholinergic innervation led to the disappearence of those TH/NPY-labeled fibers, suggesting that the two markers were present in the parasympathetic fibers. These data provided strong evidence that in response to sympathetic deafferentation, mature parasympathetic neurons are capable of expressing sympathetic activity. As a matter of fact the biochemical environment of the neurons can influence their gene activity. Another example is provided by in vitro experiments on cultured sympathetic neurons showing that the cells express either noradrenergic or cholinergic activity, depending on the culture medium (see HENDRY, 1985). The question remains to determine whether or not the newly expressed potentialities by mature neurons correspond to the derepression of immature neuronal characteristics.

With regards to the sensorimotor structures, the striatal model once again provided several illustrations for genomic plasticity. In that way, after nigrostriatal dopaminergic deafferentation, besides the increase in dopaminergic receptor expression, increase or decrease in expression of neuroactive peptides related to various intrinsic neuronal populations have been reported to occur (BANNON et al., 1986; NORMAND et al., 1988; YOUNG et al., 1986). In addition, we have contributed to provide evidence for change in the phenotype of certain striatal intrinsic neurons. This finding has emerged from an immunohistochemical study of NPY-containing neurons in the rat striatum (KERKERIAN et al.,1988; 1990). NPY is to be found in a population of medium-sized aspiny neurons that may represent interneurons. From an animal to another, the number of striatal NPY-immunoreactive neurons is stable so that we have been able to determine the density and the topographical distribution of these neurons in basal conditions on striatal sections. Interestingly, we have shown that unilateral lesion of the nigrostriatal dopaminergic pathway results in an increase in the number of NPY-immunodetectable cells in the ipsilateral striatum as compared to control values. A detailed topographical analysis revealed that increases in the density of NPY neurons are restricted to the medioventral striatal areas. One interpretation of these results, in view of the limits of detection of the immunohistochemical methods, was that neurons containing NPY, but in too low levels to be visualized in basal conditions, become detectable after deafferentation because of an increase in intraneuronal NPY levels. An alternative proposal was that in response to dopaminergic denervation certain striatal neurons normally non expressing NPY have newly get the capacity to produce NPY, which suggest the involvement of genomic derepression processes. Recent data from in situ hybridization studies have conforted this last proposal by showing that similar denervations result in an increase in the number of striatal neurons producing messenger RNAs encoding NPY precursor (LINDEFORS et al., 1990). These finding indicated that lesion of the nigrostriatal dopaminergic pathways resulted in an activation of NPY gene expression in many non or low expressing neurons, which suggested at first sight that dopamine basally inhibit NPY expression in a topographically defined striatal neuronal population. To verify this hypothesis we have examined the effects of non lesional impairment of striatal dopaminergic transmission, by blocking dopamine production or dopamine receptors, on striatal NPY immunostaining. Surprisingly these treatments did not reproduce the lesion effects. This data indicated that the activation of NPY gene expression in response to nigrostriatal dopaminergic lesion did not result from the single suppression of dopamine influence on striatal target neurons. Complementary experiments led us show that NPY response to dopaminergic deafferentation was related to the previously reported development of dopamine D1 receptor supersensitivity response (HERVE et al., 1988) and was dependent on the integrity of corticostriatal fibers. Further data suggested that adaptive changes in NPY expression may result from modifications in the dynamic interaction between dopaminergic and glutamatergic receptors at striatal target neuron level.

IV - Intracerebral transplantations : another approach to brain plasticity.

Neural grafting has emerged over the last 15 years, as a viable approach to study the development and the regeneration of neuronal connections in the CNS of mammals. This approach appeared as a useful tool to investigate the mechanisms of embryonic neuronal plasticity, by looking for example at the reinnervation induced by the grafted cells, which has to resemble that of the lacking natural innervation as closely as possible. Indeed, fetal serotonergic neurons transplanted to one of their previously denervated target, the hippocampus, exhibit near to normal ultrastructural characteristics, and even normal morphological relational features with the host hippocampal population (DASZUTA et al., 1989). This suggest that the grafted neurons may be able to express a genomic program defining the characteristics of the hippocampal serotonergic innervation, and/or that target cells can contribute to regulating the reinnervation.

As regards to the restoration of function, transplant-assisted recovery in motor systems have been one of the first demonstrations of the functional effects of grafting fetal brain tissue on the adult host animal's behaviour. Indeed, grafts of fetal dopamine-rich mesencephalic tissue implanted into the dopamine-depleted striatum of rodents can improve many of the lesion-induced motor impairments (see BRUNDIN et al, 1987). In this model, transplanted dopaminergic neurons develop axonal growth and formation of synaptic contacts with the host striatal neurons. They have been shown to restore dopamine levels and dopamine release in the denervated striatum, and the degree of functional recovery was correlated with the recovery of striatal dopamine level in the grafted animals. Other experiments also indicate that transplanted neurons developed in this ectopic position are able to receive informations from the host (DOUCET et al., 1989). These data altogether, suggest a high degree of integration of the grafted embryonic cells to the host tissue. Then this model illustrates the opportunity provided by intracerebral transplantations to analyze both the capacities of adaptation of developing fetal cells, and the reactivity of mature host elements to the graft. For instance, the most successful reinnervation of the target tissue (the striatum) was obtained after a previous specific dopamine denervation of the host (DOUCET et al., 1990). The reinnervation processes occuring following neural grafting can be basically compared with homotypic sprouting and reactive synaptogenesis described above, where target neurons may play an essential role in promoting the reafferentation. The functional recovery induced by the transplant may involve the restoration of functional cellular interactions between the graft-derived terminals and the mature host neurons. Indeed, partial reinnervation of the rat striatum by transplanted dopaminergic neurons was found to be associated with changes in the reactivity of striatal peptidergic populations (DASZUTA et al., 1990). For example, the hyperactivity of NPY containing neurons in response to the dopaminergic deafferentation was found to be reversed in the grafted striata. The effect was rapidly visible and generalized to the entire striatum despite its only partial dopaminergic reinnervation. Added to the re-establishment of TH-NPY synaptic contacts, these data suggest that both dopaminergic synaptic transmission and dopamine diffuse release may contribute to this normalization. It is therefore worth considering that cellular events going considerably beyond the limits of actual structural brain reinnervation may underly the graft-related functional restorations.

Conclusion

In conclusion, numerous observations of structural reorganizations and of metabolic and genomic adaptations triggered by brain lesions are now available showing the high degree of plasticity of mature neuronal circuits. However, in most of the cases, the precise mechanisms underlying these adaptive changes, their significance and their contribution to functional recovery or to deficits remains to be determined. An interesting possibility emerging from the data reported above is that the mature brain may be still able to express disused embryonic properties to control nerve proliferation, synapse formation and stabilisation, and neuronal phenotype determination. In this respect, it is worth mentioning that glial cells may play a key role in the remodeling of brain circuits, since they represent a major source of neurotrophic factors, and since their activity is presumably controlled by neurotransmitter receptors present on their membrane. Another view is that the mature brain may present "silent" neuronal systems that may be selectively activated in response to changes in the biochemical environment of the cells. For example, expression of additional neurotransmitter by a given biochemically characterized neuronal system in response to brain injury can be assume to provide this system with new functional specificities. One might therefore wonder whether apparently adaptive changes in behavioural strategy after lesions might not be underlied by profound changes in the organization and activity in neuronal networks. The last point we wish to stress, in view of the complexity of the normally occuring cellular interaction processes and of the adaptive responses triggered by selective deafferentation, is that one can no more consider the functional deficits related to neurodegenerative diseases as resulting from the simple direct interruption of a given neuronal transmission. Behavioural impairments may likely reflect a new functional state of deprived neuronal systems which is to be taken into consideration in therapeutic approaches of human neuropathology.

Acknowledgements

The authors are grateful to K. Mattei for kindly preparing the manuscript.

References

AGID Y., JAVOY F. and GLOWINSKI J. (1973) Hyperactivity of remaining dopaminergic neurons after partial destruction of the nigrostriatal dopaminergic system in the rat. Nature 245, 150-151.

BANNON M.J., LEE J.M., GIRAUD P., YOUNG A., AFFOLTER H.U. and BONNER T.I. (1986) Dopamine antagonist haloperidol decreases substance P, substance K and preprotachykinin mRNAs in rat striatonigral neurons. J.Biol.Chem. 261, 6640-6642.

BJORKLUND A. and STENEVI U. (1979) Regeneration of monaminergic and cholinergic neurons in the mammalian central nervous system. Physiol.Rev. 59, 62-100.

BJORKLUND H., HOKFELT T., GOLDSTEIN M., TERENIUS L. and OLSON L. (1985) Appearance of the noradrenergic markers tyrosine hydroxylase and neuropeptide Y in cholinergic nerves of the iris following sympathectomy. J.Neurosci. 5, 1633-1643.

BLACK I.B., ADLER J.E. DREYFUS C.F., FRIEDMAN W.J., LAGAMMA E.F. and ROACH A.H. (1987) Experience, neurotransmitter plasticity and behavior. in:Psychopharmacology, ed. by H.Y. MELTZER, Raven Press, New York, 463-469.

BRUNDIN P., STRECKER R.E., LINDVALL O., ISACSON O., NILSSON O.G., BARBIN G., PROCHIANTZ A., FORNI C., NIEOULLON A., WIDNER H., GAGE F.H. and BJORKLUND A. (1987) Intracerebral grafting of dopamine neurons. Experimental basis for clinical trials in patients with Parkinson's disease. Ann.New York Acad.Sci., 495, 473-495.

COTMAN C.W. and NADLER J.V. (1978) Reactive synaptogenesis in the hippocampus. in: Neuronal Plasticity, ed. C.W. COTMAN, Raven Press, New York, pp.227-271.

CREESE I. and SNYDER S.H. (1979) Nigrostriatal lesions enhance striatal ^{3}H-apomorphine and ^{3}H-spiroperidol binding. Eur.J.Pharmacol., 56, 277-281.

DASZUTA A., BJORKLUND A., BRUNDIN P., CHAZAL G., DESCARRIES L., KALEN P. and STRECKER R. (1989) Morphological and functional analysis of raphe transplants in adult rat hippocampus. Restaur.Neurol.Neurosci. 1, 45-46.

DASZUTA A., MOUKHLES H. and VUILLET J. (1990) Morphological and functional analysis of cell interactions between graft-derived dopaminergic fibers and NPY striatal population in the rat. Europ.J.Neurosci., Suppl. 3, Abst. n°1343, p.79.

DOUCET G., BRUNDIN P., DESCARRIES L. and BJORKLUND A. (1990) Effect of prior dopamine denervation on survival and fiber outgrowth from intrastriatal fetal mesencephalic grafts. Eur.J.Neurosci., 2, 279-290.

DOUCET G., MURATA Y., BRUNDIN P., BOSLER O., MONS N., GEFFARD M., OUIMET C.C. and BJORKLUND A., (1989) Host afferent into intrastriatal transplants of fetal ventral mesencephalon. Exp.Neurol. 106, 1-19.

FENTON H.M., LESZCZAK E., GERHARDT S., LIEBMAN J.M. (1985) Evidence for heterogeneous rotational responsiveness to apomorphine, 3-PPP and SKF 38393 in 6-hydroxydopamine-denervated rats. Eur.J.Pharmacol., 106, 363-372.

FEUERSTEIN C., DEMENGE P., CARON P., BARRETTE G., GUERIN B., MOUCHET P. (1981) Supersensitivity time course of dopamine antagonist binding after nigrostriatal denervation : evidence for early and drastic changes in the rat corpus striatum. Brain Res., 226, 221-234.

FISONE G., WU C.F., CONSOLO S., NORDSTROM O., BRYNNE N., BARTFAI T., MELANDER T. and HOKFELT T. (1987) Galanin inhibits acetylcholine release in the ventral hippocampus of the rat : histochemical, autoradiographic, in vivo and in vitro studies. Proc.Natl.Acad.Sci. USA, 84, 7339-7343.

FLOHR H. (1988) Post-lesion neural plasticity. Springer, Berlin.

GAGE F.H., BJORKLUND A. and STENEVI U. (1983) Reinnervation of the partially deafferented hippocampus by compensatory collateral sprouting from spared cholinergic and noradrenergic afferents. Brain Res., 268, 27-37.

GIRAULT J.A., HALPAIN S. and GREENGARD P. (1990) Excitatory amino acid antagonists and Parkinson's disease. Trends in Neurosci. 13, 325-326.

HEBB D.O. (1949) The organization of behavior : a neurophysiological theory. John Wiley and sons. New York, pp.227-231.

HENDRY I.A. (1985) Phenotypic plasticity of sympathetic adrenergic and cholinergic neurons. Trends in Pharmacol.Sci. 165, 126-129.

HERVE D., TROVERO F., BLANC G., STUDLER J.M., THIERRY A.M., GLOWINSKI J. and TASSIN J.P. (1988) Heterologous regulation of dopaminergic D1 receptors in the rat striatum : involvement of cortico-subcortical relationships. Europ.J.Neurosci. Suppl.388, Abst. n°88-3.

HUSTON J.P. (1990) Neurobehavioral indices of recovery from unilateral sensory deprivation and lesion of the nigrostriatal system. Eur.J.Pharmacol. 183, 32.

KELLER A., ARISSIAN K. and ASANUMA H. (1990) Formation of new synapses in the motor cortex following lesions of the deep cerebellar nuclei. Exp.Brain Res. 80, 23-33.

KENNARD M. (1940) Relation of age to motor impairment in man and in subhuman primates. Arch.Neurol.Psychiatry 44, 377-397.

KERKERIAN L., DUSTICIER N. and NIEOULLON A. (1987) Modulatory effect of dopamine on high affinity glutamate uptake in the rat striatum. J.Neurochem. 48, 1301-1306.

KERKERIAN L., SALIN P. and NIEOULLON A. (1988) Pharmacological characterization of dopaminergic influence on expression of neuropeptide Y immunoreactivity by rat striatal neurons. Neuroscience 26, 809-817.

KERKERIAN L., SALIN P. and NIEOULLON A. (1990) Cortical regulation of striatal neuropeptide Y (NPY) containing neurons in the rat. Putative modulatory influence on striatal dopamine-NPY relationships. Eur.J.Neurosci. 2, 181-189.

LINDEFORS N., BRENE S., HERRERA-MARSCHITZ M. and PERSSON H. (1990) Neuropeptide gene expression in brain is differentially regulated by midbrain dopamine neurons. Exp.Brain Res. 80, 489-500.

LINDEFORS N. and UNGERSTEDT U. (1990) Bilateral regulation of glutamate tissue and extracellular levels in caudate-putamen by midbrain dopamine neurons. Neurosci.Lett., 115, 248-252.

MERILL E.G. and WALL P.D. (1978) Plasticity of connection in the adult nervous system. in: Neuronal Plasticity, C.W. COTMAN (ed.) Raven Press, New York, pp.97-111.

MURAKAMI F., KATSUMARU H., SAITO K. and TSUKAHARA N. (1982) A quantitative study of synaptic reorganization in red nucleus neurons after lesion of the nucleus interpositus of the cat : an electron microscopic study involving intracellular injection of horseradish peroxidase. Brain Res. 242, 41-53.

NIEOULLON A. (1984) Neuronal plasticity in the red nucleus and the ventrolateral thalamus of the adult cat : a biochemical approach. in: R.G. HASSLER and J.F. CHRIST (eds.) Advances in Neurology 40, Raven Press, New York, pp.107-116.

NIEOULLON A., KERKERIAN L. and DUSTICIER N. (1983) Presynaptic controls in the neostriatum : reciprocal interactions between the nigrostriatal dopaminergic neurons and the corticostriatal glutamatergic pathways. Exp.Brain Res., suppl.7, 54-65.

NIEOULLON A., VUILLON-CACCIUTTOLO G., DUSTICIER N., KERKERIAN L., ANDRE D. and BOSLER O. (1988) Putative neurotransmitters in the red nucleus and their involvement in postlesion adaptive mechanisms. Beh.Brain Res. 28, 163-174.

NORMAND E., POPOVICI T., ONTENIENTE B., FELLMANN D., PIATIER-TONNEAU D., AUFFRAY C. and BLOCH B. (1988) Dopaminergic neurons of the substantia nigra modulate preproenkephalin A gene expression in rat striatal neurons. Brain Res. 439, 39-46.

PRITZEL M., HUSTON J.P. and SARTER M. (1983) Behavioral and neuronal reorganization after unilateral substantia nigra lesions : evidence for increased interhemispheric nigro-caudate projections. Neuroscience 9, 879-888.

ROSSI F., WIKLUND L., VAN DER WANT J.J.L. and STRATA P. (1989) Collateral sprouting of climbing fibres following the subtotal lesion of the inferior olive in the adult rat. Europ.J.Neurosci., suppl.2, Abstr.90.4.

SAMUEL D., ERRAMI M. and NIEOULLON A. (1990) Localization of N-methyl-D-aspartate receptors in the rat striatum : effects of specific lesions on the ^{3}H-3-2 carboxypiperazin-4-yl-propyl-1-phosphonic acid binding. J.Neurochem. 54, 1926-1933.

SAVASTA M., DUBOIS A., BENAVIDES J. and SCATTON B. (1988) Different plasticity changes in D1 and D2 receptors in rat striatal subregions following impairment of dopaminergic transmission. Neurosci.Lett., 85, 119-124.

SCATTON B. (1982) Effect of dopamine agonists and neuroleptic agents on striatal acetylcholine transmission in the rat. Evidence against dopamine receptor multiplicity. J.Pharmacol.Exp.Ther. 220, 197-202.

TSUKAHARA N. (1981) Synaptic plasticity in the mammalian central nervous system. Ann.Rev.Neurosci. 4, 351-379.

TSUKAHARA N. and MURAKAMI F. (1983) Axonal sprouting and recovery of function after brain damage. In: Motor control mechanisms in health and disease. Edited by J.E. Desmedt. Raven Press, New York, pp.1073-1084.

UNGERSTEDT U. (1971) Postsynaptic supersensitivity after 6-hydroxydopamine induced degeneration of the nigrostriatal dopamine system. Acta Physiol.Scand. Suppl.367, 69-93.

WALL P.D. (1976) Plasticity in the adult mammalian central nervous system. Prog.Brain Res. 45, 359-379.

YOUNG S.W., BONNER T.I., BRANN M.R. (1986) Mesencephalic dopamine neurons regulated the expression of neuropeptide mRNAs in the rat forebrain. Proc.Natl.Acad.Sci., USA, 83, 9827-9831.

ZIGMOND M.J., ABERCROMBIE E., BERGER T.W., GRACE A.A. and STRICKER M. (1990) Compensations after lesions of central dopaminergic neurons : some clinical and basic implications. Trends in Neurosci. 13, 290-296.

LESION-INDUCED PLASTICITY OF THE PYRAMIDAL TRACT DURING DEVELOPMENT IN THE CAT

J. ARMAND, B. KABLY, H. JACOMY
Centre National de la Recherche Scientifique. Laboratoire de Neurobiologie
B.P. 71
13402 Marseilles Cedex 09
France

ABSTRACT. After the ablation of the sensorimotor cortex performed in cats of different ages, from birth to adulthood, the deficits and possible compensations of the forelimb movement have been correlated to the structural reorganization of corticospinal and cortico-subcortical projections from the intact hemisphere. The performances of different task-related movements indicated that in complex ones the goal could only be reached by neonatal operated animals. The performances of distal and proximal skills could only be recovered before respective critical dates of operation. The kinematic analysis of a target-reaching movement demonstrated that the motor strategy as observed in control animal was abolished, but could be replaced by a compensatory one in neonatal operated animals. In these neonatal operated animals, the anterograde transport of horseradish peroxidase injected into the intact pericruciate cortex established that new corticospinal and cortico-subcortical projections terminated in various structural substrates deprived of their innervation by the cortical lesion. The functional potentialities of this neonatal lesion-induced plasticity could accounted for the recovery of some partial functions such as motor skills, but not for the programming of the whole movement, they could allowed however the building up of compensatory motor strategies.

1. Introduction.

The search for correlations between the structural substrates and the behavioural functions of the nervous system has been exemplified at the beginning of this century by the involvement of the pyramidal tract in motor function. Its cells of origin were identified at the cortical level as the giant Betz cells (Holmes and May (1909)), and at the same time cytoarchitectonically characterized (Campbell (1905)) the excitable motor cortex (Grünbaum and Sherrington (1901)). This close relationship between structure and function however underestimated the other cortical areas at the origin of this pathway and its various corticospinal and cortico-subcortical components.

According to retrograde and anterograde transport findings, the corticospinal tract in cat has 2 components : a somatosensory and a motor one (Armand and Aurenty (1977), Armand and Kuypers (1980), Armand (1982), Armand et al. (1985)). The primary and secondary somatosensory cortices, respectively composed of areas 3 a and b, 2, 1, and 2 prae-insularis, are at the origin of only crossed corticospinal fibers terminating in the dorsal horn. The motor cortex, composed of areas 4 and 6, is at the origin of crossed and uncrossed corticospinal fibers which have 3 patterns of termination in the intermediate zone, but not in the motoneuronal cell groups. Some only crossed fibers terminate in the

J. Requin and G. E. Stelmach (eds.), Tutorials in Motor Neuroscience, 625–640.

dorsolateral part of the intermediate zone of the cervical or lumbosacral enlargement. Some crossed and uncrossed fibers also terminate to each of the spinal enlargement, but contralaterally in the dorsolateral part of the intermediate zone and bilaterally in its ventromedial part. Finally, some other crossed and uncrossed fibers terminate in the same bilateral regions of the spinal gray, but throughout the spinal cord.

Corticospinal fibers can also terminate at supraspinal levels by means of axonal collaterals in addition to specific cortico-subcortical projections following pyramidal tract fibers (Phillips and Porter (1977)). The sensorimotor cortex thus projects to cells of origin of other descending pathways, such as the red nucleus, to relay structures of ascending pathways, such as the dorsal column nuclei and the ventrobasal complex of the thalamus, and to groups of cells involved in cortico-cerebello-cortical loops, such as the pontine nuclei and the lateral reticular nucleus.

The structural substrate of the pyramidal tract, with its corticospinal and cortico-subcortical components thus includes several functional potentialities : the gating of sensory inputs at their different levels, the control of motor outputs to distal, proximal, and axial muscles through supraspinal structures and spinal interneurones, the input-output coupling within internal loops.

The pyramidal tract function has been extensively studied on the basis of the behavioural deficits observed after interruption of its fibers at different levels of their trajectory (for review, see Wiesendanger, 1981). The results obtained varied with the level of the lesion, that is with the affected components of the pyramidal tract. All these results however emphasized the functional role of the pyramidal tract in the control of distal extremity muscles. Most of these studies have been conducted in adult animals, some of them however were done in infant ones, and concluded at a better functional recovery in the latter than in the former. This infant-lesion effect (Kennard (1936, 1938, 1940, 1942)) has been recently reinvestigated in cat after hemispherectomy (Villablanca et al. (1986), Burgess and Villablanca (1986), Burgess et al. (1986)) or sensorimotor cortical ablation (Leonard and Goldberger (1987a)).Such study of the better functional recovery in animals operated earlier requires that the behavioural analysis is adjusted to the reorganized structural substrate. The correlation between the structural organization of the pyramidal tract and its functional potentialities can thus be searched also for a lesion-induced reorganization of this pathway.

For this purpose, sensorimotor cortical ablation was realized at different postnatal ages and at adulthood in order to destroy one complete pyramidal tract. At adult age, the behavioural effects of this lesion have been analysed using a forelimb movement. This analysis has favoured motor skills and the execution of the movement because of the partial and integrated functional potentialities of the pyramidal tract. Finally, in these animals operated at different ages, an eventual reorganization of the intact pyramidal tract has been searched at both spinal and supraspinal levels. In such a way we hope to answer the questions, whether a specific post-lesion induced plasticity of the pyramidal tract exists during development, and if so, whether it could be responsible of some behavioural recoveries.

2. Materials and Methods

2.1. SUBJECTS

The experiments have been carried out in 22 cats, 17 neonatal operates from 1 to 75 postnatal days (PND), 1 adult operate, and 4 control animals littermates of the neonatal operates.

2.2. CORTICAL ABLATION

In animals operated during the first postnatal month, the cortical ablation was made under Alfatesine anesthesia (1.2 ml/Kg, I.M.), and in animals operated later, it was done under Pentobarbital (30 mg/Kg, I.P.).

In all cases, the cortical ablation was made on the left hemisphere by suction under a binocular dissecting microscope, and involved the precruciate, postcruciate, lateral pericruciate regions, as well as the depth of the cruciate sulcus.

2.3. ANATOMICAL PROCEDURE

2.3.1. *Cortical Injection.* At adult age, a solution of free horseradish peroxidase (30%, HRP Boehringer) was injected in the right pericruciate cortex, in all operated animals and in control ones. Several penetrations of a glass micropipette were realized. For each penetration, 2 micro-injections (0.1μl) were made under and above cortical layer V, where the cells of origin of the pyramidal tract are located. A total amount of 4-5 μl of HRP was injected during a period of 2 hours.

2.3.2. *Perfusion.* After a survival time of 3 days, the animals were intracardially perfused under deep Pentobarbital anesthesia (60 mg/Kg, I.P.). The perfusion consisted of 2 l washing solution (NaCl 0.9%, 4 g procaïne, 4 ml heparine), followed by 4 l fixative solution (paraformaldehyde 1%, glutaraldehyde 1.25% in 0.2 M phosphate buffer) and 2 l postperfusion solution (sucrose 30%). The brain and spinal cord were removed and blocked.

2.3.3. *Histology.* In some cases, the lesioned and intact frontal lobes, and in the other cases only the lesioned one, were postfixed in formalin during several weeks, dehydrated, and embedded in paraffin. Parasagittal sections (10 μm thickness) were stained with Cresyl violet in order to reconstruct the extent of the lesion.

The other histological blocks were stored overnight in a buffer sucrose (30%) Transverse sections (40 μm thickness) were cut on a freezing microtome and reacted using a tetramethylbenzidine histochemical procedure (Mesulam (1982)).

In order to compare the density of labelling in control and operated animals, photomicrographs were taken with dark-field illumination and polarized light, the negatives of which were digitized. The density of labelling in different regions of the same section could thus be measured and compared.

2.4. BEHAVIOURAL ANALYSIS

The effects of sensorimotor cortical ablation on the forelimb movement have been studied at adult age, by using the four operant conditioning tests first designed by Sybirska and Gorska (1980). The animal had to retrieve pieces of raw meat, inaccessible with mouth, by performing a forelimb movement in two simple and two compex tasks (Figure 1). In order to test alternatively the two forelimbs, the inactive one was covered by a shoe-like cuff to prevent its use. In these different tasks, the context varied both by the posture of the animal and by the position of the target.

Among the simple tasks, in test I, the animal in a sitting position had to retrieve a piece

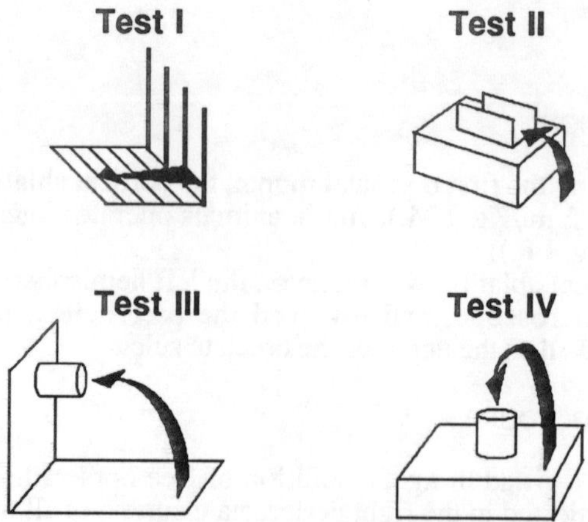

Figure 1. Schematic representation of the 4 experimental set-up. In order to retrieve a piece of meat, the cat had to perform a forelimb movement (arrow).

of meat placed on the ground, from behind vertical bars. In test II, the animal also in a sitting position had to retrieve food from a rectangular groove placed at its mouth level.

Among the complex tasks, in test III, sitting on its hindquarters, the animal had to retrieve food from inside a horizontal tube placed at its shoulder level. In test IV, the animal in quadrupedal position, its forelimb placed on a platform, had to retrieve food from inside a vertical tube.

2.4.1.*Experimental sessions*. During a trial, the execution of the movement could be subdivided into three phases (Figure 2). A reaching phase, during which the limb was lifted towards the target. A grasping phase, during which the meat was caught hold by combined plantar flexion, flexion of the claws, and adduction of the paw. A wrist rotation phase, during which the meat was brought to the mouth by means of wrist supination and flexion.

Figure 2. The 3 phases of the execution of the movement in test III. 1. reaching, 2. grasping, 3. wrist rotation.

The experimental sessions lasted for 32 weeks. There was 3 sessions of 5 trials each per week. So for each limb and each task, there was 15 trials per week.

2.4.2.*Behavioural data*. Three levels of analysis of the forelimb movement have been

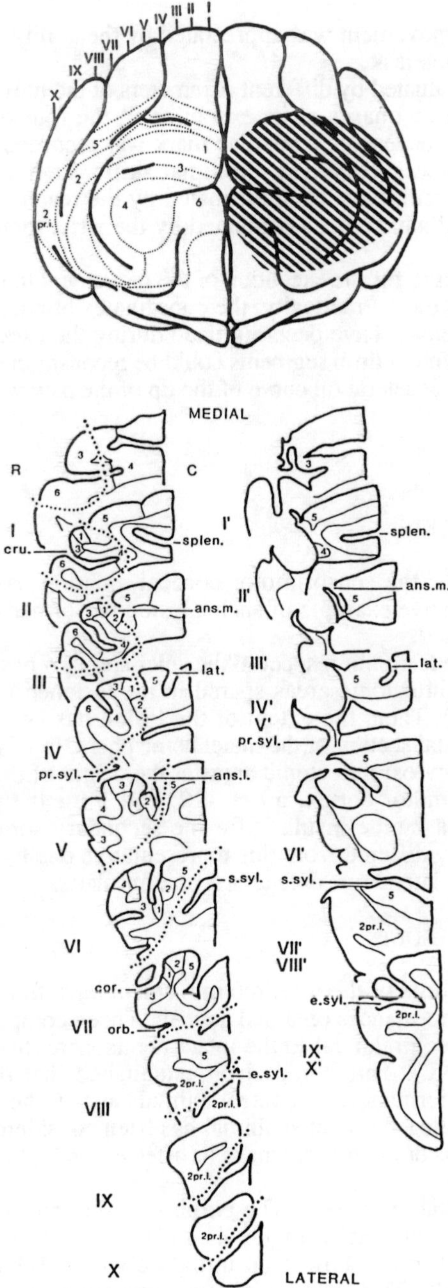

Figure 3. Reconstruction of the cortical ablation (hatched area). The extent of the lesion (left) is reported (dotted line) on each parasagittal section (I-X) on the intact hemisphere (right) in respect to the sulci and the cytoarchitectonic areas (arabic numbers).

distinguished.

The purpose of the movement was appreciated by the ability to retrieve food in 100% of trials per week, in each task.

Motor skills were evaluated by different parameters of the movement. These parameters were defined by the three phases of its execution in the four different tests. If a given phase was unique and successful, a positive mark was recorded, but if it was performed by several attempts, it was discarded. Proximal skills, such as the amplitude and the precision of the movement, have been estimated by the reaching phase in tests I and II respectively. Distal skills have been evaluated by the wrist rotation and the grasping in tests III and IV.

The motor strategy used for the execution of the movement in test III has been analysed by means of kinematic data. Practically, the coordinates of each joint, shoulder, elbow, wrist, and the tip of the paw have been digitized during the execution of the movement. The positions of the different limb segments could be reconstructed from the ground to the horizontal tube, and the acceleration curve of the tip of the paw was plotted.

3. Results

3.1. CORTICAL ABLATION

In all operated animals, the sensorimotor cortical ablation has been reconstructed at adulthood. A representative example of such a lesion performed in a 6-days-old kitten is shown in Figure 3.

The definition of the lesion in respect to the sulci and gyri has been complemented by identifying the cytoarchitectonic areas spared in the lesioned hemisphere (Hassler and Mühs-Clement (1964)). Then, the extent of the lesion has been reconstructed on each corresponding parasagittal section of the intact hemisphere. In this representative case, the lesion included all the cytoarchitectonic areas at the origin of the pyramidal tract in cat, areas 4 and 6 for the motor cortex, areas 3, 2, and 1 for the primary somatosensory cortex, and part of area 2 prae-insularis for the secondary somatosensory cortex. The other cases only slightly differed from this representative one by the peripheral extent of the lesion, especially its lateral one onto area 2 prae-insularis.

3.2. BEHAVIOURAL RESULTS

The three levels of analysis of the forelimb movement have first been applied to control animals. The motor performances obtained have then been compared in operated animals with those of the limb contralateral to the intact hemisphere, that we will refer to as the non-affected limb (NAL). This comparison established that there was no difference between the motor performances in control animals and in the NAL of operated ones. Afterwards, the NAL of each operated animal has been considered as an intra-individual reference for the analysis of the movement of its affected limb (AL).

3.2.1. *The Purpose of the Movement.* The purpose of the movement was appreciated in two simple and two complex tasks (Figure 4). In simple tasks, such as tests I and II, the performances of the AL only showed an initial deficit of 2-3 weeks in respect to the performances of the NAL. This short initial impairment was found in neonatal as well as adult operated animals. Thus, the AL performances in simple task-related movements quickly allowed all operated animals to achieve their purpose.

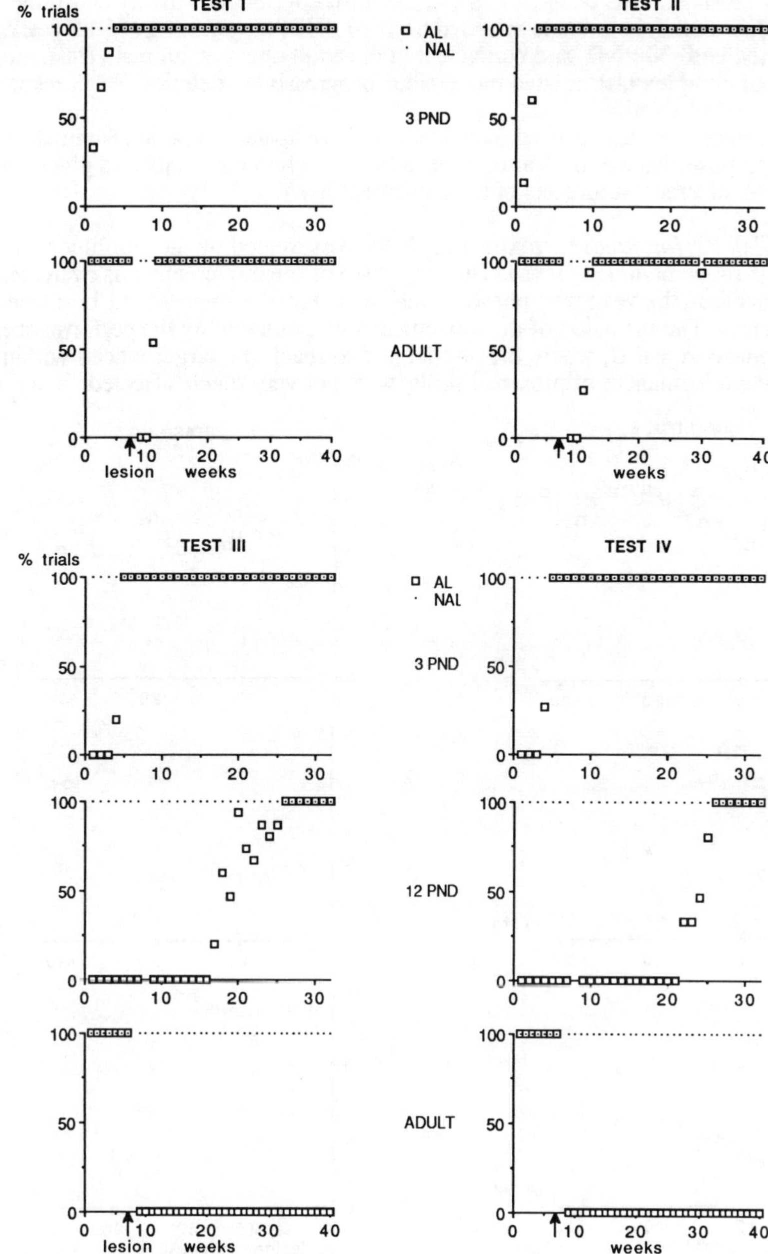

Figure 4. Simple and complex task-related movements. The performances of the affected limb (AL) and non-affected limb (NAL) were compared in simple tasks (tests I and II) and complex ones (tests III and IV) in animals operated at different postnatal days (PND) and at adulthood.

In complex tasks, such as tests III and IV, the initial deficit was always absolute, but short-lasting (3-4 weeks) in animals operated until 10 PND, long-lasting (15-25 weeks) in animals operated until 30 PND, and permanent in the adult operated animal. Thus, the AL performances of complex task-related movements progressively deteriorated in respect to the age of the cortical ablation.

These performances of task-related movements only indicated if operated animals could achieve their purpose. Further insight into the deficits or compensations was given by the performances of different parameters of the movement itself.

3.2.2. *Motor Skill Performances*. Proximal skills were estimated by the amplitude and the precision of the movement (Figure 5). The amplitude of the movement was evaluated by the performances of the reaching phase in test I, where the meat could be placed at different distances. The precision of the movement was evaluated by the performances of the reaching phase in test II, where the animal had to reach the target placed within the groove. These performances of proximal skills were not very much affected in animals

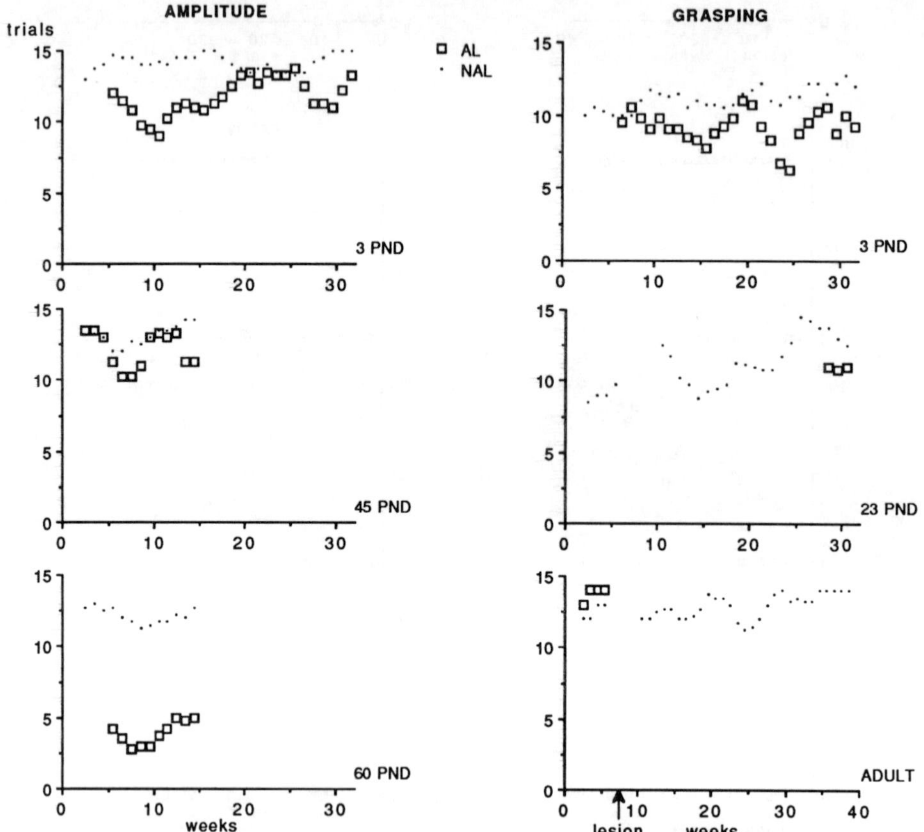

Figure 5. Performances of proximal and distal skills. The performances of the amplitude were compared in 3 neonatal operated animals (3, 45, 60 PND). The performances of the grasping were compared in 2 neonatal (3, 23 PND) and 1 adult operated animals.

operated before 60 PND, but in animals operated after their deficit was very pronounced.

Distal skills were evaluated by the performances of the wrist rotation and the grasping phases (Figure 5) in tests III and IV. These performances of distal skills could be compensated in animals operated before 30 PND, but in animals operated after their deficit remained absolute.

In respect to proximal or distal skills different critical dates for the sensorimotor cortex lesion could be determined : 60 PND was a critical date for proximal skills, and 30 PND for distal ones.

The AL performances of the animal operated at 23 PND (Figure 5) could not be evaluated during the first 28 weeks of tests, because these distal skills were not yet inbuilt in a purposeful movement. It is during this stage that the compensatory process occured in neonatal operated animals. This process built up new motor strategies, unsuccessful first, but progressively more adjusted to the task, to finally culminate in a compensatory strategy which could be successful.

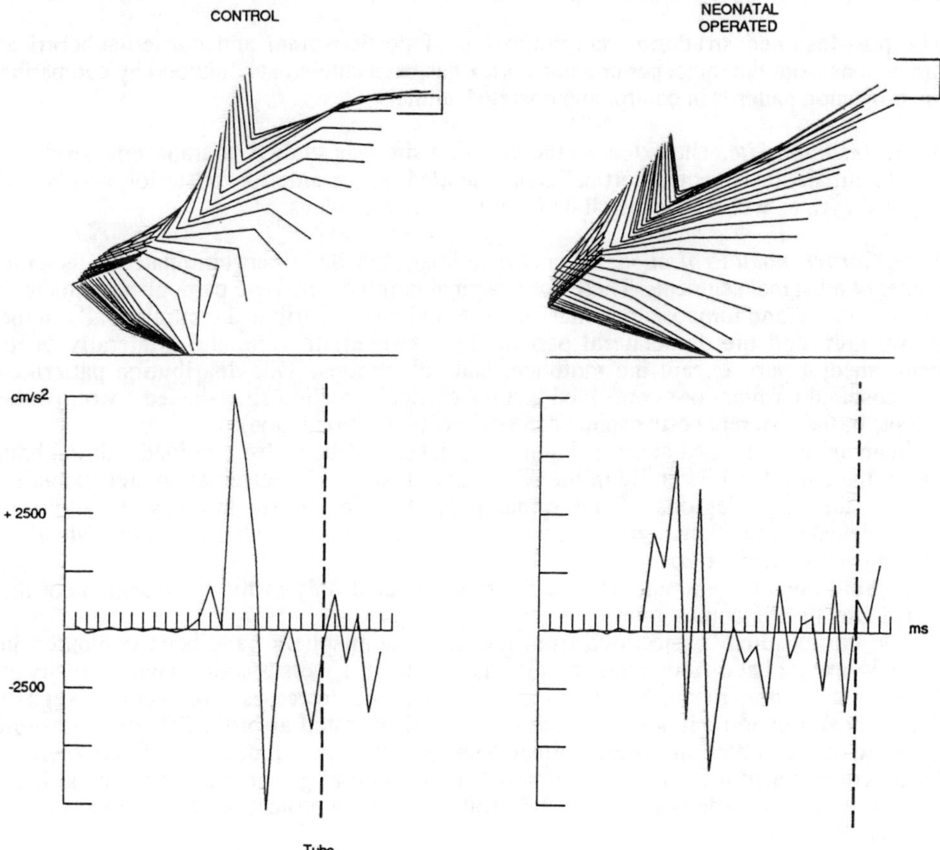

Figure 6. Execution of the movement in test III by a control and a neonatal operated animals. Top : stick diagram of the positions of the limb segments. Bottom : acceleration curves of the tip of the paw. Time base : 20 ms.

3.2.3. *Motor Strategy*. The description of the motor strategy has been made in test III by means of kinematic analysis (Figure 6).

In control animals, the acceleration curve of the tip of the paw was composed, from the ground to the tube level of a continuous movement with an acceleration and a deceleration,followed by a discontinuous movement within the tube with a succession of accelerations and decelerations. The ballistic movement of the reaching was followed by a terminal adjustement onto the target.

In neonatal operated animals, the reaching movement from the ground to the tube level was built up of consecutive accelerations and decelerations indicating that the movement could be permanently controlled by sensory feedbacks. This new strategy however could be successful.

In contrast, in adult operated animal the ballistic movement remained alone. This strategy was deficient.

3.3. ANATOMICAL RESULTS

The post-lesioned structural reorganization of corticospinal and cortico-subcortical projections from the intact pericruciate cortex has been studied at adulthood by comparing the projection patterns in control and operated animals.

3.3.1. *Injection Site*. The extent of the injection site was studied on transverse sections. In all animals, the injected cortical area extended on the anterior, posterior, and lateral sigmoid gyri, as well as the rostral depth of the cruciate sulcus.

3.3.2. *Corticospinal Projections*. In control animals, labelled fibers terminated in the gray matter of all spinal segments. These corticospinal terminations were particularly numerous in the cervical and lumbosacral enlargements, and were distributed contralaterally in the dorsal horn and the dorsolateral part of the intermediate zone and bilaterally in its ventromedial part, except the motoneuronal cell groups. This distribution pattern of corticospinal terminations confirmed that the cortical injection site labelled a worthwhile sample of the different corticospinal components (see Introduction).

In an animal operated at birth (Figure 7A), labelled fibers also terminated throughout the spinal cord, but bilaterally in the whole gray matter. The termination area extended thus in the same regions of the spinal gray than in control animals. In addition, corticospinal terminations were also found in the ipsilateral dorsal horn and dorsolateral part of the intermediate zone.

In adult operated animal, labelled fibers terminated only in the same regions of the spinal gray than in control ones.

The corticospinal projections from the intact hemisphere have been compared in animals operated at different ages on the basis of the ipsi- versus contra-lateral density of labelling in the gray matter of the segment T1. This ratio decreases in respect to the age of the cortical ablation : it was 78% for the animal operated at birth, 56% for the adult operated one, and 35% in control animal. Moreover this overall decrease did not seem to be concomittant in the different regions of the ipsilateral gray matter. The dorsal horn projections began to decrease at 15 PND, followed by the projections to the intermediate zone at 30 PND.

3.3.3. *Cortico-subcortical Projections*. In neonatal operated animals bilateral projections from the intact pericruciate cortex were also found to a great number of subcortical structures, that were only unilateral in control and adult operated animals. All of these

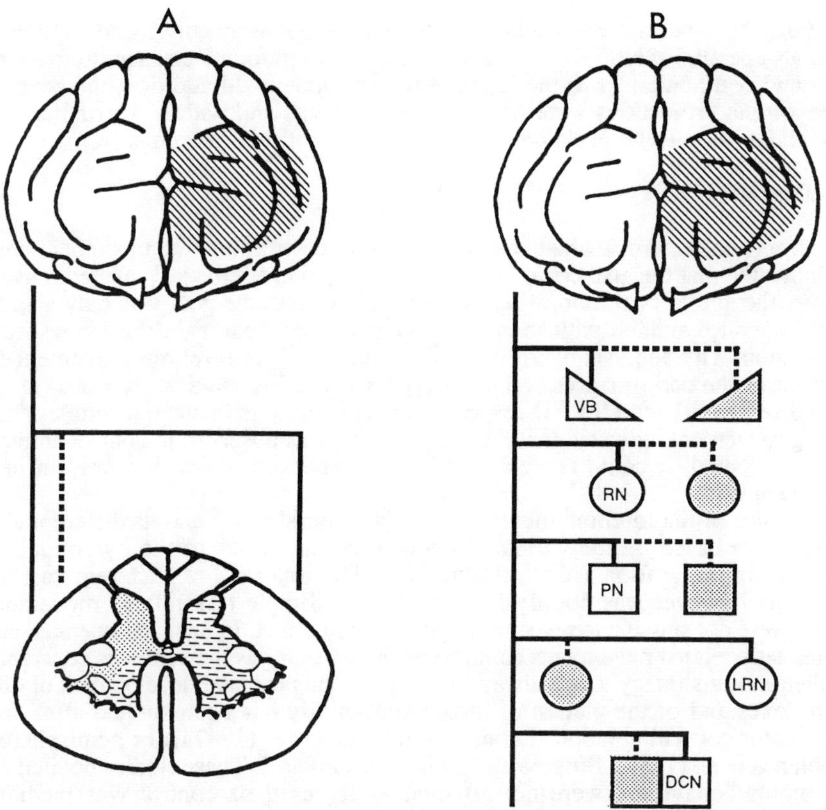

Figure 7. Reorganization of the corticospinal (A) and cortico-subcortical (B) projections in an animal operated at birth. The projections as seen in control animal (thick lines and horizontal lines area) are complemented by new projections (dotted lines and dotted areas).

structures had in common that they were deprived of their innervation by the sensorimotor cortical ablation.

Three examples of cortico-subcortical projections have been selected to be compared in control and neonatal operated animals, projections to cells of origin of other descending pathways, projections to relay structrures of ascending pathways, and projections to groups of cells involved in internal cortico-cerebello-cortical loops (Figure 7B).

In control animals, corticorubral projections were mainly found ipsilaterally. In animals operated at birth also, corticorubral projections were found ipsilaterally. In addition, new and numerous contralateral projections could be observed.

In control animals, cortical projections were found mainly contralaterally to the dorsal column nuclei and ipsilaterally to the ventrobasal complex of the thalamus. In animals operated at birth, corticocuneate and corticogracile projections were found bilaterally, as were found the cortical projections to the ventrobasal complex.

In control animals, cortical projections to the pontine nuclei were mainly found ipsilaterally, whereas those to the lateral reticular nucleus were only found contralaterally. In animals operated at birth also, cortical projections were found ipsilaterally to the pontine nuclei, and contralaterally to the lateral reticular nucleus. In addition, the contralateral corticopontine projections were much more numerous and widely distributed, and the ipsilateral lateral reticular nucleus received abundant cortical projections.

4. Discussion.

After sensorimotor cortical ablation in cat, the analysis of a forelimb movement presented here, indicated that the goal could be quickly achieved in simple task-related movements, whatever the age of operation, whereas in complex tasks the goal was only achieved in neonatal operated animals with an increasing duration of the initial deficit in respect to the age-at-lesion. The analysis of different parameters of the forelimb movement further indicated that the performances of distal and proximal skills could be recovered in animals operated before 30 and 60 PND respectively. The analysis of the execution of a target reaching movement indicated that the motor strategy as observed in control animals was always abolished, it could be replaced by a compensatory one but only in neonatal operated animals.

The choice of the forelimb movement as behavioural index and its different levels of analysis were based on the various functional potentialities of the pyramidal tract in sensorimotor integration (see Introduction). The analysis of different task-related movements however was not always meaningful. Simple task-related movements for instance were not suitable to reveal any age-at-lesion effect. In neonatal operated animals, complex task-related movements could be achieved as far as the goal was concerned, but as to their motor strategy it was always destroyed. The postnatal development of different limb reflexes and of the pattern of locomotor activity has been studied after neonatal sensorimotor cortical ablation (Leonard and Goldberger (1987a)) or hemispherectomy (Villablanca et al. (1986), Burgess and Villablanca (1986)). These studies pointed out that these motor behaviours were not affected as far as their control was mediated by subcortical levels, but became impaired when cortical control occured in normal development. This 'growing into a deficit' exemplified the corticalization process. This aspect of neurological development is not directly connected to the functional potentialities of the lesioned pyramidal tract nor to the reorganized intact one, that we wanted to put forward.

Different critical dates during development have been determined by performing the sensorimotor cortical ablation at several postnatal ages. Each critical date concerned the performances of a given motor skill. In animals operated before this date, the performances of the AL could reach the reference level of the NAL, and in animals operated after they could not. During development, critical dates determine the threshold between compensation and deficit of the performances of motor skills. Such critical dates have been suspected to exist during development, but the two ages at lesion examined were restrained to the day of birth and adulthood, without any comparison between animals operated at several ages (Leonard and Goldberger (1987a)). In animals operated at a mean age of 8 days (range 5-25) and in adult operated ones, the performances of distal skills, as observed in a paw attitude-movement test, were never recovered (Burgess and Villablanca (1986)). The present work is the first attempt to determine critical dates of sensorimotor cortex ablation during development in cat.

The execution of a target reaching movement involves several neural substrates, among which the sensorimotor cortex builds up the final motor command. In control animal, the motor sequence consisted of a ballistic phase followed by a terminal adjustment onto the target. After ablation of the sensorimotor cortex, performed at early and adult age, this motor strategy was always destroyed. In adult operated animals, the motor strategy was resumed to the ballistic phase, whereas in neonatal operated ones it consisted of a discontinuous movement. Such motor strategy composed of consecutive accelerations and decelerations has been observed in different experimental conditions. During the training period of a step-tracking task in monkey, the arm was transported by a discontinuous movement (Brooks and Watts (1988)). This was interpreted as a mode of motor control characterisitic of this learning period and involving sensory feedback. Afterwards, the transport of the limb was achieved by a quick and precise ballistic phase followed by a terminal adjustment. After deafferentation, the motor program shifted from the optimal, feed-forward, mode of motor control to a much less efficient mode based on peripheral feedback and the overall performance deteriorated (Jeannerod (1988)). After neonatal sensorimotor cortical ablation in cat, the discontinuous movement of the reaching could have been considered as a learning period, but its duration largely exceeded a normal learning. It could also have been considered as a deficit, but in fact it allowed the animal to reach the goal. Thus we have considered this new motor strategy in neonatal operated animals as a compensatory one, deficient in respect to the one observed in control animal, but successful in respect to that of adult operated one. A compensatory strategy has been reported to occur in cat after sensorimotor cortical ablation performed at both neonatal and adult age. It consisted of an increased velocity of the flexion during swing in order to compensate for the time lost due to the increase number of mistakes made during conditioned locomotion (Leonard and Goldberger (1987a)). The data presented here indicate that in neonatal operated animals proximal or distal skills can be recovered, whereas the motor strategy of execution of the movement itself is always destroyed. In respect to the various functional potentialities of the pyramidal tract, its partial functions can be recovered before some critical dates of operation, its integrated function however is always abolished, but can be replaced by new compensatory strategies.

After neonatal sensorimotor cortical ablation the intact pyramidal tract is deeply reorganized at spinal as well as supraspinal levels. At spinal level, corticospinal terminations were bilateral in the whole gray matter, except the motoneuronal cell groups. At supraspinal level, cortical projections to diencephalic, mesencephalic, pontine, and medullary nuclei which were only unilateral in control animal, were found bilateral in animal operated at birth.

Ipsilateral corticospinal terminations have only been described in the base of the dorsal horn and at the cervical enlargement after neonatal hemispherectomy, which were not observed in control animal (Gomez-Pinilla et al. (1986)). In the present study, ipsilateral corticospinal terminations have been observed in the whole dorsal horn and the dorsolateral part of the intermediate zone, and in addition throughout the spinal cord. In contrast, the corticospinal tract originating from the intact sensorimotor cortex of neonatal operated cat has been reported to maintain a normal distribution (Leonard and Goldberger 1987b)). In the latter study however, the sensorimotor cortical ablation did not seem to involve all the cytoarchitectonic areas at the origin of the pyramidal tract.

After neonatal sensorimotor cortical ablation, the structural reorganization of the cortical projections to supraspinal levels has been scarcely observed. In neonatal operated animals bilateral cortical projections, which were only unilateral in control animal, have only been reported to the red nucleus, the dorsal column nuclei, and the thalamus.

The contralateral corticorubral projections reported here in animals operated at birth confirm the observations of various authors. According to anatomical (Naus et al. (1985), Leonard and Goldberger (1987b), Murakami and Higashi (1988), Villablanca et al. (1988)) and electrophysiological (Tsukahara et al. (1983), Murakami et al. (1988))

findings, corticorubral projections were found contralaterally after neonatal sensorimotor cortical ablation, in addition to the ipsilateral projections observed in control animal. These new contralateral projections however, as studied by the retrograde double labelling technique (Naus et al. (1985)), appear not to be axonal collaterals of ipsilateral corticorubral fibers.

The new ipsilateral cortical projections to both gracile and cuneate nuclei reported here in animals operated at birth have not been mentioned previously. After neonatal hemispherectomy, the anterograde transport of labelled amino acids, injected into the intact motor cortex and sparing the somatosensory one, only revealed ipsilateral corticogracile projections in addition to the controlateral ones observed in control animal (Gomez-Pinilla et al. (1986)).

The controlateral cortical projections to the ventrobasal complex of the thalamus observed in neonatal operated animals but not in control ones have not been established previously. Other authors have described new corticothalamic projections but never in respect to the various nuclei of the thalamus. After neonatal hemispherectomy, where the thalamus could also be encroached upon, the contralateral thalamic projections as a whole increased in respect to those seen in control animal (Villablanca and Gomez-Pinilla (1987)). After neonatal sensorimotor cortical ablation, the contralateral corticothalamic projections increased more in medial nuclei of the thalamus, which still received bilateral projections in control animal, than in nuclei of the ventral group (Leonard and Goldberger (1987b)).

The intense reorganization of corticospinal and cortico-subcortical projections from the intact pericruciate cortex after neonatal ablation of the opposite one seems to include functional potentialities. For instance, some new corticospinal terminations and the rubrospinal terminations from the red nucleus receiving new cortical projections overlapped in the dorsolateral part of the spinal intermediate zone. This region contains propriospinal neurons projecting to motoneurons of the distal extremities contralateral to the cortical ablation. These anatomical data are in keeping with the behavioural recovery of distal skills in neonatal operated animals. In contrast, despite this intense plasticity the programming of the whole movement was never recovered but could be replaced by a compensatory strategy in neonatal operated animals.

Abbreviations

C : caudal, R : rostral. *Sulci* : ans. l. : lateral branch of the ansate, ans. m. : medial branch of the ansate, cor. : coronal, cru. : cruciate, e. syl. : ectosylvian, lat. : lateral, orb. : orbital, pr. syl. : presylvian, splen. : splenial, s. syl. : suprasylvian. *Nuclei* : DCN : dorsal column nuclei, LRN : lateral reticular nucleus, PN : pontine nuclei, RN : red nucleus, VB: ventrobasal complex of the thalamus.

References

Armand, J. (1982) 'The origin, course and terminations of corticospinal fibers in various mammals', in H.G.J.M. Kuypers and G.F. Martin (eds.), Descending Pathways to the Spinal Cord, Progr. Brain Res. 57, Elsevier Biomedical, Amsterdam, New York, Oxford, pp.329-358.

Armand, J. and Aurenty, R. (1977) 'Dual organization of motor corticospinal tract in the cat', Neurosci. Letters 6, 1-7.

Armand, J., Holstege, G. and Kuypers, H.G.J.M. (1985) 'Differential corticospinal projections in the cat. An autoradiographic tracing study', Brain Res. 343, 351-355.

Armand, J. and Kuypers, H.G.J.M. (1980) 'Cells of origin of crossed and uncrossed corticospinal fibers in the cat', Exp. Brain Res. 40, 23-34.

Brooks, V.B. and Watts, S.L. (1988) 'Adaptive programing of arm movements', J. Mot. Behav. 20, 17-132.

Burgess, J.W. and Villablanca, J.R. (1986) 'Recovery of function after neonatal or adult hemispherectomy in cats : II. Limb bias and development, paw usage, locomotion and rehabilitative effects of exercise', Behav. Brain Res. 20, 1-18.

Burgess, J.W., Villablanca, J.R. and Levine, M.S. (1986) 'Recovery of function after neonatal or adult hemispherectomy in cats : III. Complex functions : open field exploration, social interactions, maze and holeboard performances', Behav. Brain Res.20, 217-230.

Campbell, A.W. (1905) Histological Studies on the Localization of Cerebral Function, Cambridge Univ. Press.

Gomez-Pinilla, F., Villablanca, J.R., Sonnier, B.J. and Levine, M.S. (1986) 'Reorganization of pericruciate cortical projections to spinal cord and dorsal column nuclei after neonatal or adult cerebral hemispherectomy in cats', Brain Res. 385, 343-355.

Grünbaum, A.S.F. and Sherrington, C.S. (1901) 'Observations on the physiology of the cerebral cortex of some of the higher apes', Proc. Roy. Soc. 69, 206-209.

Hassler, R. and Mühs-Clement, K. (1964) 'Architektonischer Aufbau des sensomotorischen und parietalen Cortex der Katze', J. Hirnforsch. 6, 377-420.

Holmes, G. and May, W.P. (1909) 'On the exact origin of the pyramidal tracts in man and other mammals', Brain 32,1-43.

Jeannerod, M. (1988) The Neural and Behavioural Organization of Goal-Directed Movements, Oxford Psychology Series 15, Clarendon Press, Oxford.

Kennard, M.A. (1936) 'Age and other factors in motor recovery from precentral lesions in monkeys', Am. J. Physiol. 115, 138-146.

Kennard, M.A. (1938) 'Reorganization of motor function in the cerebral cortex of monkeys deprived of motor and premotor areas in infancy', J. Neurophysiol. 1, 477-496.

Kennard, M.A. (1940) 'Relation of age to motor impairment in man and in subhuman primates', Arch. Neurol. Psychiat. (Chicago) 44, 377-397.

Kennard, M.A. (1942) 'Cortical reorganization of motor function. Studies on series of monkeys of various ages from infancy to maturity', Arch. Neurol. Psychiat. (Chicago), 48, 227-240.

Leonard, C.T. and Goldberger, M.E. (1987a) 'Consequences of damage to the sensorimotor cortex in neonatal and adult cats. I. Sparing and recovery of function', Dev. Brain Res. 32, 1-14.

Leonard, C.T. and Goldberger, M.E. (1987b) 'Consequences of damage to the sensorimotor cortex in neonatal and adult cats. II. Maintenance of exuberant projections', Dev. Brain Res. 32, 15-30.

Mesulam, M.M. (1982) Tracing Neural Connections with Horseradish Peroxidase, I.B.R.O. Handbook Series, Methods in Neurosciences, Wiley and Sons, New York.

Murakami, F. and Higashi, S. (1988) 'Presence of crossed corticorubral fibers and increase of crossed projections after unilateral lesions of the cerebral cortex of the kitten : A demonstration using anterograde transport of Phaseolus vulgaris leucoagglutinin', Brain Res. 447, 98-108.

Murakami, F., Oda, Y. and Tsukahara, N. (1988) 'Synaptic plasticity in the red nucleus and learning', Behav. Brain Res. 28, 175-179.

Naus, C.C., Flumerfelt, B.A. and Hrycyshyn, A.W. (1985) 'An anterograde HRP-WGA study of aberrant corticorubral projections following neonatal lesions of the rat sensorimotor cortex', Exp. Brain Res. 59, 365-371.

Phillips, C.G. and Porter, R. (1977) Corticospinal Neurones. Their Role in Movement, Monographs of the Physiological Society 34, Academic Press, London, New York, San Francisco.

Sybirska, E. and Gorska, T. (1980) 'Effects of red nucleus lesions on forelimb movements in the cat', Acta Neurobiol. Exp. 40, 821-841.

Tsukahara, N., Fujito, Y. and Kubota, M. (1983) 'Specificity of the newly-formed corticorubral synapses in the kitten red nucleus', Exp. Brain Res. 51, 45-56.

Villablanca, J.R., Burgess, J.W. and Olmstead, C.E. (1986) 'Recovery of function after neonatal or adult hemispherectomy in cats : I. Time course, movement, posture and sensorimotor tests', Behav. Brain Res. 19, 205-226.

Villablanca, J.R. and Gomez-Pinilla, F. (1987) 'Novel crossed corticothalamic projections after neonatal hemispherectomy. A quantitative autoradiography study in cats', Brain Res. 410, 219-231.

Villablanca, J.R., Gomez-Pinilla, F., Sonnier, B.J. and Hovda, D.A. (1988) 'Bilateral pericruciate cortical innervation of the red nucleus in cats with adult or neonatal cerebral hemispherectomy', Brain Res. 453, 17-31.

Wiesendanger, M. (1981) 'The pyramidal tract. Its structure and function', in A.L. Towe and E.S. Luschei (eds), Handbook of Behavioral Neurobiology, 5. Motor Coordination, Plenum Press, New York, pp. 401-491.

ON THE ROLE OF THE VENTROLATERAL THALAMUS IN MOTOR RECOVERY AFTER BRAIN DAMAGE.

M. FABRE-THORPE and F. LEVESQUE.
Neuroscience Institut (C.N.R.S. - U.P.M.C.)
Department of Comparative Neurophysiology.
9 Quai Saint Bernard, 75230 Paris cedex 5
France

ABSTRACT. Lesions of the ventro-lateral thalamic nucleus (VL) have been found to have little effect on the execution of well learned movements, whereas they affect motor learning. In the present study we investigated the role of VL in the phase of cerebral reorganization that takes place after damage to the latero-posterior thalamic nucleus (LP) (relay of the visual extrageniculate pathway) had perturbed the execution of a previously learned movement. Cats were trained to perform a reaching movement towards a moving target-spot. They underwent bilateral brain lesions after performance had stabilized. A VL lesion induced a very transient increase of reaction time. A LP lesion severely disrupted accuracy and reaction time but was followed by full functional recovery. However, when both lesions were performed together, the deficits, although similar to those induced by LP lesion, were more pronounced and postoperative training did not lead to full compensation of the visuo-motor deficit. Furthermore, the recovery of accuracy crucially depended upon regular testing. These results support the involvement of VL in motor learning or re-learning.

1. Introduction

When learning a new motor skill, movement execution becomes more and more automatic. The importance of attentional constraints linked to the performance decreases progressively whereas the salient sensory information is processed more and more efficiently. Considering that different sensory-motor strategies are used before and after motor acquisition, one can postulate that the brain pathways involved in learning are distinct from those involved in the performance of the well-learned skill. Thus, a structure crucially involved in the "dynamic" motor learning phase of a new skill may progressively lose its importance during acquisition. One such structure could be the thalamic nucleus ventralis lateralis (VL), a main subcortical relay for messages originating in the basal ganglia and cerebellum (Angaut, 1970; Angaut & Bowsher, 1970; Hendry et al., 1979) on their way to the motor cortex (Strick, 1970; Rispal-Padel et al., 1973; Asanuma et al., 1974). Beside the crucial anatomical position of VL, cellular recordings in the behaving animal have shown that the firing of VL neurons is tightly linked to motor and sensori-motor events (Strick, 1976; Neafsey et al., 1978; Smith et al., 1978; Schmied et al., 1979; Macpherson et al., 1980). However, bilateral surgical lesions have little effect on the execution of well learned motor acts (Ranish & Soechting, 1976; Bénita et al., 1979; Fabre & Buser, 1980). On the other hand, we have previously shown that a bilateral VL lesion severely perturbs the acquisition phase of a reaching movement directed towards a moving target (Fabre & Buser, 1980). In the present study we attempted to show that VL is also a necessary structure for learning how to compensate a motor deficit after another brain damage. This plasticity period during which the visuo-motor system has to "relearn" how to perform the required

J. Requin and G. E. Stelmach (eds.), Tutorials in Motor Neuroscience, 641–647.

movement with the remaining intact cerebral pathways, could be likened to a new phase of motor acquisition. To induce the motor deficit we used a lesion of the thalamic nucleus lateralis posterior (LP) previously shown to produce a deficit of a known strength and duration (Fabre-Thorpe et al., 1986). Thus, we shall first report briefly on the behavioral effects of either a LP or a VL lesion in our task. Then we shall analyse the additionnal visuo-motor impairments induced when both lesions are performed simultaneously.

2. Material and Methods

2.1. TASK AND SET-UP

Cats were trained (Fig 1) to perform a reaching movement towards a spot of light that appeared and moved randomly on a tilted screen in front of the animal (Fabre-Thorpe et al., 1984). A trial started with a 500 msec tone in response to which cats had to stay immobile until the target appeared on the screen. They were then allowed three attempts without real time constraint (10 seconds) to reach the spot of light. Out of the various parameters evaluated we will consider in the present report: (i) an accuracy index based on the proportion of trials for which the first pointing attempt was accurate and (ii) two indices of movement speed, the reaction time (RT) which refers to the delay between the target's appearance and the lift of the cat's paw, and the movement time (MT), namely delay between the paw lift and the contact with the target when the first pointing attempt was correct.

Figure 1: *Testing Situation*. Ongoing reaching movement towards the moving spot of light that has appeared on the screen.

After the level of performance had stabilized seven cats underwent a bilateral lesion of either VL alone (2 cats), LP alone (3 cats) or both VL and LP (2 cats). The scores obtained after lesion were compared to preoperative ones and the functional recovery was analyzed over a few months.

2.2. SURGERY

Lesions were performed using a thermode stereotaxically lowered in the brain to the appropriate coordinates. The tip temperature was raised to 75-80°C for 30-90 seconds depending of the required lesion extent. The extent of the lesions were determined on serial frontal frozen sections. They usually involved most of the target nuclei and could slightly encroach on neighbouring structures.

3. Results

3.1. EFFECTS OF LESIONING VL AND LP ON THEIR OWN.

3.1.1 *Lesioning VL* (figure 2) had no effect on general behaviour. The analysis of visuomotor performances showed that accuracy and movement time were not impaired whereas movement onset was consistently delayed for both cats. The increase of RT reached 45% for one cat and 64% for the other (corresponding respectively to 104 and 127 msec). However, as cats were tested 36 hours after lesion, postoperative trauma can explain the severe impairment recorded during the first testing session after lesion. This increase of movement latency was transient and after a week visuo-motor performances were back to their preoperative level.

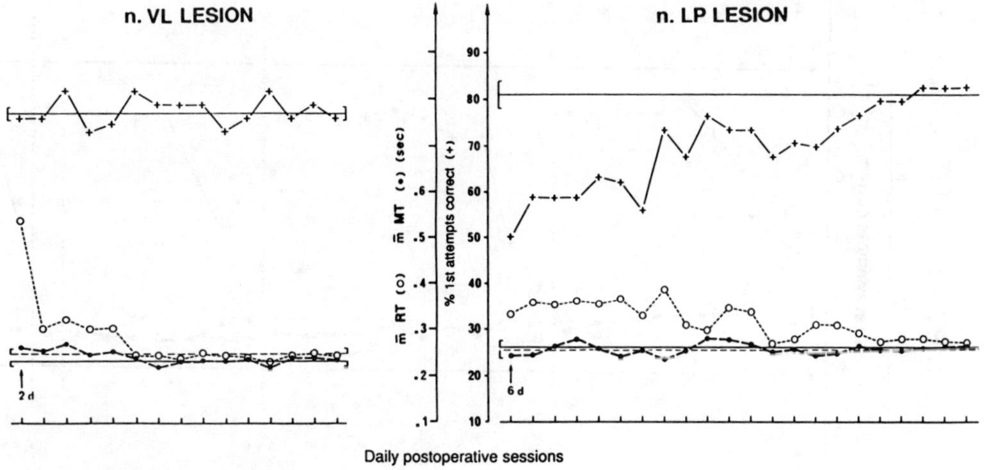

Figure 2: *Effects of lesioning either n.VL or n.LP*. Postoperative performance is shown for a cat with a VL lesion (left) and for a cat with a LP lesion (right). For each postoperative testing session (along the x axis), the percentage of accurate reachings at the first attempt is shown by a cross, whereas the corresponding mean RT and mean MT are respectively plotted with an open circle and a black dot. These postoperative scores have to be compared to the mean preoperative performance indicated for each cat by three baselines: upper full line is the proportion of first accurate attempts, bottom full line is the mean RT and broken line is the mean MT. Note the unimpaired accuracy and the transient increase of RT after VL lesion, whereas the LP lesion induces a longlasting impairment of both accuracy and RT.

This result confirms that whereas lesioning VL may delay movement onset it does not affect the execution of a well learned motor skill.

3.1.2 *Lesioning LP* (figure 2), which is the thalamic relay of the visual extrageniculate pathway, induces a deficit in general spatial orientation and in attention abilities. Visuo-motor performance was strongly impaired. These deficits had already been evaluated on another task that required pointing towards a moving target (Fabre-Thorpe et al., 1986). However we re-evaluated the LP induced deficit on three cats using the task described above. Visuo-motor performance was severely impaired. The accuracy level dropped by 27-48%, and the movement onset was delayed by 32-45% (corresponding to an RT increase ranging from 85 to 100 msec). However the time of execution (MT) stayed unaffected. The induced deficits were longlasting and 4 to 5 weeks of postoperative training were necessary to complete full recovery.

3.1.3 *Effects of Simultaneously Lesioning LP and VL* (figure 3). This combined lesion was followed by deficits in spatial orientation, and attention abilities. A new feature was that the animal had difficulty when shifting posture.

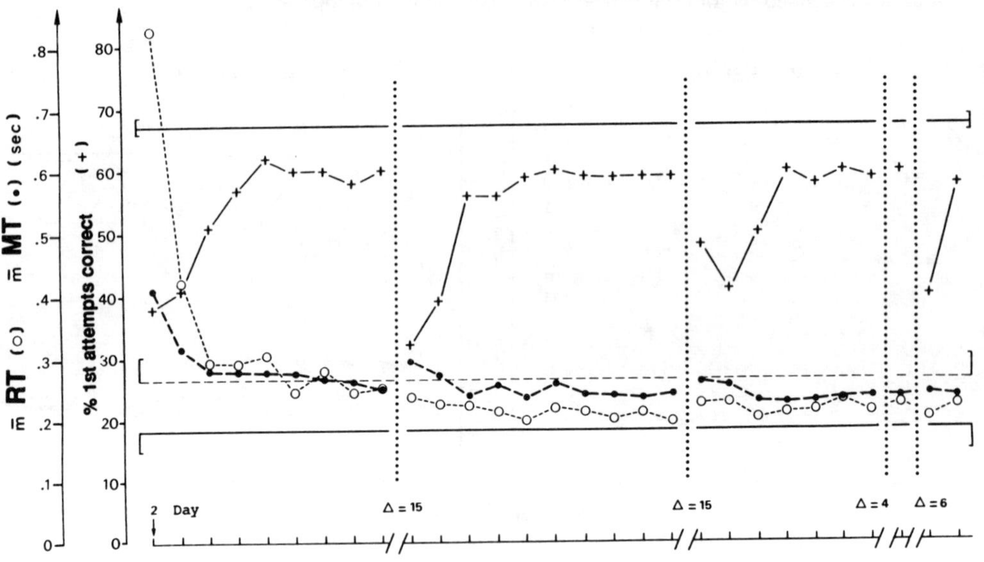

Figure 3: *Effects of simultaneously lesioning n.VL and n.LP*. Pre and postoperative performance of one cat are plotted. The three baselines represent the preoperative performances (mean and standard deviations). Upper full line is the proportion of first accurate reachin attempts, bottom full line is the mean RT and broken line is the mean MT. For each postoperative testing session along the x axis, the percentage of first accurate reaching attempts is represented with a cross, whereas the mean RT and mean MT are respectively plotted with an open circle and a black dot. The 4, 6 or 15 days training interruptions are indicated by dotted vertical lines. Note the dramatic effect on the accuracy level when postoperative training is stopped for 6 days or more.

Visuo-motor performance was also severely impaired. The accuracy, showed a dramatic decrease. The movement onset was more delayed than after any of the individual thalamic lesions: the increase of RT could reach 120% (corresponding to 230 msec). The movement time was only slightly and transiently affected.

Additional impairments appeared. Firstly, cats were unable to compensate their deficit: the two to three months of postoperative training did not induce total recovery. The accuracy level stabilized after 2 weeks, well under the preoperative level and did not show any sign of further improvement. The latency of the movement onset decreased progressively, but RT stabilized after 6 weeks 20% above its preoperative level corresponding to 30 msec. Interestingly, the MT unaffected initially was found to decrease to stabilize around 15% under its preoperative level. When postoperative scores had stabilized, the accuracy was still impaired, the initiation of the reaching movement was delayed and its execution was faster, conversely the number of timing errors decreased whereas the number of spatial errors increased. The pointing movement was thus performed with a strategy that was clearly different from those used before the combined lesion and even from those used after full recovery from VL or LP lesions when performed on their own. Another feature of the combined lesion was the crucial importance of the regular training to maintain the newly acquired level of accuracy; when testing was stopped for more than six days, the accuracy deficit re-appeared with all its initial strength. Such a drop of accuracy was never observed on normal cats or after recovery from VL or LP lesions when performed on their own.

Thus the VL plays a major role during the recovery phase following a LP damage. When VL and LP were both lesioned, the recovery lead to a new motor strategy that needed regular training to be preserved.

4. Discussion

After a VL lesion, the animals could not compensate for the motor deficits induced by a simultaneous LP lesion, which may indicate an unability to build a new motor program, or to consolidate a new "motor engram". Moreover, despite daily training no consolidation could really occur, since the new motor program was found to be unstable (drop of accuracy level) when training was stopped. Our results strongly support a role for the VL in motor learning (or re-learning). Considering that VL is the thalamic relay for cerebellar information on its way to the motor cortex, and that this cerebello-thalamo-cortical pathway has been shown to modulate its activity during motor learning (Sasaki & Gemba, 1983), it seems that this circuit could thus be crucially involved during the dynamic period of acquisition, and progressively lose its importance as the new motor skill becomes more automatic. The cerebello-rubro-spinal pathway appears then to be the most likely to take over the motor commands (Ito, 1984; Massion 1988) since neurotoxic lesions of the red nucleus lead to severe perturbation of the well learned motor skill that can be compensated by further training (Levesque et al., 1990). In conclusion we wish to propose the sequential involvement of two brain circuits; the cerebello-thalamo-cortical pathway being involved during learning, with the cerebello-rubro-spinal circuit taking charge when the movement is well learned. However it has to be stressed that beside cerebellar messages, VL receives messages originating in the basal ganglia that may also play an important role in motor plasticity (Buchwald et al., 1979; Canavan et al., 1989).

References

Angaut, P. (1970) The ascending projections of the nucleus interpositus posterior of the cat cerebellum: an experimental study using silver impregnation methods. Brain Research, 24, 377-394.

Angaut, P. & Bowsher, D. (1970) Ascending projections of the medial cerebellar (fastigial) nucleus: an experimental study in the cat. Brain Research, 24, 49-68.

Asanuma, H., Fernandez, J., Scheibel, M.E. & Scheibel A.B. (1974) Characteristics of projections from the nucleus ventralis lateralis to the motor cortex in the cat: an anatomical and physiological study. Experimental Brain Research, 20, 315-331.

Bénita, M., Condé, H., Dormont, J-F. & Schmied, A. (1979) Effects of cooling the thalamic ventrolateral nucleus of cats on a reaction time task. Experimental Brain Research, 34, 435-452.

Buchwald, N.A., Hull, C.D. & Levine, M.S. (1979) Neuronal Activity of the basal ganglia related to the development of "behavioral sets". In Brazier M.A.B. (Ed.), Brain Mechanisms in memory and learning (pp 93-103): from the single neuron to man. New-York: Raven Press.

Canavan, A.G.M., Nixon, P.D. & Passingham, R.E. (1989) Motor learning in monkeys (Macaca fascicularis) with lesions in motor thalamus. Experimental Brain Research, 77, 113-126.

Fabre, M. & Buser, P. (1980) Structures involved in acquisition and performance of visually guided movements in the cat. Acta Neurobiologica Experimentalis, 40, 95-116.

Fabre-Thorpe, M., Viévard, A., André, C., Fuzellier, J. & Buser, P. (1984) Visually guided movements in the cat: a test using a randomly moving target. Behavioral Brain Research, 11, 11-19.

Fabre-Thorpe, M., Viévard, A. & Buser, P. (1986) Role of the extra-geniculate pathway in visual guidance: II - Effects of lesioning the pulvinar-lateral posterior thalamic complex in the cat. Experimental Brain Research, 62, 596-606.

Hendry, S.H.C., Jones, E.G. & Graham, J. (1979) Thalamic relay nuclei for cerebellar and certain related fiber systems in the cat. Journal of Comparative Neurology, 185, 679-714.

Ito, M. (1984) The cerebellum and Neural control. Raven press (New-york).

Levesque, F. & Fabre-Thorpe, M. (1990) Motor deficit induced by red nucleus lesion: re-appraisal using kaïnic acid destructions. Experimental Brain Research, 81, 191-198.

Macpherson, J.M., Rasmusson, D.D. & Murphy, J.T. (1980) Activity of neurons in "motor" thalamus during control of limb movement in the primate. Journal of Neurophysiology, 44, 11-28.

Massion, J. (1988) Red nucleus: past and future. Behavioral Brain Research, 28, 1-8.

Neafsey, A.J., Hull, C.D. & Buchwald, N.A. (1978) Preparation for movement in the cat. II - Unit activity in the basal ganglia and thalamus. Electroencephalography and Clinical Neurophysiology, 44, 714-723.

Ranish, N. A. & Soechting, J.F. (1976) Studies on the control of simple motor tasks. Effects of thalamic and red nuclei lesions. Brain Research, 102, 339-345.

Rispal-Padel, L., Massion, J. & Grangetto, A. (1973) Relations between the ventrolateral nucleus and motor cortex and their possible role in the central organization of motor control. Brain Research, 60, 1-20.

Sasaki, K.& Gemba, H. (1983) Learning of fast and stable hand movement and cerebro-cerebellar interactions in the monkey. Brain Research, 277, 41-46.

Schmied, A., Bénita, M., Condé, H. & Dormont, J-F. (1979) Activity of ventrolateral thalamic neurons in relation to a simple reaction time task in the cat. Experimental Brain Research, 36, 285-300.

Smith, A.M., Massion, J., Gahéry, Y. & Roumieu J. (1978) Unitary activity of ventrolateral nucleus during placing movement and associated postural adjustment. Brain Research, 149, 329-346.

Strick, P.L. (1970) Cortical projections of the feline thalamic nucleus ventralis lateralis. Brain Research, 20, 130-134.

Strick, P.L. (1976) Activity of ventralolateral thalamic neurons during arm movement. Journal of Neurophysiology, 39, 1032-1044.

AUTHOR INDEX